Insects
in Relation to
Plant Disease

Insects
in Relation to
Plant Disease

SECOND EDITION

Walter Carter

Entomologist, and Senior Scientist Emeritus
Pineapple Research Institute of Hawaii

Professor of Entomology, Emeritus
University of Hawaii

a Wiley-Interscience Publication

JOHN WILEY & SONS

New York　　　London　　　Sydney　　　Toronto

Copyright © 1962, 1973, by John Wiley & Sons, Inc.

All rights reserved. Published simultaneously in Canada.

Library of Congress Cataloging in Publication Data:

Carter, Walter, 1897–
Insects in relation to plant disease.

"A Wiley-Interscience publication."
Includes bibliographies.
1. Insects as carriers of plant diseases. 2. Plant diseases. I. Title.

SB931.C35   1973   632   73-4362
ISBN 0-471-13849-5

Printed in the United States of America

10 9 8 7 6 5 4 3 2 1

Preface

A considerable measure of selectivity has been necessary in the preparation of this second edition. It does not presume to impinge on the subject matter of the highly specialized virology texts and reviews which have been numerous and which no doubt will continue to appear frequently. The introduction to plant virology *per se* which was essential to the first edition has been retained, but in a somewhat reduced form.

The objectives of this book remain the same, i.e., to provide a textbook for graduate students and a reference source for the whole field of the transmission of plant diseases by insects.

There has been an increased interest in bacterial and fungal diseases with which insects and a few other animals are concerned.

New cases of phytotoxemia, some representing quite new expressions of this condition, have been reported. While the number of examples is steadily increasing, a basic research need is for the characterization of the toxic principle, although some encouraging progress can be reported.

This edition combines the identification, naming, and classification aspects and notes significant developments in this controversial field. It is unlikely that any individual's proposals will be acceptable and the committee process, with its inevitable compromises, will find unanimity of acceptance difficult to achieve.

Virus transmission sections have been brought up to date, with emphasis on unusual situations. Experience has shown that the virus-vector tabulations have proved more useful than the vector-virus list and the latter has been omitted. New cases have been listed in a combination table.

Virus-vector relationships have proved to be a fruitful source of new material, and while the debate on the stylet-borne aphid-transmitted viruses has slackened off somewhat, it is still essential to know to which category a virus belongs.

Ecology as an essential facet has continued to generate interesting data significant to epidemiology and control. Ecology is an important tool in the development of integrated control programs; in fact, in its proper perspective, integrated control is simply applied ecology.

v

There is evidence that control of virus diseases, using all the available methods, is making slow but steady progress.

The initial statement on mycoplasmas introduces the latest development in the field of insect transmitted diseases of plants. If the etiological significance of these microorganisms is confirmed, as one paper appears to indicate, a fairly large and economically important group of plant diseases will cease to remain in the field of plant virology unless interrelationships are discovered. They will, however, remain in the field of insect transmission of plant diseases and the vectors, which have not changed, and will still present control problems which require solution. Initially, the emphasis in control studies will be on the application of antibiotics of the tetracycline group to the diseased plants. Whether or not the treatments actually control the disease on an economic level, they will no doubt be extensively used in diagnosis, since reaction to the tetracyclines seems to be characteristic of the mycoplasmas in plants.

ACKNOWLEDGMENTS

Once more I am indebted to Dr. R. Duncan Carter for critical reading of the manuscript and to my wife for her careful proof reading. I am also much indebted to my colleagues for providing new illustrations; they are acknowledged in the appropriate places in the text.

Walnut Creek, California   *Walter Carter*

Preface to the First Edition

It is now more than twenty years since Leach wrote the first and only inclusive text on the subject of insects in relationship to plant diseases, although there have been several adequate treatises on that section of the subject that is limited to the plant viruses. This is understandable, because developments in the plant virus field during the last twenty years have been so extensive that frequent summaries have been appropriate and necessary. Furthermore, the number of separate disciplines now perforce related has added to the desirability of occasional review treatments, not only for the benefit of the as-yet unspecialized biologist, but also for the restricted specialist who desires to be kept oriented to the subject as a whole.

Insect transmission of phytopathogenic bacteria has received little attention during recent years, but fungus diseases with which insects are concerned have become of increasing importance. There also has been a growing awareness of the toxicogenic insect as a phytopathological agent, and increasing attention to this section of the subject can confidently be expected. Leach wrote from the standpoint of the phytopathologist, but with a keen sense of the essential unity of the field of inquiry. This book is written in the same spirit, stressing the entomological and ecological aspects only insofar as seems desirable for better understanding.

When a subject involves as many disciplines as this does, it is manifestly impossible for a single worker to deal with all of them with equal comprehension. In an effort to reduce inadequacies and omissions, I have leaned heavily on the writings of specialists for some of the more theoretical aspects; for those shortcomings that remain, I must take the blame.

October, 1962   *Walter Carter*

Contents

1. Introduction   1

PART ONE   PLANT PATHOGENS TRANSMITTED BY INSECTS

2. Bacterial Pathogens   9

GENERALIZED TRANSMISSION OF BACTERIAL
   PATHOGENS
Fire Blight, Raspberry Blight, Walnut Blight, Bacterial
Heart Rot of Celery, Halo Blight of Beans, Pod Twist of
French Beans, Ring-Rot Disease of Potatoes, Bacterial Rot
of Apples, Bacterial Necrosis of Giant Cactus, Bacterial
Wilt of Bluggoe Banana, Bacterial Disease of Sugarcane,
Salmonella montevideo in Wheat, Citrus canker and Leaf-
miners
SPECIALIZED TRANSMISSION OF BACTERIAL
   PATHOGENS
Bacterial Wilt of Cucurbits, Bacterial Wilt and Stewart's Leaf
Blight of Corn, Potato Blackleg, Soft Rot of Rutabaga,
Bacterial Infection of Wheat and Maize, The Water-Marked
Disease of Willows, Olive-Knot Disease
SPECIALIZED VECTOR-PATHOGEN RELATION-
   SHIPS

3. The Mycoplasmas   51

CULTURE IN VITRO
CLASSIFICATION

4. Fungal Pathogens   69

INFECTION INCIDENT TO POLLINATION
Endosepsis of Figs

INFECTION THROUGH TRAUMATIC INJURY
Azalea Flower Spot
INTERNAL AND EXTERNAL CONTAMINATION
OF THE INSECT AS A RESULT OF FEEDING ON
FUNGAL MASSES
Ergot, Witch's Broom of Hackberry, Powdery Mildews,
Insects and Diploidizing of Rust Fungi
FUNGI DEVELOPING ON INSECT EXUDATES
Sooty molds
INFECTION THROUGH FEEDING AND OVIPOSI-
TION WOUNDS
Persimmon Wilt, Black and Yellow Molds of Corn, Root and
Basal Stem Rots of Cereals and Grasses, Anthracnose,
Gleosporium musarum in Musa balbiniana, Stem or Bole
Rot of Sisal, Boll Rot of Cotton, Aspergillus flavus in
Cotton, Aspergillus flavus in groundnut, Red Rot of Sugar-
cane, Taproot Rot of White Clover, Fruit Spoilage Diseases,
Brown Rot of Stone Fruits, Brown Rot of Apples, Needle
Blight and Fall Browning of Red Pine, Beech Bark Disease,
Calonectria rigidiuscula and its Pathogenicity to Cacao,
Cushion Gall of Cacao, Tree Cricket Canker of Apples,
Thielaviopsis paradoxa in Cacao Pods, Alternaria brassi-
cola in Cabbage, Phytophora in Blackpepper Foot Rot,
Fungus Gnats and Fusarium in Alfalfa and Red Clover
INFECTION THROUGH FEEDING PUNCTURES
Alternaria Blight of Rubber Trees, Cabbage Phomosis,
Sclerospora on Corn and Sorghum, Leaf Spot of Diffen-
bachia, The Pine Spittle Bug and Burn Blight of Pine Trees,
Bud-Blast Disease of the Rhododendron Caused by the
Fungus Pynostyanus, Internal Boll Disease of Cotton, Coffee
Bean Rot, Yeast Spot in Soybean, Nematospora in Rice,
Lint Rot of Cotton
INFECTION RESULTING FROM SYMBIOTIC
ASSOCIATION BETWEEN INSECT AND FUNGUS
The Woolly Apple Aphid and Perennial Canker, Phthiriosis
of Coffee and Vine Roots, Pediculopsis Mites and Fungal
Diseases of Plants, Oak Wilt
BIOLOGICAL ASPECTS OF TRANSMISSION
Bark Beetles and Blue-Stain Fungi, Ceratocystis Infection
of Stone Fruits and Cacao, Canker-Stain Fungus in Plane
Trees, Dutch Elm Disease, Hemileia Vastatrix on Coffee,
White Rot of Softwoods

PART TWO   THE TOXICOGENIC INSECT AND PHYTOTOXEMIA

5. **The Feeding Processes of Hemipterous Insects and Their Salivary Secretions**   **139**

   THE ANATOMY OF HEMIPTEROUS MOUTH
     PARTS AND THE FEEDING PROCESS
   THE SALIVARY SECRETIONS
   CRITERIA FOR THE DETERMINATION OF A
     PHYTOTOXEMIA
   THE PHYTOTOXEMIAS AND THEIR SYMPTOMS

6. **Localized Phytotoxic Effects**   **171**

   LOCAL LESIONS AT THE INSECT'S FEEDING
     POINT
   *Leaf Spotting by Leafhoppers, Leaf Spotting by Aphids,
   Leaf Spotting by Acarina, Leaf Spotting by Pentatomidae*
   LOCAL LESIONS WITH DEVELOPMENT OF
     SECONDARY SYMPTOMS
   *By Coccidae, By Pentatomidae, By Aphididae, By Miridae
   (Capsids), Coreidae*

7. **Primary Malformations**   **209**

   PLANT TISSUE MALFORMATIONS
   *By Aphididae, By Coccidae, By Thysanoptera, By
   Miridae, By Lygeidae, By Coreidae, By Cercopidae, By
   Membracidae, By Pentatomidae, By Cicadellidae, By Dip-
   tera, By Acarina*
   INSECT GALLS
   *Origin of the Stimulus, Nature of the Stimulus, Effect of
   Causal Stimulus*
   FRUIT MALFORMATIONS

8. **The Systemic Phytotoxemias**   **259**

   TOXEMIC SYMPTOMS INDICATING TRANSLOCA-
     TION OF THE CAUSAL ENTITY
   Limited Translocation of the Toxin

SYSTEMIC TOXIC EFFECTS
Of Aphid Toxins, Of Leafhopper Toxins, Mealybug Stripe
of Pineapple, White Streak of Pineapple Plants, Froghopper
Blight of Sugarcane, Psyllid Yellows of Potatoes and Toma-
toes, Mealybug Wilt of Pineapple

PART THREE   THE PLANT VIRUSES

9.   The Plant Virus as an Entity   **317**

HISTORICAL
DEFINITION OF A VIRUS
THE ORIGIN OF VIRUSES
The Novogenesis of Plant Virus Diseases
THE LATENT VIRUS
THE BIOLOGICAL APPROACH TO PLANT VIRUS
   PROBLEMS
IDENTIFYING, NAMING, AND CLASSIFYING
VIRUSES AND VIRUS DISEASES,
CLASSIFICATION

10.   The Clinical Aspects of Plant Virus Diseases   **351**

EXTERNAL SYMPTOMS
Color Changes, Teratological Symptoms, Variability of
Symptoms, Virus Complexes, Local Lesions
FACTORS AFFECTING EXTERNAL SYMPTOM EX-
   PRESSION
Temperature, Light Intensity, Genetic Factors, The Incu-
bation Period for Symptom Expression, Acute and Chronic
Symptoms
INTERNAL SYMPTOMS
In Leaf Parenchyma, In Phloem, In Xylem, Nuclear
Changes Induced by Virus Infection, Intracellular Inclu-
sions
PHYSIOLOGICAL SYMPTOMS
CHEMICAL AND PHYSICAL EVIDENCE OF INFEC-
   TION
By Color Tests, By Ultraviolet Radiation, By Paper
Chromatography and Electrophoresis

*SEROLOGICAL EVIDENCE FOR SPECIFIC INFEC-
TION*
*DIAGNOSIS AND DIFFERENTIATION OF VIRUS
STRAINS*
*By Interference Phenomena, By Physical and Chemical
Properties*

11. Modes of Plant Virus Transmission   **435**

TRANSMISSION THROUGH SEED
Virus Inhibitors, Genetic Aspects
TRANSMISSION THROUGH SOIL
TRANSMISSION THROUGH DODDER
TRANSMISSION BY GRAFTING
*Cleft Grafts, Natural Root Grafts, Tongue Grafts, Excised
Leaf Grafts, Budding, Bark Grafts, Core Grafts*
MECHANICAL TRANSMISSIONS
*Methods of Transmission, Factors Affecting Effectiveness of
Mechanical Transmission*
*THE INDICATOR PLANT AS AN AID IN VIRUS
TRANSMISSION*

**12. Plant Virus Transmission by Arthropods and Other
Animals**   **479**

TRANSMISSION BY ARTHROPODS
*Historical, Aphididae, Miridae; Piesmidae, Aleyrodidae,
Coccoidea, Thysanoptera, Mandibulate Insects, Mites*
*TRANSMISSION BY ANIMALS OTHER THAN
ARTHROPODS*
The Virus-vector Lists

13. Virus-Vector Relationships   **539**

THE FEEDING PROCESS OF VECTOR INSECTS
The Mechanism of Feeding
*VECTOR SPECIFICITY IN TRANSMISSION OF
PLANT VIRUSES*
*Stage or Species Specificity, Area Specificity, Group
Specificity, Regional Specificity, Vector Nonspecificity,
Factor of Ingestion, Insect Blood as a Virus Reservoir,
Complex Transmissions, Cross Protection*

*GENETIC VARIABILITY AS A FACTOR IN
  SPECIFICITY*
TRANSOVARIAL TRANSMISSION
*MULTIPLICATION OF VIRUSES IN INSECT
  VECTORS*
PERSISTENCE OF VIRUSES IN INSECTS
The Latent Period, The Persistent and Nonpersistent Virus
*THEORIES ON THE MECHANISM OF TRANSMIS-
  SION OF THE NONPERSISTENT VIRUS*
*The Inactivator Hypothesis, So-Called Biological Trans-
mission, Role of Stylets*
*THE TRANSMISSION OF PERSISTENT VIRUSES
  BY APHIDS*
CYCLICAL TRANSMISSION
*THE HOST PLANT IN RELATION TO TRANSMIS-
  SIBILITY AND VECTOR EFFICIENCY*
EFFECTS OF VIRUSES ON INSECT VECTORS

14. Ecological Aspects of Plant Virus Transmission   621
*VECTOR MOVEMENT, DISPERSAL AND MIGRA-
  TION*
*METHODS OF MEASURING VECTOR POPULA-
  TIONS AND THEIR MOVEMENTS*
*Quantitative Methods, Aphid Traps, Height of Insect
Vector Flights*
*PHYSICAL FACTORS OF THE ENVIRONMENT IN
  RELATION TO VECTOR MOVEMENTS*
*Humidity & Light Intensity, Wind Velocity, Temperature
and Rainfall, Cataclysmic Weather Conditions*
HOST PLANT SEQUENCE
Intrahost Movement, Interhost Movement, Plant Succession
THE VIRUS SOURCE
Susceptible Plant Hosts, In Hibernating Insects
PHENOLOGY
*THE SIZE OF VECTOR POPULATIONS AND THEIR
  MOVEMENTS*
Incidence of Disease
ESTIMATES OF VECTOR POPULATIONS
Advisory Services, Statistical Correlations
PHYSIOLOGICAL AND BEHAVIORAL FACTORS
Flight Behaviors

15. The Control of Viruses and Virus Diseases of Plants   **675**

CULTURAL METHODS OF CONTROL
Isolation, Barrier Crops, Density of Stand, Planting and Harvesting Dates, Avoidance of Infection
HEAT THERAPY OF VIRUS-DISEASED PLANTS
Inactivation of Viruses by Heat
INDEXING: A METHOD FOR PRODUCTION OF VIRUS-FREE PLANTS
Certification Programs, Serological Methods, Apical and Meristem Cultures
CHEMICAL CONTROL OF VIRUSES
Antibiotics, Inhibitors, Miscellaneous Chemical Treatments
CHEMICAL CONTROL OF VECTORS
Potato Diseases, Sugar Beets, Sugar Beet Curlytop, Miscellaneous Crops, Nematode Vector Control
VECTOR-HOST-PLANT CONTROL
Replacement Control
RESISTANT AND TOLERANT VARIETIES
Genetics of Resistance, Sources of Resistant Genes, Use of Resistance as a Commercial Control Measure
CROSS PROTECTION
THE FUTURE OF VIRUS DISEASE CONTROL

Index   **749**

Introduction

The subject of insects in relation to plant diseases is extremely broad, for it includes in its scope the insect, pathogen, and host plant. Each of these can be studied independently, but they are inseparable when the disease that is caused is under consideration.

In this era of specialization, which begins much too early in the training of scientific workers, appreciation of this unity usually comes only after a degree of maturity in one specialized field has been achieved. The approach to the insect-pathogen relationship is, therefore, from the standpoint of one or another specialized field, either entomology, pathology, or biochemistry. It should be possible, however, at least in graduate schools, to so orient the student that he completes his formal training with an understanding of all the disciplines involved.

The tendency for a field of inquiry to "splinter" when it becomes unwieldy is pointed up by the development of virology as a separate discipline, with plant and animal virologists in uneasy partnership. The basic disciplines of entomology, plant pathology, and biochemistry will remain the essentials, however the subdivision may be designated.

Since this book is concerned only with insects and other animals in relation to plant diseases, interrelationships between insects and plants, such as the pollination of flowers by bees and other insects, the curious adaptations of entomophagous plants which permit them to trap and absorb the captured insects, and the symbiosis of insects with bacteria and fungi are considered only if plant pathogenicity is a result of the relationship.

The symbiosis between insects and microorganisms is so complex and so frequently encountered that, at least in some instances, the microorganisms might be expected to be plant pathogenic. This is true only in rare cases, and even then not where the symbiosis has evolved to the stage where the microorganism is host-bound and hereditarily transmitted.

In the sucking insects, principally concerned in the transmission of pathogens, this type of symbiosis is the rule. One poorly explored field is the relationship of biological strains of insects to specific microsymbionts, for it can be presumed that the symbiont will mutate more readily than will its host.

Examples of symbiosis in which plant pathogens are involved are important, however, and are treated more fully in later chapters.

Proofs of pathogenicity of fungi and bacteria were provided by the work of De Bary, Pasteur, Koch, and Burrill in the last half of the nineteenth century, and experimental proof of the role of bees and wasps in transmitting a bacterial disease was obtained by Waite in 1881. By 1911, the subject had achieved its first review article when Norton (1911) wrote on "The Health of Plants as Related to Insects" and suggested certain categories, as follows:

1. Direct injuries
   a. Physiological troubles following injury
   b. Diseases caused by the presence of insects or their products
2. Diseases indirectly caused by insects (by dissemination of the pathogens)

Other reviews followed (Heald, 1913; Gardner, 1918; Rand and Pierce, 1920; Rand *et al.,* 1921). Leach's comprehensive review (1935) was followed by the first full textbook treatment of the subject (Leach, 1940). Carter (1939, 1952) reviewed the subject of the toxic effects of insect feeding on plants, and there is a steady stream of review articles, somewhat repetitious, on various aspects of virus transmittal. Special note should be made, however, of the inclusive although uncritical compilation of insect vectors of virus diseases by Heinze (1959).

Classification of plant diseases with respect to their dissemination by insects has helped to define useful categories that, however, have no phylogenetic basis. Leach (1940) considered that there is some evolutionary significance in the association of insects and plant pathogens, and he draws parallels between the development of insect-plant associations, such as those on which pollination depends, and some insect-pathogen associations. It is worthy of note that all his examples of the latter type are of fungus-insect associations.

The earliest attempt at classification after Norton's appears to be that of Rand and Pierce (1920), which is as follows:

1. External transmission of parasitic microorganisms
   a. External dissemination and direct inoculation by the insect vectors
   b. External dissemination and accidental infection without direct inoculation
2. Infection through insect wounds without dissemination by the wounding agent
3. Internal transmission of parasitic microorganisms
   a. Mechanical internal transmission
   b. Biological internal transmission

Böning's (1929) objection to the Rand and Pierce classification, on the ground that the distinction between external and internal was an artificial one, resulted in an oversimplification, as follows:

1. Dissemination without wounding
2. Dissemination with wounding
   a. Insect obligatory
   b. Insect not obligatory

Leach's classification mirrors his primary concern with bacterial and fungal pathogens, and is as follows:

1. Wounding of the plant by an insect without dissemination of a pathogen
   a. Insect toxicogenic (psylid yellows, mealybug wilt of pineapple, etc.)
   b. Insect not toxicogenic (white grubs and crown gall)
2. Dissemination of a pathogen by an insect without wounding the plant
   a. Mechanical dissemination only (sphinx moths and anther smut, flies and ergot)
   b. Biological dissemination (rare)
      (1) Insect not obligatory (flies and fire blight)
      (2) Insect obligatory (theoretically possible but no examples known)
3. Dissemination of a pathogen by an insect with wounding of the plant
   a. Mechanical dissemination only (bark beetles and blue stain, bark beetles and Dutch elm disease)

   b. Biological dissemination
      (1) Insect not obligatory (dipterous insects and bacterial soft
          rot, *Dacus oleae* and olive knot)
      (2) Insect obligatory (cucumber beetles and bacterial wilt of cu-
          curbits, apple maggot and bacterial rot of apples)

Huff's (1931) classification, as given by Leach (1940), is shown below
and is appropriate in connection with the virus diseases of plants, although it is more generally applicable to disease of animals.

   1. Biological
      a. Cyclopropagative (organisms undergo cyclical change and mul-
         tiply within the vector)
      b. Cyclodevelopmental (organisms undergo cyclical change but do
         not multiply within the vector)
      c. Propagative (organisms undergo no cyclical change but multi-
         ply within the vector)
   2. Mechanical (organisms undergo neither cyclical change nor
      multiplication within the vector)

Austwick (1957) classified the fungal diseases transmitted by insects,
along the lines of Böning and Leach, as follows:

   1. Insect feeding primarily on plant or animal host
      a. Transmission of disease without wounding of host, e.g., anther
         smut of campions by bees, etc.
      b. Transmission of disease with wounding of host
         (1) Wounding by transmitting insect, e.g., stigmatomycosis of
             cotton by stainer bugs
         (2) Wounding by another agency, e.g., epizootic lymphagitis by
             pasture flies, oak wilt by beetles, etc.
   2. Insect feeding primarily on fungus or its metabolic products
      a. Direct transmission, e.g., ergot, by flies feeding on honeydew
      b. Indirect transmission, e.g., diploidization of rust mycelia by flies

A classification modified from that of Leach and including the concepts of Huff is presented below as an outline of the subject matter of
this book.

A Classification of Plant Pathogens Transmitted
by Arthropods and Other Animals

   1. Pathogens disseminated without wounding of the plant
      a. Mechanical dissemination only
      b. Biological dissemination

      (1) Insect not obligatory
      (2) Insect obligatory
2. Pathogens disseminated with wounding of the plant
  a. Mechanical
  b. Biological
      (1) Insects not obligatory
      (2) Insects obligatory
3. Pathogens disseminated by injection
  a. Fungi
  b. Toxins
  c. Viruses
      (1) Biological—passing through hemocoele
         (a) No multiplication in vector
         (b) Propagative, multiplying in vector
      (2) Mechanical

More detailed divisions of the toxin and virus categories are found in later chapters. The toxins fall into various categories depending on the nature and extent of the symptoms; the classification of the plant viruses presents a formidable problem that is still largely unsolved.

The increasing importance of plant diseases with which insects are concerned has grave implications for the world's food supply. This is serious enough in the Western world, where agricultural research and phytosanitary technology are highly developed, but it is calamitous in the more tropical, underdeveloped countries. A few examples will illustrate. Cassava is the staple starchy food of a considerable section of the world's underfed population, but cassava mosaic is rampant, and only in East Africa does it appear to have been ameliorated by the development of resistant varieties (through the efforts of European scientific workers). Still unsolved is the problem of the decline of coconut trees, which appears to be complex in its ecology and worldwide in its distribution, from Cape St. Paul wilt in Togo and Ghana to cadang-cadang disease in the Philippines. The peasant farmers of northern Nigeria suffer losses from a toxin-secreting mealybug that twists and distorts their guinea corn.

There are some bright spots. Capsids in West Africa, formerly a limiting factor in cocoa production, are under control by modern methods, and there are tolerant varieties available to lessen the impact of the swollen-shoot virus diseases.

What has been done, however, is only an infinitesimal fraction of what needs to be done, not only in basic research, but also in the vastly greater practical problem of developing control methods.

REFERENCES

Austwick, P. K. C., "Insects and the spread of fungal disease," in *Biological Aspects of the Transmission of Disease*, Oliver and Boyd, London, 1957, pp. 73–79.

Böning, K., "Insekten als Übertrager von Pflanzenkrankheiten," *Z. Angew. Entomol.*, 15, 182–206 (1929).

Carter, W., "Injuries to plants caused by insect toxins," *Botan. Rev.*, 5, 273–326 (1939).

Carter, W., "Injuries to plants caused by insects toxins, II.," *Botan. Rev.*, 18, 680–721 (1952).

Gardner, M. W., "Mode of dissemination of fungus and bacterial diseases of plants," *Ann. Rept. Mich. Acad. Sci.*, 20, 357–423 (1918).

Heald, F. D., "The dissemination of fungi causing disease," *Trans. Am. Microscop. Soc.*, 32, 1–29 (1913).

Heinze, K., *Phytopathogene Viren und Ihre Überträger*, Duncker & Humblot, Berlin, 1959.

Huff, C. G., "A proposed classification of disease transmissions by arthropods," *Science*, 74, 456–457 (1931).

Leach, J. G., "Insects in relation to plant diseases," *Botan. Rev.*, 1, 448–466 (1935).

Leach, J. G., *Insect Transmission of Plant Diseases*, McGraw-Hill, New York, 1940.

Norton, J. B. S., "The health of plants as related to insects," *J. Econ. Entomol.*, 4, 269–274 (1911).

Rand, F. V., and Pierce, W. D., "A coordination of our knowledge of insect transmission in plant and animal diseases," *Phytopathology*, 10, 189–231 (1920).

Rand, F. V., Ball, E. D., Caesar, L., and Gardner, M. W., "Insects as disseminators of plant diseases," *Phytopathology*, 12, 225–240 (1922).

Waite, M. B., "Results from recent investigations in pear blight," *Botan. Gaz.*, 16, 259 (1891).

PLANT PATHOGENS TRANSMITTED BY INSECTS

2

Bacterial
Pathogens

Bacterial diseases of plants are widespread and often destructive, but with few exceptions they are not entirely dependent on insects for their dissemination or for the inoculation of the pathogen into the plant tissue. There is also an almost complete lack of specific relationship between insect vector and pathogen if we exclude the few cases where the insect is a more or less obligatory agent in transmission.

The fact that the plant pathogenic bacteria are unable to penetrate plant tissue without a court of entry has led to some vagueness as to the significance of the role of insects in transmission. Stomata, water pores, lenticels, and flower nectaries are all natural courts of entry. The insect contributes through feeding and oviposition wounds, as a mechanical carrier of the organism on its body, and in some cases, by virtue of a mutualistic relationship between the organism and the insect which insures a continuing association between pathogen, insect, and host plant. There are, however, certain limitations on plant pathogenic bacteria, as shown by Crosse (1958). The majority form no resting spores or structures comparable to those of fungi, and they must survive unfavorable environmental periods and periods when susceptible hosts are absent, either in association with perennial hosts, or in the soil, or in hibernating insects. Furthermore, dispersal is accomplished by a variety of agencies—man, wind, and rain, as well as by insects—and many pathogens require a combination of these agencies to accomplish transmission.

GENERALIZED TRANSMISSION OF BACTERIAL PATHOGENS

Westcott (1950) has classified the plant pathogens, of which over 200 species are known, into seven genera, namely, *Agrobacterium, Bacterium, Corynebacterium, Erwinia, Pseudomonas, Streptomyces,* and *Xanthomonas,* following the seventh edition of *Bergey's Manual of Determinative Bacteriology.* The bacterial diseases were placed in three general categories, depending on the nature of the injury to the host plant: (1) blights, rots, and leaf spots where the parenchyma tissues are killed, (2) wilts due to the invasion of the vascular system or water-conducting tissue, and (3) hyperplasia as found in galls.

Since Leach's adequate presentation in 1940, very few new cases of insect-transmitted bacterial diseases of plants have been recorded; the more significant contributions have perhaps been along the lines of disease control.

Westcott's three categories, being based on the plant tissues subject to attack, do not lend themselves to consideration based primarily on the transmission by insects. Actually, most of the cases for consideration fall in the first category, there being only four in the second category involving vascular tissues, and two in the third, where gall formation occurs. The discussion of these cases which follows includes all those for which adequate evidence for an insect association is available.

Fire Blight (Erwinia amylovora (*Burrill*) *Winslow* et al.)

Fire blight has a wide host range, including some 90 species of orchard trees and some ornamentals, but it is primarily a disease of apple and pear trees (Fig. 2.1–2.4). It also occurs on a number of noncultivated species. This classic disease of orchard trees is unique in that it was the occasion for the first demonstration that bacteria could cause plant diseases (Burrill, 1881) and that insects were involved in transmission (Waite, 1891).

On apples and pears, blossoms and young twigs are initially affected. On larger branches, destructive cankers are formed and trunks suffer from "collar blight." The pathogen overwinters in the cankers of the branches and trunks of infected trees, and in early spring a bacterial exudate oozes from these cankers. Leach (1940) has discussed in detail the

FIGURE 2.1. Fire blight. Group of leaves with marginal browning. This is one of the first obvious signs of the disease in an orchard (Photograph: East Malling Research Station, Crosse *et al.*, 1959).

various theories relating to the role of the holdover cankers as sources of the organism for the initiation of primary infections, for as he says, "The sources of primary inoculum in the spring and the relative importance of the different means of dissemination of the pathogen constitute the crux of the fire-blight problem."

Thomas and Ark (1934) provided experimental proof that flies and ants initiated primary infection by carrying the bacteria from holdover cankers to blossoms. These same authors (1934a) questioned the conclusions of several workers that rainfall was an important agent of dissemination but related increase of infection in rainy weather to nectar concentration. In dry weather, the nectar is too concentrated for bacterial growth. This conclusion was later supported by Ivanoff and Keitt (1941).

FIGURE 2.2. Fire blight. Tree naturally infected with fire blight (Photograph: East Malling Research Station, Crosse *et al.*, 1959).

The nectar concentration of blossoms is equally important for the transmission of the organism by bees. Keitt and Ivanoff (1941) demonstrated that when bees were allowed to sip nectar containing 2–12% sugar from inoculated blossoms and then from healthy blossoms containing 0–35% sugar, 49% transmission occurred. On the other hand, no infection occurred when the inoculated blossoms contained 10–14, 42–56, or 48–75%

FIGURE 2.3. A small branch canker originating from an infected summer blossom at the end of a shoot (Photograph: East Malling Research Station, Crosse *et al.,* 1959).

sugar, and the corresponding healthy blossoms, 10–18, 44–47, and 46–70%, respectively. It would appear from these data that the nectar concentration of the previously infected blossoms was the more significant. Transmission by tarnished plant bugs, wasps, aphids, bark beetle, codling moth, and yellow jackets, in addition to flies, ants, and bees, has been reported on the basis of experimental evidence. The extensive range, taxonomically and biologically, of these agents of dissemination indicates the improbability of success of control measures directed against the insect vectors or the host range of the pathogen. While much of this transmission is no doubt accidental, being due to contamination of insect bodies on contact with active pathogens, there are some data available to indicate a closer relationship. Ark and Thomas (1936) showed that the pathogen could survive in the intestinal tract of three species of flies, *Drosophila melanogaster* Meigen, *Musca domestica* L, and *Lucilia sericata* (Meigen), and that in the case of *M. domestica* the eggs were externally

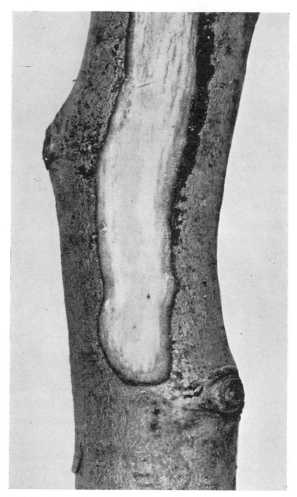

FIGURE 2.4.   Part of a newly diseased branch with leaves removed to show internal appearance of an active canker. The wood and deeper layers of the bark are still white, but they are overlaid by a thin brown layer of diseased bark. The surface of the bark in the uncut area is gray-black in color (Photograph: East Malling Research Station, Crosse *et al.,* 1959).

contaminated with the pathogen. The persistence of the pathogen through pupation of *D. melanogaster* and *M. domestica* is even more important and suggests a relationship between vector and pathogen which is more than casual. Leach's suggestion that the pathogen may overwinter in insect puparia does not seem to have received the attention it merits.

The occurrence of the disease in the United States is sporadic and varies both in incidence and intensity from year to year, being most severe in the warm southern United States (Gilgut, 1953). The development of the bacterium is favored by warm humid weather, and rapidly growing succulent tissue is especially susceptible. Hildebrand's study (1954) of isolates of *E. amylovora* from the United States, Canada, and New Zealand may well provide an explanation for the variable behavior of the blight organism, and, in turn, the extreme variation in blight incidence in the field. Hildebrand determined both virulent and avirulent forms, with morphological differences in cell and flagella length. Organisms isolated from old dormant cankers were low in virulence and grew rough colonies. Passage through susceptible tissue restored the virulence of the bacterium. When isolated from young infections, all the colonies were smooth and high in virulence. Whether or not virulence is changed by variable nectar concentration or by passage through the dipterous intestinal tracts might be fruitful speculation.

During 1958, the disease was first identified in Great Britain (Crosse *et al.*, 1958) as attacking pears, and there have been serious losses reported since. By the end of 1958, 15 outbreaks had been reported, 14 in Kent and one in Worcestershire. In Kent, the distribution is generally consistent with spread, probably by pollinating insects, from three centers (Lelliot, 1959).

In 1941, Marshall, Childers, and Brody indicated that leafhoppers were active carriers of the pathogen affecting apple trees. The leafhoppers became contaminated by feeding upon or crawling over infected tissue. Later, they inoculated healthy apple twigs during the feeding process. While earlier authors (Tullis, 1929; Miller, 1929) questioned the importance of sucking insects as vectors of the pathogen, it seems more than likely that this method of feeding, involving the contamination of mouth parts, might be an effective method of distributing the pathogen. According to Leach, twig blight of pear most commonly follows inoculation by sucking insects. The blossom stage of orchard trees is the most susceptible to blight infection, and there seems no doubt that epidemics originating in the main blossom period are largely due to dissemination by insects. Even then, however, control measures have perforce been directed at the organism itself.

Since humidity increases normally at blossom time and susceptibility therefore increases with the dilution of the sugar content of the nectaries, control has been directed to control of the organism by preventing its growth in the nectaries. This has been accomplished in California by spraying with copper compounds during blossom time and repeating the sprays as new blossoms open. In Colorado, the standard method of control of fire blight in pears and apples is with Dithane Z-78 (zinc ethylene-bisdithiocarbamate) fungicide spray, and with additional control measures such as sanitary pruning and treatment of holdover cankers (Thomas and Henderson, 1952). An objection to copper sprays is that they sometimes cause russetting of the fruit.

The latest methods of control are based on the use of antibiotics to which the pathogen is sensitive. Effective control with antibiotic sprays has been reported by Ark (1953), Heuberger and Poulos (1953), Winter and Young (1953), Goodman (1954), and Clayton (1955). Ark obtained an 83.3% control of pear blight in California using bentonite dust containing 240 ppm of streptomycin without fruit russetting. Heuberger and Poulos showed that Dithane Z-78 Purizate fungicides were highly effective; of the antibiotics, thiolutin and streptomycin sulfate at 30, 60, and 120 ppm controlled fire blight, but terramycin was ineffective. Winter and Young controlled experimentally infected Jonathan apple fire blight in Ohio with three applications (early, late blossom, and petal-fall periods) of streptomycin at concentrations of 60–120 ppm. Goodman and Clayton used streptomycin nitrate and Agrimycin (a mixture of streptomycin and terramycin). Concentrations of 50–100 ppm gave good control although foliage chlorosis occurred temporarily on heavily sprayed trees. The sodium and zinc salts of 2-pyridinethione-1-oxide were more effective control agents than a number of antibiotics or Bordeaux mixture, according to Hamilton and Szkolnik (1957).

Breeding resistant varieties as a control measure has been rather successful in the case of pears, but has not yet been satisfactory for apples. Westcott (1950) lists the variety Richard Peters as immune, and Orient, Hood, and Pineapple as highly resistant. Some varieties of apple are classed as more or less resistant, e.g., Jonadel (Brooks and Olmo, 1958). When hundreds of tree seedlings are required to be tested for resistance, the mechanics of the operation become formidable. Dunegan, Moon, and Wilson (1953) developed a rapid method for mass inoculating pear trees with an aqueous suspension of *E. amylovora* by power sprayer. The extreme virulence of the pathogen is illustrated by the fact that in two years only 200 trees had survived out of 600 under test. Working with seedling crosses of *Pyrus* pears, Carpenter and Shay (1953) found no appreciable resistance in seedlings smaller than the 8-leaf stage. Resistance increased with age, but there was a steady elimination of susceptibles. Crosse *et al.* (1958) found that Bartlett pear was virtually unaffected in the field although showing little resistance in inoculation experiments. This is an example of "field" as opposed to "true" resistance.

Raspberry Blight (Erwinia amylovora f. rubi f. nov. *Folsom*)

First noted by Folsom in Maine in 1947, later by Starr *et al.* (1951), and again by Folsom in 1954, raspberry blight resembles fire blight of apple and pear. The disease has also been reported from North Carolina. Up to 1951 all young shoots showing girdling by the raspberry cane maggots (*Phorbia rubivora* Coq., *Hylemyia rubivora* Coq., and *Pegomya rubivora* Coq.) were removed from one plantation and incubated in a humid atmosphere in the laboratory. Incubated plants showed purpling, purpling plus milky-colored bacterial drops, or no symptoms at all. In the field where no roguing was done, the maggot-injured plants not only showed symptoms but the nearby, rapidly growing canes were also infected.

Bacteriological examination revealed that the pathogen was identical with the fire blight organism except in its lack of pathogenicity to apples. Folsom therefore designated the isolate as *E. amylovora f. rubi f. nov.* Roguing in 1951 eliminated the disease on one plantation in 1952. The following year's infection was associated only with maggot injury occurring that year and not with the location of blight canes the previous year.

Starr *et al.* (1951) reported that the disease can start in inflorescences, or in leaves and stems. It would seem, therefore, that while the activities of the maggots perhaps provide the most important courts of entry of the disease, they are not the only possibility.

Walnut Blight (Xanthomonas juglandis (*Pierce*) Bergey et al.)

Walnut blight is a disease of English walnut, black walnut, butternut, and Siebold walnut. The symptoms are black, dead spots on young nuts, green shoots, and leaves. Nuts are infected early, fall prematurely; others mature with husk, shell, and kernel, although they are badly damaged.

The pathogen overwinters in mummified nuts but the method of dispersal is not established finally. The most likely agent is the walnut erinose mite (*Eriophyes tristriatus* Nalepa var. *erinea* Nalepa). Rudolph (1943), reporting experimental work, found this mite hibernating beneath the scales of a healthy dormant bud of an English walnut tree. From one of the mites, transferred to beef agar, the blight organism developed along with other bacteria in a mixed colony. The identity of the blight organism isolated from the mite was established by needle puncture into stems, young shoots, and nuts. Suspensions of the bacteria also produced numerous lesions when atomized on unwounded nuts.

Since bacterial slime oozes from blight lesions, it may well be that transmission also occurs by the agency of casual visitors in the same manner as seems to be established in the case of fire blight. The fact that the erinose mite is an obligate parasite of the English walnut would seem to indict that animal as guilty by association, even though, on Rudolph's own evidence, only a very small percentage of the mites inhabiting the tree may be infected. The high humidity of the microenvironment normally inhabited by these mites would seem to be ideal for the development of the pathogen, and, by the same token, ideal for the efficiency of the mites' transmission of the organism. Michelbacher (1954) reported on the association of walnut blight with codling moth infestation. The first generation larvae infest the nuts at the blossom end and aid in inoculating the tree with the blight organism. The second brood of caterpillars appears to favor damaged nuts, but probably only because the caterpillars can enter them more easily.

According to Ark (1944), diseased catkins are an important reservoir of infection. The catkins are highly susceptible organs present throughout the year, and, when infected, can provide a constant source of waterborne infection. The pollen from these catkins carries viable blight bacteria with the obvious implication that many casual insect visitors to the catkins may be involved in the spread of the disease.

Rudolph (1946) found that penicillin and streptomycin inhibited and killed the pathogen *in vitro*. However, when injected into the trees at a concentration of 1 ppm they did not control the disease, presumably because of dilution by the sap of the tree. Aspitarte, Bollen, and Miller (1953) tested some of the newer antibiotics against the organism *in vitro*. Aureomycin, streptomycin, terramycin, crystal violet, and puratized spray gave various degrees of inhibition, whereas penicillin, bacitracin, choloromycetin, sulfanilamide, and Actidione were ineffective in inhibiting the development of the pathogen.

In 1953, Miller reported that walnut blight was widely distributed in Oregon in 1951 and 1952, causing up to 35% loss in unsprayed areas. He recommended Bordeaux mixture, ammoniacal copper carbonate (Salcopper), and copper oxide (yellow cuprocide) as pre- and post-blossom sprays to control blight, with more frequent applications in seasons of heavy rainfall during blossoming. However, in 1959 Miller modified his recommendation to include the possible use of Agrimycin-Cu dust at 10-day intervals starting at early prebloom stage. Cu+lime+S+oil+dust (15–30–10–1.5) was also effective and was used with only a trace of injury.

Bacterial Heart Rot of Celery (Erwinia carotovora (*Jones*) Bergey et al.)

Bacterial heart rot of celery is of especial interest because of the evidence that environmental conditions govern the activity of insects that open courts of entry. Leach (1940) described how the insects, two species of leafminers, *Scaptomyza gramineum* Fall and *Elachiptera costata* Loew, are influenced by moisture and are therefore influenced in their role in transmission of the pathogen. The insects normally deposit their eggs where the relative humidity is high and in periods of wet weather they usually deposit them on the outer, older leaves. These leaves are normally stripped off at harvest so that loss is light from infection of outer leaves. In periods of dry weather, however, the flies lay their eggs in the hearts of the plants where the humidity is high at the feeding points of the larvae; the tender heart leaves are very susceptible and the terminal bud is killed, resulting in a complete loss. No detailed studies of the relationship between the pathogen and the insect were available to Leach, and there have apparently been none since.

Halo Blight of Beans (Pseudomonas medicaginis *var.* phaseolicola)

Halo blight of beans is caused by *Pseudomonas phaseolicola,* originally termed *Phytomonas medicaginis* var. *phaseolicola* Burkholder. The accidental and mechanical transmission of bacteriosis (halo blight) of beans by thrips in the greenhouse was referred to by Leach (1940). According to Buchanan (1932), the disease lesions were always associated with feeding punctures of *Heliothrips femoralis* Reut. Although this thrips species has not been associated with the field incidence of the disease, the importance of the disease in areas of high rainfall and humidity and the suitability of such climatic conditions for *H. femoralis* would indicate the desirability of further study. Currently, control measures are standard; the use of seed from blight-free areas, preferably from dry climates, and the avoidance of rotations which include susceptible crops. The use of antibiotics (Mitchell, Zaumeyer, and Preston, 1954) was not promising, even though some reduction in severity of the disease was obtained when treatments were made early in the infection stage.

Pod Twist of French Beans (Pseudomonas flectens *Johnson*)

Although no evidence obtains for field transmission by thrips of halo blight of beans, just discussed, it would seem that thrips feeding, involving as it does the maceration of epidermal tissue, should provide an efficient court of entry for bacterial pathogens.

Pod twist, recorded from Queensland in 1951, has been described by Johnson (1956) as being caused by a species of *Pseudomonas* (*P. flectens*). In this case, the evidence for association with thrips was obtained from field observations. Wild *Phaseolus lathyroides* L was found to be naturally infected, and the infections were widespread. When a bean crop growing near infected *P. lathyroides* itself became infected, attention was drawn to the presence of large thrips populations on both the wild and

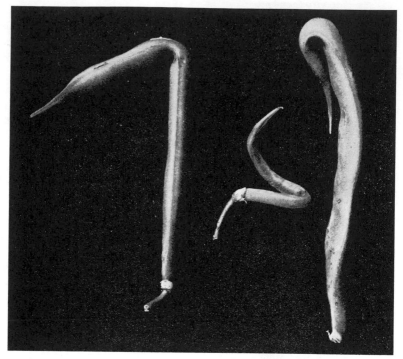

FIGURE 2.5. Pod twist of bean. Natural infection showing twisting about the infected area (Photograph: Johnson, 1956).

the cultivated hosts. *Taeniothrips nigricornis* (Schmutz) was found to be commonly present on both hosts, and transmission tests using the thrips as the vector were positive. *P. lathyroides* appears to be an important reservoir for the pathogen, providing a continuous source of infection throughout the summer.

The symptoms, as described by Johnson, are quite striking. Young pods less than 1 ½ inches long, from which the blossom has just fallen, are most likely to show infection. Many wither and drop off while others continue to enlarge in spite of considerable affected surface. From this point, the characteristic bending and curling of the pods occurs (Fig. 2.5), because the infection is usually confined to one side of the pod, extending lengthwise but not to the dorsal structure of the pod. Apparently there is some natural inhibitory process involved, because after a short time within the host the bacteria cease to be active, and the invaded area does not become necrotic but takes on a darker green appearance. As is typical of so many bacterial diseases, however, the water-soaked invaded area may exude transparent white droplets which dry to leave a shiny encrustation over the surface.

FIGURE 2.6. Pod twist of bean. Symptoms produced by confining infective thrips to bean plants during the flowering period. Note twisted beans and the completely blighted uppermost truss (Photograph: Johnson, 1956).

The thrips concerned in transmission, *T. nigricornis,* produces a lumpiness and distortion of the bean pod as a result of its feeding, but this symptom can be clearly differentiated from those induced when the insect is a vector and transmits the bacterial disease (Fig. 2.6).

Ring-Rot Disease of Potatoes (Corynebacterium sepedonicum (Spieckermann and Kotthoff) Skaptason and Burkholder)

Ring-rot disease of potatoes was reported by Brentzel and Munro (1940) as having been present in North Dakota since 1930; it was found in Maine in 1932 and a year later in eastern Canada. By 1940 it was found throughout the potato-growing states. Belova (1952) reported a severe outbreak of hollow rot and ring-rot in Russia from 1942 to 1944. Tomato and eggplant are also susceptible and the pathogen is seed-transmitted (Larson, 1944). The disease has also been reported in Greece (Sarejanni *et al.,* 1956), in Sweden (Granhall, 1956), and in Pakistan (Map 20, Commonwealth Mycological Institute).

Rapid spread of a disease of this type indicates that insects could be carriers, and this case is no exception. Brentzel and Munro observed that grasshoppers were abundant in the northern prairies of the United States, but experiments with four species of *Melanoplus* grasshoppers failed to give evidence of transmittal. The authors concluded that other factors such as infected seed, contaminated implements, and storage conditions were responsible for the spread of the disease, rather than insects.

However, List and Kreutzer (1942) were able to transmit ring-rot disease to healthy potato and tomato plants by using a grasshopper (*Melanoplus differentialis* Thomas), the black blister beetle (*Epicauta pennsylvanica* (De Geer)), and the potato beetle (*Leptinotarsa decemlineata* (Say)), which fed on the diseased plants and tubers. They concluded, while the evidence could not be taken as proof, that insects are responsible for rapid spread of ring-rot disease in the field, the insect relationship being clearly indicated. Differences in the results by these two groups of workers can only be accounted for by some local environmental condition restricting the development of the disease, on the one hand, or by more effective techniques, on the other.

Symptoms appear only when the plant is nearly full grown, with one or two stems wilting and with chlorosis of the lower leaves. Tuber infections start at the stem end with a characteristic brown vascular ring. When the stem is cut, a cream mass exudes.

Immunity is very rare, but Baribeau (1948) reports that the French variety Furore has never shown any trace of infection, and the variety Teton appears to be highly resistant. There is evidence for genetic resistance. Bonde *et al.* (1959) described a method for testing seedling progenies and reported resistance up to 50%. The variety Merrimack is highly resistant (Anon., 1959).

There are some practices, besides potato-seed certification, that have reduced infection. Bins, crates, and barrels can be treated with copper sulfate, and machines can be disinfected with formaldehyde. MacLachlan (1958) found the quaternary ammonium compounds superior to other disinfectants of equipment. Semesan bel (1 lb per 10 gal) and mercuric chloride (1:1000 or 1:500) were best for disinfection of cutting knives. Starr and Cinnamon (1953) found that nuclear radiation was the only effective use of radiation to sterilize burlap bags; ultraviolet and high- and low-voltage rays were unsuccessful. MacLachlan (1959) used electronic heating. Terramycin showed promise, being absorbed by the roots of actively growing plants (MacLachlan and Sutton, 1957). Copper and boron increased resistance and increased yields (Malenev, 1956).

Bacterial Rot of Apples (**Pseudomonas melophthora Allen and Riker**)

The apple maggot is probably one of the most familiar of insects in the larval stage, and the rot associated with its feeding in apples was presumed to be quite secondary until Allen (1931) described the pathogenic organism involved and related its activity to the activity of the apple maggot, suggesting that it could be carried by the adult flies (*Rhagoletis pomonella* Walsh) (Fig. 2.7 and 2.8). In a later paper, Allen and Riker (1932) reported that two stages of the insect appeared to assist in the dissemination of the bacteria: the adult fly, by depositing contaminated eggs beneath the cuticle of the apple; and the larvae, by aiding the movement through the fruit by their burrowing habit. Isolations were made from the surface of adult flies and from the adult alimentary canal, from larval burrows, and from the surfaces of young larvae.

The pathogen was described as *Phytomonas* (reclassified as *Pseudomonas*) *melophthora* by Allen, Pinckard, and Riker (1943), who expanded their studies with a detailed investigation of the association of the organism with the various stages of the insect. Oviposition punctures showed a high percentage of infection, and egg-surface washings were positive as were egg shells from which the larvae had emerged. There was no evidence of internal transmission through puparia, but overwintering of the organism on the surface of puparia is likely.

Although Leach (1940) called attention to the anatomical similarities between *R. pomonella* and *Dacus oleae* Rossi as revealed by a comparison of the work of Petri (1909, 1910) and Dean (1933, 1935), no further study seems to have been made which might indicate a bacterial symbiosis such as Petri described for *D. oleae*. However, the fact that the pathogen is transmitted exclusively, as far as is known, by the apple maggot, lends some support to a bacterial symbiosis hypothesis. Control of bacterial rot of apples is aided by heavy insecticide applications during

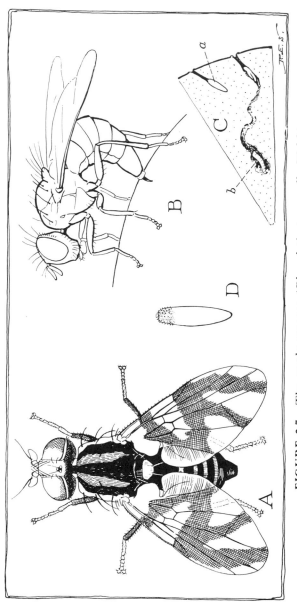

FIGURE 2.7. The apple maggot (*Rhagoletis pomonella* Walsh). A: adult female; B: female fly puncturing the skin of an apple preparatory to depositing an egg; C: section of apple showing an egg inserted at *a*, and a young maggot tunneling into the pulp at *b*; D: an egg (greatly enlarged) (Drawing after Snodgrass).

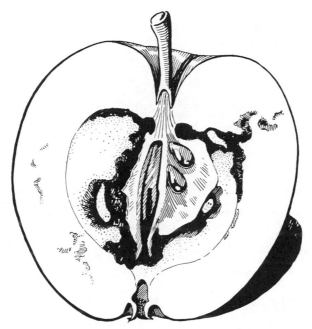

FIGURE 2.8. An infested apple cut open to show the tunnels of the apple maggot through the flesh (Drawing after Snodgrass).

July to destroy the adult flies before they lay their eggs. Dean (1951, 1954) reported variable results with methoxychlor. In 1954, a formulation of Perthane (diethyldiphenyl dichlorethane) gave the best control. Garman (1954) obtained better control with a methoxychlor-TDE combination than with DDT-parathion or lead arsenate.

Bacterial Necrosis of Giant Cactus (Erwinia carnegieana Lightle et al.)

The rapid increase of necrosis of giant cactus (*Carnegiea giganta* Britt and Rose and *Cereus giganteus* Englem.) in the United States, principally in Arizona, was first noted and described by Lightle, Standring, and Brown (1942). The initial symptom is a small, circular, light-colored spot, usually with water-soaked margins. Later, the underlying parenchymatous tissue rots and a brown liquid exudes. The rotted tissues dry and fall to the ground, leaving the woody stelar strands bare (Fig. 2.9). The pathogen was recovered from the surfaces of the eggs and larvae, from the intestinal tract of the larvae, and from the adults.

Two pathogens, *Bacterium cactivorum* Pasin et Buzz.-Tr. and *Bacillus*

(a)

(b)

FIGURE 2.9. Bacterial rot of giant cactus. (a) Three branches showing necrosis; in background, black lesions at junction of lower branch. (b) Foreground, terminal; in background, three branches infected and drooping (Photographs: U.S. Department of Agriculture).

cacticidus Johnson and Hitchcock, can cause rots of cacti belonging to other genera. The isolate from the giant cactus did not resemble either of the two and was named by the authors *Erwinia carnegieana*.

Boyle (1949) confirmed that the disease was transmitted to healthy plants through closely associated root systems and through root wounds, but from the evidence she also concluded that the rapid spread of the disease was largely due to insect transmission. The insect vector is *Cactobrosis fernaldialis* (Hulst.) and has one generation a year. The larval period lasts from November to the following April and is spent tunneling in the giant cactus. The pupal period lasts from 28 to 33 days. The adult, a nocturnal moth, is short-lived, laying its eggs and dying within three days.

It is appropriate to note here that *B. cacticidus* is associated with the collapse of prickly pear following infestation with *Cactoblastis cactorum* Berg. and the larvae of longicorn beetles (*Moneilema* spp.) in Australia. The insects are not only agents of ingression by the bacteria, but also aid in dissemination through tissues other than the parenchyma to which the bacteria are normally restricted.

Bacterial Wilt of Bluggoe Banana (Pseudomonas solanacearum E.F.Sm.

An insect-spread epiphytotic of bacterial wilt of Bluggoe banana occurred in Honduras in 1962. Infection occurs when bacterial ooze is carried by insects (5% of 700 bees and wasps were found to be infected) from infected peduncles to moist cushions on the peduncles of healthy plants from which the male flowers had recently abscissed. Infection does not occur if the bud is broken off before the first cushion of a male flower is exposed. The distance between small patches of Bluggoe banana is a factor limiting local spread, but there are instances known where the disease has been insect-borne over a distance of 60 miles from a small disease source which had been known for five years previously. The strain of *Pseudomonas solanacearum* responsible can be distinguished morphologically from other strains affecting bananas and causing moko disease, so the Bluggoe-infecting strain can be easily traced (Buddenhagen and Elasser, 1962).

Bacterial Disease of Sugarcane (Xanthomonas sp.)

White flies (Aleyrodidae) are unusual vectors of pathogenic bacteria but two cases are recorded: a bacterial disease of sugarcane caused by a species of *Xanthomonas* is transmitted by *Aleurobolus borodensis* M., and

Trialeurodes vaporariorum (West.) transmits (*X. pelargonii* to *Pelargonium hortorum*. The organisms were recovered from insects which had fed on infected plants (Bugbee, 1962). With both species of white fly involved with the same genus of bacteria, there is a suggestion here of biological factors other than surface contaminants.

Salmonella montevideo *in Wheat*

The rice weevil (*Sitophilus oryzae* (L), after being in wheat which was contaminated with (*S. montevideo* for seven days, proved to be carrying the organism both internally and externally for one week. If kept in the contaminated wheat for 14 or 21 days, the bacteria were retained in the insect for five weeks and contamination of the clean wheat by the weevils was greater when the exposure was for seven days (Husted *et al.,* 1969)

Citrus Canker (Xanthomonas citri *(Hasse) Dowson*) *and Leafminers*

Spread of citrus canker is aided by leafminer attack, which in India is often widespread. The canker attacks the insect-affected leaves more freely than it does the healthy ones (Sohi and Sandhu, 1968).

SPECIALIZED TRANSMISSION OF BACTERIAL PATHOGENS

Under this heading are to be considered those bacterial diseases of plants for which a specialized relationship has developed between the causative pathogen and the transmitting insect. In one of these, bacterial wilt of cucurbits, the specialization appears to be as much between insect and host plant as between insect and pathogen. In both the cucurbit disease and bacterial wilt of corn (Stewart's disease), the specialization appears to lie in the fact that the pathogen overwinters in the insect's body. Actually, this renders the insects ideal reservoirs and disseminators of the pathogen but does not involve any high degree of specialization. The species of insect concerned—all Coleoptera—overwinter as adults, and if the last plants on which they feed before low temperatures in the fall overtake them are diseased, the pathogen will remain in the intestinal tract during the insect's hibernation. In the spring, when the insect resumes activity, the pathogen is viable. The relationship between insect and pathogen in the transmission of potato blackleg, soft rot of crucifers and olive knot appears to be of a different order, with the last named the most highly developed.

Bacterial Wilt of Cucurbits (Erwinia tracheiphila (*Erw. Smith*) *Winslow* et al.)

This strictly vascular parasite causes a serious disease of cucumber and muskmelon. It is serious in the middle west, and in the north central and Northeastern United States, but it is rare in the southern and western states. The disease is also known in Europe, South Africa, and Japan. Other cucurbits, such as pumpkin and squash, are susceptible to infection, but they are rarely damaged severely.

The disease is entirely dependent on the two insect vectors—the striped cucumber beetle (*Diabrotica vittata* Fabr.) and the twelve-spotted cucumber beetle (*D. duodecimpunctata* Oliv.)—for inoculation, dissemination, and overwintering of the pathogen. Plants become infected after the overwintering beetles, infested with the bacteria, feed on them. Only a small proportion of the overwintering beetles harbor the organism, but these are sufficient to establish centers for secondary spread. As the adult beetles feed, they introduce the bacteria into the tissues, which multiply and spread through the vascular system of the plant.

The first symptom of infection is the wilting of a single leaf, which remains green. Eventually all the leaves wilt and the plant dies. A stringy white viscid ooze emerges from freshly cut stems. According to Leach (1940), this last symptom is the best for diagnostic purposes.

Rand and Cash (1920a) demonstrated experimentally that feces from infective beetles may contain viable bacteria and serve as inoculum, but only if dropped into fresh wounds. Dew and rain facilitated this secondary spread. Gould (1944) reported that the striped cucumber beetle as found in Indiana has two overlapping generations. The adults are polyphagous (61 plants belong to 20 families). The overwintered adults fed on plants belonging to the Rosaceae and on wild cucumber in the spring, but later migrated to cucurbit seedlings. This period of feeding before the cucumbers are attacked makes it unnecessary to assume that the mass of inoculum in the intestines of the overwintering beetles must be considerable in order to survive the cleansing effect of feedings after emergence from hibernation. A more likely occurrence is that the wild cucumbers and perhaps other hosts are first infected, so that when the cultivated cucurbits are attacked, the insects are by then carrying fresh inoculum in their intestines.

There are no commercial resistant varieties known, and the control of cucurbit wilt is directed chiefly at the insects. Sprays and dusts of calcium arsenate, cryolite, and, more recently, the chlorinated hydrocarbons have been used (Fronk and Peterson, 1954).

Bacterial Wilt and Stewart's Leaf Blight of Corn (Xanthomonas stewartii (Erw. Smith) Bergey et al.)

The earlier literature referred only to bacterial wilt of corn, used synonymously with the term Stewart's disease after the original describer who first observed it on Long Island in 1895. The disease has also been reported from Russia (Cheremisinov, 1956) and from Poland (Rojecka, 1957). In the most recent comprehensive treatment, however, Robert (1955) separates them under the two distinct names, on the basis of symptomology and development in sweet and dent corn; in other words, as varietal symptom expressions of infection by a single species of pathogen.

The two combined are among the most important diseases with which the corn grower has to contend. Bacterial wilt is most severe on sweet corn, particularly the early yellow varieties, but will also attack dent, flint, flour, and popcorn varieties (Fig. 2.10 and 2.11). Stewart's leaf blight is primarily found in dent corn and dent corn hybrids, although

FIGURE 2.10. Rows of sweet corn showing varieties susceptible (*left*) and resistant (*right*) to bacterial wilt (Photograph: U.S. Department of Agriculture, Robert, 1955).

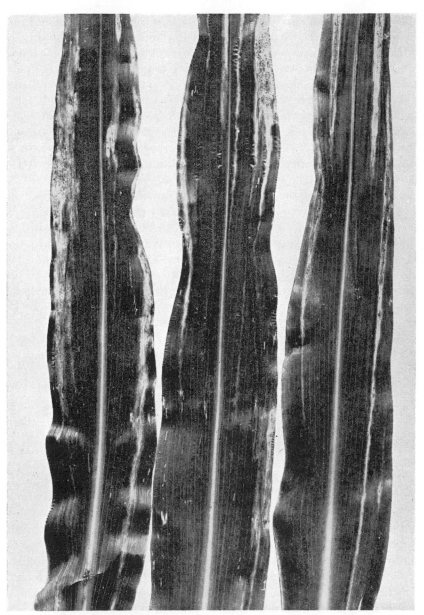

FIGURE 2.11. Susceptible Golden Bantam sweet corn leaves showing streaking of leaves caused by bacterial wilt (Photograph: U.S. Department of Agriculture, Robert, 1955).

not all are equally susceptible. Some sweet corn hybrids, resistant to early infection with bacterial wilt, are susceptible to Stewart's leaf blight.

The pathogen is a vascular parasite which may multiply when inoculated into a leaf, filling the water-conducting tissues of the veins and causing dead streaks in the leaf. When these streaks are extensive, the result is known as Stewart's leaf blight. If the bacteria spread into the stalk, clogging the water-conducting vessels, the whole plant is deprived of water, and bacterial wilt results. Yellow bacterial ooze from cut stems is diagnostic for bacterial wilt, but since infection rarely spreads from the leaves in Stewart's leaf blight infection, cut stalks will not generally show bacterial oozing.

Both expressions of the disease have the same relationship with their insect vectors. Seed infection can occur, but only 2% of the plants develop bacterial wilt when infected seed is planted. According to earlier authors (Rand and Cash, 1920), the development of the pathogen in germinating seed is conditioned by injury by the southern corn root-worm (*Diabrotica undecimpunctata howardi* Baker), the northern corn root-worm (*D. longicornis* (Say)), the western corn root-worm (*D. virgifera* Lec.), and other agents such as the larvae of white grubs (*Phyllophaga* spp.), since the pathogen is unable to invade the embryo unless such injury occurs.

The disease organism is chiefly disseminated by the feeding activities of the corn flea beetle (*Chaetocneme pulicara* Melsh.) and the toothed flea beetle (*C. denticulata* Ill.). Elliott and Poos (1934) demonstrated that the pathogen overwinters in the intestinal tract of the flea beetle, and infection follows the beetle's feeding in the spring.

As has been suggested in the case of bacterial wilt of cucurbits, wild host plants could well bridge the gap between the insects' emergence from hibernation and the appearance of freshly growing corn. Elliott (1935) reports natural infection with *P. stewartii* on teosinte (*Euchlaena mexicana* Schrad.), and Poos (1939) extended possibilities along this line when he was able to show that the pathogen was present in Kentucky blue grass, orchard grass, and wheat in the absence of wilt symptoms, although symptoms do appear on Job's tears (*Coix lacrymijobi* L) and yellow bristle grass (*Setaria lutescens* (Weigel) F. T. Hubb).

Strains of *P. stewartii*, of different virulence, are known (McNew, 1940; McNew and Braun, 1940; Braun and McNew, 1940). It should be a fruitful field for investigation to determine the effect on virulence of overwintering in the intestinal tract of the insect and of passage through wild host plants, since this might well account for some of the variables noted in field incidence and severity of the disease in corn.

Poos (1945) obtained good control of bacterial wilt of sweet corn by applying DDT on the seedlings to protect them against flea beetles. Seed treatment is ineffective. The use of resistant varieties or hybrids, however, is recognized as the only practicable method of controlling bacterial wilt, but the annual forecasts of disease prospects issued by the Illinois Natural History Survey, Urbana, are of great value in determining the necessity for their use in any particular year. In the early 1930's the early yellow varieties of sweet corn were popular, but susceptibility to bacterial wilt led to their abandonment in favor of the resistant white varieties. Yellow sweet corn hybrids were developed, however, particularly for the canning industry. The first to be released for general use—Golden Cross Bantam—was available in 1932 and 1933, both epidemic years. The work of the plant breeder continued so that by 1954 the situation was well in hand as far as sweet corn was concerned.

Dent corn hybrids show differences in susceptibility to Stewart's leaf blight although specific breeding for resistance to the disease has not received much attention. There are, however, a number of moderately resistant hybrids that are being grown successfully throughout the United States.

Potato Blackleg (Erwinia carotovora (*Jones*) *Winslow* et al.)

The organism causing potato blackleg is the soft-rot pathogen which is found in most cultivated soils. Leach's summary (1940) includes discussion of all the important research, mostly performed by Leach himself, on this major disease of the potato.

Although the bacteria survive in soil, their entry into the potato seed-piece is difficult without insect intervention if conditions are suitable for the formation of a cork layer on the cut surface of the potato. The microenvironment is important. In wet soils with a low oxygen supply, cork formation is poor but the organism, being a facultative anaerobe, can grow well. Wet, heavy, poorly drained soils are therefore associated with the greater prevalence of blackleg. The seedcorn maggot, *Hylemyia cilicrura* (Rond.), can, however, be a direct agent of entry for the organism. The adults lay their eggs in the soil in the spring. The emerging larvae are possessed of the mouth hooks typical of dipterous larvae with which they can penetrate the cut surface of the potato. This presents an ideal court of entry for the bacteria which may have contaminated the surface of the egg when it was deposited, or which may have been picked up by the larvae from the soil or from the contaminated surfaces of the seed-pieces (Fig. 2.12).

Leach has developed the concept of a mutualistic symbiosis between *E. carotovora* and the seed-corn maggot (1931, 1933). He demonstrated that the intestinal tracts of both adult flies and larvae contain several species

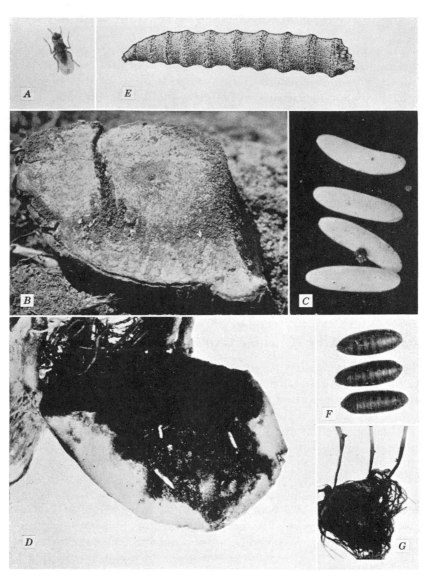

FIGURE 2.12. Stages of the life cycle of the seed-corn maggot (*Hyelmyia cilicrura* Rond.). A: adult female fly, approximately x1½; B: eggs of the seed-corn maggot deposited on the surface of a seed-piece planted in the soil; C: four eggs, approximately x 50; D: seed-piece showing three maggots and type of injury caused by their feeding; E: larva greatly enlarged; F: three puparia, approximately x 3; G: blackleg plant resulting from seed-piece infested with maggots in the laboratory (Photographs: A, B, C, and F: Leach, 1940; D, E, and G: Bonde, 1950).

of physiologically similar bacteria among which *E. carotovora* is frequently found. These bacteria can survive pupation, which means that adult flies, on emergence, can contaminate eggs as they are laid.

According to Leach, survival during pupation, which would seem to be the most significant part of the relationship between the insect and the bacteria, occurs in three different locations in the puparium—in the cast-out linings of the fore and hind intestines and in the lumen of the mid-intestine of the pupa (Fig. 2.13). The first two mentioned are, of course, defunct larval tissues with no organic connection with the pupa, and the survival of bacteria on these tissues must be entirely saprophytic. However, even in the lumen of the mid-intestine, which is an area commonly inhabited by bacteria, only a nonpathogenic form has been found in the few cases where specific identity has been established.

Earlier, Leach (1926, 1931) had indicated that bacteria contributed to the nutritional requirements of the larvae by breaking down the starch of the seed-piece, but there is no evidence in Leach's summary to indicate that the nonpathogenic bacteria surviving in the pupae of the seed-corn maggot are an essential link in the chain. Actually, the opportunities for contamination of the surface of the seed-piece in the soil by bacteria and the entry of these following the scraping action of the newly hatched larvae must be very great.

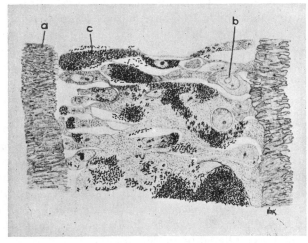

FIGURE 2.13A. A camera-lucida sketch from a stained slide showing the contents of the mid-intestine of a pupa of the seed-corn maggot: *a*, the embryonic cells of the new mid-intestine wall; *b*, the epithelial cells of the old larval mid-intestine being sloughed; *c*, living bacterial cells, approximately x 250 (After Leach, 1940).

FIGURE 2.13B. A camera-lucida sketch from a stained slide showing a section of the hind-intestine of a freshly emerged imago of the seed-corn maggot: *a,* wall of hind-intestine; *b,* residue of disintegrated epithelial cells. The bacteria which have survived through metamorphosis are multiplying on the disintegrated epithelial cells. Approximately x 250 (After Leach, 1940).

According to Bonde (1930), *Hylemyia trichodactyla* Rond appears to function as a disseminator in the same manner as does *H. cilicrura* (Fig. 2.14). In the United States, the control of blackleg and seed-piece decay are serious problems in potato production. The sporadic occurrence of blackleg raises the question of source of infection. Efforts are made by growers to get seed from disease-free areas, but there are many other factors involved. Bonde (1950) stressed the necessity for allowing the seed-pieces to suberize the cut surfaces before planting. Storing the cut seed-pieces at 78–80% relative humidity favors healing of cut surfaces, which reduces the attractiveness of the potatoes to insects. Delaying planting time until maggots of the first generation enter pupation also reduces infection, but planting time for the farmer is established by many factors of local weather and soil conditions and can rarely be altered practically, except in a very narrow range. Furthermore, such a recommendation requires a close watch on fly generations, which involves more quantitative surveys than can normally be offered by entomologists.

Bonde (1955) experimented with dips of Agrimycin as well as streptomycin nitrate dips. Both reduced the amount of seed-piece decay and blackleg, and increased crop yield. Crossan (1959) used Captan 50W to protect seed-pieces.

Leach (1940) considers that the soft-rot of crucifers associated with the

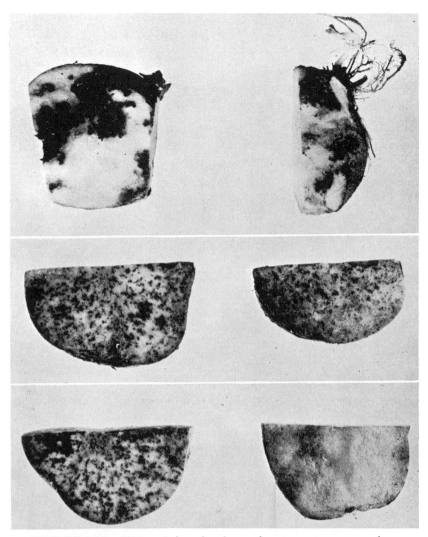

FIGURE 2.14. *Upper:* Injury by the seed-corn maggot on unsuberized seed-pieces planted in soil from South Carolina containing pathogenic bacteria. *Lower and Middle:* Primary bacterial lesions on freshly cut seed-pieces planted in nonsterilized soil. The young seed-corn or potato maggots enter through such lesions and may destroy the seed-pieces. Healthy seed-piece on lower right was suberized and was not attacked by soil bacteria and the maggots (Photographs: Maine Agricultural Experimental Station, Bonde, 1950).

feeding of the cabbage maggot, *Hylemyia brassicae* (Bouché), is in the same general category as potato blackleg, with similar symbiotic relationships between the pathogen and the insect vectors.

Soft Rot of Rutabaga Erwinia amylovora (*Burrill*) Winslow et al.

Erwinia amylovora persists for at least 72 hours in the apple aphid (*A. pomi* DeGeer) when the latter is fed through a membrane on a synthetic medium in which the bacteria are growing. Other *Erwinia* spp. found in the eggs, pupae, and adult flies of the cabbage maggot fill the same role in the transmission of soft rot of rutabaga (swede turnip) as the seed-corn maggot does in the case of potato blackleg (Deane and Chapman, 1964). Included in this same general category is bacterial soft rot of carrots and potatoes. There is a relationship between *Drosophila buskeii* (Coquillet). The organisms have been isolated from the surface of the insect's eggs and from larvae and pupae after surface sterilization (Tamimi and Banfield, 1969).

Bacterial Infection of Wheat and Maize (Pseudomonas *spp.*)

Symbiotic bacteria of the genus *Pseudomonas* in two species of chloropid flies, *Oscinella frit* (L) and *O. pusilla* Meig., were found in numerous organs of the insects, even in the haemolymph. During larval feeding the bacteria pass into the mid-gut in a continuous flow; they can also be found in the sloughed cuticle and on body surfaces. Pure cultures of the organisms can be prepared from the salivary glands, and when inoculated into growing points of maize and wheat, local infections similar to those produced by the feeding of infective *Oscinella* larvae (Ryzkova, 1962) result. Secondary smut infection in maize is also favored by larval feeding (Shapiro, 1963).

The Water-Marked Disease of Willows (Erwinia salicis (*Day*) Chester)

The water-marked disease of willows is a serious problem in the cricket-bat industry in East Anglia, England. The pathogen as determined by Dowson (1937) is *Bacterium* (reclassified as *Erwinia*) *salicis* (Day) Chester. Diseased trees under three years old are rarely found in nature. Metcalf (1940, 1941) described three distinct regions occurring in a diseased branch, viz., "a narrow black outer zone extending irregularly around the annual ring; a red-brown waterlogged area between this and the initial parenchyma; and a dark brownish-green sodden inner zone involving the whole of the wood with the pith." The bacterial population of the water-marked wood changes with time, *E. salicis* being replaced after one year by three associated organisms which are simply designated as A, B, and C.

Gray (1940) studied the relationship of the willow wood wasp *Xyphydria prolongata* (Geoffr.) to the disease. He was able to isolate an organism from the ovum, larva, pupa, and imago which suggested a hereditary transmission similar to that described for potato blackleg. The organism, however, was not the actual pathogen but a closely allied form associated with it in diseased wood. Since Metcalf reported that infection of the outer elements of the new annual ring takes place in late summer by way of holes bored by the insect larvae, and that the complete symptom picture is the result of the activities not only of *E. salicis* but of the associated organisms, it suggests that the isolation of a primary pathogen from an insect is not a *sine qua non* for the establishment of an insect-disease association.

Olive-Knot Disease (Pseudomonas savastonoi (Erw. Smith) Bergey et al.)

This is an example of the third of Wescott's categories, where infection results in hyperplasia or gall formation (Fig. 2.15). Olive-Knot disease is prevalent in the olive-growing areas of the Mediterranean countries and has been known in California since 1898 (Wilson, 1935). A variety of the fungus causes a similar disease of ash in Russia (Oganova, 1957). In California, where the olive fly, *Dacus oleae* Rossi, is not known, rain water is considered the most important agency of dissemination, infection being through courts of entry provided by wounds due to freezing and through frost cracks and leaf scars. In Russia a bark beetle, *Leperisinus fraxini* Panz., is implicated; in Italy the organism was isolated from minute tumors at the end of a gallery of an undetermined species of leafminer (Martelli, 1958). However, in the Mediterranean region, *D. oleae* was implicated through the studies of Petri (1909, 1910), who described a highly developed symbiotic relationship between the insect and the bacterium.

The intestinal tract of the fly is constantly contaminated with bacteria, *P. savastonoi* usually being present (Fig. 2.16 and 2.17). Some of these bacteria are passed on with the feces. The loci of the bacterial infestations are in sac-like evaginations at the junction of the anal tract and the vagina, which have a common opening to the exterior. The bacteria are smeared over the eggs as they pass the opening of the evaginations. The bacteria find their way through the micropyle of the egg and are found

FIGURE 2.15. Olive knot. Galls arising on young terminals (*left*) and a heavily infected branch (*right*) (Photographs: Plant Pathology Department, University of California).

in the body of the developing larvae. The larvae are therefore internally contaminated before they emerge from the egg. The bacteria established in the ceca of the intestinal tract survive pupation, so that when the fly emerges as an adult, it is always internally contaminated. In addition to the intestinal contamination of the egg, oviposition also results in the deposition of bacteria into the oviposition wound. According to Petri, this method of infection is responsible for much of the olive knot found in Italy.

Two significant differences can be noted between the case of *D. oleae* and the other specialized transmissions thus far noted. One is the entry of the bacteria into the egg through the micropyle, which distinguishes it from a casual surface contamination process, and the other is the development of the diverticulum of the oesophagus in the pupa. This diverticulum acts as a reservoir for the bacteria from which the intestinal tract of the adult fly becomes contaminated.

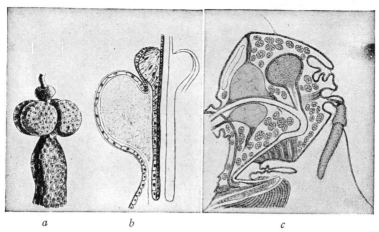

FIGURE 2.16. *a,* The anterior portion of the mid-intestine of the larva of *Dacus oleae,* in perspective, showing the caeca filled with bacteria; *b,* the same as seen in section; *c,* a longitudinal section of the head of an imago of *Dacus oleae* showing the bulbous diverticulum of the esophagus. During metamorphosis, the bacteria from the intestinal tract accumulate in this organ, from which the entire intestinal tract is later recontaminated (After Petri).

Control of the disease in California consists of removing the galls where and whenever possible with tools disinfected with 1/1000 mercuric chloride, spraying trees with Bordeaux mixture in November, December, and March in heavily infected areas, and avoiding the use of infected nursery stock.

Bottari and Spina (1953) described the seven varieties of olive trees grown in Sicily. These varieties, although highly resistant to knot disease, are susceptible to other diseases under certain environmental conditions. The *Oliarola messenese* (or *Castriciana*) variety was highly resistant to knot disease but was, however, subject to defoliation caused by *Cydoconium oleaginum* Cast, if grown under wet conditions. The *Moresca* (*Morghetana*) and *Cerasuola* were very susceptible to the olive-knot pathogen when grown in hilly localities subject to frost and hail, agreeing with the California experience in this regard, but otherwise highly resistant.

Marcovitch (1954) reported good control of the olive fly in Israel using DDT, methoxychlor, and dieldrin. Quite apart from the problem of olive knot, the olive fly is one of the most important insects affecting the economic well being of farmers in the Mediterranean region, and a concerted effort is being made toward the development of control measures.

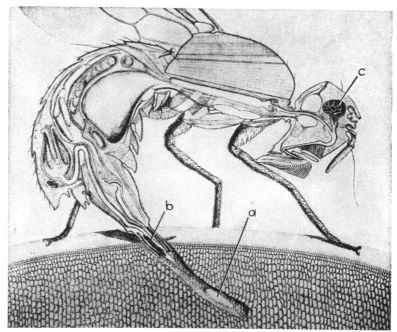

FIGURE 2.17. A diagrammatic sketch showing how the eggs of *Dacus oleae* are deposited and how the tissues of the plant are inoculated with bacteria: *a,* eggs; *b,* saclike evaginations of the anal tract filled with bacteria; *c,* diverticulum of esophagus filled with bacteria (After Petri).

SPECIALIZED VECTOR-PATHOGEN RELATIONSHIPS

Leach (1940) devoted a chapter to the subject of symbiosis between insects and microorganisms which skillfully summarized this field of inquiry up to 1940. Using Buchner's classification (1930) of symbiotic associations into two main divisions of ectosymbiosis and endosymbiosis, the latter was recognized as the more complex type, with four subdivisions:

1. The microsymbiote develops free in the lumen of the intestinal tract or in intestinal ceca.
2. The microsymbiote develops within the epithelial cells of the intestinal tract.
3. The microsymbiote develops in the region of the mesoderm but retains its connection with the epithelium.

4. The microsymbiote develops within special cells (mycetocytes) or special tissues (mycetomes) without any connection with the intestinal tract.

The pathogens responsible for bacterial wilt of cucurbits and bacterial wilt and Stewart's leaf blight of corn would presumably fall in subsection one of the above classification, if symbiosis is to be invoked to explain the insect-pathogen relationship. It is doubtful whether such is valid in either case. The vectors of both are biting insects and gross tissue feeders. It follows that any microorganisms present in the tissue on which they feed would enter and pass through the intestinal tract. This provides an excellent method of dissemination of the pathogens which might accidentally be included, but dissemination alone does not connote any symbiotic association. The passage of weed seeds through a sheep is a comparable phenomenon.

The cases of *E. carotovora* and the seed-corn maggot fall into a slightly different category. The survival of bacteria throughout pupation, and particularly the growth of bacterial colonies in the pupa before adult fly emergence, as Leach has shown, suggest a symbiotic relationship. The relationship cannot be considered a parasitic one because the tissues on which the bacteria develop are cast-off tissues. There is, however, no specificity associated with those pupation-surviving bacteria, and the actual pathogen is sometimes, but not always, found. So while it is true that bacteria are in apparent symbiotic relationship with the insect, there is no evidence that *E. carotovora* has any specific or even favored position. The relationship between *D. oleae* and symbiotic bacteria is set apart from those cases just discussed by the presence of an organ of the insect apparently specially adapted for the transmission of the bacteria from the adult, throught the egg, to each succeeding generation. Here again, however, the relationship between the insect and specific microorganisms does not appear to be established. The specialized organs do infer a more highly developed relationship than previously encountered, but in the absence of specificity, the insect-*pathogen* relationship would appear to be more coincidental than predetermined. Furthermore, the California experience shows clearly that neither *D. oleae* nor any other insect vector is essential to transmission of the organism causing the disease. In Italy, the infection is active before the seasonal activity of the insect develops. This casual or coincidental association between bacterial plant pathogens and insects is at most a primitive and not highly organized symbiosis. Buchner (1930) defines endosymbiosis as "a well regulated association between two partners of different kinds where one—usually the more highly organized partner—harbors the other, and where mutual adaptation is of so intimate a character that we are justified in supposing the arrangement to be useful to the host." But most of the examples from Buchner's monumental study are of cases where the relationship is so intimate that the symbionts are host-bound and specific. It is granted that all symbionts, however closely host-bound they now are, must have come originally from "outside," but until the plant-pathogen-insect symbiotic relationship becomes specifically fixed, then it must remain as a coincidental relationship.

Studies on the function of the symbionts have gone a long way since Leach called attention to the work of Koch (1930), showing among others that symbionts serve as a source of vitamins that the insects cannot synthesize. This has been completely confirmed, and many cases are known wherein the symbionts make the insect independent of vitamin-containing nourishment. Furthermore, symbionts are able to decompose metabolic end products such as uric acid. They can also produce amino acids in pure galactose culture media. Buchner (1965) has reviewed this progress in detail. But these phenomena occur in "closed" biological systems comprised of host-bound symbionts and their insect hosts. There is no pathogenicity involved.

REFERENCES

Allen, T. C., "Bacteria producing rot of apple in association with the apple maggot, *Rhagoletis pomonella,*" *Phytopathology,* **21**, 338 (1931).

Allen, T. C., and Riker, A. J., "A rot of apple fruit caused by *Phytomonas melophthora,* n. sp., following invasion by the apple maggot," *Phytopathology,* **22**, 557–571 (1932).

Allen, T. C., Pinckard, J. A., and Riker, A. J., "Frequent association of *Phytomonas melopthora* with various stages in the life cycle of the apple maggot, *Rhagoletis pomonella,*" **24**, 228–238 (1934).

Anon., "New vegetable varieties, List 5," *Proc. Am. Soc. Hori. Sci.,* **71**, 591–600 (1958).

Ark, P. A., "Pollen as a source of walnut bacterial blight infection," *Phytopathology,* **34**, 330–334 (1944).

Ark., P. A., "Use of streptomycin dust to control fire blight," *Plant Disease Reptr.,* **37**, 404–406 (1953).

Ark., P. A., and Thomas, H. E., "Persistence of *Erwinia amylovora* in certain insects," *Phytopathology,* **26**, 375–381 (1936).

Aspitarte, T. R., Bollen, W. B., and Miller, P. W., "Effects of some antibiotics and bactericides on growth of *Xanthomonas juglandis* and *X. corylina,*" *Plant Disease Reptr.,* **37**, 263–265 (1953).

Baribeau, B., "Bacterial ring rot of potatoes," *Am. Potato J.*, **25**, 71–82 (1948).

Belova, O. D., "Determination of infection of potato by hollow and ring rot," *Orchard & Garden*, **4**, 62–64 (1952).

Bonde, R., "Some conditions determining seed-piece-decay and blackleg induced by maggots," *Phytopathology*, **20**, 128 (1930).

Bonde, R., "Factors affecting potato blackleg and seed-piece decay," *Bull. Maine Agr. Exptl. Sta.*, 482 (1950).

Bonde, R., "Antibiotic treatment of seed potatoes in relation to seed-piece decay, blackleg, plant growth and yield rate," *Plant Disease Reptr.*, **39**, 120–123 (1955).

Bonde, R., Akeley, R., and Merriam, D., "A method for testing seedling progenies of potato for ring-rot resistance," *Plant Disease Reptr.*, **43**, 198–200 (1959).

Bottari, V., and Spina, P., "The varieties of olive cultivated in Sicily," *Ann. Sper. Agrar. (Rome)*, **7**, 937–978 (1953).

Boyle, A. M., "Further studies of the bacterial necrosis of the giant cactus," *Phytopathology*, **39**, 1029–1052 (1949).

Braun, A. C., and McNew, G. L., "Agglutinin absorption by different strains of *Phytomonas stewartii*," *Botan. Gaz.*, **102**, 78–88 (1940).

Brentzel, W. E., and Munro, J. S., "Bacterial ring rot of the potato. Investigation of possible dissemination by grasshoppers," *Bull. N. Dakota Agr. Sta. Bull.*, No. 295 (1940).

Brooks, R. M., and Olmo, H. P., "Register of new fruit and nut varieties, List 13," *Proc. Am. Soc. Hort. Sci.*, **72**, 519–541 (1958).

Buchanan, D., "A bacterial disease of beans transmitted by *Heliothrips femoralis Reut.*," *J. Econ. Entomol.*, **25**, 49–53 (1932).

Buchner, P., *Tier und Pflanze in Symbiose*, Gebruder Borntraeger, Berlin, 1930.

Buchner, P., *The Endosymbiosis of Animals with Plant Microorganisms*, Wiley-Interscience, 1965.

Buddenhagen, I. W., "An insect-spread bacterial wilt epiphytotic of Bluggoe banana" *Nature*, **194** (4824) 164–165 (1962).

Bugbee, W. M., "Whitefly transmission of *X. pelargonii*," *Phytopathology*, **52**, 1 (1962).

Burrill, T. J., "Anthrax of fruit trees; or the so-called fire-blight of pear, and twig blight of apple trees," *Proc. Am. Assoc. Advance. Sci.*, **29**, 583–597 (1881).

Carpenter, T. R., and Shay, J. R., "The differentiation of fire-blight resistant seedlings within progenies of interspecific crosses of pear," *Phytopathology*, **43**, 156–162 (1953).

Cheremisinov, N. A., "Michurin theory for the control of maize disease," *Trans. Voronezh. Agr. Inst.*, **26**, 130–146 (1956).

Clayton, C. N., "Streptomycin for the fire blight control on apple in North Carolina," *Plant Disease Reptr.*, **39**, 128–131 (1955).

Crossan, D. F., "Control of potato seed piece decay," *Plant Disease Reptr.*, **43**, 543–545 (1959).

Crosse, J. E., "Dispersal of bacterial plant pathogens," in *Biological Aspects of the Transmission of Disease,* Oliver and Boyd, London, 1958.

Crosse, J. E., Bennett, M., and Garrett, C. M. E., "Fire blight of pear in England," *Nature,* **182,** 1530 (1958).

Crosse, J. E., Bennett, M., and Garrett, C. M. E., "Fire blight disease of pear," *Ann. Rept., East Malling Research Sta., Kent,* **1958,** 1–4 (1959).

Dean, R. W., "Morphology of the digestive tract of the apple maggot fly, *Rhagoletis pomonella* Walsh," *N. Y. Agr. Expt. Sta. (Geneva, N. Y.), Bull.,* No. 215 (1933).

Dean, R. W., "Anatomy and postpupal development of the female reproductive system in the apple maggot fly, *Rhagoletis pomonella* Walsh," *N. Y. Agr. Expt. Sta. (Geneva, N. Y.), Bull.,* No. 229 (1935).

Dean, R. W., "Evaluation of some new insecticides for apple maggot control," *J. Econ. Entomol.,* **44,** 147–153 (1951).

Dean, R. W., "Further studies of insecticides for apple maggot control," *J. Econ. Entomol.,* **47,** 479–485 (1954).

Deane, J. F., and Chapman, R. K., "The relation of the cabbage maggot, *Hylemyia brassicae* (Bouche) to decay in some cruciferous crops," *Entomologia Exp. Appl.,* **7,** 1–8 (1964).

Dowson, W. J., "*Bacterium salicis* Day, the cause of the watermark disease of the cricket-bat willow," *Ann. Appl. Biol.,* **24,** 528–544 (1937).

Dunegan, J. C., Moon, H. H., and Wilson, R. A., "A rapid method for mass inoculation of pear trees with the blight organism," *Plant Disease Reptr.,* **37,** 14–15 (1953).

Elliott, C., "Dissemination of bacterial wilt of corn," *Iowa State Coll. J. Sci.,* **9,** 461–480 (1935).

Elliott, C., and Poos, F. W., "Overwintering of *Aplanobacter stewartii,*" *Science,* **80,** 289–290 (1934).

Folsom, D., "Bacterial twig and blossom blight of raspberry in Maine," *Plant Disease Reptr.,* **31,** 324 (1947).

Folsom, D., "Bacterial fire blight of raspberry associated with raspberry cane maggot," *Plant Disease Reptr.,* **38,** 338–339 (1954).

Fronk, W. D., and Peterson, L. E., "Some effects of treating muskmelons with insecticides," *J. Econ. Entomol.,* **47,** 807–811 (1954).

Garman, P., "Spray combinations for control of apple pests in Connecticut," *J. Econ. Entomol.,* **47,** 731–734 (1954).

Gilgut, C. J., "Summary of disease incidence in 1952 at the New England fruit spray conference," *Plant Disease Reptr.,* **37,** 116–117 (1953).

Goodman, R. N., "Fire blight control with sprays of Agrimycin, a streptomycin-terramycin combination," *Plant Disease Reptr.,* **38,** 874–878 (1954).

Gould, G. E., "The biology and control of the striped cucumber beetle," *Bull. Indiana Agr. Expt. Sta.,* **490,** 28 (1944).

Granhall, I., "Ring rot (*Corynebacterium sepedomicum*) in potatoes," *Rept. 10th Congr. Scand. Agr. Sci. Soc., Stockholm* (1956).

Gray, E., "The willow wood wasp and water-marked disease of willows," *Vet. J.*, **96**, 370–373 (1940).

Hamilton, J. M., and Szkolnik, M., "Omadine, a promising new organic fungicide for the control of blossom blight of pears," *Plant Disease Reptr.*, **41**, 301–302 (1957).

Heuberger, J. W., and Poulos, P. L., "Control of fire blight and frog-eye leaf spot (black rot) diseases of apples in Delaware in 1952," *Plant Disease Reptr.*, **37**, 81–83 (1953).

Hildebrand, E. M., "Relative stability of fire blight bacteria," *Phytopathology*, **44**, 192–197 (1954).

Husted, S. R., Mills, R. B., Foltz, V. D., and Crumrine, M. H., "Transmission of *Salmonella montevideo* from contaminated to clean wheat by the rice weevil," *J. Econ. Entomol.*, **62**, 1489 (1969).

Ivanoff, S. S., and Keitt, G. W., "Relations of nectar concentrations to growth of *Erwinia amylovora*, and the fire blight infection of apple and pear blossoms," *J. Agr. Research*, **62**, 733–743 (1941).

Johnson, J. C., "Pod twist: a previously unrecorded bacterial disease of French bean (*Phaseolus vulgaris* L.)," *Queensland J. Agr. Sci.*, **13**, 127–158 (1956).

Keitt, G. W., and Ivanoff, S. S., "Transmission of fire blight by bees and its relations to nectar concentration of apple and pear blossoms," *J. Agr. Research*, **62**, 745–753 (1941).

Koch, A., "Über das verhalten symbionten freier sitodrepa larven," *Biol. Zentr.*, **53**, 199–203 (1930).

Larson, R. H., "The ring rot bacterium in relation to tomato and eggplant," *J. Agr. Research*, **69**, 309–325 (1944).

Leach, J. G., "The relation of the seed-corn maggot (*Phorbia fusciceps* Zett.) to the spread and development of potato blackleg in Minnesota," *Phytopathology*, **16**, 149–176 (1926).

Leach, J. G., "Further studies on the seed-corn maggot and bacteria with special reference to potato blackleg," *Phytopathology*, **21**, 387–406 (1931).

Leach, J. G., "The method of survival of bacteria in the puparia of the seed-corn maggot (*Hyelmyia cilicrura* Rond.)," *Z. Angew. Entomol.*, **20**, 150–161 (1933).

Leach, J. G., *Insect Transmission of Plant Diseases*, McGraw-Hill, New York, 1940.

Lelliot, R. A., "Fire blight of pears in England," *Agriculture (London)*, **65**, 564–568 (1959).

Lightle, P. D., Strandring E., and Brown, J. G., "A bacterial necrosis of the giant cactus," *Phytopathology*, **32**, 303–313 (1942).

List, G. M., and Kreutzer, W. A., "Transmission of the casual agent of the ring-rot disease of potatoes by insects," *J. Econ. Entomol.*, **35**, 455–465 (1942).

MacLachlan, D. S., "Machinery and warehouse disinfection in potato ring rot control," *Potato Handbook,* 31–32 (1958).

MacLachlan, D. S., "Electronic heating as a means of sterilizing potato bags contaminated with bacterial ring spot," *Am. Potato J.,* **36,** 52–55 (1959).

MacLachlan, D. S., and Sutton, M. D., "The use of antibiotics in the control of potato ring rot," *Rept. Quebec Soc. Protect. Plants,* **1956,** 76–82 (1957).

Malenev, F. E., "The influence of boron, copper, manganese and zinc on the resistance of potatoes to *Phytophthora* and other diseases," *Sbornik Microelem. Agr. Med. Riga. Acad. Sci. Latv. S. S. R.,* 429–436 (1956).

Marcovitch, S., "The insect pest situation in Israeli agriculture," *J. Econ. Entomol.,* **47,** 19–32 (1954).

Marshall, G. E., Childers, N. F., and Brody, A. W., "Leafhoppers can weaken apple trees and reduce the crop," *Proc. Ohio Hort. Soc.,* **74,** 61–66 (1941).

Martelli, G. P., "Leaf infections of olive knot disease in galleries of mining insects," *Inform. Fitopatol.,* **4,** 2 (1958).

McNew, G. L., "Factors influencing attenuation of *Phytomonas stewartii* cultures," *J. Bacteriol.,* **39,** 171–196 (1940).

McNew, G. L., and Braun, A. C., "Agglutination test applied to strains of *Phytomonas stewartii,*" *Botan. Gaz.,* **102,** 64–77 (1940).

Metcalf, G., "The watermark disease of willows. I. Host parasite relationships," *New Phytologist,* **39,** 322–332 (1940).

Metcalf, G., "The watermark disease of willows. II. Pathological changes in the wood," *New Phytologist,* **40,** 97–107 (1941).

Michelbacher, A. E., "Control of codling moth in walnuts," *J. Econ. Entomol.,* **38,** 347–355 (1945).

Miller, P. W., "Studies of fire blight of apple in Wisconsin," *J. Agr. Research,* **39,** 579–621 (1929).

Miller, P. W., "Nut diseases in Oregon in 1952," *Plant Disease Reptr.,* **37,** 18 (1953).

Miller, P. W., "A preliminary report on the comparative efficiency of copper-lime and Agrimycin dust mixtures for the control of walnut blight in Oregon," *Plant Disease Reptr.,* **43,** 401–402 (1959).

Mitchell, J. W., Zaumeyer, W. J., and Preston, W. H., Jr., "Absorption and translocation of streptomycin by bean plants and its effect on the halo and common blight organism," *Phytopathology,* **44,** 25–30 (1954).

Oganova, E. A., "On bacterial canker of ash," *Soobshcheniya Inst. Lesa, Akad. Nauk S. S. S. R.,* **8,** 81–89 (1957).

Petri, L., "Ricerche sopra i batteri intestinali della Mosca Olearia," *Mem. Reale Staz. Pathol. Vetetable Roma* (1909).

Petri, L., "Utersuchungen uber die Darmbakterien der Olivenfliege," *Zentr. Bakteriol. II,* **26,** 357–367 (1910).

Poos, F. W., "Host plants harboring *Aplanobacter stewartii* without showing external symptoms after inoculation by *Chaetocnema pulicaria*," *J. Econ. Entomol.*, **32**, 881–882 (1939).

Poos, F. W., "DDT to control corn flea beetle on sweet corn and potato leafhopper on alfalfa and peanuts," *J. Econ. Entomol.*, **38**, 197–199 (1945).

Rand, F. V., and Cash, L. C., "Some insect relations of *Bacillus tracheiphilus* E. F. Smith," *Phytopathology*, **10**, 133–140 (1920).

Rand, F. V., and Cash, L. C., "Bacterial wilt of cucurbits," *U. S. Dept. Agr., Tech. Bull.*, No. 828 (1920a).

Robert, A. L., "Bacterial wilt and Stewart's leaf blight of corn," *U. S. Dept. Agr. Farmers' Bull.*, **2092** (1955).

Rojecka, N., "White bacteriosis (*B. stewartii*) of maize," *Postepy Nauk Rolniczej*, **4**, 69–72 (1957).

Rudolph, B. A., "A possible relationship between the walnut erinose mite and walnut blight," *Science*, **98**, 430–431 (1943).

Rudolph, B. A., "Attempts to control bacterial blights of pear and walnut with penicillin," *Phytopathology*, **36**, 717–725 (1946).

Ryzhkova, E. V., "Phytopathogenic symbionts of the Swedish flies *Oscinella frit* (L) and *O. pusilla* Mg., (Diptera, Chloropidae) and their practical use," *Rev. Ent. U. S. S. R.*, **41**, 788–795 (1962).

Sarejanni, A. J., Demetriades, S. D., Zachos, D. G., and Papaioannou, A. J., "Rapports sommaires sur les principales maladies des plantes cultivees observees en Grece au cours des annees 1953, 1954, 1955." *Ann. Inst. Phytopathol. Benaki*, **8**, 77–83 (1954); **9**, 3–9 (1955); **10**, 3–8 (1956).

Shapiro, I. D., "Oscinella frit on maize," *Zasch. Roast. 1963*, pt. 6, (in Russian) (1963).

Sohi, G. S., and Sandhu, M. S., "Relationship between citrus leaf-miner (*Phyllocnistic citrella* Stainton) injury and citrus canker (*Xanthomonas citri* (Hasse) Dowson incidence on citrus leaves," *J. Res. P. A. U.*, **5**, 66 (1968).

Starr, G. H., and cinnamon, C. A., "Effect of ultra-violet rays, X-rays, and nuclear radiation on ring-rot bacteria infecting burlap potato bags," *Phytopathology*, **43**, 439–440 (1953).

Starr, M. P., Cardona, C., and Folsom, D., "Bacterial fire blight of raspberry," *Phytopathology*, **41**, 915–919 (1951).

Tamini, K. M., and Banfield, W. M., "Transmission of bacterial soft rot by fruit flies," *Phytopathology*, **59**, 403 (1969).

Thomas, H. E., and Ark, P. A., "Fire blight of pears and related plants," *Calif. Agr. Expt. Sta. Bull.*, No. 586 (1934).

Thomas, H. E., and Ark, P. A., "Nectar and rain in relation to fire blight," *Phytopathology*, **24**, 682–685 (1934a).

Thomas, W. D., Jr., and Henderson, W. J., "Spray experiments for the control of fire blight on apples and pears in 1947–1950," *Plant Disease Reptr.*, **36**, 273–275 (1952).

Tullis, E. C., "Studies on the overwintering and modes of infection of the fire blight organism," *Mich. State Univ. Agr. Expt. Sta. Tech. Bull.*, No. 97 (1929).

Waite, M. B., "Results from recent investigations in pear blight," *Botan. Gaz.*, **16**, 259 (1891).

Wescott, C., *Plant Disease Handbook*, Van Nostrand, New York, 1950.

Wilson, E. E., "The olive knot disease: its inception, development, and control," *Hilgardia*, **9**, 233–264 (1935).

Winter, H. F., and Young, H. C., "Control of fire blight of apples in Ohio," *Plant Disease Reptr.*, **37**, 463–464 (1953).

The
Mycoplasmas

The mycoplasma as organisms were recognized by the classical bacteriologists in the group Mycoplasmatales. The first mycoplasma to be cultured was that of bovine pleuropneumonia (Nocard *et al.*, 1898). In the ensuing 40 years, dogs, rodents, goats, and sheep, as well as poultry, were found to be subject to diseases caused by mycoplasma. Finally, a mycoplasma pathogenic to man was cultured (Channock *et al.*, 1962).

Interest in mycoplasmas as disease agents in plants is more recent. Transmission of plant viruses by leafhoppers and other arthropods, and the rough grouping of many of these as "yellows" or "witch's broom" diseases has been accepted for many years, but recent work has raised questions as to the true etiology of these diseases. Now it is considered likely that, rather than being virus diseases, they are the result of infection by mycoplasmas.

Doi *et al.* (1967) first described mycoplasma-like bodies in the sieve tubes and occasionally from phloem parenchyma in plants infected with mulberry dwarf, potato witch's broom, aster yellows, and Paulonia witch's broom. Since then, other so-called virus diseases have been added to the list, e.g., sweet potato and legume witch's broom of *Cryptolaemia japonica*.

The Japanese workers described these bodies from plant tissues and

51

later (Kawakita and Ishiie, 1970) from vector tissues, as specific, pleo-morphic, spherical to irregularly ellipsoidal in shape and 80–800 mμ in diameter, enclosed in a 2-layered limiting membrane about 8 mμ thick rather than by a cell wall. In gross morphology and structure, similarities were drawn between these bodies and those of mycoplasma species or chlamydia of the psittacosis-lymphogranuloma-trachoma complex (Fig. 3.1).

In a concurrent paper, Ishiie *et al.* (1967) describe partial inhibition of mulberry dwarf and plant recovery following treatment with antibiotics of the tetracycline group. The mycoplasmas disappeared in plants follow-ing treatment, a phenomenon not expected to occur with plant virus dis-eases. Davis *et al.* (1968) applied 1000 ppm chloretracycline through roots and by sprays every three days to plants showing aster yellow disease symptoms, and these were suppressed for two to four weeks. Infection of vectors with aster yellows with the same dosage of the antibiotic resulted in failure of the insects to transmit. They concluded that the causal entity of aster yellows might be a mycoplasma-like or bedsonia-like agent, but Davis and Whitcomb (1971) now prefer the designation chlamydia-like; they include bedsoniae, PLT group, psittacosislike organisms in the chla-mydiae.

The constant association of the organisms in the phloem of diseased but not of healthy plants, coupled with the failure to demonstrate true virus particles in these same damaged tissues and the reaction of the dis-eased tissues to antibiotic treatments, were all indicative of a previously unrecognized etiology of these diseases.

The two papers from the Japanese workers set off a tremendous surge of activity in the redescription of known materials in mycoplasmic terms. There were two approaches, depending on facilities available. One of these was the testing of the tetracyclines against infected plants; these

FIGURE 3.1. (*a*) Mycoplasma-like organisms (MLO) and mito-chondria in a sieve cell of white clover affected with aster yellows. X51,000. (*b*) MLO in the sieve cells and phloem parenchyma cells of a potato plant affected with witch's broom. X4,400. (*c*) MLO in sieve cells of *Cryptotaenia japonica* affected with witch's broom. X17,600. (*d*) MLO located at both sides of a sieve pore of a potato plant affected with witch's broom. X18,600. (*e*) MLO in a sieve cell of morning glory affected with sweet potato witch's broom. X17,400. (*f*) MLO in sieve cells of sweet potato affected with witch's broom. Note slender bodies near the sieve pores. X18,600. (*g*) MLO in a sieve tube of a potato plant affected with witch's broom. Note fine structure. X73,000. (*h*) MLO in sieve cells of cosmos affected with legume witch's broom. X18,600 (Photograph Professor H. Asuyama).

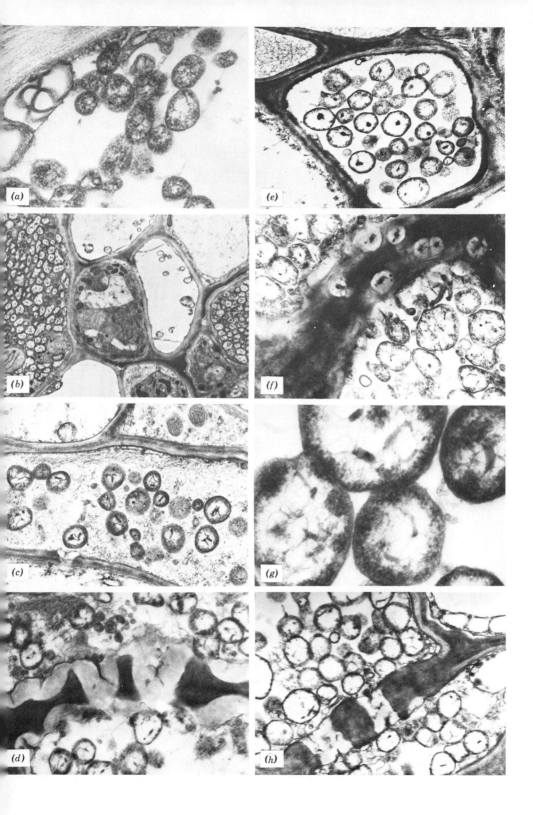

tests were easy to perform. The second approach was the use of the electron microscope on ultrathin sections, first of plant tissue and later of the insect vectors.

There has been a plethora of reviews already published, somewhat repetitious, as could be expected, with a span of only three or four years work available for review (Caspar, 1969; Shikata *et al.*, 1969; Maramorosch *et al.*, 1970; Bos, 1970; Hull, 1970; Whitcomb and Davis, 1970; and Davis and Whitcomb, 1971). The last two reviews are essential references. The 1970 review with its 510 references includes consideration of the phytarboviruses primarily, but emphasizes the parallelism in biological cycles between them and the mycoplasmas. The latest review (1971) is specifically concerned with mycoplasmas, Rickettsiae, and Chlamydiae as possible plant and insect disease agents.

Shortly after the publications by the Japanese workers, the European yellows-type diseases were investigated thoroughly by electron microscopy by Ploaie *et al.* (1968). The study included Crimean yellows, European clover dwarf, stolbur, and parastolbur, and without exception, electron microscope studies revealed the presence of mycoplasma-like bodies. See also Ploaie and Maramorosch (1969).

In the United States, aster yellows has been the subject of intensive study, largely perhaps, because the disease and its vector-virus relationships have been studied over many years; in fact, it is one of the classical cases of leafhopper transmission of the agent of a plant disease. Now the term "agent" is replacing the old term "virus," which is simply indicative of the parlous state of the etiology of the disease.

However, the number of plant diseases, formerly known as virus diseases, in which the infected plants have been shown to contain mycoplasma-like bodies, has increased materially. The currently recognized list is shown in Table 3.1, and it is now clear that the phenomenon is not limited to those diseases previously known under the generic titles of yellows and witch's broom, although diseases in these two groups are still the favored candidates for study.

In the review by Maramorosch *et al.* (1970), which includes consideration of 32 of the examples listed in Table 3.1, an analysis reveals that in 22 cases there are clear data on the presence of the mycoplasma-like bodies in the infected plants. In six of these cases, they are also present in the insect vectors. Remission with tetracycline treatment is recorded in 15 cases. Nine are considered dubious.

A characteristic of the effects of treatment with antibiotics in many cases is that the symptoms reappear after varying perids of time. One case of complete remission has been reported (Story and Halliwell, 1969) for papaya bunchy top; acromycin (tetracycline) and aureomycin (chlortetra-

TABLE 3.1. *Plant Diseases in Which Mycoplasma-like Bodies Have Been Shown to Occur in Infected Plant Tissue*

Alfalfa witch's broom	Molasses dwarf
Apple chat fruit	Oat sterile dwarf
Apple proliferation	Papaya bunchy top
Apple rubbery wood	Parastolbur (potato)
Ash witch's broom	Paulownia witch's broom
Aster yellows	Peach Western X
Blueberry stunt	Peach yellows
Broadbean virescence	Phloem necrosis of elm
Cactus witch's broom	Phormium yellow leaf
Cassava witch's broom	Pear decline
Cherry buckskin	Potato purple top
Citrus greening	Potato stolbur
Citrus Likubin	Potato witch's broom
Citrus stubborn	Rice giallume yellows
Clover dwarf	Rice "paddy jantan"
Clover phyllody	Rice yellow dwarf
Clover Stolbur	*Rubus* stunt
Coconut lethal yellowing	Sandalwood spike
Corn stunt	Strawberry green petal
Cotton virescence	Sugarcane white leaf
Cranberry false blossom	Sweet potato witch's broom
Crimean yellows	Sunflower phyllody
Cryptolaenia japonica witch's broom	Tobacco yellow dwarf
	Tomato big bud
Current reversion	Tomato purple top
Eggplant little leaf	Wallflower virescence
Flavescens doree (grape)	Witch's broom of groundnut
Lavender yellow decline	Witch's broom of legumes
Legume little leaf	Yellow leaf of *Areca* palm
Little peach	Yellow wilt of sugarbeet
Mal azul (tomato)	*Vinca rosea* yellows (stolbur)

cycline) were the antibiotics used. Shikata *et al.* (1969) reported complete freedom from white leaf symptoms in sugarcane and absence of mycoplasma-like bodies for a period of three months, after treatment for 72 hours with oxytetracycline.

The Davis *et al.* paper (1968) has already been mentioned. Freitag and Smith (1969) have shown that the tetracyclines, acromycin and aureomycin will reduce the titer of the aster yellows agent in some plants, with remission of symptoms in others. Treatment reduced the acquisition capacity of the leafhopper vectors as well as transmission by preinoculated leafhoppers (Fig. 3.2A–C).

The use of the tetracyclines has been developed by some without regard for the need for controlled conditions used by Davis *et al.* (1968); but Klein *et al.* (1971) have reiterated the necessity for control of buffers, pH, molarity, and mode of application, since these factors affect the efficiency of antibiotic therapy of aster yellows and should be specified.

Two additional groups of antibiotics, tylosin tartrate and chloramphenicol, have demonstrated efficacy against the yellows disease agents (Davis and Whitcomb). The findings of Paddock *et al.* (1971), which suggest that control of leafhopper-bone plant disease agents might be due to direct action of systemic biocides on the mycoplasmas, will be significant if confirmed. The chemicals they used were systemic organophosphorus compounds and oxime carbamates.

Mycoplasma-like bodies in insect vectors were reported very shortly after they had been discovered in infected plants. Grenados *et al.* (1968) showed, for the first time, that the organisms could be found in both infected plants and vectors of corn stunt. Kawakita and Ishiie (1970) found the organisms in viruliferous leafhoppers transmitting mulberry dwarf disease. Shikata and Maramorosch (1969) found the organisms in great numbers in yellows-infected plants and in the fat body of *M. fascifrons* and *D. maidis,* vectors of aster yellows and corn stunt respectively. Elongated bodies were seen in sieve-tube pores, and it was suggested that this is a means of distributing the organisms throughout infected plants. Pleomorphic forms were found in the phloem of aster-yellows-infected plants and in *Nicotiana;* they were limited to the sieve tubes and adjacent parenchyma (Worley, 1970).

The morphology of the mycoplasma-like organisms associated with peach Western X disease was described by Huang and Nyland (1970). The organisms occur only in mature sieve tubes of leaves showing symptoms. Cell to cell movement is probably through the pores since the organisms have been seen extending through pores of adjacent sieve tubes; they are up to 5400 mμ long and 120–130 mμ wide. Hirumi and Maramorosch (1969) found mycoplasma-like bodies in the salivary glands of insect vectors of the aster yellows disease agent (Fig. 3.3 1 and 2).

Pear decline, associated with the psyllid, *Psylla pyricola,* was first believed to be a phytotoxemia and then considered a virus disease; it is now included in the list of suspected mycoplasma-induced diseases. A brown-leaf-vein symptom was recognized as part of the syndrome in 1966 (Tsao *et al.,* 1966), and recently, Hibino and Schneider (1970) have reported that mycoplasma-like bodies are to be found in the sieve tubes of the swollen brown veins of old leaves, but only sparsely in new leaves not yet showing the symptom.

FIGURE 3.2. (*a*) Aster-yellows-infected plants treated with tetracyclines. Effect of acromycin on aster: (*left*) treated plant showing some floral development; (*right*) untreated plant showing no recovery from symptoms. (*b*) Aster-yellows-infected plants treated with tetracyclines. Effect of aureomycin on common plantain; (*left*) treated plant showing recovery from symptoms; (*right*) untreated plant showing typical symptoms. (*c*) Aster-yellows-infected plants treated with tetracyclines. Effect of aureomycin on celery; (*left*) treated plant showing recovery; (*right*) untreated plant control (Photographs Freitag and Smith, *Phytopathology,* 59, 1969).

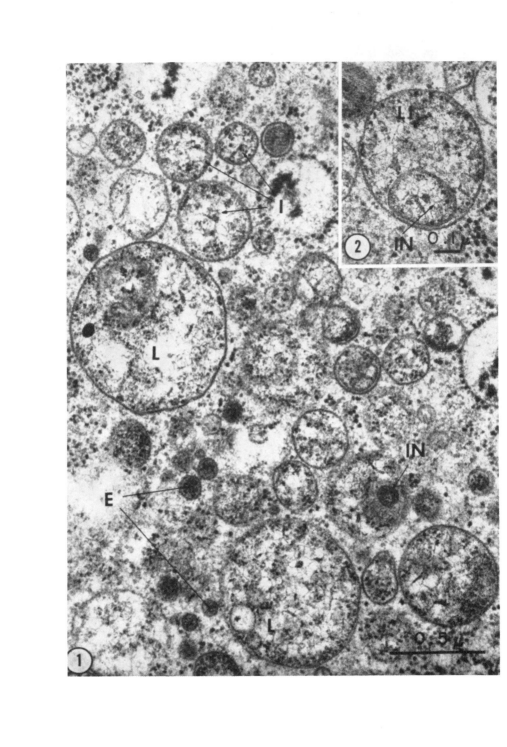

Phormium yellow leaf has a somewhat similar history. After some confusion with the phytotoxic effect of the insect's feeding, it was determined that it was a virus disease. Now, however, Ushiyama *et al.* (1969) have shown the presence of mycoplasma-like bodies in both infected plants and in the vector *Oliavarus atkinsoni* Myers.

Citrus greening, vectored by the psyllid *Trioza erytreae,* the citrus psylla, is suspected of being caused by a mycoplasma-like organism and has been shown to have a temperature-dependent agent-vector relationship. There is a high + correlation between symptom development and symptom expression with temperatures below 25 °C (Schwarz and Green, 1970).

Dodder transmission of mycoplasma-like bodies of witch's broom of brown ash to *Vinca rosea* was accomplished by using *Cuscuta camprestris* Yunker and *Fraxinus americana* Dur. and Hilg. by Hibben and Wolanski (1971). The extreme abundance of the bodies in the tissue of the dodder indicates that reproduction takes place in that plant. Dale and Kim (1969) had come to the same conclusion with respect to the transmission of aster yellows agent by *Cuscuta* spp.

A report by Amin and Jensen (1971) provides some indirect evidence of the etiological significance of the mycoplasma-like bodies in Western X disease. Acquired either by feeding or infection, the antibiotic reduces transmission, extends the incubation period of Western X disease in the leafhopper vector, and extends the insect's life. The treated leafhoppers laid more eggs than did the controls, indicating reduction of the pathogenic effects of the disease agent in the insect (see also Nasu *et al.,* 1970).

Apart from indirect evidence, however, there is abundant evidence of the association of mycoplasma-like bodies with certain plant diseases, as well as the remission of symptoms after treatment with antibiotics. But

FIGURE 3.3, 1 & 2. 1. Mycoplasma-like bodies of aster yellows in the cytoplasm of a salivary gland cell of an infective leafhopper vector, *Macrosteles fascifrons,* 31 days after acquisition by feeding. Various forms of the bodies bounded by unit membranes; small compact body (E), intermediate bodies (I), large body (L), and inclusion (IN). Ribosomes of the bodies are smaller than those of the host cells. DNA-like strands are seen in paler nuclear areas of the bodies, except in the small compact body. Magnification X49,000. 2. Large mycoplasma-like body (LI) with inclusion (IN). The inclusion, which is bounded by a unit membrane, also contains DNA-like strands and small ribosomes. Magnification X66,000 (From Hirumi, H., and Maramorosch, K., *J. Virology,* 3, 83 (1969). With the permission of American Society of Microbiology).

this is all circumstantial evidence and does not constitute proof of pathogenicity.

The *sine qua non* is the cultivation of the organisms on solid media and the satisfaction of Koch's postulates. There are some inherent difficulties in achieving a positive Koch test for pathogenicity. The process of acquisition and transmission by the vector is presumably the same as for the virus diseases, and the large body of evidence on the vector-virus relationships is still applicable. The organisms can only be detected in plants by electron microscopy; in pure culture, free of particulate matter, they can be seen in the dark field of the light microscope or by phase contrast. There is the possibility, indicated by the long history of work on the animal mycoplasmas beginning with Nocard *et al.* (1898) and their culture of the mycoplasma of bovine pleuro-pneumonia, that the organisms found in plants may also be cultured *in vitro*.

In the case of the animal and human mycoplasmas of the pleuro-pneumonia group (PPLO), their definition includes colony structure on solid media (Marmin, 1967). Marmin's article, which is a comprehensive review of mycoplasmas with special reference to those pathogenic to man, contains a section and an appendix on laboratory techniques including culture media and procedures for use with the electron microscope.

CULTURE IN VITRO

Whether or not the culture media established for animal mycoplasmas will serve equally well for the plant mycoplasma-like bodies remains to be seen. In a conservative approach to the matter, Davis *et al.* (1970) reported on the viability of the aster yellows agent in cell-free media. Slight modifications of the medium increased the longevity of the aster yellows infectivity up to 48 hours. Storage of viable aster yellows bodies was suggested; the medium might be enriched for eventual culture of the organisms.

Reports of successful culture of the mycoplasma-(chlamydia)-like bodies are confusing. Hull *et al.* (1969) failed to culture the bodies from sandal spike disease although remission of the symptoms was obtained by treatment with antibiotics. Nayer (1971) on the other hand reported culture of the bodies from yellow leaf of *Areca* palm using the same technique as that used by Hull. She did not report pathogenicity data. The study by Hampton *et al.* (1969) has not been accepted.

Chen and Granados (1970) cultured small pieces of ataclosteles from corn infected with corn-stunt disease. The cultures were centrifuged, the pellet used for subcultures and injecting the insect vector. Infectivity from cultures and subcultures 5, 8, 14, and 50 days old was shown. All contained mycoplasma-like bodies (PPLO). The authors considered that their study established a positive Koch test, but they did not, apparently, culture on solid media.

Mycoplasma-like bodies were found in tissue of white-leaf-diseased sugarcane plants by Lin and Lee (1968). Lin *et al.* (1970) immersed cuttings from infected plants for 48–72 hr in tetracycline; plants arising from these cuttings showed no symptoms for 10 months or longer, and when the symptoms did appear, they were in the form of narrow stripes rather than the completely white leaves of untreated controls. The treatment times were 48 and 72 hr for the cases showing a long period of symptom remission; at 24 hr, there was only a slight delay and the full white leaf symptoms appeared.

The mycoplasma-like bodies were cultured in Morton's PPLO medium, both liquid and agar, according to the *Difco Manual* 9th ed. Buds were introduced into the liquid media and transfers made to solid agar on which the typical "fried egg" colonies developed in two out of six attempts, using 30 buds. Inoculaion of buds resulted in only a few plants showing symptoms, but the controls were completely free. The meticulous detail with which this study was reported makes for confidence in the results, but unfortunately, efforts to repeat this study in the same laboratory have been a failure (personal communication with Taiwan).

Gianotti *et al.* (1971) appears to be the latest report to date. It concerns the cultivation of a mycoplasma-like body from clover phyllody-infected plants. Pathogenicity was retained after at least four subcultures in broth and produced "fried egg" colonies on agar.

While the isolation of mycoplasma-like bodies from plants and insects appears to have been abundantly confirmed, there appears to be a pronounced lack of pathogenicity data, with the exception perhaps of the Gianotti report.

CLASSIFICATION

If these mycoplasma-like bodies are in fact mycoplasmas, they would be placed in the Mycoplasmatales, or rather in a separate Class Mollicutes, or in the Bertonellacae (Ricksettsia). The antibiotic spectra indicate that the yellows-disease agents are induced by wall-free procaryotes, which taxonomically are limited to those two groups.

However, since the known mycoplasma species are usually extracellular, and the suspected agents of the yellows disease are intracellular, Davis and Whitcomb (1971) have suggested that another group of intracellular procaryotic parasites deserve consideration, i.e., those in the genus *Chlamydia* in the Chlamydiacae.

Before any conclusion is possible, however, the suspected disease agents must meet certain conditions. Edward *et al.* (1967) reported the ground rules laid down by the 12th International Congress of Microbiology. If these rules are followed they will effectively prevent premature naming. Currently, naming, except by reference to the disease, is impossible.

The requirements are exacting. No new specific names will be acceptable unless accompanied by an adequate description that will allow laboratory identification supported by serological comparisons with known species. Before such a description can be written, the organism must be cultured on solid agar and satisfy Koch's postulates for pathogenicity.

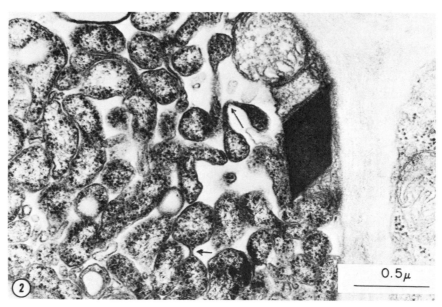

FIGURE 3.4.   Mycoplasma of aster yellows in a phloem cell from a flower stalk of a China aster plant (*Callistephus chinensis*) infected with aster yellows. Noninvaded cell at right has retained its normal appearance. Note apparent binary fission (arrows) of several mycoplasmas inside the plant cell. The microorganisms are bounded by unit membranes with DNA strands in the nuclear regions, and with RNA-containing ribosome granules at the periphery (From Shikata, E., Maramorosch, K., and Ling, K. C., *FAO Plant Prot. Bull.,* **17,** 121–128 (1969). With the permission of FAO, Rome).

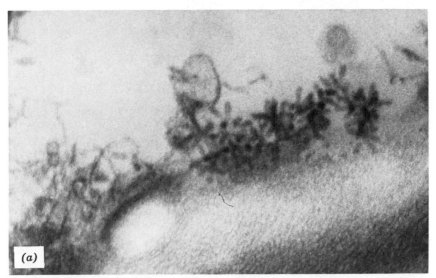

FIGURE 3.5. (a) Numerous virus-like particles in free distribution in the rest of the cytoplasm of phloem cell of *Vinca rosea* infected by clover dwarf mycoplasma. X60,000.

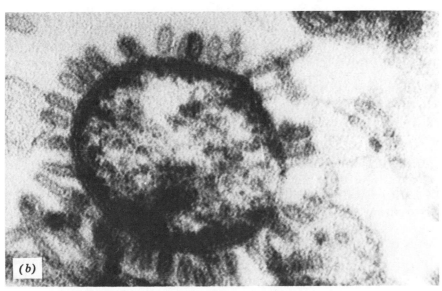

FIGURE 3.5. (b) Numerous rod-shaped virus-like particles fixed on unit membrane of mycoplasma cell. X120,000 (Photographs P. G. Ploaie, Bucharest).

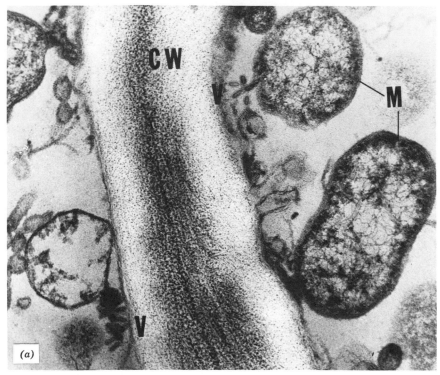

FIGURE 3.6. (*a*) Mycoplasma-like bodies in the phloem of periwinkle infected with clover dwarf together with rod-shaped virus-
like particles. X56,320.

Since no *Mycoplasma* species has been described from plants, the subject of classification remains an open field of inquiry. A further complication was revealed recently when Ploaie (1971) showed that particles resembling rod-shaped viruses can be detected free in the phloem cells or
fixed on mycoplasma cells which are presumed to be the etiologic agent
of clover dwarf disease. He presents the hypothesis of the existence of a
virus infecting the mycoplasma or transmitted by mycoplasma. His electron micrographs all show the particles attached to the mycoplasma or
free in the cytoplasm of the phloem cells (Fig. 3.4, 3.5 and 3.6). It is
known that viruses infect mycoplasma recognized as belonging to the Mycoplasmatales; 11 of these are now known (Lisa and Maniloff, 1971) (see
also Gourley, 1970).

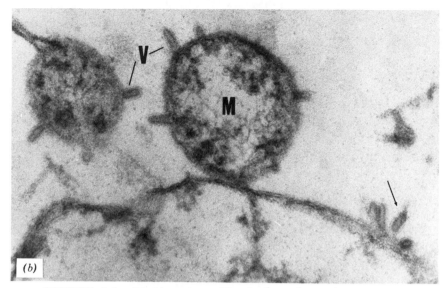

FIGURE 3.6. (*b*) Mycoplasma-like body and membranous structure with virus-like particles attached. X90,000 M=mycoplasma-like bodies; V=virus-like particles; C=cell wall (Photographs P. G. Ploaie).

REFERENCES

Amin, P. W., and Jensen, D. D., "Effects of tetracycline on the transmission and pathogenicity of the Western X-disease agent in its insect and plant hosts," *Phytopathology*, **61**, 696 (1971).

Bos, L., "Mycoplasmas een nieuw hoofdstuk in de Plantezeitenkinde?" *Overduk it Gewasbesch*, **1**, 45–54 (1970).

Casper, R., "Mykoplasmen als erreger von Planzenkheiten," *Nachricht. Deutsch. Pflanzenschutz*, **12**, 177–182 (1969).

Channock, R. M., Heyflick, M. L., and Barile, M. F., "Growth on artificial medium of an agent associated with atypical pneumonia and its identification as a PPLO," *Proc. Nat. Acad. Sci.*, **48**, 41–49 (1962).

Chen, T. A., and Granados, R. R., "Isolation, maintenance and infectivity studies of a plant pathogenic mycoplasma, the causal agent of corn stunt disease," *Phytopathology*, **60**, 573 (Abstr.) (1970).

Dale, J. L., and Kim, K. S., "Mycoplasma-like bodies in dodder parasitizing aster yellows-infected plants," *Phytopathology*, **59**, 1765 (1969).

Davis, R. E., and Whitcomb, R. F., "Spectrum of antibiotic sensitivity of aster yellows disease in insects and plants," *Phytopathology*, **59**, 1556 (Abstr.) (1969).

Davis, R. E., and Whitcomb, R. F., "Mycoplasmas, Rickettsiae and Chlamydiae; possible relation to yellows diseases and other disorders of plants and insects," *Ann. Rev. Phytopathology,* **9**, 119–154 (1971).

Davis, R. E. and Worley, J. F. "Spiroplasma; motile, helical organism associated with corn stunt disease" *Phytopathology,* **63**, 403–408 (1973).

Davis, R. E., Whitcomb, R. F., and Purcell, R., "Viability of the aster yellows agent in cell-free media," *Phytopathology,* **60**, 573 (Abstr.) (1970).

Davis, R. E., Whitcomb, R. F., and Steere, R. L., "Remission of aster yellows disease by antibiotics," *Science,* **161**, 793–794 (1968).

Doi, Y., Teranaka, M., Yora, K., and Asuyama, H., "Mycoplasma- or PLT group-like microorganisms found in the phloem elements of plants infected with mulberry dwarf, potato witch's broom, aster yellows or Paulonia witch's broom," *Ann. Phytopath. Soc. Japan,* **33**, 259–266 (1967).

Edward, D. J. *et al.,* "Recommendations on nomenclature of the Order Myco-plasmatales," *Science,* **155**, 1694–1696 (1967).

Freitag, J. H., and Smith, S. H., "Effects of tetracyclines on symptom expression and leafhopper transmission of aster yellows," *Phytopathology,* **59**, 1820 (1969).

Gianotti, J. *et al., C. Rond. Ser. D,* **272**, 1776–1778 (1971).

Goheen, A. C., Nyland G., and Lowe, S. K. "Association of a rickettsialike organism with Pierce's disease of grapevines and alfalfa dwarf and heat therapy of the disease in grapevines" *Phytopathology,* **63**, 341–345 (1973).

Gourley, R. N., "Isolation of a virus infecting a strain of *Mycoplasma laidla-wii,*" *Nature,* **225**, 1165 (1970).

Granados, R. R., Maramorosch, K., and Shikata' E., "Mycoplasma: suspected eti-ologic agent of corn stunt," *Proc. Natl. Acad. Sci.,* **60**, 841–844 (1968).

Hampton, R. O., Stevens, J. O., and Allen, T. C., "Mechanically transmissible mycoplasma from naturally infected peas," *Plant Disease Reptr.,* **53**, 499 (1969).

Hibben, C. R., and Wolanski, B., "Dodder transmission of a mycoplasma from ash witch's broom," *Phytopathology,* **61**, 151 (1971).

Hibino, H., and Schneider, H., "Mycoplasma-like bodies in sieve tubes of pear trees affected with pear decline," *Phytopathology,* **60**, 499 (1970).

Hirumi, H., and Maramorosch, K., "Mycoplasma-like agents of aster yellows in the salivary glands of leafhopper vectors," *Phytopathology,* **59**, 399 (1969).

Huang, J. and Nyland, G., "The morphology of a mycoplasma associated with peach X-disease," *Phytopathology,* **60**, 1534 (Abstr.) (1970).

Hull, R., "Mycoplasma-like organisms in plants," *Rex Appl. Mycol.,* **50**, 121–130 (1971).

Hull, R., Horne, R. W., and Nayer, R. M., "Mycoplasma-like bodies associated with spike sandal disease," *Nature,* **224**, 1121, 5224 (1969).

Ishiie, T., Doi, Y., Yora, K., and Asuyama, H., "Suppressive effects of antibiotics of tetracycline group on symptom development of mulberry dwarf disease," *Ann. Phytopath. Japan,* **33**, 267–275 (1967).

Kawakita, H., and Ishiie, T., "Electron microscope studies on the mulberry dwarf disease. (1) The presence of mycoplasma-like organisms in viruliferous leafhoppers, *Hishimonus sellatus* Uhler and *Hishimonoides sellatiformis* Ishihara," *J. Sericult. Sci. Japan*, **39**, 413–419 (1970).

Klein, M., Frederick, R. J., and Maramorosch, K., "Buffers affecting tetracycline chemotherapy," *Phytopathology*, **61**, 130 (Abstr.) (1971).

Lin, S. C., and Lee, C. S., "Mycoplasma or mycoplasma-like organisms in white leaf disease of sugarcane," *Ann. Rep. Taiwan Sugar Exp. Stn.*, 1967–1968 (1968).

Lin, S. C., Lee, C. S. and Chiu, R., "Isolation and cultivation of, and inoculation with a mycoplasma causing white leaf disease of sugarcane," *Phytopathology*, **60**, 795 (1970).

Liss, A., and Maniloff, J., "Isolation of Mycoplasmatales viruses and characterization with MVL1, MVL52, MVG51," *Science*, **173**, 725–727 (1971).

Marmin, B. P., "The Mycoplasmas," in *Recent Advances in Medical Bacteriology*, A. P. Waterson, Ed., Churchill Ltd., London, 1967.

Maramorosch, K., Granados, R. R., and Hirumi, H., "Mycoplasma disease of plants and insects," *Adv. Virus Res.*, **16**, 135–193 (1970).

Nasu, S., Jensen, D. D., and Richardson, J., "Electron mycroscopy of mycoplasma-like bodies associated with insect and plant hosts of peach Western X-disease," *Virology*, **41**, 583–595 (1970).

Nayer, R., "Etiological agent of yellow leaf disease of *Areca cathecu*," *Plant Disease Reptr.*, **55**, 170–171 (1971).

Nocard, E., Roux, E. R., with Borrel, Salimbeni, and Dujardin-Beaumetz, "Le microbe de la peripneumonie," *Ann. Inst. Pasteur, Paris*, **12**, 240 (1898).

Paddock, R. G., French, F. L., and Turner, P. L., "Control of leafhopper-borne viruses possibly due to direct action of systemic biocides on mycoplasmas," *Plant Disease Reptr.*, **55**, 291 (1971).

Ploaie, P. G., "Particles resembling viruses associated with mycoplasma-like organisms in plants," *Rev. Roum. Biol.-Botanique, Bucharest*, **16**, 3–6 (1971).

Ploaie, P., Granados, R. R., and Maramorosch, K., "Mycoplasma-like structures in periwinkle plants, with Crimean yellows, European clover dwarf, stolbur and parastolbur," *Phytopathology*, **58**, 1063 (Abstr.) (1968).

Ploaie, P., and Maramorosch, K., "Electron microscope demonstration of particles resembling Mycoplasma or Psittacosis-Lymphogranuloma-Trachoma group in plants infected with European yellows-type diseases," *Phytopathology*, **59**, 536–544 (1969).

Schwarz, R. E., and Green, G. C., "Citrus greening and the citrus psylla, *Trioza erytreae*, a temperature-dependent causative agent /vector," *Z. Pflanzenkrankh Pflanzenschutz.*, **77**, 490 (1970).

Shikata, E., and Maramorosch, K., "Mycoplasma-like bodies in sieve tubes of yellows diseased plants and in fatbody cells of two insect vectors," *Phytopathology*, **59**, 1559 (Abstr.) (1969).

Shikata, E., Teng, W. S., and Matsumota, T., "Mycoplasma-like or PLTA-like microorganisms detected in leaves of sugarcane plants infected with white-leaf disease and the suppression of the disease symptoms by the antibiotics of tetracycline group," *J. Facul. Agr. Hokkaido Univ.,* **56,** 79–90 (1969).

Story, G. E., and Halliwell, R. S., "Association of a mycoplasma-like organism with the bunchy top disease of papaya," *Phytopathology,* **59,** 1335 (1969).

Tsao, P. W., Schneider, H., and Kaloostian, G. H., "A brown leaf-vein symptom associated with greenhouse grown pear plants infected with pear decline virus," *Plant Disease Reptr.,* **50,** 270–274 (1966).

Ushiyama, R., Bullivant, S., and Matthews, R. E. F., "A mycoplasma-like organism associated with *Phormium* yellow leaf disease," *N. Z. J. Botany,* **7,** 363–371 (1969).

Whitcomb, R. F., and Davis, R. E., "Mycoplasma and phytarboviruses as plant pathogens persistently transmitted by insects," *Ann. Rev. Entomology,* **15,** 405–464 (1970).

Worley, J. F., "Possible replicative forms of a mycoplasma-like organism and their location in aster-yellows diseased *Nicotiana* and aster," *Phytopathology,* **60,** 284 (1970).

Fungal
Pathogens

Wind, water, man, and animals spread pathogenic fungi as effectively as do insects, and relatively few fungal diseases depend on insects for the dissemination of the fungus spores. Spores that are adapted to dissemination by insects, particularly those that are "wet" and produced in sticky masses, have a better chance of reaching a suitable substrate because they are usually transported directly to the susceptible plant and are often deposited in wounds where infection occurs immediately. Wind and water dispersal are more wasteful of inoculum. Man and animals assist in dissemination by transporting infected material or by wounding susceptible plants and thus opening courts of entry. Ingold (1953), in his monograph on dispersal in fungi, contributes a chapter on dispersal by insects. Austwick (1958) has estimated that some 45 different fungal diseases of 34 plant species involving over 100 species of insect and 66 species of fungi of all classes are found in more or less close association. Many of these associations are casual, and some are inferred rather than established; the latter have been generally excluded from consideration in the discussion of individual cases which follows in this chapter.

Attempts at classification of diseases with which insects are associated have been described in Chapter 1, including, as far as the fungi are concerned, Austwick's (1958) rearrangement of the classifications proposed by

Leach (1940) following that of Böning (1929).

As Austwick remarks, our knowledge of the epidemiology of these diseases is not adequate for classification of the fungus-insect relationships. Actually, all such classifications are categories of convenience without phylogenetic significance, except those inferred by differences in complexity of the insect-fungus relationship. In this chapter, specific cases are considered under the following headings:

1. Infection incident to pollination
2. Infection through traumatic injury
3. Internal and external contamination of the insect as a result of feeding on fungal masses
4. Fungi developing on insect exudates
5. Infection through feeding and oviposition wounds
6. Infection through feeding punctures
7. Infection resulting from symbiotic association between insect and fungus

INFECTION INCIDENT TO POLLINATION

Cases of infection incident to pollination are considered to be the simplest form of transmission of pathogens by insects. Another smut of campions (*Melandrium* spp.) is caused by *Ustilago violacea* (P) Fuckel; in the diseased plant, the pollen is replaced by smut spores without affecting the petals or their attractiveness to insects, which transfer the spores to previously unaffected flowers (Brefeld and Falck, 1905). Another smut of pinks is caused by the same fungus and the phenomenon of infection is the same. A similar relationship exists between insects and blossom blight of red clover (*Trifolium pratense* L) and *Botrytis anthophila* Bond, according to Silow (1933), who demonstrated that bees are the principal vectors. Insects are not, however, essential for primary infection of these diseases since both are systemic and both are seed-borne. A more complex insect relationship is involved in the transmission of the organisms causing endosepsis of figs by the wasp *Blastophaga psenes* L, but since infection is incidental to pollination the disease is considered here rather than in a more general category of fruit spoilage diseases.

Endosepsis of Figs

Endosepsis of figs is one of the three important diseases of the fruit, manifesting itself as soon as the fruits begin to ripen. Brown streaks develop

FIGURE 4.1. Souring of figs. A fruit affected with souring. Note exudate from the "eye" (After Smith and Hansen, 1931).

down the flower stalks and progress until the entire pulp is disintegrated. Development of external symptoms depends on weather conditions, dryness being conducive to normal drying of the fig without external symptoms; under moist conditions, the fig appears sagging, wet, dripping, and deformed (Fig. 4.1 and 4.2).

The disease was first described by Caldis (1927), who named the causal organism *Fusarium moniliforme* var. *fici* n. var. Transmission occurs when the fig flower is pollinated. The fig is dioecious and insect-pollinated; the staminate and pistillate flowers are borne on different trees. The male fig, known as the caprifig, bears staminate flowers around the opening and gall flowers in the rest of the cavity; the female fig bears only pistillate flowers. Both sexes produce figs, those of the caprifig being inedible. The fruit of both is actually an aggregation of fruits lining the cavity of a hollow receptacle with an opening developing at the center of

FIGURE 4.2. The same fruit opened to show internal breakdown
(After Smith and Hansen, 1931).

the flattened distal end as the fruit begins to ripen.

The process of pollinating the pistillate flowers is known as caprification and is effected by the fig wasp. The fig wasp parasitizes the gall flowers of the male, or caprifig, by inserting its ovipositor through the style
and depositing an egg in the ovule. Although Caldis does not mention it,
the presumption is that oviposition results in the stimulus to gall formation. The egg hatches and the larva lives in the gall to maturity and pupation. The adult females are fertilized while still in the galls and emerge
through the eye of the fig, brushing against the staminate flowers which
surround the eye and carrying the pollen to the flowers of the pistillate

female fig. The wasp enters the female fig receptacle for purposes of oviposition but, owing to the length of the styles of the female fig, oviposition is not accomplished, although pollination occurs.

Transmission of the pathogen by the wasp was convincingly demonstrated by Caldis. Infection is carried externally, the insect picking up spores in the cavity of the fig after issuing from the gall and before coming out of the eye. The spores lodge among the spines with which the insect's body and appendages are covered.

Control of endosepsis seems to be largely a matter of orchard sanitation and the destruction of surplus caprifigs (Smith and Hansen, 1927). These writers also succeeded in disinfecting caprifigs containing *Blastophaga* without injuring the insect, but the method is unsuitable for large areas.

INFECTION THROUGH TRAUMATIC INJURY

Azalea Flower Spot

Although, in the very broad sense, any sort of insect feeding can be considered as causing trauma, azalea flower spot has been set in this unique category because of the mechanism of infection.

Azalea flower spot (*Ovulinia azaleae* Weiss) is spread by at least 11 species of bees, 3 species of ants, and a thrips (*Heterothrips azaleae* Hood), according to Smith and Weiss (1942). Curiously, although pollen was taken from infective insects, no infection was obtained from it, even though the pollen itself carried spores of the fungus. Infections were obtained, however, from the heads and legs of the insects.

The disease is characterized by the appearance of small specks, called the fleck stage, which enlarge to become irregular blotches spreading to one or more petals. The disease then progresses to the stage known as limp blight (Fig. 4.3). According to Smith and Weiss, insects are probably not concerned in primary infection, nor is there any evidence that bees or other insects carry the organisms over in their winter quarters. The infectivity of the insects increases later in the season at which time the insects are effective vectors, gradually dropping their spore load so that a succession of flowers is infected. The insects are considered more important in spreading the fungus to areas as far as two miles away from the source infection, rather than serving to increase damage in any one planting. The fungus, once established, is capable of re-entry without insect aid. The mechanism of infection is unusual. The delicate flowers are scratched and abraded by the spines on the insect's legs, the wounds serving as courts of entry for the spores carried on the insect's body (Fig. 4.4).

(a)

(b)

FIGURE 4.3. Azalea flower spot. (a) Azalea plant in full bloom with flowers undamaged. (b) Azalea plant with flowers in the limp blight stage (Photographs: Smith and Weiss, 1942).

FIGURE 4.4. Azalea flower spot. Extensive injury to late-season flower, after numerous visits by bees, showing the tissue clotted, purple, and collapsed (Photograph: Smith and Weiss, 1942).

Smith and Weiss concluded that disease control, rather than insect.control, was the best method of approach since the insects are apparently not involved in primary infection. Gill (1953) has reviewed recent work on control of the fungus and reports that Nabam (Dithane D-14 and Parzate liquid) and zineb (Dithane Z-78 and Parzate) have been outstanding fungicides against petal blight. Brierley (1950) used chemicals applied to the soil in order to prevent formation of apothecia. Calcium cyanamide applied to the soil around azalea plants at the rate of 400 lb/ac was successful, but there is danger to surrounding plants. For such controls to be worthwhile they would need to be used on a community-wide basis.

INTERNAL AND EXTERNAL CONTAMINATION OF THE INSECT AS A RESULT OF FEEDING ON FUNGAL MASSES

In this category belong ergot, witch's broom of hackberry, certain powdery mildews, and the rust diseases of the Gramineae.

Ergot

Ergot, a disease common on grasses and cereals, occurs most commonly on rye. *Claviceps purpurea* (Fr.) Tul. is the principal fungus involved. The imperfect (*Sphacelia*) stage of the fungus develops on the flowers, which exude honeydew, a sweet and sticky substance in which the conidia of the fungus are found. As early as 1847, Leunis noted the insect visitors to this honeydew and suggested their possible role in transmission of the fungus, but it was not until many years later that experimental proof became available. The most common insects visiting this fungus-bearing sticky mass are Diptera and a few Coleoptera. Since it has been shown that the honeydew of ergot contains a large number of amino acids, its attractiveness and usefulness to insects can be readily understood. Mercier (1911) showed that the spores of the fungus were found externally on a fungus gnat (*Sciara thomae* L) and were passed through the intestinal tract in a viable condition. Langdon and Champ (1954) confirmed this also for the green fly *Pyrellia coerulea* (Wied.) transmitting *Claviceps paspali* Stev. and Hall to *Paspalum dilatatum* Poir in Queensland. Primary infection is not dependent on insects, since the ascospores can be disseminated by wind before the development of the sticky honeydew, but insects appear to be the principal agents of secondary infection, and Atanasoff (1920), quoting Stager, lists 37 species of flies and beetles observed visiting the honeydew.

Langdon and Champ (1954) have separated the vectors of ergot into two groups: (1) those which carry inoculum on the external surfaces of their bodies, and (2) those which ingest conidia. In the first group are the chance visitors *Paspalum dilatatum,* such as thrips and aphids, which normally are found on flowers and which, therefore, can be important in carrying the fungus to healthy flowers. The second group includes a number of insects which are strong fliers. These are more important in long-range spread and could well be the principal agents of pandemics.

Witch's Broom of Hackberry

Witch's broom of hackberry and its relationship to eriophyd mites was recently investigated by Snetsinger and Himelick (1957). Earlier references go back over fifty years—to Kellerman and Swingle (1888, 1889) and Salmon (1900)—who described the fungus (*Sphaerotheca phytopto-phila* Kell. and Swingle) and discussed the relationship between it and the mite.

This more recent paper, while making no claim to either confirming or refuting the earlier work, actually does confirm the association between a mite of the genus *Aceria* and a powdery mildew corresponding to the earlier species described. The cleistophecia of the fungus were never found unless the mite was present, and hackberry trees free from witch's broom were not infested with the mite. What is lacking is knowledge of the mechanism of transmission.

Powdery Mildews

An association of thrips with several species of powdery mildew on grape in California, described by Yarwood (1943), is of quite a different character. In this case, the thrips feed on the mildew and can destroy established mildew colonies on rose and grape leaves. While no evidence for transmission is presented, it would seem extremely likely that this occurs when adult thrips fly off to uninfested plants to feed in their normal manner on the leaf tissue.

Insects and Diploidizing of Rust Fungi

The attraction of insects to the spermagonia of rust fungi was noted long ago by Persoon (1801) and Rathay (1882). The latter author recorded 139 species of insects visiting these pycnia, as they are now called, and actually determined the presence of the spermatia (pycniospores) of *Endophyllum* in the wet footmarks of flies on a glass window. The significance of this was emphasized by Craigie (1927), who discovered that these spores are the diploidizing agents of the rust fungi, which enable the crossing of different physiological races. The resultant new races may be extremely virulent to cereals, and the plant breeder has found it necessary to maintain a constant program of breeding resistant varieties to meet the ever-arising new threats.

FUNGI DEVELOPING ON INSECT EXUDATES

This category of fungi developing on insect exudates includes the sooty molds, and perhaps it should also include phthiriasis of coffee, vines, and citrus. This latter is, however, a more specific case which involves a symbiotic relationship, and for that reason it is included in the last category in this chapter.

Sooty Molds

The sooty mold fungi characteristically appear as a black layer on the surface of the leaves of many trees, particularly those in tropical and subtropical areas. Citrus is especially susceptible, but many other trees and shrubs are affected. In Hawaii, gardenias often present a completely blackened appearance, and most of the local limes and other citrus are badly disfigured. These fungi are not parasitic but live on the honeydew deposited by many species of insect. Since honeydew has been found to contain many amino acids (Gray, 1952; Maltais and Auclair, 1952; Mittler, 1953) as well as sugars, it offers an excellent growth medium. Apart from the unsightly appearance of affected trees, photosynthesis is interfered with; trees sprayed to eliminate the insects will show severe chlorosis of the leaves when the black mold drops off, but they soon recover. *Coccus viridis* Green is a principal offender among the insects in Hawaii, but aleyrodids, scales, and mealybugs as well as aphids have all been implicated (Webber, 1897).

Yamamoto (1951) has shown that many insects, other than those that secrete honeydew, disseminate spore and hyphal fragments of sooty molds. Fragments of the fungi pass through the intestinal tracts of these insects and emerge in a viable condition. Included in this group are flies, wasps, bees, and ladybird beetles; since honeydew is an excellent source of food for these insects, transfer of the sooty molds can easily occur.

Bonnemaison and Missonier (1956) described the toxic damage caused by the feeding of *Psylla pyri* L, but they also reported that the sooty mold that develops on the honeydew excreted by the insects reduces photosynthesis.

The Comstock mealybug, *Pseudococcus comstocki* (Kuw.), secretes sticky honeydew on the leaves of the catalpa tree on which sooty mold grows, giving the foliage a dirty, distressed appearance (Hough, 1925; Cutright, 1939). When the insect is destroyed, the condition disappears.

INFECTION THROUGH FEEDING AND OVIPOSITION WOUNDS

This category of infection through feeding and oviposition wounds is a broad one as far as the range of diseases caused by fungi entering through feeding wounds is concerned. Most of the insects involved are those which feed grossly and, with two exceptions, are mandibulate insects. The two exceptions, capsids on cacao, and the beech scale, although sucking insects, nevertheless affect tissue grossly.

Persimmon Wilt

Persimmon wilt was first discovered in Tennessee in 1937 and proved to be the fastest-killing disease of vigorous and healthy trees on record (Crandall and Baker, 1950). The fungus is *Cephalosporium diospyri* Crandall, and it gains entrance through wounds, including those made by wind, browsing cattle, and insects. The authors made a thorough search for insect species causing wounding and settled on two beetles as the most likely suspects. One was the powder-post beetle, *Xylobiops basilaris* (Say) and the other, the twig girdler, *Oncideres cingulatus* (Say). The powder-post beetle normally attacks dead and dying trees but will, in the absence of more suitable host material, attack healthy trees. Only a very small percentage of such attacks is successful, however, due to the reaction of the healthy tree to the wound. A mass of gum is thrown out by the tree, filling the tunnel and forcing the insect to vacate. When ground moisture is deficient, however, the gumming process is sometimes long delayed and the tunnels remain exposed to spore entrance. The wounds made by *Oncideres* are also often sealed over with gum, but there is wide variation in the rate at which it happens. The fungus was not recovered from the beetles' bodies but was cultured from wood taken from immediately behind wounds made by the insects, with a higher incidence of infection in *Oncideres* wounds.

Black and Yellow Molds of Corn

Black mold (*Aspergillus niger* van Tiegh) and yellow mold (*A. flavus* Link) of corn invariably start growth on maize at the point where an insect, especially the corn earworm (*Heliothis zeae* Boddie) has penetrated the husk into the ear. The moist excrement of the insect provides an excellent medium for the fungus. In young, tender ears the fungus spreads rapidly, but in partly ripened ears infection will remain localized within the feeding area of the insect. Thoroughly ripened ears are practically immune from either earworms or fungus. The insect carries spores of the fungi on its body. Both organisms are also frequently found on cotton bolls that have been attacked by *H. zeae*.

According to Taubenhaus (1920), the disease is most prevalent in dry seasons when the insect is most prevalent. Control of the corn earworm is described as the remedy, with worm-resistant varieties recommended. Shank, stalk, and ear rots are important diseases of corn in the midwestern states, and spread of the European corn borer has increased the incidence of stalk rot. The greater the infestation of the corn borer, the greater the amount of stalk rot. Varieties of corn normally resistant to the rot develop considerable injury when infested with the borer.

"The corn-borer aids in the development of rots in several ways; it provides an entry for fungi; the boring of the larvae distributes organisms inside the plant; the frass within larval tunnels serves as a medium for the rapid increase of the invading organisms and the injuries caused by the borer weaken adjoining tissues, thereby making them more subject to fungus attack." (Christensen and Schneider, 1950).

There are many species of pathogens associated with infested corn tissue. *Fusarium* is the most common, but Taylor (1952) listed *Cephalosporium acremonium* Corda, *Fusarium moniliforme* Sheldon, bacteria and yeasts generally, *Gibberella zeae* (Schwabe) Petch, *Nigrospora oryzae* (B. and Br.) Petch, and *Mucor* species occasionally, and *Diplodia zeae* (Schwabe) Lev. rarely, as associated with corn borer and corn rots. Ullstrup (1953) described several ear rots of corn but ascribed susceptibility to hybrid characters and spread to wind and rain. Only in the case of some widely distributed fungi, not specialized in their parasitism, was insect intervention noted by this author.

Root and Basal Stem Rots of Cereals and Grasses

Root and basal stem rots of cereals and grasses are inconspicuous but destructive diseases which attack the roots and the basal portions of the culms of many cereals and grasses. They are caused by many species of fungi which live in or on seed, in the soil, and in dead plant tissue. Some of these fungi are relatively weak parasites and can cause damage only when the plant is weakened or injured by insects, nematodes, or mechanical damage (Christensen, 1953).

Although several kinds of insect are involved, the most important appear to be the billbugs, of which Satterthwait (1932) records 18 species attacking cereals and grasses in the United States. Of these, *Calendra parvula* (Gyllenhal) is the most common and destructive.

Hanson *et al.* (1950) have discussed the role of the insect in excellent detail, and this account is drawn largely from their discussion. The insect attacks the lower internodes of the stems and roots of grasses and cereals, and because of this is an important factor in the development of root and basal stem rots. The insect weakens the plant by its feeding and therefore predisposes the plant to attack by pathogenic fungi, including many that require wounds as courts of entry. In addition, the insect carries the pathogens both on and in its body. Decay starts around wounds produced by the adult when feeding and ovipositing and then follows the path of the larvae inside the culm. The larval frass is also a favorable medium for the organisms. Fortunately, the short-lived cereals are not attacked early enough in the season to prevent maturing the seed, even though they may be killed by rots toward the end of the season. The earlier the infestation, other factors being equal, the greater is the severity of rots and damage to the plant on the perennial grasses; both insects and rot have a greater opportunity to develop and to continue their activities until late in the fall. Loss of hay and pasture, and premature failure of sod, sometimes occur.

Control is aimed at the control of the insect, which is possible with good cultural conditions and suitable crop rotations using immune crops such as alfalfa, clover, soybeans, flax, or Irish potato.

Most of the fungi isolated from infected cereals and grasses were *Fusarium* spp. and the predominance of this group increased as the season progressed. On some hosts *Helminthosporium* was prevalent early in the season. Another species of billbug, *Anacentrus deplanatus* (Casey), has been reported associated with root and basal stem rot of barnyard grass (*Echinochloa crusgalli* Beauv.) by Hanson and Milliron (1942). This species of weevil has been definitely reported only from barnyard grass and sugarcane, but it is unlikely that it is so limited in its host range.

Boosalis (1954) reported that the Hessian fly (*Phytophaga destructor* (Say)) has become an important factor in the increase of crown and basal stem rot of wheat. Since the larvae of the Hessian fly attack the culms, the role of this insect is essentially the same as that of the billbugs.

Insect involvement in the development of these diseases of cereals and grasses complicates the work of the plant breeder in developing resistant varieties because some plants showing resistance may develop severe infection when injured by insects.

Anthracnose

Collectotrichum lagenarium (Pass.) Ell. and Halst., the anthracnose fungus, often occurs at or near the feeding sites of the spotted cucumber beetle (*Diabrotica undecimpunctata howardi* Barber). Experiments by Libby and Ellis (1954) demonstrated that the beetle transmitted the fungus even though the number of positive plants was low.

Gleosporium musarum *in* Musa balbiniana

Gleosporium musarum causing anthracnose of *Musa babiniana,* an important dietary staple in Costa Rica, is transmitted by *Polybia occidentalis* Oliv., *Synoeca surinama* (L), and *Trigona* spp. The floral parts of the *Musa* are the sites of infection and when the floral parts are dropped while still fresh (a peculiarity of the *Musa* species) the sweet deposit attracts the vectors which are Hymenoptera (Cardenos-Barrigal, 1963).

Stem or Bole Rot of Sisal

Aspergillus niger, already implicated in corn stem rots and fruit spoilage, is also responsible for stem or bole rot of sisal (*Agave sisalensis*). The sisal weevil (*Scyophophorus interstitialis* Glyh.) feeds on infected plants; spores of the fungus are ingested by the weevil and retain viability and pathogenicity after passage through the alimentary tract. Although not considered a primary agent of spread the weevil may be contributory (Wienk, 1965).

Boll Rot of Cotton

Boll rot of cotton can be initiated by several insect species, e.g., through feeding and oviposition wounds of *Anthonomus grandis* (Boh.), *Heliothis zea* (Boddie), and *Lygus lineolaris* (P. de B.). Nectar feeding by *Drosophila melanogaster* (Meig.) and *Trichoplusia ni* (Hb.) caused infection in the inner and outer involucral nectaries, leading to boll infection. The fungi involved are *Fusarium moniliforme* and *Alternaria tenuis* (Bagga and Laster, 1968).

Aspergillus flavus *in* Cotton

Infection of cotton bolls by *A. flavus* and consequent contamination of the seed with aflatoxin is predisposed by the presence of the boll weevil.

The fungus apparently enters through the exit holes of the mature larvae, but not, however, through the small holes made by the first instar larvae when they first enter the boll (Ashworth *et al.*, 1971).

Aspergillus flavus *in Groundnut*

Mites (*Sancassania* and *Tyrophagus*) can be maintained on pure diets of the fungus or of groundnut. They are heavily infested with spores and can disseminate the spores when they penetrate the groundnut pods and feed on the kernels. Aflatoxin is produced in the kernels (Aucamp, 1969).

Red Rot of Sugarcane

Red rot of sugarcane has a common court of entry for the fungus *Colletotrichum falcatum* Went. through the tunnels of the moth borer *Diatraea saccharalis* (Fab.). The resulting disease is known as red rot, a serious disease in the southern United States and in many tropical and subtropical sugarcane-growing areas. Losses are in seed-pieces, resulting in faulty stands or sometimes in total loss; reductions of stands in the following or ratoon crops; and losses in sucrose in the mature cane following infection in the stalks attacked by the moth-borer. Fungus lesions on the leaf midribs are not seriously injurious to the plant, but they are important as sources of spores that cause stalk infection. These lesions are dark or blood-red in color. Infection after harvest of the underground parts of the stem, on which the second or ratoon growth depends, may occur through the tunnels of the sugarcane weevil (*Anacentrinus subnudus* Buchanan).

Effective control depends on the development of resistant varieties; according to Abbott (1953), some progress has been made in this direction. The use of seed cane free of borer infestation is desirable.

Taproot Rot of White Clover

Insects and fungi are related to taproot survival of white clover. The taproots of white clover (*Trifolium repens* L) are subject to diseases which kill off the roots and reduce the stand. Kilpatrick and Dunn (1958) sampled taproots at monthly intervals. The May samples were severely damaged and girdled by larvae of the clover root curculio (*Sitona* spp.). Later, adults of *S. hispidula* (F.) and *S. flavescens* (Marsh) were found in clover fields. Root rot developed rapidly, and *Fusarium oxysporum* Schlecht. was the major fungus isolated.

Fruit Spoilage Diseases

Fruit spoilage diseases, with the exception of endosepsis of fig, already discussed, are initiated by insects in many cases. Souring of figs can best be observed on the fruit of parthenocarpic varieties of fig that do not require caprification and, therefore, are not subject to the more specific endosepsis. Caldis (1930) restricts the term "souring" to cover spoilage by fermentation organisms. The organisms concerned are three species of yeast which Caldis describes. An affected fig first shows a change in the color of the pulp which loses its pink color and becomes watery. A pink liquid exudes from the eye, dropping on to the leaves or jellying at the eye. In the latest stages the figs shrivel and dry and may remain attached to the twigs.

Caldis used the findings of Phillips *et al.* (1925) to initiate studies on insect transmission of the fermenting yeasts, and tested the dried-fruit beetle, *Carpophilus hemipterus* L, and the vinegar fly, *Drosophila melanogaster* Low. Both of these species, the most commonly found on figs still on the tree according to Phillips *et al.*, were found to transmit the organisms, both internally and externally.

"Smuts and Molds" is the designation of one of the spoilage grades in dried figs, according to Hansen and Davey (1932). This designation includes the black dusty type of smut (*Aspergillus niger* van Tiegh) as well as the variously colored growths of other fungi. Hansen had earlier (1929) reported the presence of thrips carrying cryptogamic organisms into green figs of parthenocarpic varieties, and in 1931 Smith and Hansen reported predaceous mites in the same role. Hansen and Davey showed that mites and thrips entered the figs long before the eye scales began to loosen, and that these insects were carrying the organisms of smut and mold with them. The presence of larger insects such as *C. hemipterus* and *D. melanogaster* was not necessary for the initiation of the disease in the fig. Davey and Smith (1933) also found that the dried-fruit beetle was not an important factor in the transmission of smut and mold; predaceous mites, and, to a lesser extent, thrips, were the only insects to which transmission could be attributed.

Under the general designation of "fig rot," Schewket (Sevket) (1934) recorded the association of *Lonchacea aristella* Beck with rot in Smyrna figs. The flies oviposit in the profici figs in May, and a month later the new generation of flies emerges. Oviposition follows in the young fruits of cultivated figs and losses of 25–30% of the crop can occur.

Brown Rot of Stone Fruits

Brown rot of stone fruits is extremely destructive and infection is greatly facilitated by wounding. The plum curculio (*Conotrachelus nenuphar*

Herbst.) both feeds and oviposits on the fruit, and the wounds so made are portals of entry for the fungus (*Sclerotinia laxa* Aderh. and Ruhl). Scott and Ayres (1910) reported that 93% of fruit infection could be traced to curculio wounds, although brown rot is often present in the absence of curculios. Ogawa (1957) showed that *C. hemipterus* was contaminated with the spores of *Sclerotinia fructicola* (Wint.) Rehm. by feeding on diseased peaches and was able to transmit the organism to naturally wounded peaches. *Carpophilus* beetles are known in Hawaii as "souring beetles." They are attracted to overripe and rotten fruit, and the ripe bases of pineapple fruits are sometimes black with them. They are responsible for the introduction of fermenting yeasts into the fruits, but they also serve a useful purpose in the decay of the starchy stumps of the pineapple plant when a field is knocked down in preparation for a new planting. Schmidt (1935) has described the life history of these insects in this connection in adequate detail.

Black mold and brown rot (*Rhizopus nigricans* Ehr. and *Monilinia fructicola* [Wint.] Rehm) are greatly increased in peaches by *Drosophila* spp. infestations. Willison and Dustan (1956) found that the stage of maturity that fruit reached during the holding period after picking is of prime importance. In the early stages of ripening, the flies appear to play an active part by implanting inoculum in injured areas. The amount of stem-end injury therefore governs the extent of brown rot wastage. With increase in susceptibility as the fruit ripens further, the activities of the flies become more important. Black mold apparently cannot make entry until the fruit is nearly ripe, but the flies are efficient vectors of the inoculum.

The association of the weevil *Rhyncites baccus* (L) with the conidia of *Monilia* spp. has been pointed out by Stojanovic (1956). Conidia of the fungus were found on the legs and lower parts of the abdomen of the weevil and it is presumed that, since the weevils are fruit-feeders, they may play an important part in the spread of the fungus.

Brown Rot of Apples

Brown rot of apples is caused by *Sclerotinia fructigena* Aderh. and Ruhl. Its entry is facilitated by the feeding of the earwig, *Forficula auricularia* L, which bores circular holes about 4–6 mm in diameter and hollows out a shallow cavity beneath the skin of the apple. Brown rot follows rapidly, and it is not unusual to find 10% brown rot under these conditions. Grassing-down orchards contributes to the problem since earwigs are present in orchard grass the year round, and they become precursors of the brown rot in the fall when the fruit begins to ripen. Measures recommended for control are those designed to control the insect, namely, banding trees with sacks soaked in insecticide or coating the ground below the trees with residual insecticides in the early summer (Croxall, Collingwood, and Jenkins, 1951).

Needle Blight and Late Fall Browning of Red Pine

Haddow and Adamson (1939) and Haddow (1941) report that since 1932, plantations of red pine (*Pinus resinosa* Lamb) have suffered from two abnormalities of the current season's foliage. One of these is needle blight, which affects the lower part of the needles and causes them to become hooked and drop. The other is a browning of the foliage which occurs in the late fall and gives the tree the appearance of having been damaged by fire. Both types of injury are associated with infestation by a cecidomyid midge. The adults emerge in spring or early summer and oviposit in the sheath of scales at the base of the needles. The larvae hatch in July and form cavities at the base of the needles. Needle blight was found to be due to invasion of the bases of the shoots by the fungus *Pullularia pullulans* (de Bary) Berkl., which is commonly present on the sheath scales. The exact methods of transmission of the fungus are not known. Late fall browning of the foliage was found to be due directly to injury by the larvae.

Beech Bark Disease

The fungus *Nectria coccinea,* which follows the attack of the beech scale *Cryptococcus fagi* (Baer), has become a serious menace to beech trees in Canada and the United States. Fruiting bodies develop in areas on the stem and branches previously occupied by the scale (Ehrlich, 1934). As infestation progresses, the foliage and twigs wither and die; whole branches cease to put forth leaves; and large areas of bark on the trunk crack, loosen from the wood, and fall away. In general, the fungus is able to enter and extensively infect only bark on which the insect has been present for a year or more. According to Hawboldt (1944), no appreciable fungus injury follows a light infestation of the insect, but a heavy one is followed by the death of patches of bark and sometimes of the cambium.

The fungus enters through the tissue lesions caused by the insects' feeding and grows rapidly through the bark, phloem, cambium, and peripheral sapwood. The general health of the tree is greatly impaired by the scale infestation, and the tree then becomes more susceptible to the fungus. The fungus and the egg and larval stages of the scale are disseminated by wind; the fungus spores are also disseminated by rain. Control on ornamental trees can be effected by control of the scale. In forest stands, infected and recently killed timber should be removed.

Calonectria rigidiuscula (*Berk. and Br.*) *Sacc. and its Pathogenicity to Cacao* (Theobroma cacao L)

The effect of this fungus on cacao and the etiology of the disease bear a fairly close parallel to that of *Nematospora* on cotton in that the feeding of a plant bug provides the initial court of entry. Two capsid species, *Sahlbergella singularis* Hagl. and *Distantiella theobroma* Distant, are major pests of cacao in West Africa. They are both extremely toxic, and, in addition, their feeding lesions are ideal courts of entry for the fungus *Calonectria rigidiuscula*. Crowdy (1947) has described the symptoms and the association of the fungus with capsid feeding lesions while Owen (1956) reported that capsid lesions particularly favor the development of the fungus. A. L. Wharton, in an unpublished note, describes the relationship as follows: "In single capsid lesions on young shoots and twigs, the fungus becomes occluded in infected lesions and so far there is no evidence that such infections ever break out. The fungus, however, does remain viable and can, in fact, be isolated from the center of old tree trunks and branches from lesions which must have originated when the trees were seedlings or saplings. Massive capsid attack and/or other debilitating factors result in *Calonectria* infections which are not occluded. The fungus continues unchecked in the debilitated tissue causing a chronic stage of infection—canopy and branch die-back. When the original cause

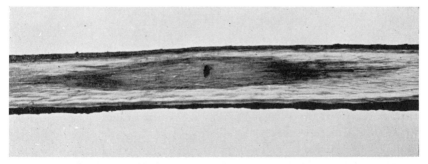

FIGURE 4.5. Fungal attack (probably *Calonectria rigidiuscula* (Berk. and Br.) Sacc. and/or *Botryodiplodia theobromae* Pat.) in the wood of a 1-yr-old cacao seedling following gallery excavation by *Xyleborus morstatti* Hag. (Coleoptera, Scolytidae). (Photograph West African Cacao Research Institute, Nigeria).

of ill health is removed or eliminated, e.g., by chemical control of capsids, the trees may regenerate from sound tissues below the die-back areas. The fungus is checked when it reaches the new vigorous growth and eventually the old dead tissues containing the *Calonectria* are shed. After a number of years it is difficult to detect any pre-history of capsid canker and *Calonectria* die-back on such regenerated trees."

There is some variation in susceptibility and some upper Amazon varieties may be resistant. Since these latter are now being used extensively in breeding of cacao varieties in the West African region, some particularly resistant types may accrue. *Botryodiplodia theobromae* Pat. is next in importance to *Calonectria* in its relationship to die-back (Fig. 4.5).

Capsids are not the only insects concerned. Scolytid borings also provide opportunity for invasion of the fungi.

Cushion Gall of Cacao

Cushion gall of cacao in Ghana has also been shown to be caused by infection with *C. rigidiuscula* (Brunt and Wharton, 1960). A naturally occurring gall was found to be infested with the mealybug *Pseudococcus njalensis* Laing. When these were transferred to half-beans, galls developed in a high percentage of cases. The organism in culture is extremely infectious; merely painting the half-bean with a suspension of spores is sufficient to infect. The epidemiology of this disease is extremely interesting: in Central America cushion gall is epidemic and can sterilize a tree; in Ghana it is of rare occurrence. Clearly, the mere presence of *P. njalensis,* universal as it is in Ghana cacao, is not enough. Perhaps a rare strain of the fungus is involved; or *P. njalensis* may not be an efficient vector, or even a vector at all under field conditions. A comparative study of the epidemiology of this disease in Ghana and in Central America is clearly called for (Fig. 4.6).

Tree Cricket Canker of Apples

Although feeding and oviposition wounds most frequently occur together, and separation of these as separate courts of entry is not justified, the case of tree cricket canker appears to be one in which the oviposition wound is a specialized court of entry for the organism causing a canker of apple trees, *Leptosphaeria coniothyrium* (Fuckel) Sacc. This fungus also causes blight of raspberry and is associated with attacks of the raspberry cane midge, *Thomasiniana theobaldi* Barnes (Brooks, 1928). The tree crickets, *Oecanthus niveus* DeG. and *O. angustipennis* Fitch, are gross feeders, using soft-bodied insects and decaying plant tissue, along with

(a)

(b)

FIGURE 4.6. Cushion gall of cacao. (a) Chupon, terminal gall (natural infection). Contained *Pseudococcus njalensis* colonies from which first transmission to beans obtained. (b) Example of first transmission to half-beans after feeding *P. njalensis* from tree gall at the rate of 10/bean (Photographs: West African Cacao Research Institute, Ghana).

other organic materials, as food. Their bodies are liberally contaminated with viable spores of many species of fungi both internally and externally.

The insects oviposit in the bark of the tree, first chewing a hole and then depositing the egg. The hole is then plugged with a pellet of fecal matter, in the case of *O. niveus,* and with a pellet of chewed wood in the case of *O. angustipennis.* When infected feces (or diseased wood) are used for this purpose, the introduction of the pathogen is accomplished efficiently (Parrot and Fulton, 1914; Parrot *et al.,* 1915; Gloyer and Fulton, 1916).

Thielaviopsis paradoxa *in Cacao Pods*

An unusual relationship between birds and fungal infection was noted in Costa Rica where white-tipped brown jays puncture cacao pods, permitting entry of *T. paradoxa* causing pod rot. 97% of infected pods had been punctured (Hansen, 1964).

Alternaria brassicola *in Cabbage,* Phytophora *in* Blackpepper Foot Rot

Slugs and snails serve as vectors of fungi. *A. brassicola* is carried by slugs to cabbage (Hasan and Vago, 1966). Snails were found to be carrying zoospores and sporangia of *Phytophora* causing foot rot of black pepper (*Piper nigrum*) in Sarawak. The snail provides a means of continuing spread during the dry season; peak spread is during the wet season (Turner, 1964).

Fungus Gnats and Fusarium *in* Alfalfa and Red Clover

There is an interaction between these fungi and the larvae of fungus gnats (*Bradysia* spp.; *Sciaridae* spp.) in greenhouses. The larvae feed on the roots of seedlings and thereby predispose the plants to fungal attack. Plants slightly injured by the larvae are killed at a much higher rate than are uninjured plants (Leath and Newton, 1969).

INFECTION THROUGH FEEDING PUNCTURES

By definition, this group of diseases is one with which sucking insects are concerned. The feeding lesions of sucking insects are usually restricted to specific tissues—the actual external feeding lesion is very small—yet these characteristics lend themselves to effective transmission of those fungi specifically adapted to entry under these conditions.

Alternaria *Blight of Rubber Trees*

Martin (1947) described a blight of rubber trees caused by an apparently new *Alternaria* species, which results in defoliation early in the development of the leaf cycle. Infected leaves which develop to maturity may have brown, concentrically zonate spots in between the veins of the leaf, or somewhat elongate spots along the main veins. One clone could be infected by the fungus without insect or mechanical injury, but in the presence of punctures of the insect, *Tomaspis inca* Guer., the disease was much more severe. Another clone showed no symptoms in the absence of the insect. However, typical, although light, blight occurred when the insect punctures were present.

Since no data are available on the function of the insect, other than the association of severe disease with insect-punctured leaves, the presumption is that the feeding punctures are courts of entry.

Cabbage Phomosis

Cabbage phomosis caused by *Phoma lingam* (Tode) Desmaz. is severe in wet seasons in the USSR, although first-year plants, weakened by the absence of sufficient moisture, are also susceptible. By laboratory experiments, Buryhina (1950) showed that inoculated plants contracted the disease only when infested with the harlequin cabbage bug, *Murgantia histrionica* Hahn.

Sclerospora *on Corn and Sorghum*

Sclerospora sorghi infects corn and sorghum. When infected plants are kept at 21°C day and 7°C night temperatures, loss of fungal mycelium occurs. When these plants are fed on by *Rhopalosiphon maidis* and *Schizaphis graminum* and then transferred to healthy plants, variously shaped and sized nuclear bodies found in the source plants are transmitted. The bodies could be found in the recipient plants before a mosaic-like symptom appeared, 1 wk after feeding (Naqvi and Futrell, 1970). This preliminary datum will no doubt be amplified in due course; it is, without doubt, unique.

Leaf Spot of Dieffenbachia

The citrus mealybug, *Pseudococcus citri* (Risso), often infests *dieffenbachia,* and it is responsible for most of the wounds required by the fungus for entry into the tissue. The disease was described by Linn in 1942 and the fungus named *Cephalosporium dieffenbachia* Linn. Infection can occur through both dorsal and ventral surfaces of the leaf, but wounding is required except in the case of convolute or recently unrolled leaves. *P. citri,* in common with other mealybug species, deposits a track in the tissues when its stylets enter a leaf. Since the track is tubular, it would be of interest to know what species of fungus can enter the outer opening and grow through the track into the plant tissue after the insect's stylets have been withdrawn and the track vacated.

The Pine Spittle Bug and Burn Blight of Pine Trees

Aphrophora parallela Say, when it becomes abundant, renders Scots pine trees liable to mass infection with the fungus *Diplodia pinea* (Desm.) Kickx. Symptoms of attack are in three phases: (1) a twig and tip blight of various conifers, which is rather rare; (2) a tip blight of Scots pine seedlings, which is not of particular importance; and (3) a serious and lethal bark disease of twigs, branches, and stems of Scots pine, which is epidemic. Trees over 15 years old develop a massive crown infection that often kills the whole tree.

It is this last expression of the fungus infection which is dependent on association with the cercopid. Under conditions of mass infection, a 24-ft tree bore 1500 frothy clusters and 2000 nymphs in early June. Penetration of the stylets is often deep and may reach the cambium when the insect feeds on young internodes. Heavy infestations of the insect can occur without the tree suffering any apparent gross injury; but when invasion of the fungus occurs on heavily infested trees, the effects are often lethal (Haddow and Newman, 1942). Another spittle insect, *A. Saratogensis* Fitch, is associated with burn blight on jack and red pine. According to Gruenhagen *et al.* (1947), the disease was epidemic over 6500 acres in Wisconsin and in several areas in Michigan and the Cass Lake region of Minnesota in 1945. The fact that the distribution of the disease does not yet coincide fully with the area occupied by the insect suggests that spread can be expected to continue.

Nectria cucurbitula Sacc. has been identified as the causal agent. Symptoms are characteristic, a diseased tree having the appearance of having been singed by fire. The symptom sequence begins with the insect puncture which is visible in the cambium when the bark is peeled away. From there, spreading necrosis girdles the branch and advances both up and down. The disease usually starts in the twigs at the upper part of the tree and then works towards the main stem. Red pine appears to be a more favorable host to the insect but more resistant to the fungus than is the jack pine.

The ingress of the fungus is clearly related to the feeding punctures of the insect. The fungus produces vast numbers of spores which are abundant on the foliage of the trees where they can be easily picked up by crawling insects. Gruenhagen *et al.* (1947) have presented adequate evidence on transmission, and they suggest that in addition to acting as vectors, the insects also serve as agents to weaken the twigs and render them more susceptible to fungus entry.

Bud-Blast Disease of the Rhododendron Caused by the Fungus Pycnostyanus

Pycnostyanus (*Sporocybe*) *azaleae* (Peck) Mason may be directly associated with the leafhopper *Graphocephala coccinea* Forst. Medium-sized buds appear to be most susceptible both to the insects and subsequent disease attack. A number of species and a lesser number of hybrids were resistant to the disease. Baillie and Jeppson (1951) recommended multiple Bordeaux sprays for the fungus and 0.1–0.2% DDT sprays for the insect.

Internal Boll Disease of Cotton

Earlier described as "stigmatomycosis" by Ashby and Nowell (1926), the more general and perhaps more descriptive term of internal boll disease of cotton seems to have become better established. The disease is apparently common wherever cotton is grown, and it is universally associated with the feeding of several species of plant bug, principally belonging to the genus *Dysdercus* and collectively known as cotton stainers. These insects are ranked as one of the three great pests of cotton, the other two being the Mexican boll weevil and the pink bollworm.

The disease derives its name from the fact that no external symptoms are seen. Originally, it was believed that the staining of the cotton was due to the crushing of *Dysdercus* and other sucking insects in the cotton gins and to the excrement of these insects. Actually, the staining is now known to be due to the extrusion of sap from the injured seed and to the introduction of various fungi at the feeding point of the insect.

Frazer (1944) has written the most complete account of the processes involved. The fungi concerned are *Nematospora gossypii* A. and N. and allied species (Fig. 4.7 and 4.8). Spores and mycelium of the fungus were found present as external contaminants on the mouth parts and internally in the deep stylet pouches of *Dysdercus* species. Spore germination is possible in the pouches. The fungus is cast off with the exuviae during molting but recontamination occurs. The most likely explanation for the presence of the fungus spores in the stylet pouches is that it occurs because of leakage during feeding. The spores apparently do not remain

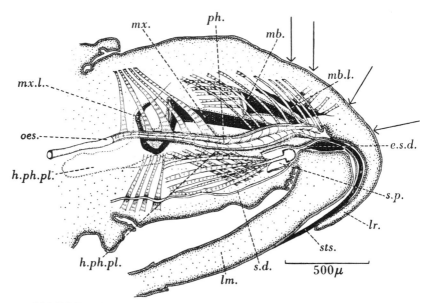

FIGURE 4.7. Partly diagrammatic view of the head of *Dysdercus* to show relative positions of feeding organs. The arrows indicate the directions of the sections shown in Figure 4.8, A–D. Key to abbreviations: *e.s.d.*, exit duct from the salivary pump; *h.ph.pl.*, hypopharyngeal plate (forming inner margin of maxillary stylet pouch); *lm.*, labium; *lr.*, labrum; *mb.*, mandible; *mb.l.*, lever to articulate mandible; *mx.*, maxilla; *mx.l.*, lever to articulate maxilla; *oes.*, esophagus; *ph.*, pharynx; *s.d.*, salivary duct from salivary glands to pump; *s.p.*, salivary pump; and *sts.*, stylets (Reproduced by permission from *The Annals of Applied Biology*, Frazer, 1944).

FIGURE 4.8. A: Section through head of fourth-stage nymph, showing junction of food channel with pharynx; B: section through head of fifth-stage nymph, showing junction of salivary duct with salivary channel; C: transverse section through head of fourth-stage nymph posterior to junction of pharynx and food channel; the positions in which *Nematospora* spores accumulate are indicated by arrows; D: section farther back than C, with each stylet in its separate

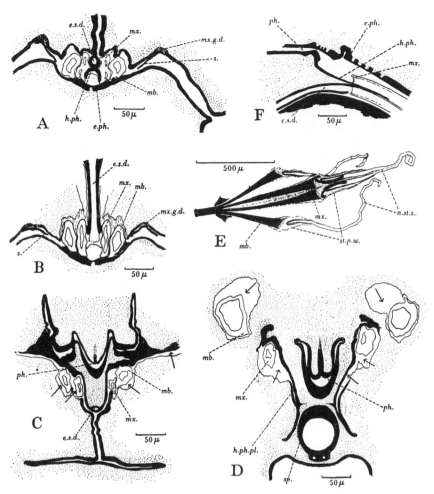

pouch; the positions of *Nematospora* spores are indicated by arrows; E: basal parts of stylets in exuviae, showing sheaths from new stylets and the chitinous lining of stylet pouches; the positions of *Nemato-spora* spores are indicated by arrows; F: median longitudinal section of fifth-stage nymph head at junction of pharynx and salivary duct with food and salivary channels. Key to abbreviations: *e.ph.*, epipharynx; *e.s.d.*, exit duct from salivary pump; *h.ph.*, hypopharynx; *h.ph.pl.*, hypopharyngeal plate (forming inner margin of maxillary stylet pouch); *mb.*, mandible; *mx.*, maxilla; *mx.g.d.*, maxillary gland ducts; *n.st.s.*, sheath from new stylets; *ph.*, pharynx; *s.*, secretion from maxillary glands; *s.p.*, salivary pump; and *st.p.w.*, chitinous wall of stylet pouch (Reproduced by permission from *The Annals of Applied Biology,* Frazer, 1944).

viable in the intestine. Since the disease develops only when the fungus is introduced through the wall of the cotton boll, the presumption is that the inoculum must be conveyed via the proboscis. Infections occur entirely at random; transmission is purely mechanical; and the insect is an obligatory agent solely in its function as a syringe to inject the fungus, which is otherwise unable to reach a suitable substrate.

Time of infection is important in determining the development of the disease. Pearson (1947) has shown that since the bolls reach their maximum size during the first four weeks of development, damage must occur during that time if the size of the boll is to be affected. If the boll is inoculated when one or two weeks old, growth is almost completely arrested. Water relationships are evidently important, since there is a close correlation between the moisture content of the developing lint and the degree of staining following infection. Introduction of sterile water by needle inoculation can produce some staining near the puncture.

The life cycle of *Dysdercus* is from 30 to 45 days. There is a large alternate host list whereby the insect is able to maintain itself until the favored cotton crop is available. Climatic factors influence the distribution and migration of the insect. Wet years are heavy stainer-damage years; in dry seasons, the insect will desert cotton and move to other crops in moister regions.

Mendes (1956) reported that three species of stainers of the genus *Dysdercus* and three species in related genera have been recorded as attacking bolls and doing some—but not significant—damage to fibers in the state of São Paulo, Brazil, although the insects induced rotting in 43 of 107 tests. He recommends clean cultivation and thorough clearing of fields after harvest as a prophylactic control measure. These control measures are in line with older recommendations. Hargreaves and Taylor (1938) recommended reducing the amount of seed litter around gins, reduction of sowing period, frequent picking of open bolls, and reduction of brush shade around cotton plots, in addition to thorough cleaning of the fields after harvest.

Cleaning up old cotton fields completely and maintaining a cotton free period is fairly standard practice in many African native communities. Since the process involves a consideration of tribal organization and local mores, the agricultural officer responsible must combine efficiency with infinite patience and tact.

Coffee Bean Rot

Nematospora spp. are also concerned in a serious disease of coffee berries in Kenya. The insects concerned are species of *Antestia, A. lineaticollis*

Stål, and *A. faceta* Germ. LePelley (1942) has reported in detail on the food and feeding habits of these species (See also Chapter 5).

The berries are the chief food of the insect; in fact the large green berry is essential to the normal development of the insect. It is during feeding on both large and small green berries that the fungus is introduced. Two species of *Nematospora* are involved—*N. corylii* Peglion and *N. gossypii* A. and N. These develop only in the fruits and cause a blackening, followed by rot. Blackened berries are separated by flotation during washing, but infected berries must be removed by handpicking.

Experimental evidence implicating *Antestia* is presented by LePelley; he found a very close correlation between the number of *Antestia* and the number of damaged beans. Some idea of the efficiency of the insect is seen in the conclusion that as high as 97% of the berries have been found rotted when *Antestia* is numerous (343 to a tree); but even very small populations, as low as 4 to a tree, can cause economic damage.

Yeast Spot in Soybean

The stinkbugs (Pentatomidae) transmit yeast spot disease fungi in soybeans. Since wounding or artificial puncturing gave negative results, it is concluded that the relationship between the insect and pathogen (*Nematospora corylii*) is an intimate one (Daugherty, 1967). It is clearly not an insect-species specific case, however, since six species of stinkbug were involved. *N. corylii* was later isolated from the green stinkbug (*Acrosternum hilare* (Say)) by Foster and Daugherty (1969) who located the organisms throughout the salivary system of the insect. Spores in viable state were found in the insect's body. Frazer (1944) found the same organism and *N. gossypii* in the stylet pouches of the cotton stainer (*Dysdercus*), but this apparently did not occur in the stinkbug.

The problem was investigated in depth by Clarke and Wilde (1970, 1970a, 1971). The organism is retained in the insect's head for 90 days—less frequently in the thorax and abdomen. It was isolated from feces but not from salivary secretions. Infected insects transmitted the yeast to soybeans in 60% of their feedings. They were also able to acquire the organism after feeding on dogwood (*Cornus drummondi*) and on soybeans which had been fed on by infected insects. In their third (1971) paper, these authors reported no significant damage to oil or protein content when the soybeans were infected later in their development, and damage was not increased significantly in the presence of *N. corylii*. These last findings are important to exporters of soybeans.

Nematospora *in Rice*

The rice stinkbug, *Oebalus pugnax* (F), discolors rice by its feeding and reduces the milling quality. The rice becomes infected with *N. corylii* which remains viable in the rice after as much as a year's storage (Daugherty and Foster, 1966).

Lint Rot of Cotton

Siteroptes reniformis is the vector of *Nigrospora oryzae* which causes lint rot of cotton. Infection can occur in a low percentage in the absence of the mites, but 100% infection occurred when the bolls were infested with mites and fungus together. Young female mites carry 1–2 spores on the ventral surface of their bodies. Bolls infested with 1–3 mites carrying 2 spores each produced 100% infection (Laemmlen, 1969).

INFECTION RESULTING FROM SYMBIOTIC ASSOCIATION BETWEEN INSECT AND FUNGUS

Cases included in this category vary from the relatively simple mutualism exhibited in the interrelationship between the woolly apple aphid and perennial canker to the highly developed bark beetle-fungus associations as typified by Dutch elm disease. The order in which these cases are discussed represents an attempt to place them in ascending order of specialization without ascribing either rigid distinction or phylogenetic significance to the arrangement.

The Woolly Apple Aphid and Perennial Canker

The woolly apple aphid (*Eriosoma lanigerum* (Hausm.)) was first described in 1802. In the eastern United States, the American elm (*Ulmus americanus* L) is a primary host; apples and related trees are secondary hosts. In the western United States and in other parts of the world where the American elm does not exist, the insect has adapted itself to a single host, the apple. The insects are bark feeders, feeding on the bark under the top two or three inches of soil, or rarely, on the true roots.

The aphid produces warty galls and the damage tends to be accumulative in that young warty galls are very suitable places for young aphids to establish. Zeller and Childs (1925) described a fungus, which they named *Gloeosporium perennans,* that causes a disease of apple trees which ex-

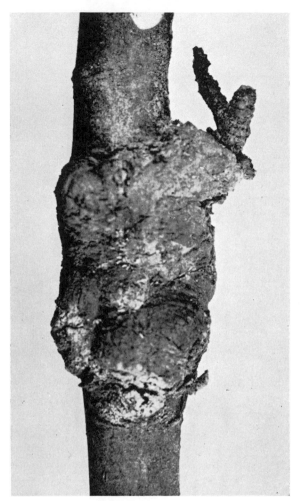

FIGURE 4.9. A large perennial canker of apple, showing the numerous annually formed callus rings. After each year's growth of callus, the pathogen is completely walled off, and reinfection is dependent upon the feeding wounds of the woolly aphis (Photograph: Oregon State University Agricultural Experiment Station).

presses itself as fruit rot and branch canker. The term "perennial canker" is actually a misnomer since new cankers are formed each year (Childs, 1929; McLarty, 1933) (Fig. 4.9 and 4.10).

There is a close relationship between the fungus and aphid. The fungus requires wounds through which to enter the tissue. These are provided when the warty galls produced by the insect burst upon freezing.

FIGURE 4.10. A perennial canker, similar to that in Figure 4.9, with the dead bark removed to show the woolly aphis to which the canker affords both food and shelter (Photograph: Oregon State University Agricultural Experiment Station).

The fungus, living saprophytically on the old canker, enters these broken galls and develops a fresh canker during the winter, roughly parallel to the old one. In the summer, the canker heals, and when the new callus is being formed the fungus is sealed off. The aphids establish themselves on the new callus and the process is repeated.

Control measures have developed along two lines: (1) the control of the aphid by spraying and clearing up old canker tissue, and (2) by breeding varieties resistant to gall formation by the aphids (See Chapter 5).

Leach (1940) and Austwick (1958) consider the relationship between the fungus and the insect as a type of mutualism.

Phthiriosis *of Coffee and Vine Roots*

There is some doubt as to the coidentity of the fungi involved in phthiriosis of coffee and vine roots. The species from coffee has been described as *Polyporus coffeae,* but Maublanc and Roger (1935) refer it to the genus *Bornetina,* leaving open the question of its specific identity with *B. corium* affecting vines in Palestine.

In both cases the fungi are not parasitic but develop on the sap that exudes from the feeding punctures of coccids living on the roots and on the honeydew excreted by the insect. Growth of the fungus forms an impermeable sheath around the roots and losses can be serious.

Steyaert (1946) reported the same fungus on coffee in the Belgian Congo, where it develops on the exudate from *Pseudococcus citri* Risso. The insect lives in cells made of the mycelium which forms a brown sheath around the roots. In Java, *P. deceptor* is the coccid involved. Donk (1948) refers the fungus to the genus *Dicanthoides* to which genus it was transferred from *Daedalea. Bornetina corium* is the imperfect stage of the fungus.

A similar coccid-fungus association is found on citrus growing in the vicinity of Rio de Janeiro. There, *Pseudococcus comstocki* (Kuw.) develops colonies on the roots of citrus which are surrounded by the mycelial crust of what is believed to be *Boletus tropicus* Rick. The appearance of the sporophores of this fungus on the surface of the ground under the trees is a sure sign of injury, which can be severe when the mycelial crusts are numerous. There is a close relationship between the coccid and the fire ant *Solenopsis saevissima* var. *moelleri* Forel. Goncalves (1940) considers that the ant is the agent chiefly responsible for the mutualistic association between coccid and fungus. The ants carry the mealybugs to the interior of the ant nest or the mealybugs enter through the superficial openings of the galleries. These mealybugs establish themselves on uncovered roots, and establish colonies which are soon covered by the mycelial crust. In the absence of *Solenopsis* nest construction, the crusts are rarely present and never cause severe damage.

Pediculopsis *Mites and Fungal Diseases of Plants*

A disease of wheat in the province of Aragon, Spain, attributed to the joint intervention of the mite *Pediculopsis graminum* Reut. and fungus

Nigrospora oryzae (Berk. and Br.) Petch, was recorded by Alfaro (1946). The spikelets were killed in the incipient stage of their development. Within the leaf enclosing the ear, and on its basal node, were some whitish hyphae with an abundance of black conidia, among which the mites were swarming. The globose, white masses, situated between the disorganized spikelets, especially near the basal node, were found to be gravid females of *P. graminum*. The association between the mite and the fungus aggravates the damage inflicted by either partner alone. The nongravid females transport the conidia of the pathogen in their abdominal sacs, while the mite, in turn, feeds on the fungus and the decomposed host tissues. Peyronel (1950) reported a similar situation in Italy, where the wheat spikelets were covered by the mycelium of a *Fusarium* species. Embedded in the mycelium were whitish or pinkish balls which were identified as the abdomens of gravid female mites. A similar association between *P. graminum* and *F. poae* (Peck) Wr. was observed in abortive and rotting ears of corn in the vicinity of Turin; the mites were observed to be disseminating the conidia of the fungus. The observations of Alfaro were confirmed and, in addition, infection of corn by *N. oryzae* in association with a species of *Pediculopsis* was also noted.

Oak Wilt

Oak Wilt has been recognized within relatively recent years. First records, as is usual in such cases, are a matter for speculation, but the first formal recognition of the disease came in 1943 when Henry and Moses noted that "a rapid dying of trees in the red oak group has been severe in Wisconsin during the past decade." A fungus, tentatively ascribed to the genus *Chalara,* was described as the cause and specifically described by Henry (1944) as *Chalara quercina* n. spp. Four important reviews followed in quick succession: McNabb (1954), Hepting (1955), Barnett and True (1955), and True *et al.* (1960).

The disease had spread to 616 counties in 20 States by 1965 (Rexrode and Lincoln, 1965). A forecast of the status of oak wilt in Pennsylvania and West Virginia estimated that if present trends continue, 1% of the oaks in Pennsylvania will be infected in 50 years; in West Virginia, 1% in 25 years and 50% in 40 years (Merrill, 1967). In either case the forecast should stimulate the breeders of resistant oaks.

The perithecial stage of the fungus was described by Bretz (1951), who at the same time suggested the possibility of heterothallism, the condition in which the fusion of two unlike mycelia is necessary for sexual reproduction. Bretz (1952) described the perfect stage of the fungus as *Endoconidophora fagacearum.* Hunt (1956) transferred the species to the genus *Ceratocystis,* to join that infamous company which includes *C. fim-briata* Ellis and Halst, the pathogen causing tapping panel disease of rubber, and *C. pilifera* (Freis) Moreau, causing southern pine blue stain.

The symptoms on the leaves are described as crinkling, bronzing, and finally browning (Henry and Moses, 1943). Dark streaks in the current year's wood have also been described. These symptoms are apparently caused by tyloses blocking water movements from the roots to the leaves. Beckman *et al.* (1953) concluded that "tylose formation was a response of the host to some action of the pathogen and that the tyloses were the cause rather than the result of the reduced transpiration stream." White and Wolf (1954) isolated two toxins associated with oak wilt: one, an alcohol-insoluble compound; the other, an alcohol-soluble component. The former was suspected of causing leaf wilting; the latter, the development of areas of necrosis. Barnett and True (1955) suggested that toxins and polysaccharides might be involved in the expression of wilt symptoms. They also noted differences in susceptibility of oak species to the disease; those in the red oak group are killed in one season whereas the white oaks may last several years. This observation is important in connection with dispersion, since, obviously, the white oaks, surviving longer with active fungi available to vectors, would pose a more difficult problem than would the red oaks. Merek and Fergus (1954) reported that under forest conditions the pathogen could persist for only 3 wk in twigs but at least a year in the roots. Survival in stumps, depending on the fungicide used, varied from 15 to 44 wk. Yount (1958) recovered the fungus from the roots of dead trees 3 yr after the trees had died.

Inoculation is through wounds or through the agency of natural root grafts. Since Guyton (1952) first recorded the disease in oaks blazed to mark a projected road, evidence has been accumulating that wounding is responsible for overland spread (Craighead and Morris, 1952; Dorsey *et al.,* 1953; Norris, 1953; Griswold, 1953; Craighead *et al.,* 1953). Norris (1955) reported overland spread occurring in 5 out of 6 experimental areas in which trees had been wounded by cutting T-shaped or axe marks in the bark. Direct inoculation with tools is possible (Jones and Bretz, 1955) but not important.

A factor limiting the efficiency of wounds as infection courts is the age of the wound. Zuckerman (1954) found that the inoculum must arrive at the wound site within 24 hr for most efficient ingress, although infection did occur as late as 124 hr after stems were broken to produce wounds. Inoculation through root grafts is by the conidial form of the fungus, which can also be found on other parts of the tree, although not in acorns (Bretz and Buchanan, 1957).

The movement of the pathogen from the roots of one tree to those of

(a)   (b)   (c)

FIGURE 4.11. Oak wilt. Nitidulid beetles, the most important vectors of oak wilt. (a) *Colopterus morio* (Er.); (b) *Colopterus semitectus* (Say); (c) *Glischrochilus quadrisignatus* (Say) (Photographs: West Virginia University Agricultural Experimental Station).

another is evidently rather slow. Studying the time interval between deaths of trees where natural root grafts were apparently the means of spread, Jeffery (1954) found the time varied between less than 1 yr to as long as 6 yr. Yount (1958) inoculated red oak trees each year for 4 yr, using trunk inoculations as checks on root inoculations made at varying distances from the trunk. In each annual series, root inoculations showed a longer incubation period than did trunk inoculations, but there was no relationship between distance from trunk and incubation period.

Distance or overland dispersal requires vectors, and the role of insects has now been well established. Insects visiting both wounds in healthy trees and mycelial mats first came under suspicion. The former criterion appears to be the more important. Morris *et al.* (1955) used radio-active isotopes to demonstrate that various kinds of wounds caused by blazing, stripping, and crushing were attractive to insects. Boyce and Garren (1953) made the suggestion, later confirmed by Bretz and Jones (1957), that injured flowers could also serve as infection courts.

Nitidulids were implicated as vectors by Norris (1953), who observed these souring beetles breeding in mycelial mats. Dorsey *et al.* (1953) and other authors as indicated, suggested the following species as possible vectors (see Fig. 4.11 and 4.12): *Carpophilus lugubris* Mur.; *Colopterus niger*

FIGURE 4.12. Oak wilt. Adults and larvae of nitidulids exposed by removing the bark from over a mat (Photographs: West Virginia University Agricultural Experimental Station).

Say (Yount *et al.,* 1955); *Colopterus semitectus* (Oliv.); *Cryptarcha ample* Er. (Dorsey and Leach, 1956); *Eupurea* spp. (Yount *et al.,* 1955); *Eupurea corticina* (Norris, 1953); *Glischrochilus confluentus* (Say); *Glischrochilus fasciatus* (Oliv.); *Glischrochilus sanguinolenta* (Oliv.); *Glischrochilus quadrisignatus* Say (Himelick and Curl, 1955).

External contamination is not the only method of transmission; Jewell (1954) and Bart and Griswold (1955) demonstrated the presence of viable conidia in nitidulid feces. Most of the free-flying insects of these species do not carry spores either externally or internally (Thompson *et al.,* 1955). Yount *et al.* (1955) confirmed this observation even though artificially contaminated beetles, refrigerated for three months, still contained infective material at the end of the period. The work of these authors suggests that in nature chance plays a considerable part in the acquisition of viable spores by insects visiting mycelial mats. Stambaugh and Fergus (1956) collected dead or dormant nitidulid beetles from the duff layer of the forest floor and found that conidia survived on or in the insects for 94 days, ascospores for 151 days; they concluded that spores could survive for the entire period of hibernation of the insect (Fig. 4.13). Three species of scolytid beetles have also been implicated as vectors, viz., *Monarthrum fasciatum* (Say) by Dorsey *et al.* (1953), *Pseudopityophthorus minutissimus* (Zimm.) by Griswold (1956), and *P. pruinosus* (Eic.) by Griswold and Bart (1954). Oak bark beetles carry the spores of the oak wilt fungus. The beetles commonly breed in diseased trees, and, emerging during the same year as symptoms develop in the tree, are contaminated at that time. They also overwinter in diseased trees and emerge the following spring when oaks are highly susceptible. Some of the beetles carry the fungus internally; beetle excreta contains propagules of the fungus (Rexrode and Thomas, 1971).

Buchanan (1957) reported that a brentid beetle, *Arrhenodes minuta* (Drury) will feed on cultures of the oak wilt fungus and ingest the small globules of exudate that form on the surface of the fungus. Furthermore, the insect is attracted to fresh blazes made in May and June; it bores into exposed xylem where the females oviposit, and it could easily become contaminated with the fungus if the tree were infected. The buprestid *Agrillus bilineatus* Web. may spread conidia through an infected tree in the course of its burrowing (Parmeter *et al.,* 1956).

Himelick and Curl (1955) used the flat-headed borer *Chrysobothris femorata* Oliv. in greenhouse experiments. The insect transmitted the fungus to 13 of 54 oak seedlings. Since the insect is often found associated with wilt-killed oaks in nature, it may well be implicated in natural spread of the fungus. These same authors, however, concluded that nitidulids were the important vectors on the basis of field experiments. A mite, *Germania bulbicola* Oudemans, in the same greenhouse experiments, transmitted the fungus to artificially made wounds on 11 of 32 oak seedlings.

Although Jewell (1956) reported negative results with pomace flies, Griswold (1953) was able to show that *Drosophila* species could move spores from fungus mats to wounds. Later (1956) Griswold showed that *D. melanogaster* Mg. transmitted the fungus after being allowed to feed on the moisture at the tips of hyphae on sporulating mats of the fungus and then on fresh artificial wounds in the xylem of healthy oaks, with resulting infection in all the plants used. Even collembolans may infect a wounded oak tree up to 16 hr after feeding on a mycelial mat (Curl, 1956).

Ceratocystis fagacearum, the organism causing oak wilt, has been clearly associated with many insect species, especially Nitidulidae. Rexrode *et al.* (1965) refers to the bark beetle *Pseudopityaphorus pruinosus* (Eic.) as a vector, but according to the same author (Rexrode, 1968), no insect has ever been conclusively proved to be a vector of the fungus from

(a)

(b)

FIGURE 4.13. Oak wilt. (a) Short cracks in the slightly bulging bark of affected oak trees are the first indication of the presence of fungus mats below. (b) Sporulating mats of *Ceratocystis fagacearum* showing the pressure cushions. Observe the duplex nature of the cushion with one portion attached to the wood (*left*) and the other attached to the bark (*right*) (Photographs: West Virginia University Agricultural Experimental Station).

diseased to healthy trees. The explanation for this might lie in the definition of healthy; previous authors have agreed that wounds are a primary attraction for the insect vectors.

Insect families with species transmitting the oak wilt fungus can be ranked roughly in the order of their probable importance as follows: Nitidulidae, Scolytidae, and Drosophilidae. Members of the families Curculionidae, Brentidae, Buprestidae, Elateridae, Histeridae, Ostomidae, and the Collembola may also be vectors, but it seems unlikely that either the Membracidae or the Formicidae are significantly involved.

Biological Aspects of Transmission

With such a large list of insect species implicated, it is a matter of interest to consider the biology of the transmitting insects and of the fungus to determine if anything other than a casual relationship exists.

Dorsey and Leach (1956) have provided the best-documented study as far as the nitidulids are concerned. It is quite clear that fresh wounds, oozing sap, are attractive to a large number of insect species, of which the nitidulids appear to be the most consistent and numerous. This attraction continues for two or three weeks or until the sapwood under the disturbed bark dries. It is during this period that the conditions for transmission of the fungus to these oozing wounds are most favorable.

At this point, the biology of the fungus becomes important. Hepting *et al.* (1952) showed that the fungus was hermaphroditic, self-sterile, and interfertile. It produces two kinds of spores, endoconidia and ascospores, the latter being produced as a result of a sexual process. There are two compatibility types (A and B), and the endoconidia perpetuate only the compatibility type producing them. The two compatibility types may exist on the same tree but are usually separate. Actually, according to Yount (1954), most naturally infected trees contain only one type, although in an oak wilt area, defined as a group of wilt-infected trees, both types were present.

Each type is bisexual but self-sterile. A mycelium of either compatibility type produces endoconidia and also perithecial elements with hyphae, which are the receptive female elements. A mycelial mat of type B may form under the bark of one tree with specialized "pressure cushions" (Zuckerman, 1954) which eventually crack the bark. Before ascospores can be formed on this mat, it must be fertilized by endoconidia from a type A mat, and vice versa. Since both types exist in an area, but rarely on the same tree, some method of transportation from one type mat to the other must be available. The endoconidia are sticky and not transportable by wind. Insects are the ideal agents, especially those that feed on the mycelial mat, for these latter carry spores not only externally but internally (Fig. 4.14).

The transportation of endoconidia can occur by the agency of any of the numerous insects to be found on mycelial mats, but nitidulids and perhaps *Drosophila* are particularly important. Both groups feed and breed in the mycelial mats; both are strongly attracted to fermenting

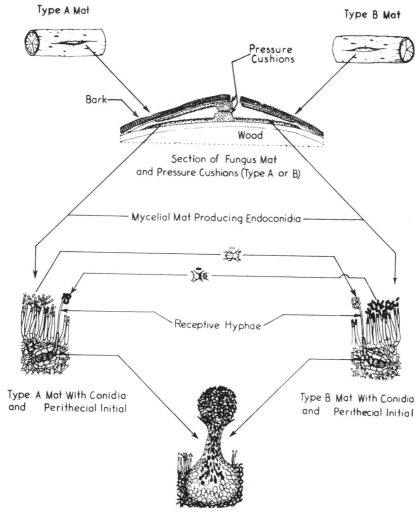

FIGURE 4.14. A diagrammatic representation of the role of insects in the spermatization of *C. fagacearum* (Drawing: West Virginia University Agricultural Experimental Station).

odors. Boyce (1954) showed that marked nitidulids, removed from fresh mycelial mats on a wilting oak tree, returned to them from distances of 100–500 ft. It is not surprising then that nitidulids are important vectors and important in aiding the fertilization process which produces ascospores. It is true that the disease can be spread without the production of ascospores, but ascospores are more durable and more resistant to local environmental conditions than are conidia (Stambaugh and Fergus, 1956).

Control measures are currently in their infancy. There is a difference in susceptibility between the red and white oak groups, and perhaps resistant segregants could be found in the less-susceptible white oak group. Chemotherapy of sick trees, which might prevent the formation of mycelial mats, might be feasible but not economic. Cutting-out programs as a single measure of control are rarely completely successful, and for forest trees, impractical. The prospects for the red oak group seem gloomy, and for the white oak group, only slightly less so. There is also the possibility that other species may be threatened, especially if more virulent strains of the fungus develop. Bart (1956) tested 17 species of trees native to Ohio and found several of them susceptible, and even the apple varieties Jonathan and Delicious.

The poorer the growth, the greater the incidence of the disease (Cones and True, 1967). Rexrode *et al.* (1965) reported that if oaks killed by the fungus are girdled to the heartwood and the bark below removed to ground level, the saprophytic life of the fungus is shortened due to the accelerated drying up of the wood; the production of mycelial mats is reduced. Another attempt at control was made by Himelick and Fox (1961). Belts of poisoned trees around infestation centers were established to reduce local distribution.

Bark Beetles and Blue-Stain Fungi

Diseases caused by the blue-stain fungi, most commonly of the genus *Ceratostomella* (or *Ceratocystes,* depending on whose nomenclature is favored), are known principally among the conifers (Fig. 4.15). According to Leach (1940) the fungi are more frequently found in felled timber where the discoloration reduces the timber's value. When living trees are attacked, however, death ensues, sometimes rapidly (Craighead, 1928). Several species of bark beetles are associated with blue-stain fungi. Von Schrenk (1903) suspected that *Dendroctonus ponderosae* Hopk. was disseminating the fungus, but he was unable to isolate it from the insect. Craighead (1928) did not believe that direct beetle attack could be responsible for the rapid death of southern pine and suggested, instead,

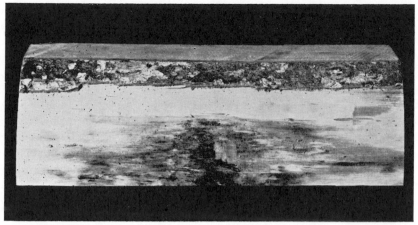

FIGURE 4.15. Blue stain in Maracaibo boxwood caused by *Lasiodiplodia theobromae* (Pat.) Griff. et Maubl. and associated with the attacks of longicorn beetles and pinhole borer beetles (Photograph: Savory. Department of Scientific and Industrial Research, Forest Products Laboratory, London).

that the blue-stain fungus was primarily responsible. This was confirmed by Nelson and Beal (1929) and Nelson (1934).

Leach *et al.* (1934) presented experimental proof of the association between two species of bark beetles (*Ips pini* Say and *I. grandicollis* Eich.) and the fungus. These authors considered that the association was one of true mutualistic symbiosis on the grounds that the fungus is dependent on the insect for dissemination and penetration into the bark, and that the fungus, in turn, improves the microenvironment for the developing brood of insects. Several other species of bark beetle are implicated. *Ips emarginatus* (LeC.), *I. integer* (Eich.), and *I. oregoni* (Eich.) are all associated with *Ceratostomella ips,* the fungus described by Rumbold (1931) as responsible for blue stain (Rumbold, 1936).

The very clear illustration in Leach (1940, Fig. 100) was of great diagnostic value to the writer in the state of Pernambuco, Brazil, in 1946. There the rapid dying of mango trees was called to his attention, together with the evidence of the presence of bark beetles. Typical dark areas were obvious in the cut cross-sections of the trunks of these trees. Chaves-Batista addressed himself vigorously to this problem, with the result that the relationship between another species of the family Ipidae and blue-stain fungi was revealed. This example from the tropics is of considerable interest, not only in extending the range of this type of disease beyond the conifers, but in demonstrating that it is not limited to the temperate regions (Chaves-Batista, 1947). Chaves-Batista named the new disease of mango "Recife disease." The first symptom is gum exudation from the insect's entrance holes; later the tree wilts. Death occurs quickly if the disease begins in summer, but after a longer period if onset occurs in autumn. The timber, is discolored brown internally, and this discoloration is radial, as in the conifers. At other times, the xylem is stained black. The fungus was described by Chaves-Batista as *Diplodia recifensis* n. spp., and the beetle is *Xyleborus affinis* Eich. Chaves-Batista also found a relationship between the number of infection courts and the rapidity of development of symptoms. In nature, these are sufficient to ensure death of the tree. Out of 20,868 mango trees surveyed, 27.03% were diseased, an appalling percentage of a total tree population in a single area. As with other species of this family, the insect has a direct mutualism with the ambrosial fungus (Fig. 4.16).

Francke-Grosmann (1952) described the relationship between blue-stain fungi and bark beetles as a symbiotic ambrosia culture. The beetles concerned are *Myelophilus* (*Blastophagus*) *minor* (Htg.) and *Ips acuminatus* Gryll. *Trichosporium tingens* L. L. and M. is constantly associated with *M. minor* and is regularly found in its breeding places; a similar relationship exists between *T. tingens* var. *macrosporum* n. var. and *I. acuminatus*. The host tree is pine. The emergence of a strain of the fungus associated with a specific beetle may be of some significance. It is possible that the two species of beetles may differ sufficiently in their attack on the pine to favor the emergence of a fungus mutant. This relationship, if true, and continued, could well develop into a separate and distinct bio-coenose.

Short-leaf pines are also subject to a fatal combination of the bark beetle *Dendroctonus frontalis* Zimm. and fungi, which are the cause of heavy losses among pine stands in the southern United States. According to Bramble and Holst (1940), heavy fungus infection invariably accompanies infestation of living pines by the beetle. The tunneling by the beetles is extensive, going through the corky bark and resulting in long, winding egg galleries in the phloem which may extend to the sapwood. The larvae extend the tunneling through the phloem, and, on becoming full-grown, bore to the outer bark where they pupate.

The complex of microorganisms found in infected sapwood was cultured from beetles and found both on and in the insects. *Ceratocystis pini* Munch., *Dacryomyces* spp., *Saccharomyces pini* Holst, and *Monilia* spp. were all found on both egg and adult stages, and, except for the first- and last-named, on larvae also. The invasion of this complex is accomplished within a short time after beetle infestation, some of the organisms acting as primary invaders. The chief visible effect of fungi ac-

(a)

(b)

(c)

(d)

FIGURE 4.16. Recife disease of mango. (a) Galleries of *Xyleborus affinis* Eichoff formed in the trunk of a diseased mango. Note also a discoloration of the wood. (b) Radial discoloration of the wood of a mango tree which is caused by the fungus *Diplodia recifensis* Bat. (c) Gum exuding from the apertures made by *Xyleborus affinis* Eichoff. (d) The final stage of the disease (Photographs: Chaves-Batista, 1947).

companying beetle attack is stoppage of conduction, which hastens the death of the tree. The drying that follows (Nelson, 1934) may also be important in the development of the beetle.

The wood-staining fungi of spruce are associated with another species of *Dendroctonus, D. engelmanni* Hopk. (Davidson, 1951, 1954). Although four species of fungi are concerned, *Leptographium* spp. is the most consistent associate of the beetle. This fungus causes conspicuous gray stain in the sapwood of attacked trees. *Endoconidiophora coerulescens* Munch is sometimes present in beetle galleries, causing stain. Two *Ophiostoma* species have been isolated from galleries and adult beetles. All four species of fungus are well adapted for insect transmission, but their exact role in the life cycle of the beetle is not known.

Blue staining in pine logs can be controlled on a satisfactory basis by using an insecticide-fungicide spray comprising 2% Santobrite, 2% borax, and 0.75% gamma BHC. It is essential, however, to spray whole logs immediately after felling (Savory, 1964).

Ceratocystis *Infection of Stone Fruits and Cacao*

Ceratocystis fimbriata, the causal organism of canker of stone fruits, is vectored by several species of insect, *Carpophilus ligneus* Murray, *C. hemipterus* (L), and a species of *Liturgus*. All of these are attracted to bark wounds and all can live and complete their cycle on cultures of the fungus (Moller and DeVay, 1966). *Ceratocystis* wilt of cacao is associated with *Xyleborus* beetles, but there are other possibilities. Mites infest the beetles and nematodes are attached to the mites. Actually *Fusarium* spp. are cultivated by the beetles for food in preference to *Ceratocystis* (Anon., 1965).

Canker-Stain Fungus in Plane Trees

Beetles were collected from plane trees (*Platanus* spp.) suffering from stem canker caused by *Ceratocystis fimbriata* and caged over knife wounds on healthy plane trees. Transmission of the fungus was effected in 2 of 7 trials by *Cryptarcha ampla* Erichs and *Laemophloeus biguttatus* (Say). Beetles collected from healthy trees were allowed to feed for 30 min on a culture of the fungus and were then confined on healthy trees. Transmission was obtained in 6 of 11 trials with *Coleopterus (Colastus) unicolor* (Say), *C. morio* (Erichs), and *Carpophilus lugugbris* Murr (Crone and Bachelder, 1961).

Dutch Elm Disease

Discovered in Holland in 1919 and described by Spierenberg in 1922, Dutch elm disease has attracted considerable attention throughout Europe and the United States (Fig. 4.17). As long ago as 1937, a bibliography of no less than 678 references was issued by Goss and Moses. Shortly after its original discovery, the disease was recognized throughout central and southern Europe. A single tree was found infected at Totteridge, Herts., England, in 1927 (Wilson and Wilson, 1928); in the next ten years the disease had spread over all England and part of Wales. In the United States, the exact date of its arrival is not certain, but there is some evidence to indicate that it was in 1929. May (1934) mentioned 3 trees that were examined which showed discoloration in the 1929 rings. According to Beattie (1934) the disease was first discovered in Cleveland and Cincinnati, Ohio, in 1930, when 3 sick trees were found in the former and 1 in the latter city. In 1931, 4 sick trees were found in Cleveland and were removed. In 1932 no new cases were found and it was thought that the disease had been exterminated. Although it was known (Knull, 1934) that the elm bark beetles had been present in the United States for some years, the manner in which the disease organism had been transported to the United States was a matter of conjecture until inspectors of the Bureau of Entomology and Plant Quarantine discovered that burl elm logs, arriving from Europe at Baltimore, Norfolk, New York, and New Orleans in August of 1933, were all diseased and infested with the elm bark beetles, *Scolytus multistriatus* (Marsh) and *S. scolytus* Fabr.

However, since interceptions at quarantine can reasonably be described as samples of infestations that get through unnoticed, it is not surprising that before long new records of infected trees began to accumulate. By 1935, Readio was able to report more than 7600 infected trees within a 40-mi radius of New York City. The distribution of the disease in that area was correlated with the fact that New York was the only area where *S. multistriatus* had been previously established. Since then, there has been extensive spread of the disease, as shown by Fig. 4.18, in which Holmes (1961) has mapped the recorded distribution as of 1959. This map omits a Colorado record (Nance, 1949) to permit the use of a larger scale. However, I have been informed that *S. multistriatus* had been known in the Ohio Valley for several years before the disease was reported in the United States or in Europe.

The first symptom of the disease is a wilting of the leaves in one or more of the limbs in the upper crown, with a color change to yellow or brown in the leaves on the affected branches. Shriveling of the leaves follows and there is premature shedding. When the leaves have fallen, a di-

FIGURE 4.17. Dutch elm disease. An early stage of the disease (Photograph: Canadian Department of Agriculture, Forest Biology Division).

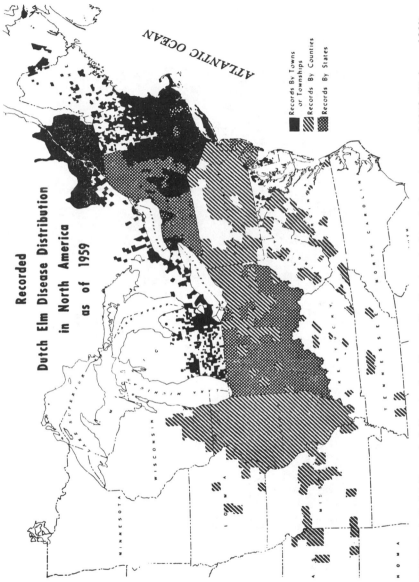

FIGURE 4.18. (*a*) Recorded Dutch elm disease distribution in North America as of 1959 (Holmes, 1961).

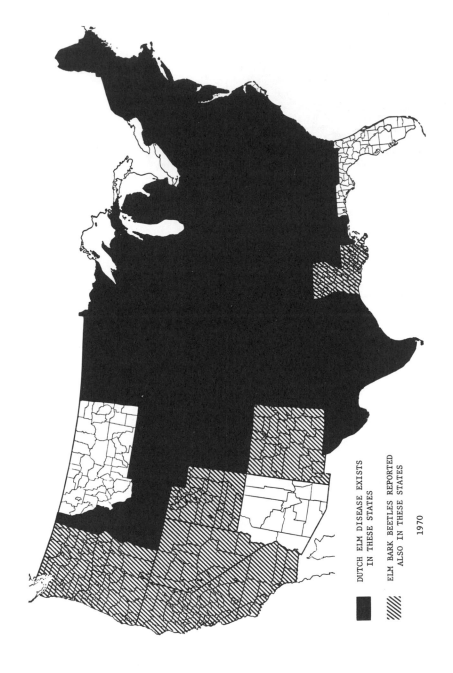

DUTCH ELM DISEASE EXISTS
IN THESE STATES

ELM BARK BEETLES REPORTED
ALSO IN THESE STATES

1970

FIGURE 4.18. (b) Distribution of the Dutch elm disease and the smaller European elm beetle as of 1970. This map is still (1972) current (Photograph: J. H. Barger, USDA, ARS).

agnostic aid is the appearance of the affected branches, which assume a downward, hooklike curve, appropriately referred to as "shepherd's crook." Internally, in diseased branches, a brown ring shows in the outer sapwood when viewed in cross-section, and as long discontinuous streaks in longitudinal-section. These symptoms are not specific to Dutch elm disease, being closely approximated by a disease caused by *Verticillium* wilt and *Cephalosporium* wilt, but the causal fungi can be readily separated in laboratory cultures (Fig. 4.19).

The fungus responsible for Dutch elm disease is *Ceratostomella ulmi* (Buism.) C. Moreau, and it is primarily a parasite of water-conducting tissue. Apart from some clogging of these tissues, a further lethal factor is a toxin produced by the fungus. According to Feldman *et al.* (1950), this toxin exists in two fractions: one a polysaccharide, which causes the upward curling of the leaves with marginal withering; and the other, an alcohol-soluble fraction that is responsible for the interveinal necrosis. With the death of the host tissue, the fungus continues to grow saprophytically. Rabkin *et al.* (1941) found the fungus as a saprophyte in dead elm trees, and Tyler and Parker (1945) reported that the fungus could survive at least two years in diseased wood which was not subject to very high temperatures and rapid dehydration.

The insect vectors of the fungus are species of elm bark beetles, of which two—*Hylurgopinus rufipes* (Eich.), the native elm bark beetle, and *S. multistriatus* (Marsh), the smaller European elm bark beetle—are the most important. Because of its more aggressive habits, the latter crowds out the native species and, within its established range, becomes the principal vector.

Both species are limited to species of elm for feeding and breeding, but there are significant differences in their life history. *S. multistriatus,* on emergence from its brood galleries, feeds first on twigs and crotches of healthy trees before building its own brood galleries in sickly trees, freshly cut elm wood, or storm-broken limbs. On the other hand, Kaston and Riggs (1938) showed that *H. rufipes* prefers freshly cut logs for feeding and produces shallow feeding tunnels which the authors consider analogous to the twig and crotch feeding of the European species (Fig. 4.20).

The native species overwinters in both adult and larval stages; the European species, only in the larval stage. Both species produce two generations a year. The first generation to emerge is probably the more important as far as dissemination of the fungus is concerned, because successful inoculation of the fungus into living elms depends upon the spores entering the long, functioning-conducting vessels of the spring and early-summer wood. During this period it is easier for the adults to introduce

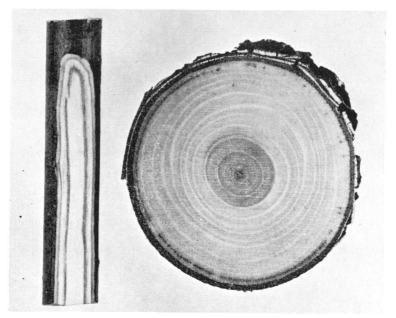

FIGURE 4.19. Dutch elm disease. Diseased elm branch in cross- and longitudinal sections showing the brown ring and the streaks (Photograph: Canadian Department of Agriculture, Forest Biology Division).

spores since the conducting vessels are then closer to the surface of the wood. Later in the season, the vascular system consists of much shorter and relatively compressed conducting vessels; spore movement into them is restricted, and only localized infection occurs. Parker *et al.* (1941) reported that late season infection was rarely successful.

When the adult insects move from their food trees and construct brood

FIGURE 4.20. Dutch elm disease. (*a*) Adult of the native elm bark beetle, *Hylurgopinus rufipes* (Eichh.), X18. (*b*) Tunnels of the adult and larval galleries of the native elm bark beetle on the inner bark surface of elm, about X0.6. (*c*) Feeding scars made by the adults of the native elm bark beetle of the bark of healthy elm, about X0.6. (*d*) Adult of the smaller European elm bark beetle *Scolytus mulfistriatus* (Marsh.), X19. (*e*) Tunnel of the adult and larval galleries of the smaller European elm bark beetle on the inner bark surface of elm, about X0.6. (*f*) Adult of the smaller European elm bark beetle feeding, and feeding scars in the twig crotches of healthy elm, about X1.5 (Photographs: Canadian Department of Agriculture, Forest Biology Division).

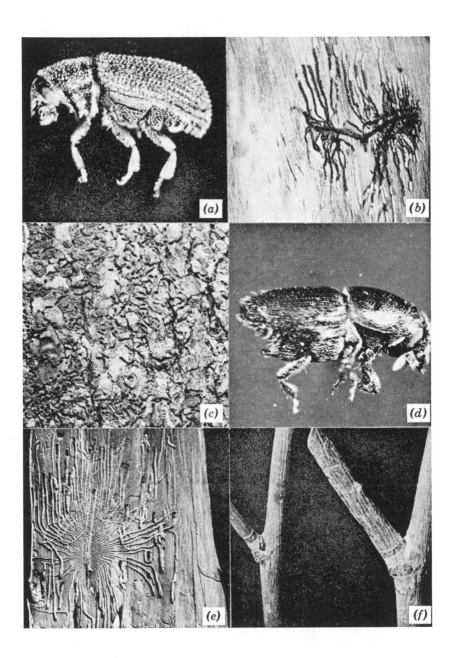

galleries, they carry the spores of the fungus on their bodies. The fungus then grows saprophytically within the galleries, producing spores which are sticky. When the new adults emerge in the spring, their bodies are contaminated externally (and sometimes internally) with the spores so that when they feed on twigs and crotches, or in the case of *H. rufipes,* on the bark, they can infect the healthy trees and continue this infection into the new brood galleries. The combination of insect life history and capacity of the fungus to grow saprophytically in wood eminently suitable for the insect's breeding is so effective that only a resistant species or variety of elm could expect to survive. However, the introduction of the fungus into the insect's feeding points does not necessarily mean that true infection is established. Parker *et al.* (1941) showed that the fungus was established in many feeding scars without true infection occurring and that the fungus could be isolated from such scars on uninfected trees as long as three years later. Furthermore, the fungus sometimes kills only an infected twig, and it sometimes dies itself in the tree.

The problem of dispersal of the fungus has become of major importance in connection with control measures. Hoffman and Moses (1940) found that a single *S. multistriatus* was capable of spreading the fungus spores to a number of brood galleries in the process of going from one to the other in order to mate. Within a small area, this could be important. These authors were able to determine that even under conditions of extremely low winter temperatures, as low as $-10°F$, viable spores were found in the intestinal tracts of overwintering beetles, thus assuring an active supply of inoculum in the critical early spring and summer months.

Rankin *et al.* (1940) found that the annual incidence of the disease depended on the rate of effective infection, which is a function of the volume of material in which the beetle currently breeds and which is in turn conditioned by all the factors that produce the death of elm wood.

Wolfenbarger and Jones (1943) found that twig-crotch injuries caused by *S. multistriatus* were more frequent around dispersal points than around breeding points, and that the average rate of injury decreased with the increase in distance from the point of dispersion. They found that the beetles may travel as far as four miles from the point of emergence to the place where they construct brood galleries.

Wadley and Wolfenbarger (1944) examined the population densities of *S. multistriatus* with respect to distance from centers of dispersion. They concluded that in dispersions radiating from a center (1) the density decreased as the circumference increased, (2) the trend was modified by the tendency of the beetles to stop when they found satisfactory feeding places, and (3) the density would decrease to some low level, characteristic of the region, but not to zero. Zentmeyer *et al.* (1944), in a statistical analysis of the problem, confirmed Wadley and Wolfenbarger's work when they concluded that the probability of infection decreased with the log of the distance from the focus of infection, and that wind currents affected the probability of influencing the direction of movement of emerging adults.

Dispersal to a point as far west of the main centers of infection as Colorado can probably only be accounted for on the basis of transport of infected and infested wood, but discontinuous local foci of infection are, most likely often the result of serious wind storms which not only cause physical damage to the trees, thus developing favorable breeding sites, but also aid in the dispersal of the insect vectors.

Mention should be made of several other vectors which have been recorded. *Scolytus scolytus* Fabr. is comparable in importance with *S. multistriatus* and *H. rufipes*. Tyler *et al.* (1939) found that adults of *Saperda tridentata* L, emerging from moribund and diseased trees, were contaminated with *C. ulmi* spores and were able to inoculate them into feeding wounds. Collins (1941) listed *Xylosandrus germanus* Blfd., *Xylobiops basilara* Say, and *Conotrachelus anaglypticus* Say as only rarely contaminated and hence of little importance.

In Europe, several other bark beetles of the genus *Scolytus* are known to be important vectors—*S. affinis* Egg., *S. laevis* Chap., *S. pygmaeus* Fabr., and *S. sulcifrons* Rey.—with *S. sulcatus* LeC. an important suspect (Goidanich and Goidanich, 1934; Kalandra and Pfeffer, 1935).

The possibility that mites are also concerned in transmission was pointed out by Jacot (1934) and Fransen (1939). Jacot found that oribatid and tyroglyphid mites were always present, and they were constantly in and out of the tunnels of *H. rufipes* and the *Scolytus* spp. and had ample opportunity to become contaminated from the coremia heads.

Control of the disease cannot be expected to be complete, but spread can be reduced to 1–2% a year (Anon., 1958). Responsibility for control calls for close cooperation between local and federal authorities and private owners. There are two main divisions of the actual control program, and both are essentially prophylactic measures, since once a tree has become infected there is practically no hope of saving it.

The first of these methods is the reduction in population of bark beetles and the prevention of their bark feeding by spraying during the dormant season, either in the fall or early spring. An annual spray of DDT will provide protection.

The second method is sanitation and consists of the removal of all possible breeding wood, such as freshly cut logs, dying and recently·dead trees, broken branches, and the stumps of recently felled trees. If this material is debarked and the bark burned, the wood can be used for its intended purposes.

The control of the disease is also being sought through resistance as well as by more direct methods, although it is a long term program with the established procedure taking 11 yr. One yr-old seedlings are repeatedly infected annually and the resistants selected after four yr. Callus cuttings from these are greenhouse tested and again, the most resistant are planted out for further testing (Tohernoff, 1966). Heybroek (1963) reported that a resistant hybrid, a clone of *Ulmus hollandica,* bred from an English *U. glabra* crossed with the French *U. carpinifolia,* was released in 1963.

As an aid to timing spray programs, Kapter (1967) recorded the phenological events associated with the spring emergence of the small European elm bark beetle. A flowering sequence of four plant species which bloom prior to the emergence of the beetles was established. This use of phenological data is long overdue.

The prophylactic action of Benomyl against Dutch elm disease has been shown by Biehn and Dimond (1970). Soil injection reduced foliar symptoms an average of 79% in 1969 when Benomyl 50% (520 lb/ac) plus Surfactant F 819 gal/ac was used. In 1970, with an increase in the Benomyl 50% to 810 lb/ac, and the addition of Tween 20, the reduction in foliar symptoms was increased to 97%. The fungi-toxicant was present in elm shoots, twigs, and leaves. Hock *et al.* (1970) reported the suppression of Dutch elm disease in American elm seedlings by Benomyl. The presence of *C. ulmi* in uninoculated trees was found to be 80%, but all the treated trees were negative. They concluded that Benomyl showed marked fungi-toxic activity when applied to the sand in which the seedlings were growing, as a soil drench. However, Hock and Schreiber (1971) found no fungi-toxicant effect following soil and trunk infections of trees growing in a highly organic soil.

Hemileia Vastatrix *on Coffee*

Hemileia vastatrix is a serious pest of coffee in East Africa and it has recently involved Brazil. One aspect of the transmission is unique. Midge larvae eat the fungus spores; two hymenopterous parasites of the midge larvae emerge carrying inoculum which might be sufficient to permit infection of coffee in areas remote from other coffee areas, depending on the mobility of the adult parasite (Crowe, 1963).

White Rot of Softwoods

The siricid woodwasps, *Sirex gigas* L and *S. cyaneus* Fabr., live in close symbiotic association with the fungus *Stereum sanguinolentum* (A. & S.)

Fr., which causes a white rot of softwoods. The association appears to be a true and highly adapted symbiosis. The fungus is present in the intersegmental sacs situated at the anterior end of the ovipositor of adult females. The egg becomes infected with fungus oidia at the start of its passage down the ovipositor. When the egg has been deposited in timber, mycelial growth commences and the fungus subsequently precedes the larva in its boring. Modifications of the larval integument on the posterior aspect of the first abdominal segment have been discovered. The structures so formed, termed hypopleural organs, contain a fungus that has been isolated and identified as *S. sanguinolentum*. The organs are found only in some of the larvae, which suggests that they occur in female larvae only. No fungus could be detected in the pupae (Parkin, 1942).

REFERENCES

Abbott, E.V., "Red rot of sugarcane," *Yearbook Agr. U.S. Dept. Agr.,* 536–539 **1953**.

Alfar, A., "El Acaro *Pediculopsis graminum* Reut. y el hongo *Nigrospora oryzae* (Berk. et Br.) Petch en association parasitario sobre Tigos aragoneses)," *Bol. Patol. Vegeta y Entomol. Agr.,* **16,** 321–334 (1946).

Anon., "Dutch elm disease," *Can. Dept. Agr. Publ.,* No. 1010 (1958).

Anon., "Ceracystis Wilt," in *Annual Reports on Cacao Research, I.C.T.A.,* Washington D.C., 1965.

Ashby, S. F., and Nowell, W., "The fungi of stigmatomycosis," *Ann. Botany (London),* **40,** 69–83 (1926).

Ashworth, L. J. Jr., Rice, R. E., McCleans, J. L., and Brown, C. M., "The relationship of insects to infection of cotton bolls by *Aspergillus Flavus,*" *Phytopathology,* **61,** 488 (1971).

Atanasoff, D., "Ergot of grains and grasses," *U.S. Dept. Agr., Bur. Plant Ind.,* **1920,** 1–27 (mimeographed).

Aucamp, J. L., "Role of mite vectors in the development of aflatoxin in groundnut," *J. Stored Prod. Res.,* 5, 245–249 (1969).

Austwick, P. K. C., "Insects and the spread of fungal disease," *Biological Aspects of the Transmission of Disease,* Oliver and Boyd, London, 1958.

Bagga, H. S., and Laster, M. L., "Relation of insects to the initiation and development of boll rot of cotton," *J. Econ. Entomol.,* **61,** 1141 (1968).

Baillie, A. F. H., and Jepson, W. F., "Bud blast disease of the rhododendron in its relation to the leaf-hopper *Graphocephala coccinea* Forst.," *J. Roy. Hort. Soc.,* **76,** 355–365 (1951).

Barger, J. H., and Hock, W. F., "Distribution of Dutch elm disease and the smaller European elm bark beetle in the U.S. as of 1970," *Plant Disease Reptr.,* **55,** 271 (1971).

Barnett, H. L., and True, R. P., "The oak wilt fungus, *Endoconidiophora faga-cearum," Trans. N. Y. Acad. Sci.,* 17, 552–559 (1955).

Bart, G. J., "Host-parasite relationships of the oak wilt fungus," *Dissertation Abstr.,* 16, 2008–2009 (1956).

Bart, G. J., and Griswold, C. J., "Recovery of viable spores of *Endoconidi-ophora fagacearum* from excrement of insects used in disease transmission stud-ies," *Phytopathology,* 43, 466 (1953).

Beattie, R. K., "The new outbreak of the Dutch elm disease," *J. Econ. Ento-mol.,* 27, 569–572 (1934).

Beckman, C. H., Kuntz, J. E., Riker, A. J., and Berbee, J. G., "Host responses associated with the development of oak wilt," *Phytopathology,* 43, 448–454 (1953).

Biehn, W. N., and Dimond, A. E., "Prophylactic action of Benomyl against Dutch Elm disease," *Phytopathology,* 60, 571 (1970).

Boning, J., "Insekten als Uebertrager von Pflanzenkrankheiten," *Z. Angew. En-tomol.,* 15, 181–206 (1929).

Bonnemaison, L., and Missonnier, J., "Le psylle du poirer (*Psylla pyri* L) mor-phologie et biologie. Methodes de lutte," *Ann. Inst. Natl. Recherche Agron. Ser. C,* 7, 263–331 (1956).

Boosalis, M. G., "Hessian fly in relation to the development of crown and basal stem rot of wheat," *Phytopathology,* 44, 224–229 (1954).

Boyce, J. S., Jr., "Nitidulid beetles released 500 feet away reinfest oak wilt fun-gus mats," *Plant Disease Reptr.,* 38, 212–213 (1954).

Boyce, J. S., Jr., and Garren, K. H., "Compatibility types of the oak wilt fungus in 23 Appalachian trees," *Phytopathology,* 43, 644–645 (1953).

Bramble, W. C., and Holst, E. C., "Fungi associated with *Dendroctonus fron-talis* in killing short leaf pines and their effect on conduction," *Phytopathology,* 30, 881–899 (1940).

Brefeld, O., and Falck, R., "Die Bluteninfektion bei den Brandpilzen und die naturliche Verbreitung der Brandkrankheiten," *Untersuchungen aus dem Ge-sammtgebiete der Mycol.,* 13, 1–74 (1905).

Bretz, T. W., "A preliminary report on the perithecial stage of *Chalara quer-cina* Henry," *Plant Disease Reptr.,* 35, 298 (1951).

Bretz, T. W., "The ascigerous stage of the oak wilt fungus," *Phytopathology,* 42, 435–437 (1952).

Bretz, T. W., and Buchanan, W. D., "Oak wilt fungus not found in acorns from diseased tree," *Plant Disease Reptr.,* 41, 546 (1957).

Bretz, T. W., and Jones, T. W., "Oak flowers may serve as infection courts for the oak wilt disease," *Plant Disease Reptr.,* 41, 545 (1957).

Brierley, P., "Eradicant fungicides of possible value against the azalea petal blight fungus, *Ovulinia azaleae," Phytopathology,* 40, 153–155 (1950).

Brooks, F. T., "Disease resistance in plants," *New Phytologist,* 27, 85–97 (1928).

Brunt, A. A., and Wharton, A. L., "Transmission of a gall disease of cacao," *Na-ture,* 187, 80 (1960).

Buchanan, W. D., "Brentids may be vectors of the oak wilt disease," *Plant Disease Reptr.,* 41, 707 (1957).

Buryhina, E. K., "Cabbage phomosis and its control," *Sad i Ogorod,* 1, 51–52 (1950); *Rev. Appl. Mycol.,* 31, 157–158 (1952).

Caldis, P. D., "Etiology and transmission of endosepsis (internal rot) of the fruit of fig," *Hilgardia,* 2, 287–328 (1927).

Caldis, P. D., "Souring of the figs by yeasts and the transmission of the disease by insects," *J. Agr. Research,* 40, 1031–1051 (1930).

Cardenosa-Barrigal, R., "La antracnosis del Platana "Cachaco" en el Tolema," *Turrailba,* 13, 88–95 (1963).

Chaves-Batista, A., "Mal do Recife (Grave doenca da Manguiera)," *Escola Super. Agr. Pernambuco, Recife,* 1947, 1–106.

Childs, L., "The relation of woolly apple aphis to perennial canker infection with other notes on the disease," *Oregon State College, Agr. Expt. Sta. Bull.,* No. 243 (1929).

Christensen, J. J., "Root rots of wheat, oats, rye, barley," *Yearbook Agr., U.S. Dept. Agr.,* 1953, 321–328.

Christensen, J. J., and Schneider, C. L., "European corn borer (*Pyrausta nubilalis* Hbn.) in relation to shank, stalk, and ear rots of corn," *Phytopathology,* 40, 284–291 (1950).

Clarke, R. G., and Wilde, Gerald E., "Association of the green stink bug and the yeast spot disease organism of soybeans. I. Length of retention, effect of molting, isolation from feces and saliva," *J. Econ. Entomol.,* 63, 200–204 (1970).

Clarke, R. G., and Wilde, G. E., "Association of the green stink bug and the yellow spot disease organism of soybeans. II. Frequency of transmission to soybeans, transmission from insect to insect, isolation from field populations," *J. Econ. Entomol.,* 63, 355–357 (1970a).

Clarke, R. G., and Wilde, G. E., "Association of the green stink bug and the yeast-spot organism of soybeans. III. Effect on soybean quality," *J. Econ. Entomol.,* 64, 222–224 (1971).

Collins C. W., "Studies of elm insects associated with Dutch elm disease fungus," *J. Econ. Entomol.,* 34, 369–372 (1941).

Cones, W. L., and True, R. P., "Oak wilt incidence related to site productivity in Hampshire County, W. Va.," *Phytopathology,* 57, 806 (Abstr.) (1967).

Craighead, F. C., "Interrelation of tree-killing bark beetles (*Dendroctonus*) and blue stains," *J. Forestry,* 26, 886–887 (1928).

Craighead, F. C., and Morris, C. L., "Possible importance of insects in transmission of oak wilt—a progress report," *Penn. Forest and Waters,* (Nov.–Dec. 1952).

Craighead, F. C., Morris, C. L., and Nelson, J. C., "A preliminary note on the susceptibility of wounded oaks to natural infection by the oak wilt fungus," *Plant Disease Reptr.,* 37, 483–484 (1953).

Craigie, J. H., "Discovery of the function of the pycnia of the rust fungi," *Nature,* 120, 756–767 (1927).

Crandall, B. S., and Baker, W. L., "The wilt disease of American persimmon caused by *Cephalosporium diospyri*," *Phytopathology*, 40, 307–325 (1950).

Crone, L. J., and Bachelder, S., "Insect transmission of canker stain fungus," *Phytopathology*, 51, 576 (Abstr.) (1961).

Crowdy, S. H., "Observations on the pathogenicity of *Calonectria rigidiuscula* (Berk. and Br.) Sacc. on *Theobroma cacao* L.," *Ann. Appl. Biol.*, 34, 45–59 (1947).

Crowe, T. J., "Possible insect vectors of the uredespores of *Hemeleia vastratrix* in Kenya," *Trans. British Mycol. Soc.*, 46, 24–26 (1963).

Croxall, H. E., Collingwood, C. A., and Jenkins, J. E. E., "Observations on brown rot (*Sclerotinia fructigena*) of apples in relation to injury caused by earwig (*Forficula auricularia*)," *Ann. Appl. Biol.*, 38, 833–843 (1951).

Curl, E. A., "Experimental transmission of *Endoconidiophora fagacearum* by Collembola," *Plant Disease Reptr.*, 40, 455–458 (1956).

Cutright, C. R., "Comstock's mealybug, *Pseudococcus comstocki*, on apple in Ohio," *J. Econ. Entomol.*, 32, 888 (1939).

Daugherty, D. M., "Pentatomidae as vectors of yeast-spot disease of soybeans," *J. Econ. Entomol.*, 60, 147–152 (1967).

Daugherty, D. M., and Foster, J. E., "Organism of yeast spot disease isolated from rice damaged by rice stink bug," *J. Econ. Entomol.*, 59, 1282–1283 (1966).

Davey, A. E., and Smith, R. E., "The epidemiology of fig spoilage," *Hilgardia*, 7, 523–551 (1933).

Davidson, R. W., "A deterioration problem in beetle-killed spruce in Colorado," *Phytopathology*, 41, 560 (1951).

Davidson, R. W., "Species of Ophiostomataceae associated with Engelmann spruce bark beetle," *Phytopathology*, 44, 485 (1954).

Donk, M. A., "Notes on Malesian fungi I," *Bull. Jard. Botan. Buitenz.*, Ser. III, 17, 473–482 (1948).

Dorsey, C. K., and Leach, J. G., "The bionomics of certain insects associated with oak wilt with particular reference to the Nitidulidae," *J. Econ. Entomol.*, 49, 219–230 (1956).

Dorsey, C. K., Jewell, F. F., Leach, J. G., and True, R. P., "Experimental transmission of oak wilt by four species of Nitidulidae," *Plant Disease Reptr.*, 37, 419–420 (1953).

Ehrlich, J., "The beech bark disease, a *Nectria* disease of *Fagus* following *Cryptococcus fagi* (Baer)," *Can. J. Research*, 10, 593–692 (1934).

Feldman, S. W., Caroseli, N. E., and Howard, F. L., "Physiology of toxin production by *Ceratostomella ulmi*," *Phytopathology*, 40, 341–354 (1950).

Foster, J. E., and Daugherty, D. M., "Isolation of the organism causing yeast spot disease from the salivary system of the green stink bug," *J. Econ. Entomol.*, 62, 424–427 (1969).

Francke-Grosmann, H., "On the ambrosia culture of the two pine bark beetles, *Myelophilus minor* Htg. and *Ips acuminatus* GYLL," *Medd. Skogsforskn Inst. Stockholm*, 41, 1–52 (1952).

Fransen, J. J., *Iepenziekte, Iepenspintkevers en Beider Bestrijding,* Dissertation, Landbouwhoogeschool, Wageningen, 1939.

Frazer, H. L., "Observations on the methods of transmission of internal boll disease of cotton by the cotton-stainer bug," *Ann. Appl. Biol.,* **31,** 271–290 (1944).

Gill, D. L., "Petal blight of azalea," *Yearbook Agr. U.S. Dept. Agr.,* **1953,** 578–582.

Gloyer, W. O., and Fulton, B. B., "Tree crickets as carriers of *Leptosphaeria coniothyrium* (Fckl.) Sacc. and other fungi," *N. Y. State Agr. Expt. Sta. (Geneva, N. Y.(Bull.,* No. 50 (1916).

Goidanich, S., and Goidanich, G., "*Lo Scolytus sulcifrons* Rey (Coleoptera-Scolytidae) nella diffusione del Pirenomicete *Ceratostomella (Graphium) ulmi* (Schwarz) Buis, nell Emilia," *Boll. Lab. Entomol. Regio 1st Supl. Agrar. (Bologna),* **7,** 145–163 (1934).

Goncalves, C. R., "Observacoes sobre *Pseudoccus comstocki* (Kuw.) attacando citrus na baixada fluminense," *Rodriguesia,* 13, 121–129 (1940).

Gray, R. A., "Composition of honeydew excreted by pineapple mealybugs," *Science,* 115, 129–133 (1952).

Griswold, C. L., "Transmission of the oak wilt fungus by the pomace fly," *J. Econ. Entomol.,* **46,** 1099–1100 (1953).

Griswold, C. L., "Transmission of the oak wilt fungus by *Pseudopityophthorus minutissimus* (Zimm.)," *J. Econ. Entomol.,* **49,** 560–561 (1956).

Griswold, C. L., and Bart, G. J., "Transmission of *Endoconidiophora fagacearum* by *Pseudopityophthorus pruinosus,*" *Plant Disease Reptr.,* **38,** 591 (1954).

Gruenhagen, R. H., Riker, A. J., and Richards, C. A., "Burn blight of jack and red pine following spittle insect attack," *Phytopathology,* 37, 757–772 (1947).

Guyton, T. L., "An unusual occurrence of oak wilt in Pennsylvania," *Plant Disease Reptr.,* 36, 386 (1952).

Haddow, W. R., "Needle blight and late fall browning of red pine (*Pinus resinosa* Ait.) caused by a gall midge (Cecidomyiidae) and the fungus (*Pullularia pullulans* (de Bary) Berkhout)," *Trans. Roy. Can. Inst.,* 23, 161–189 (1941).

Haddow, W. R., and Adamson, M. A., "Note on the occurrence of needle blight and late fall browning in red pine (*Pinus resinosa* Ait.)," *Forestry Chronicle,* 15, 107–110 (1939).

Haddow, W. R., and Newman, F. S., "A disease of the Scots Pine (*Pinus Sylvestris* L.) caused by the fungus *Diplodia pinea* Kick associated with the pine spittle-bug (*Aphrophora parallela* Say). I. Symptoms and etiology," *Trans. Roy. Can. Inst.,* 24, 1–17 (1942).

Hansen, A. J., "Role of the brown jay in epiphytotics of *Thielaviopsis* pod rot cacao," *Phytopathology,* 54, 886 (Abstr.) (1964).

Hansen, H. N., "Thrips as carriers of fig-decaying organisms," *Science,* 69, 356–357 (1929).

Hansen, H. N., and Davey, A. E., "Transmission of smut and molds in figs," *Phytopathology,* 22, 247–252 (1932).

Hanson, E. W., and Milliron, H. E., "The relation of certain weevils to root rot and basal rot of cereals and grasses," *Phytopathology*, **32**, 22 (1942).

Hanson, E. W., Milliron, H. E., and Christensen, J. J., "The relation of the blue grass billbug, *Calendra parvula* (Gyllenhal), to the development of basal stem rot and root rot of cereals and grasses in north-central United States," *Phytopathology*, **40**, 527–543 (1950).

Hargreaves, H., and Taylor, T. H. C., "Report on investigations of cotton stainers for 1936," *Rept. Dept. Agr. Uganda*, 1936–37 (1938).

Hasan, S., and Vago, C., "Transmission of *Alternaria brassicola* by slugs," *Plant Disease Reptr.*, **50**, 764–767 (1966).

Hawboldt, L. S., "History of spread of the beech scale, *Cryptococcus fagi* (Baerensprung), an insect introduced into the Maritime Provinces," *Acadian Naturalist*, **1**, 137–146 (1944); *Rev. Appl. Entomol.*, **33**, 353–354 (1945).

Henry, B. W., "*Chalara quercina* n. sp., the cause of oak wilt," *Phytopathology*, **34**, 631–635 (1944).

Henry, B. W., and Moses, C. S., "An undescribed disease causing rapid dying of oak trees," *Phytopathology*, **33**, 18 (1943).

Hepting, G. H., "The current status of oak wilt in the United States," *Forest Sci.*, **1**, 95–103 (1955).

Hepting, G. H., Toole, E. R., and Boyce, J. S., Jr., "Sexuality in the oak wilt fungus," *Phytopathology*, **42**, 438–442 (1952).

Heybroek, H. M., "De Iep 'Groeneveld' (The elm groeneveld)," *Ned Bosch. Tijdschr.*, **35**, 370–374 (1963).

Himelick, E. B., and Curl, E. A., "Experimental transmission of the oak wilt fungus by caged squirrels," *Phytopathology*, **45**, 581–584 (1955).

Himelick, E. B., and Fox, H. W., "Experimental studies on control of oak wilt disease," *Bull. Illinois Agric. Exp. Sta.*, No. 680 (1961).

Hock, W. K., Schreiber, L. R., and Roberts, B. R., "Suppression of Dutch elm disease in American elm seedlings by Benomyl," *Phytopathology*, **60**, 391 (1970).

Hock, W. K., and Schreiber, L. R., "Evaluation of Benomyl for the control of Dutch elm disease," *Plant Disease Reptr.*, **55**, 58 (1971).

Hoffman, C. H., and Moses, C. S., "Mating habits of *Scolytus multistriatus* and the dissemination of *Ceratostomella ulmi*," *j. econ. Entomol.*, **33**, 818–819 (1940).

Holmes, F. W., "Recorded Dutch elm disease distribution in North America as of 1959," *Plant Disease Reptr.*, **45**, 74–75 (1961).

Hough, W. S., "Biology and control of Comstock's mealybug on the umbrella catalpa," *Virginia Agr. Expt. Sta., Tech. Bull.*, No. 29, 1–27 (1925).

Hunt, J., "Taxonomy of the genus Ceratocystis," *Lloydia*, **19**, 1–58 (1956).

Ingold, C. T., *Dispersal in Fungi*, Oxford Univ. Press, London, 1953.

Jacot, A. P., "Acarina as possible vectors of the Dutch elm disease," *J. Econ. Entomol.*, **27**, 858–859 (1934).

Jeffery, A. R., *The Oak Wilt Survey and Eradication Program in Pennsylvania,* Penn. Department of Agriculture, Harrisburg, Penn., 1954.

Jewell, F. F., "Viability of the conidia of *Endoconidiophora fagacearum* Bretz in the fecal material of certain Nitidulidae," *Plant Disease Reptr.,* 38, 53–54 (1954).

Jewell, F. F., "Insect transmission of oak wilt," *Phytopathology,* 46, 244–257 (1956).

Jones, T. W., and Bretz, T. W., "Transmission of oak wilt by tools," *Plant Disease Reptr.,* 39, 498–499 (1955).

Kalandra, A., and Pfeffer, A., "Ein Beitrag zum Studium der Ulmengraphiose," *Lesnicka Prace,* 14, Reprint, 17 pp. (1935); *Rev. Appl. Mycol.,* 14, 536 (1935).

Kapter, J. E., "Phenological events associated with the spring emergence of the small European elm beetle in Dubuque, Iowa," *J. Econ. Entomol.,* 60, 50 (1967).

Kaston, B. J., and Riggs, D. S., "On certain habits of elm bark beetles," *J. Econ. Entomol.,* 31, 467–469 (1938).

Kellerman, W. A., and Swingle, W. T., "New species of Kansas fungi," *J. Mycol.,* 4, 93–95 (1888).

Kellerman W. A., and Swingle, W. T., "Branch knot of the hackberry," *Kansas Agr. Sta. Rept.,* 1889, 302–315.

Kilpatrick, R. A., and Dunn. G. M., "Observations of insects and fungi associated with taproot survival of white clover in New Hampshire," *Plant Disease Reptr.,* 42, 819–820 (1958).

Knull, J. N., "Scouting for elm scolytids," *J. Econ. Entomol.,* 27, 865–866 (1934).

Laemmlen, F. F., "The association of the mite *Siteroptes reniformis* and *Nigrospora oryzae* in Nigrospora lint rot of cotton," *Phytopathology,* 59, 1036 (1969).

Langdon, R. F. N., and Champ, B. R., "The insect vectors of *Claviceps paspali* in Queensland," *J. Australian Inst. Agr. Sci.,* 20, 115–118 (1954).

Leach, J. G., *Insect Transmission of Plant Diseases,* McGraw-Hill, New York, 1940.

Leach, J. G., Orr, L. W., and Christensen, C., "The interrelationships of bark beetles and blue-staining fungi in felled Norway pine timber," *J. Agr. Research,* 49, 315–342 (1934).

Leath, K. Y., and Newton, R. C., "Interaction of a fungus gnat, *Bradysia* sp. (Sciaridae) with *Fusarium* spp. on alfalfa and red clover," *Phytopathology,* 59, 257–258 (1969).

LePelley, R. H., "The food and feeding habits of *Antestia* in Kenya," *Bull. Entomol. Res.,* 33, 71–89 (1942).

Leunis, J., "Insects as spreaders of ergot," *Synopsis des Pflanzenkunde Hahn,* 522 (1847).

Libby, R. R., and Ellis, D. E., "Transmission of *Colletotrichum lagenarium* by the spotted cucumber beetle," *Plant Disease Reptr.,* 38, 200 (1954).

Linn, J. B., *"Cephalosporium* leaf spot of *Dieffenbachia,"* Phytopathology, **32,** 172–175 (1942).

Maltais, J. B., and Auclair, J. L., "Occurrence of amino acids in the honeydew of the crescent marked lily aphid, *Myzus circumflexus* (Buckt.)," *Can. J. Zool.,* **30,** 191–193 (1952).

Martin, W. J., "Alternaria leaf blight of Hevea rubber trees," *Phytopathology,* **37,** 609–612 (1947).

Maublanc, A., and Roger, L., "La phthiriose du Cafeier au Cameroun," *Rev. Appl. Botan.,* **15,** 25–32 (1935).

May, C., "Outbreaks of the Dutch elm disease in the United States," *U.S. Dept. Agr. Circ.,* No. 322 (1934).

McLarty, H. R., "Perennial canker of apple trees" *Can J. Research,* **8,** 492–507 (1933).

NcNabb, H. S., "The status of oak wilt in Iowa," *Proc. Iowa Acad. Sci.,* **61,** 141–148 (1954).

Mendes, L. O. T., "Podridao interna dos capulhos do algodoeiro obtida por meio de insetos," *Bragantia,* **15,** 9–11 (1956).

Mercier, L., "Sur le role des insects comme agents de propagation de l'ergot des graminees," *Compt. Rend. Soc. Biol.,* **70,** 300–302 (1911).

Merek, E. L., and Fergus, C. L., "Longevity of the oak wilt fungus in diseased trees," *Phytopathology,* **44,** 328 (1954).

Merrill, W., "The oak wilt epidemics in Pennsylvania and West Virginia; an analysis," *Phytopathology,* **57,** 1206–1210 (1967).

Mittler, T. E., "Amino acids in phloem sap and their excretion by aphids," *Nature,* **172,** 207–208 (1953).

Moller, W. J., and DeVay, J. E., "The role of insect vectors in the *Ceratocystis* canker disease of stone fruit trees," *Phytopathology,* **56,** 891 (Abstr.) (1966).

Morris, C. L., Thompson, H. E., Hadley, B. L., and Davis, J. M., "Use of radioactive tracer for investigation of the activity pattern of suspected insect vectors of the oak wilt fungus," *Plant Disease Reptr.,* **39,** 61–63 (1955).

Nance, N. W., "New or unusual records and outstanding features of plant disease development in the U.S. in 1948," *Plant Disease Reptr.,* Suppl. no. 184, 179–206 (1949).

Naqvi, N. Z., and Futrell, M. C., "Aphid transmissible material produced by *Sclerospora sorghi* in corn and sorghum plants," *Phytopathology,* **60,** 586 (Abstr.) (1970).

Nelson, R. M., "Effect of bluestain fungi on southern pines attacked by bark beetles," *Phytopathol. Z.,* **7,** 327–353 (1934).

Nelson, R. M., and Beal, J. A., "Experiments with bluestain fungi in southern pines," *Phytopathology,* **19,** 1101–1106 (1929).

Norris, D. M., Jr., "Insect transmission of oak wilt in Iowa," *Plant Disease Reptr.,* **37,** 417–418 (1953).

Norris, D. M., Jr., "Natural spread of *Endoconidiophora fagacearum* Bretz to wounded red oaks in Iowa," *Plant Disease Reptr.*, 39, 249–253 (1955).

Ogawa, J. M., "The dried fruit beetle disseminates spores of the peach brown rot fungus," *Phytopathology*, 47, 530 1957).

Owen, H., "Further observations on the pathogenicity of *Calonectria rigidiuscula* (Berk. and Br.) Sacc. to *Theobroma cacao* L.," *Ann. Appl. Biol.* 44, 307–321 (1956).

Parker, K. G., Readio, P. A., Tyler, L. J., and Collins, D. L., "Transmission of the Dutch elm disease pathogen by *Scolytus multistriatus* and the development of infection," *Phytopathology*, 31, 657–663 (1941).

Parkin, E. A., "Symbiosis and siricid wood wasps," *Ann. Appl. Biol.*, 29, 268–274 (1942).

Parmeter, J. R., Jr., Kuntz, J. E., and Riker, A. J., "Oak wilt development in bur oaks," *Phytopathology*, 46, 423–436 (1956).

Parrott, P. J., and Fulton, B. B., "Tree crickets injurious to orchard and garden fruits," *N. Y. State Agr. Expt. Sta. (Geneva, N. Y.) Bull.*, No. 64 (1914).

Parrott, P. J., Gloyer, W. O., and Fulton, B. B., "Some studies on the snowy tree cricket with reference to an apple bark disease," *J. Econ. Entomol.*, 8, 535–541 (1915).

Pearson, E. O., "The development of internal boll disease of cotton in relation to time of infection," *Ann. Appl. Biol.*, 34, 527–545 (1947).

Persoon, D. H., *Synopsis methodica fungorum*, Dieterich, Gottingen, 1801.

Peyronel, B., "A mutualistic association between mites of the genus *Pediculopsis* and some fungus parasites of plants," *Atti Accad. Sci. Torino, Classe Fis. Mat. e Nat.*, 84, Reprint, 9pp. (1950).

Phillips, E. H., Smith, E. H., and Smith, R. E., "Fig. smut," *Calif. Univ. Agr. Expt. Sta. Bull.*, No. 387:1925).

Rankin, W. H., Parker, K. G., and Collins, D. L., "Dutch elm disease fungus prevalent in bark beetle infested elm wood," *J. Econ. Entomol.*, 34, 548–551 (1941).

Rathay, E., *Untersuchungen uber die Spermogenien der Rostpilze*, Denkschr. Akad., Wien, 1882.

Readio, P. A., "The entomological phases of the Dutch elm disease," *J. Econ. Entomol.*, 28, 341–353 (1935).

Rexrode, C. O., "Tree wounding insects as vectors of the oak wilt fungus," *Forest Science*, 14, 181–189 (1968).

Rexrode, C. O., Kulman, H. M., and Dorsey, C. K., "Bionomics of the bark beetle, *Pseudopityophorus pruinosius* with special reference to its role as a vector of oak wilt, *Ceratocystis fagacearum*," *J. Econ. Entomol.*, 58, 913–916 (1965).

Rexrode, C. O., and Lincoln, A. C., "Distribution of oak wilt," *Plant Disease Reptr.*, 49, 1007–1010 (1965).

Rexrode, C. O., and Jones, T. W., "Oak bark beetles carry oak wilt fungus in early spring," *Plant Disease Reptr.*, 55, 108–111 (1971).

Rumbold, C. T., "Two blue-staining fungi associated with bark beetle infestation of pines," *J. Agr. Research*, 43, 837–873 (1931).

Rumbold, C. T., "Three blue-staining fungi, including two new species, associated with bark beetles," *J. Agr. Research*, 52, 419–437 (1936).

Salmon, E. S., "A monograph of the Erysiphaceae," *Torrey Botan. Club Mem.*, 9, 1–292 (1900).

Satterthwait, A. F., "How to control billbugs destructive to cereal and forage crops," *U.S. Dept. Agr. Farmers' Bull.*, No. 1003 (1932).

Savory, J. G., "Control of blue-stain in homegrown pine logs," *Suppl. Timber Trades J.*, 19–20 (June 13, 1964).

Schewket (Sevket) N., "Die Feigeninekten und die wesentlichsten Urachen der Feigenfruchfaule," *Anz. Schadlingskunde*, 10, 118–119 (1934).

Schmidt, C. T., "Biological studies of the Nitidulid beetles found in pineapple fields," *Ann. Entomol. Soc. Amer.*, 28, 475–511 (1935).

Scott, W. M., and Ayres, T. W., "The control of peach brown rot and scab," *U. S. Dept. Agr. Bur. Plant Ind. Bull.*, No. 174 (1910).

Silow, R. A., "A systemic disease of red clover caused by *Botrytis anthophila* Bond," *Trans. Brit. Mycol. Soc.*, 18, 239–248 (1933).

Smith, F. F., and Weiss, F., "Relationship of insects to the spread of azalea flower spot," *U.S. Dept. Agr. Tech. Bull.*, No. 798, 1–44 (1942).

Smith, R. E., and Hansen, H. N., "The improvement of quality in figs," *Calif. Univ. Agr. Expt. Sta. Circ.*, No. 311 (1927).

Smith, R. E., and Hansen, H. N., "Fruit spoilage disease of figs," *Calif. Univ. Agr. Expt. Sta. Bull.*, No. 506 (1931).

Snetsinger, R., and Himelick, E. B., "Observations on witches' broom of hackberry," *Plant Disease Reptr.*, 41, 541–544 (1957).

Spierenburg, D., "Een Onbekende ziekte in de Iepen. II," *Verslag. Mededel. Plantenziektenkundigen Dienst Wageningen*, 24, (1922); *Rev. Appl. Mycol.*, 2, (1933).

Stambaugh, W. J., and Fregus, C. L., "Longevity of spores of the oak wilt fungus on overwintered Nitidulid beetles," *Plant Disease Reptr.*, 40, 919–922 (1956).

Steyaert, R. L., "Plant protection in the Belgian Congo," *Sci. Monthly*, 63, 268–280 (1946).

Stojanovic, D., "Jabucni svrdlas kao prenosilac konidija *Monilia spp.* (*Prethodno saopstenje*). (*Rhynchites bacchus* as a vector of the conidia of *Monilia* spp.)," *Plant Protect. (Belgrade)*', 34, 63–65 (1956).

Taubenhaus, J. J., "A study of the black and yellow moulds of ear corn," *Texas Agr. Expt. Sta. Bull.*, No. 270 (1920).

Taylor, G. S., "Stalk rot development in corn following the European corn borer," *Phytopathology*, 42, 20–21 (1952).

Tchernoff, V., "Methods for screening and for rapid selection of elms for resistance to Dutch elm disease," *Acta. Bot. Neerl.*, 14, 409–452 (1966).

Thompson, H. E., Hadley, B. L., and Jeffrey, A. R., "Transmission of *Endoconidiophora fagacearum* by spore infested Nitidulids caged on wounded healthy oaks in Pennsylvania," *Plant Disease Reptr.*, **39**, 58–60 (1955).

True, R. P., Barnett, H. L, Dorsey, C. K., and Leach, J. G., "Oak wilt in West Virginia," *West Va. Univ. Agr. Expt. Sta. Bull.*, No. 448T (1960).

Turner, G. J., "Transmission by snails of the species of *Phytophora* which causes foot rot of *Piper nigrum* L in Sarawak," *Nature*, **202**, 1133 (1964).

Tyler, L. J., and Parker, J. G., "Factors affecting the saprogenic activities of the Dutch elm disease pathogen," *Phytopathology*, **35**, 675–687 (1945).

Tyler, L. J., Parker, K. G., and Pechuman, L. L., "The relation of *Saperda tridentata* to infection of American elm by *Ceratostomella ulmi*," *Phytopathology*, **29**, 547–549 (1939).

Ullstrup, A. J., "Several ear rots of corn," *Yearbook Agr.*, *U.S. Dept. Agr.*, 1953 390–392.

Von Schrenk, H., "The 'Bluing' and the 'Red Rot' of the western yellow pine, with special reference to the Black Hills Forest Reserve," *U.S. Dept. Agr. Bur. Plant Ind. Bull.*, No. 36 (1903).

Wadley, F. M., and Wolfenbarger, D. O., "Regression of insect density on distance from center of dispersion as shown by a study of the smaller European elm bark beetle," *J. Agr. Research*, **69**, 299–308 (1944).

Webber, H. J., "Sooty mold of the orange and its treatment," *U.S. Dept. Agr. Div. Veg. Physiol. Pathol. Bull.*, No. 13 (1897).

White, I. G., and Wolf, F. T., "Toxin production by the oak wilt fungus, *Endoconidiophora fagacearum*," *Phytopathology*, **44**, 334 (1954).

Wienk, J. F., "The transmission of *Aspergillus niger* by adult sisal weevils," *Ann. Rep. Sisal Res. Sta. Mlingano* (1964–1965).

Willison, R. S., and Dustan, G. G., "Fruit flies and fungal wastage in peaches," *Can. J. Agr. Sci.*, **36**, 233–240 (1956).

Wilson, M., and Wilson, M. J. F., "The occurrence of the Dutch elm disease in England," *Gardeners' Chronicle*, **83**, 31–32 (1928).

Wolfenbarger, D. O., and Jones, T. H., "Intensity of attacks by *Scolytus multistriatus* at distances from dispersion and convergence points," *J. Econ. Entomol.*, **36**, 399–402 (1943).

Yamamoto, W., "Studies on the dissemination of sooty moulds by insects," *Mem. Hyogo Univ. Agr. (Phytopathol. Ser. No. 1)*, **1**, 1–50 (1951) (Japanese; English summary).

Yarwood, C. E., "Association of thrips with powdery mildews," *Mycologia*, **35**, 189–191 (1943).

Yount, W. L., "Identification of oak wilt isolates as related to kind of inoculum and pattern of disease spread," *Plant Disease Reptr.*, **38**, 293–296 (1954).

Yount, W. L., "Results of root inoculations with the oak wilt fungus in Pennsylvania," *Plant Disease Reptr.*, **42**, 548–551 (1958).

Yount, W. L., Jeffrey, A. R., and Thompson, H. E., "Spores of *Endoconidiophora fagacearum* on the external surfaces of the body of Nitidulids," *Plant Disease Reptr.,* 39, 54 (1955).

Zeller, S. M., and Childs, L., "Perennial canker of apple trees. A preliminary report," *Oregon State College Agr. Expt. Sta. Bull.,* No. 217 (1925).

Zentmeyer, G. A., Wallace, P. P., and Horsfall, J. G., "Distance as a dosage factor in the spread of Dutch elm disease," *Phytopathology,* 34, 1025–1033 (1944).

Zuckerman, B. M., "Relation of type and age of wound to infection by *Endoconidiophora fagacearum* Bretz," *Plant Disease Reptr.,* 38, 290–292 (1954).

THE TOXICOGENIC INSECT AND PHYTOTOXEMIA

CHAPTER **5** CHAPTER

The Feeding Processes of Hemipterous Insects and Their Salivary Secretions

The great majority of insect species responsible for phytotoxic effects on plants belong to the Order Hemiptera; important exceptions occur in the Acarina (mites). Gall-forming insects are also found in other insect Orders. The Hemiptera are sucking insects and derive their food by piercing the plant tissue and sucking up the plant sap; in so doing they inject salivary secretions into the plant.

Basically there are two types of feeding process employed by plant bugs; in one the insects insert their stylets, and by moving them to and fro, lacerate a number of cells. A flush of saliva then enables them to take up some of the released cell contents; some of this saliva gells to form irregular aggregations. This group of insects belongs to the Heteroptera and the most important economic examples are found in the mirids (capsids).

In the second type of feeding process, the insect's secretions contain a viscous material that gells as it leaves the stylets and builds up a sheath around them, whatever route they take in the plant tissue. When suitably stained, this stylet track is seen to penetrate, inter- or intracellularly. Oc-

139

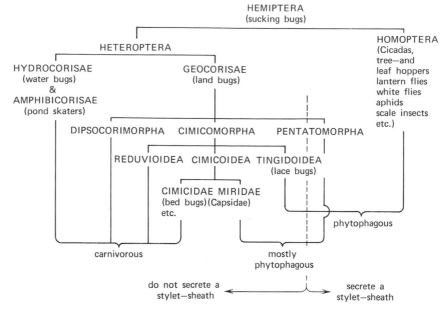

FIGURE 5.1. The feeding habits of Hemiptera in relation to their taxonomy (From Miles *Annual Review of Phytopathology* (1968). Annual Reviews Inc.).

casionally some hypertrophy occurs adjacent to the track, but gross lesions are not normally present. The Homoptera—leafhoppers, treehoppers, aphids, scale insects, lantern flies, etc.—all belong to this second group. One heteropterous group, the Pentamorpha, which includes the stinkbugs, is also a sheath depositor.

The feeding habits of the Hemiptera in relation to their taxonomy are shown in Fig. 5.1.

THE ANATOMY OF HEMIPTEROUS MOUTH PARTS AND THE FEEDING PROCESS

In general, the structure of the mouth parts of the Hemiptera follow a similar pattern for all but some carnivorous species. The principal parts of the hemipterous feeding mechanism are the clypeus, or labrum; the labium, a long slender rigid organ with a deeply concave anterior surface forming the channel of the beak; and the mandibular and maxillary stylets which lie in this channel. The maxillary stylets are grooved along the inner sides, which in turn are divided into parallel grooves by a ridge.

When the two maxillary stylets are in apposition, two channels are formed. One of these is the salivary canal, the other, the suction canal. The two pairs of stylets, the apposed maxillary and the mandibular, lying in the groove of the labium, form a compact bundle or fascicle which constitutes the piercing organ.

Two other organs are concerned in the ingestion process, the sucking pump and the salivary syringe. The first connects with the food canal between the maxillary stylets, is strongly muscled, and provides the mechanism whereby food solutions are drawn up into the pump chamber and expelled into the anterior part of the stomach. The salivary syringe is a small, hollow, cup-shaped organ, connected at its distal end with a duct from the salivary glands. Again, by the contraction of its muscles, the salivary fluid is drawn into the chamber through the salivary duct and expelled through the salivary canal formed by the maxillary stylets. In the Coccidae, the stylets are very long, often longer than the insect's body; when not in use, they are looped. How these slender hairlike structures can be driven into hard woody tissue and controlled after their entry is somewhat of a mystery. Weber's explanation (1930; 1933) is shown diagrammatically in Fig. 5.2. The piercing organ enters the plant tissue usually through the epidermis, rarely through a stoma. There is great variability even within the Hemiptera both as to the tissues sought and the manner of reaching them.

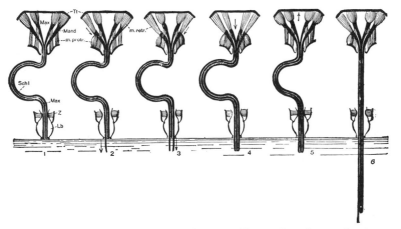

FIGURE 5.2. A diagrammatic drawing illustrating the mechanism of penetration of plant tissue by Hemipterous insects having long setae stored in a coil when retracted. 1, position of beak before penetration; 2–6, successive stages of insertion (From Snodgrass after Weber).

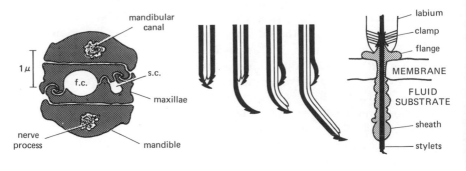

FIGURE 5.3. The stylets and stylet sheath of phytophagous Hemiptera. A. Semi-diagrammatic cross-section of the stylet bundle of an aphid. Key: f.c.=food canal; s.c.=salivary canal. Each nerve process contains two axoplasmic filaments (After Parrish (111)). B. Diagram showing sequence of movements of the tips of the stylets in a solid substrate resulting in direction of the bundle to one side and thence straight on. Note that the barbs on the tips of the mandibles (black) serve as anchors when located at the sides of the maxillae (white) (After Miles). C. Formation of the stylet-sheath (Based on diagrams from Nault and Gyrisco). (Drawing Miles, P. W. *Annual Review of Phytopathology*, **6**, 1968, Annual Reviews Inc.).

Two diagrammatic representations of hemipterous mouth parts and their penetration into the plant tissue are shown in Fig. 5.2 and 5.3.

THE PLANT TISSUES SOUGHT. The most easily demonstrable evidence of the tissues which are sought by the insect after penetration of the tissue is the stylet sheath, although stylet track is the more generally applicable term.

As long ago as 1891, Busgen classified the route of the stylets of the Aphididae into three categories: (1) an intercellular course to the phloem, (2) an intracellular course to the parenchyma, and (3) an intracellular course to the phloem.

Tate's study of 18 species of aphid (1937) showed that the vascular bundles were the objective in all cases. His three categories are intercellular, intra- or intercellular, and direct. The last-named is surely synonymous with intracellular. K. M. Smith (1926) found phloem, xylem, and tracheids tapped by *Myzus persicae* (Sulz.) with the course of the stylets intercellular. Roberts (1940) made a quantitative study of three aphid species with respect to the course of the stylets. *M. persicae* and *Macrosiphum gei* Koch tracks were intercellular in more than 50% of the cases. Roberts also determined the length of time necessary for penetration to

FIGURE 5.4. *Longistigma carvae* Harris feeding on *Tilia americana* L. (Photograph: Zimmermann, by permission from *Science, 133*, 3446, 1961).

the phloem and found very few penetrations within 15 minutes of feeding time by the insect. Some insects were still feeding on nonvascular tissue after 24 hours. An unusually clear delineation of aphid feeding and the feeding tracks is shown in Fig. 5.4 and 5.5.

The spotted alfalfa aphid, *Therioaphis maculata* (Buckton), feeds on mesophyll parenchyma and phloem, its stylets following an intercellular course. However, many tracks end in the parenchyma. This is in contrast to the greenbug, *Toxoptera graminum* (Rond.), which appears to be highly selective in seeking phloem tissue (Diehl and Chatters, 1956). The penetration of sugar beet leaves by the stylets of *Myzus persicae* (Sulz.) is shown in Fig. 5.6 (Esau *et al.*, 1961). It cannot be assumed that because the objective as determined by the end of the stylet track is a specific tissue, the insect's feeding is limited to that tissue. This is particularly true in the case of intracellular penetration, for here the insect's stylet tips are passing through an almost continuous medium of cell sap. The variation

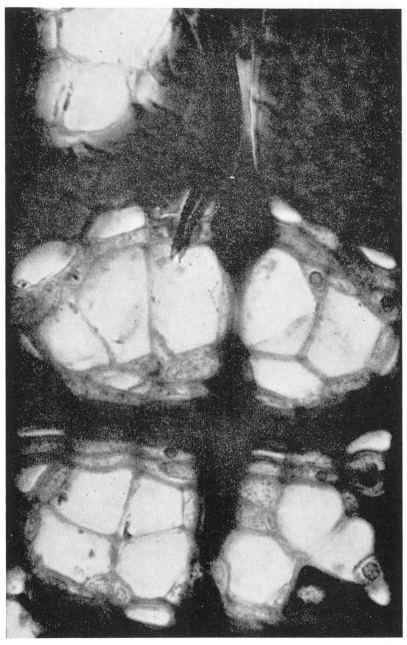

FIGURE 5.5. Stylet tips in an individual sieve element. Two of the three tips are in focus (X950).

FIGURE 5.5 (Continued). The path of the stylets through the bark of linden (*Tilia americana* L.) shown in four subsequent transverse sections mounted to form the correct sequence (Photographs: Zimmermann, by permission from *Science,* 133, 3446, 1961).

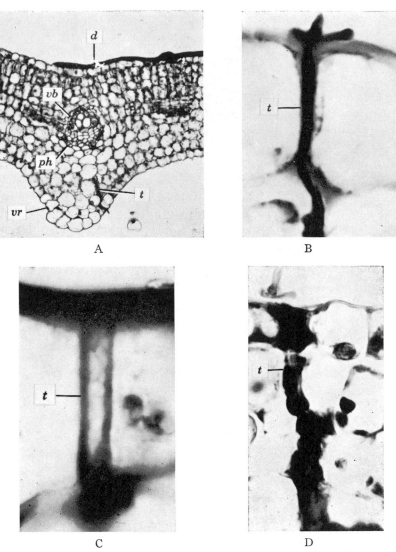

A   B

C   D

FIGURE 5.6. Penetration of sugar beet leaves by stylets of *Myzus persicae*. A: Transection of leaf showing the track *t* of saliva sheath formed during an approximately 4-min penetration with the parenchyma of the vein rib *vr*. The track is directed toward the vascular bundle *vb,* the phloem of which is located at *ph.* Part of the India ink dot *d* appears on the upper surface of the leaf. B: Part of a saliva track *t* in intercellular position in the epidermis, shown in sectional view of the wall. C: Part of saliva track *t* in intercellular position in the epidermis, shown in surface view of the wall. D: Saliva track *t* that passed through an epidermal cell.

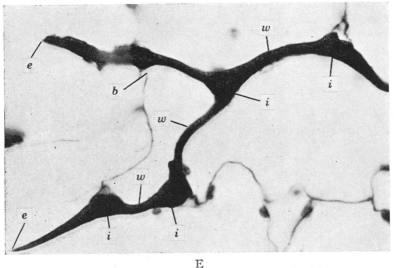

E

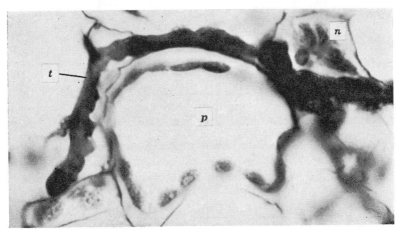

F

FIGURE 5.6 (*Continued*). E: Part of a saliva track within vein-rib parenchyma. It shows swellings in the intercellular spaces *i* and is smooth within the cell wall *w*. At left, above, one arm of the track has broken the wall of a parenchyma cell at *b*. Ends of the two arms of the track are indicated by letters *e*, one within cell wall (below), the other in the lumen of a cell (above). F: Part of a saliva track *t* in mesophyll. It is intracellular at *n*, and mostly in inter-cellular space elsewhere. The pierced cell at *n* is partly necrosed, the parenchyma cell at *p*, in contact with the track, is normal. (Photographs: Esau *et al.*, 1961).

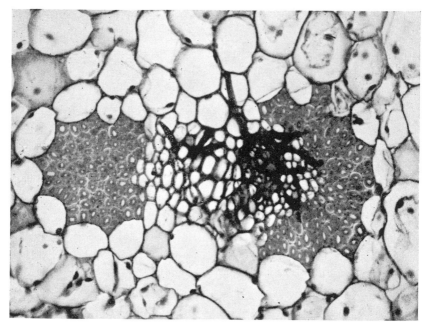

FIGURE 5.7. Section through green spot on a pineapple leaf caused by the feeding of *Dysmicoccus neobrevipes.* Feeding tracks in phloem and xylem. There was collapse of phloem cells and degeneration of chloroplasts. Note also deeply stained chloroplasts around bundle. (Photograph: Esau *et al.,* 1961).

in relative frequency of inter- and intracellular penetration by a single species of aphid, and the feeding in nonvascular tissue by an insect "normally" a phloem feeder, as shown by Roberts (1940), clearly indicates that the insect's dietary requirements and the tissues liable to be affected by its feeding processes are not limited to the "objective" tissue.

Mittler (1957) presents an interesting hypothesis on the actual feeding process of *T. salignus.* The phloem sieve tube pressure is high and forces sap up the insect's food canal at the rate of 1–2 mm³ / hr. The pressure necessary to achieve this was calculated to be 20–40 atm. The aphid uses the phloem sieve tubes as feeding sites because that tissue has the best capacity for maintaining sap supply. This hypothesis relegates the sucking pump to a swallowing function only while the mouth parts are in phloem.

While it is reasonable to assume that the natural turgor pressure of the host plant is an aid to the insect's feeding, it is not a *sine qua non.* Sucking insects have been fed on liquid artificial media (Carter, 1927, 1928, 1945; Hamilton, 1935; Bennett, 1934; Maltais, 1952) at atmospheric pres-

(a)   (b)

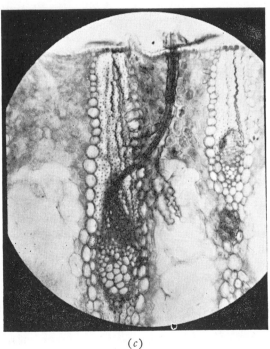

(c)

FIGURE 5.8. (a) *Oliavarus atkinsoni* parasitized by entomogenous fungus and anchored to leaf. (b) Mouth parts of *O. atkinsoni* embedded in leaf and reaching the phloem. (c) Stylets and probe tracks of *O. atkinsoni* in *Phormium* leaf (Photographs: Cumber, New Zealand Department of Scientific and Industrial Research).

sure. Furthermore, not all the tissues sought by the insect have the turgor pressure to offset the necessity for a sucking apparatus. For example, *D. brevipes* stylets reach the fibrovascular bundles but equally consistently move over to the fiber bundles.

The Coccidae vary in the tissues sought. *A. hederae* Vallot apparently only reaches the cortical cells but *P. adonidum* feeds on phloem. *D. neobrevipes* invariably reaches the vascular tissue (Fig. 5.7), but its stylet tracks leading to the fiber bundles were difficult to explain until Krauss (1949) described a starchy sheath surrounding these bundles. It is necessary to postulate some powerful enzyme secreted by the mealybug to account for the use of this starch in the feeding process. All three coccid species have intracellular penetration. Aleyrodidae show little evidence of a stylet track; according to K. M. Smith (1926) contra Hargreaves (1915), many of the stylets enter stomata; their course is intercellular.

The Capsidae, according to Smith, also show no distinct stylet track, although saliva is deposited freely. The insects repeatedly plunge the stylets into the tissue within a circumscribed area. The tarnished plant bug, *Lygus lineolaris* (Palisot de Beauvois) (*L. oblineatus* (Say)), feeds by short rapid thrusts, with frequent withdrawal. During this process, considerable damage is done to tissue. Cell walls are punctured by the stylets which travel both between and through the cells without following any regular pattern (Flemion *et al.*, 1954).

The Cicadellidae vary greatly in the tissues sought. Some are mesophyll feeders, others feed on the phloem. Stylet tracks are often poorly defined (Smith and Poos, 1931), but there is considerable deposit of saliva, often in rough masses. Fig. 5.8 illustrates the feeding of *Oliavarus atkinsoni* V. (Cixiidae) in *Phormium* Myers.

THE SALIVARY SECRETIONS

The nature of the material deposited in the tissue has been the subject of some speculation. Horsfall (1923) considered it a wound reaction on the part of the plant and therefore, of plant origin. Petri (1908, 1911) considered that the tracks laid down by *Phylloxera* were composed of callose and insoluble calcium pectate, and that tannin was deposited (presumably as a wound reaction) after the wall of the sheath was formed.

F. F. Smith (1933) investigated the nature of the material in the feeding tracks produced by two species of leafhoppers, *Empoasca fabae* (Harris) and *Stictocephala festiva* (Say). Sheath material could be differentiated from plant tissue by four color reactions. Adamkiewicz, Sakaguchi, quinone solution, and Benedict's modification of the Hopkins-Cole

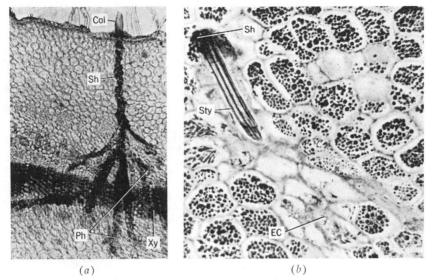

(a)   (b)

FIGURE 5.9. Feeding tracks and types of secretion of *Oncopeltus fasciatus* (Dall.). (a) Hand-cut section of milkweed stem showing the stylet, Sh, and the external collar of sheath material, Col, produced by *Oncopeltus*. The tracks end in the phloem bundles, Ph, on either side of the xylem, Xy. (b) Microtome section of a peanut showing the regions of empty cells, EC, which result from the feeding of *Oncopeltus* and a fragment of the stylets, Sty, *in situ*. The stylets are surrounded at the proximal end by a thick safranin-stained mass of sheath material, Sh, which diminishes toward the center of the peanut and is absent from the region of empty cells. Starch grains fill the other cells. (Photographs: Waite Institute, Miles, 1960).

reaction. All tests for plant reserve, wound response, and plant cell-wall substances were negative except the color test for pectin. Other color tests indicated that the sheath material was proteinaceous. From these data, Smith concluded that the sheath is largely of insect origin and contains no plant substance.

Miles (1959) found evidence that the sheath material of *Oncopeltus fasciatus* (Dall.) is a salivary secretion (Fig. 5.9 and 5.10), and later (Miles, 1960), that the sheath material contains a lipoprotein, is rich in tyrosine, and contains traces of tryptophan. Chitin and carbohydrate appear to be absent. Properties of unfixed sheath material are shown in Table 5.1. Miles (1960a) then went to the salivary glands (Table 5.2) and showed that the sheath material originates as two distinct secretions produced re-

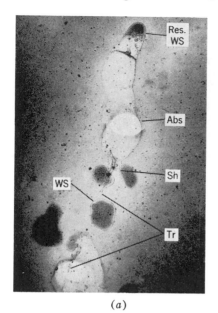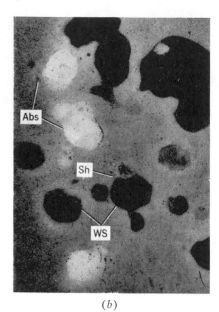

(a)   (b)

FIGURE 5.10. Feeding tracks and types of secretion of *Oncopeltus fasciatus* (Dall.). Films of sucrose-agar gel on which *Oncopeltus* has fed, fixed in formol and stained (a) with safranin and light green and (b) with toluidine blue. Abs, areas where the gel has been absorbed; Res. WS, partially resorbed watery saliva; Sh, mounds of sheath material; Tr, partially formed "track" of sheath material; WS, deposits of watery saliva. (Photographs: Waite Institute, Miles, 1960a).

spectively in the anterior and posterior lobes. Table 5.2 shows the histochemical and stain reactions of the glands.

Evidence that the deposition of salivary material is achieved independently of plant tissue or plant reaction is found in Bennett's report on finding salivary deposits by *Circulifer* (*Eutettix*) *tenellus* (Baker) in a liquid medium (1934); Carter's finding that *Dysmicoccus brevipes* deposited feeding tracks in plain agar (1945); Nickel and Sylvester's work with *M. persicae* (1959) (Fig. 5.11); and the work of Van Hoof (1958) (Fig. 5.12). The mealybug tracks were straight and presented the appearance of a string of beads with a cleanly drilled internal channel. In the process of penetration, which was observed in detail, the mealybug ejected a continuous series of clear liquid globules which apparently solidified quite rapidly. It cannot be assumed, although it often is, that the secretion which solidifies to form the stylet track is the only component of the secretion. As Carter (1945) showed by feeding mealybugs on radioactive solutions

TABLE 5.1. *Properties of Unfixed Sheath Material* [a]

Reagent or test	Reaction [b]	Inference
Amino acid and protein tests		
Millon's	+	Tyrosine present
Morner's	+	Tyrosine present
Tetrazonium after performic acid	+	Tyrosine present
Hopkins-Cole-Adamkiewicz	±	Tryptophan (trace) present
Proteinase	Dissolves	Protein present
Benzoquinone	Tans	Protein present
Carbohydrate tests		
Molisch	0	Carbohydrate absent
PAS	+	Carbohydrate or lipid present
PAS after acetylation	+	Carbohydrate absent?
Lipid tests		

	Pyridine extraction		
	Before	After	
Sudan stains	±	±	
PFAS	0	0	
Chlorated HNO_3	Oily droplets	0	Lipid present
Liebermann-Burchardt	Pink, turns green	0	Steroid (?) present
Other tests			
Chitosan	Dissolves in KOH		Chitin absent
Ruthenium red (Johansen technique)	0		Pectic substances absent

[a] After Miles (1959; 1960).
[b] +, positive; ±, weak or variable; 0, negative.

and then transferring them to agar, the radioactive component diffused far beyond the area in which the feeding tracks were deposited. There are many cases where the effects of the salivary secretions are evident in plant tissues farther removed from the stylet track itself. Lawson *et al.* (1954) demonstrated this with P^{32} injected by the green peach aphid into tobacco. Kloft (1960) provided confirmation using other species of Hemiptera.

Miles (1959, 1959a) has provided the experimental evidence on the secretions themselves, using *O. fasciatus, Dysdercus fasciatus* Sign., and *Aphis craccivora* Koch. All three species of insect were found to secrete two types of saliva: sheath material which coagulates rapidly and forms a lining in the path of the stylets; and a watery and water-soluble saliva.

TABLE 5.2. *Reactions of the Salivary Gland Contents and Salivary Secretions of Oncopeltus* [a]

Test or treatment	Anterior lobe	Lateral lobe	Posterior lobe	Accessory gland	Sheath material	Watery saliva
Isotonic saline*	Coagulates	Swells, coagulates	Soluble		Insoluble	Soluble
Conc. HCl	Dissolves	Resistant	Dissolves		Resistant	Dissolves
Millon's reagent	+	++	+		++	+
Tetrazonium	+	++	±		+	±
Benzoquinone	++	+	++		++	0
Schiff's reagent	+	+ Fades	0		+ Fades	0
PAS	++	0	+		+	+
PAS after acetylation	±	0	0		+	+ Bluish
Toluidine blue	Pale blue-green	0	Dark blue	Violet duct pink	Pale blue-green	Dark blue to pink
Fat stains [b]	0	+	0		±	0
Argentaffin	0	±	++		+	0
G-Nadi*	± [e]	± [e]	0	++ [d]	± [e]	
Aniline blue in lactophenol	On standing +	On standing ++	0		+	0
Isoelectric point	5.4	5.4	5.4		5.4	

Key: ++, strongly positive; +, positive; ±, weakly positive or variable; 0, negative.
[a] After Miles (1960). The reactions of the cell contents of the accessory gland have been noted only where they differed from those of the walls of the principal gland. Tests marked with an asterisk were made on fresh material; the others on formol-fixed material.
[b] Sudan black B, Sudan III, Sudan IV.
[c] Not inhibited by cyanide.
[d] Inhibited by cyanide but not by azide.

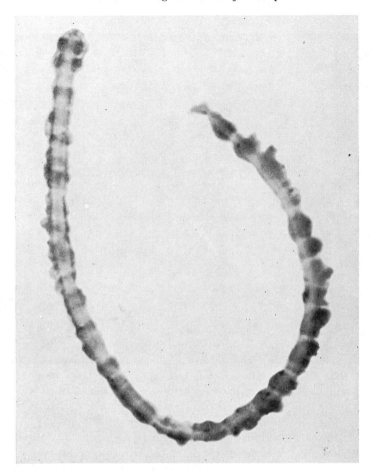

FIGURE 5.11. Salivary sheath deposited by *Myzus persicae* when the insect fed on a watery medium. (X1520). Photograph: Nickel and Sylvester, 1959).

This watery saliva is secreted and sucked back into the food canal even when no sheath material is being discharged (Fig. 5.13). The watery saliva of *Oncopeltus* originates in the posterior lobe of the salivary glands but the accessory gland is believed to be the source of the mucoids which are often found in the watery saliva.

It will be clear during the consideration of individual cases, which follows, that the effect of the salivary secretions on plant cells cannot be harmonized with the concept of a single solidifying component of uncertain function. This does not necessarily mean that the material in the stylet track is not affected by the plant sap in the cells through which it passes

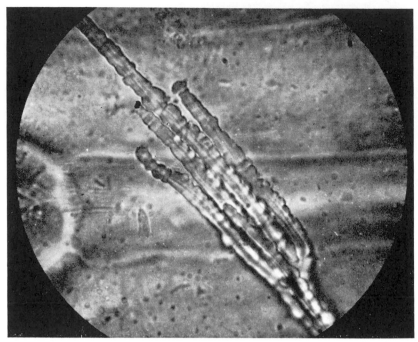

FIGURE 5.12. (a) Salivary sheath of *Aphis fabae* Scop. formed in the liquid above the stripped epidermis of hyacinth through which the insect fed.

or as it dissolves the middle lamella. The stylet tracks of *Tuberolachus salignus* (Gmelin) are beaded in spaces but are smooth in cells (Mittler, 1957). The act of solidification involves reaction, and it could well be that separation of soluble from insoluble components occurs at that time.

The work of Miles (1959, 1959a, 1960, 1960a) in separating the major components of the secretions has resulted in considerable progress in characterizing their nature. The compounds forming the sheath, the precursors, are elaborated in the anterior lobe of the salivary glands.

Miles (1965) fed Homoptera and Pentamorpha species on solutions, using paraffin membranes. The sheaths formed in these solutions were treated with indicators and by this means the presence of disulphide and hydrogen bonds, and sulphydril groups, was determined; they are secreted with a polyphenol oxidase. The sheath has a pH of 6. The watery secretion is secreted by the posterior lobe of the salivary gland. This component contains many soluble substances, including enzymes and metabolites, and has a pH of 8 or slightly higher.

FIGURE 5.12. (*b*) Salivary sheath of *Aphis fabae* Scop. photographed with electron microscope at X5000 (Photographs: Institute of Phytologic Research, Wageningen, Netherlands, Van Hoof, 1958).

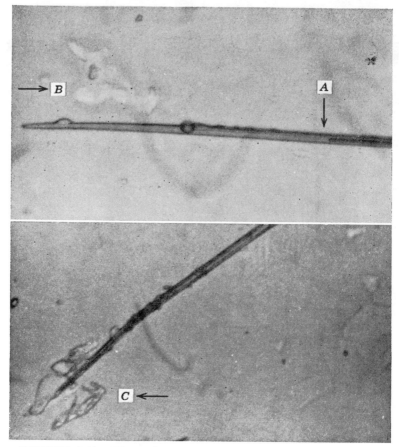

FIGURE 5.13. Photomicrographs of the stylet bundle of *Aphis craccivora*. A. The meniscus of a watery secretion which has been discharged from the salivary canal and sucked back into the previously empty food canal. B. Sheath material secreted on to an underlying glass slide. C. Sheath material being extruded from the tip of the stylet bundle (Photographs: Waite Institute, Miles, 1959).

ENZYMES IN THE SECRETIONS.   Nuorteva (1954, 1956, 1965a) found enzymes in mirids and leafhoppers which he considered were definitely implicated as factors in plant growth. In *Miris dolobratus* (L) proteases occur, but only in the older nymphs and adult females. In *Capsus ater* (L) proteases could be found when the insects were taken from oats, but not when they were taken from grasses. In a comparison of the salivary glands of seven spp. of leafhoppers, Nuorteva found proteases in four of the seven spp. with varying degrees of activity, but in the remaining three, none at all.

There are specific differences between leahoppers with respect to the presence or absence of proteases or amylases depending on whether they are phloem or mesophyll feeders (Saxena, 1954). In the phloem feeders, *Tettigella spectra* (Sing.) and *Parabolocratus porrestus* Walk., the two enzymes are absent; but both are present in the mesophyll feeders, *Empoasca kerri* Pruthi and *E. devastana* Distl. Herford (1935) showed that *Empoasca solana* DeL. was able to inject diastase into a feeding solution and secrete an enzyme capable of inverting pure sucrose. Diastase was also found in the salivary glands of three *Psylla* spp.: *P. pyri* (L); *P. pyricola* Frost; and *P. pyrisuga* (Frost) (Pussard 1939). Fier and Beck (1961) determined that the digestive enzymes—anylase, invertase, protease, and lipase—contained in the salivary glands of *O. fasciatus,* originate in the salivary glands of the insect and are neither regurgitated nor transported from the lumen of the mid-gut. Esterases were found in the saliva of the capsid bugs attacking cocoa (Goodchild, 1952) and a cellulose-hydrolyzing component in aphid saliva was reported by Adams and Drew (1963).

Sogawa (1968) found phenolases in the salivary glands of the rice leafhopper, and Miles (1969) considers that salivary phenolases are secreted by most sucking insects. He generalized on the digestive enzymes found in the saliva of the Hemiptera: those following an intercellular path mostly secrete a pectin-hydrolyzing enzyme; the phloem and xylem feeders may have, in addition, proteinase and lipase, among others.

Nuorteva and Laurema (1961, 1961a) reported on the occurrence of pectin polygalacturonase in salivary glands of Heteroptera and Homoptera Auchenorrhyncha; its occurrence in *Lygus hesperus* Knight was shown by Strong and Kruitwagen (1968). Pectinase (pectin polygalacturonase) has been shown to be present in extracts of whole aphids (McAllen and Cameron, 1956); in the saliva of *Myzus persicae* (Sulz) (Adams and McAllan, 1956); and in a large number of other aphid species as well as four other species representing other families of sucking insects (Adams and McAllan, 1958). The last named authors showed differences between alate and apterous forms of the same species and differences between the same species when fed on different hosts. The suggestion was made that the function of the enzyme might be to dissolve the middle lamella of the plant cell wall.

FREE AMINO ACIDS.   These were found, from 4 to 14, in salivary glands (Anders, 1958; Kloft, 1960; Schaller, 1963; Miles, 1964). Miles (1967a) determined the presence of 18 amino acids in the anterior and posterior lobes of *Eumecopus australaspiae* Don. and believes that amino acids in variable concentration occur in most if not all plant bugs. Anders and Kloft both believe that these compounds are responsible for the effect of the insects' feeding on the physiology of the plant. Ingested free amino acids occurring in the hemolymph, as well as such compounds as glucose and glycerol, are transferrable unchanged to the watery salivary component. Nuorteva has shown (1962) that amino acids fed artificially to *Calligypona pellucida* (F) reduced the growth of oats to which the insects were transferred. Protein molecules of about 40,000 mol wt can be transferred from the laemolymph to saliva via the accessory gland (Miles and Slovak, 1970).

GROWTH REGULATING COMPOUNDS IN THE SECRETIONS. The hormones have long been suspected of disarrangement by insect feeding. Allen (1947) reviewed the relationship between insect damage and horomones. He suggested that insects may affect the activity of, and response to, growth substances, either by infection of materials or the withdrawal of them. The latter carries the corollary that the term "mere withdrawal" (of cell sap) may be an empty phrase. The relationship, according to Allen, is indicated by (1) the similarity in growth changes in plants caused by hormones and those caused by insect feeding, (2) the similarity in enzyme activity, and (3) the suppression of insect damage in certain plants after treatment with hormones. Abscission caused by the feeding of the tarnished plant bug is prevented by the application of ANA (a-naphthalene acetic acid), so that possibly the insect, in feeding, produces an auxin inactivator or interferes in some other way with the normal food supply. Fisher *et al.* (1946) have shown the presence of an auxin inhibitor in the salivary glands of *Lygus oblineatus* Say.

Plant-growth-inhibiting substances were found by Nuorteva (1956) in a species of mirid, *Miris dolopratus* (L); a pentatomid, *Dolycorus baccarum* (L); and an aphid, *Cinara piceae* (Panz.). This same author (1956a) found no growth-stimulating substances in salivary glands, but was able to show that β-indolylacetic acid could be transferred in an active form from a synthetic medium to the salivary glands of *Stenodema calcaratum* Fall. He did, however, suggest the possibility that plant-growth-promoting substances may be present in, and therefore, transferable from, the salivary glands if auxins are present in them, especially in great amounts, as was suggested by Nuorteva. His suggestion that at least some of the toxic substances in the salivary glands of toxicogenic insects originate from the host plant, rather than from the insect, has the important corollary that the physiological status of the plant could condition the content of the oral secretions of the insect feeding on it. Even though certain compounds may be picked up by the insect and passed with unchanged activity through the mid-gut epithelium to reach the hemolymph and finally the salivary glands, the actual oral secretions of the insect injected into the plant must be complex, and the action of the various fractions will be either complementary or antagonistic in their action on the plant tissues.

An example can be taken from Nuorteva's paper. Indoletriacetic acid was not transferred from nutrient to the salivary glands of *Dolycoris baccarum* L in active form, possibly because of auxin inhibitors in the salivary glands.

That components of the watery saliva may be derived from synthesis by the salivary glands was shown by Miles and Lloyd (1967) to occur in aphids by the conversion of tryptophan to β-indolylacetic acid (IAA), or by transfer of the ingested IAA by a mirid and a leafhopper to salivary glands. These same authors infected phenylamine as a precursor into pentatomids, and the tryptophan was converted into IAA. When both tryptophan and the precursor were inserted into the abdomens and the insect then fed on seedlings, swelling and splitting of nodes, epinasty and crooked stems—all symptoms of localized application of IAA—were produced.

Schaller (1965) attempted to relate the content of IAA in the saliva of aphids to phytopathogenicity. IAA was found in saliva in variable concentrations, but these were not relatable to feeding effects. Schaller comments that for a specific instance, quantity may not necessarily be of overriding importance. However, this is contrary to the case in many phytotoxemias which are clearly influenced by quantitative considerations. Qualitative considerations could of course predominate if a component of the salivary secretions has a catalytic function. The mechanisms of toxicity of the oral secretions of *L. hesperus* have been the subject of a detailed study by Strong (1970). He found no evidence of IAA synthesis in salivary glands during salivation. He also showed that bugs fed on various plant hormones showed no effect on abscission of buds when the insects were transferred to cotton plants after five days of feeding on the diets. Gibberellic acid could be found in the insect's body but not in the excised salivary glands. No evidence was found for specific abscissors or for ethylene production, which, according to Pratt and Goeschl (1969), is produced by abscission accelerators. The only plant-physiologically active compound in the saliva of *L. hesperus* is polygalacturonase, according to Strong. On this basis, he set up a hypothesis to account for the various plant reactions to the digestive action of the enzyme which causes hormonal imbalance in the plant. Strong does not attempt to explain systemic effects or evidence of the toxic element, and in the light of the results reported by other workers, if the saliva of *L. hesperus* contains only the physiologically-active compound, it is probably unique.

Miles' recent review (1968) is inclusive and timely. Included is a discussion of plant responses to injected saliva. Miles comments on the lack of direct evidence for the influence of virus infection on the elaboration of nonviral toxins, "direct" in this case meaniing the isolation and identification of the toxins. By this definition, the following would be considered examples of indirect evidence. There is one instance recorded of the production of a highly toxic nonviral substance by *Calligypona pellucida* (F), fed on oat plants infected with oat sterile dwarf virus (OSDV). When injected into oat stems, it produced symptoms closely resembling those of OSDV and different from those caused by *C. pellucida* from virus-free plants (Virtanen *et al.,* 1961).

The converse was noted when Staples (1968) reported that *Paratrioza cockerelli* (Sulc.) lost its phytotoxicity when fed on potato plants infected with haywire virus, a virus for which the insect is not a vector. A perhaps-related example, also, is that the presence of a latent virus is necessary in pineapple plants to permit the development of mealybug wilt toxin in the mealybugs (Carter, 1963).

Miles said (1969), ". . . when any sucking bug feeds on a growing plant, all the cells immediately surrounding those actually pierced will be to some extent subject to the influence of all the components that can be found in any of the salivary secretions," and it can be added that some of these components are translocated. Although perhaps for no insect species are the data complete, the total result is often profound disturbance of the plant's physiology, resulting in clearly defined symptoms of disease. The disease syndrome is known as phytotoxemia. An insect whose salivary secretions induce symptoms of disease is termed toxicogenic and the term applies to the species. If the insect does not induce disease under certain conditions, the active state is described as toxiniferous (Carter, 1936).

CRITERIA FOR THE DETERMINATION
OF A PHYTOTOXEMIA

The establishment of the criteria for the determination of a phytotoxemia has always presented difficulties which even now are not entirely resolved. Although one of the earliest cases known—hopperburn of potato —is now generally considered to be the result of a phytotoxic secretion, a good case was made earlier for the influence of mechanical interference with translocation as a result of the insect's feeding.

Where only local lesions are involved, determination is not complicated. Usually, a single insect will produce the lesions. The effect on the plant tissue is localized at the insect's feeding point; the lesion is specific to that species of insect; the capacity to produce it is inherent in the insect; and no new lesions arise after the removal of the insect from the plant.

Where there is secondary development of symptoms from the local lesions, the sequence of symptoms will relate the local lesion to it. Here it is necessary to recognize the factors of secondary infection with pathogenic fungi and bacteria, which may occur and which may materially change the syndrome. With the tissue malformations, the criteria are modified somewhat. The number of insects involved becomes important as a satisfactory criterion to use.

The translocated and systemic toxic effects are the most difficult to establish; the more complex they are, the more closely akin to virus diseases they become. Leach (1940) modified an earlier attempt by Carter (1939) to tabulate differences between the systemic phytotoxemias and virus diseases, and then proceded to indicate, quite correctly, how narrow some of these differences were. During the intervening years there has been no widening; on the contrary, the differences have narrowed even further.

The most widely used criteria for establishing the toxicogenicity of an insect are (1) that the capacity to induce symptoms is inherent and (2) that it is characteristic of the insect species, whatever the developmental stage or the host plant to host plant-feeding sequence, and symptoms follow feeding by the insect on the first host plant on which feeding occurs after the insect hatches. There are no examples known where the incidence and severity of symptoms are not directly related to the number of insects feeding, although some species are so toxic that a single insect may induce symptoms on a small plant. On the other hand, in some cases, increasing the number of insects beyond a certain limit produces no further additive effect. Recovery, once the insects are removed, is usual, but a disease can progress to the point of no return if the attack is sustained long enough. On the other hand, after recovery, a new attack may result in a repetition of the symptoms.

If these criteria can be met, the conclusion that a phytotoxemia is involved is reasonably certain, but rarely is a case so simple. A species of insect, inherently toxicogenic, may not always be toxiniferous. A nymphal stage may be more or less toxicogenic than other nymphal stages, and the adult may or may not be capable of inducing the symptoms. The feeding sequence of the insect may affect its capacity to induce disease. Nuorteva (1959, 1959a) has confirmatory evidence for this in his studies with *Calligypona pellucida* (F.). This leafhopper damages oats, and leafhoppers collected from the area of oat damage have more toxic substances in their salivary glands than do specimens collected outside this area. The explanation offered is not, as has been suggested, that there are two physiologic strains of the leafhopper, one being restricted to the area of oat damage. It is rather that this area coincides with the area devoted to the production of timothy seed. Years of bad oat damage are years of intensive timothy seed cultivation; the leafhopper occurs in the timothy seed fields and migrates from there to oats.

The toxic effect of an insect's feeding can be made manifest months after the insect has been removed and growth of the plant has continued. The mechanism for this phenomenon is completely obscure. That a complex chemical entity can remain stable *in vivo,* suffer the dilution that plant growth would involve, and later cause symptoms or affect the plant's susceptibility to infection, or to other closely related toxins, is purely speculative. There is also the equally speculative possibility that the action of the toxin is a sensitizing process, analogous, perhaps, to allergy.

There is also some evidence that a latent virus, or something akin to it, can change the physiology of a plant in such a manner that it becomes a suitable substratum on which the insect can develop a toxic secretion. Finally, there is the possibility that toxic feeding by insects may provide the necessary stress to force the expression of unfavorable genetic factors, resulting in symptoms associated with the genetic factor and quite atypical of those normally associated with the insect's feeding.

THE PHYTOTOXEMIAS AND THEIR SYMPTOMS

If the symptoms as described in the literature are assembled, it is possible to group them in certain broad categories, necessarily with ill-defined limits, which serve, nevertheless, to "classify" them on a basis of complexity. There can be four of these categories, which we include with their symptoms. These categories will be used as a basis for discussion in this and succeeding chapters.

Categories of Phytotoxemias Based on External Symptoms

1. Local lesions at the insect's feeding point: Leaf-spotting, leaf-stippling, with or without local diffusion

2. Local lesions with development of secondary symptoms: Cankers, cork formation, fruit scabs, pitted seed and aborted pods, split lesions, leaf-drop, premature nut fall
3. Tissue malformations: Leaf-curling, savoying, rosetting, witch's brooms, formation of adventitious buds, shortening of internodes and petioles, hopperburns, phyllody, cat-facing, and gummosis
4. Symptoms indicating translocation of the causal entity:
   a. Limited—Chloroses, leaf-banding, leaf-striping, vein-clearing, wilt
   b. Systemic—Growth depressions, killing of roots, wilting, abortive tubers, susceptibility to freeze injury, phloem necrosis, leaf-streaking

REFERENCES

Adams, J. B., and Drew, M. E., "A cellulose-hydrolyzing factor in aphid saliva" *Can. J. Zool.,* **43,** 489–496 (1963).

Adams, J. B., and McAllan, J. W., "Pectinase in the saliva of *Myzus persicae* (Sulz.) (Homoptera: Aphididae)," *Can. J. Zool.,* **34,** 541–543 (1956).

Adams, J. B., and McAllan, J. W., "Pectinase in certain insects," *Can. J. Zool.,* **36,** 305–308 (1958).

Allen, T. C., "Suppression of insect damage by means of plant hormones," *J. Econ. Entomol.,* **40,** 814–817 (1947).

Anders, F., "Aminosauren als galleneregende Stofee der Reblaus (Viteus) (Phylloxera) vitifolii Shimer," *Experimentia,* **14,** 62–63 (1958).

Bennett, C. W., "Plant-tissue relations of the sugar-beet curly-top virus," *J. Agr. Research,* **48,** 665–701 (1934).

Busgen, M., "Der Honigtau," *Jena. Z. Naturw.,* **25,** 339–428 (1891).

Carter, W., "A technique for use with homopterous vectors of plant diseases, with special reference to the sugar-beet leaf hopper, *Eutettix tenellus* (Baker)," *J. Agr. Research,* **34,** 449–451 (1927).

Carter, W., "An improvement in the technique for feeding homopterous insects" *Phytopathology,* **18,** 246–247 (1928).

Carter, W., "The oral secretions of the pineapple mealybug," *J. Econ. Entomol.,* **38,** 335–338 (1945).

Carter, W., "The toxicogenic and toxiniferous insect," *Science,* **83,** 522 (1936).

Davidson, J., "Biological studies of *Aphis rumicis* Linn. The penetration of plant tissues and the source of the food supply of aphids," *Ann. Appl. Biol.,* **10,** 35–54 (1923).

Diehl, S. G., and Chatters, R. M., "Studies on the mechanics of feeding of the spotted alfalfa aphid on alfalfa," *J. Econ. Entomol.,* **49,** 589–591 (1956).

Esau, K., Namba, R., and Rasa, E. A., "Studies on penetration of sugar beet leaves by stylets of *Myzus persicae*," *Hilgardia*, **30**, 517–527 (1961).

Feir, D., and Beck, S. D., "Salivary secretions of *Oncopeltus fasciatus* (Hemiptera: Lygaeidae)," *Ann. Entomol. Soc. Amer.*, **54**, 316 (1961).

Fisher, E. H., Riker, A. J., and Allen, T. C., "Bud blossom and pod drop of canning string beans reduced by plant hormones," *Phytopathology*, **36**, 504–523 (1946).

Flemion, F., Ledbetter, M. C., and Kelley, E. S., "Penetration and damage of plant tissues during feeding by the tarnished plant bug *(Lygus lineolaris)*," *Contribs. boyce Thompson Inst.*, **17**, 347–357 (1954).

Glass, E. H., "Feeding habits of two mealybugs *Pseudococcus comstocki* (Kuw.) and *Phenacoccus colemani* (Ehrh.)," *Virginia Agr. Expt. Sta. Tech. Bull.*, No. 95 (1944).

Goodchild, A. J. P., "Digestive system of the West African cacao capsid bugs (Hemiptera Miridae)," *Proc. Zool. Soc. London*, **122**, 543–572 (1952).

Hamilton, M. A., "Further experiments on the artificial feeding of *Myzus persicae* (Sulz.)," *Ann. Appl. Biol.*, **22**, 243–258 (1935).

Hargreaves, E., "The life-history and habits of the greenhouse white fly (*Aleyrodes vaporariorum* Westd.)," *Ann. Appl. Biol.*, I, 303–334 (1915).

Herford, G. V. B., "Studies on the secretion of diastase and invertase by *Empoasca solana* Delong (Rhynchota, Homoptera, Jassidae)," *Ann. Appl. Biol.*, **22**, 301–306 (1935).

Horsfall, J. L., "The effects of feeding punctures of aphids on certain plant tissues," *Penn. State Univ. Agr. Expt. Sta. Bull.*, No. 182 (1923).

King, W. V., and Cook, W. S., "Feeding punctures of mirids and other plant-sucking insects and their effect on cotton," *U.S. Dept. Agr. Tech. Bull.*, No. 296 (1932).

Kloft, W., "Wechselwirkungen zwischen planzensaugenden Insekten und den von ihnen besogenen Planzengeweben," *Z. Angew. Entomol.*, **45**, 337–381; **46**, 42–70 (1960).

Krauss, B., "Anatomy of the vegetative organs of the pineapple II. The leaf," *Botan. Gaz.*, **110**, 333–404 (1949).

Lawson, F. R., Lucas, G. B., and Hall, N. S., "Translocation of radioactive phosphorus injected by the green peach aphid into tobacco plants," *J. Econ. Entomol.*, **47**, 749–752 (1954).

Leach, J. G., *Insect Transmission of Plant Diseases*, McGraw-Hill, New York, 1940.

Maltais, J. B., "A simple apparatus for feeding aphids asceptically on chemically defined diets," *Can. Entomologist*, **84**, 291–294 (1952).

McAllan, J. W., and Cameron, M. L., "Determination of pectin polygalacturonase in four species of aphids," *Can. J. Zool.*, **34**, 559–564 (1956).

Miles, P. W., "The salivary secretions of a plant-sucking bug, *Oncopeltus fasciatus* (Dall.) (Heteroptera; Lygaeidae) I: The types of secretion and their roles during feeding," *J. Insect Physiol.*, **3**, 243–255 (1959).

Miles, P. W., "Secretion of two types of saliva by an aphid," *Nature,* **183,** 756 (1959a).

Miles, P. W., "The salivary secretions of a plant-sucking bug, *Oncopeltus fasciatus* (Dall.) (Heteroptera: Lygaeidae) II. Physical and chemical properties," *J. Insect Physiol.,* 4, 209–219 (1960).

Miles, P. W., "III. Origins in the salivary glands," *J. Insect Physiol.,* 4, 271–282 (1960a).

Miles, P. W., "Studies on the salivary secretions of plant-bugs; the salivary secretions of aphids," *J. Insect Physiol.,* 11, 1261 (1965).

Miles, P. W., "Studies on the salivary physiology of plant bugs," *J. Insect Physiol.,* 13, 1787–1802 (1967a).

Miles, P. W., "Studies on the salivary physiology of plant-bugs; the chemistry of formation of the sheath material," *J. Insect Physiol.,* 10, 147–160 (1964).

Miles, P. W., "The physiological division of labor in the salivary glands of *Oncopeltus fasciatus* (Dall.) (Heteroptera; Lygaeidae)," *Aust. J. Biol. Sci.,* 20, 785 (1967).

Miles, P. W., "Insect secretions in plants," *Ann. Rev. Phytopathology,* 6, 137–164 (1968).

Miles, P. W., "Interaction of plant phenols and salivary phenolases in the relationship between plants and Hemiptera," *Ent. Exp. Applic.,* 12, 736–744 (1969).

Miles, P. W., and Lloyd, J., "Synthesis of a plant hormone by the salivary apparatus of plant sucking bugs," *Nature,* 213, 801–802 (1967).

Miles, P. W., and Sloviak, D., "Transport of whole protein molecules from blood to saliva of a plant bug," *Experimentia,* 26, 611 (1970).

Mittler, T. E., "Studies on the feeding and nutrition of *Tuberolachnus salignus* (Gmelin) (Homoptera, Aphididae). I. The Uptake of Phloem Sap," *J. Exptl. Biol.,* 34, 334–341 (1957).

Nickel, J. L., and Sylvester, E. S., "Influence of feeding time, stylet penetration, and developmental instar on the toxic effect of the spotted alfalfa aphid," *J. Econ. Entomol.,* 52, 249–254 (1959).

Nuorteva, P., "Die bedeutung mechanischer Schadigung des Weizenkorns durch Wanzen fur das Korn und fur die Backfahigkeit des Mehles," *Ann. Entomol. Fennici,* 19, 29–33 (1953).

Nuorteva, P., "Studies on the salivary enzymes of some bugs injuring wheat kernels," *Ann. Entomol. Fennici,* 20, 102–124 (1954).

Nuorteva, P., "Notes on the anatomy of the salivary glands and on the occurrence of proteases in these organs in some leafhoppers (*Hom. Auchenorrhyncha*)," *Ann. Entomol. Fennici,* 22, 103–108 (1956).

Nuorteva, P., "Studies on the effect of the salivary secretions of some Heteroptera and Homoptera on plant growth," *Ann. Entomol. Fennici,* 22, 108–117 (1956a).

Nuorteva, P., "Om bollnassjukans natur," *Nord. Jordbrugsforskn.,* 41, 25–31 (1959).

Nuorteva, P., "Bollnassjukan i Finland," *Notulae Entomol.*, **39**, 119–126 (1959a).

Nuorteva, P., "Studies on the causes of the phytopathogenicity of *Calligypona pellucida* (F) (Hom. Araeopidae)," *Ann. Zool. Soc. Zool. Bot. Fennicae "Venana" Tom.*, **23**, 3–58 (1962).

Nuorteva, P., and Laurema, S., "On the occurrence of pectin polygalacturonase in the salivary glands of Heteroptera and Homoptera Auchenorrhyncha," *Ann. Entomol. Fennici*, **27**, 89–93 (1961).

Nuorteva, P., and Laurema, S., "Observations on the activity of salivary proteases and amylases in *Dolycoris baccarum* (L)," *Ann. Entomol. Fennici*, **27**, 93–97 (1961a).

Petri, L., *Centr. Bakteriol., II*, **21**, 375–379 (1908).

Petri, L., *Rend. Accad. Naz. Lincei*, **20**, 57–65 (1911).

Pratt, H. K., and Goeschl, G. G., "Physiological roles of ethylene in plants," *Ann. Rev. Plant Physiology*, **20**, 541 (1969).

Pussard, R., "Contribution a l'etude de la nutrition des psyllides. Presence d'une anylase dans les glandes salivaires de quelques psyllides adults," *Compt. Rend. Soc. Savantes*, **71**, 291–292 (1939).

Roberts, F. M., "Studies on the feeding methods and penetration rates of *Myzus persicae* Sulz., *M. circumflexus* Buckt. and *Macrosiphum gei* Koch," *Ann. Appl. Biol.*, **27**, 348–358 (1940).

Saxena, K. N., "Feeding habits and physiology of digestion of certain leafhoppers, Homoptera, Jassidae," *Experientia*, **10**, 383–384 (1954).

Schaller, G., "Papierchromatographische analyse der Aminosauren und Amide des Speichels und Honigtaus von 10 Aphidenarten mit unterscheidlicher Phytopathogenitat," *Zool. Jahrb. Abt. Allgem, Zool. Physiol. Tiere*, **70**, 399–406 (1963).

Schaller, G., "Investigation of the beta indolylacetic acid content of the saliva with various phytopathogenicity," *Zool. Jahrb. Abt. Allgem. Zool. Physiol. Tiere*, **71**, 385–392 (1965).

Smith, F. F., "The nature of the sheath material in the feeding punctures produced by the potato leaf hopper and three-cornered alfalfa hopper," *J. Agr. Research*, **47**, 475–485 (1933).

Smith, F. F., and Poos, F. W., "The feeding habits of some leafhoppers of the genus *Empoasca*," *J. Agr. Research*, **43**, 276–286 (1931).

Smith, K. M., "Investigation of the nature and cause of the damage to plant tissue resulting from the feeding of capsid bugs," *Ann. Appl. Biol.*, **7**, 40–55 (1920).

Smith, K. M., "A comparative study of the feeding methods of certain Hemiptera and of the resulting effects upon the plant tissue, with special reference to the potato plant," *Ann. Appl. Biol.*, **13**, 109–139 (1926).

Snodgrass, R. E., *Principles of Insect Morphology*, McGraw-Hill, New York, 1935.

Sogawa, K., "Studies on salivary glands of rice leafhoppers III. Salivary phenolases," *Appl. Ent. & Zool.*, **3**, 13 (1968).

Staples, R., "Gross protection between a plant virus and potato psyllid yellows," *J. Econ. Entomol.*, **61**, 1378–1380 (1968).

Strong, F. E., "Physiology of injury caused by *Lygus hesperus*," *J. Econ. Entomol.*, **63**, 808 (1970).

Strong, F. E., and Kruitwagon, E. C., "Polygalacturonase, in the salivary apparatus of *Lygus hesperus* (Hemiptera)," *J. Insect Physiol.*, **14**, 1113–1119 (1968).

Tate, H. D., "Method of penetration formation of stylet sheaths and source of food supply of aphids," *Iowa State Coll. J. Sci.*, **11**, 185–206 (1937).

Van Hoof, H. A., "Oonderzoekingen over de Biologische Overdracht Van een Nonpersistent Virus," *Dissertation der Landbouwhogeschool te Wageningen*, Van Putten & Oortmeijer, Aikmaar, 1958.

Virtanen, A. I., Hietala, P. L., Karvonen, P., and Nuorteva, P., "Inhibition of growth and panicle formation with an extract of the leafhopper, *Calligypona pellucida* (F)," *Physiol. Plantarum*, **14**, 683–696 (1961).

Weber, H., *Biologie der Hemipteren*, Springer, Berlin, 1930.

Weber, H. *Lehrbuch der Entomologie*, Fischer, Jena, 1933.

CHAPTER **6** CHAPTER

Localized
Toxic Effects

LOCAL LESIONS AT THE INSECT'S
FEEDING POINT

Local lesions at the insect's feeding point are the simplest expressions of the toxic effects of insect feeding. There are no apparent systemic effects, and cellular disturbance, being limited to a small area, can be well defined.

Leaf Spotting by Coccidae

Mealybug feeding results in various types of leaf spotting. *Pseudococcus adonidum* L on pineapple plants produces a circular chlorotic spot with discrete margins, quite different in character from the irregular chlorotic areas produced by *Dysmicoccus brevipes* (Ckl.).

In the latter case, the effects are due to colony feeding within a circumscribed area. Cell walls are thickened in the chlorotic area, the thickening being more pronounced in the center of the area and reduced at the periphery. Within the area of pronounced cell-wall thickening, the cells are full of a homogeneous, nongranular substance which takes a slightly less dense stain than the thickened cell wall. The chloroplasts are degener-

171

ated (Carter, 1933, 1939). *P. adonidum* spotting, on the other hand, re-corded from Queensland (Carter, 1942) and frequently encountered in Hawaii during the relatively rare and seasonal occurrence of the insect on pineapple plants, is clearly the result of a single insect's feeding, and is visible two or three days after the insect's feeding begins, although the destruction of the chloroplasts is similar to that by *D. brevipes*. Some individuals of *P. adonidum* are highly efficient spotters. A single insect produced 50 spots of a soft-leaved pineapple seedling in 10 days.

Green spotting by *D. neobrevipes* Beardsley (Beardsley, 1959), which insect previously was considered a strain of *D. brevipes* (Ito, 1931), is in quite a different subcategory. In its most typical form, it is bizonate, with a deep green central zone and a lighter green outer zone, the outer zone being, however, deeper green than the surrounding nonspotted area of the leaf (Fig. 6.1). This type of spot, which is undoubtedly a strictly localized secretion effect, is probably due to a complex of secretions, one of which diffuses farther than the other. There is no evidence of thickening of cell walls, emptying of cell contents, or dissolution of chloroplasts; on the contrary, the typical condition is a hyperplasia, which increase in the number of chloroplasts. This can be seen most clearly in a green spot which is being formed at or near the junction of the white and green tissue at the base of the leaf. There, the chloroplasts are clearly formed within the green spotted area, while the normal white tissue surrounding it contains no visible chloroplasts. A green spot may retain its identity when all the surrounding tissue has dried brown, with the cells in the green spotted area apparently alive. No study of this phenomenon has ever been recorded by a cytologist or a specialist in plant pathological anatomy, but Miles (1968) has suggested that increases, or prolongation of photosynthesis in isolated patches, or green islands, are perhaps related to a localized increase in concentration of growth hormones in plants due to the action of salivary amino acids.

Those produced by *Melanaspis bromeliae* may represent another case in point. *Melanaspis bromeliae* Leonard was reported as occurring on pineapple plants in Brazil by Carter (1949). Here again there is evidence for more than one component of the insect's secretion. Underneath the scale, the color of the leaf is darker than the normal green of the leaf. This darker green spot is circular. A much larger chlorotic ring forms a halo around this spot. Development of the dark central ring begins shortly after establishment of the crawler on the leaf. The chlorotic ring develops later and becomes depressed, leaving the green center as an elevated island.

A diaspine scale was observed on pineapple plants in Fiji (Carter, 1942). The chlorotic spot, centered by the insect, reached a diameter of

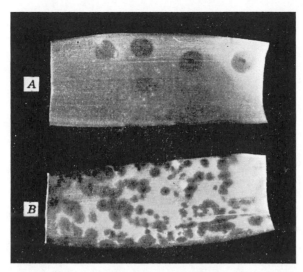

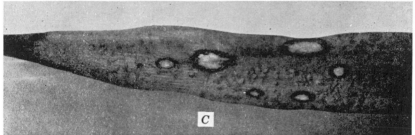

FIGURE 6.1. Green spotting of pineapple by *Dysmicoccus neo-brevipes* Beardsley. (*A*) Large green spots showing dark centers. (*B*) Chlorotic leaf showing bizonate spots remaining as green islands. (*C*) Leaf with raised green spots in median area; flat spots at margins which have lost the dark centers; areas of secondary necrosis (After Carter, 1933).

almost 1 cm. Such a feeding effect is obviously associated with a diffusible component of the insect's secretion, since the circular area involved is beyond that which could possibly be directly tapped by the insect's piercing organ.

Lecanium coryli L causes severe damage or death to ash trees. Only adult females' feeding brings about symptoms. From histological studies, it has been shown that the saliva becomes apparent around the tips of the stylets; later the cells along the stylet track become brown. The cell walls are then destroyed, presumably by the action of cellulase in the saliva which then penetrates the adjoining cells. Cork tissue is formed through

FIGURE 6.2. Geometric mosaic pattern of leaflet of coconut palm (After Carter, 1967).

the cambium; in heavy infestations this results in cutting off the vascular tissue with resulting die-back of twigs and branches. The histological evidence in this case is important, since otherwise, the corking of the cambium, actually a very secondary symptom, would probably have been ascribed to mechanical damage only (Komarek, 1946).

Geometric mosaic, a curious example of localized effect, was recorded by Carter (1967) as occurring on coconut leaflets. The area involved coincided with the feeding area of a colony of the scale insect *Xanthophthalma concinna* (Ckl. and Parrott); the actual feeding tracks were not determined. The symptom was one of dark, scattered areas of the leaflet which have, invariably, perfectly straight margins bounded by the veins and, depending on their length, appear as squares or rectangles (Fig. 6.2).

Leaf Spotting by Leafhoppers

Leaf spotting and stippling by leafhoppers (Cicadellidae) is specific for each of five species of *Empoasca: E. bifurcata* DeL., *E. erigeron* DeL., *E. filamenta* DeL., *E. abrupta* DeL., and *E. maligna* Walsh (Fig. 6.3). All are mesophyll feeders, and the result of their feeding is a definite stippling on the upper surface of the leaves which is characteristic for each species; whitish spots are produced on the upper surface of the leaves that are fed on from the lower epidermis. The cell walls of the spongy mesophyll and the palisade layers are torn, the cells are emptied of their contents, and in some cases even larger portions of the cell walls disappear. The type of feeding was typical for each of the species on all host plants tested (Smith and Poos, 1931). F. F. Smith (1940) found an interesting relationship between stippling leafhoppers, their mesophyll-feed-

(a)   (b)

FIGURE 6.3. (a) Typical stippling produced by *Empoasca bifurcata* on *Solidago* spp., one of its preferred hosts. Note the normally colored area in the center of the many spots; also the slightly coarser stippling than in (b). (b) Typical stippling produced by *E. erigeron* on *Solidago* spp., one of its preferred hosts. Compare with (a). The dark area near the tip of the leaf was produced by infection of mildew (Photographs: Smith and Poos, 1931).

ing habit, and the appearance of the fecal deposits. All excreted dark-colored material; that of the phloem-feeding, nonstippling species consisted of clear droplets. The stipple design and fecal deposits were characteristic of the species (Fig. 6.4).

When intact cells were found devoid of contents in the area of the white spots typical of leafhopper feeding, K. M. Smith (1926) concluded that chlorophyll destruction had occurred by the diffusion of the salivary secretion through the cell wall (See also Horne and LeFroy, 1915; DeLong, 1938).

Eutettix strobi Fitch produces characteristic crimson spots on pigweed (*Chenopodium album* L), but these spots are produced only by the nymphs and not by adults. They appear not less than 24 hours after feeding has started. The size of the spot depends on the length of time the insect remains feeding in one place, and the ability to spot leaves increases with

FIGURE 6.4. Scattered stippling by *Typhlocyba pomaria* McA. (Photograph: Smith, 1940).

the age of the nymph (Fenton, 1925; Carpenter, 1928). When, as in this case, nymphal and adult feeding effects differ, the conclusion is reached that, not only do the adult and nymphal secretions differ, but that these are developed by the insect and not, as Nuorteva suggested, picked up from the plant.

Graminella nigrifrons Forbes produces chlorotic spots on rice plants, but this appears to be an enhancement of a condition which arises in control plants, but less frequently and later than when the plants are infested. The condition is noninfectious (Deemsteegt *et al.*, 1969). What may be an analagous situation was reported by Carter (1963) who found that mealybug feeding induced or enhanced the expression of a genetically induced symptom called black freckle in pineapple plants.

Vidano (1967) considers that leafhopper depigmentation, by speices of Typhlocibid leafhoppers, is the simplest form of pathological symptom induced by sucking insects. *Erythroneura comes* Say, *E. tricincta* Fitch, *E. ziczac* Walsh, *Zyginia rhamni* Ferrari, and *Arboridia dalmatia* Novak-Wagner are examples from a study of insects affecting the vine (Fig. 6.5). Vidano's paper is well illustrated and has a useful English summary. More complex foliar disorders of vines are referred to in Chapter 7.

Leaf Spotting by Aphids

Aphididae-induced phytotoxemia usually falls into a more complex category than that of simple leaf spotting, but Wadley (1929) records the occurrence of pale yellow spots with red centers on oats and other graminaceous plants caused by *Toxoptera graminum* Rond. due to the injection of some enzymelike substance. Jensen (1954) found that two species of *Myzus*, *M. ornatus* Lang and *M. persicae* (Sulz.), produced chlorotic spots on *Cymbidium* buds and that these persisted sometimes when the flowers were open.

Leaf Spotting by Acarina

The toxic feeding of the mites has attracted attention in view of recent work implicating this group in virus transmission. It is necessary to separate the symptoms of toxic feeding from those of virus transmission.

Yellow spots on raspberry, for example, are difficult to ascribe to either cause, being valid symptoms of both toxic feeding and virus disease.

Van Dinther (1951) found that quantitative counts of the mites present were a useful indicator if the developing symptom picture was closely followed. Chlorotic fleck of Myrobalan plum is due to the feeding activity of a toxicogenic eriophyid mite, *Vasates fockeui* (Nal. and Trt.) (Gilmer and McEwen, 1958). Symptoms appear approximately 14 days after the establishment of the mite population on immature, not on mature, tissue, and consist of more or less well-defined chlorotic areas ranging in size from pinpoint to 1–2 mm in diameter. The larger flecks may show indications of concentric rings. The authors concluded that no virus disease was involved when they failed to induce chlorotic fleck, either by bud inoculation or propagation from infected seedlings; field infested seedlings failed to develop symptoms on new growth in the greenhouse. Control of mites was achieved, and the development of fleck was prevented with 3 applications of Demeton applied at 2-wk intervals.

FIGURE 6.5. Folia depigmentation on *Vitis* due to *Zygina rhamni* Fabricus, A, and section of affected leaf, B (Photograph: Carlo Vidano).

A similar abnormality of peach, known as yellow spot, and resulting from the feeding of a closely related mite, *Vasates cornutus* (Banks), was described earlier (1952) by Wilson and Cochran. The period for symptom development is essentially the same as that reported by Gilmer and McEwen, but a feeding period of 24 hr was sufficient to incite symptom development 12 days afterwards. Wilson and Cochran also concluded that the symptom was the result of injection of toxic substances.

Flock and Wallace (1955) found it necessary to develop a colony of virus-free *Aceria ficus* (Cotte) in order to distinguish between symptoms of mite feeding leaf-distortion, russetting, and slight chlorosis, from those of fig mosaic.

Whether the effects of chlorotic fleck are strictly local is questioned by the work of Hildebrand (1945) and Brase and Parker (1955), which showed poor growth and certain plum variety budlings when propagated on Myrobalan seedlings. The latter authors suggested that the decline of Stanley prune trees may be associated with chlorotic fleck in the Myrobalan seedlings used as root stocks. If this is true then a most important corollary would be that toxic injections by the mite had permanently altered the physiology of the Myrobalan plum, and, to carry the matter further into the speculative field, had done so by altering the somatic tissue.

Lepra explosiva was described by Vergani (1944) as a result of a mite-injected toxin. Infestations of healthy oranges with the mite produced the first lesions after 2–2½ months. The symptoms were local. It was not transmitted by grafting, by juice inoculating, by inoculation of the juice of macerated mites, or by juice expressed from lesions. The disease was diminished in direct proportion to the degree with which the mites were controlled. However, over 20 years later, the etiology of this disease is still confused (Knorr, 1968). The current situation with respect to lepra explosiva as a virus disease is treated later in Chapter 12.

Knorr and Denmark (1970) have considered the role of *Brevipalpus phoenicis* (Geijkes) in inducing phytotoxemia. One expression of this is *phoenicis* blotch, a diffuse chlorotic spotting of orange trees which resembles the early stages of lepra explosiva but does not lead to gumming of affected areas. Single nymphs reared from eggs will produce blotch. A single spraying of wettable sulphur (5–10 lb/100 gal water), or chlorobenzilate (¼ pt 45.5% liquid in 100 gal water) will control the mite.

Halo scab presents a unique case of a combination of fungus and *Brevipalpus phoenicis* attack. Scab, caused by the fungus *Elsinore fawcetti* Bitancourt and Jenkins, attacks sour orange, but there is no leaf drop. When the scab lesions are infested with the mites, halo scab develops two weeks after infestation, with heavy leaf drop following (Knorr and Denmark, 1970) (Fig. 6.6). The chlorotic areas are phoenicis blotch.

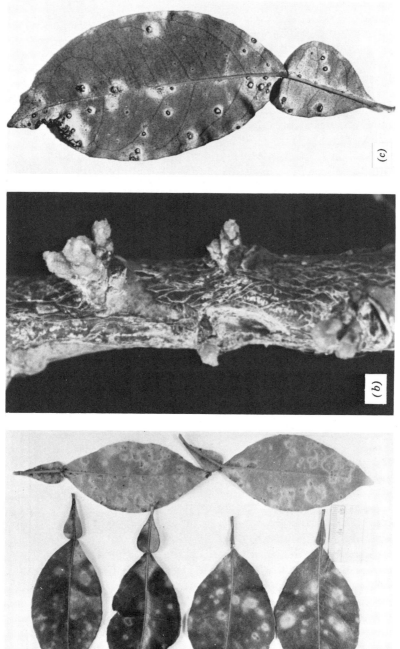

FIGURE 6.6. (a) Phoenicis blotch resulting from a field infestation of *Brevipalpus phoenicis*, Tice, Florida. (b) Early stage of gall proliferation due to infestation of *Brevipalpus phoenicis*. (c) Haloscab in association with *Brevipalpus phoenicis* on sour orange (Photographs: L.C. Knorr).

Leaf Spotting by Pentatomidae

The stinkbug *Euschistus variolarius* (P. deB.) produces symptoms on pears which differ depending on the variety of pear attacked. Small white cottony pockets, extending ⅛ in. into the flesh, occur at the feeding points of the insect on d'Anjou pears, and the feeding points may be depressed. External symptoms do not appear, however, on Bartlett pears, and they can only be seen when the skin is removed (Wilks, 1964).

LOCAL LESIONS WITH DEVELOPMENT OF SECONDARY SYMPTOMS

By Coccidae

The feeding punctures of *Matsucoccus feytaudi* Duc. induce two types of feeding puncture, although the processes are similar. A single puncture will induce local cellular hypertrophy and hyperplasia; multiple punctures will induce a spread of the production of resin which is excreted outwards. Death of the organ or tree follows. Carle *et al.* (1970) describe the condition as a toxemia with slight systemic symptoms and abnormal resinogenesis.

By Pentatomidae

The pentatomid, *Bathycoelia thalissima* (H-S), feeds on cocoa pods, causing premature ripening. Development of pods is inhibited if feeding occurs on them when the pods are 10–17 mm long. The affected pods resemble those called "cherelle wilt," but injured beans inside the pod are diagnostic for the pentatomid injury (Lodos, 1967).

By Aphididae

The chlorotic spotting of black raspberry induced by the feeding of *Amphorophora rubitoxica* Knowlton properly belongs here. The chlorotic spot appears at the feeding point of the insect and can develop after only 15 min of feeding, although one hr of feeding is more adequate. The ability to produce spots is inherited by the insect, but in a succession of feedings an occasional negative is found. This is also true for the green spotting caused by *D. neobrevipes;* green spots are not produced by every feeding. When *A. rubitoxica* has been feeding in large numbers on a plant for some time, the spots may coalesce to form a chlorotic blotch, and, in some instances, there is vein clearing to form a diffuse, netlike appearance (Stace-Smith, 1954).

By Miridae (Capsids)

By far the most important group of insects in this category are species of mirids. The damage they do to economic crops is enormous, particularly in the tropics where they are often a limiting factor to agricultural development, whether by peasant proprietor or organized plantation. The genus *Helopeltis* contains at least seven species which have been recorded as injurious to economic crop plants in India, Java, Ceylon, and the tropical African countries of Nigeria, Tanganyika, Nyasaland, French Equatorial Africa, and the Congo.

The typical lesion caused by the feeding of these insects is a circular spot, often red in color. If no further development takes place, as is true on tea if feeding occurs for a short time only (Smee, 1928), or on cacao where a slight sinking and blackening occurs, the lesions would fall in the first category of stippling and spotting. It is the secondary effects of *Helopeltis* feeding, however, with which we are now concerned.

HELOPELTIS.   On tea, the condition known as gnarled stem canker has been shown to be due to the feeding of *Helopeltis bergrothi* Reut., the tea mosquito bug (Leach and Smee, 1933). While the insect is still feeding on the stem, a water-soaked area develops around the point of entry of the stylets, and the area becomes well-defined within an hour. Within 24 hours, the canker shows a uniform dark brown-purple, oblong, ovate discoloration. There is a definite relationship between the development of cankers and the penetration of the stylets to the pericycle parenchyma. Penetration to phloem and xylem is only very rarely followed by canker formation, and not at all when only cortex and pith are penetrated. Cork bark grows over the canker so that the latter is hidden until later, when a small sunken area marks its position. The development of gall wood around each canker gives the lesions the gnarled appearance resulting from many cankers arising from single feeding points. Fungi and bacteria are not found in the cankers until the collapse of the cortex, indicating saprophytic or weakly parasitic status. The cankers are in such close proximity, however, that fungi can spread from one to the other.

Leach and Smee concluded that the effect of the insect's feeding was not due to mechanical injury, on several grounds: (1) the effect of the stimulus was felt at a considerable distance from the lacerated tissues (up to 12 mm), (2) the collapse was too rapid for fungi or bacteria to be primarily concerned, (3) neither fungi nor bacteria have been isolated from the earliest stage of the disease, and (4) they have not been seen in the collapsing cells.

Helopeltis also cause extensive damage to tea irrespective of canker formation. The production of tea depends on the growth of fresh new shoots, which the insects especially favor. The extreme rapidity with which lesions develop means that a small population of *Helopeltis,* as few as 4–6 (Smee, 1928), can destroy 1 lb of green leaves during their developmental cycle, and 1 bug can destroy 2 shoots or 3 leaves and a bud in a day. According to Smee and Leach (1932), the damage varies with the developmental stage of the bug, the damage increasing as the insect reaches maturity.

Environmental factors affecting outbreaks have been recognized. Seasons of long drought are associated with low *Helopeltis* populations. Smee (1928) found a measure of correlation between outbreaks and seasons of many rainy days, particularly in November and February. However, the advent of rain, especially after drought, is the signal for flushing of fresh new leaves, and the low populations of *Helopeltis* may be directly correlated with suitable food supply. De Jong (1935) inferred this when he suggested that changes in the plants during the wet season favored the insect's reproduction. High winds with low humidity are limiting factors to the development of *Helopeltis* in eastern Sumatra, where the insect develops only in well-protected, fairly dense shade.

Altitude, species, and damage are correlated, according to Leefmans (1916, 1916a). Under 4000 ft, *H. antonii* Sign injured both tea and cinchona. Above that altitude, the darker variety of *antonii* is found which prefers cinchona but does no serious damage to it. Cinchona grown near tea below the 4500-ft line, however, provides a dangerous neighbor for tea, in view of the capacity of the insect to reproduce on cinchona.

The system of alternate pruning, and the Pegat system of interspersed green manuring crops, were interesting practical applications of data on the response of the insect to environmental conditions. When tea plantations are pruned in alternate strips of approximately 10 ft in width, the development of the insects in the pruned areas is very much reduced; these pruned areas also constitute a barrier through which *Helopeltis* will not move freely from foci of infestation. The Pegat system of control consists of planting green manure plants between the tea rows. Its purpose was also to attempt the isolation of foci of infestation, and at the same time provide a source of green manure (Boode, 1922; Garretsen, 1922, 1923; Van Hooff, 1922, 1922a; Hart, 1923; Feuilletau, 1924).

Factors affecting attractiveness of the tea plant to *Helopeltis* attack have received considerable attention. There is general agreement that soil conditions affect susceptibility to an extreme degree; the more unsuitable they are to the plant, the greater the attractiveness is (Smee, 1928). Light (1930) concurred that serious defoliation only occurs in plants in poor condition.

Water-logged areas, slopes with underlying water pans, and sites surrounded by bush, are all primary attraction factors, as is malnutrition
when the original topsoil has been lost (Anstead and Ballard, 1922; Smee
and Leach, 1932). The ratio of available potash to available phosphoric
acid in the soil is a determining factor in attraction to *Helopeltis* attack,
according to Andrews (1914, 1919, 1921, 1923). Andrews injected tea
bushes with potash and quickly freed the plants from *Helopeltis*. New arrivals abandoned efforts to feed and moved over to untreated bushes. Soil
applications, as might be expected, gave irregular results. In a study of
three districts, Andrews reported: "There is a correlation between the
ratio of available potash to available phosphoric acid; between the percentage availability of the phosphoric acid; between the value of the ratio
of available potash to available phosphoric acid for a given percentage
availability of phosphoric acid and between the soil acidity of the soils of
the three districts, and the varying degree in which the gardens situated
in the districts suffer from attack. No relationship could be traced between the total quantities of potash and phosphoric acid in the soil, and
the incidence of the pest." Andrews' contribution goes beyond the specific
problem in establishing that the nutritional status of a plant affects the
ability of insects to feed on it.

Other information on the factor of nutrition is available from De Jong
(1934), who showed that fewer eggs were laid by *Helopeltis* on high sugar
content leaves than on those of lower content, and that attractiveness of
the tea plant rises as the proportion of carbohydrate falls, or as the relative quantity of albumens increases (De Jong, 1935). On cacao, *Helopeltis*
confines its attack to maturing pods but cannot survive on pods with a
high sugar content (Cottrell, 1927).

The injury caused by the feeding of *H. bergrothi* can simulate fungal
attack on mangoes, as shown by Leach (1935). A stem canker, an angular
spot, a fruit scab, and a fruit rot of mango were all shown to be the direct result of *H. bergrothi* feeding. In all four cases, the primary lesions
are recognizable before the insect has withdrawn its stylets from the
sucked tissue. "No organism has been seen or isolated from the freshly affected tissue and pieces of this tissue applied to the cut or uncut surfaces
of healthy young stems, leaves, or fruit have not transmitted the disease."

Fruit scab lesions always turn black, due perhaps to the resin content
of the skin. The initially affected area contains small parenchymatous
cells, with numerous resin ducts and small vascular bundles. It is Leach's
opinion that the toxin introduced through the insect's stylets can diffuse
only slowly through this mass and cannot penetrate to the large thin-
walled cells of the flesh before the plant reacts by forming scab tissue.

Fruit rot lesions are associated with scab lesions. Whether or not the former develop apparently depends on whether the insect penetrates deeper than usual to reach the larger parenchymatous, thin-walled cells of the inner skin where the toxin can diffuse rapidly. This is the only one of the four diseases mentioned here where the rot spreads in a manner similar to that of a fungus or bacterium. Leach suggests that the clear liquid in the hollow rotted area possibly includes the toxin in solution.

Schmitz (1958) has provided a complete monograph on *Helopeltis* affecting cotton in Central Africa. Along with *Lygus vosseleri* Poppius, *Helopeltis schoutedeni* Reuter is a principal enemy of cotton in that area. Damage to cotton is shown in Fig. 6.7. Figure 6.8 shows *Helopeltis* damage to cashew (Swaine, 1959). Initially, stem canker lesions are similar to those already described on tea. The blackening of the lesions is due to the invasion of secondary fungi and bacteria which can easily enter through the bead of sap that is often exuded from the puncture of a fresh lesion. Hyperplasia originates, after a week or ten days, from unaffected cells in or near the border of the affected area. It develops in the parenchyma, pushing out the cortex to form a slight swelling. This cortex may become necrotic due to the invasion of secondary microorganisms, or it may remain healthy throughout the development of the canker. Hyperplasia of the pith often occurs, that of cambium rarely; but when it does, the canker may form an open wound with the development of wound wood. Final symptoms of angular leaf spot depend on weather conditions. In wet weather, secondary organisms will blacken the spots quickly; in dry weather, there is a slow darkening of a light brown with darker brown edges. Hyperplasia occurs if the spot is next to a midrib.

THE CACAO MIRIDS.   The vast forest that stretches along the west coast of Africa is the area in which the great bulk of the cocoa beans of commerce are produced. Introduced as an exotic, *Cacao theobroma* L quickly adapted itself to life under the high forest canopy and became the principal economic crop of the area. In due time, however, transfer of forest insects to the cacao tree occurred; two of these, the mirids *Sahlbergella singularis* Hagl. and *Distantiella theoborma* Distant, are responsible for what is probably the most widespread and economically significant case of injury by this group of insects (Fig. 6.9). It was possible before control measures were developed to travel for hundreds of miles along the forest roads and never be out of sight of evidence of mirid damage.

Voelcker and West (1940) described *Sahlbergella* blast, as they called it, as a major contributing cause of die-back in cacao. Nicol (1945) summarized the data up to 1945 for the Cocoa Conference in London. Since

(a)

(b)

FIGURE 6.7. Damage to cotton caused by *Helopeltis schoutedeni* Reuter is illustrated in (a) and (b) (Photographs: Schmitz, 1958).

(a)

(b)

FIGURE 6.8. (a) Cashew stem and leaves damaged by *Helopeltis*. Bagamoyo, June, 1955. (b) Malformation of 2-yr-old cashew tree caused by *Helopeltis*. Bagamoyo, June, 1955.

(c)

FIGURE 6.8. (c) *Helopeltis* damage to developing cashew fruit. Bagamoyo, June, 1955 (Photographs: Commonwealth Institute of Entomology, Swaine, 1959).

that time, progress reports have appeared regularly in the annual reports and technical papers of the West African Cocoa Research Institute and the Department of Agriculture of Ghana.

The first symptoms are very similar to those of *Helopeltis* feeding, except that the cacao mirids are much larger insects, and the water-soaked area developing around the insect's feeding point while it is still feeding is more extensive. The initial gray, water-soaked area rapidly changes in color to the typical brown or blackish spot (Fig. 6.10). This primary damage, which is not restricted to the point of insertion of the stylets, has been shown by Goodchild (1952) to be due to the histolytic effect of esterases in the saliva injected into the wound. No organic pathogens have been found in the saliva (Squire, 1947). On a green shoot, the toxic effect of the salivary secretions plus mechanical damage may be sufficient to

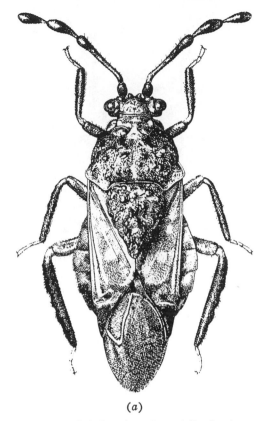

(*a*)

FIGURE 6.9.   (*a*) Adult female, *Distantiella theobroma* (Dist.).

cause death. Numerous punctures on the leaves may involve death of both leaves and shoot.

The relationship between mirid damage and the fungus *Calonectria rigidiuscula* (Berk. and Br.) has already been referred to (See Chapter 4). Provided that infection by the fungus does not take place, the lesion heals about three days after puncturing, and no cankering or bark roughening takes place. If infection with the fungus does take place, subsequent damage is not confined to the medullary rays in the peripheral tissues, but it spreads to the phloem and xylem. The infected lesion is characterized by a deep-seated canker extending into the xylem and by the roughened bark which appears as the lesion finally becomes occluded (Fig. 6.11).

Taylor (1954), from whose report this summary is largely drawn, describes three fairly distinct forms of damage on mature cacao. The first is seasonal and has little permanent effect on the trees. It is due to the mir-

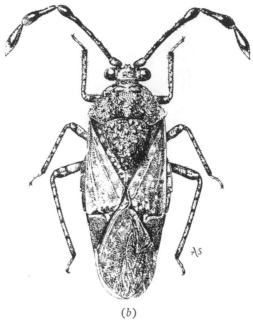

(b)

FIGURE 6.9. (b) Adult female, *Sahlbergella singularis* Hagl. (Drawings: West African Cacao Research Institute).

ids feeding on the fan shoots after the main crop of pods has been removed. The shoots are killed, the leaves turn brown and remain hanging on the tree for some time. This condition has been named "blast." With persistent attack, whole areas of trees become unthrifty and heavily cankered, with much reduced leaf growth. Cacao in this condition is described as "stag-head." The third type of damage is acute, the canopy branches are completely lost, and vigorous chupon (sucker) growth takes place. These new chupons are ideal for mirid reproduction and lead to the condition known as "capsid pockets." These pockets develop where the canopy has been broken, either as a result of adverse water relations or from damage caused by falling forest trees. Conditions of shade influence both the incidence of attack and the severity of damage (Williams, 1953, 1953a). The situation is complex because both excessive shade and reduced shade are factors rendering the cacao either susceptible to attack or reducing the capacity of the tree to regenerate.

Squire (1947) reported that "pockets" occurred in cacao grown under the forest shade. In Nigeria and the Ivory Coast, where there is little or no shade, damage is more diffuse. The observation clearly indicates either that the insect is attracted to more open spaces, where these are

FIGURE 6.10. Capsids on cacao. Primary lesions developing around feeding punctures (Photograph: West African Cacao Research Institute, Taylor, 1954).

available, or that the chupons on well-lit trees are more liable to attack and more susceptible to damage. Mirids attacking cocoa throughout the world have been listed by Entwistle, and control methods have been described (1965, 1966). Seasonal population changes of cocoa capsids were studied by Gibbs *et al.* in Ghana (1968). Build-up of cocoa capsid populations coincides with the development of the main crop from July or August to October. After harvest, *S. singularis* moves to vegetative tissues; *D. theobroma* moves to more local changes of feeding site. The phenology of cocoa and its associated insects have been investigated by Gibbs and Leston (1970).

Leston (1970), in a review of the insects affecting cocoa, has provided a detailed summary of the cocoa capsids. There are 11 named species plus

(a)   (b)

FIGURE 6.11. (a) Infested capsid lesions on cacao becoming occluded by the growth of callus tissue. (b) Roughened bark of cacao which appears after the infected lesions have become occluded (Photographs: West African Cacao Research Institute, Taylor, 1954).

12 spp. of *Helopeltis* and 12 *Monalonion* spp. whose systematics are inadequately known. Losses occur of not less than 20% of the cocoa crop in Ghana and up to 60% in New Britain.

Environmental factors, such as water-balance restoration after a spell of dessication, may materially affect subsequent damage on new flush and shoot production; these are preferred sites for capsid increase. On the other hand, water stress can result in humidities too low to permit survival, particularly of the larvae of *Distantiella*.

The inter-relationships between ant species and capsids are significant. The presence of *Oecophylla* is estimated to protect up to 15% of Ghana cocoa. The role of the other species is more complex; the state of the systematics of the ant species is a negative factor.

Control of both *Helopeltis* and capsids was extremely difficult and unsatisfactory until the advent of the chlorinated hydrocarbons. One of the earliest uses of DDT was that by Nicol (1945), who had some success in controlling the insects in young cacao by painting the crotch with DDT emulsion. This was based on the tendency of the insect in both nymphal and adult stages to rest in the crotch during the day. The method was not satisfactory for mature trees.

The control program was summarized by Hammond (1957), although there was little in his factual account to indicate the magnitude of the task. The insect is not difficult to kill and it was early learned that the gamma isomer of BHC (benzene hexachloride) applied at the rate of 4 oz active toxicant/ac in 5 gal (Imperial) of emulsified spray was adequate. The problem was primarily one of application, and this has been solved by the development of mist sprayers carried in knapsack fashion and run by a small engine. Even so the problem of covering 150,000 acres of forest presented some extremely difficult logistics. The results have been such, however, that all 4 million acres of Ghana cacao forest will in due time be treated. These results, summarized by Hammond, appear in Table 6.1.

TABLE 6.1.   Control of Cacao Mirids with Insecticides[a]

Average yield per acre during the first season of spraying	No. of plots	Increase in average yield per acre (%)	
		Second year of treatment	Third year of treatment
Under 100 lb/ac	2	224	623
100–200 lb/ac	8	112	244
200–300 lb/ac	4	78	156
300–400 lb/ac	2	19	68
400–600 lb/ac	2	15	120
Over 600 lb/ac	1	4	43

[a] Summary of yield increases over three years (based on main crop yields).

The methods described are those found necessary in an area where land tenure and traditional methods preclude the organization of commercially operated plantations. In the Congo, plantations are laid out in orderly rows and crops are grown under efficient technical supervision. Trees are pruned to develop a low spreading canopy, easy to inspect and easy to reach with small hand-dusting equipment. Regular inspections are made, and damaged or diseased branches and infected pods are removed. If mirids are found, a spot treatment with an insecticidal dust that contains 2.2% gamma isomer of BHC is applied. A total consumption of ½ lb dust ac/mo is adequate (Nicol and Taylor, 1954).

The most widely used insecticide in recent years has been an emulsified concentrate of Lindane but pockets of resistant capsids have developed.

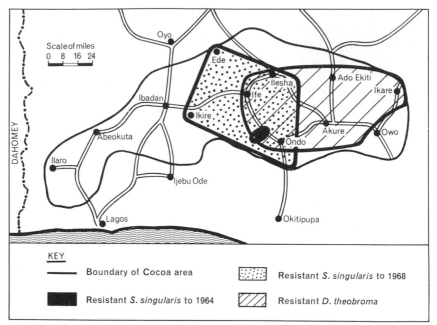

FIGURE 6.12.   Map showing the distribution of *S. singularis,* resistant to lindane in Nigeria (as of March, 1968) in relation to the main cocoa area (Photograph: Youdeowei, 1971, University of Ibadan, Nigeria).

How extensive these are throughout the country is not known, but a recent paper by Youdeowei (1971) records the data on the distribution of resistant capsids in Nigeria and methods for their detection. Recorded spread in the main cocoa growing areas in Nigeria is shown in Fig. 6.12. Booker (1970) has established three well defined categories of resistance as determined by a standard laboratory test. These are: (1) resistant—where the rate of mortality was similar to that of the controls (less than 60%: (2) partly-resistant—where the mortality rate after 12 hr was in the range of 65–80%: (3) susceptible—where mortality was 100% in the 12 hr period.

Progress reports on capsid control can be found in the annual reports of the cocoa research institutes in Ghana (Tafo) and Nigeria (Ibadan). A report from the International Capsid Research Team has been summarized after six years' work by a team of five workers (Anon., 1971). It is an important contribution.

LYGUS.   *Lygus* injury to lima bean has been shown to produce symptoms very closely resembling those reported as caused by *Nematospora*, in-

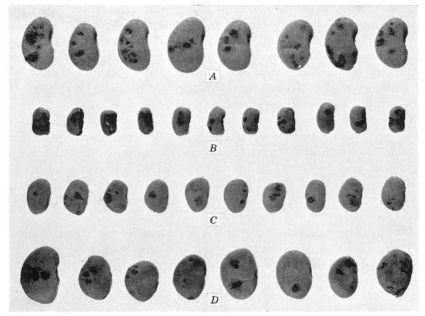

FIGURE 6.13. *Lygus* pitting of field-grown lima bean and cowpea seed. (*a*) Ventura vining-type lima bean. (*b*) Blackeye cowpea. (*c*) Baby lima bean. (*d*) Fordhook bush lima (Photograph: Baker *et al.*, 1946).

troduced by the green stinkbug *Nezara hilaris* Say (Leach, 1940). Baker *et al.* (1946) showed that in the absence of the pathogen, the toxic feeding of *Lygus hesperus* Knight, *L. elis* Van Duzee, and perhaps other species caused pitting of bean seed in the pod, in addition to the shedding of blossoms and of pods up to 2 in. long. This is one of the few instances where the action of pathological primary or secondary organisms has been eliminated as a prelude to the conclusion that a toxin is involved (Fig. 6.13). *Lygus hesperus* and *Lygus elisus* feeding on developing carrot seed reduce both yield and seed, as well as germination rates. The insect derives an advantage when feeding on floral parts although after 24 hr feeding, the number of florets is greatly reduced. Abscission is limited to buds attacked when less than 7 mm long, but new buds appear in direct proportion to the extent of debudding so that bud production is prolonged. Suitable feeding sites are therefore available for a lengthened period (Strong, 1968).

Scott (1969, 1970) compared growth rates and yield of plants grown from carrot and bean seed which had been fed on by *Lygus* bugs, with those from *Lygus*-free plants. While the fed-on seed was lighter than that produced by the controls, plants produced from that affected seed grew faster and outyielded the controls. Scott presents three hypotheses to account for this: (1) destruction of a growth inhibitor in the seed by the salivary secretions of the *Lygus* bugs, (2) injection of a plant auxin into the developing seed as has been shown to occur (Miles and Lloyd, 1967), and (3) a change in the carbon-protein in the seed as result of the feeding. Scott favors the second hypothesis.

Resistance to the feeding of *L. hesperus* and *L. elisus* on a developing carrot seed has been shown (Scott, 1970a). Three varieties were used, Nantes, Imperida, and Royal Chantenay. Resistance was found in all three varieties, but Royal Chantenay possessed the highest degree of resistance. *Lygus lineolaris* (P. deB) is the most important of the plant bugs causing apical seediness in strawberries. Effective insecticides confirmed the role of the insect (Schaefers, 1966).

The split lesion on cotton plants is a developmental symptom resulting from the feeding of any of several species of mirid. Other effects of these insects' feeding habits more properly belong in the next category to be discussed.

The local lesion is a swelling which later may develop into the typical split lesion. All the feeding punctures do not result either in an externally visible swelling or in a split lesion, but microscopic examination reveals that internally the tissues are affected to a lesser degree (King and Cook, 1932). The swellings do not affect the tissue for more than 5 mm from the point of puncture (Ewing, 1929; Painter, 1930). If the swellings are numerous enough to be seen over the whole plant, the damage appears superficially to be systemic. Painter also obtained typical split lesions by injections of macerated cotton fleahoppers (*Psallus seriatus* Reut.), and somewhat the same effect with injection of diastase. King and Cook (1932) were not able to repeat Painter's experiments with the same degree of certainty, and rightly pointed out that no sterile needle is as fine a piercing instrument as the stylets of the insect, and that it almost always causes a reaction in the plant because of mechanical damage.

The development of the swelling has been described by these same authors. The early reaction is an enlargement of the nuclei and the cells at some point in the feeding area. Cell division is stimulated. The milder lesions may show only evidence of increased cell activity and more or less distortion, but in the more severe cases, areas of broken down tissue develop. External swellings are usually visible by the second or third day.

No sheath material was found about the punctures of the mirids, so that clogging and mechanical obstruction as described by Heriot (1944) were definitely not factors in the development of the lesions, although some species of Homoptera, with well-defined sheath material in their feeding tracks, were also efficient producers of both swellings and split lesions.

ADELPHOCORUS LINEOLATUS *(Goeze)*. *Adelphocorus lineolatus* (Goeze), the alfalfa plant bug, feeds on the buds, flowers, and immature pods of alfalfa. Bud blast, flower fall, and pod injury result. Necrosis is at first localized around the feeding puncture, but it later spreads to other parts of the individual flower. Cell disintegration in ovary and ovules was evident 18 hours after mirid feeding. Hughes (1943) believed that mechanical injury was minor and that the effect was primarily one of phytotoxicity.

Coreidae

PSEUDOTHERAPTUS. The work of Way (1953) provides a detailed study of a species of *Theraptus,* later named *Pseudotheraptus wayi* Brown, and the damage it causes to coconuts in East Africa. Injury to the male flowers is superficial, and consists of a small brown spot surrounding each feeding puncture. The later instars and adults prefer to feed on the immature and the mature female flowers and the young nuts. The insertion of the stylets into the bracts and through to the ovary or young nut causes a shrunken necrotic area which is not due to the invasion of pathogenic organisms, but is a direct effect of the insect's feeding.

A single puncture of a later instar nymph will destroy flowers and young nuts up to 8 wk old. Single feeding punctures may cause premature nut-fall of 8–12 wk-old nuts. With additional feeding punctures, nut-fall will continue up to 12 wk, but after that the nut continues to grow and the necrotic area caused by the punctures appears from beneath the bracts as a ring of lesions. Masses of gummy material exude from these lesions. The disease was known as "gumming disease" before the relationship of the insect to it was known. The number of insects necessary to damage a plantation severely is very small; the maximum number rarely exceeds 2/tree and frequently none is found.

There is a steady increase in the number of necrotic areas made in young coconut fruits as the insect passes through its five instars to the adult stage. Way's data are shown in Table 6.2.

Vanderplank (1958, 1959) attempted to estimate populations by direct methods, but hand collection, trapping, knock-down sprays, and the use of radioactive tracer techniques were all unsuccessful. Instead, a method of estimating damage was developed by counting lesions on freshly fallen young nuts known locally as *vidaka,* the damage being expressed as a percentage—the Vidaka Damage Ratio (VDR). Inversely, the VDR is a good measure of expected yield.

TABLE 6.2. *Results of Insectary Experiments Showing*
the Mean Numbers of Necrotic Areas Made
in Young Coconut Fruits by each Nymphal
Instar and by a Pair of Adult Theraptus

	Instar					Adult (male and female)
	1st	2nd	3rd	4th	5th	
Mean no. of nuts damaged	0.4	5	8	9	12	124
Mean no. of necrotic areas	0.6	13	24	26	34	346

On the other hand, Yeo and Foster (1958) found that assessing damage had its limitations in determining the effect of control measures because of the lapse of time between application and damage records. They found that searching the palms systematically and counting the total nymphal populations gave fairly reproducible results.

AMBLYPELTA. Two species of *Amblypelta, A. cocophaga* China and *A. theobromae* Brown, are now known to cause damage to cacao in New Guinea and British Solomon Islands essentially similar in nature to that already described for the Miridae (Fig. 6.14–6.17). Brown's (1958) illustrations, however, suggest that damage to pods, especially distortion as shown in his Fig. 4, Plate XX, is much more severe than that which occurs following cacao mirid attack (Fig. 6.18 and 6.19).

Attack on the terminal shoots causes a die-back, but this is restricted to the tip and occurs with a population of less than one insect, in the feeding stage, per tree. Brown also describes the damage done to a number of other host plants of these insects. These vary, from the damage to fruit already described, to stem injuries which may involve wilting of terminal shoots or the development of cankerous swellings as on cassava (*Manihot esculenta* Crantz) and injuries to petioles which result in downward drooping.

Premature nut-fall and blossom drop are secondary developments following puncturing by coreids. Young macadamia nuts fall prematurely when the feeding puncture of *Amblypelta luctescens* Dist. is visible only as a watery white spot. *A. cocophaga* China is responsible for premature nut-fall of coconuts in the Solomon Islands. One or two insects per tree are sufficient to render the plantation nonbearing (Phillips, 1940; Brimblecombe, 1948).

The relationship of ants to immature nut-fall is significant. Species of the genus *Oecophylla* are antagonistic to the bugs causing nut-fall, in as

FIGURE 6.14. The left-hand cacao pod has been caged with *Amblypelta cocophaga* for 27 days (Photograph: Commonwealth Institute of Entomology, Brown, 1958).

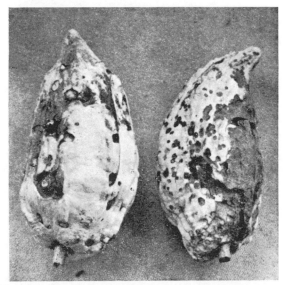

FIGURE 6.15. Cacao pods damaged by *Amblypelta cocophaga malaitensis* at Auki, Malita (Photograph: Commonwealth Institute of Entomology, Brown, 1958).

FIGURE 6.16. (*a*) The right-hand pod has been caged with *Amblypelta* for 8 days, and the cage removed for photographing the pod. The scale is in cm.

FIGURE 6.16. (*b*) The same pods 13 days later. Feeding scars have coalesced to form larger necrotic areas (Photographs: Commonwealth Institute of Entomology, Brown, 1958).

FIGURE 6.17. *Amblypelta cocophaga cocophaga* China, male, from Guadalcanal. Scale in mm alongside (Photograph: Commonwealth Institute of Entomology, Brown, 1958).

FIGURE 6.18. Larva of *Amblypelta cocophaga cocophaga* China on young coconut at Honiara, Guadalcanal (Photograph: Commonwealth Institute of Entomology, Brown, 1958).

201

FIGURE 6.19. Damage to coconuts by *Amblypelta* spp. at Sangara, near Popoendetta, New Guinea (Photograph: Commonwealth Institute of Entomology, Brown, 1958).

widely separated areas as East Africa and Melanesia. When this ant is dominant, there is no problem with nut-fall, but when *Pheidole* and *Iridomyrmex* are dominant, neither of which is antagonistic to the bugs, nut-fall becomes a serious problem (Lever, 1961).

REFERENCES

Andrews, E. A., "A note on the relation between the tea mosquito (*Helopeltis theivora*) and the soil," *Quart. J. Sci. Dept. Indian Tea Assoc., Calcutta,* 4, 31–35 (1914).

Andrews, E. A., "A preliminary note on the present state of the mosquito-blight enquiry," *Quart. J. Sci. Dep. Indian Tea Assoc. Calcutta,* 4, 119–129 (1919).

Andrews, E. A., "Some notes on attempts to produce immunity from insect attack on tea," *Rept. Proc. 4th Entomol. Meeting, Pusa, Calcutta,* 56–59 (1921).

Andrews, E. A., *Factors Affecting the Control of the Tea Mosquito Bug (Helopeltis theivora* Waterh.), Indian Tea Assoc., London, (1923).

Anon., *International Capsid Research Team, Cocoa Growers' Bulletin,* 17, Cadbury Limited, Bournville, England (Sept., 1971).

Anstead, R. D., and Ballard, E., "Mosquito blight of tea," *Planters' Chron., Coimbatore,* 17, 443–447, 453–455 (1922).

Baker, K. F., Snyder, W. C., and Holland, A. H., "*Lygus* bug injury of lima bean in California," *Phytopathology,* 36,493–503 (1946).

Beardsley, J., "On the taxonomy of pineapple mealybugs of Hawaii, with a description of a previously unnamed species (Homoptera: Pseudococcidae)," *Proc. Hawaiian Entomol. Soc.,* 17, 29–37 (1959).

Boode, F. J. C., "Van Hooff's Om- De-Andere-Rij-Snoeisysteem tegen *Helopeltis*," *Mededeel. Proefsta. Thee,* 81, 47–49 (1922).

Booker, R. H., "Resistance of *Sahlbergella singularis* Hagi (Hemiptera, Miridae) to the cyclodiene insecticides in Nigeria," *Bull. Ent. Soc. Nigeria,* 2(1970).

Brase, K. D., and Parker, K. G., "Decline of Stanley prune trees," *Plant Disease Reptr.,* 39, 358–362 (1955).

Brimblecombe, A. R., "Fruit-spotting bug *(Amblypelta lutescens)* as a pest of the Macadamia or Queensland nut," *Queensland Agr. J.,* 67, 206–211 (1948).

Brown, E. S., "Injury to cacao by *Amblypelta Stål* (Hemiptera, Coreidae) with a summary of food-plants of species of this genus," *Bull. Entomol. Res.,* 49, 543–554 (1958).

Carle, P., Carde, J. P., and Bonlay, C., "Feeding behaviour of *Matsucoccus feytaudi* Duc., Histological and histochemical characterization of the tissue disorganization caused in the host plant *(Pinus pinaster* Ait. var *mesogeenis,"* *Ann. Sci. Forestry,* 27, 89 (1970).

Carpenter, I. P., "Study of the life history and spotting habits of *Eutettix chenopodii* (Homoptera Cicadellidae)," *Univ. Kansas Sci. Bull.,* 18, 457–483 (1928).

Carter, W., "The spotting of pineapple leaves caused by *Pseudococcus brevipes,* the pineapple mealybug," *Phytopathology,* 23, 243–259 (1933).

Carter, W., "The toxicogenic and toxiniferous insect," *Science,* 83, 522 (1936).

Carter, W., "Injuries to plants caused by insect toxins," *Botan. Rev.,* 5, 273–326 (1939).

Carter, W., "The geographical distribution of mealybug wilt with notes on some other insect pests of pineapple," *J. Econ. Entomol.,* 35, 10–15 (1942).

Carter, W., "Insect notes from South America with special reference to *Pseudococcus brevipes* and mealybug wilt," *J. Econ. Entomol.,* 42, 761–766 (1949).

Carter, W., "Mealy wilt; a reappraisal," *Ann. N. Y. Acad. Sci.,* 13, 741–764 (1963).

Carter, W., "Geometric mosaic, a new expression of phytotoxaemia," *J. Econ. Entomol.,* 60, 603 (1967).

Cottrell, G. S., "Pests of cacao in the Gold Coast," *Proc. 1st W. African Agr. Conf., Ibadan, Nigeria,* 1, 98–112 (1927).

Damsteegt, V. D., Webger, A. J., and Graham, C. L., "A chlorotic leaf spot of rice; insect induction," *Phytopathology,* 59, 1556 (Abstr.), (1969).

DeJong, J. K., "*Helopeltis* in cacaotuinen," *Bergcultures,* 8, 658–667 (1934).

DeJong, J. K., "De Voedselopname van *Helopeltis,"* *Bergcultures,* 9, 292–294 (1935).

DeLong, D. M. "Biological studies on the leafhopper *Empoasca* fabae as a bean pest," *U.S. Dept. Agr. Tech. Bull.,* No. 618, 1–60 (1938).

Del Rivero, J. M., "La roseta de los agrios," *Bol. Patol. Vegetal y Entomol. Agr.,* 20, 193–210 (1955).

Entwistle, P. F., "Cocoa mirids; I. A world review of biology and ecology; II Control," *Cocoa Growers' Bull.,* (I) 5, (1965); (II) 6, (1966).

Ewing, K. P., "Effects on the cotton plant of the feeding of certain Hemiptera of the family Miridae," *J. Econ. Entomol.,* 22, 761–765 (1929).

Fenton, F. A., "Notes on the biology of the leafhopper *Eutettix strobi* Fitch," *Proc. Iowa Acad. Sci.*, 31, 437–440 (1925).

Feuilletau, De B. W., "*Helopeltis*-bestrijding," *Mededeel. Proefst. Thee*, 86, 1–14 (1924).

Flock, R. A., and Wallace, J. M., "Transmission of fig mosaic by the eriophyid mite *Aceria ficus*," *Phytopathology*, 45, 52–54 (1955).

Garretsen, A. J., "Het snoeien om de andere rij ter bestrijding van *Helopeltis*," *Mededeel. Proefst. Thee*, 81, 36–39 (1922).

Garretsen, A. J., "Groenbemesters en *Helopeltis*," *De Thee*, 4, 57–58 (1923).

Gibbs, D. G., and Leston, D., "Insect phenology in a forest cocoa-farm locality in West Africa," *J. Appl. Ecology*, 7, 519–548 (1970).

Gibbs, D. G., Pickett, A. D., and Leston, D., "Seasonal population changes in cocoa capsids (Hemiptera, Miridae) in Ghana," *Bull. Entomol. Res.*, 58, 279–291 (1968).

Gilmer, R. M., and McEwen, F. L., "Chlorotic fleck, an eriophyid mite injury of Myrobalan plum," *J. Econ. Entomol.*, 51, 335–337 (1958).

Hammond, P. S., "Capsid control on mature cocoa," *New Gold Coast Farmer*, 1, 109–115 (1957).

Hart, S. J. G., "De lamtoro in verband met de *Helopeltis*," *Algem. Land-bouw-weekblad Ned.-Indie*, 8, 477–478 (1923).

Heriot, A. D., "Comparison of the injury to apple caused by scales and aphids," *Proc. Entomol. Soc. Brit. Columbia*, 41, 13–15 (1944).

Hildebrand, E. M., "Myrobalan mottle and asteroid spot," *Phytopathology*, 35, 47–50 (1945).

Horne, A. S., and Lefroy, H. M., "Effects produced by sucking insects and red spider upon potato foliage," *Ann. Appl. Biol.*, Nos. 3, 4, 370–386 (1915).

Hughes, J. H., "The alfalfa plant bug *Adelphocoris lineolatus* Goeze and other Miridae (Hemiptera) in relation to alfalfa-seed production in Minnesota," *Minn. Univ. Agr. Expt. Sta., Tech. Bull.*, No. 161 (1943).

Ito, K., "Studies on the life history of the pineapple mealybug *Pseudococcus brevipes* (Ckll.)," *J. Econ. Entomol.*, 31, 291–298 (1938).

Jensen, D. D., "The effect of aphid toxins on *Cymbidium* orchid flowers," *Phytopathology*, 44, 493–494 (1954).

Knorr, L. C., and Denmark, H. A., "Injury to citrus by the mite *Brevipalpus phoenicis*," *J. Econ. Entomol.*, 63, 1966 (1970).

Komarek, J., "The physiological damage upon the ash tree made by the scale insect *Lecanium coryli* L,." *Acta Soc. Zool. Cechoslov.*, 10, 156–165 (1946).

Leach, R., "Insect injury simulating fungal attack on plants. A stem canker, an angular spot, a fruit scab and a fruit rot of mangoes caused by *Helopeltis bergrothi* Reut. (Capsidae)," *Ann. Appl. Biol.*, 22, 525–537 (1935).

Leach, R., and Smee, C., "Gnarled stem canker of tea caused by the capsid bug (*Helopeltis bergrothi* Reut.)," *Ann. Appl. Biol.*, 20, 691–706 (1933).

Leefmans, S., "Over *Helopeltis* in theetuinen," *Mededeel. Proefsta. Thee*, No. 46 (1916).

Leefmans, S., "Bijdrage tot het Helopeltis-vraagstuk voor de Thee," *Mededeel. Proefsta. Thee*, No. 50 (1916a).

Leston, D., "Cocoa insects," *Ann. Rev. Entomol.*, 15, 270 (1970).

Lever, R. J. A. W., "Immature nutfall of coconuts," *World Crops*, 13, 60–62 (1961).

Lewis, H. C., "Lime zinc spray as a repellent for leafhoppers on citrus," *J. Econ. Entomol.*, 35, 362–364 (1942).

Light, S. S., "*Helopeltis* in Ceylon," *Tea Quart.*, 3, 21–26 (1930).

Lodos, N., "*Pseudotheraptus* sp. on cocoa," (note) *Rept. Cocoa Res. Inst. Ghana*, for 1963–65 (1966).

Lodos, N., "Studies on *Bathycoelia thalassina* (H-S) (Hemiptera; Pentatomidae) the cause of premature ripening of cocoa pods in Ghana," *Bull. Entomol. Res.*, 57, 288–299 (1967).

Miles, P. W., "Insect secretions in plants," *Ann. Rev. Phytopathology*, 6, 137–164 (1968).

Miles, P. W., and Lloyd, J., "Synthesis of a plant hormone in the salivary apparatus of plant sucking bugs," *Nature*, 213, 801–802 (1967).

Nicol, J., "The present position of research on capsid pests in West Africa," *Rept. Cocoa Conf. London* (1945).

Nicol, J., and Taylor, D. J., "Capsids and capsid control in the Belgian Congo," *W. African Cacao Research Inst. Tech. Bull.*, No. 2 (1954).

Painter, R. H., "A study of the cotton flea-hopper *Psallus seriatus* Reut. with especial reference to its effect on cotton plant tissues," *J. Agr. Research*, 40, 485–516 (1930).

Phillips, J. S., "Immature nutfall of coconuts in the Solomon Islands," *Bull. Entomol. Res.*, 31, 295–316 (1940).

Schaefers, G. A., "The reduction of insect-caused apical seediness in strawberries," *J. Econ. Entomol.*, 59, 698–906 (1966).

Schmitz, G., "*Helopeltis* du contonnier en Africa centrale," *Publ. Inst. Natl. Etude Agron. Congo Belge. Ser. Sci.*, 71, (1958).

Scott, D. R., "*Lygus* bug feeding on developing carrot seed; effect on plants grown from that seed," *J. Econ. Entomol.*, 62, 504–505 (1969).

Scott, D. R., "Feeding of *Lygus* bugs on developing carrot seed and bean seed increase growth and yields of plants grown from that seed," *Ann. Entomol. Soc. Amer.*, 63, 1604–1608 (1970).

Scott, D. R., "*Lygus* bugs feeding on developing carrot seed; plant resistance to that feeding," *J. Econ. Entomol.*, 63, 959 (1970a).

Smee, C., "Tea mosquito bug in Nyasaland (*Helopeltis bergrothi* Reut.) and notes on two potential pests of tea, (1) the tea leaf weevil (*Dicasticus mlanjensis* Mishl.) and (2) the bean flower capsid (*Callicratides rama* Kirby)," *Bull. Dept. Agr. Nyasaland*, 4, 1–10 (1928).

Smee, C., and Leach, R., "Mosquito bug the cause of stem canker of tea," *Bull. Dept. Agr. Nyasaland,* 5, 1–7 (1932).

Smith, F. F., "Certain sucking insects causing injury to rose," *J. Econ. Entomol.,* 33, 658–662 (1940).

Squire, F. A., "On the economic importance of the Capsidae in Guinean Region," *Rev. Entomol. Rio de Janeiro,* 18, 219–247 (1947).

Stace-Smith, R., "Chlorotic spotting of black raspberry induced by the feeding of *Amphorophora rubitoxica* Knowlton," *Can. Entomologist,* 86, 232–235 (1954).

Strong, F. E., "The selective advantage accruing to *Lygus* bugs that cause blasting of floral parts," *J. Econ. Entomol.,* 61, 315–316 (1968).

Swaine, G., "A preliminary note on *Helopeltis* spp. damaging cashew in Tanganyika Territory," *Bull. Entomol. Res.,* 50, 171–181 (1959).

Taylor, D. J., "A summary of the results of capsid research in the Gold Coast," *W. African Cacao Research Inst., Tech. Bull.,* No. 1 (1954).

Vanderplank, F. L., "Studies on the coconut pest, *Pseudotheraptus Wayi* Brown (Coreidae), in Zanzibar. I. A method of assessing the damage caused by the insect," *Bull. Entomol. Res.,* 49, 559–584 (1958).

Vanderplank, F. L., "Some data on the yields of coconuts in relation to damage caused by the insect," *Bull. Entomol. Res.,* 50, 135–150 (1959).

Van Dinther, J. B. M., "*Eriophyes gracilis* Nal., als verwekker van glle bladviekken op framboos," *Tijdschr. Plantenziekten,* 57, 81–94 (1951).

Van Hooff, H. W. S., "Snoeien en *Helopeltis,*" *Mededeel. Proefsta. Thee,* 81, 26–31 (1922).

Van Hooff, H. W. S., "De op Tjiboengoer genomen Maatregelen tegen *Helopeltis,*" *Mededeel Proefsta. Thee,* 81, 40–44 (1922a).

Vergani, A. R., "Transmission y naturaleza de la "Lepra explosiva" del naranjo," *Ministerio Agr. Buenos Aires, Ser. A,* 1, 3 (1945).

Vidano, Carlo, "Sintomalogia esterna ed interna da insetti fitomizo su vitis," Publication 116 del Centro di Entomologia alpina e forestale del Consiglo Nazionale delle Ricerche 1967.

Voelcker, O. J., and West, J., "Cacao die-back," *Trop. Agr. (Trinidad),* 17, 27–31 (1940).

Wadley, F. M., "Observations on the injury caused by *Toxoptera graminum* Rond. (Homoptera: Aphididae)," *Proc. Entomol. Soc. Wash.,* 31, 130–134 (1929).

Way, M. J., "Studies on *Theraptus* sp. (Coreidae); the cause of the gumming disease of coconuts in East Africa," *Bull. Entomol. Res.,* 44, 657–667 (1953).

Wilks, J. M., "The spined stink bug; a cause of cottony spot in pear in British Columbia," *Canadian Entomologist,* 96, 1198–1201 (1964).

Williams, G., "Field observations on the cacao mirids *Sahlbergella singularis* Hagl., and *Distantiella theobroma* (Dist.) in the Gold Coast," *Bull. Entomol. Res.,* 44, 101–119 (1953).

Williams, G., "Field observations on the cacao mirids *Sahlbergella singularis* Hagl., and *Distantiella theobroma* (Dist.) in the Gold Coast. Pt. II," *Bull. Entomol. Res.*, 44, 427–437 (1953a).

Wilson, N. S., and Cochran, L. C., "Yellow spot, and eriophyid mite injury on peach," *Phytopathology*, 42, 443–447 (1952).

Yeo, D., and Foster, R., "Preliminary note on a method for the direct estimation of populations of *Pseudotheraptus wayi* Brown on coconut palms," *Bull. Entomol. Res.*, 49, 585–590 (1958).

Youdeowei, A., "Resistance in cocoa capsids to insecticides in use in Nigeria," *Cocoa Growers' Bull.*, (Cadbury Ltd., Bournville, England), 16, 15–20 (1971).

Primary
Malformations

PLANT TISSUE MALFORMATIONS

This category includes toxemias that are made evident by the development of primary, gross tissue malformations. Local lesions are sometimes associated, but the malformations are not secondary developments from the local lesions.

By Aphididae

Many species of aphid are notorious for the leaf-curling, rolling, and other distortions caused by their feeding. Tate (1937) was able to generalize, on the basis of study of 15 species, that plants showing such symptoms were found to exhibit marked histological abnormalities, principally in the form of pseudovascular tissue. The extent of this development probably depends upon such factors as: (1) severity of infestation, (2) duration of feeding activity, and (3) specific plant reaction to the stimulus of the insect secretion. *Myzus convolvuli* (kalt.) feeding on lily (*Lilium longiflorum* Thumb.) produces symptoms simulating those of the virus disease known as lily rosette. There is downward curling and stunting of young leaves, with irregular mottling of the older leaves. The

209

FIGURE 7.1. Virus diseaselike injury to *Antirrhinum* spp. caused by the feeding of *Myzus persicae*. *a, A. Majus* with stunted and distorted lower leaves and normal terminal growth produced after removal of aphids. *b,* Apical leaves of *A. Majus* showing distortion and discoloration. *c, A. Majus* with development of growing point completely checked by aphids and normal new growth arising laterally following removal of aphids (Photograph: Baker and Tompkins, 1942).

old injury persists, but new growth emerging after the aphids are removed is normal. Smith and Brierley (1948) reported that the aphid causes somewhat similar symptoms on many of its host plants. It is not a vector of lily rosette.

A similar case of simulation of virus symptoms caused by *M. persicae* (Sulz.) was noted by Baker and Tompkins (1942). On snapdragon, apical growth is checked, the leaves are reduced in size and laterally distorted. The damage is limited to leaves fed upon, but it occurs with the feeding of very small numbers of aphids. The authors took the precaution of using a nonviruliferous line of *M. persicae* which had been maintained for many generations on turnip (Fig. 7.1).

There is specificity, both with respect to aphid species and host plant. Neither *M. persicae* nor *Aphis* (*Cerosipha*) *gossypii* Glover produces visible lesions when feeding on *Commelina nudiflora* L, but *M. solanifolii* (Ashmead) causes severe leaf curl with yellowed vein margins. Severin and Freitag (1938) found *A. gossypii* curling leaves of celery, while *M. volvuli* caused only white spotting. The foxglove aphid (*M. solani*) can cause very severe russetting on potato foliage with only slight infestations (Wave *et al.,* 1965).

By Coccidae

The Coccidae include a few species of mealybugs which cause plant malformation. *Pseudococcus citri* Risso collected from jack fruit (*Artocarpus heterophyllus* Lam.) killed the growing points of young cacao seedlings and caused adventitious budding. Single mealybugs caged on young leaves next to the growing point produced a witch's broom appearance of the bud. The leaves abscised after emerging to a length of 1–2 cm (Carter, 1949). Since *P. citri* was found on other cacao plants in the Rio Botanical Gardens that exhibited no evident symptoms, the experiment indicated that *P. citri* from the jack fruit had injected specific oral secretions.

A species of *Pseudococcus* was found associated with savoying of both terminal and stem leaves of kapok seedlings in Ceylon. Experimental cagings of the insect of healthy seedlings resulted in typical symptoms. New tissue produced after the insects were removed was normal in appearance (Carter, 1956).

Mealybugs cause an injury to strawberry plants which bears a strong resemblance to symptoms of a virus disease known as crinkle. Spotting is a symptom when the insect feeds on the young curled leaves, but mottling and dwarfing develop as the leaves grow older; the plants become

FIGURE 7.2. Distortion of guineacorn (*Sorghum vulgare*) caused by a mealybug, *Heterococcus nigeriensis* Williams, in Northern Nigeria (also see Figures 7.3 and 7.4.). Field symptoms. Plant on right shows an early stage of distortion; healthy plant of same age on left (Photograph: Harris, 1961).

flattened as a result of dwarfing and shortening of the petioles (Hildebrand, 1939).

Matsucoccus vexillorum Morrison and *M. bisetosus* Morrison, belonging to the closely related family Margarodidae, cause "flagging" or needle blight of pines. The effect is one of injury to twigs and branches, weakening of the crown, and deformation of the younger trees. There is

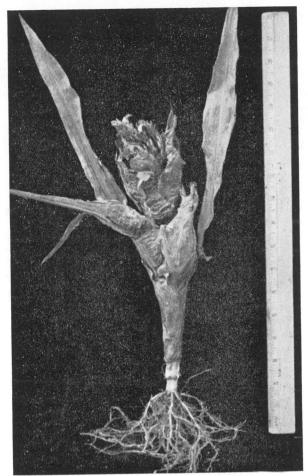

FIGURE 7.3. Distortion of maize, field symptoms. Plant at harvest time photographed beside an 18 in. rule (healthy plants were about 6 ft high). The oldest leaves are normal (Photograph: Harris, 1961).

abnormal formation of resin (McKenzie *et al.,* 1948; McKenzie, 1941).

Harris (1961) has reported a malformation of guineacorn caused by the feeding of *Heterococcus nigeriensis* Williams (Fig. 7.2–7.5), which is a striking example of the effects of toxic feeding. Symptoms appear very shortly after single insects have fed for a short time, although regular growth is resumed when the insects have been removed. An unusual feature is the occurrence of distortion symptoms on a number of other hosts. Evidently two strains of the mealybug are involved: one, a yellowish-pink form, feeds on the meristematic tissue and is responsible for distortion; the other strain, a gray form, feeds on mature, expanded leaves or normal

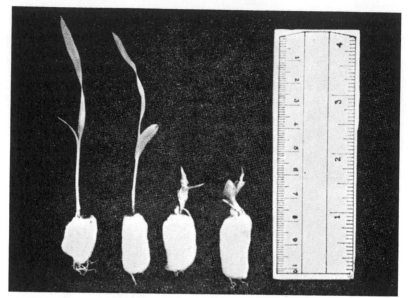

FIGURE 7.4. Distortion of guineacorn seedlings, experimental symptoms. Four 7-day-old seedlings (two controls on the left) photographed 6 days after single mealybugs had been placed on the two seedlings on the right (PFT of 24 hr) (Photograph: Harris, 1961).

tillers and does not normally cause distortion. There is an interesting parallel here with the gray and pink forms of the pineapple mealybug which were finally (after thirty years) separated taxonomically. Harris regards the effect of the insect's feeding as analogous to gall formation rather than to a toxic effect, on the grounds that no wilting or death of cells occurs. These latter criteria are not essential, and, in any case, the stimulus to gall formation is a toxicity even though the toxic entity may be hormonal.

By Thysanoptera

The rarity of reports on the toxic feeding of thrips is perhaps related to their limited capacity to feed in any but the mesophyll tissues (Wardle and Simpson, 1927), but Kratochvil and Farksy (1942) reported malformation of young shoots resulting from thrips feeding, a condition which had previously been attributed to mites, other insects, fungi, and viruses.

By Miridae

The Miridae have already been discussed at some length in Chapter 4, but many of the toxic effects of the feeding of this group involve more

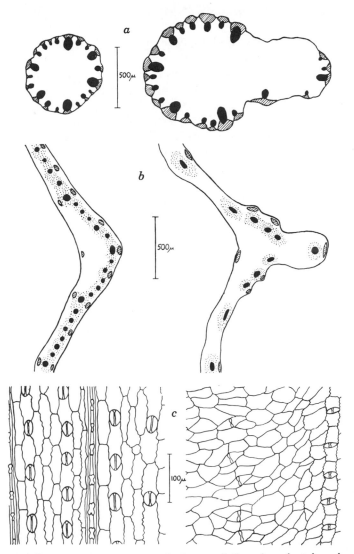

FIGURE 7.5. The anatomy of plants of *Cynodon dactylon* showing symptoms of distortion associated with the presence of a mealybug. *Heterococcus nigeriensis* Williams. *a,* Diagrammatic cross-sections of normal and distorted flower stems; *b,* diagrammatic cross-sections of normal and distorted leaves (midrib in center). *c,* lower epidermis of normal and distorted leaves (camera-lucida drawing). In all cases the normal material is on the left. In *a* and *b* the black areas represent vascular bundles, the shaded areas, sclerenchyma; and the dotted areas, tissue containing chloroplasts (After Harris, 1961).

than local lesions and secondary development of symptoms from them.

Lygus vosselri (Simoni) is a good example. This insect is a pest of cotton in Uganda and its feeding effects on the cotton plant have been described by Taylor (1945); he stocked cotton plants with 25 *Lygus* nymphs and repeated this at intervals of two weeks. His comparisons were then drawn between these artificially infested plants and naturally infested plants that were considered severely damaged. The differences between these two sets of plants, representing two types of infestation, were sufficiently pronounced to indicate that the effect was more severe in the case of the artificially infested plants from either the total population or the regular increments of fair-sized populations. These differences are shown in Table 7.1. The immediate reaction of tissues of *Lygus* attack is in contrast to the effects of prolonged infestation. Black spots appear within three hours after feeding, and the cells in these spots are dead. Prolonged infestation, particularly when repeated, results in symptoms which involve the whole plant and which are closely analogous to those resulting from hormonal growth-inhibiting substances.

Taylor concludes that the damage is primarily mechanical and secondarily toxic, but the extreme growth malformations he reports suggest that

TABLE 7.1.   Symptoms Resulting from Heavy Artificial Infestations Compared with Field Incidence of Lygus *Bugs on Cotton* [a]

Heavily infested at intervals	Field incidence considered severe
Columnar in shape, owing to shortness of all branches	Conical; the lower branches, including the lower sympodia, being much longer
Very ragged in appearance	Much more normal
Primary sympodia averaged 2.1 nodes	Primary sympodia averaged 3.1 nodes
Internodes on main stem and branches often very short so that leaves bunched	Internodes mostly normal
Many abnormal secondary, tertiary, and auxillary branches, all weak and useless	Comparatively few abnormal branches
Number of bolls that persisted was 6 or 4 each	Number of bolls that persisted was 21 each
All bolls unhealthy and useless; crop completely destroyed	At least 10 bolls per plant appeared healthy and likely to mature

[a] After Taylor (1945).

mechanical damage by stylet probing simply permits the rapid diffusion of the toxic secretions. This is perhaps characteristic, to a greater or lesser degree, of mirid damage to plants.

A mechanical localized damage to alfalfa buds is caused by the feeding punctures and lacerations of *Lygus,* but the rapid disintegration of buds following the traumatic injury is believed to be due to toxic secretions. Rosetting, characterized by development of racemes of buds near the tips of main stems and branches into disk-like clusters, follows heavy *Lygus* infestations. On the vegetative buds there is retardation of growth and a tendency to extensive branching. Evidence of recovery becomes noticeable within ten days after removal of the insects (Carlson, 1940). *Lygus hesperus* Kgt., feeding on guayule, reduces both the weight and viability of the seed; when feeding occurs on the current season's shoots, both subsequent growth and flowering are inhibited (Romney *et al.,* 1945). Anatomical studies on plants infested with 25 of the insects showed that branch tips were withered and gray, with the stem apex and surrounding young leaves severely shrunken. There was complete collapse of tissues near the stem apex; below, areas of collapsed cells were interspersed with areas of intact cells (Addicott and Romney, 1950). Other references to *Lygus* and other mirids in relation to seed production are to be found in Sorensen (1946); Flemion *et al.* (1949); Flemion and Olson (1950); Kho and Braak (1956).

Lygus bug injury to alfalfa has been studied quantitatively by MacLeod and Jeppson (1942) and histologically by Jeppson and MacLeod (1946). They reported that although disintegration of the tissue was limited to a small area around the stylet punctures, there were large areas of cell disintegration in terminal and lateral bud primordia, indicating retardation of growth of the lateral bud in the process of substitution for the injured terminal region.

Adelphocorus lineolatus (Georze), the alfalfa plant bug, induces the same type of damage as does the potato leafhopper, and differences in symptoms cannot be visually distinguished. Although it should be possible to develop alfalfa varieties with resistance to alfalfa bug, since there are clonal differences in damage sustained, it will be essential to distinguish damage by the two species of insect if resistant varieties to either pest are to be developed. It is probable that injury previously attributed to leafhoppers may be due in part to the alfalfa plant bug (Radcliffe and Barnes, 1970).

Citrus false exanthema is induced by the feeding of the mirid *Platytylus bicolor* (Lep. & Serv.) which causes blisters on the bark, gum exudations, and open lesions in leaves, with secondary chlorosis and leaf drop. A single nymph or adult will induce 10 lesions in 24 hours (Muller and Coster, 1965).

By Lygeidae

Nysius raphanus Howard increases to epizootic proportions on weed hosts and migrates into cultivated fields when the weeds dry. On cotton, dried scorched appearance of stems, petioles, and leaves results from their feeding (Tappan, 1970). Leigh (1961) described similar symptoms on a long list of alternate host plants. The insect has assumed pest status in southern California vineyards, and Barnes (1970) described three reasons: (1) the introduction of London rocket (*Sisymbrium irio* L, a cruciferous weed which is favored by current cultivation methods) (2) the strong attraction of the weed to overwintering adults, and (3) the exhaustion or destruction of the weed leading to migration.

Control is achieved by timing of cultivation to destroy weeds before grape vines leaf out. If the weed persists after leaf-out, they should be sprayed with Malathion before destroying the weeds. The feeding of the insect results in rapid wilting, suggesting a toxin; nymphs are more toxic than adults.

By Coreidae

The coreid, *Ambypelta lutescens* Dist., has already been referred to in Chapter 4 in connection with fruit spotting; but on papaya very small populations, one or two per tree, will distort the growing points within a month after feeding has started. Suppression of normal growth results in a dense bunch of leaves with short, distorted stalks and very short internodes. Nymphs appear to be more injurious than adults but this may be a matter of the relative mobility of the two forms (Sloan, 1946).

By Cercopidae

Prosapia bicinta (Say), the two-lined spittle bug, causes a phytotoxemia of coastal Bermuda grass (a cultivar of *Cynodon dactylon*) characterized by stippling and streaking of the leaves, and by browning and death of the entire aerial portion of the plant. Although both nymphs and adults feed in the xylem, only adult feeding induces the symptoms. A single adult feeding for one hour will induce symptoms; their intensity is correlated with the degree of infestation and the length of the feeding time. There is evidence that the toxin is translocated (Byers and Wells, 1966). The same insect has been found to seriously damage sweet corn. The damage is proportionate to numbers of insects used in cage experiments to verify field observations (Janes, 1971).

By Membracidae

The buffalo treehopper, *Ceresa bubalus* Fabr., attacks vines in Italy, occasionally resulting in successive downward leafroll. Necrosis of both phloem and cambium occurs (Vidano, 1967) (Fig. 7.6).

By Pentatomidae

The pentatomids, *Antestia* spp., cause a drop of green berries of coffee which is a response on the part of the plant to interruption of normal growth caused by the insect's feeding. Other injuries are scarring and distortion of the leaves, nonsetting of flowers, and multiple branching which may result in bunchy and matted growth (LePelley, 1942).

By Cicadellidae

The Cicadellidae, particularly species of the genus *Empoasca,* are notorious for their capacity to induce plant malformation as a result of their feeding. The most notable example is that of hopperburn, primarily associated with the feeding of *Empoasca fabae* Harris. The disease has been known since 1895, when L. R. Jones described it as tipburn, ascribing it to conditions of high temperature, bright sunlight, and low humidity. Osborn (1896) described the injury on potato due to *E. fabae,* but it was not until Ball (1918) established the relationship between the insect and disease that the situation was clarified. As Leach (1940) quite appropriately notes, Ball's findings focused attention on the significance of the toxicogenic insect in inducing plant disease (Fig. 7.7).

The symptoms of hopperburn on potatoes were described by Ball (1918, 1919), Parrott and Olmstead (1920), and Leach (1922); it is from these authors that the following has been summarized. The first sign is the appearance of brownish areas at the tip of the leaf and occasionally on the margins of the leaflets. These areas coalesce as the burning progresses until the entire margin of the leaf is brown and more or less curled. The burned margin increases in width until only a narrow strip along the midrib remains. In severe cases all the leaflets curl and dry up, the petioles wither, and defoliation occurs with only slight disturbance of the plant.

While these symptoms are those of "burn," they are not actually the first symptoms. As Leach has pointed out, the first symptom consists of shortening of the petioles with a crowding of leaflets, and this symptom precedes that of burn. The burn symptoms develop very rapidly. Even in

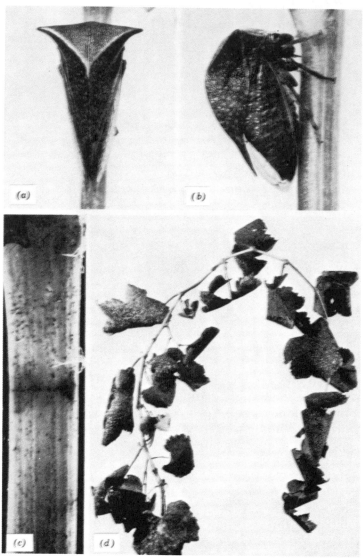

FIGURE 7.6. A–B, Adult buffalo treehopper, *Cereas bubalus* Fabr.; C, traces of feeding punctures in grapevine shoot; D, caulinar alterations and successive downward leafroll (Photograph: Carlo Vidano).

a   b

FIGURE 7.7. Hopperburn of potato. *a*, Initial burn symptoms. *b*,
Advanced burn symptoms (Photographs: A. G. Peterson).

the partial shade of a cage, leafhoppers can burn and roll the leaves in
three days. A single leafhopper can destroy a leaf.

Ball also recorded similar damage to other hosts of the same insect. On
dahlias, marginal burning and distortion of the midribs and veinlets
occur as on potatoes. The water sprouts and fast-growing tips of box
elder trees and other nursery stock are similarly affected, as are the upper
leaves of fast-growing raspberry canes. Other workers have noted additional symptom manifestations on other crops and have used other
names, e.g., alfalfa yellow, to designate them, but hopperburn seems to
be generally applicable and should be used (Poos and Westover, 1934).

On apples, discoloration of the leaf margins is followed by curling of
the tips and distal margins of the leaves. On legumes, yellowing and
dwarfing are additional symptoms; the number of flowers is reduced, and
the root system is weakened. On alfalfa, there is mottling and blotching,
fusing to form striated discolorations between lateral veins; deep yellow
to orange discoloration with bronzing of leaf edges; stunting and shortening of internodes; and rosetting and proliferation of new dwarfed shoots
(Granovsky, 1928; Hollowell *et al.,* 1967; Jones and Granovsky, 1927;
Monteith, 1968).

On eggplant, there is a distinct cupping of the younger leaves with the margins turning light green and yellowish; further cuppings result in extreme irregularity of the leaves. "Peanut pouts" was the term used by Metcalf (1937) to designate *E. fabae* injury on peanuts, although Poos (1941) ascribes it to thrips. The nature of the injury has been investi gated rather thoroughly but complete agreement has not been achieved. Symptoms, resembling those of hopperburn, have been produced by the injection of macerated adults and nymphs (Fenton and Ressler, 1921, 1922; Eyer, 1922). The influence of needle wounds was avoided by Fenton and Ressler when they conducted a potometer experiment with positive results. Granovsky (1926) used macerated insects and sap from injured leaves in injection experiments on apple, with positive results. All these workers ascribed the damage to toxic substances injected by the leafhopper.

Carbohydrate accumulation was shown to occur in potato foilage by Granovsky (1930); he considered that enzymic secretions by the insect were responsible for the disorganization of the phloem of affected tissues. Johnson (1934, 1938) also observed the phenomenon, but he stated that it was due to the clogging of the food-conducting elements and consequent improper translocation.

Feeding by *E. fabae* on potatoes increases the respiration rate of the plant, and reduces both true and apparent photosynthesis within 24 hours after feeding starts. Reduction in photosynthesis is quantitatively related to the length of the feeding period and the developmental stage of the insect (Ladd and Rawlins, 1965). The effects of *E. fabae* on rose are shown in Fig. 7.8.

Medler (1941) compared the histological effects of the feeding punctures of *E. fabae* with two other leafhoppers, neither of which produces the symptoms of hopperburn on alfalfa. With *E. fabae,* there was hypertrophy of affected cells characterized by nuclear enlargement and prominent safranin staining nucleoli; with the other two species, neither was present. External symptoms of chlorosis and reddening, noted by previous workers and ascribed to interference with translocation, were shown to be secondary symptoms.

Since the insect is a phloem feeder and deposts sheath material, it is possible that secondary symptoms are conditioned by interference with translocation. As Heriot (1944) points out, sheath material is left behind when the insect moves to another feeding site. However, Leach's earlier observation was that the first external symptom is a shortening of the petioles with a crowding of leaflets. The positive results with insect macerations and the conclusions of Medler would seem to throw the weight of evidence on the insect's toxicogenicity as the cause of primary symptoms.

There is agreement as to the relative ability of the various stages of the

FIGURE 7.8. Injury on Butterfly rose by *Empoasca fabae* (Harris). Tip killed, lower leaves wilted, stem with brownish necrotic patches. New lateral developed after removal of leafhoppers (Photograph: F. F. Smith).

insect to produce symptoms. All nymphal stages produce the disease, but the first and second instars produce no effect in large numbers; the older the nymph, the more injurious it is (Fenton, 1921). The last three instars are more toxic than either the first or second, or the adults. It would be difficult to account for greater interferences with translocation by an older nymph than by an adult, but in fact the adults are not as effective as the older nymphs.

Injury is in direct proportion to the number of leafhoppers feeding. Peterson and Granovsky (1950) showed these quantitative relationships between number of *E. fabae* and potato yields. Nymphal density and per-cent of hopperburn can be expressed as a straight line relationship on a logarithmic scale. Relatively great potato-yield reductions resulted from low leafhopper densities and with further increases in leafhopper popula-tions the yield reductions became proportionately less great.

There are varietal differences in susceptibility to the effects of *E. fabae* attack. McFarlane (1942) showed that the early maturing varieties in gen-eral were not susceptible to hopperburn, but stage of maturity is not the effective factor. Cross of resistant crossed with resistant produced about 90% of the seedling progeny in the lowest damage class, and segregation of susceptible and resistant was clear in the F generation.

Peterson and Granovsky (1950a) compared the relative feeding effects of *E. fabae* on resistant and susceptible potato varieties. Varietal differ-ences in reaction to leafhopper feeding may involve both morphological and physiological characters. The resistant variety Sequoia has thicker-walled collenchyma, and its phloem is more extensive and is often at a greater distance from the lower epidermis, thus reducing the interference with translocation and the accumulations of carbohydrates as compared with the susceptible Cobbler. There was also evidence of a varietal differ-ence in physiological response to the toxin injected by the leafhopper during the course of feeding.

Sleesman (1940) noted differences in susceptibility to attack by *E. fabae* among 12 species of *Solanum* in Ohio. Four of these species were highly resistant, if not immune. McFarlane and Rieman (1943) tested 27 varie-ties of snap and lima beans for resistance to *E. fabae*. In general, the early maturing varieties were more susceptible and the late maturing ones tended to be resistant.

Other species of *Empoasca* are also involved in hopperburn, although at the time of Ball's work no other species were implicated. *E. libyca* de Bergevin is a phloem feeder and is responsible for hopperburn on cotton in the Sudan. The first symptoms on cotton appear as a paling of the leaf edges. "This rapidly extends to the tissue between the main veins, pro-ducing reddening and yellowing and eventual death of the leaf tissue from the periphery inwards, as if scorched. When the attack is heavy the young leaves fail to expand fully and remain small and curled with brown, dry margins. Growth is brought to a premature standstill and the apical buds and young bolls are shed" (Cowland, 1947). The same species has been reported by Klein (1948) as damaging a wide variety of plants in Palestine. His description of symptoms leaves no doubt that they are typical of hopperburning species.

Empoasca libyca has also been reported from vines in Italy (Vidano, 1967), the attack of the insect causing vein browning, downward rolling and marginal burning (Fig. 7.9). *E. flavescens* Fab. prevented the estab-

FIGURE 7.9. A. *Empoasca libyca* Bergevin; B. Grapevine leaves showing vein browning; C. Showing downward rolling and marginal burning (Photograph: Carlo Vidano).

lishment of cotton growing in Luzon. The species caused wilting and drying of the leaves, and infested plants appeared stunted and unthrifty (Cendena and Baltazar, 1947). Other host plants were less susceptible to damage. However, this same species is also toxic to castor bean. The leaf symptoms are characteristic of hopperburn and there is poor capsule formation during the population increase period which is between November and January (Jayaraj and Basheer, 1964). In studies of resistance, Jayaraj (1966, 1967) found that the leafhopper infestation increased the respiration of the susceptible varieties more than it did that of the tolerant and resistant varieties; the increase in the susceptibles was linked with increased accumulation of sucrose, total nitrogen, and total amino acids. The susceptibles also developed higher populations of the insect.

E. flavescens is also known as a pest of vines in Italy. It causes vein browning and interveinal chlorosis (Vidano, 1967).

Empoasca elongatus DeL. causes die-back of *Quercus ilex* (holly oak). Feeding on midribs or the large veins of young leaves results in necrotic spotting and death of the youngest growth. Stem blisters with a necrotic core develop if the shoots are 2–3 in. long before being attacked, and a gnarled appearance and splitting follow. The plant is not a breeding host and the insect is a night feeder; its breeding hosts are unknown (Loos, 1965).

In Hawaii, *E. solani* causes typical hopperburns on lettuce and celtuce. Martin and Pemberton (1942) concluded from experimental evidence that a toxic secretion was involved. The degree of injury varied directly with the size of the insect population, and the plants recovered after the insects were removed. On the basis of symptoms developing in cages containing watermelon and castor bean plants, the same species caused hopperburn on these plants (Carter, 1939) (Fig. 7.10).

It appears safe to generalize that phloem-feeding *Empoasca* spp. are responsible for hopperburn; the mesophyll-feeding species are not.

Control of *E. fabae* is essential in potato culture since the degree of hopperburn damage varies directly with the populations of the insect. Older methods have been superseded by use of the newer organic insecticides.

The chlorinated hydrocarbons appear to be favored currently. DDT was naturally the first one used; Granovsky (1944, 1944a) reported effective control and greatly increased potato yields following its use. This was confirmed by many later studies, the reduced populations of *E. fabae* and the correlated higher yields giving striking confirming evidence of the importance of the insect as a potato pest (Peterson and Granovsky, 1950). Medler and Fisher (1953) reported yield increases of alfalfa of from 30 to 300% following the use of methoxy-DDT (methoxychlor) at the

FIGURE 7.10. Hopperburn of cucurbits. An advanced symptom on watermelon following the feeding of *Empoasca solani* (Carter, 1939).

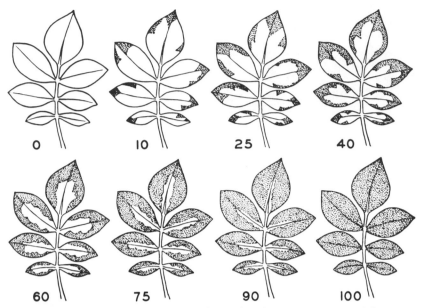

FIGURE 7.11. Diagrammatic chart showing percentages of hopperburn on potato leaves (After Granovsky and Peterson, 1954).

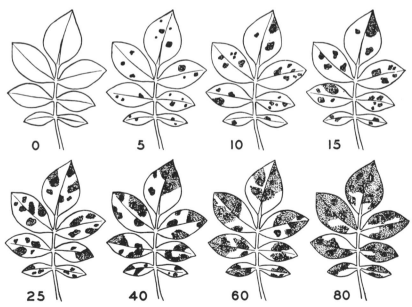

FIGURE 7.12. Diagrammatic chart showing percentages of early blight injury on potato leaves (After Granovsky and Peterson, 1954).

rate of 1 lb technical/ac applied in low volume. Ethyl-DDT and methoxy-DDT gave 88 and 80% control, respectively, of *E. fabae* on snap beans. These were applied as dusts of 25 lb of 5% material/ac (Brett and Brubaker, 1956). However, DDT and other chlorinated hydrocarbons have been outlawed in some areas, perhaps without an adequate balance having been struck between their disadvantages and advantages. In the meantime, the growing of affected crops in North America will become more difficult and more expensive to produce.

Evaluation of injury caused by leafhoppers on potato, in the presence of other factors, calls for the development of quantitative methods of diagnosis if insecticide programs are to be properly evaluated. Granovsky and Peterson (1954) essayed this on the basis of leaf-symptom estimation of damage by potato flea beetles, potato leafhopper, and early blight of potatoes. Diagrammatic charts for two of these are shown in Fig. 7.11 and 7.12.

By Diptera

Dipterous insects, apart from the gall formers, are not frequently associated with plant malformations, but the midge *Contarinia nasturtii* Kieff, which feeds in the larval stage on many crucifers, causes a twisting of the heart leaves of young cabbage plants. Frickhinger (1943) ascribes the damage to a checking of the young growth where the larvae feed, and also probably to the secretion of toxins by the larvae.

By Acarina

The Acarina, particularly the Eriophyidae, includes many species, apart from the gall formers, which cause plant malformation. The blistering of apples and pears is the result of feeding by *Eriophyes pyri* Pgst. Typical blisters appear first on the leaves as pale greenish spots, which later turn brown and fall off. Injury to fruit is generally outgrown (Venables and Heriot, 1937) (Fig. 7.13 and 7.14). The blisters are caused by enlargement of the loose mesophyll cells which expel intercellular air, thus closing the stomata and raising the epidermis from the underlying tissue (Heriot, 1935). Wilson (1915) states that *E. pyri* oviposits in parenchyma and that the larvae extend the cavity in all directions, resulting in a mass of blisters over the whole leaf.

Killing of fruit buds in winter, weakened flowers, and resulting misshapen fruit, due to a species of blister mite morphologically indistinguishable from *E. pyri,* but differing from it in biology and the damage it causes, has been described by Borden (1932).

FIGURE 7.13. Blister mite injury to pear tree foliage (Photograph: H. F. Madsen).

Blister mite on cotton (*E. gossypii* Banks) is a major pest in some places. Injury is seen in crumpled leaves, distorted growth, and lack of fruiting branches (Fife, 1937). The development of types of sea island cotton immune to *E. gossypii* was reported on as long ago as 1919 (Harland). F_2 generations segregated into immune and nonimmune; in the F_3 generation, the immune bred true, and the nonimmune segregated further into immune and nonimmune.

The blackberry mite (*Aceria essigi* Hassan) causes the disease known as redberry disease of the blackberry (Essig, 1925), which is characterized by the presence of berries that may be entirely or partly red and that never ripen. The disease is caused by the mites entering the flowers and feeding

FIGURE 7.14.   Blister mite of pear (Photograph: H. F. Madsen).

near the bases of the immature droplets (Hanson, 1930, 1933; Mote and Wilcox, 1931). The same species of mite has since been reported as the causal agent of the same disease in New Zealand, where it has become serious since 1946 (Hamilton, 1949).

Big bud of black currants, not to be confused with "reversion" disease, is caused by *E. ribis* Nal., which stimulates the growing point to irregular development (Lees, 1918). On red currant big buds are not produced but instead a somewhat swollen bud, or a dense growth of buds, which do not develop normally (Massee, 1928).

The effect of the mite *Phytopus* (*Eriophyes*) *ribis* (Nalapa) on black currant bushes has been clarified by Thresh (1965), thus resolving the confusion caused by the combined action of mites *per se* and reversion virus disease. Infested apical buds become rounded and produce characteristically malformed leaves; infestation of axillary buds is only recognizable by microscopic examination. Virus-infected bushes provide a more favorable environment for mites (Thresh, 1964).

Somewhat similar symptoms follow the attack of *A. mangiferae* Taher Sayed on mango buds. The inflorescence becomes rounded in form, and, along with terminal buds, dies. The terminal buds are replaced by lateral ones which also become infested so that the stem appears stunted and deformed (Hassan, 1944; Taher Sayed, 1946). A related eriophyid mite, *Vasates mangiferae* Attiah, produces the same type of symptom on mango terminal buds and inflorescences, but there is no production of laterals such as occurs after attack by *A. mangiferae* (Attiah, 1955).

Red leaf of grape vines in California is controlled by controlling the mite *Tetranychus pacificus* McG. The disease is characterized by development of bronze-red color in the leaf tissues and partial to complete defoliation. Fruit often fails to mature and may sunburn, or the berries may shrivel up beginning at the tips of the clusters. Control measures must be instituted before the injury is far advanced; if it is left until red coloring has developed, further development of the injury cannot be checked (Jacob *et al.*, 1941).

Symptoms closely resembling those produced by viruses on garlic have been shown by Smalley (1956) to be the result of feeding by an eriophyid mite, *Aceria tulipae* Keifer. Early symptoms include twisting and streaking of the leaves and dwarfing of the plants. Frequently large areas of the leaves become devoid of chlorophyll in marked contrast to the mild streak of garlic mosaic. Later, a tangled mass of interlooped leaves develops (Fig. 7.15).

The separation of the mite feeding symptoms from those of virus disease on peaches presented some difficulty to Wilson and Cochran (1952) when they demonstrated that *Vasates cornutus* (Banks) produced spotting and chlorosis of vein-bordering tissue of peach foliage. The question of a virus involvement was settled by means of index grafting and by reinfesting seedlings which (1) had been previously infested, (2) had shown symptoms, and (3) had produced clean, symptom-free tissue following the removal of the mites. The second infestation produced a second cycle of symptoms.

The probable relationship between mite feeding and phyllody, or the change of floral organs into vegetative ones, was suggested by Breakey and Batchelor (1950), who observed stunting of the internodes and phyllody in chrysanthemum plants infested with *Paraphytopus chrysanthemi* Keifer. Proeseler (1968) differentiated virus-like symptoms from injuries due to gall mite feeding from those due to virus infection.

Eriophyid mites (*Aceria medicaginis* Sayed) induce a virus-like witch's broom proliferation of lucerne. Stubbs and Meagher (1965) consider this due solely to the mites' feeding and no virus is involved. Recovery takes place when a miticide (Thimetron) is used, and symptom severity is directly related to density of the mite population. It is interesting to note that an earlier worker (Edwards, 1935) reported graft transmission of the disease. Stubbs and Meagher suggest that witch's broom and big bud might have been confused, but it is also possible that Edwards material had not been completely cleaned of mites, since in 1935 methods for achieving this were not highly developed. A new disease of barley and other cereals resulted from infestations of *Tetranychus* mites and was ascribed to the toxic effects of the mites' feeding (Wallace and Sinha, 1961). Extreme malformation of Douglas fir seedlings occurs as a result of eriophyid mite feeding (Lavender *et al.*, 1967).

The wheat curl mite (*Aceria tulipae* Keifer), vector of wheat streak

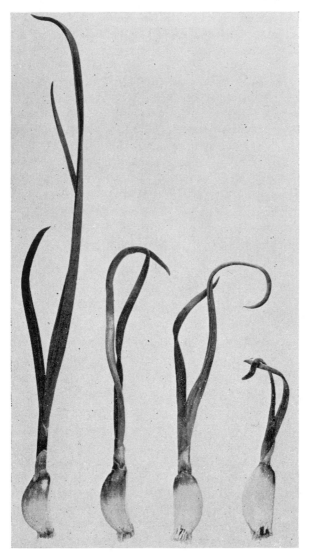

FIGURE 7.15. Mite injury to garlic (Photograph: Smalley, 1956).

mosaic (WSMV), is responsible for a phytotoxemia known as red kernel disease of corn. The disease is found in areas where WSMV does not occur. WSMV infection may, however, predispose the plant to heavier mite infestations (Nault *et al.,* 1967) as has been clearly shown in the case of reversion disease of black currant (Thresh, 1964). Yew trees have a reputation for growing slowly but this in fact may be referable to mite infestation. There is no virus known to be involved, and big bud and dry bud, with severe stunting of terminal buds and excessive lateral growth, occur on yew trees as a result of infestation by *Cecidophyopsis psilaspis* (Nal.) (Hodgman, 1966). Slykhuis *et al.* (1968) confirmed the relationship between red kernel disease of corn and *Aceria tulipae* (K) from studies in Ontario.

Tetranychus sinhai Baker causes a syndrome resembling net blotch of barley. The leaves darken, become yellow, and wilt from the bend of the leaf to the tip. Symptom severity parallels the size of the mite populations. Of 165 varieties of barley, 15 were resistant, 15 susceptible; the balance were intermediate in susceptibility. In general, the resistants were those of arid area origin (Sinha and Wallace, 1963).

The citrus bud mite, *E. sheldoni* Ewing, is known in southern California, Queensland, New South Wales, and Hawaii. Blasted buds, multiple buds, bunched growth, and deformed twigs, leaves, blossoms, and fruits result from injury to the buds. Blackening of the rind tissue beneath the button of the fruit results from their feeding in that location. Lemon is the more seriously affected (Boyce and Korsmeier, 1941; Caldwell, 1945; Di Martino, 1952). An unidentified species causes the same types of damage in Sicily.

INSECT GALLS

The insect galls, or zoocecidia, are the oldest known toxic effect of insect feeding on plants. They were recognized in ancient times and used for medicinal purposes and in the dyeing industry. They have also attracted the attention of systematists who have been concerned with the classification of galls and of the gall-forming insects. Gall producers occur in many different insect orders with the largest number in the families Itonididae and Cynipidae. There are a large number to be found among the mites, chiefly in the genus *Aceria* (*Eriophyes*). Systematists have classified very large numbers. From the Dutch East Indies 1536 separate and distinct galls are listed (Docters, 1926); from Central and South America, a single paper lists 1341 galls in 98 plant families with only 229 of the insect species known (Houard, 1933). Wells (1916) described 17 distinct galls from a single species of plant (*Celtis occidentalis* L). A single insect family may contain gall formers causing bud galls, flat circular galls on leaf laminae, galls folding or rolling the leaf, and stem galls (Burkill, 1930). Felt described over 400 species of Cecidomyidae forming galls on nearly 200 genera of plants (1913–1925). The fundamental biological problems involved, however, are concerned with the origin, nature, site of action, and the effect of the causal stimulus.

Origin of the Stimulus

The origin of the stimulus to gall formation has been the subject of speculation since early times. Malpighi (1687) suggested that it was of animal origin. Since then, and up to the end of the nineteenth century, four main hypotheses had been presented: (1) the injection of a stimulating fluid at the time of oviposition, (2) the physical irritation set up by the presence of a foreign body, (3) the secretion of an active saliva, and (4) the excretion of metabolic products. These were effectively analyzed by Plumb (1953), from whose paper this discussion is largely drawn.

The initial stimulus to gall production could come from the ovipositing insect injecting the stimulating material along with the egg, but for the complete development of the gall, a continuing stimulus is necessary. Mere puncture, either by ovipositor or stylets, will not stimulate a typical growth. As Horsfall (1923) has pointed out, many species of aphid infesting plants that are susceptible to galling do not produce zoocecidia. The eggs of some gall-producing insects are laid on the leaf, and gall formation does not begin until the larva has hatched and migrated to the place of gall formation.

Adler (1881) and Beijerinck (1883) are credited with establishing the chemical nature of the stimulus. Laboulbene (1892) was essentially calling attention to the specificity of this material when he wrote: "The same cause stimulates whether the liquid is secreted by the cells of a gland opening itself into the mouth with or without organs of suction, or it oozes through the very walls of the body of a larva, a helminth, or finally, whether it is provided by a galligenous bacterium. It is, therefore, neither a puncture, nor an incision, nor a foreign body which is able to produce a lasting plant outgrowth, a tree gall; it is the soluble material, elaborated by the animal or plant cells, and these liquid substances have a special action, necessary, indispensable."

Experimental injections of the dried Malpighian tubules, and of water in which gall-forming larvae were crushed, or of fragments of the same larvae, have given some indications of the process of gall forming (Laboulbene, 1892; Rossig, 1904).

The excrement of *Pontania salicis* Christ, is capable of producing cell division (Kuster, 1911) as is that of *P. pomum* Walsh (Cosens, 1912). LaRue (1935, 1937) observed outgrowths around the fecal pellets which were in contact with the mesophyll in poplar leaves tunneled by leaf-mining larvae. Injection of 0.00005% heteroauxin in the mesophyll induced a mass of cell outgrowths twice as thick as the original leaf.

Plumb (1953) injected extracts of *Adelges abietis* L, a gall former, on Norway spruce in the spring about the time of swelling of the buds, and caused the initiation of gall formation in five buds. The transfer of gallicolae-producing eggs to the injected buds permitted the completion of gall formation in one case.

Parr (1939) injected extracts of salivary glands of *Matsucoccus gallicolus* Morrison into pine needles and induced symptoms typical of the actual damage done by the insect. Another gall-making coccid, *Asterolecanium variolosum* Ratz., was studied by the same author (Parr, 1940). Using the same technique of macerating salivary glands and injecting them by means of a microinjection bulb, similar results were obtained.

Nature of the Stimulus

As to the real nature of the stimulus, Plumb quotes Wells (1920): "I cannot find a single structural fact produced by any one in the study of the initial stages of Hemiptera galls that throws any light whatever upon the profound problem of the nature of the highly specific stimulus applied by the insect to the embryonic plant tissue."

Whatever may be the nature of the stimulating material, stimulation is nevertheless a cardinal principle in gall production, and the production of plant galls is dependent on stimulating of plant cells in a meristematic or plastic condition. Following Kuster's classification (1903) they can be separated into two main classes: (1) the kataplasias, or those forms of heteroplasias whose cells and tissues did not vary widely from the normal; and (2) the prosoplasias, or those forms of heteroplasias whose cells, and particularly whose tissue forms, differ fundamentally from those of the normal parts. According to Cook (1923), all insect galls arise from the meristematic cells and are kataplasias first, many becoming prosoplasias with fibrous and sclerenchymatous tissues prominent. Cell activity ceases when the gall larvae reach maturity and in this respect differ from fungal, slime mold, and bacterial galls.

Boysen-Jensen (1948) concluded from experiments on larval secretions that they had two effects, an embryonic growth with cell division, and cell elongation, and that these effects were similar to those induced by growth hormones.

Nysterakis (1946) went further with the suggestion that root and leaf galls due to *Phylloxera* feeding are the result of the injection of indole-3-acetic acid. He also ascribed the resistance of American vines and their hybrids to their relatively low sensitivity to the heteroauxin secreted by the insect. Modification of this sensitivity of environmental conditions probably accounts for variable results recorded.

Martin (1942) injected extracts of two species of leafhopper and a mealybug into sugarcane and induced stem gall formation. Since none of these insects is capable of inducing these symptoms when fed on the cane, the presumption is that growth-promoting substances were freed from internal organs not intimately connected with the salivary glands. Arrillaga (1949) obtained very similar stem galls by the injection of indole acetic acid, 2,4-D, and naphthalene acetic acid.

Weidner (1950) suggested that the stimulating substance may be regarded as an antiauxin activating the auxins of the plant cells.

Lewis and Walton (1947, 1958), working with the aphid *Harmaphis hamamelidis* Fitch, concluded that the stimulating material was separate from the salivary secretion and is injected by a different process. Their later paper contains the unique suggestion that the cecidogen is to be found in many organs and cells of the insect—salivary glands, cross-cells, mycetomes, symbionts, and germ cells—and is transmitted through the egg. It can be recognized by staining reactions, by its structure, and by its diagnostic crystals, and is identical in plant and insect. The suggestion that the cecidogen is a virus will no doubt be received with some reserve. If the cecidogen is transmitted through the egg, then all stages of the insect should be capable of gall making, but only the stem mother is able to do so. Again, repeated stinging is necessary for normal gall formation, a difficult thing to reconcile with the concept of a multiplying virus. Furthermore, if it is a virus, then the cecidogen is probably the largest virus known and with a unique developmental cycle, capable of being clearly illustrated at magnifications of 300–660.

Anders (1957) considered that amino acids provided the stimulus to gall formation, and this opinion reduced interest in IAA. He challenged the hypothesis that IAA could not be found in the saliva. As Miles (1968) points out, however, Anders' standard of what constituted adequate concentrations was based on quantities necessary when applied to plants externally, and he ignored the possibility that IAA can be produced in the saliva after injection into the plant. Miles (1968a) was able to show that IAA was produced during salivation of a plant bug, and a normally noncecidogenic insect could produce galls temporarily when injected with excessive amounts of metabolic precursors of IAA. Miles' conclusion was that IAA could be a universal cause of cecidogenesis.

The specificity of gall structure gave Miles a little trouble. He reconciled specificity and the universality of IAA as the stimulating agent by suggesting two modifying factors: (1) the type and stage of development of the plant tissue attacked, and (2) the behavior of the insect once feeding has begun. However, when several gall formers attack the same plant species, the second modifying factor is likely to be determinative.

The sawfly *Pontania pacifica* appears to contain low molecular weight materials in its accessory glands, and these promote gall production. Whether those compounds are also responsible for gall initiation is left as an open question. The most rapid and sustained growth of galls was obtained when a mixture of indole-3-acetic acid (10/mg/1), adenine (50 mg/1), and kineton (10 mg/1) was injected (McCalla *et al.*, 1962).

Japiella medicaginis, the cecidonyid gall midge, causes simple leaf galls, and microscopic examination showed that the layer of cutin secreted by the epidermal cells of the plant begins to disappear once the larva has hatched and is soon lost from all the epidermal cells lining the larval chamber. This destruction of the cuticular layer presumably makes the epidermis permeable to the passage of solutes (Heath, 1961).

Effect of Causal Stimulus

Parr (1940) has described the formation of the gall of *A. variolosum*. The sequence of events is as follows. The cells through which the stylets pass, or those immediately surrounding them, are either killed immediately by the salivary fluids or by extraction of the cell contents. They may be inhibited from proliferation if not killed. Collenchyma of the twig is stimulated and begins to proliferate. The gall grows in size as the larva grows. The plant tissues surrounding the parasite proliferate radially by both hyperplastic and hypertrophic activity. In the vicinity of Rio de Janeiro, *A. pustulans* Ckll. behaves similarly. Parr (1939) also described the gall-forming process by a coccid, *Matsucoccus gallicolus* Morrison. Collapse and disintegration of plant tissue around the base of the stylets is followed by the formation of a depression into which the first instar of the insect feeds. The second instar becomes completely covered by plant tissue, which forms a gall.

Gall formation stimulated by the eriophyids shows a very intimate mite-host relationship. Keifer (1946) has reviewed the economic species and described the damage. This includes warty deformations, leaf blisters, hairy deformations, terminal bud galls, formation of erineum patches on the leaf endodermis, stunting and deformation of leaves, formation of pouch galls on upper leaf surfaces, yellowing and dropping of pine needles, formation of woody galls around buds, conspicuous cell-like galls on leaves, and hairy growths on petioles.

Galls on economic plants are relatively few in number compared with the number described; some are useful articles of commerce (Wiesner, 1927). The mite galls, as Keifer has shown, are of considerable economic importance. More recently, Nishida and Holdaway (1955) have described the damage to *Litchi chinensis* Sonn. caused by *Aceria* (*Eriophyes*) *litchii* Keifer. This insect produces a typical "felt" gall on the lower surfaces of the young leaves and a general curling of the foliage. The most severe damage occurs when the tree is in the flush of new growth. Control measures must begin at the beginning of the new flush and be continued for the growth period. Wettable sulfur at 5 lb/100 U.S. gal applied five times during the year gave effective control.

Knorr and Denmark (1970) reported an injury to citrus by the mite *Brevipalpus phoenicis* (Geijskes). One aspect of this injury is designated as *Brevipalpus* gall. It occurs in the field only after the citrus have been defoliated previously by cold, drought, or disease. Duplicating this condition in the laboratory, it was found that noninfested plants produced normal shoot growth, whereas infested plants showed complete growth suppression. The repeated killing of emerging shoots leads to a cushion gall formation. The condition is of economic importance in Venezuela where 60% destruction of seedlings has been reported (Fig. 7.16).

The woolly apple aphid has already been referred to in connection with perennial canker (see Chapter 3). The warty galls are produced by stimulation of the cambium layer and the dissolution of the middle lamella, which in the sclerenchyma results in the formation of separated medullary rays (Staniland, 1924). The damage tends to be accumulative in that young warty galls are very suitable places for young aphids to become established. The insect is primarily a bark-feeder, feeding either on the bark under the top two or three inches of soil or, rarely, on the true roots. Leaf galls are occasionally produced on the American elm (*Ulmus americanus* L) in the eastern United States, and on apples in the western United States and in other parts of the world where the elm is not found. Greenslade (1936) surveyed the already extensive literature (Fig. 7.17 and 7.18).

The difficulty of artificial control measures has led to emphasis on the factors of resistance and immunity of apple stocks. Immunity, not to the aphid but to gall formation, exists in the variety Northern Spy, and breeding for immunity has been carried on with the object of combining the resistance of Northern Spy with more suitable horticultural varieties. The immunity is carried over to many of the seedlings. Jancke (1931) recorded tests with 103 varieties of edible apple and 79 wild apples and hybrids, which indicate that a considerable degree of resistance to the aphid is found in hybrids.

Resistance and susceptibility have been defined by Underhill and Cox (1938): resistant varieties being those on which the aphids establish temporary infestations but do not produce galls; susceptible varieties being those on which the aphids establish permanent infestations and stimulate the formation of galls. Both appear to vary with geographical location, and this has been attributed to soil type. This in turn, however, must be reflected in the physiology of the plant. If aphids can establish only temporarily on a plant, the nutritional substratum provided must be inade-

FIGURE 7.16. *Brevipalpus* gall resulting from an infestation of
B. phoenicis (Photograph: L. C. Knorr).

quate. Since galls are not formed under these circumstances, the gall-
forming secretion is absent and the insect, while toxicogenic, is not
toxiniferous under these conditions of nutrition (Fig. 7.19).

Indications that the factor of resistance is correlated with the chemical
composition of the bark have been found by Roach and Massee (1931)
and by Greenslade *et al.* (1934). Bramstedt (1938) found a positive correlation between the degree of immunity of resistance of apple seedlings to
the woolly apple aphid and the characteristic changes which the insect
caused in the attacked tissues. This was true not only of the susceptible
gall-forming varieties but also of those which showed immunity.

Phylloxera of the vine presents many aspects similar to those of the
woolly aphid problem, but there is great need for a fully documented account of what is apparently a very complex matter. The insect's life history varies with location, and much of the literature is concerned with
these differences (Topi, 1924, 1926). Damage is due to the formation of
galls on the small feeder roots, but whether the decline and final death of

FIGURE 7.17. Woolly apple aphid. Soft galls on Lord Lamborne apple shoots. Aphids eaten and galls damaged by long-tailed tits (Photograph: Massee, 1928).

the vine is entirely due to this damage will remain moot until the problem of court-noué, which includes concepts of virus etiology, is clarified.

Leaf galls are not found universally, and in some locations many years have elapsed between the appearance of root galls and those on leaves (Froggatt, 1922; Borner and Schilder, 1933; Veitch, 1933). The question of biological races of *Phylloxera*, raised by Borner (1922, 1923) and the cause of considerable controversy, presents many problems which may not be easily, if at all, soluble. Whether the biotypes have arisen by sepa-

FIGURE 7.18. Galls that developed on apple roots as a result of woolly aphid infection (Photograph: G. D. Jancke).

ration from mixed stocks imported from America, or whether they represent mutation as a result of their new environment, and whether they can be differentiated morphologically, are all of interest and some importance. That biotypes do exist and that they react differently on different varieties of vine is of most significance here.

Phylloxera perishes on the roots of immune vines owing to a reaction of the cells around the puncture; galls develop on the leaves of vines with susceptible roots (Stellwaag, 1924).

Vines resistant to *Phylloxera* are characterized by having larger titers of phenolic compounds, or lower ones of quinone reductase (Henke, 1963). According to Denisova (1965), changes in the biosynthesis of phenolic compounds in vines infested with *P. vitifoliae* are related to resistance. Gall formation is inhibited by the phenolics from an immune variety; those in the susceptible varieties stimulate gall formation. In the immune variety, the tannins are strictly localised in a dense layer which appears to prevent the aphid secretions from reaching the meristematic tissue. Galls form even on the resistant tissues but do not disrupt root

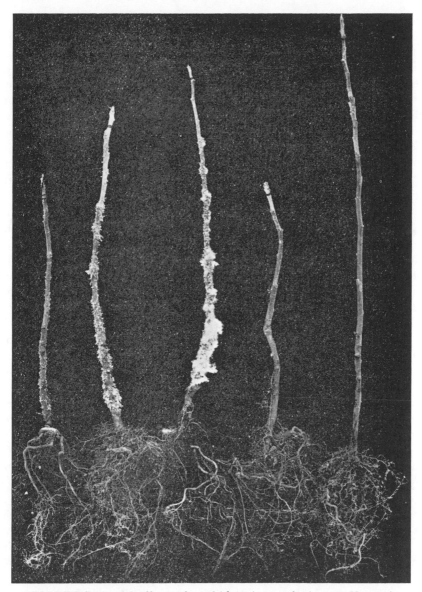

FIGURE 7.19. Woolly apple aphid (*Eriosoma lanigerum* Hausm.). Apple seedlings tested to determine their resistance to woolly aphid attack. Three on left, susceptible; two on right, resistant (Photograph: Massee, 1928).

functioning. *Phylloxera* bred on European or Asiatic vines for 20–30 years loses its ability to form galls on American vines and recovers it only after a long period of development on the American vine roots (Printz, 1937). The ontogeny of the *Phylloxera* gall on grape leaves has been described by Sterling (1952). In initiating leaf galls on *Vitis vulpina* L, the *Phylloxera* nymphs may suck from either side of the young lamina. Only those galls started from the adaxial surface develop into typical pouchlike cecidia. Mitotic activity is depressed in laminar cells adjacent to the proboscis and stimulated at some distance from the organ. An upwalling growth on the adaxial surface produces a cuplike enclosure for the feeding insect. The great volume of adaxial gall tissue is formed by the activity of a cambium-like meristem.

Phylloxera on pecans are extremely rapid gall-formers, the galls appearing on twigs, petioles, and leaves 3–5 days after infestation (Whitehead, 1933). Sakimura (1947) has reviewed the literature on gall-forming thrips. They are minor and rather primitive gall-formers, with the shape of the galls usually specific to the host plant and to the thrips species producing them. The same species of thrips produces different shapes of galls on different species of plants, while different species of thrips produce different shaped galls on the same species of plant.

Phylloxera on the roots of vines have been controlled by combinations of soil fumigation (DD) and Hexachlorobutadiene. Some fractions of the DD were more effective than others (Slonovskii, 1963).

The balsam woolly aphid (*Adelges piceae* (Ratz)) produces tumors and compression wood. The effects were duplicated by applying IAA in lanolin to scarified bark surfaces. These effects included swelling and distortion of twigs, inhibition of bud growth, and dying back of branches, all comprising the syndrome known as gout. Gout appears to be an upset of the normal balance of growth regulators in the tree, and it is suggested by the authors (Balch *et al.,* 1964) that the insect's saliva contributed to this imbalance by synergistic or exzymatic action.

There are gall-formers among the Psyllidae. *Apsylla cistellata* Buckton is responsible for extensive losses in the mango crop in India. The buds are changed into hard, conical galls by developing scale leaves which imbricate the central axis. The disease is most severe and common on the grafted varieties (Singh, 1954).

Psylla isitis Buckton causes the psylla disease of indigo. Curling of leaflets and the knotting of the apical shoot occur only as a result of nymphal feeding. Citrus leaves are pitted by the nymphs of *Trioza (Spanioza) erythrease* Del. G. The pits enlarge with the growth of the nymphs (Harris, 1936). Nymphs of *Phytolyma lata* Wlk., on the other hand, produce galls on *Chlorophora excelsa* Benth. and Hook which enclose the insect. With heavy infestations, large multiple galls are formed.

The toxic effects of leafhopper feeding are very varied, and when these are present in virus infected plants, complications of interpretations of the syndrome can result. One recently clarified is that which is part of a complex involving leaf galls and enanismo virus disease of small grains. The leaf galls are a reaction to the toxic secretions of the leafhopper vector *Cicadulina pastusae* (Ruppel & DeL.) (Galvez-E., 1965).

Empoasca fabae, the potato leafhopper, is a prime example of an insect inducing various symptoms on different host plants. On alfalfa, the insect feeding causes the development of amorphous tumors on the leaves. Punctures made with pin points dipped in water containing macerated insects induced tumors in the same period as did nymphs or male and female adults feeding normally. Varietal resistance to the insect's feeding was expressed as slightly raised leaf areas only. The pinprick technique was successful even after heating to 89–90°C or the addition of formaldehyde to the water containing the macerated insects. Clearly no virus was involved (Barnes and Newton, 1963). *E. fabae* fed on tissue-cultured plants derived from broad bean callus tissue caused injury manifest at the cellular level by cell hypertrophy (Hollenborne *et al.,* 1966).

Sheffield (1968) reported that a single leafhopper (species undetermined) produced galls on sugarcane which were ascribed to a toxin rather than to Fiji disease, and which arose from excessive production of ground tissue, not from proliferation of phloem. Only one specimen of the insect was available, and it was fed on 3 leaves, but 500 galls were produced on 11 leaves remote from the 3 leaves fed upon; in one case, galls appeared on the leaves of a different tiller. Galls appeared on 11, 14, and 18 days after feeding, but none on later growth after the insect was removed. The galls are not sett-borne. This is the first case of a translocated cecidogen known to the author, and while it is a pity that the insect could not have been determined to species, to develop such a complete case from a single specimen was a remarkable achievement.

Maramorosch *et al.* (1961) have added *Cicadulina bipunctella* China to the list of leafhoppers inducing leaf galls (amorphous tumors) by toxic secretion. If all the leafhoppers of a single species population induce the symptom, the presumptive evidence is strong, not only for the involvement of a toxin, but also for a very stable toxin which is probably independent of a specific feeding substratum. This latter point could be experimentally determined by using a series of host plant species both as sources and as test plants. A field sample considered adequate to indicate that all the individuals induce a phytotoxic symptom may not, however, be sufficient. Wallaby ear is a case in point. Grylls has worked on this problem for a number of years, but has not published until he could resolve the etiological problem. He has been kind enough to provide the following statement on the present status of his findings: "As a result of early studies with the maize wallaby ear disease, Grylls (unpublished) considered that the disease was caused by a virus. Field collections of the vector, *Cicadulina bimaculatus* (Evans), resulted in transmission of the disease either by the adults collected or by their progeny. Individual adults and nymphs failed to transmit the disease, but when these were mated some of the progeny transmitted the disease. Grylls interpreted this as transovarial transmission, but Maramorosch interpreted it as evidence that the disease was caused by an insect toxin (Maramorosch, 1959, Maramorosch *et al.*, 1961, Granados, 1969). Work in progress has demonstrated virus particles in the salivary glands of the vector and wallaby ear infected maize sap, but not in healthy maize sap. It has yet to be demonstrated that the virus particles found are the causal agent of the maize wallaby ear disease" (Grylls private communication).

FRUIT MALFORMATIONS

The injuries that represent malformation following feeding of mirids, and, rarely, some other insect families, are important in causing losses because of development of unmarketable fruit. As early as 1916, Fryer reported damage to apples by capsid bugs. The feeding of the bug results in a check to the surrounding tissues so that as the fruit grows, some parts develop more rapidly than others and distortion results. Cat-facing of peaches by the tarnished plant bug, *Lygus lineolaris* (P. de B.), was described by Rice (1938). This also appears to be a wound reaction, but its development seems to be specific. Specific wound reactions would seem to indicate a reaction to specific substances. Chandler (1955) defines cat-facing of peaches as "an early injury to the fruit, [which] checks the growth at that point and results in a distorted and frequently lopsided peach, with a lack of pubescence in the area affected." Chandler implicated *Lygus,* a number of species of stinkbugs, and the plum curculio. Injury caused by the last-named could be duplicated by pin pricks, and the injury caused by *Lygus* is indistinguishable from that caused by stinkbugs (Fig. 7.20) (See also Bobb, 1970).

Fetola disease of orange develops from the punctures of an *Empoasca* species. The fruits are disfigured by small discolored skin areas, ranging in size from a few millimeters to 15–20 mm (Dulzetto and Muscatello, 1939; Vittoria, 1940).

Gummosis, while probably generic as to cause, is definitely associated with wounds caused by insect feeding. *Thecla basiliodes* in its larval stage is a serious pest of pineapple fruits in Brazil and north to the Caribbean. Affected fruits develop large masses of brown gum, and the fruit

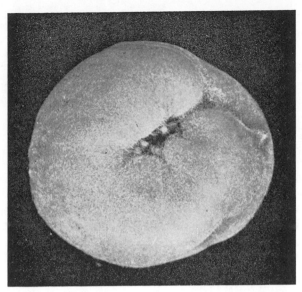

FIGURE 7.20. Cat-facing of peaches (Photograph: Illinois Natural History Survey, Chandler, 1955).

is worthless. Pineapple gummosis in Puerto Rico is associated with another lepidopterous insect, *Batrachedra* spp. (Perez, 1957).

A capsid bug, *Zmegacoelum modestum* Distant, attacking stone fruits, causes severe gumming on Elberta peach (Pescott, 1940). The meadow spittle bug, *Philaenus leucophthalmus* (L), is associated, in Delaware, with extensive gumming of peaches, the gum exuding in droplets from minute punctures on the fruit (Stearns, 1956).

REFERENCES

Addicott, F. T., and Romney, V. E., "Anatomical effects of *Lygus injury to guayule, Botan. Gaz.,* **112,** 133–134 (1950).

Adler, H., "Ueber den Generationswechsel der Eichengallwespen," *Z. Wiss. Zool.,* **35,** 151–245 (1881).

Anders, F., "Uber die gallenerregenden Agenzien der Reblaus (*Viteus (Phylloxera) vitifolii* Shimer," *Vitis,* **1,** 121–124 (1957).

Arrillaga, J. G., "Formation of galls in stems and leaves of sugar cane in response to injections of growth regulating substances," *Phytopathology,* **39,** 489–493 (1949).

Attiah, H. H., "A new eriophyid mite on mango from Egypt (Acarina)," *Bull. Soc. Roy. Entomol. Egypt,* **39,** 379–383 (1955).

Baker, K. F., and Tompkins, C. M., "A virosis-like injury of snapdragon caused by feeding of the peach aphid," *Phytopathology*, **32**, 93–95 (1942).

Balch, R. E., Clark, J., and Bonga, J. M., "Hormonal action in production of tumors and compression wood by an aphid," *Nature*, **202**, 721–722 (1964).

Ball, E. D., "Leaf burn of the potato and its relation to the potato leafhopper," *Science*, **48**, 194 (1918).

Ball, E. D., "The potato leafhopper and its relation to the hopperburn," *J. Econ. Entomol.*, **12**, 149–155 (1919).

Barnes, M. M., "Genesis of a pest; *Nysius raphanus* and *Sisymbrium irio* in vineyards," *J. Econ. Entomol.*, **63**, 1462–63 (1970).

Barnes, D. K., and Newton, R. C., "Amorphous tumors induced in alfalfa by potato leafhoppers," *Nature*, **199**, 95–96 (1963).

Beijerinck, M. W., "Beobachtung uber die ersten Entwicklungsphasen einiger Cynipidengallen," *Verhand. Akad. Wetenschap. Amsterdam*, **22**, 1–198 (1883).

Bobb, M. L., "Reduction of cat-facing injury to peaches," *J. Econ. Entomol*, **63**, 1026 (1970).

Borden, A. D., "The pear leaf blister mite as a cause of fruit-bud injury," *Calif. Agr. Expt. Sta. Circ.*, No. 324 (1932).

Borner, C., "Gibt es eine oder zwei Reblausarten amerikanischer Herkunft?" *Weinbau W. Kellerwirtsch.*, **1**, 245–249 (1922); *Z. Pflanzenkrh. W. Gallenk.*, **33**, 136–137 (1923).

Borner, C., "Neue Aufgaben der Reblausforschung," *Z. Schadlingsbekampf.*, **1**, 32–38 (1923).

Borner, C., and Schilder, F. A., "Ueber das bisherige Aufreten der Blattgallenreblaus in Deutschland," *Arb. Biol. Reichsanstalt Land-u. Forstwirtsch.*, **20**, 325–346 (1933).

Boyce, A. M., and Korsmeier, R. B., "The citrus bud mite, *Eriophyes sheldoni* Ewing," *J. Econ. Entomol.*, **34**, 745–756 (1941).

Boysen-Jensen, P., "Formation of galls by *Mikiola fagi*," *Physiol. Plantarum*, **1**, 95–108 (1948).

Bramstedt, F., "A histological method of determining the immunity of apple varieties to woolly aphis," *Z. Pflanzenkrh.*, **48**, 480–488 (1938).

Breakey, E. P., and Batchelor, G. S., "Phyllody of chrysanthemum and the eriophyid mite, *Paraphytopus chrysanthemi* Keiffer," *Ann. Entomol. Soc. Amer.*, **43**, 492–494 (1950).

Brett, C. H., and Brubaker, R. W., "Potato leafhopper control on snap beans," *J. Econ. Entomol.*, **49**, 571 (1956).

Burkill, H. J., "Some notes on midge galls," *Entomologist (London)*, **63**, 59–61 (1930).

Byers, R. A., and Wells, H. D., "Phytotoxemia of coastal bermuda grass caused by the two-lined spittle bug, *Prosapia bicincta* (Hom. Cercopidae)," *Ann. Entomol. Soc. Amer.*, **59**, 1067–1071 (1966).

Caldwell, N.. E. H., "The citrus bud mite," *Queensland Agr. J.*, **60**, 228–230 (1945).

Carlson, J. W., "*Lygus* bug damage to alfalfa in relation to seed production," *J. Agr. Research*, **61**, 791–815 (1940).

Carter, W., "Injuries to plants caused by insect toxins," *Botan. Rev.*, **5**, 273–326 (1939).

Carter, W., "Insect notes from South America with special reference to *Pseudococcus brevipes* and mealybug wilt," *J. Econ. Entomol.*, **42**, 761–766 (1949).

Carter, W., "Notes on some mealybugs (Coccidae) of economic importance in Ceylon," *Plant Protect. Bull., FAO*, No. 4 (1956).

Cendaña, S. M., and Baltazar, C. R., "A biological study of *Empoasca flavescens* Fabricius (Cicadellidae, Homoptera)," *Philippine Agriculturist*, **31**, 1–17 (1947).

Chandler, S. C., "Biological studies of peach catfacing insects, in Illinois" *J. Econ. Entomol.*, **48**, 473–475 (1955).

Cook, M. T., "The origin and structure of plant galls," *Science*, **57**, 6–14 (1923).

Cosens, A., "A contribution to the morphology and biology of insect galls," *Trans. Royal Can. Inst.*, **9**, 297–387 (1912).

Cowland, J. W., "The cotton jassid (*Empoasca libyca* Berg.) in the Anglo-Egyptian Sudan, and experiments on its control," *Bull. Entomol. Res.*, **38**, 99–115 (1947).

Denisova, R. V., "The phenolic complex of vine roots infested by Phylloxera as a factor in resistance," *Vest. Sel. Sk. Nauki*, **10**, 114–118 (1965) (In Russian).

DiMartino, E., "Abnormalities of development in citrus plants caused by an eriophyd," *Ann. Sper. Agrar. (Rome)*, **1**, 241–248 (1952).

Docters, J. van Leeuwen-Reijnvaan, and van Leeuwen Docters, W. M., *The Zoocecidia of the Netherlands East Indies*, Monograph, *Botanic Gardens*, Buitenzorg (1926).

Dulzetto, F., and Muscatello, G., "Fetola disease in oranges and its probable causes," *Ital. Agr.*, **76**, 586–693 (1939).

Edwards, E. T., "Witch's broom—a new virus disease of lucerne," *J. Aust. Inst. Agric. Sci.*, **1**, 31–32 (1935).

Essig, E. O., "The blackberry mite, the cause of redberry disease of the Himalaya blackberry, and its control," *Calif. Agr. Expt. Sta. Bull.*, No. 399, (1925).

Eyer, J. R., "Preliminary note on the etiology of potato tipburn," *Science*, **55**, 180–181 (1922).

Felt, E. P., "A study of gall midges," *N. Y. State Museum Bull.*, No. 165, 175, 180, 257 (1913, 1915, 1916, 1925).

Fenton, F. A., "Progress report on the season's work on the production of potato tipburn," *J. Econ. Entomol.*, **14**, 71–83 (1921).

Fenton, F. A., and Ressler, I. L., "Artificial production of tipburn," *J. Econ. Entomol.*, **14**, 510 (1921).

Fenton, F. A., and Ressler, I. L., "Artificial production of tipburn," *Science,* **55,** 54 (1922).

Fife, L. C., "Damage to sea island cotton by the West Indian blister mite (*Eriophyes gossypii* Banks) in Puerto Rico," *J. Agr. Univ. Puerto Rico,* **21,** 169–177 (1937).

Flemion, F., and Olson, J., "*Lygus* bugs in relation to seed production and occurrence of embryoless seeds in various umbelliferous species," *Contribs. Boyce Thompson Inst.,* **16,** 39–46 (1950).

Flemion, F., Poole, H., and Olson, J., "Relation of *Lygus* bugs to embryoless seeds in dill," *Contribs. Boyce Thompson Inst.,* **15,** 229–310 (1949).

Frickhinger, H. W., "The midge causing deformed heart and its control," *Kranke Pflanze,* **20,** 65–68 (1943).

Froggatt, W. W., "Leaf galls of *Phylloxera* at Howlong," *Agr. Gaz. N. S. Wales, Australia,* **33,** 360 (1922).

Fryer, J. C. F., "Capsid bugs," *J. Board Agr.,* **22,** 950–958 (1916).

Galvez-E, G. E., "Toxin from *Cicadulina pastusae,* a vector of enanisme virus causing galls on leaves of small grains," *Phytopathology,* **55,** 1059 (Abstr.) (1965).

Granovsky, A. A., "Studies on leafhopper injury to apple leaves," *Phytopathology,* **16,** 413–422 (1926).

Granovsky, A. A., "Alfalfa yellow top and leafhoppers," *J. Econ. Entomol.,* **21,** 261–267 (1928).

Granovsky, A. A., "Differentiation of symptoms and effect of leafhopper feeding on histology of alfalfa leaves," *Phytopathology,* **20,** 121 (1930).

Granovsky, A. A., "Tests of DDT for the control of potato insects," *J. Econ. Entomol.,* **37,** 493–499 (1944).

Granovsky, A. A., "The value of DDT for the control of potato insects," *Am. Potato J.,* **21,** 89–91 (1944a).

Granovsky, A. A., and Peterson, A. G., "Evaluation of potato leaf injury caused by leafhoppers, flea beetles and early blight," *J. Econ. Entomol.,* **47,** 894–902 (1954).

Greenslade, R. M., "Horticultural aspects of woolly aphis control together with a survey of the literature," *Imp. Bur. Fruit Prods. Tech. Comm.,* **8** (1936).

Greenslade, R. M., Massee, A. M., and Roach, W. A., "A progress report of the causes of immunity to the apple woolly aphis (*Eriosoma lanigerum* Hausm.)," *Ann. Rep. East Malling Research Sta. Kent,* **21,** 220–224 (1934).

Grenados, R. R., "Maize viruses and vectors," in *"Viruses, Vectors and Vegetation,"* Wiley-Interscience, New York, 1969.

Hamilton, A., "The blackberry mite (*Aceria essigi*)," *New Zealand J. Sci. Technol., A,* **31,** 42–45 (1949).

Hanson, A. J., "The redberry disease of blackberries," *Proc. Wash. State Hort. Assoc.,* **26,** 199–201 (1930).

Hanson, A. J., "The blackberry mite and its control (*Eriophyes essigi* Hassan)," *Bull. Wash. Agr. Expt. Sta.*, No. 279 (1933).

Harland, S. C., "The inheritance of immunity to leaf-blister mite (*Eriophyes gossypii* Banks) in cotton," *West Indian Bull.*, **17**, 162–166 (1919).

Harris, E., "Distortion of guineacorn (*Sorghum vulgare*) caused by a mealybug, *Heterococcus nigeriensis* Williams, in Northern Nigeria," *Bull. Entomol. Res.*, **51**, 677–684 (1961).

Harris, W. V., "Notes on two injurious psyllids and their control," *East African Agr. J.*, **1**, 498–500 (1936).

Hassan, A. S., "Notes on *Eriophyes mangiferae* S. N. (Acarina)," *Bull. Soc. Fouad I^{er}Entomol.*, **28**, 179–180 (1944).

Heath, G. W., "An investigation into leaf deformation in *Medicago sativa* caused by the gall midge *Japiella medicaginis* (Cecidomyidae)," *Marcellia Suppl.*, **30**, 185–199 (1961).

Henke, O., "Über den stoffwechsel reblausanfalliger und unanfalligar Reben," *Phytopath. Zeit.*, **47**, 314–326 (1963).

Heriot, A. D., "Notes on the blister made by *Eriophyes pyri* Nal.," *Proc. Entomol. Soc. Brit. Columbia*, **31**, 41–42 (1935).

Heriot, A. D., "Comparison of the injury to apple caused by scales and aphids," *Proc. Entomol. Soc. Brit. Columbia*, **41**, 13–15 (1944).

Hildebrand, A. A., "Notes on mealybug injury on strawberry and its resemblance to crinkle," *Can. J. Research, C*, **17**, 205–211 (1939).

Hodgman, M. G., "The big bud mite on yew," *J. Royal Hort. Soc.*, **91**, 133–134 (1966).

Hollenborne, J. E., Medler, J. T., and Hildebrandt, A. C., "The feeding of *Empoasca fabae* (Harris) on broad bean callus in tissue culture," *Can. Entomologist*, **98**, 1259–1264 (1966).

Hollowell, E. A., Monteith, J., Jr., and Flint, W. P., "Leafhopper injury to clover," *Phytopathology*, **17**, 399–404 (1927).

Horsfall, J. L., "The effect of feeding punctures of aphids on certain plant tissues," *Penn. State Univ. Agr. Expt. Sta. Bull.*, No. 182 (1923).

Houard, C., *Les Zoocecidies des Plantes de l'Amerique du Sud et de l'Amerique Centrale,* Hermann et Cie., Paris, 1933.

Jacob, H. E., Hewitt, W. B., and Proebsting, E. L., "Red leaf of grapevines in California prevented by controlling mites," *Proc. Am. Soc. Hort. Sci.*, **39**, 285–292 (1941).

Jancke, O., "Beitrage zur innertherapeutischen schadingsbekampfung. I. Mitteilung," *Z. Angew. Entomol.*, **18**, 276–318 (1931).

Janes, Melvin, J., "Two-lined spittle bug adults severely damage sweet corn seedlings," *J. Econ. Entomol.*, **64**, 976 (1971).

Jayaraj, S., "The effect of leafhopper infestation on the respiration of castor bean varieties in relation to their resistance to *Empoasca flavescens* (F) (Homopt. Jassidae)," *Experimentia*, **22**, 445 (1966).

Jayaraj, S., "Studies on the resistance of castor plants (*Ricinus communis* L.) Z. *Angew Ent.*, **59,** 117–126 (1967).

Jayaraj, S., and Basheer, M., "Biological observations on the castor leafhopper *Empoasca flavescens* Fab. (Jassidae, Homopt.)," *Madras Agric. J.*, **51,** 89 (1964).

Jeppson, L. R., and MacLeod, G. F., "*Lygus* bug injury and its effect on the growth of alfalfa," *Hilgardia*, **17,** 165–188 (1946).

Johnson, H. W., "Nature of injury to forage legumes by the potato leafhopper," *J. Agr. Research*, **49,** 379–406 (1934).

Johnson, H. W., "Further determinations of the carbohydrate-nitrogen relationship and carotene in leafhopper-yellowed and green alfalfa," *Phytopathology*, **28,** 273–277 (1938).

Jones, F. R., and Granovsky, A. A., "Yellowing of alfalfa caused by leafhoppers," *Phytopathology*, **17,** 39 (1927).

Jones, L. R., "Potato blights and fungicides," *Vermont Agr. Expt. Sta. Bull.*, No. 49 (1895).

Keifer, H. H., "A review of North American economic eriophyid mites," *J. Econ. Entomol.*, **39,** 563–570 (1946).

Kho, Y. O., and Braak, J. P., "Reduction in the yield and viability of carrot seed in relation to the occurrence of the plant bug *Lygus campestris* L.," *Euphytica*, **5,** 146–156 (1956).

Klein, H. Z., "Notes on the green leafhopper *Empoasca libyca* Berg. (Hom. Jassid) in Palestine," *Bull. Entomol. Res.*, **38,** 579–584 (1948).

Knorr, L. C., and Denmark, H. A., "Injury to citrus by the mite *Brevipalpus phoenicis*," *J. Econ. Entomol.*, **63,** 1996 (1970).

Kratochvil, J., and Farsky, O., "Das Absterben der diesjahrigen terminalen Larchentriebe," *Z. Angew. Entomol.*, **29,** 177–218 (1942).

Kuster, E., *Die Gallen der Pflanzen*, Hirzel, Leipzig, 1911.

Kuster, E., *Pathologische Pflanzenanatomie*, 3rd ed., Fischer, Jena, 1925.

Laboulbene, A., "Essai d'une theorie sur la production des diverses galles végétales," *Compt. Rend., Hebd. Acad. Sci. Paris*, **114,** 720–723 (1892).

Ladd, T. L., Jr., and Rawlins, W. A., "The effects of feeding of the potato leafhopper on photosynthesis and respiration in the potato plant," *J. Econ. Entomol.*, **58,** 623–628 (1965).

LaRue, C. D., "The role of auxin in the development of intumescences on poplar leaves; in the production of cell outgrowths in the tunnels of leaf-miners; and in the leaf-fall in Coleus," *Am. J. Botan.*, **22,** 908 (1935).

LaRue, C. D., "The part played by auxin in the formation of internal intumescences in the tunnels of leaf miners," *Bull. Torrey Botan. Club*, **64,** 97–102 (1937).

Lavender, D. P., Nagel, W. P., and Doerksen, A., "Eriophyid mite damage to Douglas fir seedlings," *J. Econ. Entomol.* **60,** 621 (1967).

Leach, J. G., "Leafhopper injury of potatoes," *Phytopathology*, **12,** 37 (1922).

Leach, J. G., *Insect Transmission of Plant Diseases,* McGraw-Hill, New York, 1940.

Lees, A. H., "Reversion and resistance to 'big bud' in black currants," *Am. Appl. Biol.,* **5,** 11–27 (1918).

Leigh, T. F., "Insecticidal susceptibility of *Nysius raphanus,* a pest of cotton," *J. Econ. Entomol.,* **54,** 120–122 (1961).

LePelley, R. H., "The food and feeding habits of *Antestia* in Kenya," *Bull. Entomol. Res.* **33,** 71–89 (1942).

Lewis, I. F., and Walton L., "Initiation of cone gall of witch hazel," *Science,* **106,** 419–420 (1947).

Lewis, I. F., and Walton, L., "Gall-formation on *Hamamelis virginiana* resulting from the material injected by the aphid *Hormaphis hamamelidis,*" *Trans. Am. Microscop. Soc.,* **77,** 146–200 (1958).

Loos, C. A., "Leafhopper found cause of shoot dieback in holly oak," *Calif. Dept. Agric.,* **54,** 148 (1965).

MacLeod, G. F., and Jeppson, L. R., "Some quantitative studies of *Lygus* injury to alfalfa plants," *J. Econ. Entomol.,* **35,** 604–605 (1942).

Malpighi, M., *Opera Omnia, Partis Secundae,* de Gallis, Londini, 1687.

Maramorosch, K., *et al.,* "Further studies on the maize and rice leaf galls induced by *Cacadulina bipunctella,*" *Ent. Exp. Appl.,* **4,** 86–89 (1961).

Martin, J. P., "Stem galls of sugar cane induced with insect extracts," *Science,* **96,** 39 (1942).

Martin, J. P., and Pemberton, C. E., "Disease symptoms in lettuce and celtuce caused by the bean leafhopper *Empoasca solana* (*SOLANI*) Del.," *Hawaiian Planter's Record,* **46,** 111–118 (1942).

Massee, A. M., "The black currant gall mite on red currants," *Ann. Rept. East Malling Research Sta.,* **14,** 151–152 (1928).

McCalla, D. R., Genthe, M. K., and Hovanitz, W., "Chemical nature of an insect gall growth factor," *Plant Physiology,* **37,** 98–103 (1962).

McFarlane, J. S., "Breeding for hopperburn resistance in the Irish potato," Ph.D. Thesis, Univ. Wis., Madison (1942).

McFarlane, J. S., and Rieman, G. H., "Leafhopper resistance among the bean varieties," *J. Econ. Entomol.,* **36,** 639 (1943).

McKenzie, H. L., "*Matsucoccus bisetosus* Morrison, a potential enemy of California pines," *J. Econ. Entomol.,* **34,** 783–785 (1941).

McKenzie, H. L., Gill, L. S., and Ellis, D. E. "The Prescott scale (*Matsucoccus vexillorum*) and associated organisms that cause flagging injury to Ponderosa pine in the Southwest," *J. Agr. Research,* **76,** 33–51 (1948).

Medler, J. T., "The nature of injury to alfalfa caused by *Empoasca fabae* (Harris)," *Ann. Entomol. Soc. Amer.,* **34,** 439–450 (1941).

Medler, J. T., and Fisher, E. H., "Leafhopper control with methoxychlor and parathion to increase alfalfa hay production," *J. Econ. Entomol.,* **46,** 511–513 (1953).

Maramorosch, K., "An ephemeral disease of maize transmitted by *Dalbulus elimatus*," *Ent. Exp. Appl.*, **2**, 169–170 (1959).

Metcalf, Z. P., "Peanut 'pouts'," *Science*, **86**, 374 (1937).

Miles, P. W., "Studies on the salivary physiology of plant bugs; experimental induction of galls," *J. Insect Physiology*, **14**, 97–106 (1968a).

Monteith, J., Jr., "Leafhopper injury of legumes," *Phytopathology*, **18**, 137–138 (1928).

Mote, D. C., and Wilcox, J., "Redberry mite of the blackberry," *Proc. Wash. State Hort. Assoc.*, **27**, 203–207 (1931).

Muller, G. W., and Costa, A. S., "Citrus false exanthema induced by feeding of a mirid," *FAO Pl. Prot. Bull.*, **12**, 97–104 (1965).

Nault, L. R., Briones, M. L., Williams, L. E., and Barry, B. D., "Relation of the wheat curl mite to kernel red streak of corn," *Phytopathology*, **57**, 986–989 (1967).

Nishida, T., and Holdaway, F. G., "The erinose mite of lychee," *Hawaii Agr. Expt. Sta. Circ.*, No. 48 (1955).

Nysterakis, F., "Resistance of American vines to *Phylloxera* and the sensitiveness of their cells to phytohormones," *Compt. Rend. Acad. Agr. France*, **10**, 444–446 (1946).

Osborn, H., "A new pest of potatoes," *Iowa Agr. Expt. Sta. Bull.*, No. 33 (1896).

Parr, T. J., "*Matsucoccus* sp., a scale insect injurious to certain pines in the northeast (Hemiptera-Homoptera)," *J. Econ. Entomol.*, **32**, 624–630 (1939).

Parr, T. J., "*Asteroleconium variolosum* Ratzeburg, a gall-forming coccid, and its effect upon the host trees," *Yale Univ., School Forestry Bull.*, No. 46 (1940).

Parrott, P. J., and Olmstead, R. D., "The work of *Empoasca mali* on potato foliage," *J. Econ. Entomol.*, **13**, 224–226 (1920).

Perez, M. E., "Pineapple gummosis in Puerto Rico and its control," *Puerto Rico Univ. Agr. Expt. Sta. Tech. Paper, No. 21 (1957)*.

Pescott, R. T. M., "A capsid plant bug attacking stone fruits," *J. Australian Inst. Agr. Sci.*, **6**, 101–102 (1940).

Peterson, A. G., and Granovsky, A. A., "Relation of *Empoasca fabae* to hopperburn and yields of potatoes," *J. Econ. Entomol.*, **43**, 484–487 (1950).

Peterson, A. G., and Granovsky, A. A., "Feeding effects of *Empoasca fabae* on a resistant and susceptible variety of potato," *Am. Potato J.*, **27**, 366–371 (1950a).

Plumb, G. H., "The formation and development of the Norway spruce gall caused by *Adelges abietis* L.," *Conn. Agr. Expt. Sta. Bull.*, No. 566 (1953).

Poos, F. W., "On the causes of peanut 'pouts'," *J. Econ. Entomol.*, **34**, 727–728 (1941).

Poos, F. W., and Westover, H. L., " 'Alfalfa yellows'," *Science*, **79**, 319 (1934).

Printz, Y., "Zur Frage der Virulenzveranderung der Reblaus-Biotypen," *Plant. Protect. Fasc.*, **12**, 137–142 (1937).

Proeseler, G., "Gall mites causing virus-like damage," (Translated title). *Nach Br. PflSchutzdienst*, **22**, 48 (1968).

Radcliffe, E. B., and Barnes, D. K., "Alfalfa plant bug injury and evidence of plant resistance in alfalfa," *J. Econ. Entomol.*, **63**, 1995 (1970).

Rice, P. L., "Cat-facing of peaches by the tarnished plant bug *Lygus pratensis* L.," *Trans. Peninsula Hort. Soc.*, *1937*, 131–136 (1938).

Roach, W. A., and Massee, A. M., "Preliminary experiments on the physiology of the resistance of certain rootstocks to attack by woolly aphis," *16th-18th Ann. Rept. East Malling Research Sta., Kent, 1928–1930*, **2**, 111–120 (1931).

Romney, V. E., York, G. T., and Cassidy, T. P., "Effect of *Lygus* spp. on seed production and growth of guayule in California," *J. Econ. Entomol.*, **38**, 45–50 (1945).

Rossig, H., "Von welchen Organen der Gallwespenlarven geht der Reiz zur Bildung der Pflanzengalle aus?" *Zool. Jahrb. System*, **20**, 19–90 (1904).

Sakimura, K., "Thrips in relation to gall-forming and plant disease transmission. A review," *Proc. Hawaiian Entomol. Soc.*, **13**, 59–95 (1947).

Severin, H. H. P., and Freitag, J. H., "Western celery mosaic," *Hilgardia*, **11**, 495–558 (1938).

Sheffield, F. M. L., "Gall on sugar cane leaves due to a toxicogenic insect and not to Fiji disease," *Ann. Rept. East Africa Agric. Forestry Orgn. for 1967* (1968).

Singh, S. M., "Studies on *Apsylla cistellata* Buckton causing mango galls in India," *J. Econ. Entomol.*, **47**, 563–564 (1954).

Sinha, R. C., and Wallace, H. A. H., "*Tetranychus sinhai* Baker (*Acarina, Tetranychidae*) a new pest of cereals; varietal resistance of barley," *Canadian Entomologist*, **95**, 588 (1963).

Sleesman, J. P., "Resistance in wild potatoes to attack by the potato leafhopper and the potato flea beetle," *Am. Potato J.*, **17**, 9–12 (1940).

Sloan, W. J. S., "The fruit spotting bug," *Queensland Agr. J.*, **62**, 229–233 (1946).

Slonovskii, I. F., "New fumigants for the control of Phylloxera," *Kishinev Akad. Nauk. Mold. SSR Inst. Zool.*, 85–95 (1963).

Slykhuis, J. T., Mortimer, C. G., and Gates, L. F., "Kernel red streak of corn in Ontario and confirmation of *Aceria tulipae* (K) as the causal agent," *Canadian J. Plant Science*, **48**, 411 (1968).

Smalley, E. B., "The production on garlic by an eriophyid mite of symptoms like those produced by viruses," *Phytopathology*, **46**, 346–347 (1956).

Smith, F. F., and Brierley, P., "Simulation of lily rosette symptoms by feeding injury of the foxglove aphid," *Phytopathology*, **38**, 849–851 (1948).

Sorenson, C. J., "Mirid bug injury as a factor in declining alfalfa seed yields," *Faculty Assoc. Utah State Agr. Coll.* (1946).

Staniland, L. N., "The immunity of apple stocks from attacks of woolly aphis.

(*Eriosoma lanigerum* Hausmann). Part II. The causes of the relative resistance of the stocks," *Bull. Entomol. Res.,* 15, 157–170 (1924).

Stearns, L. A., "Meadow spittlebug and peach gummosis," *J. Econ. Entomol.,* 49, 382–385 (1956).

Stellwaag, F., "Die Grundlagen fur den Anbau reblauswiderstandsfahiger Unter-lagsreben zur Immunisierung verseuchter Gebiete," *Monograph. Angew. Entomol.,* 7, 1–88 (1924); *Z. Angew. Entomol.,* 10, 1–88 (1924).

Sterling, C., "Ontogeny of the *Phylloxera* gall of grape leaf," *Am. J. Botany,* 39, 6–15 (1952).

Stubbs, L. L., and Meagher, J. W., "A virosis-like proliferation (witch's broom) of lucerne (*Medicago sativus*) caused by an eriophyid mite (*Aceria medicaginis* Keifer)," *Aust. J. Agric. Res.,* 16, 125–129 (1965).

Taher Sayed, M., "*Aceria mangiferae* nov. spec. (*Eriophyes mangiferae* Hassan MS) (Acarina-Eriophyidae)," *Bull. Soc. Fouad I^{er}Entomol.,* 30, 7–10 (1946).

Tappan, William, B., "*Nysius raphanus* Howard, attacking tobacco in Florida and Georgia," *J. Econ. Entomol.,* 63, 658 (1970).

Tate, H. D., "Method of penetration, formation of stylet sheaths and source of food supply of aphids," *Iowa State Coll. J. Sci.,* 11, 185–206 (1937).

Taylor, T. H. C., "*Lygus simonyi* Reut. as a cotton pest in Uganda," *Bull. Entomol. Res.,* 36, 121–148 (1945).

Thresh, J. M., "Increased susceptibility to the mite vector (*Phytopus ribis* Nal.) caused by infection with black currant reversion virus. Association between black currant reversion virus and its gall mite vector," *Nature,* 202, 1028, 1085–1087 (1964).

Thresh, J. M., "The effect of gall mite on the leaves and buds of black currant bushes," *Plant Pathology,* 14, 26–30 (1965).

Topi, M., "Ancora sulla esistenza di diverse specie di filossera della vite e sulla attaccabilita delle viti americane da parte della filossera," *Atti Accad. Naz. Lincei Rend. Classe Sci. Fis. Nat.,* 33, 528–530 (1924).

Topi, M., "Sulle probabili cause del diverso comportamento della filossera, specialmente gallecola, in rapporto ai vari vitigni americani," *Monit. Zool. Ital.,* 37, 74–84 (1926).

Underhill, G. W., and Cox, J. A., "Studies on the resistance of apple to the woolly apple aphid, *Eriosoma lanigerum* (Hausm.)," *J. Econ. Entomol.,* 31, 622–625 (1938).

Veitch, R., "The grape *Phylloxera,*" *Queensland Agr. J.,* 39, 79–83 (1933).

Venables, E. P., and Heriot, A. D., "The blister mite of apple and pear," *Can. Dept. Agr. Publ.,* No. 577 (1937).

Vidane, C., "Internal and external symptomology from plant sucking insects on vitis" (Trans.), *Annali. Accad. Agric. Torino,* 109, 117–136 (1967).

Vittoria, A., "The cause of fetola disease of citrus and suggestions for control," *Ital. Agr.,* 77, 355–357 (1940).

Wallace, H. A. H., and Sinha, R. N., "Note on a new disease of barley and other cereals," *Canadian J. Plant Sci.,* **41,** 871 (1961).

Wardle, R. A., and Simpson, R., "The biology of *Thysanoptera* with reference to the cotton plant. 3. The relation between feeding habit and plant lesions," *Ann. Appl. Biol.,* **14,** 513–528 (1927).

Wave, H. E., Shands, W. A., and Simpson, G. W., "Biology of the foxglove aphid in the northeastern United States," *USDA Tech. Bull.,* 1138 (1965).

Weidner, H., "The effects of the gall-insects on their host plants (Die Wirkung der Gallinsekten auf ihre Wirtspflanzen)," *Naturw. Rundschau,* **3,** 364–368 (1950).

Wells, B. W., "The comparative morphology of the Zoocecidia of *Celtis occidentalis,*" *Ohio J. Sci.,* **16,** 249–290 (1916).

Wells, B. W., "Early stages in the development of certain *Pachypsylla* galls on Celtis," *Am. J. Botany,* **7,** 275–285 (1920).

Whitehead, F. E., "Preliminary report on the pecan *Phylloxera,*" *Rept. Oklahoma Agr. Expt. Sta., 1930–1932,* 265–267 (1933).

Wiesner, J. von, in P. Krais, ed., *Die Rohstoffe des Pflanzenreichs,* Vol. I, 4th ed., Brehmer, Leipzig, 1927.

Wilson, H. F., "Injurious gall mites," *2nd Biennial Crop Pest and Hort. Rept. Oregon Agr. Expt. Sta. 1913 and 1914,* 123–126 (1915).

Wilson, N. S., and Cochran, L. C., "Yellow spot, and eriophyid mite injury on peach," *Phytopathology,* **42,** 443–447 (1952).

CHAPTER **8** CHAPTER

The Systemic
Phytotoxemias

TOXEMIC SYMPTOMS INDICATING
TRANSLOCATION OF THE CAUSAL ENTITY

This category includes some cases in which the toxic effect is apparent some distance from the feeding site, indicating a limited translocation of the toxic entity, and others in which the effect is clearly systemic. The symptoms of the first-named are generally chloroses, vein-banding, and chlorotic streaks and stripes, but a limited degree of wilting may also occur. Systemic action is demonstrable either by involvement of the whole plant, or by the expression of symptoms on new growth after the toxicogenic insects have been removed.

Translocated toxic effects are extremely difficult to duplicate by mechanical inoculation methods. Although this is frustrating to many workers who would like to use the standard methods of plant virology, it is really not surprising. The enzymic or growth-inhibiting nature of the toxic secretions—as determined by their effects on plants, their activity at extremely low concentration, the small quantity of saliva injected, and their inability to reproduce in the plant—presents formidable technical difficulties which thus far have remained unsolved. Any method of extraction, whether it be of salivary glands or the collection of saliva by feeding through membranes into artificial solutions, is subject to neutralizing action, either of tissue mixtures or of microbial contamination. There is then the problem of introducing the extracted toxin, modified as it no doubt would be by the process of extraction, into the appropriate receptor cells of the test plant in sufficient concentration. Perhaps the solution will come through the development of bioassay methods, similar in sensitivity to the *Avena* test.

Limited Translocation of the Toxin

Limited translocation of the toxic entity is known in a few cases. Anasa wilt of cucurbits has resulted in the abandonment of squash and cucurbit growing in many parts of Utah. The disease is caused by the feeding of the squash bug, *Anasa tristis* DeG. Above the feeding point, whether the petiole or the main stem, wilting results in 1–16 days depending on the age of the plant, the number of insects feeding, and the season at which attack occurs. If wilting is not complete when the insects are removed, the plants recover (Robinson and Richards, 1931).

Wilting of peach seedlings after feeding by the leafhopper, *Macropsis trimaculata* Fitch, is another example. Ten adults feeding for a week caused sudden wilting of the plant. Death is avoided if the insects are removed; otherwise, the tree dies from the tips downward. Wilting may occur, however, several days after the insects have been removed, suggesting translocation of the toxin from the feeding point. This is the first known case of an insect-transmitting a toxin independent of virus transmission by the same insect (Kunkel, 1933).

Lygus pratensis L causes an aggravated form of tarnished plant bug injury under greenhouse conditions. Two weeks' feeding by the insect on the stems of potato plants will produce a characteristic wilt. When the bugs are removed after the 2 weeks' feeding, a marked chlorosis appears in the terminal leaves 3 weeks later. This chlorosis precedes wilting by 10 days. The Miridae are normally so violently toxic that local or secondary lesions from the feeding point are the rule. In this case, however, diffusion of the toxic principle is indicated, since the insects were allowed to feed only on the stems (Leach and Decker, 1938).

SYSTEMIC TOXIC EFFECTS

Of Aphid Toxins

Systemic toxic effects of aphid feeding are rare, but Severin and Tompkins (1950) have documented a study of aphid feeding on ferns which provides satisfactory evidence of the systemic toxic effect of two species,

Myzus solani (Kaltenbach) and *Idiopterus nephrelepidis* Davis. The foxglove aphid, *M. solani,* feeding on the fronds of bird's-nest fern, first causes circular chlorotic areas, which later become spindle shaped, resembling beads and extending laterally from the midrib. These often fuse so that chlorosis extends along the midrib.

The extent of the symptom development depends on the number of aphids used. A single aphid produces a noticeable blanching effect along the midrib, as well as lateral stripes; 5 aphids increase the blanching to a chlorosis and a fusing of the lateral stripes. Increasing the number of aphids to 10 and 20 intensified these symptoms. These symptoms are not local; when 100 aphids were fed on the youngest frond, as many as 4 newly developing fronds might show the symptoms several months after removal of the aphids. When the aphids were confined to one of the oldest fronds, mild symptoms appeared on the youngest leaflet of the frond and sometimes on 1 or 2 of the newly developing fronds.

The same species of aphid feeding on holly ferns induced similar chlorotic areas or streaks resembling beads. When all of the fronds were cut off, symptoms appeared on the newly developing fronds. The same is true of the fern aphid *I. nephrelepidis* on bird's-nest fern. The symptoms differ, being dark green interveinal areas and dark green vein-banding, rather than chlorosis; but while the oldest fronds on which the aphids have fed are severely distorted—a local lesion with secondary development—symptoms appear on the newly developing fronds when all the older fed-on fronds have been cut off.

The mechanism of systemic toxicity of this kind has not been studied, nor is there any speculative hypothesis available. The toxic material is evidently translocated, and when a larger quantity of it is available, e.g., from 100 aphids, effects are apparent in parts of the plant distant from the feeding site. The time lapse of 4 months between feeding and the appearance of symptoms on newly developing fronds, however, suggests the possibility that the toxic material is "stored" and reacts only when the cells involved are in a condition to be reactive; it could involve meristematic tissue. Again, this situation calls for the active interest of workers in abnormal plant physiology.

Aphid yellows of celery (List and Sylvester, 1954) appears from 10 to 35 days after infestation with *Aphis heraclella* Davis, the time depending on the initial aphid population and the rate of colony increase. The discoloration begins above the feeding point of the colony and is followed by a systemic light green chlorosis, succeeded by stunting of the heart leaves, twisting, splitting and recurvature of the pithy petioles, and prostration. There is some varietal difference in symptom expression, bronzing in the pascal varieties, and hydrangea pink in the self-bleaching varieties (Fig. 8.1–8.4).

FIGURE 8.1. Aphid yellows of celery. Two celery plants of the Pascal type showing prostrate growth (left) due to diseased condition in comparison with normal plant. Plants of the same age from the same planting (Photograph: List and Sylvester, 1954).

FIGURE 8.2. A diseased leaflet (left) showing complete chlorosis as compared with a healthy plant (right) (Photograph: List and Sylvester, 1954).

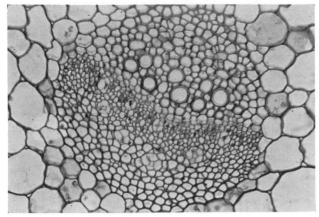

FIGURE 8.3. A transverse section of a healthy celery petiole. The dark round bodies in some of the phloem and xylem cells are safranin deposits (Photograph: List and Sylvester, 1954).

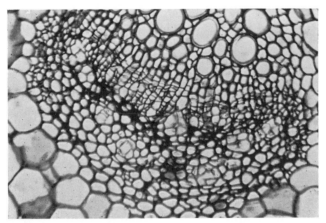

FIGURE 8.4. A transverse section of a vascular bundle of a diseased celery petiole showing the collection of material between the cells of the phloem (Photograph: List and Sylvester, 1954).

The capacity to produce the disease is inherent in the insect; diseased plants recover if the insects are removed, but reinfestation will again produce disease symptoms in a recovered plant. The insect is a phloem feeder but there is no evidence of plugging of the phloem. Control of the disease is achieved by the use of standard aphicides.

Another type of systemic injury by aphids is seen in the effects on the physiology of wheat, barley, and oats as measured by increases in susceptibility to freeze injury. The capacity for recovery was measured by tiller development. When two aphids were compared, *Toxoptera graminum* (Rond.) and *Rhopalosiphum fitchii* (Sand.), the former was found to be much more effective in causing susceptibility to freeze injury, although both species reduced tiller development (Kantack and Dahms, 1957).

Of Leafhopper Toxins

The leafhopper *Xerophloea vanduzeei* Law. secretes an active toxic principle which induces systemic symptoms in sugar beets and China asters (Severin *et al.*, 1945). On asters, symptoms by this insect resemble those of aster yellows, and on sugar beets, those of curly top; they run the whole gamut of what are normally virus disease symptoms—clearing of veins, swelling of veinlets on the lower surface of young leaves, and protuberances on the upper surfaces of the leaves. Yellowing, necrosis, and malformation of leaves also occur on sugar beets (Fig. 8.5–8.9).

On China aster, clearing of veins and veinlets appears first. The pale yellow banding, indistinguishable from that caused by the aster yellows virus, develops later. Small, interveinal, green, blisterlike elevations may occur on the youngest leaves, which may also be asymmetrical.

Symptoms on flowers are striking. The flower buds may be reduced in size with petals twisted in corkscrew fashion. Virescence or greening of the flowers may occur, sometimes in part only. Breaking in color of the petals occurs when the leafhoppers are caged on the plants in the late bud stage. The breaking consists of white streaks alternating with the normal color of the flowers. Removal of the leafhoppers after the feeding period does not prevent the development of this symptom. Severin considered three possibilities to account for the symptoms: (1) a virus disease, (2) a mechanical injury, and (3) the effect of toxic secretion.

Sugar beet seedlings showing symptoms induced by this leafhopper were fed upon by noninfective beet leafhoppers which were then transferred to healthy beet seedlings. No symptoms resulted. Similarly, noninfective short-winged and long-winged aster leafhoppers failed to produce symptoms on healthy asters. The evidence for a virus disease was therefore negative.

Mechanical inoculations using the carborundum method also failed, but inoculation of sap from sugar beets showing symptoms into the crown of a sugar beet between the bases of the petioles resulted in yellowing and necrosis of the outer leaves; no symptoms developed on the younger leaves. The same yellowing and necrosis of outer leaves occurred when the salivary glands of *Xerophloea* nymphs were used to inoculate in the same manner. Noninfective *Xerophloea* nymphs were transferred

FIGURE 8.5. China aster (*Callistephus chinensis*). *a* and *b*, Cleared veins and veinlets induced by toxic secretion of *Xerophloea vanduzeei; c,* leaf from healthy check or control plant; and *d* and *e,* cleared venation caused by the aster-yellows virus (Photograph: Severin *et al.,* 1945).

singly; the cleared vein symptom appeared on the youngest leaf in 4–12 days.

Confining the insects to outer leaves resulted in symptoms appearing on the youngest leaves, and when from 1 to 5 nymphs were caged onto the petiole of a base leaf, it was possible to determine by the appearance of symptoms on various parts of the plant the rate at which the inciting agent moved. On asters, the average time for development of symptoms was less for all the nymphal instars than for the adult, and longer for three color forms of the overwintering females. The toxic effects of feeding by phlepsid leafhoppers are shown in Fig. 8.10.

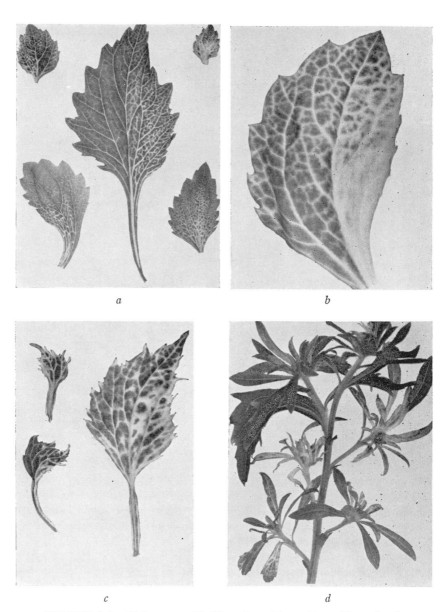

a *b*

c *d*

FIGURE 8.6. China aster (*Callistephus chinensis*). *a*, Cleared veins and veinlets, asymmetrical leaves; *b*, cleared venation with yellow vein-banding; *c*, left, two leaves showing asymmetry and pale yellow vein-banding; right, small, green, interveinal, blisterlike elevations; and *d*, axillary shoots showing pale yellow vein-banding or chlorotic involucre bracts and intermediate leaves (Photograph: Severin *et al.*, 1945).

FIGURE 8.7. China aster (*Callistephus chinensis*). *a*, Axillary shoot showing pale yellow vein-banding and curved or twisted linear bracts; *b*, axillary shoot showing chlorotic involucre bracts and intermediate leaves near the basal region; *c*, small flower buds with small, twisted petals; and *d*, flower with petals curled outward or twisted in a spiral (Photograph: Severin *et al.*, 1945).

LEAFHOPPER DAMAGE TO OATS. The leafhopper, *Calligypona* (*Delphacodes*) *pellucida* (F.), is the principal cause of severe damage to oats in a restricted area in Finland (Fig. 8.11), although the insect is common throughout the country (Kanervo *et al.*, 1957). Leafhoppers collected from the area of damage are more injurious to oats than are individuals collected from outside this area, as shown by laboratory and isolation cage experiments. Nuorteva (1958) compared the two leafhopper populations from the standpoint of the contents of their salivary glands and found that plant-growth-inhibiting substances are detectable in the salivary glands of insects from the area of damage, but they are absent from the glands of those outside this area. In discussing the nature of the in-

a   *b*

c   *d*   *e*

FIGURE 8.8. Sugar beet (*Beta vulgaris*). *a*, Younger or inner leaf showing cleared veins and veinlets; *b*, swelling of veinlets, with protruding midrib and lateral veins on the lower surface of the blade; *c*, interveinal protuberances on the upper surface of the blade; *d*, inward rolling of the margin of the youngest leaf and cleared veinlets; and *e*, inward rolling of the margin and curved midrib (Photograph: Severin *et al.*, 1945).

FIGURE 8.9. Sugar beet (*Beta vulgaris*). *a, b,* and *c,* Yellowing of basal and apical portions of blade and parts along both sides of the midrib; *d,* necrosis of midrib and veins; *e,* dried basal and middle portion of blade, with apical part still green (Photograph: Severin *et al.,* 1945).

a

b

c

d   *e*   *f*

jury, Kanervo *et al.* (1957) suggested two possibilities: (1) that there is a virus disease which overwinters in the leafhoppers and which they spread, or (2) that there are two leafhopper population strains which differ from each other in their genetic characters. One of these strains has salivary secretions that are more toxic to oats than are the salivary secretions of the other strain; or the secretions of the first contain some toxic enzymes that are lacking in the other. There is evidence that the toxic population spreads out from the area of damage. Nuorteva (1958), reviewing this discussion, agrees with the previous authors' conclusion that the known viruses, at least, are not involved. He furthermore cites the effects of larger numbers, as well as the length of feeding time, in increasing symptoms, as evidence against the virus hypothesis. Nuorteva does not accept the hypothesis of two physiological strains of the leafhopper as proved; he is more impressed by the possibility that differences in the insect's food may result in the observed differences in the secretions elaborated by the insect. Unless, however, the area of damage is one in which oats are physiologically different from oats in the area of no damage, which seems unlikely, it would be necessary to postulate an alternate host with restricted but expanding distribution from which the leafhopper moved onto the oat crop.

A fully documented study of the phytopathogenicity of the leafhopper was presented by Nuorteva (1962) in a follow up of his previous studies. The two oat viruses (OSDV and WSMV) differ symptomologically from the phytotoxemia. The latter differs from both viruses in that males are not phytotoxic, but both males and females are vectors. The intensity of symptoms of phytotoxemia is governed by the numbers of insects and the length of their feeding times. Oats growing in proximity to timothy are most seriously damaged, because timothy is the host on which the largest populations develop and from which they migrate to oats. However, in

FIGURE 8.10. Symptoms induced by the feeding of phlepsid leafhoppers. *a,* Celery (*Apium graveolens* L var *dolce*) leaflets showing cleared veins and veinlets with yellow vein-banding, induced by the feeding of *Texananus lathropi* or by *T. latipex; b,* numerous small, green islands surrounded by yellow areas caused by *T. lathropi* or *T. latipex; c,* cleared venation, white vein-banding and mottling induced by *T. pergradus; d,* outward rolling, curling, and chlorosis of celery leaflets induced by *T. spatulatus; e,* large yellow areas and necrosis of sugar beet leaf induced by *T. spatulatus;* and *f,* small yellow areas of blade and egg punctures in petiole of sugar beet leaf induced by *T. spatulatus* (Photographs: Severin *et al.,* 1945).

a

b

c

d

e   *f*

FIGURE 8.11. The toxic effect of *Calligypona pellucida* (F.) on oats. *a*, Isolated control; *b*, two leafhoppers/plant outside the area of damage; *c*, five leafhoppers/plant outside the area of damage; *d*, one leafhopper/plant from the area of damage; *e*, two leafhoppers/plant from the area of damage; and *f*, five leafhoppers/plant from the area of damage (Photographs: Aulis Tinnila).

cage tests, leafhoppers from oats were more damaging than were those from timothy and clover. Nuorteva's paper clearly shows that diet affected the phytopathogenicity of the leafhoppers to oats.

A DISEASE OF MAIZE. Maramorosch (1959) described a disease of maize induced by the feeding of the leafhopper *Dalbulus eliminatus* DeL. and W. The presenting symptom is the development of vein swellings which extend for a considerable distance along the leaf, but these disappear in about 25 days and no other symptoms develop. An intervarietal sequence appears to be involved, since the only clear results obtained were by transferring the insects from corn to the hybrid sweet corn, Golden Bantam. This probably accounts for the inability of the insects to repeat the performance when transferred from Golden Bantam to that same variety. The inclusion by Maramorosch of this latter characteristic as a criterion for a toxin is unexplainable.

LEAFHOPPER-BORNE VIRUS DISEASE.   In 1953 Maramorosch described a disease occurring on clover that is transmitted by *Euscelis plebejus* spp. *plebejus* (Fall.). It has since been shown, however, that what Maramorosch was dealing with was a complex of a phytotoxemia and a virus (De Fluiter *et al.*, 1954; Evanhuis, 1957). According to Evanhuis, the first symptoms, principally consisting of deep ridges on the upper side of the leaflets, are caused by the feeding of the insect introducing a toxic principle into the plant which spreads through the plant through the leaf stalk. The capacity to induce this symptom is inherent in insects hatched from isolated eggs on filter paper in Petri dishes, and Evanhuis found it impossible to develop a strain of leafhoppers unable to induce the symptoms. Separation of the phytotoxemia from the virus was accomplished on the basis of the time required for the development of symptoms and the presence of a long latent period, from 6 to 7 weeks, for the virus in the insect.

In this case, the toxic effect could be separated readily from that of the virus, but there are no doubt many instances where this is not so easy. So many vectors of viruses are toxic in their own right that the influence of this toxicity on virus transmission could be a matter for serious consideration.

Mealybug Stripe of Pineapple

Localized symptoms, resulting from the translocation of a toxic agent, are seen in mealybug stripe of pineapple which can be caused by the feeding of *Dysmicoccus brevipes* (Ckl.), *D. neobrevipes* Beardsley, and *Pseudococcus adonidum* Ferris. The first examples noted and described (Carter, 1944) showed evidence of feeding in the affected area in the shape of typical green spots. It is now known, however, that the green spotting was caused by the feeding of the mealybugs in the striped zone after the latter had developed (Carter, 1952). The stripe symptom varies in intensity but is characterized by a short, chlorotic-streaked zone on a number of leaves. Besides the chlorosis, the water-storage tissue is affected and in severe cases becomes necrotic. Unless necrosis occurs, the chlorotic area reverts to a normal green color and the symptom disappears. There is no extension of the striping if the mealybug colony is left on the plant, but if the colony is sprayed out and the plant is later infested a second time, typical symptoms occur a second time in a fair proportion of cases, but in a lower proportion on previously striped leaves than on new growth (Fig. 8.12).

When a green-spotting colony of mealybugs is used and sprayed out 10 days afterwards, the green spotting appears in the zone of feeding. Strip-

FIGURE 8.12. Mealybug stripe of pineapple leaves. The white streak at the top is a crack in the leaf (After Carter, 1944).

ing occurs on the inner leaves well below the zone of feeding on the youngest leaf fed upon, and on the next leaves to emerge. This indicates that the toxic agent moves down to meristematic tissue, which is susceptible. The symptoms do not recur unless the plants are reinfested, and vegetative progeny of striped plants have continued to grow without any apparent effect on later growth.

The stripe symptom is seen most frequently in hybrids after infestation with large numbers of mealybugs. The standard commercial variety is not susceptible, with the exception of one vegetative clonal selection. Small seedlings are more susceptible than larger plants. A host plant transfer from pineapple to pineapple appears to be necessary; the most successful transfers have been those using clonal selections from Cayenne as source plants, which are themselves not susceptible. Some variety-hybrid clones are both good sources and at the same time susceptible. The symptoms were also recorded in Brazil by Carter (1949).

The best evidence for the toxic nature of the causal agent is that with gross infestation more severe symptoms develop, and in a shorter time, than when smaller numbers are used, although symptoms on small seedlings have resulted when 5, 10, and 50 bug lots were used. The time required for symptom expression appears to be the time required for the causal agent to be translocated down to the meristem, plus the time required for the leaves to grow out far enough for the symptoms to be recognized. The original concentration first introduced into the leaf will affect the first-named period; the vigor and rapidity of growth, the second. A gross infestation of a small seedling can result in severe symptoms in a month; with smaller numbers of mealybugs, milder symptoms result in 7–8 weeks. Still obscure is the reason for the results from some transfers of mealybugs between clones. For example, mealybugs grown on clone A and transferred to the susceptible clone B will induce mealybug stripe in a large percentage of cases; but mealybugs from clone C transferred to clone B will be entirely negative, even though both source clones, A and C, are positive sources when seedlings are used as the test plants. This curious condition, of no economic significance, requires more study before its etiology is fully understood.

White Streak of Pineapple Plants

White streak of pineapple is caused by the feeding of *Phenococcus solani* Ferris (Carter, 1960) (Fig. 8.13). The insect is a mesophyll-feeder, and staining reveals cell-wall thickening and diffuse staining of the area contiguous to the feeding track. Toxic effects of the insect's feeding are visible 7 to 10 days after a single *P. solani* has been allowed to feed. There may be small chlorotic spots visible, but the most frequent and clearly defined symptom is a white streak which broadens as it extends in a proximal direction from the feeding point. The larger the number of insects used, the more rapid is the onset of symptoms. In the streaked area, the chloroplasts are entirely destroyed. Evidence for the diffusion of the toxic principle was obtained when insects were caged in small leaf cages; streaks appeared later in the center leaves. The capacity to induce streaking is inherent in the insect since streaking developed on plants fed upon

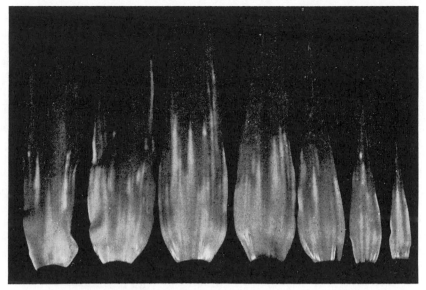

FIGURE 8.13. Toxicogenicity of *Phenococcus solani* Ferris. Seedling pineapple plant dissected to show extreme white streaking which follows feeding by *P. solani* (After Carter, 1960).

by previously unfed crawlers. Recovery from streaking occurs when the insects are removed. New tissue is streakless and the old streaks either disappear entirely or become pale green. This recovery is associated with regeneration of the chloroplasts. Streaking of the inner leaves is renewed when recovered plants are reinfested.

Froghopper Blight of Sugarcane

A blight of sugarcane caused by the feeding of *Tomaspis saccharina* Dist. has been known since 1892 and has been definitely associated with the insect since 1889 (Williams, 1921; Hardy, 1927a). Williams provides the most extensive study, and there is much incidental information in the *Minutes and Proceedings of the Froghopper Investigation Committee of Trinidad and Tobago.*

The disease expression begins with a local lesion, which is a pale area surrounding the puncture. This elongates into a long, narrow, pink streak, which later becomes brown and discolored. Necrosis follows. A single froghopper puncture may result in the destruction of ½ sq. in. of leaf tissue. The length and width of the cane internodes are both reduced, and below the area of maximum shortening of the internodes, adventitious buds develop. Adventitious roots may also be formed as high as 3 feet above the soil level. Following injury by froghoppers, the pH of the cell sap gradually falls; in fairly good soil, it rarely drops below 5.7, but on some heavy soils it may be as low as 5.4. This increase in acidity is most noticeable at the top of the cane where enzyme activity is greatest (Withycombe, 1926). The root systems of blighted cane are invariably poor and superficial. Under favorable conditions for growth, recovery takes place when the attack of froghopper has passed.

The relation of the insect to blight was proved by Nowell and Williams (1920), who showed that froghoppers could produce a definite form of blight in the absence of fungi or other insects.

Nowell (1919) records that the severity of blight in many cases does not correspond with the degree of froghopper infestation. This is unusual, in that it is safe to generalize that there is a definite correlation between the size of the insect population and the degree of resultant phytotoxemia. Pickles (1933) suggests that it may be due to the preference of froghoppers for unhealthy cane, or to different microclimatic conditions, dependent on plant growth—small or large canes, thick or sparse canes, etc.

Withycombe (1926) has described the effects of the insect's saliva on the plant tissue. The saliva has been shown to penetrate cell walls and lignified tissue. Plasmolysis of the cell contents often occurs. The saliva of the froghopper has a distinctly diastatic action upon starch, and, in addition, contains oxidases. It is slightly acid in reaction and does not invert disaccharides. Saliva is pumped into the wound; shortly afterwards there is a distinct bleaching of the chlorophyll, more longitudinally on the leaf blade than laterally. The most striking effect, noticeable even after a short feeding, is upon the contents of the border parenchyma cells, for which the saliva appears to be specific. The chlorenchyma around a froghopper puncture allows slight spreading of the insect's saliva, but in the border parenchyma the saliva passes rapidly in a longitudinal direction. Lateral extension is, however, slow, and it is usually associated with prolonged feeding. The specific effect of the saliva upon the contents of the border parenchyma is believed by Withycombe to be due to the fact that the enzymes of the froghopper's saliva exert their greatest activity at the pH of these cells. The phloem has nearly the same pH, while the chlorenchyma is slightly more alkaline, and the xylem more acid.

The main apparent effect, after short periods of feeding, is upon the plastids of the border parenchyma. With prolonged feeding, both chlorenchyma and vascular tissues become more involved. Extension of symptom expression continues after feeding has ceased and is accompanied by the appearance of a red pigment. The complex of injury set in train by the insect's feeding can be summarized as follows: direct injury and drain upon the contents of the vascular tissues, and injurious effects which extend longitudinally in the border parenchyma with removal of carbohydrates. The injury produced by the saliva is effected by several ferments in the presence of diastatic and oxidizing enzymes. The rate of respiration increases locally with oxidation. Supernormal oxidation is influenced by the action of plant oxidases, in combination with the oxidases of the insect's saliva. Local translocation is hindered and the effect is further intensified when the phloem becomes blocked. Water deficiency prevents recovery of the protoplasts in all affected cells and culminates in a drying up of the tissues. Assuming continued effect of the enzymes introduced by the froghopper, the metabolic equilibrium of the plant will continue to be upset unless a sufficiently powerful counterreaction is forthcoming.

The roots of diseased plants are invariably poor, which is not surprising in view of the metabolic upset suffered by the plant. The systemic effect of the insect's feeding in this case can therefore be ascribed, not to the translocation of a single toxic entity down to the roots, but rather to the gross physiological disturbance suffered by the plant.

With diseased roots a concomitant symptom, it would be expected that a definite relationship exists between certain types of soil and blight incidence. Canes growing in soils more acid than pH 5.5 are severely damaged in bad froghopper years (Urich and Hardy, 1927). Canes growing in markedly alkaline calcium-satisfied soils appear to be capable of resisting froghopper damage. According to Hardy (1927), alkaline, calcareous black soils support heavy crops of cane that are generally much less prone to froghopper damage than canes grown on other soils such as the acidic, pale red soils of the same district and the acidic fawn-colored soils of the great alluvial plain of the central and northern regions of Trinidad.

The mechanism whereby liming apparently confers resistance to blighting of canes growing in alkaline soils has not yet been explained. There may also be, according to Withycombe (1926), factors of soil compaction, poor physical state, oxygen deficiency, and water-logging, all of which contribute to serious froghopper damage. What appears to be quite clear is that soil conditions modify the susceptibility of the cane to the attack of a toxic insect. Pickles (1931) added the contributing factor of relative suitability of soils for froghopper oviposition, but did not question that recovery from blight may be affected by the physiological relations of the cane. Follett-Smith (1928) studied the nutrient status of soils; all the soils on which cane is susceptible possess low nutrient-supply rates as measured by the electrical conductivity of water extracts.

Variety and age of cane are both considered susceptibility factors. The leaf sap of ratoon canes at the commencement of the wet season apparently is more suitable for inducing oviposition than is the sap of plant cane or of pasture grasses. These first ratoon canes are preferred as hosts for the first-brood-adult froghoppers. Briton-Jones (1927) believes that the relative liability of ratoon and plant cane to blight is a question of condition of growth, rather than the age of the stool from which young growth arises. Plant canes, however, hardly ever suffer to the same extent as ratoons, and the percentage of blight is usually small for the plant cane crop. The question of source of froghopper infestation associated with these differences does not appear to have been studied, although Smith (1926) reported that no instances have yet been noted of blight caused by first-brood froghoppers developing from permanent pastures.

According to Urich (1927), Uba is a fairly resistant cane. It is deep rooting, covers the ground well, prevents the growth of grass, and is not trashed. Thin-type canes are less affected by blight than thick-type canes (Steven and Potter, 1928), and the degree of blight affects recovery, as the varieties least severely affected recover almost completely (Hardy and Urich, 1929).

The development of efficient insecticides has greatly reduced the emphasis on the fundamental aspects of the problem. Pickles reported excellent controls with pyrethrum and sulfur dusts (1940) and later recommended the use of Cyanogas for nymph destruction (1942). The chlorinated hydrocarbons have been used more recently (Pickles, 1946; James, 1946; Juantorena, 1950), and the current problem seems to be one of determining at what population level control measures should be instituted. This will obviously vary greatly with location, soil type, and variety of cane.

Psyllid Yellows of Potatoes and Tomatoes

The potato psyllid, *Paratrioza cockerelli* (Sulc.), has been known as a pest of solanaceous plants since 1911, when Patch (1912) took note of observations that tomatoes in Colorado had been severely damaged. This was followed in 1915 by Compere's note on injury to Jerusalem Cherry (*Solanum pseudo-capsicum* L) in California. The first mention of damage as a result of nymphal feeding appears to be that of List (1917, 1925); he reported curling of tomato leaves as a symptom.

An epidemic of disease on potatoes in Utah served to focus attention on this problem when field investigation and surveys revealed a close correlation between the presence of the insect and the disease (Richards *et al.*, 1927; Linford, 1927, 1928; Richards, 1928).

Richards (1929, 1931) and Richards and Blood (1933) established the relationship, the latter paper being inclusive and definitive up to that time. The symptoms of psyllid yellows on potato, in the field, include rolling and cupping with marginal yellowing of younger leaves, with subsequent necrosis and degeneration. Stem elongation in aerial shoots is hindered, and axillary tubers, or small rosettes of leaves malformed in witch's broom fashion, may develop at the internode. This witch's broom effect and aerial tuber formation resulted in early confusion with *Rhizoctonia* disease (Linford, 1928). Underground symptoms are expressed in an abnormally large number of stolons and abnormal root development. Tubers are small and very numerous, resulting in a great reduction in marketable tubers and sometimes in the abandonment of fields before harvest (Schaal, 1938). Internal symptoms and changes in the physiology of affected plants have been described by Eyer (1937). Injury to the border parenchyma surrounding the vascular tissue (Eyer and Crawford, 1933) is extended laterally, since necrosis of the phloem is found in both midrib and petiole. In the regions of injury, the cell proteins are broken down. Nitrate-nitrogen content of healthy potato plants is definitely higher than that of diseased plants. The same is true also of the chlorophyll and carotin content, both being decreased in the diseased plant. Schaal (1938) found an almost starch-free condition of the tubers, but a high content of starch in the aerial portions of the plant, indicating interference with translocation. Symptoms varied in different seedling varieties. In some, development of small tubers was greatly increased; in others, the converse obtained. The severity of symptom expression was related to the number of nymphs feeding, although with over 100 no further intensification occurred. Potatoes grown in highly alkaline soils (pH 8.4), or infected with fungus disease, or injured mechanically in the stems and root systems, developed symptoms of psyllid yellows with fewer insects than when the plants were growing normally.

Carry over of symptom expression into the next vegetative generation occurs and is expressed in reduction in size and yield of tubers, depending on the severity of psyllid injury to the parent plants (Edmundson, 1940). Schaal (1938) reported that tubers from diseased plants were less vigorous than those from normal tubers.

Symptoms of the disease on tomato in the field have been described by List (1917, 1935, 1938, 1939), Binkley (1929), and Blood *et al.* (1933). The consolidated descriptions used here are quoted from R. D. Carter (1954). The first symptoms to develop are a puckering of the young leaves and an upward curling of the basal portion of the older leaves. Leaf margins and veins become purple, and the basic color of the plant changes to a pale or grayish-green. The laminae become thickened and the petioles may be twisted, especially in the newer growth. Terminal leaves may develop with the petioles slender and erect and the leaflets short and narrow. Growth is greatly reduced and the plant becomes stunted, although some secondary growth may be stimulated (Fig. 8.14 and 8.15).

Fruit set is reduced or the fruit produced is very small if the plant is

FIGURE 8.14. Psyllid yellows of tomato. Healthy (*left*) and diseased leaves of seedlings of the Marglobe tomato. Interveinal puffing and marginal curling and twisting are visible in the diseased leaf (After R. D. Carter, 1954).

FIGURE 8.15. Healthy (*left*) and diseased leaflets of Marglobe tomato seedlings. Leaflet on right exhibits typical curling (After R. D. Carter, 1954).

attacked before blossoming begins. This appears to be the result of failure of blossoming and the reduction in numbers of flower clusters caused by the general stunting of the plant. When infestation occurs after blossoming begins, fruiting may be stimulated, and many small fruits may be set near the branch endings. The quality of the fruit is poor; the color is yellowish, the flesh tough and rubbery, the pulp about the seed is reduced in quantity, and the taste is poor. According to List (1935), fruit symptoms may be evident even though the psyllid infestation has not produced leaf symptoms.

Other symptoms of psyllid attack have been reported. Since the insect is clearly inherently toxicogenic, it would seem unreasonable to ascribe any other effect of the insect's feeding to a separate and distinct entity not in combination with the inherent toxicity. Spindling sprout (hair sprout) of potatoes, for example, has been experimentally demonstrated by Snyder (1946) as a phase of psyllid damage, even though the symptom occurs in areas where the psyllid is not known and is also a symptom of purple top of potatoes, which is a virus disease.

Schultz (1939) recognized that spindle sprout has been ascribed to many causes, but the evidence that it is a concomitant of psyllid yellows, and at the same time typical of hormonal effects, is significant. The permanence of such an effect may vary without vitiating the hypothesis that insect toxins and hormones have much in common. Spindling tubers apparently do not recover within at least two generations of growth, and it is possible that similar causal entities may leave permanent effects on vegetatively produced plants, without any virus being involved. The mechanism is, of course, obscure.

The problem becomes more complex when phloem necrosis of potato tubers is considered. Sanford (1952) reported this condition associated with infestations of both nymphs and adults *P. cockerelli,* contrary to the experience with typical psyllid yellows. Symptoms closely approximate those of psyllid yellows, the differences being in the fairly constant occurrence of severe internal phloem necrosis of the tubers borne on newly affected plants, and a very significant reduction of the pyramidal growth tendency typical of psyllid yellows. With low insect populations on the foliage of certain stalks of a plant, the associated tubers may fail to develop phloem necrosis.

The insect-disease relationship is complex. Early suggestions that a virus is involved (Binkley, 1929; Shapovalov, 1929; Klyver, 1931) were negated by the findings of Richards (1931) and Richards and Blood (1933). The inclusion of psyllid yellows in K. M. Smith's *Textbook on Plant Viruses* (1937) was uncritical, but correction was made in the 2nd edition, in 1957. Richards (1931) showed that psyllid nymphs with no prior access to a diseased plant could cause the disease in healthy plants. Recovery followed when all nymphs were removed 5–10 days after feeding. The extent of injury was correlated with the numbers of nymphs feeding and the time during which they fed. Feeding of up to 1000 adults to a plant did not result in symptoms, but these did develop when nymphs, the progeny of the original adults used, were present. Nymphs reared from eggs removed from healthy plants induced the disease more intensely than did nymphs of the same age which had fed on diseased plants.

The list of host plants of the insect is formidable (Knowlton and Thomas, 1934), but the preferred hosts are in the Solanaceae (Hartman, 1937). Hartman makes the interesting comment that psyllids are sometimes present in large numbers without the disease being present. Hartman says: "As soon as one field in a given community was found to be infested . . . they could be found some place within every field in the same locality . . . It was also found that after infestation by psyllids had occurred . . . some fields readily developed symptoms . . . At the time, other fields in the same community developed few or no symptoms." It would seem that there is still much to be learned concerning the ecology and genetics of the insect in relation to its toxicogenicity and to the susceptibility of its hosts.

Schaal (1938) reported successful experiments on the production of symptoms by injection of potato plants with extracts of ground-up nymphs. The more concentrated the extract, the more typical the color symptoms. It is worthy of note that curling and yellowing of leaves, normally showing 6–14 days after the first color symptoms, did not appear in the injected plants until after they started to form stolons. Once more, as in spindling sprout, there is evidence that a toxin can be injected into a plant and a long time can elapse before the effects are evident.

R. D. Carter (1954) has made the latest contribution to the problem of the insect-disease relationship. Working with tomato and some other solanaceous hosts, he found that some nymphs were never toxic; even when fed in groups, these nontoxic nymphs caused no disease. Other nymphs caused disease for a short time, then became nontoxic. The remaining nymphs were toxic up to about the time of final molting. Toxicity of nymphs was often irregular: slightly affected plants occurred alternately with severely diseased plants in a series fed on by the same nymph; some nymphs were intensely toxic to every plant in a test series, and others were only slightly toxic at any time. Similar results, from single plant tests, were reported by Daniels (1954). Pletsch (1947) reported that fifth-instar nymphs were nontoxic; and Binkley's failure to induce symptoms by transferring nymphs for feeding on healthy plants may have been the result of transitory toxicity in the nymphs.

Carter reported some carry over of toxicity from the nymphal to the adult stage, at least for the first 72 hours after emergence, suggesting that the toxic entity is stable enough to survive the metamorphosis from nymph to adult in sufficient concentration to produce effects on the host plant. This does not negate Richards' (1931) finding, for adult toxicity could be demonstrated only with teneral adults feeding on very small tomato seedlings. The only contrary data are those of Daniels (1954), who reported that as few as 5 adult psyllids, feeding for 96 hr on mo-old tomato seedlings, caused severe symptoms that persisted for 104 days; on the other hand, following 120 hr feeding by 8 nymphs on 10-day-old seedlings, no symptoms were evident after the 41st day. Individual adults were also capable of inducing disease, but symptoms were less severe than those caused by the same individual when feeding for the same length of time as a nymph.

The toxiniferous state of an inherently toxicogenic insect seems to have been well described.

R. D. Carter (1954, 1961) has found an interesting relationship between the mycetome structure of *P. cockerelli* and the nymphal and adult stages. In the nymphal stage, the mycetome is a single U-shaped body; in the adult, it is found to have divided into two separated lobes (Fig. 8.16). It would be interesting to explore the symbiont flora of the nymphal and adult mycetomes.

Mealybug Wilt of Pineapple

Mealybug wilt of pineapple is unique among the systemic phytotoxemias and is most complex, because most recent studies have led to a reappraisal of the etiology of the disease to include the concept of a latent factor which is transmissible by mealybugs but is separate and distinct from the actual wilt-inducing secretion. The first reference to this disease was by Larsen (1910), but Larsen used the word "wilt" in a generic sense to include all wilting conditions brought about by death of roots. The same generic concept is current in Australia even now; although Oxenham (1957) ascribed the primary cause to fungi, his illustrations and symptom description, however, are typical of what has been known for many years in Hawaii as mealybug wilt. Oxenham reports no experimental evidence using either fungi or fungus-soil complexes as etiological agents.

The disease is now known to occur in practically all the tropical areas of the world where pineapples are grown commercially. Carter, whose intensive work on this problem began in 1930 (Carter, 1933), has observed and recorded the incidence of the disease in Africa, Central and South America, the Caribbean, Ceylon, Malaya, Australia, and Fiji (Carter,

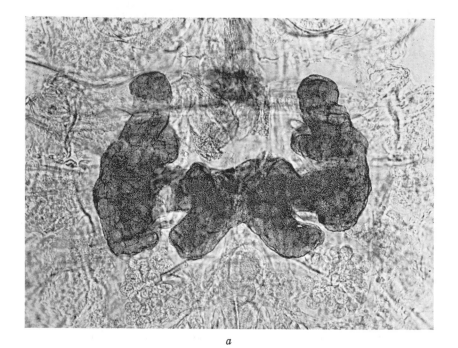

a

b

FIGURE 8.16. Anterior abdominal area of cleared, whole-mounted tomato psyllid nymphs. X77. *a*, Female, showing U-shaped mycetome and developing ovaries; *b*, male, showing mycetome and testes;

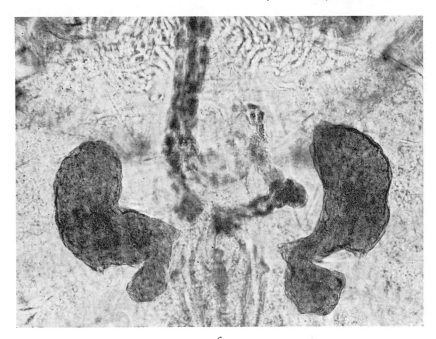

c

FIGURE 8.16. *c,* female, lobes of mycetome separated (After R. D. Carter, 1961).

1934, 1942, 1949, 1956). Jepson and Wiehe (1939) recorded the disease in Mauritius; Corbett and Pagden in Malaya (1941); Plank and Smith in Puerto Rico (1940); Westgate in southern Florida (1945); Malan in South Africa (1954); Réal (1959) and Py *et al.* (1957) in Guinea in West Africa; Takahashi (1939) and Ting-Wei Lew (1958) in Taiwan; Cook Islands 1972 (Carter unpublished). Carter (1939, 1952) summarized the available information up to 1951. Linford (1952) reported the absence of mealybug wilt in Mexico and ascribed it to the small numbers of mealybugs found, as well as to their position below the soil line.

SYMPTOMATOLOGY OF MEALYBUG WILT. Field symptoms of mealybug wilt start with a reddening of the leaves and slight inward reflexing of the leaf margins. This is designated as stage 1. In stage 2, a definite color change from red to pink occurs, and in stage 3 the reflexing of the leaves becomes pronounced. In stage 4, which is the last stage of wilting, the tips of the affected leaves have dried; in severe cases, the plant is apparently moribund (Fig. 8.17 and 8.18). A recovery stage is then recognized, characterized by the renewed growth from the center of the plant (Fig. 8.19). All these symptoms can be duplicated in the case of drought-

FIGURE 8.17. Mealybug wilt. Plants at end of rows diseased after being grossly infested with mealybugs (After Carter, 1933).

FIGURE 8.18. A ratoon field practically destroyed by mealybug wilt (After Carter, 1933).

FIGURE 8.19. Mealybug wilt. *Top:* Plants wilted 10 wk after infestation with grossly infested pieces of fruit rind. Photographed July 20, 1931. *Bottom:* The same plants photographed February 9, 1932 illustrating the type of recovery that occurs (After Carter, 1933).

stricken plants; whether this is also true of plants wilted by soil pathogens is not known, but is not likely since the pathogen complex in the soil would presumably remain viable *in situ*.

The mealybug wilt syndrome, however, clearly includes root collapse, and wilted plants invariably have poor roots. The relationship of root collapse to wilt was determined by Carter (1948), who developed a method for growing the plants in mist chambers which permitted the constant examination of the roots while the plants were growing. By this method, the first symptom was clearly shown to be the cessation of elongation of the roots, followed by collapse of the entire root system in typical cases, before leaf symptoms developed. Recovery is also clearly associated with the condition of the roots. After the roots cease to elongate and often before the old root mass collapses, new normal-appearing roots emerge from the stem above the old root mass, and this is accompanied,

or followed, by new aerial growth from the center of the plant. A comparison of the root systems of wilted and uninfested healthy plants is shown in Fig. 8.20 and data on root measurements in Fig. 8.21 and 8.22.

The period for the development of symptoms varies with the age of the plant at time of infestation. Figure 8.23 (Carter, 1945) shows this in graph form for plants grossly infested at 5 mo and 9½ mo of age with mealybugs from the rind of heavily infested fruits.

Symptom severity and relative recovery are also affected by the fre-

FIGURE 8.20.   Mealybug wilt. *Left:* A wilted plant in early recovery stage; *right:* a normal plant. Plants grown in a water-mist chamber (After Carter, 1948).

quency of infestation. When plants were infested once with a gross infestation for 10 days, 100% wilt resulted, but the plants recovered to produce fairly good-sized fruit. However, when plants in an adjacent bed were infested three times, at monthly intervals, with mealybugs from the same source, recovery was negligible (Carter, 1945).

Symptom expression, as described thus far, refers to symptoms in

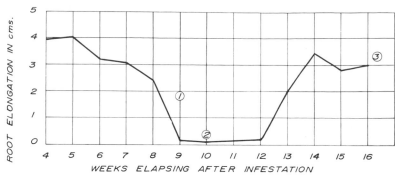

FIGURE 8.21. Graph showing root elongation in pineapple plants grown in a mist chamber and infested with mealybugs from a positive source plant. Key: (*1*) Long roots dead; median leaves reddening; (*2*) long roots dead; median leaves red all over; and (*3*) median leaves wilted, center growing (After Carter, 1948).

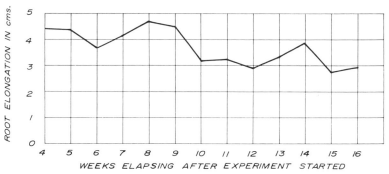

FIGURE 8.22. Means of 18 uninfested plants (After Carter, 1948).

Smooth Cayenne, the commercial variety in Hawaii. However, symptom expression varies in varieties and hybirds (Carter and Collins, 1947). The amount of anthocyanin affects the development of red and pink color in the leaves. The third-stage wilt symptom is more definite in the broader and succulent-leaved varieties than in those with narrow, stiff leaves. In varieties that are extremely low in anthocyanin and, as a consequence, pale green in color, symptoms range from a slight yellowing of a few contiguous median leaves to symptoms typical of those of Smooth Cayenne in every respect except the development of red and pink colors; the predominant color is yellow.

Seasonal variations in symptoms have been reported. Jepson and

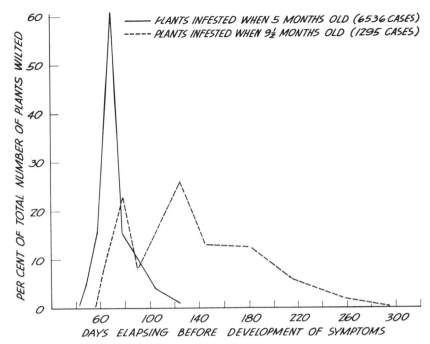

FIGURE 8.23. Period for the development of symptoms of mealy-bug wilt (After Carter, 1945).

Wiehe (1939) report that in Mauritius symptoms in summer differ from those in winter. The summer symptoms appear to be identical with those in Hawaii. In winter, the plant becomes chlorotic, the outer leaves becoming mottled. The anthocyanin coloration becomes darker, tending to a brownish color; the leaf tips become necrotic as in summer, but there is no loss of rigidity except in extreme cases; the leaves simply die back from the tips. The appearance of a field thus affected is distinctly yellow, tinged with reddish-brown, a color which large doses of nitrogen or iron sprays will scarcely alter. A high degree of recovery occurs, and commercial fruits of the highest grades are commonly obtained from these plants.

An interesting diagnostic procedure has been suggested from Malaya to separate drought wilt from mealybug wilt. If a thin strip of epidermis is removed from the upper surface of a wilted leaf, sufficient to remove the layer of cells containing anthocyanin, the underlying tissue will appear white or yellowish if mealybugs are the cause of the wilt, but the color will be green if the wilt is not due to mealybugs. This technique (Anon., 1954) was used with the variety grown in Singapore; it does not seem to apply to Smooth Cayenne. In Hawaii there is little difficulty separating wilts due to the two causes; drought wilt occurs on a localized area basis.

THE SPECIES OF MEALYBUGS CONCERNED. For almost thirty years it was presumed that only one species of mealybug, *Pseudococcus brevipes* (Ckl.), was involved, although it was recognized that if only one species was implicated, there were strains of the insect which differed biologically, even though taxonomists were agreed on the species determination. One of the strains in Hawaii was pink in color, parthenogenetic, and normally lived on the roots and crowns of grasses. This was known as the nongreen-spotting strain, since it never produced green spots (Carter, 1935) when fed on pineapple plants. The other strain, known as the green-spotting strain, is a bisexual strain (males are obligatory), and its normal habitat is on the leaves of its hosts, particularly pineapples, but also on sisal and many other plant species in many families (Ito, 1938). A further difference between the strains in Hawaii was the presence in the mycetome of the green-spotting strain of a rod-shaped symbiont, in addition to the symbiont common to both strains. Until recently, taxonomists agreed that both were the same species, but Beardsley (1960) has split the two; the nongreen-spotting strain is now known as *Dysmicoccus brevipes* (Ckl.) and the green-spotting strain as *D. neobrevipes* Beardsley. The *D. brevipes* species concept undoubtedly includes a complex of many closely related strains, or perhaps subspecies, throughout pineapple-growing areas of the world. This was illustrated by studies of symbionts in association with green spotting of pineapple leaves, which is clear in the case of *D. neobrevipes* in Hawaii, but which does not hold true in East Africa and Malaya. There, no rod symbionts were found in mealybugs on green-spotted plants (Carter, 1942). If these strains or subspecies differ biologically in any significant degree, the differences could well be reflected in the status of mealybug wilt associated with the strains or subspecies. Other species of mealybugs have been tested from time to time, but with the exception of *Pseudococcus adonidum* L, other species of mealybugs are very rare on pineapple plants or fruit. *P. adonidum* is seasonal in Hawaii. In the early spring, fair-sized colonies usually can be found in some localities on the crowns of developing fruit, and, more rarely, on 5-to 6-month-old plants. No wilt in the field has ever been associated with this species and many tests with it proved negative. Finally (Carter, unpublished), it was found that *P. adonidum* could, on occasion, cause wilt, but there are many unanswered questions as far as *P. adonidum* is concerned that require further study.

RELATIONSHIP BETWEEN MEALYBUGS AND ANTS. The recognition of the relationship between mealybugs and ants is basic to an understanding of the epidemiology of mealybug wilt. *D. neobrevipes* appears to be completely dependent on the activity of ants for its vigorous growth and reproduction, but not all ant species fill the mealybug's needs in this respect equally. Many species of ants will attend the mealybugs casually to the benefit of the insect, but in no case is the effect on the mealybug population so explosive, and the rapidity of spread so pronounced, as when *Pheidole megacephala* (Fabr.) is in attendance. *Solenopsis germinata* Forel and its varieties can, on occasion, build huge nests around colonies of mealybugs on scattered plants, but these generally are colonies of *D. brevipes,* which is normally a subterranean species, and spread is very limited. With *P. megacephala,* on the other hand, large colonies of the mealybug develop on the aerial portions of the plant, and spread can be rapid. In most of the areas in the world where the writer has observed mealybug wilt, the most typical and easily diagnosed situations have been those where *P. megacephala* was in attendance. Phillips (1934), in a study of ants in pineapple fields, traced the influx of ants into the fields, finding intense activity and nest building on the edge and where the mealybug populations were largest; a steady thinning-out of activity farther on, roughly coincident with the size of mealybug colonies, until a point was reached where scattered individual mealybugs only were found; and beyond that, foraging ants only.

EPIDEMIOLOGY OF MEALYBUG WILT. Factors affecting field incidence of mealybug wilt include climate, soil, agronomy, and pineapple variety. In general, temporary plantings in virgin soil are rarely affected, as, for example, in those areas where the jungle is cleared for a single cycle after which the land reverts to jungle again. Exceptions were noted in Surinam, where wilt was found in such plantings growing in impoverished soil which appeared to be little more than quartz sand, and in the coastal areas of Pernambuco, Brazil, where very large colonies of mealybugs were found on plants growing in poor, sandy soil.

Corbett and Pagden (1941) noted the prevalence of wilt in pineapple grown in the lateric mineral soils as compared with peat soils. Their recommendation that new plantings be restricted to peat soils was confirmed by Order in Council and has been the policy in Malaya since then. Healthy plants in Malaya, according to these authors, were to be found principally in the newly opened (i.e., virgin) peat soil areas. The incidence of heavy rains in reducing leaf populations of mealybugs and favoring establishment of subsurface colonies is also noted as a factor in minimizing wilt incidence in Malaya. Carter (1942) noted that in pineapple growing on the fringe of the plant's geographic economic range, low temperatures during the winter materially reduced the size of mealybug colonies, and that, although these colonies were quite well distributed, wilt incidence was low.

Lands rich in organic matter in Zanzibar, where small plantings on virgin soil were observed, showed only dubious cases of wilt. On the adjacent East African mainland, however, in soils deficient in organic matter and subject to seasonal rainfall and high temperatures, mealybug wilt was a limiting factor which has since caused the abandonment of the plantings. In 1938, plantings in Kenya showed only small and scattered colonies of mealybugs on small plantings on virgin land. With the development of pineapple growing, the establishment of the canning industry in that area, and the use of old sisal land for pineapple planting, wilt had increased considerably by 1956.

When Carter observed mealybug wilt in South Africa in 1937 (Carter, 1942), the disease was observed only incidentally. In 1956, on a second visit, an example of the classic syndrome was observed, i.e., invasion of *Pheidole* ants from the edge of the field and numerous young wilted plants. Private communication recently has reported that growers are now alarmed at the spread of the disease.

Jepson and Wiehe (1939) reported that in Mauritius, mealybugs were known on pineapple plants long before any development of plantations on a commercial scale, but they did not attract attention. When plantations were developed, however, mealybug populations increased and with this, serious wilt developed. In the state of São Paulo in Brazil, pineapples have been grown for over 70 years in areas of rather marginal land, and in areas where low winter temperatures occur. There has been a movement of pineapple growing from one locality to another for various reasons. *Thecla basiliodes* becomes a serious limiting factor, and the soil is so poor that without fertilization only one cycle can be grown profitably. Plantings have therefore been essentially on virgin lands. Planting material appears to be almost universally infested with a few mealybugs at the base of the plants; but on the growing plant, the mealybug colonies are for the most part confined to the base leaves under the surface of the soil. It was only in those situations where there had been movement of mealybugs up to the leaves that typical wilt was found.

In the state of Pernambuco, where more tropical conditions obtain, wilt is more common in spite of the fact that the variety grown there and known as White Paulista is more resistant than the Yellow Paulista of the south. The virgin land used there is a cleared, low scrub forest; Carter (1949) noted there a very clear comparison between virgin and second-cycle land. An area planted to pineapple, all of the same age and development, showed wilt to be much more prevalent in one-half of the field. That half proved to be a replanted section; the other half was virgin. Both sections had been planted at the same time, with planting material from the same source, and with the same agronomic treatment after planting. Virgin pineapple land in Hawaii is very rare, the crop having been grown as a monoculture for so many years. The opportunity to make comparisons there occurred only when the margins of a field were changed to include small pieces of land which up to then had been uncultivated. On these occasions, however, the same reduced susceptibility to wilt of plants grown on the virgin soil was observed. The reasons for this are, of course, obscure. Growth in virgin lands is usually more vigorous than in old cycle lands. Old cycle lands have also developed pathological root fungus complexes. It is not unreasonable to suppose that the weakening of the roots due to the toxic secretions of the mealybug will render them more liable to attack by root fungi.

In Taiwan, August, the normal planting month, is followed by a dry season during which mealybug populations multiply rapidly until the heavy rains in the following summer. With the advent of the next dry season, populations again rise rapidly and infestation becomes general. It is not until the rains begin in the third year, the beginning of the ratoon crop, that rapid collapse of the plants occurs. Ting-wei Lew (1958) ascribes this to physical degeneration of the plants due to prolonged mealybug feeding plus attacks by secondary organisms.

Localization of wilt in experimental plots has been frequently encountered (Carter, 1945) (Fig. 8.24), and in one experiment (unpublished), the accidental infection of the growing media with phythiaceous fungi materially increased both incidence and severity of wilt. The incidence of scattered wilt in an otherwise well-controlled field (Carter, 1960) could well be conditioned by pockets of fungus infection which might attack roots weakened by mealybug feeding.

In Hawaii, Fullaway (1924) observed that mealybug infestation appeared to start from the outer edges of pineapple fields and gradually work in toward the center. This observation was confirmed on a quantitative basis by Carter (1932a), who showed scattered infestations in infield blocks, but heavy infestations grouped a short distance from the edge of fields. Infestation occurred much more rapidly and was much greater in a young field contiguous to an old infested one than when the edge of the field was bordered by the wild grasses and shrubs. A major source of entry into pineapple fields was, and still is, on a much reduced scale, on planting material. In the early precontrol days, planting material was heavily and almost uniformly infested. Most of these populations disappeared in the absence of ants in the center of the field, but some remained. These scattered colonies, if they became attended by *Solenopsis* ants and developed to any size, were associated with wilting in small areas which would only develop noticeably when the plants reached the ratoon stage. Fig. 8.25 is an aerial photograph of a field in the ratoon

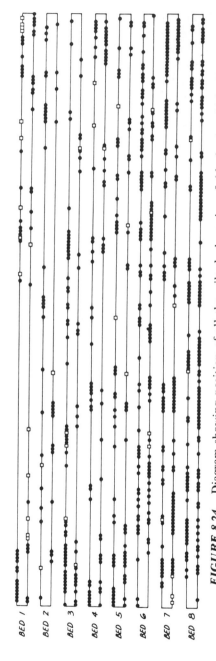

FIGURE 8.24. Diagram showing position of all the wilted plants in a field plot in which all the plants had been infested with pieces of fruit rind heavily infested with mealybugs. Note the open spaces where no plants wilted, particularly in the upper half of the diagram. Key: ● = wilt developing up to red bud stage; □ = after red bud stage (From Carter, 1945).

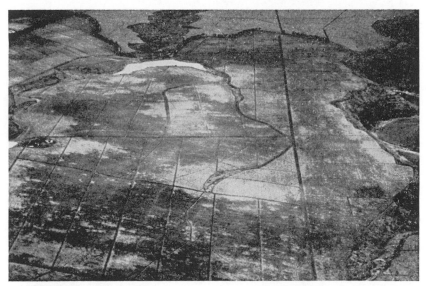

FIGURE 8.25. Aerial photograph of pineapple field showing distribution of wilt along field margins and intrafield ditches (Photograph: 11th Photo section, U.S. Air Corps., Luke Field, Oahu, Hawaii, 1932).

stage, typical of conditions throughout the pineapple industry in the early thirties. The white areas are areas of wilted plants. Note the heavy concentration along the edges and along the grass-covered drainage ditches throughout the field. It is now realized that, in all probability, the buildup of mealybug populations was mainly due to ants coming in from the edges of the field and attending mealybug colonies already introduced on the planting material when a new field was planted. An exception was when the new field was planted with only a narrow field road dividing it from an old infested planting. In such cases, both ants and mealybugs moved over as soon as the new planting began to grow.

The general field picture therefore was one which showed an early infestation along the field edges, building up to large populations that gradually thinned out toward the center of the field. Wilt followed the buildup and this in turn favored dispersal, because mealybugs leave a wilted plant. As the plants grew and became less susceptible, there would be little increase in wilt after the first flush of wilt in the young plants; but at the end of plant crop, approximately two years from planting, the field would be thoroughly infested, and wilt in the ratoon suckers would practically destroy the ratoon crop. The very earliest plants to wilt would often show a fair degree of recovery, but wilted ratoon suckers would not.

a

b

FIGURE 8.26. a, Pineapple field planted before control measures were instituted and practically destroyed by mealybug wilt in ratoon; *b,* field planted with vegetative planting material from field shown in *a,* and sprayed with the relatively inefficient oil emulsion sprays. This field gave a record yield (Photograph: Carter).

ETIOLOGY OF MEALYBUG WILT.   The hypothesis that the mealybug wilt secretion is a toxin developed by mealybugs feeding on pineapple but not on other hosts is supported by a considerable body of evidence. Planting material, taken from heavily infested fields which wilted later in ratoon, grew normally in the absence of heavy mealybug infestation of the growing plant (Fig. 8.26). The size, length of feeding time, and frequency of infestation were found to affect the number of plants wilting, symptom expression, and recovery (Fig. 8.27 and Table 8.1). Mealybugs from apparently healthy plants were effective in inducing wilt. Plants wilted a second time in a considerable number of cases, after being reinfested after recovery had taken place (Carter, 1945). Control has been achieved by application of insecticides which reduce mealybug popula-

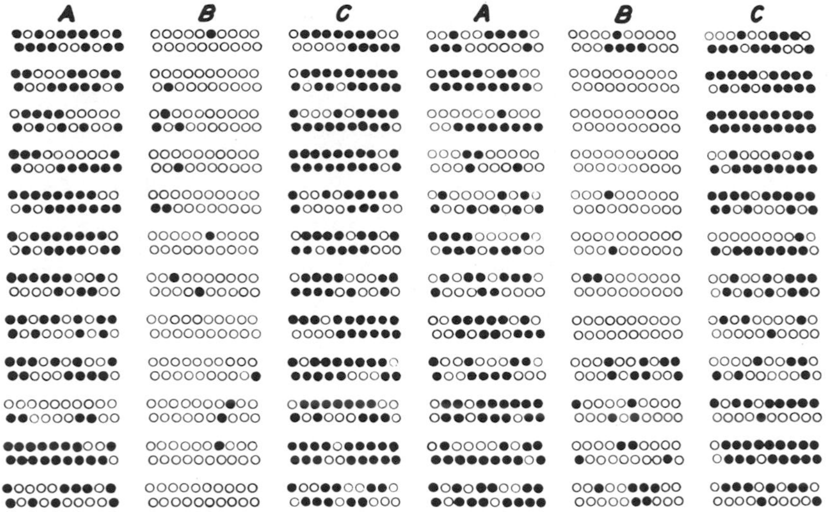

FIGURE 8.27. Distribution of wilt in field plots, each plant of which was grossly infested with mealybugs. The *A* plots were infested with mealybugs from one area and were left on the plants for 1 wk. *B* and *C* were infested with mealybugs from another area, but the feeding time was 1 wk for the *B* plots and 1 mo for the *C* plots. This result confirms that toxicity of mealybugs varies with the source field and that the length of feeding time is an important factor in wilt incidence. Comparison of *A* and *B* plots, both fed on for 1 wk, indicates the variability in toxicity between mealybugs from different source fields. Plots *B* and *C* compare the effect of 1 wk feeding (*B* plots) with 1 mo feeding (*C* plots) with mealybugs from the same source field (Photograph: Carter).

tions to low levels; but wilt once more develops if these same low populations become ant-attended and increase in size. Although all the above stiil hold true (gross field wilt no longer being demonstrable), there are some data available that called for a modification of the toxin hypothesis. These data were brought together by Carter (1951). In essence, these are (1) that some field-grown pineapple plants are negative sources for mealybug wilt, that is, mealybugs fed on them do not wilt the next plants on which they feed, and (2) that pineapple seedlings not previously infested with mealybugs are also negative sources. The involvement of an unusual type of latency was postulated to account for the lack of positive source value in field-run plants which had been grown by vegetative reproduction for over 50 years with a history of heavy mealybug infestation and accompanying wilt.

Carter (1963) has continued these studies over the last ten years. There

TABLE 8.1.   Wilt Incidence, Symptom Expression,
and Incubation Period [a]

Plant No.	Symptom stage at indicated date of observation				Symptom stage at indicated date of observation			
	Nov. 6	Nov. 13	Nov. 19	Nov. 25	Nov. 6	Nov. 13	Nov. 19	Nov. 25
	Gross infestation, Aug. 29				Infestation from 100 bugs, Aug. 29			
1	1	2	3	3		1	1	1
2	1	3	3	4				
3								
4			2	2				
5					1	2	2	2
6			1		1	2	3	
7		2	2	2	1	2	2	
8	1	1	2	3				
9	1	1	1	2				
10			1	2		1	1	1
11		1						
12		2	2	3		1	1	2
13	2	2	3	4				
14		2	3	4		1	1	1
15								
16		1	1	1				
17		1	1	2				
18						1	1	1
19		1	2	3				
20								
21		1	1	1				
	Mean symptom stage reached			2.4	Mean symptom stage reached			1.6
	Percentage wilt			71	Percentage wilt			40

[a] Result when gross colonies of mealybugs from a positive source plant were used in comparison with lots of 100 drawn from the same gross colonies.

is convincing evidence from these studies for the existence of a transmissible latent factor essential to positive source value. It is separate and distinct from the actual wilting secretion, since positive source value develops without wilt occurring. Once established, there is no doubt that the latent factor is transmitted through vegetative reproduction, although it may not be demonstrable (Carter and Ito, 1956); in the latter case it would be called "occult" in the Madison terminology (Walker *et al.*,

1958). The effect of virgin colony (a previously unfed colony) feeding on the establishment, and demonstrability of the latent factor, are important. Although virgin colony feeding prior to infection feeding (i.e., the feeding of small numbers of mealybugs from a positive source) has no effect, virgin colony feeding after infection feeding materially increases the number of plants in which the latent factor can be demonstrated. Virgin colonies derived from *D. neobrevipes* appear to be slightly more effective than those from *D. brevipes*.

The latent factor, being transmissible and evidently perpetuated in the vegetatively reproduced plant, is to be considered a latent virus, because no disease expression occurs as a result of its presence in a plant. A plant infected with the latent factor is a positive source plant, however.

Carter considered three possible relationships between the latent factor and the inherent toxicity of the insect's secretions. He accepts the hypothesis that this toxicity provides a stress factor which aids in the establishment, concentration, and distribution of the latent factor in the infected plant, but rejects the hypothesis that the latent virus is activated by the insect's secretions to become the etiological agent of mealybug wilt. Instead he presents evidence supporting the hypothesis that the actual wilt-inducing secretion is a toxin, synthesized by the insect when feeding on a plant infected with the latent virus. This hypothesis harmonizes the epidemiological and etiological aspects of the mealybug wilt problem.

OTHER TOXIC EFFECTS OF MEALYBUG FEEDING ON PINEAPPLE. Apart from the specific feeding effect which is mealybug wilt, the toxicity of both species of mealybugs expresses itself in a number of ways. Growth depression, accompanied by color changes, is a typical reaction to mealybug feeding and is especially noticeable in small seedlings. It can be demonstrated also in plants of the commercial variety grown from slips in a water mist chamber. Table 8.2 compares measurements of uninfested, infested but not wilted, and wilted plants (Carter, 1948).

A long-term carry-over effect of mealybug feeding was reported by Carter and Collins (1947) (Fig. 8.28).

More recent findings have been reported (Carter, 1963) which were obtained by the use of mealybugs from negative source plants, to ensure freedom from any complication with the wilt secretion. These are (1) premature development of fruit, (2) increased susceptibility to wilt by later infestations of mealybugs from positive sources, (3) color changes and reduced vigor related to feeding by virgin colonies, and (4) differences in vigor of growth of plants infested with mealybugs from positive sources, but which did not wilt. The differences in vigor depended on the size of the original infestation used, and they were evident for over two years

TABLE 8.2.   Comparison of Plant Measurements of Healthy
and Infested Pineapple Plants Grown
in Water-Mist Boxes

Data	Symptom class [a]			
	0	1	2	3
Number of plants	18	15	20	1
Slip weight (g)	208	206	210	194
Final plant weight (g)	563	474	386	300
Increase in plant weight (%)	170	130	87	55
Final number of roots	61	63	76	61
Total length of live roots (cm)	2696	2488	475	0
Total length of dead roots (cm)	24	14	1395	1521
Dry weight of root mass (g)	10.8	10.6	5.8	5.1
Mean length of 10 longest leaves (cm)	49	42	38	36
Mean width of 10 widest leaves (cm)	4.4	3.9	3.6	3.5
Number of green growing leaves	20	20	19	21
Total number of leaves	55	54	49	42

[a] 0 = Healthy uninfested plants showing no symptoms.
  1 = Infested plants showing no root or leaf symptoms.
  2 = Typical wilt symptoms on plants showing recovery.
  3 = Advanced third-stage wilt.

after the mealybugs had been removed by spraying. No mechanism for these prolonged carryover effects of toxic feeding can be suggested except that they must involve changes in somatic tissue, and, because of the length of time that they are observable, affect cells produced long after the initial stimulus has been delivered. Another most interesting effect is to force the expression of genetic weaknesses. Two examples of this are known (Carter, 1962).

RESISTANCE TO MEALYBUG WILT.   There is good evidence that resistance to mealybug wilt is genetically controlled (Carter and Collins, 1947; Collins and Carter, 1954), but it is not considered of any practical importance as a control measure. One reason is that resistance and positive source value appear to be associated, and there is some evidence (Carter, 1963) that resistance to mealybug wilt and susceptibility to infection by the latent virus go together. Positive source plants (plants infected with the latent virus) are also found in susceptible clones, however. Furthermore, one commercial variety, grown in a limited area in Hawaii, is a completely positive source although it is highly susceptible, and wilt was a limiting factor in that area before control measures were instituted.

FIGURE 8.28. Carry-over effect of mealybug feeding on suscepti-bility to wilt from later infestations. The plot on the right had been infested when the plants were 10 mo old; both plots were infested in the ratoon sucker stage, some 15 mo later. Severe wilt occurred only in the plot infested twice (After Carter and Collins, 1947).

FIGURE 8.29. Dipping pineapple planting material in insecticide to kill mealybugs (Photograph: Ting-wei Lew, 1958).

FIGURE 8.30. Treatment of growing plants with insecticide. A measured quantity is poured into the heart of the plant (Photograph: Ting-wei Lew, 1958).

Immunity to the latent infection would be of real value because that would mean the complete elimination of positive source plants in a plantation, but there is no evidence that such immunity exists.

CONTROL OF MEALYBUG WILT.   Control of mealybug wilt has been achieved in Hawaii on a scale of efficiency not duplicated elsewhere, although Ting-wei Lew (1958) describes the control methods used in Taiwan, as do Py *et al.* (1957) following Vilardebo (1955) for the Ivory Coast. In both these areas, dipping of the planting material is recommended, followed by spraying the growing plant with an organic phosphorous insecticide (Fig. 8.29 and 8.30). Control of ants is recommended, but it is not stressed as it is in Hawaii, where it is considered essential (Fig. 8.31). Ant species differ in their susceptibility to the chlorinated hy-

FIGURE 8.31. Two early methods for mealybug wilt control. The border planting (Carter, 1932) along the field edge delayed the entry of ants and mealybugs into the field proper. The ant fence is a 1-ft board buried halfway in the soil and kept sprayed with an oily mixture. The association of ants with wilt was recognized before the role of mealybugs was understood, and the ant fence was devised by a plantation manager.

drocarbons, but one or another of these is usually effective. The organophosphorus insecticides are used to control the residual mealybug infestations. Dosage and frequency of application are determined by local conditions. The aim should be to reduce mealybug populations as low as economically feasible, but complete elimination of mealybugs cannot be expected and is not in fact necessary to achieve mealybug wilt control.

REFERENCES

Anon., "Pests of pineapples," *Rept. 6th. Comm. Entomol. Conf., Dept. Agr. Malaya,* 270–271 (1954).

Beardsley, J. W., "On the taxonomy of pineapple mealybugs in Hawaii, with a description of a previously unnamed species (Homoptera; Pseudococcidea)," *Proc. Hawaiian Entomol. Soc.,* 17, 29–37 (1959).

Binkley, A. M., "Transmission studies with the new psyllid yellows disease of solanaceous plants," *Science,* 70, 615 (1929).

Blood, H. L., Richards, B. L., and Wann, F. B., "Studies of psyllid yellows of tomato," *Phytopathology,* 23, 930 (1933).

Briton-Jones, H. R., "Note on Mr. Edward B. Smith's suggestions for froghopper control," *Minutes Proc. Froghopper Invest. Comm. Trinidad and Tobago,* 6, 147–148 (1927).

Carter, R. D., "Toxicity of *Paratrioza cockerelli* (Sulc.) to certain solanaceous plants," Dissertation, Univ. Calif., Berkeley, 1954.

Carter, R. D., "Distinguishing sexes in nymphs of the tomato psyllid, *Paratrioza cockerelli,*" *Ann. Entomol. Soc. Amer.,* 54, 464–465 (1961).

Carter, W., "Border plantings as guard rows in pineapple mealybug control," *J. Econ. Entomol.,* 28, 1–8 (1932).

Carter, W., "Studies of populations of *Pseudococcus brevipes* (Ckl.) occurring on pineapple plants," *Ecology,* 13, 296–304 (1932a).

Carter, W., "The pineapple mealybug, *Pseudococcus brevipes,* and wilt of pineapples," *Phytopathology,* 23, 207–242 (1933).

Carter, W., "Mealybug wilt and green spot in Jamaica and Central America," *Phytopathology,* 24, 424–426 (1934).

Carter, W., "The symbionts of *Pseudococcus brevipes* (Ckl.)," *Ann. Entomol. Soc. Amer.,* 28, 60–71 (1935).

Carter, W., "Injuries to plants caused by insect toxins," *Botan. Rev.,* 5, 273–326 (1939).

Carter, W., "Geographical distribution of mealybug wilt with notes on some other insect pests of pineapple," *J. Econ. Entomol.,* 35, 10–15 (1942).

Carter, W., "A striping of pineapple leaves caused by *Pseudococcus brevipes,*" *J. Econ. Entomol.,* 37, 846–847 (1944).

Carter, W., "Etiological aspects of mealybug wilt," *Phytopathology*, **35**, 305–315 (1945).

Carter, W., "Influence of plant nutrition on susceptibility of pineapple plants to mealybug wilt," *Phytopathology*, **35**, 316–323 (1945a).

Carter, W., "The effects of mealybugs feedings on pineapple plants grown in finely atomized nutrient solutions," *Phytopathology*, **38**, 645–657 (1948).

Carter, W., "Insect notes from South America with special reference to *Pseudococcus brevipes* and mealybug wilt," *J. Econ. Entomol.*, **42**, 761–766 (1949).

Carter, W., "The feeding sequence of *Pseudococcus brevipes* (Ckl.) in relation to mealybug wilt of pineapples in Hawaii," *Phytopathology*, **41**, 769–780 (1951).

Carter, W., "Injuries to plants caused by insect toxins, II," *Botan. Rev.*, **18**, 680–721 (1952).

Carter, W., "Notes on some mealybugs (Coccidae) of economic importance in Ceylon," *Food and Agr. Organization U. N. Plant Protect. Bull.*, **4**, 49–52 (1956).

Carter, W., "A study of mealybug populations (*Dysmicoccus brevipes* (Ckl.)) in an ant-free field," *J. Econ. Entomol.*, **53**, 296–299 (1960).

Carter, W., "*Phenococcus solani* Ferris, a toxicogenic insect," *J. Econ. Entomol.*, **53**, 322–323 (1960a).

Carter, W., "Mealybug wilt: A reappraisal," *Ann. N. Y. Acad. Sci.*, **105**, 741–764 (1963).

Carter, W., and Collins, J. L., "Resistance to mealybug wilt of pineapple with special reference to a Cayenne-Queen hybrid," *Phytopathology*, **37**, 332–348 (1947).

Carter, W., and Ito, K., "Study of the value of pineapple plants as sources of the mealybug wilt factor," *Phytopathology*, **46**, 601–605 (1956).

Collins, J. L., and Carter, W., "Wilt resistant mutations in the Cayenne variety of pineapple," *Phytopathology*, **44**, 662–666 (1954).

Compere, H., "Insect note," *Monthly Bull. Calif. State Comm. Hort.*, **4**, 574 (1915).

Corbett, G. H., and Pagden, H. T., "A review of some recent entomological investigations and observations," *Malayan Agr. J.*, **29**, 347–375 (1941).

Daniels, L. B., "The nature of the toxicogenic condition [*sic*] resulting from the feeding of the tomato psyllid *Paratroza cockerelli* (Sulc.)," Dissertation, Univ. Minnesota, Minneapolis, 1954.

De Fluiter, H. J., Evanhuis, H. H., and van der Meer, F. A., "Observations on some leafhopper-borne virus diseases in the Netherlands," *Proc. 2nd Conf. Potato Virus Disease*, Lisse-Wageningen, 1954, pp. 84–88.

Edmundson, W. C., "The effect of psyllid injury on the vigor of seed potatoes," *Am. Potato J.*, **17**, 315–317 (1940).

Evanhuis, H. H., "Investigations on a leafhopper-borne clover virus," *Proc. 3rd Conf. Potato Virus Disease*, Lisse-Wageningen, 1957, pp. 251–254.

Eyer, J. R., "Physiology of psyllid yellows of potatoes," *J. Econ. Entomol.*, **30,** 891–898 (1937).

Eyer, J. R., and Crawford, R. F., "Observations on the feeding habits of the potato psyllid (*Paratrioza cockerelli* Sulc.) and the pathological history of the 'psyllid yellows' which it produces," *J. Econ. Entomol.*, **26,** 846–850 (1933).

Follett-Smith, R. R., "The nutrient status of the observation plot soils," *Minutes Proc. Froghopper Inves. Comm. Trinidad and Tobago,* **13,** 145–151 (1928).

Fullaway, D. T., "Insects affecting pineapple production," *2nd Ann. Short Course Pineapple Production, Univ. Hawaii, 1923,* 52–61 (1924).

Hardy, F. "The liming problem in Trinidad sugarcane soils," *Minutes Proc. Froghopper Invest. Comm. Trinidad and Tobago,* **7,** 202–210 (1927).

Hardy, F., "Investigations into the froghopper blight of sugarcane in Trinidad," *Minutes Proc. Froghopper Invest. Comm. Trinidad and Tobago,* **8,** 218–235 (1927a).

Hardy, F., and Urich, F. W., "Progress report. Recovery of blight canes," *Minutes Proc. Froghopper Invest. Comm. Trinidad Tobago,* **14,** 182–187 (1929).

Hartman, G., "A study of psyllid yellows in Wyoming," *Wyoming Agr. Expt. Sta. Bull.,* No. 220 (1937).

Ito, K., "Studies on the life history of the pineapple mealybug *Pseudococcus brevipes* (Ckl.)," *J. Econ. Entomol.,* **31,** 291–298 (1938).

James, H. C., "The bionomics and control of *T. flavilatera* Ur., the Demerara sugar cane froghopper," *Proc. Brit. West Indies Sugar Technologists,* **1946,** 34–79.

Jepson, W. F., and Wiehe, P. O., "Pineapple wilt in Mauritius," *Mauritius, Dept. Agr. Gen. Ser. Bull.,* No. 47 (1939).

Juantorena, J., "A note on the sugarcane froghopper," *Agricultura (Venezuela),* **15,** 46–47 (1950).

Kanervo, V., Heikinheimo, O., Raatikainen, M., and Tinnila, A., "The leafhopper *Delphacodes pellucida* (F.) (Homopt. Auchenorchyncho) as the cause and distributor of the damage to oats in Finland," *Publ. Finnish State Agr. Research Board,* **160,** 1–56 (1957).

Kantack, E. J., and Dahms, R. G., "A comparison of injury caused by the apple grain aphid and greenbug to small grains," *J. Econ. Entomol.,* **50,** 156–158 (1957).

Klyver, F. D., "California psyllids of present and potential economic importance," *Calif. Dept. Agr. Bull.,* **20,** 691–694 (1931).

Knowlton, G. F., and Thomas, W. L., "Host plants of the potato psyllid," *J. Econ. Entomol.,* **27,** 547 (1934).

Kunkel, L. O., "Insect transmission of peach yellows," *Contribs. Boyce Thompson Inst.,* **5,** 19–28 (1933).

Larsen, L. D., "Diseases of the pineapple," *Hawaiian Sugar Planters Assoc. Pathol. Physiol. Ser. Bull.,* No. 10 (1910).

Leach, J. G., and Decker, P., "A potato wilt caused by the tarnished plant bug *Lygus pratensis* L.," *Phytopathology,* **28,** 13 (1938).

Linford, M. B., "Further observations on unknown potato disease in Utah," *Plant Disease Reptr.,* **11,** 110–111 (1927).

Linford, M. B., "Plant disease in Utah in 1927," *Plant Disease Reptr. Suppl.,* **59,** 64–117 (1928).

Linford, M. B., "Pineapple diseases and pests in Mexico," *Food and Agr. Organization U. N., Plant Protect. Bull.,* **1,** 21–25 (1952).

List, G. M., "A test of lime-sulfur and nicotine sulfate for a control of the tomato psyllid and effect of these materials upon plant growth," *Colo. State Entomol. Ann. Rept.,* **9,** 40–41 (1917).

List, G. M., "The tomato psyllid, *Paratrioza cockerelli* Sulc.," *Colo. State Entomol. Circ.,* **47,** 16 (1925).

List, G. M., "Psyllid yellows of tomatoes and control of the psyllid, *Paratrioza cockerelli* (Sulc.) by use of sulphur," *J. Econ. Entomol.,* **28,** 431–436 (1935).

List, G. M., "Tests of certain materials as controls for the tomato psyllid, *Paratrioza cockerelli* (Sulc.) and psyllid yellows," *J. Econ. Entomol.,* **31,** 491–497 (1938).

List, G. M., "The potato and tomato psyllid and its control on tomatoes," *Colo. Agr. Expt. Sta. Bull.,* **454,** 1–33 (1939).

List, G. M., and Sylvester, E. S., "The relationship of aphids to a toxicogenic disease known as aphid-yellows of celery," *Colo. Agr. Expt. Sta. Tech. Bull.,* No. 50 (1954).

Malan, E. F., "Pineapple production in South Africa (with special reference to the Eastern Transvaal)," *Union S. Africa, Dept. Agr. Bull.,* No. 339 (1954).

Maramorosch, K., "A new leafhopper-borne plant disease from Western Europe," *Plant Disease Reptr.,* **37,** 612–613 (1953).

Maramorosch, K., "An ephemeral disease of maize transmitted by *Dalbulus eliminatus,*" *Ent. Exp. Appl.,* **2,** 169–170 (1959).

Nowell, W., "Investigation of froghopper pest and diseases of sugarcane," *Trinidad and Tobago Council Paper,* No. 39 (1919).

Nowell, W., and Williams, C. B., "Sugarcane blight in Trinidad: A summary of conclusions," *Bull. Dept. Agr. Trinidad and Tobago,* **19,** 8–10 (1920).

Nuorteva, P., "On the nature of the injury to plants caused by *Calligypona pellucida* (F.) (Hom. areopidae)," *Ann. Entomol. Fennici,* **24,** 49–59 (1958).

Nuorteva, P. "Studies on the causes of phytopathoegenicity of *Calligypona pellucida* (F.) (Hom. Araeopidae)," *Ann. Zool. Soc. "Venamo" Tom.,* **23,** 1–58 (1962).

Oxenham, B. L., "Diseases of the pineapple," *Queensland Agr. J.,* **83,** 13–26 (1957).

Patch, E. M., "Notes on Psyllidae," *Maine Agr. Expt. Sta. Bull.,* **202,** 215–234 (1912).

Phillips, J. S., "The biology and distribution of ants in Hawaiian pineapple fields," *Expt. Sta. Pineapple Producers Coop. Assoc. Bull.*, **15**, 1–57 (1934).

Pickles, A., "On the oviposition of *Tomapsis saccharina* Dist. (Rhynch. Cercopidae) and insect pest of the sugarcane in Trinidad," *Bull. Entomol. Res.*, **22**, 461–468 (1931).

Pickles, A., "Entomological contributions to the study of the sugarcane froghopper," *Trop. Agr.* (*Trinidad*), **10**, 222–233, 240–245, 286–295 (1933).

Pickles, A., "Control of the sugarcane froghopper by the use of pyrethrum dust," *Proc. Agr. Soc. Trinidad and Tobago,* **40**, 57–61 (1940).

Pickles, A., "A discussion of researches on the sugarcane froghopper (Homop. Cercopidae)," *Trop. Agr.* (*Trinidad*), **19**, 116–123 (1942).

Pickles, A., "Field trials of insecticides for the control of the sugarcane froghopper," *Trop. Agr.* (*Trinidad*), **23**, 9–12 (1946).

Plank, H. K., and Smith, M. R., "A survey of the pineapple mealybug in Puerto Rico and preliminary studies of its control," *J. Agr. Univ. Puerto Rico,* **24**, 49–76 (1940).

Pletsch, D. J., "The potato psyllid (*Paratrioza cockerelli* (Sulc.)) its biology and control," *Montana State Coll. Agr. Expt. Sta. Tech. Bull.,* No. 440 (1947).

Py, C., Tisseau, M. A., Oury, B., and Ahmada, F., *The Culture of Pineapples in Guinea,* Institut francais recherches fruitieres d'outre mer; Institut des fruits et agrumes coloniaux (I. F. A. C.), Paris, 1957.

Réal, P., "Le cycle annuel de la cochenille *Dysmicoccus brevipes* (Ckll.), vectrice d'un "Wilt" de l'ananas en basse Cote d'Ivoire; son determinesme," *Rev. Pathol. Vegetale Entomol. Agr. France,* **38**, 1–111 (1959).

Richards, B. L., "A new and destructive disease of the potato in Utah and its relation to the potato psylla," *Phytopathology,* **18**, 140–141 (1928).

Richards, B. L., "Psyllid yellows of potatoes," *Utah Agr. Expt. Sta. Bull.,* **209**, 50–51 (1929).

Richards, B. L., "Further studies with psyllid yellows of the potato," *Phytopathology,* **21**, 103 (1931).

Richards, B. L., and Blood, H. L., "Psyllid yellows of the potato," *J. Agr. Research,* **46**, 189–216 (1933).

Richards, B. L., Blood, H. L., and Linford, M. B., "Destructive outbreak of unknown potato disease in Utah," *Plant Disease Reptr.,* **11**, 93–94 (1927).

Robinson, L. R., and Richards, B. L., "Anasa wilt of cucurbits," *Phytopathology,* **21**, 114 (1931).

Sanford, G. B., "Phloem necrosis of potato tubers associated with infestation of vines by *Paratrioza cockerelli* (Sulc.)," *Sci. Agr.,* **32**, 433–439 (1952).

Schaal, L. A., "Some factors affecting the symptoms of the psyllid yellows disease of potatoes," *Am. Potato J.,* **15**, 193–206 (1938).

Schultz, E. S., "A non-transmissible spindling sprout of potato," *Phytopathology,* **29**, 21 (1939).

Severin, H. H. P., and Tompkins, C. M., "Symptoms induced by some species of aphids feeding on ferns," *Hilgardia,* **20,** 81–92 (1950).

Severin, H. H. P., Horn, F. D., and Frazier, N. W., "Certain symptoms resembling those of curly-top or aster yellows, induced by saliva of *Xerophloea vanduzeei,*" *Hilgardia,* **16,** 337–360 (1945).

Shapovalov, M., "Tuber transmission of psyllid yellows in California," *Phytopathology,* **19,** 1140 (1929).

Snyder, W. C., Thomas, H. E., and Fairchild, S. J., "Spindling or hair sprount of potato," *Phytopathology,* **36,** 897–904 (1946).

Smith, E. B., "Suggestions for the control of froghoppers on sugarcane estates," *Minutes Proc. Froghopper Invest. Comm. Trinidad and Tobago,* **5,** 118–123 (1926).

Smith, K. M., *A Textbook of Plant Virus Diseases,* Little, Brown, Boston, 1937; 2nd ed., 1957.

Steven, R. M., and Potter, J. A., "Cane varieties and froghopper blight," *Minutes Proc. Froghopper Invest. Comm. Trinidad and Tobago,* **13,** 162–164 (1928).

Takahashi, R., "Insect pests of pineapple, especially *Pseudococcus brevipes* (Ckl.)," *Bull. Agr. Research Inst. Formosa,* **161,** 1–17 (1939).

Ting-wei L. G., "Pineapple mealybug control in Taiwan," *Höfchen-Briefe,* **11,** 114–120 (1958).

Urich, F. W., "History of sugarcane blight in Trinidad from 1920 to 1924," *Minutes Proc. Froghopper Invest. Comm. Trinidad and Tobago,* **6,** 149–152 (1927).

Urich, F. W., and Hardy, F., "Progress report: Soil reaction and blight," *Minutes Proc. Froghopper Invest. Comm. Trinidad and Tobago,* **8,** 213–217 (1927).

Vilardebo, A., "La cochenille de l'ananas et le Wilt quelle provoque," *Fruits (Paris),* **10,** 59–66 (1955).

Walker, D. L., Hanson, R. P., and Evans, A. S., *Symposium on Latency and Masking in Viral and Rickettsial Infections,* Burgess, Madison, Wisconsin, 1958.

Westgate, P. J., "Mealybug wilt of pineapples in south Florida," *Proc. Florida State Hort. Soc.,* **58,** 194–196 (1945).

Williams, C. B., "Froghopper blight of sugarcane in Trinidad," *Mem. Dept. Agr. Trinidad and Tobago,* **1,** 1–170 (1921).

Withycombe, C. L., "Studies on the aetiology of sugarcane froghopper blight in Trinidad. I. Introduction and general survey," *Ann. Appl. Biol.,* **13,** 64–108 (1926).

THE
PLANT
VIRUSES

The Plant Virus
as an Entity

The diseases caused by the plant viruses include many that are among the most destructive known to man, and the economic losses resulting from them are incalculable, but enormous. Some may threaten the livelihood of an entire population where a single crop economy dominates, as in the case of the cocoa-producing areas of West Africa. Some may eliminate a crop plant from culture, as happened as early as 1775 when potato growing was abandoned in various parts of Europe because of losses caused by virus diseases.

In our own century, in many valleys of the intermountain region of the United States, the culture of sugar beets ceased until varieties resistant to curly top could be developed. Less spectacular, but more important in the aggregate, is the huge loss taken through the reduction of yields of many of the more important food crops. It is not, however, only the economic aspects that provide the incentives for the remarkable development in this field of study over the last thirty or forty years. Plant and animal viruses and the diseases they cause provide useful tools for the extension of knowledge and philosophical horizons as man seeks to interpret biological processes, and aided by developments in molecular biology, to control them.

317

HISTORICAL

The first definitive separation of microorganisms and viruses came with Iwanowski's discovery in 1892 that the agent of tobacco mosaic would pass through a bacterial filter. This discovery, confirmed by Beijerinck in 1898, also served to clarify terminology. Prior to Iwanowski, the term "virus" had been used in the Pasteurian sense as synonymous with bacteria.

The use of the term "filterable virus" to designate infectious filter-passing entities survived for many years, but the adjective gradually fell into disuse, aided perhaps by the finding of Mulvania (1926) and Smith and Doncaster (1935) that ability to pass through a filter was determined by filter pore size and shape of the virus particle. The word virus alone has now become current; it is a convenient term and easy to pronounce.

Even before the discoveries of Iwanowski and Beijerinck, E. F. Smith (1891, 1894) had achieved the almost *sine qua non* of modern transmission techniques when he succeeded in transmitting peach yellows by grafting scions from diseased trees on to healthy ones. Mayer (1886) had already transmitted tobacco mosaic by injecting sap from diseased plants into healthy ones.

The discovery that a leafhopper could transmit a virus disease was first announced by a Japanese worker, Takami, in 1901 (Katsura, 1936). Since this publication was in Japanese and was not readily available to the Occident, Ball's work with *Circulifer* (*Eutettix*) *tenellus* (Baker) (1906, 1917) was for a long time considered the first demonstration of its kind. It is surprising, in retrospect, that aphid vectors of plant viruses were not recognized until later. Aphids were identified as vectors in 1923 (Schultz and Folsom, Murphy), although the oldest known plant virus disease—color break in tulips—was not aphid-transmitted experimentally until 1929 (Hughes).

Many of the landmarks in the progress of this field of inquiry are associated with the development of special techniques. The number of local lesions at the point of infection was shown by Holmes (1928, 1929) to develop in proportion to virus concentration. By this method, viruses that are readily sap-transmitted could be studied quantitatively.

The discovery that plant viruses are antigenic and induce the development of antibodies when injected into animals made the methods of serology applicable to virus research, particularly in the determination of relationships between viruses which were not apparent by other criteria (Dvorak, 1927; Beale, 1928, 1929, 1931).

Bawden (1950) recognizes three other landmarks in the early work on plant viruses: (1) the discovery of the symptomless carrier by Nishimura

(1918), (2) the mutability of viruses leading to development of virus strains (Carsner, 1925; Johnson, 1926; McKinney, 1926), and (3) the phenomenon of cross protection whereby one strain of a virus will protect the plant against another strain of the same virus (Thung, 1931; Salaman, 1933).

Up to about 1935, the principal emphasis had been on description of the diseases themselves and in the search for insect vectors. A major breakthrough occurred, however, with the isolation of the tobacco mosaic virus in its paracrystalline form by Stanley (1935), its determination as a nucleoprotein by Bawden *et al.* (1936), and the purification and crystallization in true three-dimensional crystals of the virus of tomato bushy stunt by Bawden and Pirie (1937).

In the ensuing years, the early promise that all viruses would soon be known by their crystalline structure has not been realized (Fig. 9.1 and 9.2). As late as 1955, Cohen listed the viruses that had been successfully purified, and their dimensions and shape determined. There were eight of them, as shown in Table 9.1.

TABLE 9.1. *The Size, Shape, and Nucleic Acid Content of Some Plant Viruses* [a]

Virus	Dimensions (A.)	Shape	Ribonucleic acid	Crystallized
Tobacco mosaic	3000×150	Rigid rod	5.8	+
Cucumber 3 and 4	3000×150	Rigid rod	5.8	+
Potato X	4300× 98	Less rigid rod	5	0
Tomato bushy stunt	300	Sphere	14	+
Tobacco necrosis	240	Sphere	16.5	+
Southern bean mosaic	320	Sphere	21	+
Turnip yellow mosaic	218	Sphere	35	+
Alfalfa mosaic	165	Sphere	15	0

[a] From Cohen (1955).

Klug and Caspar (1960) have reviewed the structural details of the small viruses (Fig. 9.3). It is clear that the great bulk of our information comes from studies of a single virus, TMV, although there are some data on helical viruses closely related to TMV and on a few of the so-called spherical viruses.

Techniques for isolation and purification of viruses have been highly developed in the itervening years, and many more viruses have had their *in vitro* structure determined. Those that are now known are shown in the cryptograms of Gibbs (1968).

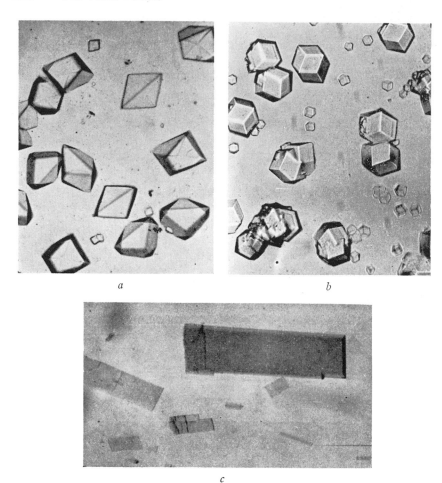

a   *b*

c

FIGURE 9.1. *a,* Crystals of tobacco necrosis virus (Tobacco II culture) produced by precipitating with ammonium sulfate; *b,* crystals of tomato bushy stunt virus produced by precipitating the virus with ammonium sulfate; *c,* crystals of a tobacco necrosis virus (Rothamsted culture) separated from a virus solution in distilled water (Photographs: Bawden, 1950).

In vivo demonstrations of the shape and size, and some indications of the internal structure, of plant viruses in both plants and vectors have been made possible by electron microscopy. Some clear examples are shown in Fig. 9.4–9.8. This technique is invaluable in revealing into which structural group a virus belongs.

The current concept of the structure of the virus particle as a single- or

FIGURE 9.2. A crystal of the rhombic type of tobacco necrosis virus in which the molecular order is unusually good. X84,000 (Photograph: Wyckoff).

double-stranded central helix of nucleic acid surrounded by a protein coat is illustrated in Fig. 9.3. An exception has been noted recently in the defining of a deficient virus in which the nucleic acid is naked, that is, without a protein coat (Sanger, 1968).

The virus nucleic acid occupies the central role in the initiation of infection. Bawden's basic premise (1960) that the nucleic acid is the infectious agent is now generally accepted; the protein fraction is not infectious. Nucleic acid contains the genetic code for replication of the virus particle although currently, the only well characterized gene product of plant viruses is the particle-protein coat, and the primary structure of the coat proteins of plant viruses has only been determined for some strains of tobacco mosaic (Gibbs, 1969).

A complication has been found in the existence of a satellite virus, that is, one which depends on another virus (tobacco necrosis) for its replication (Kassanis, 1962). Some studies on its nature and origin have been re-

FIGURE 9.3. Drawing showing a segment ¹⁄₂₀th the length of the rod-shaped tobacco mosaic virus particle. The complete virus particle is built up of 2130 identical protein molecules arranged in a helix, rather like the steps of a spiral staircase, with a single thread of ribonucleic acid winding between the turns of the protein helix. Part of the ribonucleic acid is shown uncovered to indicate the structure it takes up when it packs with the protein subunits. Each protein molecule consists of 158 amino acids linked together in a chain that is folded in a very definite way to form the elongated globular subunits illustrated here. The thread of ribonucleic acid is made up of 6280 nucleotides linked together by a very regular backbone. The infectivity of the virus is carried by the ribonucleic acid, which contains the genetic code to make more nucleic acid and protein subunits. The function of the protein subunits is principally to form a protective, stabilizing shell for the infective nucleic acid molecule (Reproduced by permission from *Advances in Virus Research* (Klug and Caspar, 1960); detailed legend courtesy of Caspar).

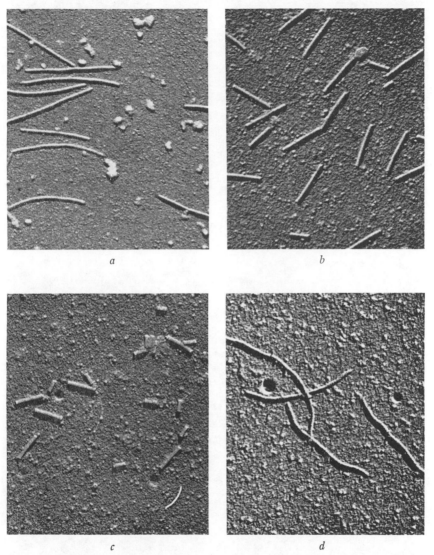

FIGURE 9.4. *a*, Pea streak virus, pd-shadowed; *b*, tobacco mosaic virus, pt-Tr-shadowed; *c*, potato stem mottle virus, pd-shadowed; *d*, potato virus Y.

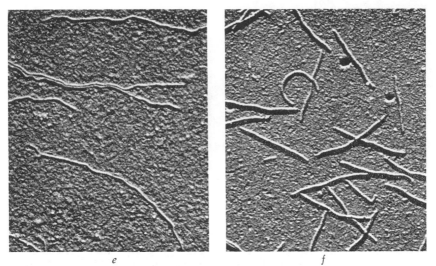

e   f

FIGURE 9.4.   *e,* sugar beet yellows virus; *f,* potato virus X (All photographs X40,000; from Brandes and Wetter, 1959).

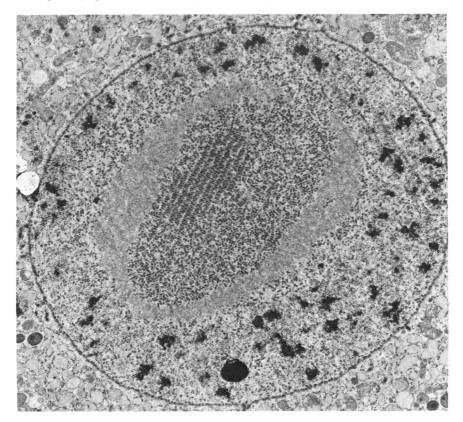

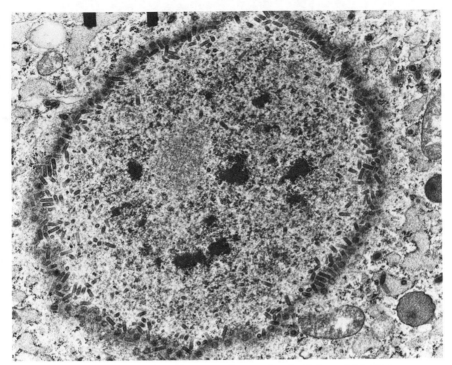

FIGURE 9.6. Nucleus from a salivary gland of the aphid, *Hypero-myzus lactucae,* infected with the sowthistle yellow vein virus in which the virus particles have accumulated at the periphery of the nucleus and the nuclear membrane is tending to disappear (From *Virology,* **42,** 1023–1042 (1970). Photograph: Sylvester and Richardson and the American Microbiological Society).

cently reported by Jennifer and Corbett (1971). However, these aspects are the primary concern of the molecular biologist and do not fall within the scope of this book; the reader is referred to up-to-date texts on plant virology. Matthews text (1970), while written from the standpoint of the molecular biologist, is written in terms which the nonspecialist can understand.

FIGURE 9.5. Nucleus from a salivary gland of the aphid *Hypero-myzus lactucae* infected with the sowthistle yellow vein virus showing a viroplasm containing and surrounded by virus particles, as well as the process of synymenosis at the nuclear membrane (From *Virology,* **42,** 1023–1042 (1970). Photograph: Sylvester and Richardson and the American Microbiological Society).

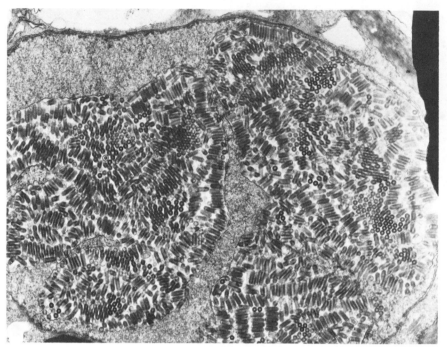

FIGURE 9.7. Thin section of diseased tissue from sowthistle, *Sonchus oleraceus,* infected with the sowthistle yellow vein virus showing perinuclear masses of bacilliform particles (From *Virology*, 35, 347–355 (1968). Photograph: Richardson and Sylvester and the American Microbiological Society).

DEFINITION OF A VIRUS

There have been many attempts to write a working definition of a plant virus (Gardner, 1931; Green, 1935; Stanley, 1938; Bawden, 1950; Lwoff, 1953, 1958). The latest definition comes from Matthews (1970) who defines it in molecular biological terms as follows: "A virus is a set of one or more nucleic acid template molecules, either DNA or RNA with the following properties, (1) they have the ability to organize their own reduplication only in a suitable intracellular environment. Intracellular components upon which virus replication is dependent include ribosomes, transfer RNA and an energy producing system. Certain viruses may also require the presence of another virus; (2) in the mature virus particle the genetic material consists of more than one nucleic acid molecule, each housed in a separate particle, or all may be housed in one. The ma-

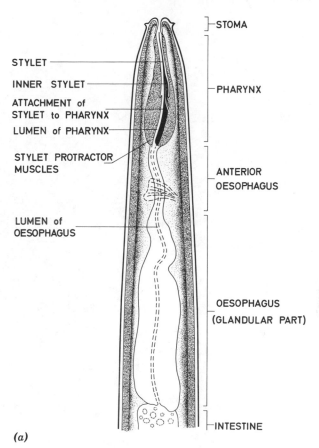

STOMA

STYLET

INNER STYLET

ATTACHMENT of
STYLET to PHARYNX

LUMEN of PHARYNX

PHARYNX

STYLET PROTRACTOR
MUSCLES

ANTERIOR
OESOPHAGUS

LUMEN of
OESOPHAGUS

OESOPHAGUS
(GLANDULAR PART)

INTESTINE

(a)

FIGURE 9.8.   (a) Diagram of the oesophageal region of *Trichodorus pachydermus* (X300).

ture virus contains no nucleic acid other than the genetic material; (3) a virus can cause disease, at least in one host under appropriate conditions."

THE ORIGIN OF VIRUSES

The early debate on the nature of viruses between the molecular hypothesis on the one hand and the organismal on the other has subsided in the light of recent developments, and there now appears to be no question. However, there has been a carryover of the organismal hypotheses into the question of the origin of viruses.

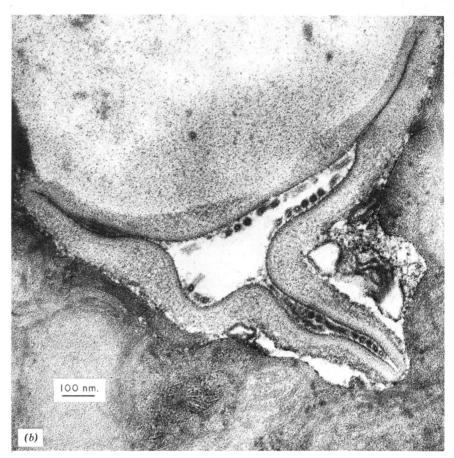

FIGURE 9.8. (*b*) Cross-section of the lumen of the pharynx, in the region where the stylet is attached, showing sectioned particles of tobacco rattle virus (Photographs: Taylor and Robertson, the *Journal of General Virology,* and the Cambridge University Press).

That viruses may represent retrogression from larger organisms has been proposed by some workers (Green, 1935; Laidlaw, 1938; Burnet, 1945), and there is indeed a continuous range of forms, from those which are undoubtedly bacteria, at the one end of the scale, to the virus nucleoproteins at the other end. The viruses would therefore, according to this view, have evolved from larger organisms by the loss of synthesizing power. Bawden objects on the grounds that the infected host cell is the functioning entity and the virus particle simply one component. Viruses could therefore have developed from cell components, and no evolutionary development either up or down from microorganisms need be postulated (Bawden, 1950).

Viruses as self-reproducing units are not unique. Chromosomes and the determinants of cytoplasmic inheritance do the same thing. The free-gene theory is a cell-constituent theory, that is, that genes have escaped from the nucleus and act independently of it (Muller, 1922; Duggar and Armstrong, 1923). Virus origin from specific cytoplasmic units has been proposed by Darlington (1944) and by DuBuy and Woods (1943).

Bawden (1964) has reviewed the various hypotheses and concludes as follows: "It might be more profitable to speculate on the likelihood that current hypotheses are wrong than to try and adapt them to accommodate awkward facts. As yet there is only a glimmering of how complex are the processes in which viruses are engaged because 'in vitro' they are not behaving like viruses." Matthews (1970) has considered the origin of viruses by critically examining three possibilities: (1) descendants of primitive cellular life forms, (2) development from normal cell constituents, and (3) degenerative microorganisms. His approach is that of the molecular biologist, and his hypothetical outline of the evolution of viruses from mycoplasma-like organisms is particularly stimulating.

The Novogenesis of Plant Virus Diseases

The novogenesis of virus diseases is not so complicated. Most of the known virus diseases of crop plants have been recognized and described in the last thirty years, and it may be significant that many more have been recognized from the New World, and even from the more newly settled areas of the New World, than from the older agriculture of Europe. Bennett (1952) has considered this as an expression of equilibrium arrived at in the Old World but not in the New. New crops are introduced into the newer agricultural areas of the world, and new vector-virus relationships are established. *Circulifer (Eutettix) tenellus* (Baker) is now known as a North African species, but curly top is not known in that area although recently reported from Turkey (Bennett, 1952). How the insect reached the western United States and extended its range far south into Mexico is not known, but once there it must have found native hosts of curly top and only needed the introduction of a new crop plant, sugar beet, to develop a devastating virus disease. Bennett's concept is of a hugh reservoir of already existing viruses awaiting transport, vectors, and suitable new nonequilibrated crop plants to achieve emergence of virus diseases. This concept does not concern itself with the novogenesis of viruses, but it could well be extended to include it.

The latent virus of dodder, when transmitted to other plants, causes severe symptoms of disease, but there is no evidence that in dodder it is a virus at all if pathogenicity is an essential part of the virus definition. The process seems clear. Dodder establishes itself parasitically on another plant, so that there is organic union between the tissues of host and parasitic plant. Some cell constituents from the parasitic dodder find receptive cells in the host and when fully integrated with these give rise to a new virus and virus disease.

There is no reason to suppose that this process cannot go on indefinitely, and no reason to postulate that the foreign cell constituents transmitted had the status of viruses in their original hosts. Transmittal is the key process, and there are no more efficient transmitters than the insects. It is a logical step, then, to the conclusion that in the novogenesis of virus diseases the insect is potentially the most efficient agent. This potentiality of the insect need not be limited to the establishment of new combinations of host plants and already defined infective cell constituents. If, as Bawden suggests, variations in normally noninfective proteins might render them infective, then the insect, particularly the toxicogenic species, because of its inherent capacity to affect plant metabolism, could well be the agent of this transformation. The increasing number of viruses now known to multiply in their insect vectors is evidence that virus synthesis can and does occur in insects (Black, 1953; Maramorosch, 1955). The preliminary transformation of noninfective to infective protein (or nucleic acid) could occur in the same manner. Nor is it essential to postulate that the insect species responsible for the initial transformation be competent as a vector of the transformed cell constituent; nor is it necessary to assume that the new infective cell constituent is anything more than latent, or with the status of a provirus in its initial host.

THE LATENT VIRUS

A number of so-called latent viruses have been described, but it may be appropriate to examine the concept of latency in viruses, and to determine, if possible, how many of these are truly latent in the light of current definitions of latency. The paracrinkle virus in King Edward potato is the classic example of a latent virus in plants, and until Bawden *et al.* (1950) showed that it could be revealed by the use of an indicator plant, its presence in King Edward potato was only determinable by the addition of other viruses causing virus complexes that resulted in disease.

A clear distinction should be made between the "masked virus" and the "symptomless carrier." Masking is, by definition, merely a covering, and as far as plant viruses are concerned, this is environmental. Hildebrand (1958) uses a much more inclusive and elaborate definition when he defines masking as "the presence of a living ultramicroscopic virus entity within the cytoplasm of living cells in the absence of distinctive symptoms"; and latency as "the presence of virus in the host with the total absence of visible symptoms over the entire range of environment to which the host or carrier is exposed." These definitions make masking a partial latency, differing only in duration, and according to Hildebrand, a long-latency virus should be practically indistinguishable from a cell component. Divergent views and some confusion of definition of terms were evident at the *Symposium on Latency and Masking in Viral and Rickettsial Infections* held at Madison, Wisconsin, and edited by Walker *et al.* (1957). Part of the difficulty stemmed from the desirability of arriving at definitions applicable to both plant and animal viruses. Latent viruses infecting animals can be activated by a number of stimuli, e.g., by sunlight and fever (herpes in man); by chilling (swine influenza); or by hormonal factors (Bittner in mice). Bawden, however, stated that "There is no evidence that any treatment of a plant has activated what I would call a truly latent virus."

Bawden preferred to use the term "commensalism" to describe the condition wherein an infected plant is free from visible lesions. This term is synonymous with "latency" and "masking" as generally understood, but Bawden considers that the term latency should be restricted to those obscure states where the lack of virulence depends on some change in the state of the virus. The symptomless carrier is best characterized by potatoes infected with virus X, and King Edward potato carrying potato paracrinkle virus. In the latter case, the symptomless carrier capacity is characteristic of the whole variety, and the virus is not found naturally in any other plant. It is, however, transmissible to tomato and other varieties of potato, still symptomless, but identifiable serologically and by electron microscopy.

A difference between "latent virus" and "latent infection" should be recognized, according to Lwoff; "latent" should be reserved to describe "infection" since a latent virus might be present in a noninfective stage. Bawden, however, considers these transmissible antigens rather than viruses, if they do not produce disease. If they cause disease, then they are viruses; if not, then they are viruslike particles. The symposium finally agreed on the following definitions:

Inapparent infections: The whole field of infections that give no overt sign of their presence.

Latent infections: Chronic inapparent infections with virus-host equilibrium.

Occult virus: Virus particles cannot be detected and the state of the virus is not ascertainable.

Provirus, vegetative virus, and infective virus are appropriate terms to use in describing developmental cycles, the infective virus being the fully formed virus particle.

Determination of the presence of a latent infection is fraught with some danger. K. M. Smith's latent virus of sugar beets appeared to be well established on the basis of the evidence (1951), but ten years later it was shown to be not a latent virus but a bacterial infection (Yarwood *et al.*, 1961).

THE BIOLOGICAL APPROACH
TO PLANT VIRUS PROBLEMS

The plant virus *in vitro* and the virus in the infected plant appear to have little in common, so it would seem that the biological approach is the only productive one for workers in the greenhouse and field. The case has been well stated by Kennedy (1951). Diagrammed, it appears somewhat as follows:

Plant————Insect————Virus————Disease

Biological Complex

Biochemical————Ecological————Physiological

Biological Adaptations

This approach relegates the isolationist test-tube attitude and the equally restrictive concept of an independent self-sufficient organism to their respective speculative niches.

The four primary components of the biological complex cannot be separated in nature, and there is abundant evidence of interrelationships. The vector may be affected by the direct physiological action of the virus on the host plant, often favorably. Populations of *Thrips tabaci* were consistently higher on diseased *Emilia sonchifolia* plants than on healthy *Emilia* (Carter, 1939). A dense, luxuriant stand of sugar beets is not a favorable environment for *Circulifer (Eutettix) tenellus* (Baker), but the insect thrives when the stand has been cut down and opened up by the action of the curly-top virus (Carter, 1930). Severin (1946) found that aster yellows in celery resulted in plants on which a number of leafhopper vector species developed faster and reproduced more than on healthy plants.

Myzus persicae (Sulz.) multiplies faster on potatoes suffering from leaf roll or severe mosaic than on healthy potatoes (Arenz, 1951).

All of Kennedy's examples refer to favorable effects of virus-diseased plants on the insect vector, but unfavorable effects of the virus itself are also now known. Jensen (1959) reported on the lethal effect of Western X virus on the leafhopper, *C. montanus* Van D.; with increasing evidence for the multiplication of plant viruses in their insect vectors, many other cases are now recognized. The mechanism of host plant selection by insects is debatable, but whatever it is, host plant selection by vectors will govern the establishment of new virus hosts. Although there is little evidence for modification of a virus by passage through specific host plants, it has been shown that it can occur (Carsner, 1925; Lackey, 1932; Bawden, 1956). Plant metabolism affects both infection and disease development. Light intensity, temperature, nutrition, and the physiological age of the plant are all factors recognized as affecting disease expression. Conversely, the virus "imposes a whole new metabolic pattern on the plant, and virus individuality becomes merged in that of the infected cell" (Kennedy, 1951). The biological approach rejects the idea that viruses, even though arising by accident or fortuitously, maintain their continuing status on any such tenuous basis. The mechanism, at least, for this continuity now appears to reside in the ribonucleic acid, but that does not vitiate Kennedy's views on the biological complex that induces disease.

One of the very specific vector-virus relationships is that between thrips and tomato spotted wilt. This is a virus disease recognized within a very short period of time in widely separated parts of the world. All the vectors are thrips species, and all have the same relationships with the virus, although regional strains of the virus exist. Wherever and however this virus originated, it is now part and parcel of a closely integrated biological complex.

The biological approach also rejects the notion that disease production is inherent in the "ultimate particle" itself. Rather, it is "the changing relations of viruses with plants, insects and the rest of the environment which produces disease" (Kennedy, 1951).

The ecological aspect is the broadest and most all-inclusive one, and it is, to use Kennedy's expression, the arena of evolutionary change. Beginning with the ecology of the healthy cell and going on to the infected cell, which Pirie is essentially doing with his biochemical approach (1955), ecology is the integrator of all the diverse and gradually ascending systems involved. Ecology will not tell us the shape and size of an ultimate particle, which, introduced into one part of a system by another, initiates a metabolism pathological to still another part, but it does lead to an understanding of that biological complex which includes plant, insect, virus, and disease. As Kunkel once remarked to the writer, virus chemistry advances notwithstanding, "We still have the diseases in the field."

IDENTIFYING, NAMING AND CLASSIFYING VIRUSES AND VIRUS DISEASES

The historical procedure has followed a standard pattern, i.e., discovery, attaching a vernacular name, and a description based on symptoms. With the development of more refined methods, descriptions became more complete. Methods of transmission, mechanical or by insects, grafting, and by dodder, became essential tools for identification. With the advent of virus purification *in vitro* and the determination of virus particle morphology, the available tools reached their present status. Individuals as well as committees of international societies have addressed themselves to the problem. Hansen, whose contributions to the problem have continued for many years (1956, 1957, 1957a, 1966, and 1968), has modified his proposals from time to time to meet objections, so that only his most recent contribution (1970) is considered in this edition.

Ross (1964) summarized the procedures for virus identification and set up the following criteria, giving them in their order of importance. He places symptomology first and virus particle characteristics last in his seven categories. The others are (B) properties in crude juice, (C) transmission characteristics, (D) interaction with other viruses, (E) response to specific genes in the host, and (F) serology.

Ranking of criteria is somewhat of a subjective matter, depending on the individual's experience with their use. However, if the Adansonian principle (Adanson, 1763) of using all known information, without ranking, is followed, it might eliminate one source of disagreement. The Provisional Committee on Nomenclature of Viruses (PCNV) set up by the International Association of Microbiological Societies in 1963 published their recommendations in 1965 (PCNV, 1965). They proposed that identification should be based only on the nucleic acid (RNA, DNA) morphology; serology is relegated to the lowest level, and symptoms are not taken into account at all except on the level of strains. A new Latin binomial nomenclature disregarding previous names was proposed, headed by the Kingdom Vira.

Prochenko (1970) proposed form, size, and structure of virus particles, chemical composition, antigenic properties, inactivation temperatures, modes of transmission, host-plant range, stability *in vitro,* and habitat, as criteria. He also added a Latin nomenclature with species, genera, and families. His 81 examples of species include 42 transmitted only mechanically.

Gibbs *et al.* (1966) have proposed using an eight-point cryptogram limited to four terms. The terms lean heavily on morphology. The example quoted is that of tobacco mosaic which would read:

$$\frac{R}{1} : \frac{2}{5} : \frac{E}{E} : \frac{S}{0}$$

which interpreted means that the virus is composed of RNA and is single stranded; molecular weight of RNA is 2×10^{-5}; 5% nucleic acid; virus particles and nucleocapsid both elongated with parallel sides and ends not rounded; host, seed plants; no vector known. However, even with an example drawn from a classical virus which is better known than any other perhaps, the cryptogram might change since Lojek (1969) has reported that if purified tobacco mosaic virus is used, *Myzus persicae* (Sulz.) can transmit the virus and retain it for five hours. With four terms, different viruses might have the same cryptograms, e.g., turnip yellow mosaic group and the squash mosaic group. The vernacular names in this case would serve to separate them. The cryptogram itself is not a name, but a shorthand method for virus identification added to the vernacular name. Since the latter is essential to the use of the cryptogram, its proponents vigorously defend the use of the vernacular name.

Gibbs (1968) has described the cryptogram in detail in an appendix to the list of plant viruses issued by the Commonwealth Mycological Institute as Phytopathological Paper No. 9. He has supplied a current cryptogram for each virus named in the list. The paucity of information on the majority of viruses is clearly demonstrated by the abundance of asterisks in the cryptograms, but the assumption is that definitive symbols will replace these in due time. In the meantime, the vernacular names will continue to serve their present purposes. Gibbs' (1969) review of the problem is inclusive (there are six pages of references) and provides a full account of the history and development of methods of naming and classifying viruses, culminating in an expansion of the use of cryptograms, with a passing reference to the possibilities of using computers. As Gibbs says, "The computer, unfettered by the well worn neuronic paths in a human mind, often produces novel and stimulating ideas." It is presumed that the neuronic paths of the human programmer will be adequate.

Hansen's basis for naming "genera" is a three-syllable cognomen derived from (1) location in the host, (2) vector-virus relationships, and (3) particle type. It is perhaps instructive to compare the symbols for identification and naming by Gibbs and Hansen in parallel columns, as shown in Table 9.2.

TABLE 9.2. *A Comparison of the Symbols used by Gibbs and Hansen for the Identification and Naming of Viruses*

Gibbs (1969)	Hansen (1970)

Four pr of symbols

Symbols for code-names of "genera"

1st. Type of nucleic acid /
    strandedness of NA
    R—RNA; D—DNA

Initial   —for site of infection
M(e)   —mechanical transmission
D(a)   —deep inoculation
G(e)   —soil transmission

2nd. Mol wt of NA (in millions)/
    % NA in infective particles

Middle   —for specific vector type

3rd. Outline of particle/
    outline of "nucleocapsid"
    Symbols for both properties
    S —essentially spherical
    E —elongated with parallel
        sides, end not rounded
    U —as E but ends rounded
    X —complex or none of above

aca   —mites
ale   —white flies
aphi   —aphids
cica   —leafhoppers
coc   —scale insects
in   —no specific vector known
het   —plant bugs
nema   —nematodes

4th. Kinds of host infected /
    kinds of vector
    Symbols for kinds of host
    A —actinomycete
    B —bacterium
    F —fungus
    I —invertebrate
    S —seed plant
    V —vertebrate
    Symbols for kinds of vector
    Ac—mite and tick
    Al —white fly
    Ap—aphid
    Au—leafhopper, plant, or treehopper
    Cc—mealybug
    Di—fly and mosquito
    Fu—fungus
    Gy—mirid, piesmid, or tingid
    Ne—nematode
    Ps —psylla
    Si —flea
    Th—thrips
    Ve—vectors known but none of above
    O —spreads without vector
    Symbols for all pairs
    * —virus properties unknown
    ( )—information doubtful

plano   —fungal zoospores
psylli   —psyllids
thy   —thrips
x(e)   —multiple vector types
Ending   —for particle type
virus   —undefined contagious sub-
    stance and collective desig-
    nation for viruses;
    Elongated particles, sides
    parallel, ends not rounded
flexus   —threadlike and flexuous
lancea   —lance shaped; threadlike,
    curved or slightly flexuous
chorda —(-chords) rigid rods,
    moderately long and thin
codex   —(stem of a tree) rigid rods,
    long, thick
pachus—(-thick) -rigid rods, rela-
    tively short and thick;
    Elongated particles with
    parallel sides, ends rounded
brevis   —(-short) bacilliform, short
    (sausage-like or eliptical)
media   —(-intermediary) bacilliform,
    relatively long (slim
    sausages)
physa   —(-vesicle) large sausages
    with a membrane, often
    one end cut

It will be seen from this comparison that while Gibbs includes host data and a category of vector to include such cases (presumably) as beetles and grasshoppers, Hansen does not, and he specifically excludes all the biting insects as not true vectors. Workers with those groups will probably object. Hansen also defines particle characteristics in much greater detail than does Gibbs and he thus opens the way for more specific objections. The greatest difference however, is in Hansen's use of Latinized code names.

GROUPINGS OF VIRUSES. Groups of viruses, once characterized, are the building blocks for future developments, and there are several of these at various levels of erudition. Bawden (1964) retains his earlier point of view (1955) that the first step in grouping is the identification of the "collective species," which are defined as groups whose members share common antigens. He suggests that serological relationships provide the best criterion, in spite of recognized limitations. Stable virus protein is an essential, and this precludes many viruses. For relationships of some of these, grouping can come from the use of interference phenomena, that is, where one virus interferes with the multiplication of another. Distinctive symptoms on a common host are essential.

However, serology as a tool for identification is perhaps most useful in identifying virus strains, or viruses so closely related that they would be considered as components of collective species. This cuts down the number of viruses recognized as distinct entities, but leaves it still necessary to retain names for purposes of identification of diseases. Van Regenmental (1966) established eight separately named viruses as strains of tobacco black-ring and four as strains of *Arabis* mosaic. Valleau (1940) drew up a key to the tobacco viruses based on the property of resistance to drying and transmissibility by plant sap. Host reactions (symptoms) were used at the species level. Although Valleau used Latin binomials, these added nothing to the value of his key. The eight viruses considered end in seven separate "genera." Over twenty years ago, Weiss (1939) published a key to typical viruses of leguminous crops. The nomenclature he used might not be acceptable now; the important contribution was the "classification" of a group of viruses attacking a major group of crop plants.

After a review of the pertinent literature on potato viruses, MacLeod (1952) set up six groups based on methods of transmission. These are:

1. *Sap inoculation but not insect transmission:* Virus F (tuber blotch), virus G (*Aucuba* mosaic), and virus X.
2. *Sap inoculation and aphid transmission:* Viruses A and Y, calico, leaf-rolling mosaic, and spindle tuber.
3. *Leafhopper or thrips transmission:* Green dwarf, tomato spotted wilt, and yellow dwarf.

4. *Aphid transmission:* Leaf roll.
5. *Dodder transmission:* Aster yellows, purple top (bunchy top), and witch's broom.
6. *Graft transmission only:* Streak and roll, etc.

MacLeod also described the techniques necessary to accomplish the separation. His paper includes a bibliography of 104 titles. Brandes and Wetter (1959) have suggested that the shape, diameter, and length of the particles of the elongated plant viruses could serve the purposes of classification of this group. Length is the most useful feature. These authors place a number of elongated viruses in twelve groups according to their normal lengths. Electron microscope photographs illustrating some of these are shown in Fig. 9.4. An even smaller group is that comprising three soil-borne viruses. Harrison and Nixon (1960) compared three viruses, all soil-borne, viz., tomato black ring, raspberry ringspot, and arabis mosaic viruses. The particle shape, mode of transmission *in vitro,* and behavior in plants suggest that these viruses should be placed in the same general group in any system of identification. Brandes and Bercks (1965) have followed up the work of Brandes and Wetter (1959) and identify some viruses on the basis of elongate morphology and serology.

Three cassava viruses, i.e., common mosaic, vein mosaic, and brown streak, can be separated by their elongate particles (Kitajima and Costa, 1964). Fifteen soil-borne viruses have been catalogued by Kristensen (1962). The crop can be a useful basis for cataloguing. Roland and Tehon's list of sugar beet viruses includes a detailed list of host plants of sugar beet yellows virus (1961). Plants tested and proved negative are also included, and there is an interesting phylogenetic tree of the host plants. It is noteworthy that with one exception, all the positive host plants are on one side of the tree. Roland and Tehon's study could well be emulated for other crops; it would be most useful for workers in the field.

Slykhuis (1967) has tabulated the cereal viruses known throughout the world in a manner which could well prove a model for similar treatments of other crops. The groupings are according to the primary natural means of transmission; soil, seed, mites, beetles, aphids (with divisions for circulative and stylet-borne), leafhoppers, and mechanically transmitted cases where the vector, if any, is not known.

Pitcher has tabulated the nematode transmitted viruses as part of a broad biological review. Thresh (1966, 1967) has illustrated his catalogues of black currant and red currant virus diseases; another useful feature is the inclusion of synonymy. Hashioka (1964) lists the known virus diseases of rice with workable descriptions. This list appears to be incomplete unless synonymy is involved. At the grass roots level, Holdeman

TABLE 9.3.   Viruses Grouped by Serological Reactions[a,b]

Group	Virus
Tobacco mosaic	Common and masked tobacco mosaic, tomato aucuba mosaic, enation mosaic and streak; Japanese petunia; Holmes' rib grass; cucumber viruses 3 and 4
Potato X	Salaman's H, G, L, S, and N strains; potato viruses B and D; potato mottle and ringspot; *Hyoscyamus* virus 4
Potato Y	Potato leaf-drop streak, rugose mosaic, and vein-banding; potato virus C; tobacco veinal necrosis; *Hyoscyamus* virus 2
Tobacco etch	Severe and mild etch; Blakeslee's Z-mosaic virus of *Datura*
Henbane mosaic	Severe and mild strains of *Hyoscyamus* virus 3
Cucumber mosaic	Price's isolates of cucumber virus 1; Valleau's delphinium virus
Soybean mosaic	
Pea mosaic	Osborn's pea viruses 2 and 3
Tulip mosaic	
Sugar beet yellows	
Tobacco ringspot	Wingard's and yellow and green tobacco ringspots; tobacco ringspot virus 2
Tobacco necrosis A	Potato, Princeton, and tobacco VI cultures
Tobacco necrosis B	Tobacco I and II cultures
Tobacco necrosis C	Rothamsted culture and bean stipple-streak
Tomato bush stunt	
Southern bean mosaic	
Turnip yellow mosaic	

[a] The viruses within each group precipitate with each other's antisera, whereas those in different groups do not.
[b] From Bawden (1955).

(1965) has developed a key for the use of field inspectors, based on symptoms of maize viruses.

Strains of a single virus, barley stripe mosaic, have been described, and their biological characteristics and evolution discussed, by McKinney and Greeley (1965). Abbott and Tippett (1966) have described the symptoms of eight strains of sugarcane mosaic on differential hosts. They concluded that to attempt to describe all the variants would have no practical value.

Gibbs (1969) has listed what he considers currently recognized groups of viruses based on their insect transmission; the aphid transmitted viruses are divided into three subgroups depending on the persistence of the virus in the insect. These groupings involve great heterogeneity, two or more (among about 100 for *Myzus persicae*) might have little in common except the vector group.

The International Committee for Nomenclature of Viruses (ICNV) and its subcommittee for the plant viruses (PVS), both reporting to the IX International Congress of Microbiology, were set up in 1966. It seems clear from the summary published in *Phytopathology News* for March, 1971 (Vol. 53) that the ICNV was primarily concerned at first (1966) with the mechanics of naming viruses. Their recommendations in 1966 were as follows:

1. The code of bacterial nomenclature shall not be applied to viruses.
2. Nomenclature shall be international.
3. Nomenclature shall be universally applied to all viruses.
4. An effort will be made towards a Latinized binominal nomenclature.
5. Existing Latinized names shall be retained whenever feasible.
6. The law of priority shall not be observed.
7. New sigla shall not be introduced.
8. No person's name shall be used.
9. No nonsense names shall be used.
10. For pragmatic purposes the species is considered to be a collection of viruses with like characters.
11. The genus is a group of species sharing certain common characters.
12. The rules of orthography of names and epithets are listed in Chapter 3, Section 6, of the proposed international code of nomenclature of names (C).
13. The ending of the name of a viral genus is "virus."
14. To avoid changing accepted usage, numbers, letters, or combinations may be accepted for names of species.
15. These symbols may be preceded by an agreed abbreviation of the Latinized name of a selected host genus, or, if necessary, by the full name.
16. Should families be required, a specific termination to the name of the family will be recommended.
17. Any family name will end by "-idae."

The debate in the March, 1971 issue of *Phytopathological News* was continued in the May, 1971 issue, and terminated, at the editor's request, in the November issue.

In 1970, at the X International Congress of Microbiology, the ICNV approved 40 group names, including 16 proposed by the PVS. Of the latter, however, the names of only 3 were acceptable: Bromovirus and Cucumovirus, because it can be imagined that they are Latin, and Nepovirus, which is already in use. The PVS did not request the status of genera for the 16 groups; the other 24 groups approved by the ICVN for bacterial, vertebrate, and invertebrate viruses were accorded that status and given Latinized names. The 16 groups proposed by the PVS are as shown in Table 9.4.

TABLE 9.4. *Groups of Viruses Approved by the ICNV*[a]

Group	Group name	Type member
1	Tobravirus	Tobacco rattle virus
2	Tobamovirus	Tobacco mosaic virus
3	Potexvirus	Potato virus X
4	Carlavirus	Carnation latent virus
5	Potyvirus	Potato virus Y
6	Cucumovirus	Cucumber mosaic virus
7	Tymovirus	Turnip yellow mosaic virus
8	Comovirus	Cowpea mosaic virus
9	Nepovirus	Tobacco ringspot virus
10	Bromovirus	Brome mosaic virus
11	Tombusvirus	Tomato bushy stunt virus
12	Caulimovirus	Cauliflower mosaic virus
13		Alfalfa mosaic virus
14		Pea enation mosaic virus
15		Tobacco necrosis virus
16		Tomato spotted wilt virus

[a] Names approved are underlined.

A major bone of contention appears to be the use of Latinized names as opposed to the vernaculars, and the ICVN, being an international body, decrees *ex cathedra* once a consensus has been obtained. Their ruling on Latinized names could be anticipated in view of the far greater number of animal virologists involved and the non-English speaking plant virologists in Continental Europe. Acceptance of their decree means that vernacular names will not be recognized; but for practicing plant virologists in the laboratory and field, and their constituents on the farms, nothing will serve to eliminate the necessity to use vernacular names. This will be true whatever language the vernacular name is written in. The cryptogram of Gibbs *et al.,* is not a system of naming, but if attached in parenthesis to a vernacular name in publication, it would provide a means of technical identification internationally, irrespective of the language of the vernacular name.

Harrison *et al.* (1971) have resubmitted the list of groups proposed by the PVS in exactly the same order, with the same sigla and the same type members. Their headings for recording data are: (A) behavior in insects, (B) vector relations, (C) particle properties, and (D) particle composition —the last named including nucleic acids, protein, lipid, and other components.

Classification

Historically, classification has been confused with naming, as is demonstrated by the systems summarized by a committee of the American Phytopathological Society in 1949. McKinney (1944) and Bawden (1955) reviewed the problem current at that time. See also Lwoff (1967).

The systems referred to are as follows:

J. JOHNSON'S SYSTEM. The designation of each plant virus consists of the common name of the host on which the virus was first found and described, followed by the term "virus" and an Arabic number corresponding approximately to the chronological order in which the virus was described on the host in question. The name is printed in italics or the equivalent. Capital letters following the number designate strains with the alphabetical order indicating the degree of attenuation. Substrains may be indicated by small letters following the capital letters. The common name of the plant host may be in other languages than English. No attempt is made to classify the viruses according to a natural system. Tobacco mosaic virus=Tobacco virus 1 Johnson.

K. M. SMITH'S SYSTEM. This system is a modification of Johnson's system. The common name of the host plant is replaced by its generic Latin equivalent. The numbers may or may not be the same as Johnson's. The type strain has no letter, and the lettering begins with strains derived from the type. The lettering does not attempt to indicate the virulence of the strain. No attempt is made at classification according to a natural system. Smith's book deals with most of the then known plant viruses. Tobacco mosaic virus=*Nicotiana* virus 1 Smith.

H. S. FAWCETT'S SYSTEM. This is a binomial system in which the first term of each name (the "genus") is derived arbitrarily from the name of the host in which the virus was first described by adding the suffix "vir" (from Latin *virus,* neuter) to the genitive form of the Latin generic name, dropping the final consonant if the genitive ends in a consonant; the second term (the "species") is a descriptive name. It is basically like the number systems but substitutes a word for the number. To avoid confusion when more than one virus is attributed to the same host, the specific name of the virus should not be based on the generic or specific name of the host. Authorities for the virus names are cited only for the specific part of the name, not for the artificial generic name. Tobacco mosaic virus might be named *Nicotianaevir commune.* (Based on the recommendations for forming names as given in the Rules of Botanical Nomenclature, the formation of these generic names might be further modified by using *i* as the connective instead of *ae* when the generic name of the host ends in *a.*)

F. O. HOMES' SYSTEM.   This system consists of an extension of the Latin binomial system of nomenclature to include viruses. An attempt is made to base the system of natural relationships among the viruses. The viruses are placed in an Order Virales. There are three suborders: Phagineae (infecting bacteria), Phytophagineae (infecting higher plants), and Zoophagineae (infecting animals). In the Phytophagineae there are six families separated on the basis of symptoms produced and means of transmission. The genera in these families are differentiated according to symptoms and insect vectors. Species separations are based on host range, symptoms, means of transmission, geographical range, and physical or chemical properties. Holmes' book deals with most of the known plant viruses as well as with bacterial and animal viruses. The plant viruses are given the rank of a division of the plant kingdom and called Viriphyta. There are two families determined on the basis of symptoms produced; the genera in these families are distinguished by their means of transmission and host reactions. Where necessary, the type species in each genus is described. The system has been applied in a description and revision of several virus species in the genera *Marmor, Fractilinea,* and *Galla.* Tobacco mosaic virus = *Marmor tabaci* (Holmes ex Valleau) McKinney.

H. H. THORNBERRY'S SYSTEM.   This proposal has been presented in outline only. Under the Plant Kingdom, Phylum Thallophyta, Class Schizomyceta, an Order Biovirales is erected in which three families are proposed, Rickettsiaceae, Phytoviraceae, and Zooviraceae. In the Phytoviraceae, five genera, *Phytovirus, Pteridovirus, Bryovirus, Thallovirus, and Bacteriophagus* are proposed. The known viruses of the higher plants would all belong to the one genus *Phytovirus.* Species would be separated according to host range, methods of transmission, and symptoms on standard hosts. Species names would be compounded from the first or more syllables of the generic name of an important host prefixed to a Latin word indicating one of the chief symptoms on a standard host. No attempt is made to classify the plant viruses according to natural relationships. Tobacco mosaic virus = *Phytovirus nicomosaicum* Thornberry.

W. D. VALLEAU'S SYSTEM.   In an application of the Latin binomial system to the viruses of tobacco, Valleau uses Holmes' terminology where possible. He creates several new genera, eliminates varietal names for type species, and favors creation of a catchall genus, probably *Marmor,* for imperfectly understood viruses. Tobacco mosaic virus = *Musivum tabaci* Valleau.

Holmes' system is an attempt at classification, and it was recognized as such by inclusion in Bergey's *Manual of Determinative Bacteriology.* Thornberry's system was later expanded in 1966 to include data available up to 1958. He used Latinized binomials based on vernacular names.

The most recent and elaborate attempt at classification to generic level is that of Hansen. Since the PVS has rejected Hansen's earlier proposals, this latest one will no doubt suffer the same fate. It is however, the most complete Latinized binominal system currently proposed, with generic names built up from the three-syllable cognomen, and specific names derived from the name of the principal host plant. Objections are likely, as with any other system, on the grounds that changes will occur in the generic name of a virus as knowledge of its characteristics is improved. An example is tobacco ringspot which has recently been shown to be transmitted by thrips (Bergeson *et al.,* 1964) which would put TRV into Hansen's Mexe-viruses. Hansen's inclusion of the mycoplasma as "virus-like" is perhaps to illustrate the flexibility of his system, but it stretches it unreasonably since there is nothing virus-like about mycoplasmas.

Currently it can be said that there is progress being made on the problem of identification and naming of viruses, but there must develop a reasonable degree of acceptance, not by decree, but by demonstration of usefulness, before meaningful progress can be made on the problem of classification. Perhaps two systems will evolve, one all inclusive of viruses wherever found, and the other a system for plant viruses for the working plant virologist in the field; the latter will inevitably retain the vernacular names.

It might, however, be wise to recall once more the advice of Andrews (1952) who said "If students of viruses take thought and time and base their classification and nomenclature on solid foundations with reference from the very beginning to type material, they can forever be free from the nightmare of change and contentiousness which bedevil nomenclature."

REFERENCES

Abbott, E. V., and Tippett, R. L., "Strains of sugarcane mosaic," *USDA Tech. Bull.,* **1340**, (1966).

Andanson, M., "Families des plantes," *Vol. 1 Vincent Paris* (1763).

Andrews, C. H., "The place of viruses in nature," *Proc. Roy. Soc. (London),* **B139**, 313–326 (1952).

Arenz, B., "Der einfluss verschiedener faktoren auf die Resistenz der Kartoffel gegen die Pfirsichblattlaus. Weitere ergebnisse über die Resistenz der Kartoffel gegen die Pfirsichblattlaus," *Z. Pflanzenbau Pflangenschutz.,* **2**, 49–62, 63–67 (151).

Ball, E. D., "The beet leafhopper (*Eutettix tenella*)," *Utah State Agr. College Expt. Sta. 16th Ann. Rept.,* 1904–1905 (1906).

Ball, E. D., "The beet leafhopper and the curly-leaf disease that it transmits," *Utah State Agr. College Expt. Sta. Bull.,* No. 155 (1917).

Bawden, F. C., *Plant Viruses and Virus Diseases,* 3rd ed., Chronica Botanica Co., Waltham, Mass., 1950.

Bawden, F. C., "The classification of viruses," *J. Gen. Microbiol.,* **12**, 362–365 (1955).

Bawden, F. C., "Reversible, host-induced, changes in a strain of tobacco mosaic virus," *Nature,* **177**, 302–304 (1956).

Bawden, F. C., "The multiplication of viruses," *Plant Pathol.,* **2**, 71–116 (1960).

Bawden, F. C., *Plant Viruses and Virus Diseases,* 3rd ed., 4th ed., Chronica Botanica Co., Waltham, Mass., 1955, 1964.

Bawden, F. C., Pirie, N. W., Bernal, J. E., and Fankuchen, I., "Liquid crystalline substances from virus infected plants," *Nature,* **138**, 1051–1052 (1936).

Bawden, F. C., and Pirie, N. W., "The isolation and some properties of liquid crystalline substances from solanaceous plants infected with three strains of tobacco mosaic virus," *Proc. Roy. Soc. llondon*), **B123**, 274–320 (1937).

Bawden, F. C., Kassanis, B., and Noxon, H. L., "The mechanical transmission and some properties of potato paracrinkle virus," *J. Gen. Microbiol.,* **4**, 210–219 (1950).

Beale, H. P., "Immunological reactions with tobacco mosaic virus," *Proc. Soc. Exptl. Biol. Med.,* **25**, 702–703 (1928).

Beale, H. P., "Immunological reaction with tobacco mosaic virus," *J. Exptl. Med.,* **49**, 919–935 (1929).

Beale, H. P., "Specificity of the precipitin reaction in tobacco mosaic disease," *Contribs. Boyce Thompson Inst.,* **3**, 529–539 (1931).

Beijerinck, M. W., "Ueber ein contagium vivum fluidum als Ursache der Fleckenkrankheit der Tabaksblatter," *Verhandel. Koninkl. Akad. Wetensch. Amsterdam Af. Natuurk.,* 6, 1–22 (1898).

Bennett, C. W., "The origin of new or little known virus diseases," *Plant Disease Reptr. Suppl.,* 211, 43–46 (1952).

Bergeson, G. B., Athow, K. L., Lavioleette, F. A., and Sister Mary Thomasine, "Transmission, movement and vector relationships of tobacco ringspot virus in soybeans," *Phytopathology,* 54, 723–728 (1964).

Black, L. M., "Viruses that reproduce in plants and insects," *Ann. N. Y. Acad. Sci.,* 56, 398–413 (1953).

Brandes, J., and Wetter, C., "Classification of elongated plant viruses on the basis of particle morphology," *Virology,* 8, 99–115 (1959).

Brandes, J., and Bercks, R., "Gross morphology and serology as a basis for classification of elongated plant viruses," *Advances in Virus Research,* 9, 1–24 (1965).

Burnet, F. M., *Virus as Organism,* Harvard Univ. Press, Cambridge, Mass., 1945.

Carsner, E., "Attenuation of the virus of sugar beet curly-top," *Phytopathology,* 15, 745–757 (1925).

Carter, W., "Ecological studies of the beet leafhopper," *U.S. Dept. Agr. Tech. Bull.,* No. 206 (1930).

Carter, W., "Populations of *Thrips tabaci,* with special reference to virus transmission," *J. Animal Ecol.,* 8, 261–276 (1939).

Cohen, S. S., "Comparative biochemistry and virology," *Advances in Virus Research,* 3, 1–48 (1955).

Darlington, C. D., "Heredity, development, and infection," *Nature,* 154, 164–169 (1944).

DuBuy, H. G., and Woods, M. W., "Evidence for the evolution of phytopathogenic viruses from mitochondria and their derivatives. II. Chemical evidence," *Phytopathology,* 33, 766–777 (1943).

Duggar, B. M., and Armstrong, J. K., "Indications respecting the nature of the infective particles in the mosaic disease of tobacco," *Ann. Missouri Botan. Garden,* 10, 191–212 (1923).

Dvorak, M., "The effect of mosaic on the globulin of potato," *J. Infectious Diseases,* 41, 215–221 (1927).

Gardner, A. D., "Microbes and ultramicrobes," *Methuen's Monographs on Biological Subjects,* London, 1931.

Gibbs, A. J., "Cryptograms," *Appendix III in Plant Virus Names, Phytopathology Papers,* No. 9, *Commonwealth Mycol. Inst.* (1968).

Gibbs, A. J., "Plant virus classification," *Advances in Virus Research,* 14, 263–328 (1969).

Gibbs, A. J., Harrison, B. D., Watson, D. H., and Wildy, P., "What's in a virus name?" *Nature,* 209, 450–454 (1966).

Green, R. G., "On the nature of filterable viruses," *Science,* 82, 443–445 (1935).

Hansen, H. P., "Correlations and interrelationships in viruses and in organisms. Classification and nomenclature of plant viruses," *Den Kgl. Vet.-Og Landbohojskoles Arskrift,* **1956,** 108–137.

Hansen, H. P., "Correlations and interrelationships in viruses and in organisms. II. The principles of the natural periodical system of plant and animal infecting viruses," *Den Kgl. Vet.-Og Landbohojskoles Arskrift,* **1957,** 31–66.

Hansen, H. P., "The natural periodical systems of organo-genes and viruses as the basis of the practical classification and nomenclature of the viruses," *Proc. 3rd Conf. Potato Virus Diseases Lisse-Wageningen,* **1957,** (1957a).

Hansen, H. P., "On the general nomenclature of viruses," *Arsskr. K. Vet.-Landbohjak.,* **75,** 191–206 (1966).

Hansen, H. P., "On the general nomenclature of viruses II," *Den. Kgl. Veterinaer-Og. Landbohjsloles Arsskrift,* **77,** 184–197 (1968).

Hansen, H. P., "Contribution to the systematic virology," *The Royal Vet. and Agric. Univ. Copenhagen,* 1–110 (1970).

Harrison, B. D., Gibbs, A. J., Hollings, M., Shepard, R. J., Valenta, V., and Wetter, C., "Sixteen groups of plant viruses," *Virology,* 45, 350–363, (1971).

Hashioka, Y., "Virus diseases of rice in the world," Reprinted from *Risso* (Dec., 1964).

Hildebrand, E. M., "Masked virus infection in plants," *Ann. Rev. Microbiol.,* **12,** 441–168 (1958).

Holdeman, Q. L., "Virus diseases of corn—tentative key for preliminary and laboratory decisions," *Plant Disease Reptr.,* **49,** 654–655 (1965).

Holmes, F. O., "Accuracy in quantitative work with tobacco mosaic virus," *Botan. Gaz.,* **86,** 66–81 (1928).

Holmes, F. O., "Inoculating methods in tobacco mosaic studies," *Botan. Gaz.,* **87,** 56–63 (1929).

Hughes, A. W. McK., "Aphis as a possible vector of 'breaking' in tulip species," *Ann. Appl. Biol.,* **17,** 36–42 (1930).

Iwanowski, D. J. V. V., "Ueber zurie krankheiten der Tabakspflanze," *Bull. Acad. Imp. Sci. St. Petersburg,* **35,** 67–70 (1892).

Jenifer, F. G., and Corbett, M. K., "Studies on the nature and relationship of satellite virus and some of its relationships to tobacco necrosis virus," *Phytopathology,* **61,** 129 (Abstr.) (1971).

Jensen, D. D., "A plant virus lethal to its insect vector," *Virology,*8, 164–175 (1959).

Johnson, J., "The attenuation of plant viruses and the inactivating influence of oxygen," *Science,* **64,** 210 (1926).

Kassanis, B., "Properties and behaviour of a virus depending for its multiplication on another," *J. Gen. Microbiology,* **27,** 477–488 (1962).

Katsura, S., "The stunt disease of Japanese rice, the first plant virosis shown to be transmitted by an insect vector," *Phytopathology,* **26,** 887–895 (1936).

Kennedy, J. S., "A biological approach to plant viruses," *Nature, 168,* 890–894 (1951).

Kitajima, E. W., and Costa, A. S., "Elongated particles found associated with cassava brown streak," *East African Agric. and For. J.,* 30,28–30 (1964).

Klug, A., and Caspar, D. L. D., "The structure of small viruses," *Advances in Virus Research,* 7, 225 (1960).

Kristensen, H. R., "Soil-borne viruses," *Tidsskr. Planteavl.,* 66, 75 (1962).

Lackey, C. F., "Restoration of virulence of attenuated curly-top virus by passage through Stellaria media," *J. Agr. Research,* 44, 755–765 (1932).

Laidlaw, P. P., *Virus Diseases and Viruses,* Cambridge Univ. Press, London, 1938.

Lojek, J., "Transmission of purified TMV by aphids," *Proc. Can. Phytopath. Soc.,* 36, 13–26 (Abstr.) (1969).

Lwoff, A., "The nature of phage reproduction," *The Nature of Virus Multiplication,* P. Fildes and W. E. Heyningen, eds., Cambridge Univ. Press, London, 1953, pp. 149–174.

Lwoff, A., *Symposium on Latency and Masking in Viral and Rickettsial Infections,* D. L. Walker, R. P. Hanson, A. S. Evans, eds., Burgess, Minneapolis, 1958, p. 186.

Lwoff, A., "Principles of classification and nomenclature of viruses," *Nature,* 215, 1314 (1967).

MacLeod, D. J., "The identification of potato viruses," *34th Rep. Quebec Soc. Protect. Plants,* 1952, 87–96.

Maramorosch, K., "Multiplication of plant viruses in insect vectors," *Advances in Virus Research,* 3, 221–250 (1955).

Matthews, R. A. F., *Plant Virology,* Academic Press, New York, 1970.

Mayer, A. E., "Üeber die Mosaikkranheit des Tabaks," *Landwirtsch. Vers.-Sta.,* 32, 451–467 (1886).

McKinney, H. H., "Virus mixtures that may not be detected in young tobacco plants," *Phytopathology,* 16, 893 (1926).

McKinney, H. H., *J. Wash. Acad. Sci.,* 34, 139 (1944).

McKinney, H. H., and Greeley, L. W., "Biological characteristics of barley stripe-mosaic virus strains and their evolution," *Tech. Bull. USDA,* 1324 (1965).

Muller, H. J., "Variation due to change in the individual gene," *Am. Naturalist,* 56, 32 (1922).

Mulvania, M., "Studies on the nature of the virus of tobacco mosaic," *Phytopathology,* 16, 853–871 (1926).

Murphy, P. A., "On the cause of rolling in potato foliage; and on some further insect carriers of leafroll disease," *Sci. Proc. Roy. Dublin Soc.,* 17, 163–184 (1923).

Nishimura, M., "A carrier of the mosaic disease," *Bull. Torrey Botan. Club,* 45, 219–231 (1918).

Pirie, N. W., "Summing-up," *J. Gen. Microbiol.,* 12, 382–386 (1955).

Pitcher, R. S., "Interrelationships of nematodes and other pathogens of plants," *Helminth. Abstracts,* 34, 1 (1965).

Provisional Virus Nomenclature Committee, "A new system of classifying viruses," *Ann. Inst. Pasteur,* 109, 625–637 (1966).

Prochenko, A. E., "A natural classification of phytopathogenic viruses," *Phytopath. Zeit.,* 68, 41 (1970).

Richardson, J., and Sylvester, E. S., "Further evidence of multiplication of sowthistle yellow-vein virus in its insect vector, *Hypermyzus lactucae,*" *Virology,* 35, 347–355 (1968).

Roland, G., and Tehon, J., "Quelque considerations sur les virus de la betterave," *Revue de l'Agriculture,* 14, 869 (1961).

Ross, A. F., "Identification of plant-viruses," *Plant Virology,* M. K. Corbett and H. D. Sisler, Ed. University of Florida Press, 1964.

Salaman, R. N., "Protective inoculation against a plant virus," *Nature,* 131, 468 (1933).

Sanger, H. L., "Defective plant viruses," *Molecular Biology,* Springer-Verlag, Berlin, 1968.

Schultz, E. S., and Folsom, D., "Transmission, variation and control of certain degeneration diseases of Irish potatoes," *J. Agr. Research,* 25, 43–117 (1923).

Severin, H. H. P., "Longevity, or life histories, of leafhopper species on virus-infected and on healthy plants," *Hilgardia,* 17, 121–137 (1946).

Slykhuis, J. T., "Virus diseases of cereals," *Rev. Appl. Mycology Contr.,* 46, 614 (1967).

Smith, E. F., "Additional evidence of peach yellows and peach rosette," *U.S. Dept. Agr. Division Veg. Pathol. Bull.,* No. 1 (1891).

Smith, E. F., "Peach yellows and peach rosette," *U.S. Dept. Agr. Farmers' Bull.,* No. 17 (1894).

Smith, K. M., "A latent virus in sugar-beets and mangolds," *Nature,* 167, 1061 (1951).

Smith, K. M., and Doncaster, J. P., "The preparation of gradocol membranes and their application in the study of plant viruses," *Parasitology,* 27, 523–542 (1935).

Stanley, W. M., "Isolation of a crystalline protein possessing the properties of tobacco mosaic virus," *Science,* 81, 644–645 (1935).

Stanley, W. M., "The reproduction of virus proteins," *Am. Naturalist,* 72, 110–123 (1938).

Sylvester, E. S., and Richardson, J., "Infection of *Hyperomyzus lactucae* by sowthistle yellow vein virus," *Virology,* 42, 1023–1042 (1970).

Takami, N., "On stunt disease of rice plant and *Nephotettix apicalis* Motsch. var. *Cincticeps* Uhl.," *J. Japan Agr. Soc.,* 241, 22–30 (1901).

Taylor, C. E., and Robertson, W. M., "Location of tobacco rattle virus in the nematode vector, *Trichodorus pachydermus* Seinhorst," *J. General Virology,* **6,** 179–182 (1970).

Thornberry, H. H., "Index of plant virus diseases," *Agric. Handbook USDA,* 307 (1966).

Thresh, J. M., "Virus diseases of black currant," *Ann. Rept. East Malling Res. Sta. for 1965* (1966).

Thresh, J. M., "Virus diseases of red currant," *Ann. Rept. East Malling Res. Sta. for 1966* (1967).

Thung, T. H., "Smetstof en plantencel bij enkele virusziekten van de Tabaksplant," *Handel. Ned.-Ind. Natuurwetensch. Congr.,* 1931, 450–463.

Valleau, W. D., "Classification and nomenclature of tobacco viruses," *Phytopathology,* **30,** 820–830 (1940).

Van Regenmental, H. V., "Plant virus serology," *Advances in Virus Research,* **12,** 207–271 (1966).

Walker, D. L., Hanson, R. P., and Evans, A. S., *Symposium on Latency and Masking in Viral and Rickettsial Infections,* Burgess, Minneapolis, 1958.

Watson, J. D., and Crick, F. H. C., "The structure of DNA," *Cold Spring Harb. Sympos. Quant. Biol.,* **18,** 123 (1953).

Weiss, F., "A key to the typical viruses of leguminous crops," *Plant Disease Reptr.,* **23,** 352–361 (1939).

Yarwood, C. E., Resconich, E. C., Ark, P. A., Schlegel, D. E., and Smith, K. M., "So-called beet latent virus is a bacterium," *Plant Disease Reptr.,* **45,** 85–89 (1961).

The Clinical Aspects of Plant Virus Diseases

Included for consideration in this chapter as aspects of virus infection are the external and internal symptoms, and their effects on the physiology of the host plant. Symptomology provides the recognition signals for the presence of virus diseases, for it is by their symptoms that they are known. However, many mineral nutrient deficiences, some insect-induced toxemias, and even some genetic conditions may produce a variety of symptoms, any or all of which have their indistinguishable counterparts within the symptom range of established virus diseases.

EXTERNAL SYMPTOMS

Color Changes

The most commonly encountered symptom is that of color change in the chlorophyll and in other colors of flowers and foliage. A primary, local lesion at the point of entry of the virus may show a range of development from a faint yellowish spot through increasing intensities to necrosis. If the virus reaches the food stream, then the yellowing becomes systemic. If the virus is in the phloem, chlorosis of cells near the fibrovascular bun-

351

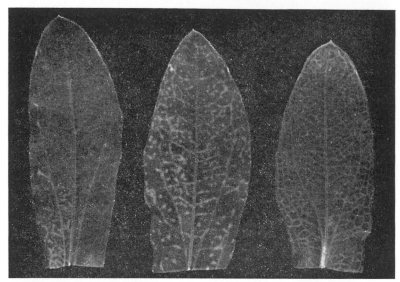

FIGURE 10.1. Three leaves of *Calendula officinalis*, infected by New Jersey yellow dwarf virus, which show clearing of veins (Photograph: L. O. Kunkel, 1954).

FIGURE 10.1A. Portions of leaf of New Jersey No. 2 hybrid corn. The leaf section on the left is from a healthy plant; that on the right shows a late stage of chlorosis caused by the virus of corn stunt disease (Photograph: L. O. Kunkel, 1954).

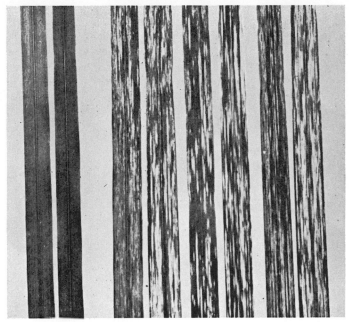

FIGURE 10.1B. Cocksfoot streak (*Dactylis glomerata*). Two healthy leaves on left (Photograph: K. M. Smith).

dles occurs, to produce the symptom known as vein-clearing (Fig. 10.1, 10.1A, 10.1B, and 10.2).

Another degree of color change is bleaching, shown in extreme development in *Thunbergia alata* Bojer following infection with New Jersey yellow dwarf of potato (Kunkel, 1954) (Fig. 10.3), but often arising by coalescence of chlorotic areas to involve the whole leaf, as with cucumber mosaic and tobacco ringspot viruses on celery (Severin, 1950).

Mottling of the mosaic type is so general a symptom that the term "mosaic" refers to a large group of diseases rather than to any specific one. Specific symptom designation depends, therefore, on characterizing the type of mosaic, or, more commonly, of associated symptoms (Fig. 10.4–10.7A).

A type of chlorosis referred to as an oak-leaf pattern is a striking symptom of virus disease infection and can occur in widely separated virus diseases. Kunkel (1954) cites an example from Turkish tobacco, caused by a strain of tobacco mosaic (Fig. 10.8), but an equally striking and distinct oak-leaf pattern is found in cacao infected with swollen-shoot disease.

Ringspot patterns of chlorosis, where there is a characteristic zoning effect, are common. When the zonated spot is a predominant symptom, it

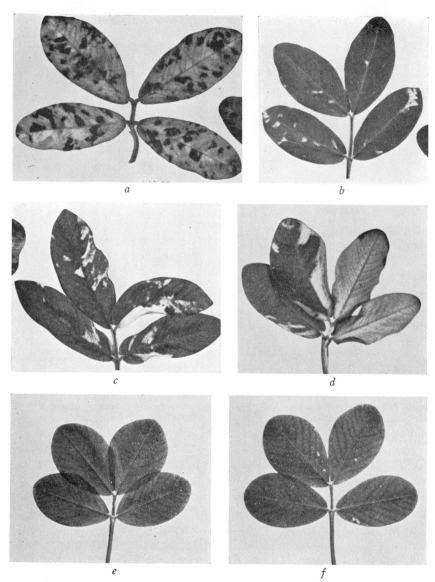

FIGURE 10.2. The range of symptoms of mosaic rosette in groundnuts. *a*, Leaf in which chlorotic areas predominate; *b*, leaf in which only small areas are chlorotic; *c*, leaf in which chlorotic areas are irregularly distributed, causing distortion. Mottle symptoms are unusually strongly marked in the "green" areas of this leaf; *d*, leaf with nearly unilateral distribution of chlorotic areas; *e* and *f*, symptoms of the mottle virus extracted from mosaic rosette; *e*, leaf showing only darkening; *f*, mottled leaf showing a few chlorotic specks (Photographs: H. H. Storey).

FIGURE 10.3. *Thunbergia alata* leaves. The leaf on the left is from a healthy plant; that on the right is from a plant infected by the virus of New Jersey yellow dwarf of potato (Photograph: Kunkel, 1954).

provides the common name for the disease (Fig. 10.9–10.14A). In some cases, the normal green color is intensified rather than becoming chlorotic. Kahn and Latterell (1955) report abnormally dark green leaves as one symptom of budblight of soy beans, a disease induced by one or the other of two ringspot viruses, those of tomato and tobacco. Vein-banding, as opposed to vein-clearing, is the term used to describe the paralleling of the veins on both sides, with a stripe of dark green color. Tompkins (1937) states that vein-banding replaces vein-clearing in cauliflower mosaic; however, this change appears to be limited to younger leaves (Caldwell and Prentice, 1942; Broadbent and Tinsley, 1953).

   Color changes in flowers (Fig. 10.15) are known as "flower-breaking." The classic example is the oldest virus disease of plants to be mentioned in the literature, namely, color breaking of tulips (van Slogteren and deB. Ouboter, 1941). Many of the bizarre patterns produced are of great beauty and are highly prized. Other examples, such as in pansies and violas (Severin, 1948), are too often associated with dwarfing and poor shape

a

b

FIGURE 10.4. *a,* Henbane mosaic on tobacco; *b,* broad bean mottle virus on *Vicia faba* (Photographs: K. M. Smith).

FIGURE 10.5. A leaf from a Turkish tobacco plant infected by ordinary tobacco mosaic virus which shows a typical mottling (Photograph: Kunkel, 1954).

FIGURE 10.6. *Left:* a typical *Abutilon striatum* leaf from a plant affected by *Abutilon* mosaic; *right:* a leaf from a healthy plant of the same species (Photograph: Kunkel, 1954).

FIGURE 10.7. Turnip yellow mosaic on Chinese cabbage (Photograph: K. M. Smith).

FIGURE 10.7A. Tobacco mosaic on tobacco var. .White Burley (Photograph: K. M. Smith).

FIGURE 10.8. Oak-leaf pattern of chlorosis caused by *Petunia aucuba* mosaic virus in a leaf of Turkish tobacco (Photograph: Kunkel, 1954).

FIGURE 10.9. Nasturtium (*Tropaeolum*) ringspot virus on tobacco var. White Burley (Photograph: K. M. Smith).

FIGURE 10.10. Cabbage black ringspot virus on young cabbage (Photograph: K. M. Smith).

FIGURE 10.11. Yellow spot in young pineapple field plant showing sideways leaning due to secondary rot on one side. This symptom gave rise to the early designation of the term "side rot" (Photograph: Linford, 1932).

FIGURE 10.12. Yellow spot in crown on immature pineapple (Photograph: Linford, 1932).

to be desirable. Similarly, color breaking in annual phlox, caused by the aster yellows virus, is also associated with distortion and virescence.

The color breaking of wallflowers and stocks can occur following infection with cabbage blackring virus, and sometimes gladioli are affected by cucumber mosaic in a similar manner. Tomato spotted wilt, one of the ringspot viruses, induces ringspot patterns in *Petunia* (Smith and Markham, 1944).

Teratological Symptoms

Teratological symptoms include leafy outgrowths called enations; crinkling, twisting, and shortening with production of secondary shoots; phyllody; leaf galls; big bud and floral gigantism; shoestringing; adventitious buds and roots; root swellings; and root tumors (Fig. 10.16–10.21). The leaf enations are outgrowths from the leaf surface and, according to Kunkel (1954), always grow downward. The inner surface of the enation represents the upper surface of the leaf in its extended growth into the enation.

Another type of leaf outgrowth is the leaf gall found in sugarcane plants infected with Fiji disease. These galls, according to Kunkel (1924), are caused by proliferation of phloem cells. Excessive growth of xylem tis-

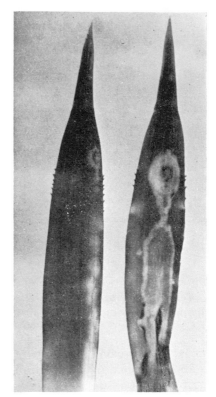

FIGURE 10.13.   Yellow spot (TSW) on pineapple seedling, showing spots, streaks, and secondary necrosis (Photograph: Linford, 1932).

sue in cacao trees infected with swollen shoot disease gives rise to the characteristic swelling along the stems, or to the clublike terminal portion of the stem (Fig. 10.22). An increase in the production of woody tissue occurs also in the wound tumor disease, particularly in the roots. This results in the typical root tumors from which the disease derives its name (Black, 1945). Curiously, tumors are stimulated by wounding of the infected plant and occur rarely, if at all, on uninjured roots.

Shoestringing, the partial or complete suppression of leaf blade development, is found in tobacco and tomato plants infected with some strains of tobacco mosaic and cucumber mosaic. Adventitious buds and roots are found in a number of virus diseases, and arise in leaves, stems, or flowers. Adventitious buds may arise in the axils of leaflets of healthy tomato plants that are grown under high levels of nutrition; but when healthy plants grown at ordinary levels of nutrition become infected by aster

FIGURE 10.14. Yellow spot symptoms on *Emilia* showing mosaic-like patterns, ringspots, and chlorotic streaks (Photograph: Linford, 1932).

yellows they regularly produce adentitious buds. Potato witch's broom virus causes the production of buds at the nodes of above-ground stems of the potato, and from these, long slender stolons develop. Adventitious buds are also developed in the stigmas of flowers infected by aster yellows virus, producing long chains of abnormal flowers. Proliferation of lateral buds gives rise to the condition known as witch's broom while shortening of internodes results in a bunched effect of top leaves.

Aerial roots are produced on plants suffering from sereh disease of sugarcane, and a similar reaction follows infection by the cranberry false blossom disease (Kunkel, 1954) (Fig. 10.23).

Wilting is only rarely found as a symptom of virus disease infection (Fig. 10.24), but Webb (1956) reported an atypical wilting in one seedling variety of potato infected with leaf roll virus, and McWhorter and Frazier (1956) describe the secondary symptom of western ringspot of beans as "quick wilt." More recently, Badami and Kassanis (1959) described a virus found in *Solanum jasminoides* Paxt. as tobacco wilt virus. Wilting due to a cacao virus is shown in Fig. 10.24A.

Excessive production of lateral roots occurs in sugar beets affected by curly top and in plants diseased by the California and New York strains

FIGURE 10.14A. Cacao necrosis virus. Transparent lesions following the veins appearing on occasional leaves 2–11 mo after grafting (Photograph: Courtesy of West African Cacao Research Institute, Ibadan, Nigeria).

of aster yellows. Malformation of fruits is a serious economic effect of some mosaic diseases; Keener (1954) described warty outgrowths and distorted fruits of cucurbits and tomato fruits resulting from mosaic of these plants.

Variability of Symptoms

Symptoms of a virus disease may be designated as "typical"; for any one host plant in a given environment, this is no doubt true. When the host range of a virus is extensive, however, disease symptoms caused by the one virus may be extremely variable. Table 10.1, taken from Holmes' in-

FIGURE 10.15. Virus-induced flower break in vanda Miss Agnes Joaquim. Healthy flower on left. (Photograph: Murakishi, Hawaii Agricultural Experiment Station).

TABLE 10.1. Symptom Range of Tobacco Mosaic Virus Disease[a]

Primary symptoms	Secondary symptoms
Necrotic and yellowish lesions	Mottling, distortion, and stunting
Starch retention	Necrosis and yellowish lesions
Chlorophyll retention	Prolonged yellowing
Abscission of inoculated leaf	Secondary lesions (demonstrable only by iodine staining)
	Tissue outgrowths
	Defoliation
	Leaf abscission
	Flower and fruit drop
	Bending of upper stem toward inoculated side
	Intensification of pigment spots on flowers
	Death by systemic necrosis

[a] After Holmes (1932).

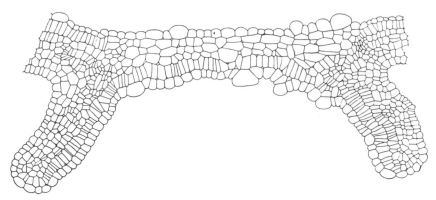

FIGURE 10.16. Teratological changes in *Nicotiana* L due to virus infection (Drawing: Boyce Thompson Institute).

FIGURE 10.17. Flax flowers affected by aster yellows virus, showing leaflike structures and sterile boll (Photograph: W. E. Sackston, Canadian Science Service).

a

b

FIGURE 10.18. *a*, Aster yellows virus symptoms on sunflower showing sector of sterile leaflike flower parts; *b*, six-spotted leafhoppers (*Macrosteles fascifrons*) feeding on sunflower leaf (Photographs: W. E. Sackston, Canadian Science Service).

FIGURE 10.19. Curly top of sugar beets showing curled leaves and vein swellings characteristic of the disease (Photograph: C. W. Bennett, United States Department of Agriculture).

a   b

FIGURE 10.19A. Symptoms of beet-leaf curl (*Piesma quadrata*). *a,* Initial symptom; *b,* advanced symptom.

FIGURE 10.20. A flower head of *Centuria imperialis* infected by aster yellows virus, with chains of flowers caused by the aster yellows disease (Photograph: Kunkel, 1954).

FIGURE 10.21. Tomato black ring virus on cucumber var. Telegraph showing enations (Photograph: Courtesy K. M. Smith).

FIGURE 10.22. Swollen shoot symptoms on infected cacao trees (Photograph: West African Cacao Research Institute, Ibadan, Nigeria).

FIGURE 10.23. Sections of stems from a Bonny Best tomato plant affected by the cranberry false blossom virus, with numerous roots that developed as a result of the disease (Photograph: Kunkel, 1954).

a   b

c

FIGURE 10.24. Purple top wilt of potatoes. *a,* Sequoia variety showing pinch rolling and top chlorosis (instead of purpling) of young leaves; *b,* sequoia variety in wilting stage; *c,* abnormal tuber production due to purple top wilt of potato (Photographs *a* and *b* from Anderson, Department of Agriculture, Victoria, Australia; Photograph *c* from Feddersen, Courtesy of Department of Agriculture, Adelaide, Australia).

FIGURE 10.24A   Cacao necrosis in a Parinari-type cacao seedling. Note complete collapse of leaves and necrosis of young axillary growth (Photograph: West African Cacao Research Institute, Ibadan, Nigeria).

vestigation (1932), exemplifies this point. Holmes used host plants of to-bacco mosaic representing twelve genera and two categories of symptoms, primary or secondary, depending on whether they occurred at the point of inoculation or some other part of the plant.

Since Holmes recorded these symptoms, many strains of tobacco mosaic have been recognized, showing various combinations of these symptoms, but it is clearly evident that with such a symptom range for a single plant, virus disease symptoms as a single unrelated criterion cannot be satisfactory for the characterization of a plant virus disease (Fig. 10.25).

a

b

c

FIGURE 10.25. Symptoms of cucumber mosaic virus on different host plants. *a,* Cucumber mosaic virus on *Nicotiana glutinosa; b,* cucumber mosaic virus on tomato; *c,* a bright yellow strain of cucumber mosaic virus on tomato (Photographs: K. M. Smith).

Virus Complexes

Complexes of more than one unrelated virus in a single plant often result in emergent symptomology, emergent being understood as "a novelty of behavior arising from the specific interaction or organization of a number of elements which therefore, constitute a whole, as distinguished from their mere sum, or resultant" (Wheeler, 1928). Many of these complexes are now well recognized. Table 10.2, which lists some of them, is derived from the *Review of Applied Mycology Supplement,* 1957.

TABLE 10.2. Virus Complexes Resulting in Emergent Symptomology in the Host Plant

Complex	Components
Raspberry mosaic	Black raspberry necrosis + rubus yellows necrosis
Rose mosaic (N.Z.)	Rose line pattern + rose vein-banding
Strawberry severe crinkle	Strawberry mottle + strawberry crinkle
Strawberry yellow edge	Strawberry mild yellow edge + strawberry mottle
Strawberry yellow edge complex	Mild yellow edge + crinkle + mottle
Strawberry mild crinkle	Strawberry vein chlorosis ± strawberry mottle
Tobacco rosette	Tobacco mottle + tobacco vein-distorting
Tomato fern leaf	Cucumber mosaic ± tobacco mosaic
Tomato mixed streak	Tobacco mosaic + potato X
Potato crinkle	Potato X + potato A
Carnation yellows	Carnation mosaic + carnation streak
Lily fleck	Lily symptomless + cucumber mosaic
Potato acropetal necrosis	Potato Y ± X
Potato rugose mosaic	Potato Y ± X
Potato mild mosaic	Potato X + A

Sometimes a virus complex is recognized even though one of the components may have been previously undescribed. Doering *et al.* (1957) have shown that tomato rosette is a disease caused by the tomato rosette strain of tobacco mosaic virus plus "shoestring," a virus previously unrecognized (Fig. 10.26–10.28).

Leaf shrivel is the term applied to a disease of tomato caused by a strain of potato virus Y. When the *Aucuba* strain of tobacco mosaic is added, the emergent symptom is known as yellow shrivel; when the added component is cucumber mosaic, it is known as fern leaf shrivel. Both the complexes result in more severe diseases and crop losses than the single component virus (Sturgess, 1956).

An extremely complex syndrome is that of sweet-potato-feathery mot-

FIGURE 10.26. Tobacco rosette—a disease caused by a virus complex (Photograph: K. M. Smith).

FIGURE 10.27. Shoestring of tomato caused by infection with tobacco mosaic and Texas carrot yellows viruses in combination, and a healthy leaf of the same age (Photograph: Kunkel, 1954).

FIGURE 10.28. Effect of virus mixtures on symptoms. Local lesions produced by virus X (strain 3XE-1) alone (*right*) and in the presence of TMV (*left*) (Photograph: Thompson, New Zealand Department of Scientific and Industrial Research).

tle, which is the result of infection by three viruses, namely, yellow dwarf, internal cork, and leafspot. Yellow dwarf is transmitted by the abutilon white fly (*Trialeurodes abutilonea*), but internal cork and leafspot are both aphid-transmitted (*Myzus persicae, Macrosiphon solanifolii, et al.*). The complexity is further increased by the existence of strains of each of the virus components (Hildebrand, 1960).

LATENT VIRUSES. Latent viruses are often important components of virus complexes. The classic example is that of potato virus X, the so-called healthy potato virus. Five of the 15 virus complexes listed in Table 10.2 have this virus as a component. The strawberry latent virus, known in three strains, produces emergent symptoms of varying degrees of severity, with host-plant varietal differences also evident. McGrew and Scott (1959) report experiments on the combination of latent A and mottle virus, and latent A combined with a mixture of latent C plus leaf curl virus. Both combinations reduced runner production, yields, and size of fruits to a greater degree than did latent A alone. The mottle complex was more severe in the Blakemore variety and the leaf curl complex more severe in the Catskill variety. The severity of symptoms due to virus complexes may be determined by a particular strain of one of the components. McKay and Loughnane (1953) showed this to be true in the case of strains of virus X with potato virus A on three potato varieties. MacNeill and Ismen (1958) found that a new strain of tobacco mosaic virus, linked with potato virus X, induced far more virulent symptoms of the tomato streak complex than when virus X was in combination with Holmes'-type strain of tobacco mosaic.

Symptomology is a major consideration in the differentiation of virus strains, even though the use of symptoms as a single criterion is not to be recommended. When, however, a wide host range is available, then differentiation of strains on the basis of symptoms produced through the host range provides a more reliable series of criteria. In spite of the limitation of the method, it remains the one most widely used.

Local Lesions

The local lesion is the term applied to the necrotic lesion which appears when a suspension of virus-infected plant juice is rubbed on a test plant leaf. When this symptom is not followed by systemic infection, as with some species of *Nicotiana,* the number of lesions can be shown to be proportional to the concentration of virus in the sample. When no obvious local lesions are produced, or when the primary symptom is only a faint chlorotic area, the method can still be used if the leaves are decolorized and stained with iodine. This process shows the local lesion either as a white spot on a dark background, or as a dark spot on a black background, depending on whether the leaf was picked in the evening or after a long, dark period (Fig. 10.29 and 10.30).

Since the original work of Holmes (1929, 1931), the development of the method occupied the attention of many workers over a period of 15 years or more, but the concern was primarily with the statistical and physiological aspects rather than with symptomology. Bawden (1950) and K. M. Smith (1951) have both reviewed these aspects of the subject.

a

b

c

d

e

f

g

FACTORS AFFECTING EXTERNAL SYMPTOM EXPRESSION

Temperature and light intensity have a profound effect on symptom expression. Below 51°F, and above 98°F, the symptoms of tobacco mosaic are masked. Although the potato grew best at 75°F, the rate of spread of this virus in the plant was most rapid between 75–85°F, being slower both above and below that range (Grainger, 1936).

Temperature

Keitt and Moore (1943) found temperature to be a sharply limiting factor in leaf symptom expression of sour cherry yellows. A night temperature of 16°C, was favorable, even if the day temperature went over 24°C. At a constant temperature of 20°C or more, the symptoms were masked. Rose mosaic expression is favored by temperatures of 15–25°C. However, when the temperature is reduced below 15°C, mosaic expression for the two varieties is the same (Baker and Thomas, 1942) even though the two varieties of rose differ in growth response to temperature.

With a soil temperature of 45°F, wheat-streak mosaic developed more severe symptoms in a shorter time at 85°F than at lower temperatures, typical stunting being slight or absent at 68 and 60°F (Sill and Fellows, 1953). Temperature may have a differential effect on two viruses in the same host plant. Cabbage blackring symptoms are increased between 16

FIGURE 10.29. Soil-borne viruses in raspberry. *a*, Outbreak of Scottish raspberry leaf curl disease in Malling Jewel raspberry caused by raspberry ringspot virus (RRSV). Most of the plants in this plantation have been killed or are severely diseased. *b*, Leaf of *Petunia hybrida* systemically infected with RRSV, showing yellow ring and line patterns; *c*, local lesions produced by RRSV in an inoculated leaf of *Chenopodium amaranticolor; d*, leaf of Malling Jewel raspberry naturally infected with the Wiltshire strain of RRSV, showing small yellowish spots; *e*, local lesions produced by RRSV in an inoculated leaf of French bean (variety Prince) in winter; *f*, leaf of *Petunia hybrida* systemically infected with RRSV and subsequently inoculated with the beet ringspot strain of tomato black ring virus. Note the local lesions caused by tomato black ring virus; *g*, Leaf of *Petunia hybrida* systemically infected with RRSV and subsequently inoculated with tobacco ringspot virus. Note the local lesions caused by tobacco ringspot virus (Photographs: J. Sunderland, *Annals of Applied Biology*, **46**, 584, 1958).

FIGURE 10.30. Leaves of *N. tabacum* var. Turkish showing local lesions of tobacco mosaic after iodine treatment. *a*, 3 days after inoculation; *b*, 4 days after inoculation; *c* and *d*, 5 days after inoculation; *e*, 7 days after inoculation; *f*, 10 days after inoculation. Lesions of the systemic disease on leaves showing vein-clearing. *g* and *h*, 6 and 10 days, respectively, after inoculation of a lower leaf of plant; *i*, leaf 10 days after inoculation of left half by rubbing with extract of mosaic plants (After Holmes, 1931).

and 28°C, but those of cauliflower mosaic are suppressed (Pound, 1945; Pound and Walker, 1945).

McKinney and Clayton (1945) found a gene-controlled temperature response mechanism which regulates the expression of the necrosis and mosaic factors. Seven species of *Nicotiana* carrying the antinecrosis (N) factor were inoculated with tobacco mosaic and cultured at several temperatures. Each was susceptible to mosaic, but only at temperatures higher than those favoring severe secondary necrosis. When the N factor was transferred to plants carrying mosaic-resistance factors, secondary necrosis was greatly reduced or eliminated, and mosaic susceptibility at high temperatures was greatly reduced.

Light Intensity

Light intensity effects are difficult to separate clearly from those of temperature, but there are both field and laboratory studies which emphasize the effect of light intensity on symptom expression. Knight (1941), working with cucumber viruses 3 and 4, found that with winter light conditions Turkish tobacco plants infected with yellow *Aucuba* mosaic (no. 4) were very close in symptom expression to plants with green *Aucuba* (no. 3). Typical symptoms developed, however, when the plants were irradiated with 500- or 1000-watt bulbs. Bawden (1950) reports seasonal differences in expression of sugar beet yellows, symptoms being strong in summer and practically lacking in winter. On the other hand, mosaic is dull in summer and bright in winter.

Tobacco necrosis and tomato bushy stunt, however, are both favored by winter conditions under glass. Plants infected with tobacco necrosis in summer show few and scattered lesions, whereas in winter (November through February), the lesions are so numerous that they coalesce. Summer symptoms of tomato bushy stunt are mild, with isolated chlorotic blotches and some petal twisting. In winter, however, the disease kills rapidly (K. M. Smith and Bald, 1935; Bawden and Roberts, 1947).

Day length was shown by McKinney (1937) to be a factor in wheat mosaic expression. Rosette symptoms appear only at an 8-hour day and at temperatures around 16°C. Wilson (1955) has related light intensity and nutrition with symptoms. Increased nitrogen reduced intensity of leaf symptoms, and there was some reduction in phloem necrosis. Phosphorus initially reduced but later increased symptoms, and phloem necrosis increased at all growth stages. Potassium resulted in slight intensification of leaf symptoms but made no difference to phloem necrosis. Low light intensity reduced symptoms. Felton (1948) showed symptom development to be delayed by high temperature, low soil moisture, and high nitrogen. A combination of these completely suppressed symptoms. Arenz (1949) working with leaf roll of potato found that nitrogen deficiency reduced time for symptom expression and increased severity; phosphorus and potassium deficiency had little effect. When nitrogen was applied to plants showing symptoms, however, vigor was restored and symptomless growth developed.

Symptoms may be artificially modified by spray treatment. Watson (1955) sprayed sugar beets with a 10% sucrose solution and increased expression of symptoms and beet carbohydrates. Low light depressed symptoms more in the unsprayed plots than in the sprayed ones. The increase in symptom expression following high light intensity was explained in terms of carbohydrate accumulation.

Light intensity effects in relation to time of inoculation were reported by Bawden and Roberts (1947, 1948). High light intensity before exposure to infection increased the number of lesions of tobacco mosaic and tomato bushy stunt if the plants had been grown in the dark for 24–72 hours beforehand. Light intensities after inoculation had little or no effect.

Altitude may affect symptom expression. MacMillan (1923) showed that potato mosaic was masked at altitudes of 8000 ft or over. The explanation was that at that altitude, short-wave length light stimulated chlorophyll production, the effect being to mask the mosaic symptoms.

Genetic Factors

Symptom expression may be governed by Mendelian factors. Zaumeyer and Harter (1943) state that bean mosaic virus 4 is a case in point. A single allelomorphic pair is involved. The dominant allele is associated with susceptibility to local lesions only, and varieties possessing this are commercially resistant. Homozygous recessives are associated with systemic expression-leaf mottling, stunting, and reduction of yield.

Boyce (1958) studied delayed symptom expression of yellow leaf disease of *Phormium tenax* Forst. and found that it was characteristic of a resistant cultivar. Diachun and Hensen (1956) found clonal differences in symptom expression of yellow bean mosaic virus. Within a single clone of beans, symptoms were similar, but between clones, symptoms varied greatly from mild mottle to severe distortion.

Genetic mottles, leaf variations, and chloroses indicating chlorophyll deficiencies are known to occur in many plants. Some have been shown to be transmitted through the cytoplasm, not the chromosomes; others through the mechanism of recessive or dominant genes contained in the chromosomes. These genetic expressions can be easily confused with virus symptoms, and when such plants are also virus infected, genetic virus disease symptom complexes may result (McWhorter and Frazier, 1956).

The Incubation Period for Symptom Expression

The time elapsing between inoculation and the appearance of symptoms has been variously described as "incubation period" and "latent period of infection." The term incubation period was current when the concept of the development of a living entity requiring a developmental period from one stage to another was struggling for dominance. The alternative term seems more appropriate since it avoids the connotation of a developmental stage, but it retains the concept of a reproductive period, which brings the concentration and distribution of the infection to the point where the physiology of the plant is so affected that visible symptoms become evident. This latent period may be a few days, or, as in the case of some tree diseases, over a year. Perhaps "presymptom period" would be a better term.

Acute and Chronic Symptoms

"Shock," "acute," and "chronic" are terms in general, but not always uniform, use. McKinney and Clayton (1943) described acute symptoms of tobacco mosaic as severe chlorosis and the early death of diseased tissue. Chronic infection exists in invaded tissue that may or may not show obvious symptoms. Death of tissues following chlorosis occurs only after a long delay and under environmental conditions that are more extreme than those obtaining for the expression of acute symptoms of chlorosis.

Berkeley (1952) has expressed the differences in more generally applicable terms. Using the stone fruit viruses as examples, Berkeley describes the first symptoms following inoculation as a severe distortion and necrosis of unfolding bud tissue. These symptoms are systemic and generally severe, so the term acute, or shock, is used to describe them. These symptoms, however, soon disappear, and the persistent or chronic phase succeeds. This may be symptomless, or nearly so, or may give rise to a chronic symptom, such as yellow leaf casting as with the cherry yellows disease.

Shock or acute symptoms result from infection of differentiating tissue, that is, tissue actively growing at the time of infection. Chronic symptoms result from meristematic tissue becoming infected before it differentiates. The acute symptom also may be described as the symptom of invasion; the chronic, as the symptom of occupation.

INTERNAL SYMPTOMS

Internal changes induced by the activity of viruses fall into two main categories, anatomical and cytological. The principal anatomical changes are (1) hypoplastic, hyperplastic, and hypertrophic changes in the mesophyll and other parenchyma, and (2) degenerative changes in the phloem, sometimes seen as sieve-tube necrosis and sometimes as growth anomalies followed by necrosis. Most of these symptoms are, of course, associated with the development of external symptoms. Cytological effects are seen in abnormal starch accumulation, the inhibition of plastid development, or the occurrence of chloroplast destruction (Esau, 1948). A more specific effect seen in the cell is the occurrence of intracellular inclusions. Apart from these, as Esau states, probably none of the symptoms mentioned is sufficiently specific that it cannot be reproduced by one or another agency besides a virus.

The order in which anatomical changes are taken up in this section is derived from Esau (1948), which in turn is a supplement to her first review (1938).

In Leaf Parenchyma

Comparing the yellow and green areas of plants suffering from mosaic, it is found that the yellow areas of the diseased leaves show hypoplasia with underdevelopment of the plastids. This is, however, a late development. In the early stages of differentiation, the yellow areas are hyperplastic, since certain phenomena of maturation occur in them earlier than elsewhere in the mesophyll (Esau, 1948). In sugar beet mosaic, the yellow areas are thinner than the green, and they show juvenile characteristics. There is no differentiation into palisade and spongy layers. Cell division is reduced in the yellow areas, giving the appearance of larger and more mature cells than in the green areas, where the mesophyll appears healthy or hyperplastic. Chloroplasts are abnormal in the yellow areas and deficient in numbers in the younger cells. Their shape may be changed, and they may fuse into amorphous masses in severely diseased cells (Esau, 1944). Abnormal or deficient chloroplast development, resulting from an inhibited development and destruction of plastids, is evidently a precursor of the chloroses and chlorotic patterns seen as external symptoms (Fig. 10.31).

When hypertrophy and hyperplasia occur in the cortical tissue of sugar beet foliage as a secondary development of primary symptoms in the phloem, mesophyll cells near the phloem are stimulated to growth and division. In these, the chloroplasts are small, sparse, pale, irregular, and

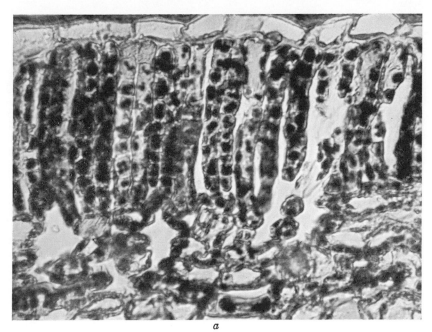

a

FIGURE 10.31. Effect of cassava mosaic virus on the anatomy of cassava leaves. *a,* TS lamina of a healthy cassava leaf showing normally developed palisade tissue and normal chloroplasts (X350; reduced ⅔ on reproduction).

degenerate, leading to the external symptom of vein-clearing. Esau (1933) limits the term vein-clearing to translucency in situations where cells divide and enlarge without an intercellular space system and with a deficiency of coloring matter.

Sheffield (1933), studying several species of Solanaceae, found that in the plant diseased by *Aucuba* mosaic, some leaf tissues were devoid of plastids. The virus inhibited the development of the plastid primordia with the result that chlorophyll was absent. Mature plastids were never affected by the virus but, according to Bawden (1950), this may be a matter of virus concentration rather than immunity of the mature plastids to virus attack.

Lackey (1954) showed that the external symptoms of sugar beet yellows —yellow thickened and brittle leaves—resulted from the production of increased numbers of parenchyma cells, so closely compacted that little intercellular space remained. Vein-clearing and pimples result from hypertrophy of cells immediately adjacent to veins.

Porter (1954) reported at length on the histological and cytological

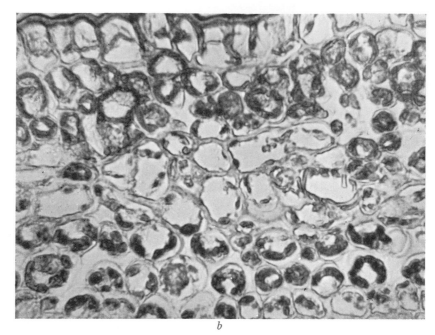

b

FIGURE 10.31. *b,* TS lamina of a chlorotic area of a cassava leaf infected with mosaic virus showing undifferentiated palisade tissue and abnormal chloroplasts (X350; reduced ⅔ on reproduction) (Photographs: Chant and Beck, 1959).

changes occurring in plants of a number of species affected by cucumber mosaic. In all cases, the initial effects originated in the mesophyll of affected leaves. Yellow areas appear hyperplastic because of suppression of cell division, premature cell vacuolation, and early development of intercellular spaces in the mesophyll. Hypoplasia develops in mature leaves because of failure to complete normal mesophyll differentiation. As in other mosaic diseases, plastid development is inhibited, resulting in reduction in number and size, or the plastids are destroyed. The dark green raised areas that occur in diseased cucumber leaves are hyperplastic, with palisade cells longer, and the spongy parenchyma cells increased in number, and with more intercellular spaces, than in healthy leaf tissue.

Secondary necrosis in Easter lily and tulip leaves is restricted primarily to the mesophyll, but in broad bean leaves it spreads rapidly to the epidermal and vascular tissues.

Parenchyma cells of sweet potato affected by the cork virus become separated as a result of dissolution of the calcium pectate intercellular binding, which Hildebrand (1956) explains is caused by enzyme action. Chloroplastids in these separated cells appear to be unaffected.

Necrosis develops in plants affected by the tomato ringspot virus; it originates in the inoculated epidermis. Spread occurs through the mesophyll to the larger veins of the leaf and progresses down the stem into the roots, but without, at first, involving the conducting tissue. The necrosis courses downward in the parenchyma tissue external to the phloem and involves the vascular tissue only in later stages (Smith and McWhorter, 1957).

In Phloem

According to Esau (1948) these symptoms fall into three general categories: (1) degeneration, beginning with a necrosis of the sieve-tubes, with or without other growth disturbances, (2) abnormal growth, which is usually followed by more or less extended necrosis of the abnormal tissue, and (3) a generalized necrosis affecting phloem cells, and, often, cells outside the phloem (Fig. 10.32).

Sieve-tube necrosis is a form of phloem degeneration common to many unrelated virus diseases. Potato leaf roll is a classical example. Phloem necrosis is restricted to the primary phloem and is initiated in the sieve tubes. Net necrosis of the tubers occurs only in tubers formed by plants infected during their growth; progeny tubers do not show net necrosis even though phloem necrosis occurs in the aerial portions of the plant. Sanford and Grimble (1944) pointed out that tuber necrosis can be induced by other agents. Net necrosis of tubers is a symptom of psyllid yellows, which is a toxemia. Davidson and Sanford (1955) relate the development of tuber phloem necrosis to varietal susceptibility and nutritional factors, but offer no explanation for the appearance of the symptom as a result of initial infection only.

There are a number of virus diseases of citrus which are characterized by sieve-tube necrosis, a condition favored by certain scion-stock relationships. Schneider and Wallace (1954) have listed four such diseases. In quick decline, sieve-tube necrosis always develops first, just below the bud union, in the sour orange rootstock. Secondary anatomical changes follow. Lemon sieve-tube necrosis, on the other hand, affects the lemon portion of the tree and is especially abundant above the union with the rootstock. Schneider (1957) described chronic decline of citrus, a bud union disorder of sweet orange on sour orange rootstock. Sieve-tube necrosis occurs below the bud union and can result in girdling effects. This disease is similar to quick decline (tristeza), in that necrosis of sieve-tubes and hypertrophy of parenchyma cells occur below the bud union. Sieve-

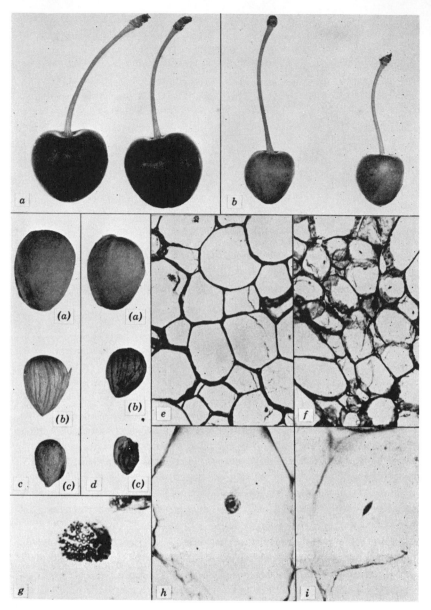

FIGURE 10.32. External and internal symptoms of Western X little cherry virus in sweet cherry trees. Collected at harvest time. *a,* Healthy Lambert fruit; *b,* severely diseased Lambert fruit; *c,* healthy (X2); (*a*) pit, (*b*) fresh seed, (*c*) air-dried seed; *d,* severely diseased (X2); (*a*) pit, (*b*) fresh seed, (*c*) air-dried seed; *e,* parenchyma in mesocarp of healthy fruit (X120); *f,* parenchyma in mesocarp of severely diseased fruit (X120); *g,* functional sieve plate in phloem of fruit spur on which all fruit were severely diseased (X1230); *h,* rounded nucleus in parenchyma cell of the mesocarp of a healthy fruit (X510); *i,* spindle-shaped nucleus in parenchyma cell of the mesocarp of a severely diseased fruit (X510).

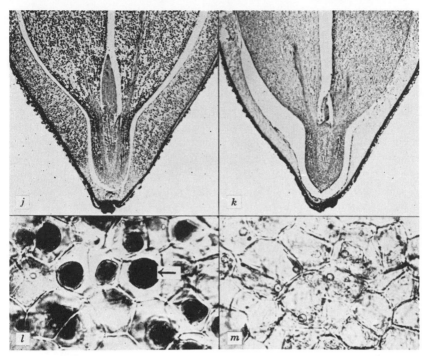

FIGURE 10.32. *j*, seed from healthy fruit (X20) *k*, seed from severely diseased fruit (X20); *l*, fresh cotyledon from healthy fruit, stained in Sudan IV (X510); large fat droplet at arrow; *m*, fresh cotyledon from severely diseased fruit, stained in Sudan V (X510) (Photographs: Leo E. Jones, Oregon State College).

tube necrosis may not always be a primary symptom, but may follow virus-induced metabolic disturbances of the plant.

The second group, as defined by Esau (1948), is best characterized by curly top of sugar beets and other crops. In this group, growth abnormalities occur in the initial stages of phloem degeneration. Necrosis in primary and secondary phloem leads to cell collapse and lesion formation. Girolami (1955) reported similar effects with both curly top and aster yellows viruses affecting flax.

Phloem degeneration occurs in Graminae affected by the barley yellow dwarf virus. The primary internal symptom appears in the phloem; secondary necrosis occurs in the xylem and parenchyma. By following the developmental changes that occurred, Esau (1957, 1957a) showed that the inception of the degeneration is related to the maturation of the first sieve-tube elements of a vascular bundle in a leaf or in the vascular cylinder of a root. From this, she concluded that the yellow dwarf is transported in the phloem.

A necrosis of *Lilium longiflorum* Thunberg affecting the vascular tissue, usually the phloem and sheath cells, occurs as a result of infection with the lily rosette virus. Since this necrosis does not occur in plants affected by other lily virus disease, it is a useful character for rapid diagnosis if freehand sections are examined in 0.05% trypan blue (McWhorter and Brierley, 1951). The symptoms do not clearly classify into any of the three groups as suggested by Esau (1948) because the necrosis appears to be more generalized than in her (1) group (sieve-tube necrosis). Although there are similarities between the lily rosette necrosis and that of curly top, there were significant differences: (1) necrotic areas were more restricted and less conspicuous than in curly top, (2) there was no evidence of necrotic products extruded into exudate channels, and (3) there was no conclusive evidence of hypertrophy or hyperplasia in any of the affected lily tissues.

Esau's third group of phloem symptoms is not as well known as are the other two. Abnormalities are much less characteristic, and necrosis does not appear to be specifically a phloem symptom and may not start in that tissue.

In Xylem

Pierce's disease virus must be introduced into the xylem, so it is probable that xylem abnormalities are primary. In psorosis of citrus, on the other hand, xylem symptoms might be secondary, since psorosis is transmitted through bark grafts. Gum formation appears to be the principal abnormality found in the xylem of diseased plants, although it is not a specific symptom. This might be in such quantity that it plugs the vessels (Esau, 1948).

Hutchins (1933) used a xylem-staining reaction for diagnosis of phony peach disease. Purplish spots appear in the xylem of diseased roots when stained with methyl alcohol acidulated with hydrochloric acid.

Lee and Black (1955) have described the development of tumor cells in petioles of virus-infected *Trifolium incarbatum* L. The ontogeny of various tissues abnormally differentiated in the region close to the xylem is diagrammed in Fig. 10.33.

Nuclear Changes Induced by Virus Infection

Malformed pollen grains and embryo sacs, causing failure to produce seed by tomato plants infected with aspermy virus, have been described

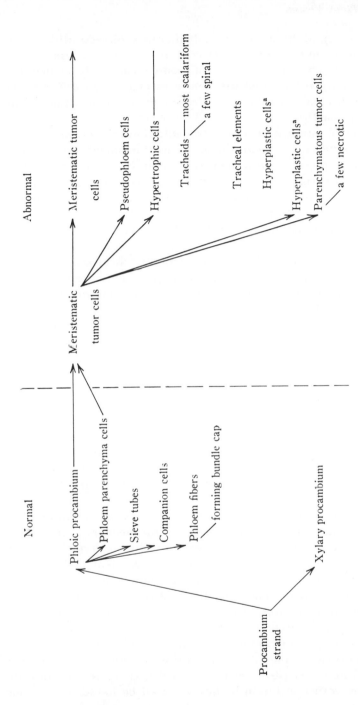

FIGURE 10.33. Diagram showing steps in the development of tumor cells in petioles of *Trifolium incarnatum* L (After Lee and Black, 1955).

[a] Show uneven thickening at the walls and necrosis.

by Caldwell (1952). Malformation of pollen grains coincides with abnormal meiosis of the microspore mother cells, none of which forms a normal pollen grain. The same sequence of events occurs with the developing ovule. No normal reduction division takes place; the nucleus of the megaspore mother cell is unable to undergo regular meiosis, and no embryo sacs are formed.

Wilkinson (1952) studied the same virus infection in *Nicotiana* species, and found the same type of meiotic aberrations. The same author (1953a) reported in detail on the meiotic aberration produced in *N. glutinosa* by aspermy virus. Abnormal multiplication of nucleolar bodies accompanied the collapse of pachytene threads of spore mother cells at the time of the first meiotic division. Sheffield (1941) reported that severe etch virus sometimes stimulates the nuclei of fully differentiated cells to divide to form binucleate cells.

In somatic division, arrested metaphase, collapse at metaphase and at spindle formation, and tendency to form giant nuclei, have been noted by Wilkinson (1953). The nucleolar material persists abnormally through anaphase, instead of dispersing during prophase.

Nuclear abnormalities in virus tumor formation were observed by Lee (1956). Normal mitotic divisions were not observed throughout tumor formation, and nuclei could not be detected in tumor cells after an early stage in tumor development.

Intracellular Inclusions

Much of the early work on intracellular inclusions, which were first recognized by Iwanowski (1903), was devoted to efforts to discover whether or not the inclusions were stages in the life cycle of causative organisms. Hoggan (1927) reviewed the early literature and concluded that in tobacco mosaic and its variants, X-bodies and striate material were constantly associated with diseased tissue but were not causal organisms. The term "X-body" had previously been applied to the amorphous type of inclusions of Iwanowski by Goldstein (1926); "striate material" refers to crystalline inclusions which become striate on treatment with acids. A third type of inclusion, the "intranuclear inclusion," has been described from severe etch disease of tobacco by Kassanis (1939) and more recently in two leguminous crop viruses by Koshimizu and Iizuka (1957).

A fourth type of intracellular inclusion, described as a "corner inclusion body" from its position in the angles subtended by the cell walls, has been shown to occur in one of two virus diseases normally found as a complex in white clover (Koshimizu and Iizuka, 1957). Amorphous, intranuclear, and corner inclusion bodies could all be found in the same cell (Fig. 10.34).

Inclusions are found throughout the plant but vary in frequency between tissues and between different diseases in specific host plants. Sheffield (1931) found them most abundant in leaf hairs, less so in epidermis, and rare in palisade and spongy tissues in *Hyoscyamus niger* L infested with *Aucuba* mosaic. In severe etch disease, the amorphous inclusions are numerous and occur in most tissues; the intranuclear inclusions are found in almost every nucleus in almost all tissues (Sheffield, 1941). In the cucumber strain of tomato ringspot disease, alveolar inclusions develop, primarily in the vascular parenchyma, in procambium, in the cells of the apical meristem, and, to a lesser degree, in the parenchyma tissues adjacent to the vascular tissues (Smith and McWhorter, 1957).

MORPHOLOGY AND FORMATION OF INCLUSION BODIES. The X-bodies appear as amorphous masses of cytoplasm that develop by the fusion of granular particles which appear shortly after the plant is infected. In shape, they are sometimes round or ovoid, sometimes irregular. When moved around in the cytoplasmic stream, they will change shape in collision with cell walls or nucleus, a characteristic which gave rise to the term "amoeboid" to describe them and which formerly lent some credence to the idea that they were pathogenic organisms.

Evidence for an enclosing membrane is indirect. Sheffield (1939) used plasmolyzing solutions, and Koshimizu and Iizuka (1957) used both plasmolyzing and deplasmolyzing techniques. The latter report recovery to initial size after deplasmolyzing, and they interpret this as evidence for the presence of an enclosing membrane.

When X-bodies lose their amorphous character and become wholly crystalline, any limiting membrane should either disappear or become strongly evident, since, in the latter case, there should be spaces between the crystals and any enclosing membrane. Bald (1948) presented a hypothesis for the formation of X-bodies from plastids (Fig. 10.35). Koshimizu (1955) and Koshimizu and Iizuka (1957) have evidence tending to confirm this, the latter authors concluding that both amorphous and corner inclusions developed from platids; intranuclear inclusions, from nucleoli.

Crystalline inclusions are colorless and transparent. When acidified, they develop striations. They are probably true hexagonal crystals; they react positively to color tests for proteins. There are no apparent differences between crystalline inclusions which develop from X-bodies and those which occur independently. Bawden (1950) also records many variants of the typical flat-plate type of crystalline inclusion, with many fibrous forms.

Rubio (1956) studied cell inclusions associated with infections by tobacco and crucifer viruses by means of both light and electron

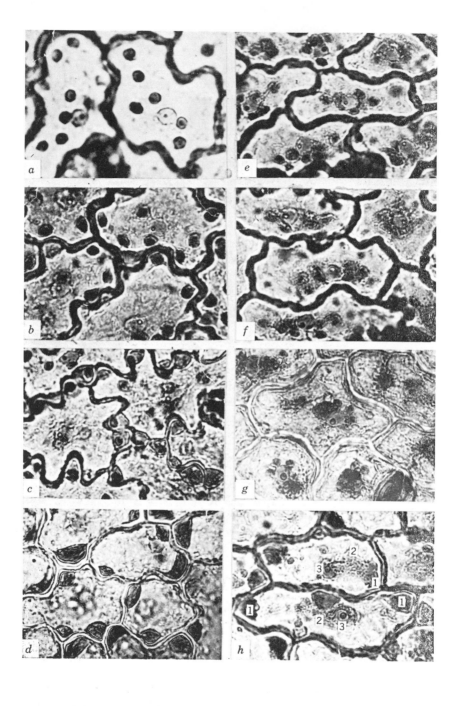

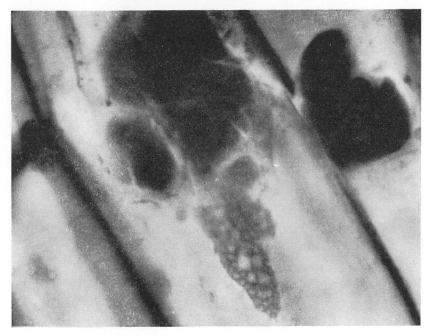

a

FIGURE 10.35. Amoeboid inclusion bodies of tobacco mosaic virus. Figures *a* and *b* are focused at different levels: *a* shows massive crystalline inclusions of tobacco mosaic; *b* shows an "amoeboid" inclusion body. Figure *a* shows clearly the stromatic structure of the amoeboid inclusion.

FIGURE 10.34. Origins and formations of intracellular inclusions. *a,* Epidermal cells of the healthy white clover (X900); *b,* an early stage of formation of the inclusion bodies. Most plastids are moved near the cell walls (X820). *c,* Plastids being enclosed in the inclusion bodies are seen (X820); *d,* fully developed inclusion bodies stained brown with IKI solution; no plastid is seen (X820). *e,* An early stage of formation of the inclusion bodies caused by the virus 2; Plastids burst into minute granules are seen (X790). *f,* Most plastids are burst and form dense masses of the minute granules (X790); *g,* these masses become more compact, forming X-bodies (X790); *h,* simultaneous occurrence of inclusion bodies of the three types: (1) "corner inclusion bodies," (2) amorphous inclusion bodies (X-bodies), and (3) intranuclear inclusions (X790) (Photographs: Koshimizu and Iizuka, 1957).

395

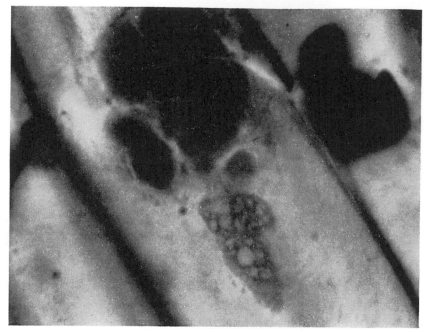

b

FIGURE 10.35.   Figure *b* shows that the inclusion is formed of two masses not yet completely fused. Four "cuboidal" bodies were visible in the whole body. One can be seen in Figure *a* and two in Figure *b*, the fourth is out of focus at both levels. X850 (Photographs: Bald, 1948).

microscopes. The formation of inclusions varied with the specific disease, a pertinent observation indeed, since it reminds us that with plant virology in its present state of development, dogmatic generalization is risky.

For example, the inclusions of turnip yellow mosaic result from aggregations of the plastids, an interesting conclusion in the light of the views of Bald (1948) and Koshimizu and Iizuka (1957); those of cabbage blackring are formed in a manner similar to those described by Sheffield (1931) for *Aucuba* mosaic, except that they have what appeared to be a distinct membrane. The inclusions of cauliflower mosaic appear to develop by increase in size but whether from "growth" or aggregation, Rubio (1956) does not make clear. Rubio and van Slogteren (1956) compared X-bodies of broad-bean-mottle virus disease with those of *Phaseolus* virus 2, using both light and electron microscopes. Those of the bean virus were granular, becoming vacuolate after two or three weeks, and, they were composed mainly of spherical virus particles. *Phaseolus* virus 2, on the other hand, gave rise to inclusions which appeared similar to those of the bean virus in the light microscope, but under the electron microscope were seen to be amorphous. A second difference is in the distribution of the inclusions in the plant tissues. Those of the bean virus are limited to the chlorotic areas of the leaf; those of the *Phaseolus* virus 2 are found in all parts of the leaf and stems. These authors did not find the isometric crystals in *Phaseolus* virus 2 as reported by McWhorter (1941); the amorphous, ovoid inclusions were not mentioned in McWhorter's paper. McWhorter (1965) has an inclusive review of the subject up to 1965 with 80 references. He decries the failure of workers to use intracellular inclusions as rapid diagnostic aids. His review established the following points: (1) various plant diseases are marked by distinctive cytological entities that are morphologically diverse, (2) the inclusion groups that characterize TMV infections should not be used as a criterion for interpreting inclusions formed by most other viruses, and (3) the great advantage for practicing plant pathologists is that the inclusions can be seen in plant tissue with the light microscope and actually do not need elaborate or specific staining techniques. Esau (1966) includes a discussion of intracellular inclusions in her review of the anatomy of plant virus infections. Inclusions contain virus in varying amounts, some mostly or entirely aggregates of virus particles; others contain noninfectious components only. McWhorter's review is pre- and Esau's post-electron microscopy.

VIRUS CONTENT OF INTRACELLULAR INCLUSIONS. Sheffield (1939, 1939a) showed X-bodies to be infective, and in 1946 she demonstrated the presence of virus particles in both X-bodies and crystalline inclusions by the use of the electron microscope (Sheffield, 1946). Bergmann and Fankuchen (1951), using X-ray diffraction techniques, concluded that the inclusion bodies most probably consist of virus protein. Steere and Williams (1953) reported that the crystalline inclusions of tobacco mosaic consist only of tobacco mosaic particles and a volatile solvent. These findings bring the debate on the nature of the inclusions full circle: from organism to by-products of cell metabolism, to carriers of virus particles, and to the virus protein itself.

THE VIRUS ITSELF AS AN INTERNAL MANIFESTATION OF INFECTION. It has already been made clear that as far as the intracellular inclusions are concerned, the actual virus particles are recognized as at least part of the intracellular inclusions. When the virus particles can be recognized either in tissue section or expressed juice, a most useful clinical symptom becomes available. It is true that very few virus diseases of plants can be characterized in this way now, but as techniques develop, more and more viruses will be recognized by the structure of the virus particle. Our knowledge of virus structure has come from the study of purified viruses, which has become so major a part of virus research that it now ranks almost as a separate discipline, separated, if not divorced, from the problems presented by the diseases which the viruses cause. Bawden (1950) devoted almost ¼ of his treatise to purification and related aspects of viruses. Burnet and Stanley's (1959) two-volume collection of papers by 21 authors includes only one contributor (Black) whose report does not depend in some measure on purified preparations.

Gold *et al.* (1954) reported on the electron microscopy of the virus of false stripe disease of barley. While it was necessary to grind tissue in order to process it for examination, so that the actual cellular relationships with the virus particle could not be determined, it was clear that virus particles could be seen in sap from leaves, embryos, endosperm, pollen, and unfertilized pistils.

K. M. Smith (1953), using ultrathin sections of tissue infected with the viruses of tomato bushy stunt and turnip yellow mosaic, was able to demonstrate the presence of spherical particles, typical of these viruses, in the plant cells. Matsui and Yamada (1958), using similar techniques, showed TMV particles in the tissues of infected tobacco.

Johnson's exudate method (1951) does not require the destruction of the plant to obtain material for electron microscope screens. He placed the plant under water pressure and obtained clear droplets of exudate from hydrathodes or cut ends of veins. Using this method, Johnson found rodlike particles typical of tobacco mosaic virus in 30 species of host plants, and in practically all tissues of these plants. Some ten other distinct viruses yielded rodlike particles by this method.

Brandes *et al.* (1959), using the exudate method, have shown the presence of virus particles of potato virus S, potato virus M, and carnation latent mosaic in infected plants; Brandes and Wetter (1959) have attempted a classification of elongate plant viruses on the basis of the morphology of the particles, determined in the same manner. Brandes (1959) examined the particles of six viruses and included data on their morphology.

PHYSIOLOGICAL SYMPTOMS

The effects of viruses on the metabolism of the host plant are an essential part of the clinical picture, even though many, perhaps most, of the studies that have been made have been directed toward an understanding of the physiology of the virus itself, rather than that of the host plant. Virus infection may affect chlorophyll, the carbohydrate content of the plant, the nitrogen balance, respiration and photosynthesis, enzyme activity, and the water content of the plant.

Reported effects on plastids and chloroplasts in infected tissue vary. Sorokin (1926, 1927) wrote on the destruction of chloroplasts in tomato mosaic during the period when the search for "organisms" was active. The first stage was the destruction of the stroma, followed by the dissolving of the entire body of the chloroplast. Esau (1933) described the destruction of chloroplasts in beets affected by curly top. Other workers have reported retardation of chloroplast development (Cook, 1930); Clinch (1932) noted that inhibited development of chloroplasts was a characteristic of chlorosis in some diseases. The chloroplasts in *Aucuba* mosaic frequently disintegrate.

Dunlap (1930) reported that with the mosaics, total nitrogen increased and total carbohydrate decreased. Carbohydrate accumulation, however, can be said to be characteristic of the group of virus diseases that fall into the general designation of "yellows" diseases. It has been ascribed to blocking of the sieve-tubes of the phloem following necrosis (Quanjer, 1913; Thung, 1928), but there is evidence to the contrary. Watson and Watson (1951) kept beet-yellows infected plants in darkness. Loss of total carbohydrates was as great in infected plants as in the healthy checks. The accumulation of reducing sugars, sucrose, and, to some extent, starch was higher in the infected plants both at the beginning and end of the dark period. Water content was reduced in the infected plant. The authors concluded that physiological changes in the plant cells were responsible rather than sieve-tube blockage.

Accumulation of starch and sugars in the leaves and stems is characteristic of tomatoes affected by curly top. According to Shapovalov and Jones (1930), the accumulation is progressive; the more severe the external disease expression, the greater the accumulation. Resistant strains of tomatoes show the same relationship to symptom expression if samples taken on the same day are compared. Stange (1953) reported fructose content of sugar beets infected with yellows disease virus to be ten times that of normal plants. Spike disease of sandal has been the subject of detailed studies by many Indian workers. Sreenivasaya and Sastri (1928) have characterized the effects of the disease on the nitrogen content of the plant as an increase in (1) total water-soluble nitrogen, (2) basic nitrogen, and (3) total amino-nitrogen. There is a decrease in the nitrate nitrogen in the disease leaves.

Commoner *et al.* (1953) showed a net increase in protein content of TMV-diseased tobacco leaf disks and a corresponding decrease in nonprotein nitrogen. These authors were concerned with the problem of virus synthesis rather than with physiological manifestations of disease. Allison (1953) studied the soluble nitrogen content of leaf roll infected potatoes, using seven varieties. Differences in glutamic acid and glutamine were most consistent. Diseased tissue contained two to three times as much glutamine as healthy tissues. In one variety, Arran Pilot, amino-nitrogen was only ⅓ that of healthy tissues. This is one of the few references to the use of such analyses in diagnosis.

Effects on respiration and photosynthesis have been studied extensively. Glasstone (1942) found that the respiration rate of diseased and healthy plants remained at the same level until the disease became systemic, when the respiration rate of the diseased tissue rose rapidly. This rise coincided with the appearance of vein-clearing. By the time mottling had developed, respiration had decreased to a level equal to that of the healthy plants (Fig. 10.36).

Whitehead (1934) showed that leaf roll infected potatoes had a higher respiration rate than had healthy plants. Under low light intensity, when additional respirable substrates were not available, the respiration of infected tissue was equal to that of healthy tissue.

Gondo (1952) measured the respiration rate of diseased tobacco plants, using whole plants. The respiration rate of the healthy plants was greater than that of the diseased plants when both were grown under the same conditions. However, the respiration of a young diseased plant showing severe symptoms was greater than that of a healthy mature plant. In cucumber-mosaic-diseased tobacco plants, the respiration index of the diseased plants was always higher than that of the healthy plants. In both cases as leaves grew older, the respiration tended to decrease, particularly in the diseased plants (Gondo, 1954).

Both respiration and photosynthesis can change very rapidly after inoculation. Owen (1955, 1957) found that in winter, 20 hours after infection, respiration rate increased in tobacco leaves infected by TMV; in summer, it decreased. Increasing light intensity before inoculation decreased respiration, but extending day length in winter had no effect. Rates began to change in less than 1 hour after inoculation. The same was true of photosynthesis. The rate was lower in infected leaves, but the effect began within 1 hour of inoculation. This result led Owen to the conclusion that the changes were not the result of virus multiplication, but rather an effect on cell metabolism. This conclusion is reasonable if the infected plant is considered as a whole organism. Infection need not reach its peak, either of concentration or distribution in the tissues, before the whole organism reacts to the infection with changes in its metabolism.

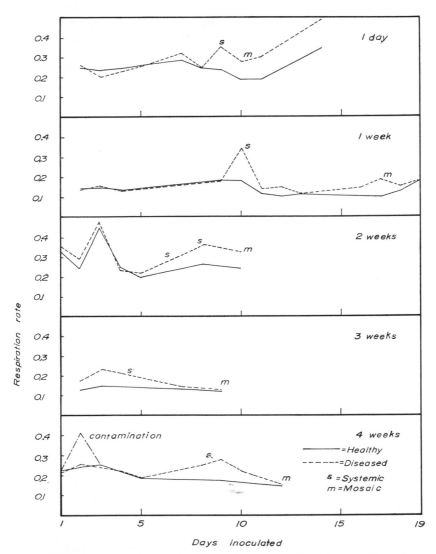

FIGURE 10.36. Respiration rates, in mg CO_2/g of fresh wt/hr, of healthy and diseased tobacco plants at different ages of inoculation (After Glasstone, 1942).

Tobacco infected with tobacco etch virus, contrary to the results with TMV, did not show respiration changes until leaves showed symptoms, when respiration rose to 40% above that of healthy leaves. Increased respiration occurred with infection at all times of the year and was maintained throughout the life of the plant. Photosynthetic activity was 20% lower in leaves showing symptoms than in healthy leaves (Owen, 1957).

Photosynthesis decreased in tobacco leaves infected with potato virus X after symptoms appeared; with TMV in *N. glutinosa*, it decreased 20% (Owen, 1958). Photosynthesis of papaya affected by leaf mosaic was reported on by Decker and Tio (1958). The mean rate for diseased leaves was 36% of that for normal leaves.

TMV in *N. glutinosa* increased respiration after local lesions appeared. The rate varied with development of lesions. This was ascribed by Yamaguchi and Hirai (1959), not to virus increase, but to a metabolic change accompanying the formation of the local lesions. The same relationship is reported by Owen (1958). The effect of virus infection with TMV on water content is to decrease it. Owen (1958a) remarks, however, that "The effects of infection with TMV on the physiology of tobacco leaves are not typical of the effects of other viruses on tobacco."

Effects on enzyme activity have been noted. Succinic dehydrogenase activity is increased and the cytochrome activity decreased in Satsuma oranges infected with dwarf disease (Yoshii and Kiso, 1957). Yoshii (1958) found no effect on oxidase activity in rice infected with the dwarf disease, and in addition he concluded that both respiration and phosphorous metabolism increased (Fig. 10.37). Auxin levels were found to be higher in plants infected with TSW and TMV (Jones, 1956). Van Duuren (1956) postulated that decreased phosphatase activity is the primary effect of yellows virus in sugar beets. Of secondary modifications, increase in reducing sugars is the most marked, there being a twentyfold increase in these in diseased tissue.

Ross and Williamson (1951) reported physiologically active emanations, presumably ethylene, from leaves inoculated with five viruses. With viruses forming numerous large necrotic lesions, ethylene production reached a maximum—a result rather than a cause of the necrosis, because production was more copious at temperatures favoring necrosis.

Physiological internal symptoms of virus infection are thus shown to be many and complex. Few generalizations are possible since it is clear that specific host reaction to specific infection occurs. The use of such symptoms as diagnostic aids has been only rarely suggested and in no case adequately established.

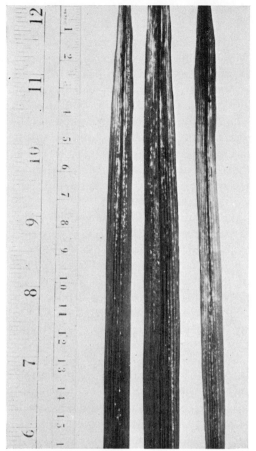

FIGURE 10.37. Leaves of rice var. Buenketan 99 infected with rice dwarf virus, showing interrupted yellowish white streaks parallel to veins and chlorotic spots or partial chlorosis on the lower parts (Photograph: Grandencio M. Reyes).

CHEMICAL AND PHYSICAL EVIDENCE OF INFECTION

The physiological-host-response characteristic of a specific disease, if known, would be of great clinical value in diagnosis. Jensen (1952) and Hirata (1958) have both reviewed progress in this field, the latter in detail, with 118 pertinent references.

By Color Tests

Chemical evidence for infection has been sought largely through color tests. Hutchins (1933) developed an acid-staining test for the presence of phony disease of peach. A cross-section of peach root or stem is placed in a test solution composed of 1–5 drops of chemically pure hydrochloric acid in 25 ml of absolute alcohol. After a few minutes of exposure to this solution, purplish dots become visible throughout the wood of diseased tissue. Any moderately strong acid can be used, according to KenKnight (1951, 1951a), who used the ingenious analogy to the development of a photographic image, the acid functioning as the developer. The acid test appears to be more specific than the phloroglucinol test for phony peach virus (KenKnight, 1952).

Lindner (1948) and Lindner *et al.* (1949, 1950, 1951) have described the development of color tests for the detection of virus diseases of stone fruit trees. Lindner's original datum was questioned by Kelly (1949) on the ground that his test was merely a modification of the Benedict test for reducing sugars. Kelly assumed that the copper sulfate in Lindner's reagent was essential to color formation, but Lindner *et al.* (1950) demonstrated that the small amount of copper sulfate added to the reagent for the colorimetric test served only to catalyze oxidation of polyphenols. The staining procedure consists of removing the chlorophyll, fixing the polyphenols, and developing blue-colored polyphenol compounds by treatment with NaOH. The test is admittedly limited to those plant species having a polyphenol system similar to that present in deciduous trees. The mechanism of accumulation of polyphenols in diseased tissue is obscure, but the authors suggest either or both of the following may be involved: (1) partial or complete blockage of the phloem, resulting in the dehydration of accumulated sugars to polyphenols; (2) mutual precipitation of virus protein and polyphenol, and by mass action, a subsequent accumulation of polyphenol-virus aggregate. Lindner *et al.* (1951) summarized the results of 70,000 tests, mainly with cherry and peach trees, to determine the most suitable tissue to use. Leaves provided the best material, but wood can be used and is useful during the dormant season. The chief limitation recognized by these workers is the inability to distinguish a specific virus disease from among several which may be present.

Sacramuzzi (1956) tested the methods of Lindner *et al.* (1951) using virus-infected almond, plum, fig, and peach trees as test material. Results were so variable that no conclusion could be reached as to the validity of the test. This seems to be a peculiarity of colorimetric tests. With a great deal of experience, workers in a restricted area, with respect both to viruses and local conditions, can evolve tests which are useful diagnostically, but these tests are difficult to apply, either to other virus infections or to other local conditions.

Accumulation of phenolic compounds in affected tissues, disturbance of the phenol and tannin metabolism of the host (Resuhr, 1942), sensitivity of necrotic phloem cells to phloroglucinol staining (Sheffield, 1943), the differential rate of starch synthesis and hydrolysis in virus-infected tissue (Holmes, 1931), the oxidation-reduction potentials and variation in dehydrogenase reactions (Wartenberg and Lindau, 1936), and the ratio between protein and reducing sugars (Hirata, 1958) have all been called on to explain the results from what were essentially empirical staining methods.

Miller and Aldrich (1951) attempted to apply the Lindner test to the detection of viruses in strawberries, and they found it useful to the extent that severely diseased plants could be separated from those less severely diseased. The occasional positive result from virus-free plants limited its usefulness as far as strawberries were concerned.

The phloroglucinol test has been applied to the detection of leaf roll in potatoes (Sheffield, 1943; Wilson, 1948); Klostermeyer (1950) applied the method in field tests to diagnose current season infection in the Netted Gem variety. The top 6 in. of the stem were cut off, the leaves removed, and the stem portion held under refrigeration; 2 mm crosssections were later cut and immersed in a 1% solution of phloroglucinol in alcohol for 1 min. and then transferred to concentrated hydrochloric acid until the xylem turned red in color. In healthy plants, only the xylem and some fibers stained red. In tissues infected with leaf roll, both inner and outer phloem tissues turned a yellowish-red color and were clearly discernible.

Childs *et al.* (1958) obtained a 98.6% correct diagnosis of exocortis virus in *Poncirus trifoliata* Raf. using the phloroglucinol-HCl test. The presence of other citrus viruses did not affect the reaction, and the test indicated infection before bark scaling occurred.

Beal *et al.* (1955) found that 2,3,5-triphenyl tetrazolium chloride could be used to detect virus infection. The test depends on the difference in the rate of color change of tissue placed in an aqueous solution of the compound at a concentration of 0.5–1%. The color change is from green to rust, to amber, and finally to blood red. Virus-free reference plants are essential; if latent viruses are involved, the test is limited in its usefulness, but the authors found that positive results accrued from seven unrelated viruses. Antoine (1958) reported the use of the same reagent in determining the presence of ratoon-stunting disease of sugarcane. Only mature cane can be used. As with Beal *et al.*, the test depends on the comparative rapidity of color change.

The reducing reaction of methylene-blue has been used by Hirata in diagnostic tests (1956). The color fading occurs more rapidly with diseased plant juice than with juice from healthy plants.

Holmes (1931) was able to determine the location of primary lesions of tobacco mosaic and later spread of the virus through the tissues by first clearing the leaves of chlorophyll and then staining with an aqueous solution of iodine as potassium iodide. This test is based on the differential rate of starch synthesis and hydrolysis in virus-infected tissue.

Muller and Munroe (1956) used rhodamine B, neutral red, and brilliant cresyl blue to determine the presence of virus Y in potato plants. All the stains were used at 5 ppm. There was a good correlation between symptom expression and the reaction of the tissues to the vital stain, although the reaction was not specific; premortal changes in leaf roll, for example, could also be detected. The stem tissues of plants which reacted only with local lesions or only with mosaic showed no reaction different from that of normal tissue. If systemic necrosis was to occur, affinity for the vital stain showed soon after inoculation and before any pathological changes in the tissue could be observed. Collenchyma reacted more readily than phloem, but staining of both tissues always preceded general necrosis.

Fluorescence microscopy was used by Keleny (1952) in a comparison of the phloroglucinol test with Bode's fluorescence test for determining the distribution of phloem necrosis of potatoes infected with leaf roll. Bode used rhodamine B and coriphosphine at 1 in 20,000, buffered at pH 2, but Keleny doubled the concentration of the dyes. The fluorescence test was superior in that it revealed a greater number of necrotic cells at an earlier stage of development. Sprau (1958) used Wasserblau 6B Extra (Bayer) as a fluorescent stain to reveal initial callose formation in primary infections.

Rawlins (1957) reported a modification of Bald's stain for viruses and cell inclusions which gave consistent results. The technique is described step by step. Although Rawlins considers that Bald's stain is the most specific biological stain for viruses yet found, it is still not completely specific. This lack of complete specificity was also found by Rawlins *et al.* (1956) to hold for the Sakaguchi histochemical method.

Van Duuren (1956) described a new chemical method for measuring the intensity of beet-yellows-virus infection, based on the use of a reagent consisting of 0.5% 3, 5-dinitrosalicylic acid dissolved in normal NaOH. There was no indication, however, that the method could be used to determine diagnostic differences between healthy and infected plants.

Hancock (1949) reported on a method for diagnosing swollen shoot of cacao, but since this method required elaborate statistical analysis of a number of observations, it lacked practicability. Tinsley and Usher (1954) modified the technique of Lindner *et al.* (1950) by using expressed sap from 10 g of mature leaves macerated in 25 ml of distilled water; 2 ml quantities of the extract were then heated to boiling point with an equal volume of normal NaOH. Infected leaves produced a red-brown to crimson color, and healthy leaves a lime-green to greenish-brown. A condition of the test is that strictly comparable leaf material be selected.

Hirata (1955) used the colloidal precipitation method to compare the reactions of sap from healthy and diseased tissue. When diseased sap was diluted two to four times with 2 ml of 30% alcohol, precipitation occurred in 3 hr. There was no reaction with healthy sap after 10 hr. This result was with radish mosaic. With potato infected with leaf roll, the best results were obtained with 1 ml undiluted sap to 1 ml $HgCl_2$ with a 1-hr observation period at 20°C.

Physical evidence of virus infection has been sought by utilizing the standard methods of physical measurement in comparing healthy with infected plant sap. Hirata (1958) lists these and finds that specific gravity, water content, osmotic pressure, temperature, turbidity, boiling, freezing point, surface tension, electromotive force, electrical conductivity, electrical resistance, electrophoresis, hydrogen ion concentration, refractive index, and ultraviolet illumination have all been used in attempts at diagnosis. For most of these, the diagnostic value is either low or entirely absent.

By Ultraviolet Radiation

Thomas *et al.* (1951) used ultraviolet to diagnose carnation mosaic. Extracts of mosaic-infected plants fluoresced a light pink color; those from healthy tissue showed no similar effects. Lindner *et al.* (1951) used ultraviolet absorption spectra to diagnose fruit tree virus infection and found that a healthy tree yielded a colorless test solution with an absorption maximum of near 260 mμ, compared with a reddish colored solution from virus-infected trees, with one or more maxima near 270 or 280 mμ. The presence of specific viruses, viz., ringspot, sour cherry yellows, and western X disease of peach, was indicated by the number and position of the maxima. Tomlinson and Woodbridge (1955) failed to confirm these results, except to a limited extent in the case of western X disease.

Kirkpatrick and Lindner (1954) used a spectrophotometric method to determine virus content of cucumber infected with a stone fruit virus

(PLM). They reported that it was more efficient than a local lesion method.

By Paper Chromatography and Electrophoresis

Paper chromatography has proved useful as a diagnostic tool. Cochran (1947) used such a method (derived from Consden *et al.*, 1944) for detecting TMV in juice from diseased Turkish tobacco leaves. Separation of virus from the normal arginine-containing proteins and chlorophyll was achieved by using water buffered at pH 4.5 or higher as the solvent (Gray, 1952).

Ragetli and van der Want (1954) used paper chromatography as a means of obtaining purified virus preparations of TMV rather than for diagnostic purposes. Loebenstein (1957) developed a quick method for detection of a sweet potato virus which combined both chromatographic and ultraviolet absorption methods. He used molar sucrose as the solvent and acid bromophenol blue as the stain. Before staining, the long "tailing" characteristic of the diseased sap absorbed ultraviolet at 265 mμ. The tailing was clearly defined by the stain.

Electrophoresis as a diagnostic method depends on the fact that plant viruses migrate in the electrophoretic cell. Vinson and Petre (1929) reported that TMV migrated to the negative electrode at pH 4.76. Wildman and Bonner (1949, 1950) compared the cytoplasmic proteins of healthy and infected plants by the electrophoretic technique. New protein components were revealed in the electrophoretic scanning pattern; in the case of TMV, the component was isolated and shown to be the virus itself. The authors used four distinct viruses and concluded that whether or not the new protein components were actually the viruses themselves, the method was a simple and effective diagnostic tool.

Hirai (1956) studied paper electrophoresis of the sap of a number of virus-infected plants by using Gray's (1952) electrophoretic technique. The proteins from these moved toward the cathode less readily than those of the healthy plants, which moved towards the anode. The same author (Hirai, 1959) reported that, using the same technique, virus-infected tulip juice failed to show virus components.

SEROLOGICAL EVIDENCE FOR
SPECIFIC INFECTION

Plant proteins are antigenic, that is, when they are injected into an animal they give rise to the production of antibodies in the blood. Antigen and antibody react together in a number of ways and these reactions have diagnostic value. It is not within the scope of this book to attempt a review of either the theory or the practice of serology, but rather to indicate its usefulness in the diagnosis of specific infection with plant viruses. Chester (1937) has given the most complete account of phytoserology as a whole, and Matthews (1957) has provided an up-to-date manual of procedures.

When crude sap from an infected plant is used as an antigen, normal host constituents will also give rise to antibodies. These are removed by absorption with sap from healthy plants, and the precipitates so formed can be removed by centrifuging. Purified viruses used as antigens do, of course, result in the production of more specific antibodies, and this method is to be preferred. The limitations of this last method will be obvious, since relatively few viruses have been purified.

When antisera derived from known viruses are available by either method, it becomes possible to test a plant to determine whether or not it is infected with a specific virus. There are some limitations involved. Matthews (1957) lists 32 viruses for which antisera have been prepared, 5 which require confirmation, and 18 which have failed to yield antisera. The virus (antigenic) content of the sap of the plant to be tested must be adequate. Without refined methods of testing, the scrological test will not differentiate between strains of the same virus, but in determining field infection, there is no particular need to get the evidence on the strain level. It is also clear that when plants that are possibly infected with more than one virus are being tested, a test with one antiserum will not reveal the mixed infection.

Goodchild (1956) confirmed, by agglutination tests, the serological activity of pea mosaic and bean mosaic viruses, two of the five listed by Matthews as requiring confirmation. Hall (1956) concluded that specific antibodies were formed only after injection with viruses relatively stable in their physical properties. It is considered unlikely that specific antisera against less stable viruses can be produced by standard methods. Moorhead (1956, 1959) obtained antisera to barley stripe, brome, and wheat mosaic viruses. The last-named contained nonvirus antibodies which were difficult to eliminate.

The development of the gel-diffusion method of serological analysis has simplified procedures and may make it easier to separate complexes of antigens. The method is simple (Bjorklund, 1952; D. H. M. van Slogteren, 1954, 1955). Agar is poured into a standard Petri dish in such a manner that pits are formed, one in the center and the others equidistant from it toward the periphery of the dish. Antiserum is pipetted into the central pit and antigens into the peripheral pits. Single lines of precipitation appear in the agar between antigen and antiserum; when two anti-gens are present with different diffusion rates, two lines will appear at some short distance from each other. Gel-diffusion tests are completed in two to five days, and the precipitation lines are remarkably stable. The time factor militates against the method for field use.

In spite of all its complexities and the limitations of the method, sero-logical diagnosis of infection is being used successfully in practice. Klecz-kowski and Watson (1944) obtained antisera to a specific antigen from sugar beet yellows and found that the precipitation test could be used for diagnosis of disease in field plants.

Bartels (1955) was able to detect virus X in potatoes and found that shoots from tubers sprouted in the dark contained more virus than those grown in the light. Zaitlin *et al.* (1954) determined the presence of color-breaking virus in *Cattleya* orchid plants by interfacial ring tests.

Probably the largest scale operation in the use of serological methods is for the detection of viruses in potatoes in connection with seed certifica-tion. Van Slogteren (1955) reports the testing of over a million mother plants a year; to provide the large quantities of antisera necessary, horses were used rather than the standard laboratory rabbits to produce the anti-sera. Similar techniques have been developed by van Slogteren to deter-mine freedom from infection of flower bulbs and other crops (Table 10.3).

Bradley (1953) developed a rapid method for testing potatoes in the field for virus X. Drops of sap are expressed from leaflets onto a glass slide and tested against dilute antiserum. Marini (1956) reports that ag-glutination reactions of the chloroplasts is a standard procedure in Italy for the diagnosis of virus X. Bercks *et al.* (1957) were able to achieve 100% positive diagnosis of infection with beet yellows virus.

DIAGNOSIS AND DIFFERENTIATION OF VIRUS STRAINS

Many variants of viruses occur, and, when these are stable enough to be differentiated, they are known as strains. Some are known to occur in na-ture but many more, notably those of tobacco mosaic, have been artifi-cially isolated from plants showing the syndrome of the type infection. McKinney (1926) selected the bright yellow areas of tobacco infected with TMV as a source of inoculum and found that this resulted in a bright yellow mottling that differed clearly from the type. Jensen (1933, 1937) used the yellow spots as a source and separated 51 "strains" by making pin punctures through the yellow spots into healthy tobacco. Norval

**TABLE 10.3. Some Viruses for Which Antisera
Have Been Prepared** [a]

Crop	Virus	Precipitation reaction	Agglutination reaction
Daffodil	Yellow stripe	+	−
	Mosaic	+	−
	Chocolate spotting	+	−
	Silver-leaf	+	−
Hyacinth	Mosaic	+	−
Tulip	Breaking	+	−
	Augustaziek o	+	−
Iris	Mosaic ×	+	−
Lily	Mosaic	+	−
Crocus	Mosaic ×	+	−
Freesia	Mosaic	+	−
Carnation	Mosaic	+	−
Chrysanthemum	Mosaic	+	−
Sugar beet	Virus-yellows	+	−
Tobacco	Mosaic +	+	+
	Rattle + +	+	−
	Necrosis o	+	−
Tomato	Mosaic +	+	+
	Aspermy	+	+
Onion	Mosaic	+	−
Potato	*Aucuba*	+	+
	X virus	+	+
	Y virus	+	+
	S virus	+	+
	Stengelbont + +	+	−
Dahlia	Mosaic	+	−
Alfalfa	Mosaic	+	−

[a] Data from E. van Slogteren (1955).

(1938) further subdivided and found 68 variants from Jensen's strains. Kunkel (1940) isolated 130 variants. The criteria used in all these examples were differential symptoms.

Local lesions may sometimes include some which are atypical. By using these as sources of inoculum, Black (1940) was able to isolate strains of potato yellow dwarf and to differentiate them by the type of local lesion they produced on *N. glutinosa*.

Naturally occurring strains are known for many viruses, and their differentiation is sometimes of great practical importance (Kunkel, 1947;

Siegel and Wildman, 1954). The *Aucuba* mosaic virus strain has distinctive symptoms from the type strain, and on *N. sylvestris* it produces only necrotic primary lesions on the inoculated leaf; except at high temperatures, there is no systemic spread (Kunkel, 1934). Holmes (1941) described a distinctive ribgrass strain of TMV from ribgrass and the broadleaf plantain (*Plantago lanceolata* L and *P. major* L). It forms necrotic ring patterns in Turkish tobacco but produces no necrotic primary lesions in tissues of *N. tabacum* and *N. sylvestris* if these have already been invaded by the type TMV.

Giddings (1944) differentiated ten strains of the curly top virus of sugar beet which were stable and could be recovered from mixtures. He used susceptible and resistant beet varieties, Turkish tobacco, and red Mexican beans, and devised a key for differentiating the strains, as is shown in Table 10.4. The reactions noted by Giddings are relative, for

TABLE 10.4. *Key for Differentiating 10 Strains of Curlytop Virus by Means of a Susceptible Variety (S. L. 842) and a Highly Resistant Variety (68) of Sugar Beet, Turkish Tobacco, and Red Mexican Bean*[a]

	Susceptible sugar beet		Resistant sugar beet		Turkish tobacco		Red Mexican bean	
Strain	Infection[b]	Severity of symptoms[c]	Infection[b]	Severity of symptoms[c]	Infection[b]	Severity of symptoms[c]	Infection[b]	Severity of symptoms[c]
1	+	+ +	+	+	+	+ +	−	−
2	+	−	+	−	0	0	−	− −
3	+	+ + +	−	−	+	+ +	+	−
4	+	−	−	−	0	0	−	− −
5	+	*+ + +	−	−	+	+ + +	−	− −
6	+	*+ + +	−	−	+	+ +	+	−
7	+	− −	0	0	0	0	0	0
8	+	+ +	−	−	+	+ +	+	+
9	+	+ + +	+	−	+	+ +	+	+ +
10	+	*+	−	−	+	−	−	− −

[a] After Giddings (1944).
[b] 0=no plants infected; − =low percentage of plants infected; + =high percentage of plants infected.
[c] − − =extremely mild symptoms; − =mild or slight symptoms; + =moderately severe symptoms; + + =severe symptoms; + + + =extremely severe symptoms; *=not much distortion of infected plants by strains 5, 6, or 10 but high mortality in case of strains 5 and 6.

later he reported (1950) that more recently developed resistant varieties and breeders stocks showed a tendency toward complete immunity against some strains, whereas in 1944, only strain 7 was in this class. Giddings (1954) also reported the discovery of strains 11 and 12. Strain 11 is characterized by extreme virulence to sugar beet, even to the resistant strains, and can kill tomato plants more rapidly than any other strain. Strain 12 is virulent to both tomato and potato.

Dean (1969) used sweet sorghum varieties for differentiation of strains of sugarcane mosaic and reported that, for the first time, a local lesion host for SMV strains could be found in Atlas sorghum.

Anderson (1954) described typical watermelon mosaic and a strain which differed from the type on the basis of symptoms which were not, however, conspicuous in warmer weather. Naturally occurring strains of tristeza virus were recorded by Olson (1956). In a later paper (1958), Olson showed that strains derived from eight sources were differentiated on infected lime and sour orange plants, as were differences in degrees of stunting of citrus varieties on various root stocks.

Dean *et al.* (1959) report a case in which there are no symptom differences incited by two strains of the common bean mosaic; differentiation is achieved by the use of snap beans, a full range of resistance and susceptibility being found among the snap bean varieties.

Allen (1957) used differences in symptom expression among 21 cereal varieties to distinguish strains of barley yellow dwarf virus, the ability to cause stunting and discoloration being the criterion. Sixteen strains were recognized from among 43 isolates.

Symptoms as a single criterion of strain differences must be used with caution. Environmental conditions such as light intensity, temperature, soil conditions, and the condition of the host plant at the time of infection can greatly modify the symptoms shown. Diachun and Henson (1956) found that virus inoculation of vegetatively propagated clones of red clover produced similar symptoms within a clone, but strikingly different symptoms from those in another clone. When a clearly defined difference between host ranges exists, as Kunkel (1932) and Severin (1934) showed for the New York and California strains of aster yellows, the case is much strengthened.

There are other criteria for the separation of strains; one of these is diagnosis by serological methods. The antisera of a type virus will react with strains of the same virus because all possess common antigens. As Bawden has pointed out, however, the strains of a virus are identical antigenically only in part.

Chester (1936) and Bawden and Pirie (1938) used the absorption technique to separate various strains of tobacco mosaic. The antigenic differences can be represented in a simple formula as follows:

Tobacco mosaic virus   w, x

Aucuba mosaic virus   w, x, y

Enation mosaic virus   w, y, z

The latter workers' results are shown in Table 10.5. It will be seen that cucumber viruses 3 and 4 are shown to be serologically related to tobacco mosaic but have few antigens in common with tobacco mosaic. The results from absorption tests are not entirely unequivocal, for different antisera made against the same virus sometimes contain different antibodies (Bawden, 1950).

Serological tests of potato virus X strains are facilitated by the use of a microslide Ouchterlony technique which utilizes a thin layer of agar on standard microscope slides. The technique is effective when intact native PVX in crude plant sap extracts is used. The precipitation lines can be enhanced by staining in 1% Buffalo Black (Black NBR) or in 1% Thiazine Red with 7.5 acetic acid in distilled water for 5 min; destaining is with 7.5 acetic acid (Ouchterlony, 1958; McCrum *et al.*, 1971). A summary of the techniques of serology can be found in a paper by Ball (1964). The latest review is that of van Regenmental (1966).

Antisera have now been developed for more than 30 viruses, but serology has failed in many cases. Serological tests, like cross-protection tests, are group- not strain-specific, nor do they reveal mixed viruses if both react to the same antiserum. All strains of a virus are serologically related, but the converse, that all serologically related viruses are strains of a single virus, can be challenged. For example, a high titer antiserum gave cross reactions between TYMV and wild CMV which have no hosts in common. Several other examples can be found in Ross's article (1964).

By Interference Phenomena

Bennett (1951) defined interference phenomena between plant viruses as "those in which a plant or plant part, after being invaded by one virus strain, is rendered immune from or resistant to infection or to invasion by a second or challenging virus or strain." He included in his definition reactions which have been variously designated as "immunization," "immune reactions," "protection," "reciprocal protection," "virus antagonism," and "mutual antagonism."

From the standpoint of diagnosis, these phenomena afford a method for distinguishing virus strains and have frequently been useful in relegating a supposedly new virus to its proper status as a strain of an already established one. An early example is that of southern celery mosaic, described by Wellman in irreproachable detail in 1934 and shown to be a strain of cucumber mosiac by Price (1935). Price infected zinnia plants

TABLE 10.5. Summarized Results of Cross-Absorption Experiments with the Tobacco Mosaic Group of Viruses[a]

Antiserum absorbed	Antigen used for absorption	Precipitation tests with absorbed sera and antigens				
		Antigen				
		Tobacco mosaic virus	*Aucuba* mosaic virus	Enation mosaic virus	Cucumber virus 3	Cucumber virus 4
Tobacco mosaic virus	TMV	−	−	−	−	−
	AMV	−	−	−	−	−
	EMV	+ +	+ +	−	−	−
	CV3	+ + + +	+ + + +	+ + + +	−	−
	CV4	+ + + +	+ + + +	+ + + +	−	−
Aucuba mosaic virus	TMV	−	+ +	+ +	−	−
	AMV	−	−	−	−	−
	EMV	+ +	+ +	−	−	−
	CV3	+ + + +	+ + + +	+ + + +	−	−
	CV4	+ + + +	+ + + +	+ + + +	−	−
Enation mosaic virus	TMV	−	+	+ +	−	−
	AMV	−	−	+	−	−
	EMV	−	−	−	−	−
	CV2	+ + + +	+ + + +	+ + + +	−	−
	CV4	+ + + +	+ + + +	+ + + +	−	−
Cucumber virus 3	TMV	−	−	−	+ + + +	+ + + +
	AMV	−	−	−	+ + + +	+ + + +
	EMV	−	−	−	+ + + +	+ + + +
	CV3	−	−	−	−	−

[a] From Bawden and Pirie (1938).
Key: − indicates that there is no precipitation; + indicates the degree of precipitation at the optimum.

with southern celery mosaic, and a specific immunity was induced against cucumber mosaic virus. Symptoms in a variety of host plants were similar for the two viruses. A second example is that of tomato aspermy virus, described by Blencowe and Caldwell from chrysanthemum (1946, 1949), but later shown by Graham (1957) and by Govier (1957) to be a strain of cucumber mosaic. On the other hand, Ross (1950) was able to show that

CMV and potato virus Y were separate viruses, not related strains of one virus.

There are three methods whereby immunizing effects of one virus strain against another can be determined. Bennett used the term "challenging virus" to denote the virus to be introduced into the already infected plant. The three methods are (1) wiping the leaves with liquids containing virus, (2) grafting, and (3) by means of an insect vector.

The wiping method is the most reliable provided the challenger virus produces local lesions, for in those cases, antagonism may be estimated quantitatively on the basis of the number of local lesions produced on healthy and diseased leaves (it might be added, provided that all the precautions necessary to standardize the local lesion technique are observed).

Inoculations by graft or by insect vector are less sensitive because both methods may involve the distribution of the challenging virus into the phloem, and thus permit it to establish and multiply in sites not already occupied by the first infection. This explanation for the insensitivity of the graft and insect-inoculation methods depends on the hypothesis that protection depends on occupancy by the challenged virus of all the sites available for virus increase by the challenger virus.

Price (1940) thought that with all viruses, one strain would protect against related strains, but Bennett (1951) suggested that later results have indicated that varying degrees of protection were to be found, since protection is not always complete.

Bennett's review (1951) of the viruses which have been tested for cross protection between strains reveals the extreme variability encountered. Although the strains of tobacco and cucumber mosaic viruses do, in general, give complete, or a high degree of, protection against other strains, there is considerable evidence that protection is not always complete. Cucumber mosaic viruses 3 and 4, while antigenically similar to TMV, do not protect against TMV; the tobacco ringspot viruses show either complete or partial protection against each other. In potato virus X, a range of degree of interference exists from protection, through merely a reduction of the severity of symptoms, to failure to protect, and equilibrium may be established between two strains. Some of the variability in results may stem from differences in methods of introducing the challenging virus. The strains of maize streak virus afford either no, or very slight, protection against each other, while in the cacao-swollen-shoot complex, there are four classes of result: (1) complete protection by attenuated strains A, and strains H and K, against A; (2) incomplete protection when strain B only delays the appearance of A; (3) no protection against strain A by ten other strains, and (4) no protection but the formation of a complex, when A is inoculated into plants already affected by strain M, which produces symptoms typical of neither strain.

Giddings (1950) found that there was no protection between strains of curly top of sugar beets. Bennett (1953) confirmed this finding, and also showed that if the challenging virus was more virulent than the one challenged, the severity of the symptoms induced by the challenging virus was increased. Gilpatrick and Weintraub (1952) found an example of protection in which the presence of the challenged virus was not essential. Two clones of *Dianthus barbatus* L reacted to carnation mosaic virus by the development of primary local lesions only. The virus could be recovered from the leaves bearing lesions, but detached symptom-free leaves from the same plants proved to be protected against further inoculation. When these symptom-free-protected leaves were used as inoculum, however, no virus lesions developed. The only plausible explanation was that there was sufficient virus present to protect, but not enough to effect transmission, even though the dilution end point of the virus lies beyond 10^{-5}. Clones of red clover were used by Diachun and Henson (1958) to identify isolates of bean yellow virus. Two such clones infected with any one of three isolates of the virus are protected against local and systemic infection by another isolate.

Thomson (1958) reported that "non-specific interference occurs not only between viruses which have similar properties . . . but also between viruses which appear to be quite distinct." Furthermore, the challenged virus might condition the plant to increased susceptibility to the challenging virus, i.e., between potato viruses X and Y. This phenomenon is considered by Thomson to rank with other factors affecting susceptibility such as host nutrition, light intensity, and so on.

A complex type of virus-strain interference has been described by Wallace (1957) in connection with psorosis disease of citrus that has been referred to as several strains of a single virus. Two different symptom manifestations are involved, one on leaves and another on bark. The first symptom appears on the leaves, but the bark symptom of psorosis A, which is a lesion, appears only on trees ten years old or older. When nonlesion inoculum of psorosis A is used to infect sweet orange seedlings, these seedlings show no immediate effect when inoculated with lesion inoculum of this same strain. When sweet orange seedlings were inoculated and infected singly with concave gum, blind pocket, crinkly leaf, and infections variegation, they all showed the same protection phenomena when reinoculated with lesion bark. They are all, therefore, considered to be related strains of the psorosis virus.

A more complicated picture is drawn by Best (1954) and Best and Gallus (1955) in describing cross protection of strains of tomato spotted wilt (TSW). A mild strain protects against a severe one, whether the severe strain is inoculated before or after the systemic symptom of the mild one appears, although the addition of the severe strain does have some effect on yield. If mixtures of the mild and severe strains were inoculated together, the same protective effects were observed, and new strains could be isolated from the plants inoculated with the mixtures. These authors account for the phenomenon by the hypothesis that there is an interchange of character determinants between the inoculated strains. This hypothesis is contrary to the generally accepted one that new strains arise only by mutation.

By Physical and Chemical Properties

NUCLEIC AND AMINO ACID CONTENT. Strains have been compared for their nucleic acid and amino acid content by a number of investigators. Knight and Stanley (1941) and Knight (1947) have shown that the nucleic acid compositions (all of the ribonucleic acid type) were essentially the same in all the strains of TMV although distinctly different from those of other viruses. The ribonucleic acids of three strains of tomato bushy stunt virus could not be distinguished from each other, but they were distinctly different from TMV nucleic acid (de Fremery and Knight, 1955). Knight (1955) concluded that since nucleic acids of strains tend to have the same composition, viruses CM 3 and CM 4 cannot be considered as strains of tobacco mosaic, since their nucleic acid is different from that found in 16 strains of TMV. Since cross-protection studies arrived at the same conclusion, the serological evidence for the relationship between CM 3 and CM 4 and TMV must be accounted for on some other, as yet, unexplained grounds.

Knight's studies indicated also that, although the nucleic acid composition of a mutant remained the same as that of the parent virus, the mutation may be accompanied by a measurable change in the amino acid composition. Different strains of TMV were found to contain different quantities of some amino acids, and a given amino acid might occur in one but not in another. Knight (1947) suggested that each mutation in a virus particle involves a change in amino acid composition, and, in general, the strains showing the most marked differences in protein composition were those which, from biological properties, appeared to be most distantly related to the parent strain. These conclusions are from studies on TMV. Since amino acid differences are not always detected in laboratory strains of TMV, Price (1954) suggested that mutations may occur which are not accompanied by changes in amino acid composition of the virus, but which nevertheless may cause profound changes in symptomology.

ELECTROPHORESIS. Singer *et al.* (1951) used electrophoresis for the differentiation of virus strains of TMV, U1 (common), and U2 (mild) in *N. tabacum*. They were also able to determine the relative amounts of each strain present and concluded that electrophoretic properties not only serve as a precise characterization of different virus strains but may also provide a means of recognizing the possible presence of strains otherwise difficult to detect.

Takahashi (1955) used electrophoresis to show that three strains of TMV—severe, mild, and Holmes' ribgrass strain—were characterized by anomalous proteins, different and specific for each strain.

Applicability of the quantitative electrophoretic technique to problems of mixed infections depends on the presence of strains with different electrophoretic mobilities. Such differences have been found in many cases, but not in all. Ginoza and Atkinson (1955) compared the isoelectric points, ultraviolet absorption spectra, and electrophoretic mobilities between pH 4.5 and 8.0 of eight strains of TMV. The infrared spectra of all strains were essentially identical, but the eight strains fell into four groups, agreeing with previous studies which had used biological inactivation by ultraviolet radiation and serological behavior. Strains within these groups, however, were distinguishable only by differences in symptom expression in *N. tabacum*.

OPTIMAL TEMPERATURE. Optimal temperature has been used to differentiate strains, since different strains of one virus may have different optimal temperatures for multiplying and causing symptoms in a given host plant. Hitchborn (1956) showed that seven strains of cucumber mosaic virus differed in their ability to multiply in plants at 37° C.

THERMAL INACTIVATION POINT. The thermal inactivation point, which is defined as the temperature at which a virus ceases to be infective, also differs between strains. The thermal inactivation point is not necessarily related to the optimum temperature for multiplication, since Holmes (1934) found that strains of TMV which had approximately equal thermal death points differed in their ability to increase at high temperatures.

Grogan and Walker (1948) used the thermal inactivation point to characterize a pod-distorting strain of the yellow mosaic virus of bean; Zaumeyer (1940) and Murphy and Pierce (1937) used it in studies of strains of pea mosaic. Bhargava (1951) differentiated cucumber mosaic strains by this method.

On the other hand, as seems to be characteristic of any method of differentiation, there are exceptions. Simons (1957) found no differences in the thermal inactivation point of three strains of cucumber mosaic virus from peppers.

DILLUTION END POINT.   Dilution end point is another method often used to differentiate strains. It involves the determination of the dilution point at which a strain is no longer infective. Along with thermal death point and optimal temperature, it is limited to viruses that can be easily transmitted mechanically or by artificial feeding of a vector.

AGING IN VITRO.   Aging *in vitro* is another method in the same category. Juice of infected plants is stored at a constant temperature of about 18° C. Inoculations are made on test plants at regular intervals, usually 24 hr. Bawden and Pirie (1938) presented a table of the properties of viruses in crude sap using the thermal inactivation point, longevity *in vitro,* and dilution end point of some 30 viruses.

X-RAY DIFFRACTION.   X-Ray diffraction has been used to determine whether the structure of strains of a virus may differ. Franklin (1956) found that three strains of TMV and CV4 were closely similar in their main features but differed in points of detail. Morphological differences that might be revealed through the use of the electron microscope have been sought, but in every case the mutant was indistinguishable from the parent strain.

REFERENCES

Allen, T. C., Jr., "Strains of the barley yellow-dwarf virus," *Phytopathology,* **47,** 481–490 (1957).

Allison, R. M., "Effect of leaf roll virus infection on the soluble nitrogen composition of potato tubers," *Nature,* **171,** 573 (1953).

Anderson, C. W., "Two watermelon mosaic virus strains from central Florida," *Phytopathology,* **44,** 198–202 (1954).

Antoine, R. "A staining technique for detecting ratoon stunting disease in sugar cane," *Nature,* **181,** 276–277 (1958).

Arenz, B., "Gefassversuche uber den Einfluss Verscheidner Ernahrungsweisen auf gesunde und blattrolkrande Kartoffeln," *Z. Pflanzenernahr.-Dung. W. bodenk.,* **47,** 114–130 (1949).

Badami, R. S., and Kassanis, B., "Some properties of three viruses isolated from a diseased plant of *Solanum jasminoides* Paxt. from India," *Ann. Appl. Biol.,* **47,** 90–97 (1959).

Baker, K. F., and Thomas, H. E., "The effect of temperature on symptom expression of a rose mosaic," *Phytopathology,* **32,** 321–326 (1942).

Bald, J. G., "The development of amoeboid inclusion bodies of tobacco mosaic virus," *Australian J. Sci. Research, Ser. B.,* **1,** 458–463 (1948).

Ball, Ellen, M., "Techniques used in plant virus research," *Plant Virology,* M. K. Corbett and H. D. Sisler, Eds., University of Florida Press, Gainesville, Fla., (1964).

Bartels, R., "The serological detection of virus X in potato sprouts exposed to light," *Nachrbl. Duet. Pflanzenschutzdienstes (Stuttgart),* **7,** 43 (1955).

Bawden, F. C., *Plant Viruses and Virus Diseases,* 3rd ed., Chronica Botanica Co., Waltham, Mass., 1950.

Bawden, F. C., and Pirie, N. W., "Plant viruses. I. Serological, chemical and physiochemical properties," *Tabulae Biol.,* **16,** 355–371 (1938).

Bawden, F. C., and Roberts, F. M., "The influence of light intensity on the susceptibility of plants to certain viruses," *Ann. Appl. Biol.,* **34,** 286–296 (1947).

Bawden, F. C., and Roberts, F. M., "Photosynthesis and predisposition of plants to infection with certain viruses," *Ann. App. Biol.,* **35,** 418–428 (1948).

Beal, J. M., Preston, W. H., Jr. and Mitchell, J. W., "Use of 2, 3, 5-triphenyl tetrazolium chloride to detect the presence of viruses in plants," *Plant Disease Reptr.,* **39,** 558–560 (1955).

Bennett, C. W., "Interference phenomena between plant viruses," *Ann. Rev. Microbiol.,* **5,** 295–308 (1951).

Bennett, C. W., "Evidence of the lack of interference between strains of the curly top virus," *Proc. Inter. Congr. Microbiol., 3rd Congr., Rome,* 387–388 (1953).

Bercks, R., Burghardt, H., and Steudel, W., "Serologische Untersuchungen zur Infektion von Beta-Ruben mit dem Vergilbungsvirus (Beta virus 4) durch *Myzodes persicae* (Sulz.) und *Doralis fabae* Scop.," *Phytopathol. Z.,* **31,** 133–138 (1957).

Bergmann, M. E., and Fankuchen, I., "X-ray diffraction studies of inclusion bodies found in plants infected with tobacco mosaic virus," *Science,* **113,** 415 (1951).

Berkeley, G. H., "Reactions of plants to virus infection," *34th Rept. Quebec Soc. Protect. Plants.,* **1952,** 101–115.

Best, R. J., "Cross protection by strains of tomato spotted wilt virus and a new theory to explain it," *Australian J. Biol. Sci.,* **7,** 415–424 (1954).

Best, R. J., and Gallus, H. P. C., "Further evidence for the transfer of character-determinants (recombination) between strains of tomato spotted wilt virus," *Enzymologia,* **17,** 207–221 (1955).

Bhargava, K. S., "Some properties of four strains of cucumber mosaic virus," *Ann. Appl. Biol.,* **38,** 337–388 (1951).

Bjorklund, B., "Serological analysis of components in hemopoietic tissue," *Proc. Soc. Exptl. Biol. Med.,* **79,** 319–324 (1952).

Black, L. M., "Strains of potato yellow dwarf virus," *Am. J. Botany,* **27,** 386–392 (1940).

Black, L. M., "A virus tumor disease of plants," *Am. J. Botany*, 32, 408–415 (1945).

Blencowe, J. W., and Caldwell, J., "A new virus disease of tomatoes," *Nature,* 158, 96–97 (1946).

Blencowe, J. W., and Caldwell, J., "Aspermy, a new virus disease of the tomato," *Ann. Appl. Biol.*, 36, 320–326 (1949).

Boyce, W. R., "Resistance to yellow-leaf disease in *Phormium tenax* Forst. and the occurrence of delayed expression of symptoms," *new Zealand J. Agr. Research*, 1, 31–36 (1958).

Bradley, R. H. E., "A rapid method of testing plants in the field for potato virus X," *Am. Potato J.*, 29, 289–291 (1953).

Brandes, J., "Elektronenmikroskopische Grobenbestimmung von acht stabchen- und fadenformigen Pflanzenviren," *Phytopathol. Z.*, 35, 205–210 (1959).

Brandes, J., and Wetter, C., "Classification of elongated plant viruses on the basis of particle morphology," *Virology*, 8, 99–115 (1959).

Brandes, J., Wetter, C., Bagnall, R. H., and Larson, R. H., "Size and shape of the particles of potato virus S, potato virus M, and carnation latent virus," *Phytopathology*, 49, 443–446 (1959).

Broadbent, L., and Tinsley, T. W., "Symptoms of cauliflower mosaic and cabbage black ring spot in cauliflower," *Plant Pathol.*, 2, 88–92 (1953).

Burnet, F. M., and Stanley, W. M., Ed., *The Viruses,* Vols. 1 and 2, Academic Press, New York, 1959.

Caldwell, J., "Some effects of a plant virus on nuclear division," *Ann. Appl. Biol.*, 39, 98–102 (1952).

Caldwell, J., and Prentice, I. W., "A mosaic disease of broccoli," *Ann. Appl. Biol.*, 29, 366–373 (1942).

Chant, S. R., and Beck, B. D. A., "Effect of cassava mosaic virus on the anatomy of cassava leaves," *Trop. Agr. Trinidad*, 36, 231–236 (1959).

Chester, K. S., "Separation and analysis of virus strains by means of precipitin tests," *Phytopathology*, 26, 778–785 (1936).

Chester, K. S., "A critique of plant serology. Part I. The nature and utilization of phytoserological procedures. Part II. Application of serology to the classification of plants and the identification of plant products. Part III. Phytoserology in medicine and general biology. Bibliography," *Quart. Rev. Biol.*, 12, 19–46, 165–190, 294–321 (1937).

Childs, J. F. L., Norman, G. G., and Eichhorn, J. L., "A color test for exocortis infection in *Poncirus trifoliata,*" *Phytopathology*, 48, 426–432 (1958).

Clinch, P., "Cytological studies of potato plants affected with certain virus diseases," *Sci. Proc. Roy. Dublin Soc.*, 20, 143–172 (1932).

Cochran, G. W., "A chromatographic method for the detection of tobacco-mosaic virus in juice from diseased Turkish tobacco plants," *Phytopathology*, 37, 850–851 (1947).

Commoner, B., Schieber, D. L., and Dietz, P. M., "Relationships between tobacco mosaic virus biosynthesis and nitrogen metabolism of the host," *J. Gen. Physiol.,* 36, 807–830 (1953).

Consden, R., Gordon, A. H., and Martin, A. J. P., "Qualitative analysis of proteins; a partition chromatographic method using paper," *Biochem. J.,* 38, 224–232 (1944).

Cook, M. T., "The effect of some mosaic diseases on cell structure and on the chloroplasts," *J. Dept. Agr. Porto Rico,* 14, 69–101 (1930).

Davidson, T. R., and Sanford, G. B., "Expression of leaf-roll phloem necrosis in potato tubers," *Can. J. Agr. Sci.,* 35, 42–47 (1955).

Dean, J. L., "A local lesion host and sorghum differential varieties for identifying strains of sugarcane mosaic," *Phytopathology,* 59, 1347, (Abstr.) (1969).

Dean, L. L., Wilson V. E., Thornton, R. E., and Agenbroad, O., "Unusual reactions of two snap bean varieties to two strains of common bean-mosaic virus," *Plant Disease Reptr.,* 43, 131–132 (1959).

Decker, J. P., and Tio, M. A., "Photosynthesis of papaya as affected by leaf mosaic," *J. Agr. Univ. Puerto Rico,* 42, 145–150 (1958).

de Fremery, D., and Knight, C. A., "A chemical comparison of three strains of tomato bushy stunt virus," *J. Biol. Chem.,* 214, 559–566 (1955).

Diachun, S., and Henson, L., "Symptom reaction of individual red clover plants to yellow bean mosaic virus," *Phytopathology,* 46, 150–152 (1956).

Diachun, S., and Henson, L., "Protection tests with clones of red clover as an aid in identifying isolates of bean yellow mosaic virus," *Phytopathology,* 48, 697–698 (1958).

Doering, G. R., Price, W. C., and Fenne, S. B., "Tomato rosette caused by a virus complex," *Phytopathology,* 47, 310–311 (1957).

Dunlap, A. A., "The total nitrogen and carbohydrates, and the relative rates of respiration, in virus-infected plants," *Am. J. Botany,* 17, 348–357 (1930).

Esau, K., "Pathologic changes in the anatomy of leaves of the sugar beet, *Beta vulgaris* L., affected by curly top;" *Phytopathology,* 23, 679–712 (1933).

Esau, K., "Some anatomical aspects of plant virus disease problems. I," *Botan. Rev.,* 4, 548–579 (1938).

Esau, K., "Anatomical and cytological studies on beet mosaic," *J. Agr. Research,* 69, 95–117 (1944).

Esau, K., "Some anatomical aspects of plant virus disease problems. II," *Botan. Rev.,* 14, 413–449 (1948).

Esau, K., "Anatomic effects of barley yellow dwarf virus and maleic hydrazide on certain Gramineae," *Hilgardia,* 27, 15–69 (1957).

Esau, K., "Phloem degeneration in Gramineae affected by the barely yellow-dwarf virus," *Am. J. Botany,* 44, 245–251 (1957a).

Esau, K., "Anatomy of plant viruses," *Ann. Rev. Phytopathology,* 5, 45–76 (1967).

Felton, M. W., "The effect of temperature, moisture, and nitrogen on the development of leaf roll symptoms in the Irish potato," *Am. Potato J.,* **25,** 50–51 (1948).

Franklin, R. E., "X-ray diffraction studies of cucumber virus 4 and three strains of tobacco mosaic virus," *Biochim. Biophys. Acta,* **19,** 201–211 (1956).

Giddings, N. J., "Additional strains of the sugar-beet curly top virus," *J. Agr. Research,* **69,** 149–157 (1944).

Giddings, N. J., "Some interrelationships of virus strains in sugar beet curly top," *Phytopathology,* **40,** 337–388 (1950).

Giddings, N. J., "Two recently isolated strains of curly top virus," *Phytopathology,* **44,** 123–235 (1954).

Gilpatrick, J. D., and Weintraub, M., "An unusual type of protection with the carnation mosaic virus," *Science,* **115,** 701–702 (1952).

Ginoza, W., and Atkinson, D. E., "Comparison of some physical and chemical properties of eight strains of tobacco mosaic virus," *Virology,* **1,** 253–260 (1955).

Girolami, G., "Comparative anatomical effects of the curly-top and aster-yellows viruses on the flax plant," *Botan. Gaz.,* **116,** 305–322 (1955).

Glasstone, V. F. C., "Study of respiration in healthy and mosaic-infected tobacco plants," *Plant Physiol.,* **17,** 267–277 (1942).

Gold, A. H., Suneson, C. A., Houston, B. R., and Oswald, J. W., "Electron microscopy and seed and pollen transmission of rod-shaped particles associated with the false stripe virus disease of barley," *Phytopathology* **44,** 115–117 (1954).

Goldstein, B., "A cytological study of the leaves and growing points of healthy and mosaic diseased tobacco plants," *Bull. Torrey Botan. Club,* **54,** 499–599 (1926).

Gondo, M., "Respiration of virus diseased tobacco plant. (I)" *Bull. Fac. Agr. Kagoshima Univ.,* **1,** 1–3 (1952).

Gondo, M., "Respiration of virus diseased tobacco plant. (II)," *Bull. Fac. Agr. Kagoshima Univ.,* **3,** 25–28 (1954).

Goodchild, D. J., "Relationships of legume viruses in Australia. II. Serological relationships of bean yellow mosaic virus and pea mosaic virus," *Australian J. Biol. Sci.,* **9,** 231–237 (1956).

Govier, D. A., "The properties of tomato aspermy virus and its relationship with cucumber mosaic virus," *Ann. Appl. Biol.,* **45,** 62–73 (1957).

Graham, D. C., "Cross-protection between strains of tomato aspermy virus and cucumber mosaic virus," *Virology,* **3,** 427–428 (1957).

Grainger, J., "Low temperature masking of tobacco mosaic symptoms," *Nature,* **37,** 31–32 (1936).

Gray, R. A., "The eclectophoresis and chromatography of plant viruses on filter paper," *Arch. Biochem. Biophys.,* **38,** 305–316 (1952).

Grogan, R. G., and Walker, J. C., "A pod-distorting strain of the yellow mosaic virus of bean," *J. Agr. Research,* **77,** 301–314 (1948).

Hall, D. H., "Studies on serology of cucurbit and bean viruses," *Dissertation Abstr.,* **16,** 219 (1956).

Hancock, B. L., "A laboratory colour test for the diagnosis of swollen shoot of theobroma cacao," *Trop. Agr. (Trinidad)*, 26, 54–56 (1949).

Hildebrand, E. M., "Cork virus leafspots on Triumph sweet potato contain separated parenchyma cells," *Science*, 123, 1034–1035 (1956).

Hildebrand, E. M., "The feathery mottle virus complex of sweetpotato," *Phytopathology*, 50, 751–757 (1960).

Hirai, T., "The diagnosis of plant virus diseases by means of paper electrophoresis," *Forsch. Pflanzenkrankh. Kyoto*, 6, 87–96 (1956).

Hirai, T., "Paper electrophoresis of the virus-infected tulip bulb," *Sonderdruck Aus Die Naturwissenschaften*, Springer, Berlin, 1959, Heft 1, S. 17.

Hirata, S., "Studies on the 'Colloidal Precipitation Method' for diagnosing virus-diseased potato, radish, turnip, and sweet potato," *Mem. Fac. Agr. Miyazaki Univ.*, 1, 137–177 (1955).

Hirata, S., "The methylene-blue method for diagnosing the virus-infected potato tubers and some other crops," *Mem. Fac. Agr. Miyazaki Univ.*, 1, 179–200 (1956).

Hirata, S., "Studies on the physical and chemical methods for diagnosing the virus-infected potatoes and some other crops," *Mem. Fac. Agr. Miyazaki Univ.*, 2, 1–87 (1958).

Hitchborn, J. H., "The effect of temperature on infection with strains of cucumber mosaic virus," *Ann. Appl. Biol.*, 44, 590–598 (1956).

Hoggan, I. A., "Cytological studies on virus diseases of solanaceous plants," *J. Agr. Research*, 35, 651–671 (1927).

Holmes, F. O., "Local lesions in tobacco mosaic," *Botan. Gaz.*, 87, 39–55 (1929).

Holmes, F. O., "Local lesions of mosaic in *Nicotiana tabacum* L.," *Contribs. Boyce Thompson Inst.*, 3, 162–172 (1931).

Holmes, F. O., "Symptoms of tobacco mosaic disease," *Contribs. Boyce Thompson Inst.*, 4, 323–357 (1932).

Holmes, F. O., "A masked strain of tobacco-mosaic virus," *Phytopathology*, 24, 845–873 (1934).

Holmes, F. O., "A distinctive strain of tobacco mosaic virus from plantago," *Phytopathology*, 31, 1089–1098 (1941).

Hutchins, L. M., "Identification and control of the phony disease of the peach," *Office State Entomol. Georgia. Bull.*, No. 78 (1933).

Iwanowski, D., "Ueber die Mosaikkrankeit der Tabakspflanzen," *Z. Pflanzenkrankh.*, 8, 1–41 (1903).

Jensen, J. H., "Isolation of yellow-mosaic viruses from plants infected with tobacco mosaic," *Phytopathology*, 23, 964–974 (1933).

Jensen, J. H., "Studies on representative strains of tobacco mosaic virus," *Phytopathology*, 27, 69–84 (1937).

Jensen, J. H., "Identification of virus infection in plant tissue," *Ann. Rev. Microbiol.*, 6, 139–150 (1952).

Johnson, J., "Virus particles in various plant species and tissues," *Phytopathology*, 41, 78–93 (1951).

Jones, J. P., "Studies on the Auxin levels of healthy and virus infected plants," *Dissertation Abstr.*, 16, 1567–1568 (1956).

Kahn, R. P., and Latterell, F. M., "Symptoms of bud-blight of soybeans caused by the tobacco-and-tomato-ringspot viruses," *Phytopathology*, 45, 500–502 (1955).

Kassanis, B., "Intranuclear inclusions in virus infected plants," *Ann. Appl. Biol.*, 26, 705–709 (1939).

Keener, P. D., "Virus diseases of plants in Arizona. I. Field and experimental observations on mosaics affecting vegetable crops," *Bull. Ariz. Agr. Exptl. Sta.*, No. 256 (1954).

Keitt, G. W., and Moore, J. D., "Masking of leaf symptoms of sour-cherry yellows by temperature effects," *Phytopathology*, 33, 1213–1215 (1943).

Keleny, G. P., "The detection of necrosis in leaf roll infected potato stems by means of fluorescence microscopy," *J. Australian Inst. Agr. Sci.*, 17, 203–206 (1952).

Kelly, S., "Interpretation of Lindner's test for plant virus diseases," *Science*, 109, 573–574 (1949).

KenKnight, G., "The acid test for phony disease of peach and its diagnostic value," *Phytopathology*, 41, 20–21 (1951).

KenKnight, G., "The acid test for phony disease of peach," *Phytopathology*, 41, 829–832 (1951a).

KenKnight, G., "Comparison of phloroglucinol test with the acid one for phony," *Phytopathology*, 42, 285 (1952).

Kirkpatrick, H. C., and Lindner, R. C., "A rapid spectrophotometric assay for some plant viruses," *Phytopathology*, 44, 525–529 (1954).

Kleczkowski, A., and Watson, M. A., "Serological studies on sugar-beet yellows virus," *Ann. Appl. Biol.*, 31, 116–120 (1944).

Klostermeyer, E. C., "The phloroglucinol test for diagnosis of leaf roll in netted gem potatoes," *Plant Disease Reptr.*, 34, 36–38 (1950).

Knight, C. A., "Effect of artificial light on the symptoms produced by two yellow mosaic viruses," *Arch. Ges. Virusforsch.*, 2, 260–267 (1941).

Knight, C. A., "The nature of some of the chemical differences among strains of tobacco mosaic virus," *J. Biol. Chem.*, 171, 297–308 (1947).

Knight, C. A., "Are cucumber viruses 3 and 4 strains of tobacco mosaic virus? A review of the problem," *Virology*, 1, 261–267 (1955).

Knight, C. A., and Stanley, W. M., "Aromatic amino acids in strains of tobacco mosaic virus and in the related cucumber viruses 3 and 4," *J. Biol. Chem.*, 141, 39–49 (1941).

Koshimizu, Y., "Leucoplasts as an origin of intracellular inclusions and site of virus reproduction," *Abstr. Ann. Meeting Phytopathol. Soc. Japan*, in Japanese (1955).

Koshimizu, Y., and Iizuka, N., "Origins and formation of intracellular inclusions associated with two leguminous virus diseases," *Protoplasma*, 48, 113–133 (1957).

Kunkel, L. O., "Histological and cytological studies on the Fiji disease of sugar cane," *Bull. Exptl. Sta. Hawaiian Sugar Planters' Assoc.*, 3, 260–267 (1924).

Kunkel, L. O., "Celery yellows of California not identical with the aster yellows of New York," *Contribs. Boyce Thompson Inst.*, 4, 405–414 (1932).

Kunkel, L. O., "Studies on acquired immunity with tobacco and *Aucuba* mosaics," *Phytopathology*, 24, 437–466 (1934).

Kunkel, L. O., "Genetics of viruses pathogenic to plants," *Am. Assoc. Advanc. Sci.*, 12, 22–27 (1940).

Kunkel, L. O., "Variation in phytopathogenic viruses," *Ann. Rev. Microbiol.*, 1, 85–100 (1947).

Kunkel, L. O., "Virus-induced abnormalities. Abnormal and pathological plant growth," *Brookhaven Symposia in Biol.*, 6, 157–173 (1954).

Lackey, C. R., "Histological changes produced by virus yellows in sugar beet leaves," *Proc. Am. Soc. Sugar Beet Technologists*, 8, 225–229 (1954).

Lee, C. L., "Virus-tumor formation in roots of *Capsella bursa-pastoris*," *Phytopathology*, 46, 140–144 (1956).

Lee, C. L., and Black, L. M., "Anatomical studies of *Trifolium incarnatum* infected by wound-tumor virus," *Am. J. Botan.*, 42, 160–168 (1955).

Lindner, R. C., "A rapid chemical test for some plant viruses," *Science*, 107, 17–19 (1948).

Lindner, R. C., Weeks, T. E., and Kirkpatrick, H. C., "A qualitative chemical test for some stone fruit virus diseases," *Phytopathology*, 39, 1059–1060 (1949).

Lindner, R. C., Kirkpatrick, H. C., and Weeks, T. E., "A simple staining technique for detecting virus diseases in some woody plants," *Science*, 112, 119–120 (1950).

Lindner, R. C., Weeks, T. E., and Kirkpatrick, H. C., "Studies on a color test for stone fruit virus diseases," *Phytopathology*, 41, 897–902 (1951).

Linford, M. B., "Transmission of the pineapple yellow-spot virus by *Thrips tabaci*," *Phytopathology*, 22, 302–324 (1932).

Loebenstein, G., "Paper chromatography of a sweet potato virus," *Nature*, 179, 1086 (1957).

MacMillan, H. G., "Potato mosaic masking at high altitudes," *Phytopathology*, 13, 39 (1923).

MacNeill, B. H., and Ismen, H., "A newly differentiated component of the tomato virus-streak complex," *Plant Disease Rept.*, 42, 898 (1958).

Marini, E., "Daignosi sierologiche con la reazione di agglutinazione dei cloroplasti," *Ital. Agr.*, 93, 695–697 (1956).

Matsui, C., and Yamada, M., "The localization of tobacco mosaic virus with diseased tobacco leaf," *Ann. Phytopathol. Soc. Japan*, 23, 75–78 (1958).

Matthews, R. F. F., *Plant Virus Serology*, Cambridge Univ. Press, London, 1957.

McCrum, R. C., Studenroth, J. C., and Olszewska, D., "Microslide Ouchterlony technique for serological detection of potato virus X," *Phytopathology*, **61**, 290 (1971).

McGrew, J. R., and Scott, D. H., "Effect of two virus complexes on the responses of two strawberry varieties," *Plant Disease Reptr.*, **43**, 385–389 (1959).

McKay, R., and Loughnane, J. B., "Effects of some single viruses and of combinations of the same viruses on three potato varieties," *Sci. Proc. Roy. Dublin Soc.*, **26**, 133–143 (1953).

McKinney, H. H., "Virus mixtures that may not be detected in young tobacco plants," *Phytopathology*, **16**, 893 (1926).

McKinney, H. H., "Mosaic diseases of wheat and related cereals," *U.S. Dept. Agr. Circ.*, No. 442 (1937).

McKinney, H. H., and Clayton, E. E., "Acute and chronic symptoms in tobacco mosaics," *Phytopathology*, **33**, 1045–1054 (1943).

McKinney, H. H., and Clayton, E. E., "Genotype and temperature in relation to symptoms caused in *Nicotiana* by the mosaic virus," *J. Heredity*, **36**, 323–331 (1945).

McWhorter, F. P., "Isometric crystals produced by *Pisum* virus 2 and *Phaseolus* virus 2," *Phytopathology*, **31**, 760–761 (1941).

McWhorter, F. P., "Plant virus inclusions," *Ann. Rev. Phytopathology*, **3**, 287–312 (1965).

McWhorter, F. P., and Brierley, P., "Anatomical symptoms useful in diagnosis of lily rosette," *Phytopathology*, **41**, 66–71 (1951).

McWhorter, F. P., and Frazier, W. A., "Analysis of the virus disease complex that reduces yields of blue lake beans in Oregon," *Oregon State Coll. Agr. Exptl. Sta., Misc. Paper*, No. 27 (1956).

Miller, P. W., and Aldrich, F. D., "Studies on the detection of viruses in the leaves of strawberries by the Lindner staining procedure," *Plant Disease Reptr.*, **35**, 131–133 (1951).

Moorhead, E. L., "Serological studies of viruses infecting the cereal crops. I. A comparison of barley stripe mosaic virus and brome mosaic isolates by means of the complement-fixation technique," *Phytopathology*, **46**, 498–501 (1956).

Moorhead, E. L., "Serological studies of viruses infecting the cereal crops. II. The antigenic characteristics of wheat streak mosaic virus as determined by the complement-fixation technique," *Phytopathology*, **49**, 151–157 (1959).

Müller, K. O., and Munro, J., "The affinity of potato virus Y infected potato tissues for dilute vital stains," *Sonderdruck aus Phytopathol. Z.*, **28**, 70–82 (1956).

Murphy, D. M., and Pierce, W. H., "Common mosaic of the garden pea *Pisum sativum*," *Phytopathology*, **27**, 710–721 (1937).

Norval, I. P., "Derivatives from an unusual strain of tobacco mosaic virus," *Phytopathology*, **28**, 675–692 (1938).

Olson, E. O., "Mild and severe strains of tristeza virus in Texas citrus," *Phytopathology,* 46, 336–341 (1956).

Olson, E. O., "Responses of lime and sour orange seedlings and four scionrootstock combinations to infection by strains of the tristeza virus," *Phytopathology,* 48, 454–459 (1958).

Ouchterlony, O., in *Progress in Allergy* Volume 5, S. Karger, New York, 1958.

Owen, P. C., "The respiration of tobacco leaves in the 20-hour period following inoculation with tobacco mosaic virus," *Ann. Appl. Biol.,* 43, 114–121 (1955).

Owen, P. C., "The effects of infection with tobacco mosaic virus on the photosynthesis of tobacco leaves," *Ann. Appl. Biol.,* 45, 456–461 (1957).

Owen, P. C., "Photosynthesis and respiration rates of leaves of *Nicotiana glutinosa* infected with tobacco mosaic virus and of *N. tabacum* infected with potato virus X," *Ann. Appl. Biol.,* 46, 198–204 (1958).

Owen, P. C., "Some effects of virus infection on leaf water contents of Nicotiana species," *Ann. Appl. Biol.,* 46, 205–209 (1958a).

Porter, C. A., "Histological and cytological changes induced in plants by cucumber mosaic virus (*Marmor cucumeris* H.)," *Contribs. Boyce Thompson Inst.,* 17, 453–471 (1954).

Pound, G. S., "Effect of air temperature on the concentration of certain viruses in cabbage," *Agr. Research,* 71, 471–485 (1945).

Pound, G. S., and Walker, J. C., "Differentiation of certain crucifer viruses by the use of temperature and host immunity reactions," *J. Agr. Research,* 71, 255–278 (1945).

Price, W. C., "Classification of southern celery mosaic virus," *Phytopathology,* 25, 947–954 (1935).

Price, W. C., "Acquired immunity from plant virus diseases," *Quart. Rev. Biol.,* 15, 338–361 (1940).

Price, W. C., "Genetic composition in relation to isoelectric point and serological reactions of strains in tobacco mosaic virus," *Trans. N. Y. Acad. Sci.,* 16, 196–201 (1954).

Quanjer, H. M., "Die Nekrose des Phloems der Kartoffelpflanze die Ursache der Blattrollkrankheit," *Mededel. Rijks. Hoogere Land.—Tuinen Boschblouwschool,* 6, 41–76 (1913).

Ragetli, H. W. J., and van der Want, J. P. H., "Paper chromatography of plant viruses," *Proc. Acad. Sci. Amsterdam, Ser. C.,* 57, 621–627 (1954).

Rawlins, T. E., "A modification of Bald's stain for viruses and for cell inclusions associated with virus infections," *Phytopathology,* 47, 307 (1957).

Rawlins, T. E., Weierich, A. J., and Schlegel, D. E., "A histochemical study of certain plant viruses by means of the Sakaguchi reaction for arginine," *Virology,* 2, 308–311 (1956).

Resühr, B., "Zur Chemie der symptombildung viruskranker Pflanzen," *Pflanzenkrankh. Pflanzenschutz,* 52, 63–68 (1942).

Ross, A. F., "Unrelatedness of potato virus Y and cucumber mosaic virus," *Phytopathology*, 40, 445–452 (1950).

Ross, A. F., "Identification of plant viruses," *Plant Pathology*, M. K. Corbett and H. D. Sisler, eds., University of Florida Press, Gainesville, Fla. 1964.

Ross, A. F., and Williamson, C. E., "Physiologically active emanations from virus-infected plants," *Phytopathology*, 41, 431–438 (1951).

Rubio, M., "Origin and composition of cell inclusions associated with certain tobacco and crucifer viruses," *Phytopathology*, 46, 553–556 (1956).

Rubio, M., and van Slogteren, D. H. M., "Light and electron microscopy of X-bodies associated with broad-bean mottle virus and *Phaseolus* virus 2," *Phytopathology*, 46, 401–402 (1956).

Sacramuzzi, G., "Saggi colorimetrici per la diagnosi rapida di malattie de virus di fruttiferi," *Ann. Sper. Agrar. (Rome)*, 10, 1417–1437 (1956).

Sanford, G. B., and Grimble, J. G., "Observations on phloem necrosis of potato tubers," *Can. J. Research, C.*, 22, 162–170 (1944).

Schneider, H., "Chronic decline, a tristeza-like bud-union disorder of orange trees," *Phytopathology*, 47, 279–284 (1957).

Schneider, H., and Wallace, J. M., "Tests on orange trees top-worked to lemons," *Calif. Citrograph.*, 39, 224, 248–252 (1954).

Severin, H. H. P., "Experiments with the aster yellows virus from several states," *Hilgardia*, 8, 305–325 (1934).

Severin, H. H. P., "Color breaking in pansies and violas," *Calif. Univ. Agr. Exptl. Sta. Circ.*, No. 377 (1948).

Severin, H. H. P., "Symptoms of cucumber-mosaic and tobacco-ringspot viruses on celery," *Hilgardia*, 20, 267–277 (1950).

Shapovalov, M., and Jones, H. A., "Changes in the composition of the tomato plant accompanying different stages of yellows," *Plant Physiol.*, 5, 157–165 (1930).

Sheffield, F. M. L., "The formation of intracellular inclusions in solanaceous hosts infected with *Aucuba* mosaic of tomato," *Ann. Appl. Biol.*, 18, 471–493 (1931).

Sheffield, F. M. L., "The development of assimilatory tissue in solanaceous hosts infected with *Aucuba* mosaic of tomato," *Ann. Appl. Biol.*, 20, 57–69 (1933).

Sheffield, F. M. L., "Some effects of plant virus diseases on the cells of their hosts," *J. Roy. Microscop. Soc.*, 59, 149–161 (1939).

Sheffield, F. M. L., "Micrurgical studies on virus-infected plants," *Proc. Roy. Soc. (London)*, B126, 529–538 (1939a).

Sheffield, F. M. L., "The cytoplasmic and nuclear inclusions associated with severe etch virus," *J. Roy. Microscop. Soc.*, 61, 30–45, (1941).

Sheffield, F. M. L., "Value of phloem necrosis in the diagnosis of potato leaf roll," *Ann. Appl. Biol.*, 30, 131–136 (1943).

Sheffield, F. M. L., "Preliminary studies in the electron microscope of some plant virus inclusion bodies," *J. Roy. Microscop. Soc.*, 66, 69–76 (1946).

Siegel, A., and Wildman, S. G., "Some natural relationships among strains of tobacco mosaic virus," *Phytopathology*, 44, 277–282 (1954).

Sill, W. H., and Fellows, H., "Symptom expression of the wheat streak-mosaic virus disease as affected by temperature," *Plant Disease Reptr.*, 37, 30–33 (1953).

Simons, J. N., "Three strains of cucumber mosaic virus affecting bell pepper in the Everglades area of South Florida," *Phytopathology*, 47, 145–150 (1957).

Singer, S. J., Bald, J. G., Wildman, S. G., and Owen, R. D., "The detection and isolation of naturally occurring strains of tobacco mosaic virus by electrophoresis," *Science*, 114, 463–465 (1951).

Smith, F. H., and McWhorter, F. P., "Anatomical effects of tomato ringspot virus in *Vicia faba*," *Am. J. Botan.*, 44, 470–477 (1957).

Smith, K. M., *Recent Advances in the Study of Plant Viruses*, 2nd ed., Blakiston, Philadelphia, 1951.

Smith, K. M., "A note on the observation of viruses in the cells of infected plants," *Biochim. Biophys. Acta*, 10, 210–214 (1953).

Smith, K. M., and Bald, J. G., "A description of a necrotic virus disease affecting tobacco and other plants," *Parasitology*, 27, 231–245 (1935).

Smith, K. M., and Markham, R., "Two new viruses affecting tobacco and other plants," *Phytopathology*, 34, 324–329 (1944).

Sorokin, H., "The destruction of the chloroplasts in tomato mosaic," *Phytopathology*, 16, 66–67 (1926).

Sorokin, H., "Phenomena associated with the destruction of the chloroplasts in tomato mosaic," *Phytopathology*, 17, 263–379 (1927).

Sprau, F., "Ein neues Farbeverfahren fur Kallose und die Uberprufung seiner Anwendung zur Diagnose Blattrollkranker Kartoffeln," *Nachrbl. Deut. Pflanzenschutzdienstes (Stuttgart)*, 10, 3–6 (1958).

Sreenivasaya, M., and Sastri, B. N., "Contributions to the study of spike disease of sandal," *J. Indian Inst. Sci.*, 11A, 23–29 (1928).

Stange, L., "Zur Frage nach den Störungen des stoffwechesls in den Blättern vergilbungskranker Zuckerrüben," *Phytopathol. Z.*, 21, 214–217 (1953).

Steere, R. L., and Williams, R. C., "Identification of crystalline inclusion bodies extracted intact from plant cells infected with tobacco mosaic virus," *Am. J. Botan.*, 40, 81–84 (1953).

Sturgess, O. W., "Leaf shrivelling virus disease of the tomato," *Queensland J. Agr. Sci.*, 13, 175–220 (1956).

Takahashi, W. N., "Anomalous proteins associated with three strains of tobacco mosaic virus," *Virology*, 1, 393–396 (1955).

Thomas, W. D., Baker, R. R., and Zoril, J. G., "The use of ultraviolet light as a means of diagnosing carnation mosaic," *Science*, 113, 576–577 (1951).

Thomson, A. D., "Interference between plant viruses," *Nature*, 181, 1547–1548 (1958).

Thung, T. H., "Physiologisch onderzoek met betrekking tot het virus der bladrolziekte van de aardappelplant, *Solanum tuberosum* L.," *Tijdschr. Plantenziekten,* 34, 1–48, 49–74 (1928).

Tinsley, T. W., and Usher, G., "A colour test for detecting swollen shoot disease in Gold Coast cacao," *Nature,* 174, 87 (1954).

Tomlinson, N., and Woodbridge, C. G., "Evaluation of a colorimetric and ultraviolet absorption test for diagnosis of plant virus diseases as applied to stone and small fruits," *Can. J. Agr. Sci.,* 35, 111–123 (1955).

Tompkins, C. M., "A transmissible disease of cauliflower," *J. Agr. Research,* 55, 33–46 (1937).

van Duuren, A. J., "De vergelingsziekte der bieten. V. Onderzoek naar de storingen in de stofwisseling van de Suikerbiet, veroorzaakt door de vergelingsziekt," *Mededel. Inst., Rationele Suikerprod.,* 25, 61–99 (1956).

Van Regenmental, H. V., "Plant virus serology," *Adv. Virus Res.,* 12, 207–271 (1966).

van Slogteren, D. H. M., "VII. Serological analysis of some plant viruses with the gel-diffusion method," *Proc. 2nd. Conf. Potato Virus Diseases, Lisse, Wageningen,* 1954, 25–29.

van Slogteren, D. H. M., "Gel diffusion of tobacco mosaic virus, demonstrated by serological analysis of its components and by electron-microscopy," *Acta Botan. Neerl.,* 4, 472–476 (1955).

van Slogteren, E., "Serological diagnosis of plant virus diseases," *Ann. Appl. Biol.,* 42, 122–128 (1955).

van Slogteren, E., and e Bruyn Ouboter, M. P., "Onderzoekingen over virusziekten in bloembolgewassen. II. Tulpen. I," *Mededel. Landbouwhogeschool Wageningen,* 45, 1–54 (1941).

Vinson, C. G., and Petre, A. W., "Mosaic disease of tobacco," *Botan. Gas.,* 87, 14–28 (1929).

Wallace, J. M., "Virus-strain interference in relation to symptoms of psorosis disease of citrus," *Hilgardia,* 27, 223–246 (1957).

Wartenberg, H., and Lindau, G., "Studien über die 'Dehydrase-Wirkung' gesunder und abbaukrander Kartoffelknollen," *Phytopathol. Z.,* 9, 297–324 (1936).

Watson, M. A., "The effect of sucrose spraying on symptoms caused by beet yellows virus in sugar beet," *Ann. Appl. Biol., 43, 672–685 (1955).*

Watson, M. A., and Watson, D. J., "The effect of infection with beet yellows and beet mosaic viruses on the carbohydrate content of sugar-beet leaves, and on translocation," *Ann. Appl. Biol.,* 38, 276–288 (1951).

Webb, R. E., "Wilting, an atypical primary symptom of infection of potatoes by the leafroll virus," *Plant Disease Reptr.,* 40, 15–18 (1956).

Wellman, F. L., "Identification of celery virus 1, the cause of southern celery mosaic," *Phytopathology,* 24, 695–725 (1934).

Wheeler, W. M., *Emergent Evolution,* Norton, New York, 1928.

Whitehead, T., "The physiology of potato leaf roll. I. On the respiration of healthy and leaf roll infected potatoes," *Ann. Appl. Biol.,* 21, 48–77 (1934).

Wildman, S. G., and Bonner, J., "Electrophoretic analysis of normal and diseased leaf cytoplasmic proteins as a means of identifying plant viruses," *Am. J. Botan.*, **36**, 830–831 (1949).

Wildman, S. G., and Bonner, J., "The electrophoretic detection of plant virus proteins," *Sci. Monthly*, **70**, 347–351 (1950).

Wildman, S. G., Cheo, C. C., and Bonner, J., "The proteins of green leaves. III. Evidence of the formation of tobacco mosaic virus protein at the expense of a main protein component in tobacco leaf cytoplasm," *J. Biol. Chem.*, **180**, 985–1001 (1949).

Wilkinson, J., "The effects of 'aspermy' virus upon nuclear behaviour in certain solanaceous plants," *Proc. Soc. Study Fertility*, **4**, 53–54 (1952).

Wilkinson, J., "Some effects induced in *Nicotiana glutinosa* by 'aspermy' virus of tomato," *Ann. Botany (London)*, **17**, 219–223 (1953).

Wilkinson, J., "Virus-induced nucleolar abnormalities in tomato," *Nature*, **171**, 658–659 (1953a).

Wilson, J. H., "The use of the phloroglucinol test for diagnosis of leaf roll in potatoes," *J. Australian Inst. Agr. Sci.*, **14**, 76–78 (1948).

Wilson, J. H., "Effects of nutrition and light intensity on symptoms of leaf-roll virus infection in the potato plant," *Ann. Appl. Biol.*, **43**, 273–287 (1955).

Yamaguchi, A., and Hirai, T., "The effect of local infection with tobacco mosaic virus on respiration in leaves of *Nicotiana glutinosa*," *Phytopathology*, **49**, 447–449 (1959).

Yoshii, H., "Studies on the nature of insect-transmission in plant viruses (IV). On the unsound metabolism in rice plant affected with the dwarf disease virus (*Oryza* virus 1)," *Lab. Plant Pathol. Coll. Agr., Ehime Univ.*, **8**, 394–405 (1958).

Yoshii, H., and Kiso, A., "Studies on the nature of insect-transmission in plant viruses (I). Some observations on the unsound metabolism in Satsuma orange affected with the dwarf disease by transmission of green broadwinged plant hopper," *Lab. Plant Pathol., Coll. Agr., Ehime Univ.*, **7**, 306–314 (1957).

Zaitlin, M., Schechtman, A. M., Bald, J. G., and Wildman, S. G., "Detection of virus in Cattleya orchids by serological methods," *Phytopathology*, **44**, 314–318 (1954).

Zaumeyer, W. T., "Three previously undescribed mosaic diseases of pea," *J. Agr. Research*, **60**, 433–452 (1940).

Zaumeyer, W. T., and Harter, L. L., "Inheritance of symptom expression of bean mosaic virus 4," *J. Agr. Research*, **47**, 295–300 (1943).

Modes of Plant
Virus Transmission

There seems little point in attempting a separation of methods of transmission into "natural" or "artificial" because, although many techniques have been devised to increase efficiency, the artificial methods all have their counterparts in natural transmission.[1]

TRANSMISSION THROUGH SEED

Transmission of viruses through seed was once considered to be rare. Smith (1951) recognized only eight cases, but either he must have intended that his list be noninclusive or he was not prepared to accept the evidence in other cases as adequate. Crowley (1957) lists 49 cases of seed transmission, of which 29 were reported prior to 1950, but he omitted four others: cherry necrotic ring spot (Gilmer, 1955); avocado sun blotch (Wallace and Drake, 1953); yellow ring spot of tobacco (Marcelli, 1955);

[1] The following abbreviations of viruses have been used throughout this chapter: TMV: tobacco mosaic virus; TSW: tomato spotted wilt; TNV: tomato necrosis virus; CMV: cucumber mosaic virus; TYDV: raspberry yellow dwarf virus; RRSV: raspberry ringspot virus; BRV: beet ringspot virus; and AMV: alfalfa mosaic virus.

and barley stripe mosaic (false stripe) in wheat (McNeal and Afanasiev, 1955). Since then Lister (1960) has added three others. Crowley's list, amended to include all the cases then known, is shown in Table 11.1. New cases reported during the last decade are shown in Table 11.2.

Although a total of about 75 seed-transmitted viruses may not appear to be a high percentage of the known viruses, seed transmission should not be considered a rare phenomenon. A virus may be seed transmitted rarely, as reference to Table 11.1 will show. Seed transmission is infrequent enough, however, to have led to considerable discussion as to the reason. According to Crowley (1957), there are four classes of viruses in which seed transmission is impossible: viz., those which kill their hosts; those which prevent flower formation; those which are limited in their distribution in the host plant; or those which are unable to tolerate the changes that take place in the seed during its maturation and dessication. However, viruses included in these classes constitute only about 10% of the known viruses, and both Bennett (1940) and Crowley have reviewed the theories to account for the lack of seed transmission in the remaining 90%.

Caldwell (1962) reviewed the hypotheses accounting for the relative infrequency of transmission of viruses through the embryo and added another. Phosphate starvation adversely affects virus multiplication (replication), and some inhibitors of respiration have the same effect. When both phenomena are functioning, high energy phosphate compounds would be reduced below the surplus levels necessary for virus multiplication in the plant cell. These surpluses are more likely to be available in the green leaf; their absence from the developing embryos may well account for the absence of virus in these tissues.

Virus Inhibitors

The presence of virus inhibitors in seed was shown by Crowley (1955), but this is not enough; virus inactivators were not demonstrable (Crowley, 1957). Viruses may induce sterility (Caldwell, 1952) and thereby prevent seed transmission, but this is known for only one or two viruses. Studies on the distribution of viruses in seeds show that embryos are rarely infected, even though the virus may be present in all the other tissues of the seed (Bennett and Esau, 1936). Crowley (1957) studied the effects of five viruses on seed infection: two viruses of bean, one seed-transmitted and the other not; TSW in *Cineraria,* reported to be 96% seed-transmitted by Jones (1944), compared with none in tomato; CMV in wild cucumber *Echinocytis lobata* Mich. T. and G., 22% seed-transmitted, according to Doolittle and Gilbert (1919), compared with 2% of

TABLE 11.1. Seed-Transmitted Virus [a]

Host	Virus [b]	% Trans.	Reference
Cannabinacae			
Humulus lupulus	Chlorotic disease of hop	27	Salmon and Ware (1935)
Caryophyllaceae	Lychnis ringspot		Bennett (1959)
Lychnius divaricata			
Chenopodiaceae			
Beta vulgaris	Beet yellows	30	Clinch and Loughnane (1948)
B. vulgaris	Beet yellows		Ernould (1950)
B. vulgaris	Beet yellows	0.1	Nikolić (1956)
Compositeae			
Cineraria spp.	Tomato spotted wilt [c]	96	Jones (1944)
Helianthus annuus	Unnamed	17–43	Traversi (1949)
Lactuca sativa	Lettuce mosaic	6	Ainsworth and Ogilvie (1939)
L. sativa	Lettuce mosaic		Couch (1955)
L. sativa	Lettuce mosaic	5	Newall (1923)
L. sativa	Lettuce mosaic	10	Ogilvie *et al.* (1935)
L. sativa	Lettuce mosaic		van Hoof (1959)
L. sativa	Mosaic	8	Grogan and Bardin (1950)
L. sativa	Yellow mosaic	30	Vasudeva *et al.* (1948)
Senecio vulgaris	Lettuce mosaic	0.5	Ainsworth and Ogilvie (1939)
S. vulgaris	Tomato black ring	3–40	Lister (1960)
Convolvulaceae			
Cuscuta campestris	Dodder latent mosaic	4.8	Bennett (1944)
Cucurbitaceae			
Cucumis melo	Muskmelon mosaic	28–94	Rader *et al.* (1947)
C. melo	Cucumber mosaic	2	Kendrick (1934)
C. melo	Cucumber mosaic	16	Mahoney (1935)
C. sativa	Cucumber mosaic	1.4	McClintock (1916)
Cucurbita pepo	Squash mosaic		Grogan (1958)
C. pepo	Muskmelon mosaic		Rader *et al.* (1947)
C. pepo	Squash mosaic	0.96	Middleton (1944)
C. pepo	Cucumber mosaic	1.5	Chamberlain (1939)
Cucumis moschata	Muskmelon mosaic	28–94	Rader *et al.* (1947)
Echinocystis lobata	Cucumber mosaic	22	Doolittle and Gilbert (1919)
Cruciferae			
Capsella bursa-pectoris	Tomato black ring	90	Lister (1960)

TABLE 11.1. *Continued*

Host	Virus [b]	% Trans.	Reference
Gramineae			
Hordeum vulgare	False stripe [d]	50–100	Gold *et al.* (1954)
H. vulgare	False stripe	86	Hagborg (1954)
H. vulgare	False stripe	58	McKinney (1951)
Triticum vulgare	False stripe	71	Hagborg (1954)
T. vulgare	False stripe		McNeal and Afanasiev (1955)
Lauracae			
Persea americana	Avocado sun blotch		Wallace and Drake (1953)
Leguminoseae			
Dolichos biflorus	Mosaic	25–40	Uppal (1931)
Glycine soja	*Arabis* mosaic	12.5	Lister (1960)
G. soja	Soybean mosaic	10–25	Kendrick and Gardner (1921)
G. soja	Raspberry ringspot	20	Lister (1960)
G. soja	Tobacco ringspot		Athow and Bancroft (1959)
G. soja	Tobacco ringspot	54–78	Desjardins *et al.* (1954)
G. soja	Tomato ringspot	76	Kahn (1956)
G. soja	Tomato black ring	91	Lister (1960)
Lathyrus odoratus	Pea mosaic		Dickson (1922)
Phaseolus limensis	Lima bean mosaic	25	McClintock (1917)
P. vulgaris	Bean mosaic [d]		Reddick and Stewart (1919)
P. vulgaris	Bean mosaic	43	Archibald (1921)
P. vulgaris	Bean mosaic		Corbett (1948)
P. vulgaris	Bean mosaic	10–30	Fajardo (1930)
P. vulgaris	Bean mosaic	20–59	Harrison (1935)
P. vulgaris	Bean mosaic		Mastenbroek (1943)
P. vulgaris	Bean mosaic		Troll (1957)
P. vulgaris	Red node	27	Thomas and Graham (1951)
Pisum sativum	Pea mosaic	0.5	Dickson (1922)
Trifolium hybridum	Pea mosaic	0.5	Dickson and McRostie (1922)
T. pratense	Pea mosaic	47	Dickson and McRostie (1922)
Vicia faba	Mosaic	1	Quantz (1953)
Vigna sesquipedalis	Asparagus bean mosaic	37	Snyder (1942)
V. sinensis	Cowpea mosaic		Anderson (1957)
V. sinensis	Cowpea mosaic	14	Gardner (1927)
V. sinensis	Cowpea mosaic	7	McLean (1941)
V. sinensis	Cowpea mosaic	11	Yu (1946)
V. sinensis	Cucumber mosaic		Anderson (1957)
Papaveracae			
Fumaria spp.	Tomato black ring	3/3	Lister (1960)
Rutaceae			
Citrus aurantifolia	Xylopsorosis	66	Childs (1956)
Malvaceae			
Abutilon spp.	Mosaic	<1	Keur (1933)

Host	Virus [b]	% Trans.	Reference
Rosaceae			
Prunus avium var.			
Mazzard	Cherry ringspot	5	Cochran (1946)
P. avium var.			
Mazzard	Cherry ringspot	56	Cation (1952)
P. cerasus	Cherry ringspot	30	Cation (1949)
P. mahaleb	Cherry yellows	7.8	Cation (1949)
P. mahaleb	Cherry yellows	41	Cation (1952)
P. mahaleb	Cherry ringspot	10	Cation (1952)
P. mahaleb	Cherry necrotic ringspot		Gilmer (1955)
P. americana	Peach ringspot		Hobart (1956)
P. persica	Peach ringspot		Wagnon *et al.* (1960)
Rubiacae			
Coffea excelsa	Coffee ringspot		Reyes (1961)
Solanaceae			
Capsicum frutescens	Tobacco mosaic [c]	22	McKinney (1952)
C. annum	Mosaic [d]		Ikeno (1930)
C. annum	Lucerne mosaic		Sutic (1959)
Datura stramonium	Q disease [d]	100	Blakeslee (1921)
Lycopersicon esculentum	Tomato streak	66	Berkeley and Madden (1932)
L. esculentum	Tobacco mosaic		Berkeley and Madden (1932)
L. esculentum	Tobacco mosaic	2	Doolittle and Beecher (1937)
L. esculentum	Cucumber mosaic	0.2	van Koot (1949)
Nicotiana rustica	Tomato black ring	1/2	Lister (1960)
N. tabacum	Tobacco ringspot	17	Valleau (1941)
N. tabacum	Tobacco ringspot		Marcelli (1955)
Petunia spp.	Tobacco ringspot	20	Henderson (1931)
P. hybrida	*Arabis* mosaic	1/5	Lister (1960)
Physalis peruviana	Tomato bunchy top	29	McClean (1948)
Solanum tuberosum	Virus Y	16	Sprau (1951)
S. tuberosum	Virus Y	14	Reddick (1936)
S. uncanum	Tomato bunchy top	53	McClean (1948)
Urticaceae			
Ulmus campestris	Elm mosaic	3	Bretz (1950)
U. campestris	Elm mosaic		Callahan (1957)

[a] Data from Crowley (1957) amended.
[b] Names of viruses used throughout are taken from the "Virus Index" in *Rev. Appl. Mycol.*, **24**, 515–56 (1945).
[c] Results not confirmed by Crowley.
[d] Also recorded to be pollen transmitted.

TABLE 11.2. Seed-Pollen Transmitted Viruses (Reported 1960–1969)

Host	Virus	% Trans.	Reference
Boraginacae			
Myosotis arvensis	Tobacco rattle	6	Cadman and Lister (1961)
Caryophilacae			
Stellaria media	Raspberry ringspot	34	Lister and Murant (1961)
Stellaria media	Tomato blackring	58	Cadman (1963)
Stellaria media	*Arabis* mosaic	55	Cadman (1963)
Chenopodiacae			
Atriplex pacifica	Sowbane mosaic	21	Bennett and Costa (1961)
Chenopodium album	Sowbane mosaic	30	Bennett and Costa (1961)
Chenopodium murale	Sowbane mosaic	45	Bennett and Costa (1961)
C. amaranticolor	Grapevine fanleaf	1.3	Dias (1963)
C. amaranticolor	Grape yellow mosaic	0.7	Dias (1963)
C. qunioa	Chlorotic leafspot	a	Cadman (1965, 1966)
Compositae			
Senecio vulgaris	*Arabis* mosaic	2	Cadman (1963)
Lactua sativa	*Arabis* mosaic	60–100	Walkey (1967)
Cruciferae			
Capsella b-pastoris	Raspberry ringspot	2–3	Cadman (1963)
Cucurbitacae			
Melon	Necrotic spot	20	Kashi (1966)
Squash	Stone fruit ringspot	b	Das and Milbrath (1961)
Cucumber	Tobacco ringspot	3–7	McClean (1962)
Graminae			
Barley	Barley yellow dwarf	—	Inouye (1966)
Barley	Barley stripe mosaic	46.5	Sutic and Tosic (1966)
Maize; sweet corn	Johnson grass str. of SCM	0.4	Shepherd and Holdeman (1965)
Leguminosae			
Alfalfa	Lucerne mosaic	—	Frosheiser (1964)
Clover, red	Viruses	—	Hamptom and Hansen (1968)
Groundnut	Chlorosis	—	Sharma (1966)
Groundnut	Bunchy top	—	Sharma (1966)
Groundnut	Ring mottle	—	Sharma (1966)
Groundnut	Stunt	Negligible	Culp and Troutman (1968)
Lupin—blue	Cucumber mosaic	6.4	Wells *et al.* (1964)
—yellow	Narrow leaf (BYM)	44.6	Blaszezak (1963)

Host	Virus	% Trans.	Reference
Phaseolus vulgare	Western bean mosaic	2–3	Scotland and Birke (1961)
Soybean	Cherry leafroll	100	Cadman (1963)
Soybean	Soybean stunt	—	Koshimiza and Iisuka (1963)
Soybean	Tobacco ringspot	100	Owusu *et al.* (1969)
Pea	Pea early browning	37	Bos and Vander Want (1962)
Pea	Pea seed-borne mosaic	10–30	Inouye (1967)
Rosaceae			
Apple and Pear	TMV	37	Gilmer (1967)
Cherry	Cherry raspleaf	b	Williams *et al.* (1963)
Cherry	Necrotic spot	93.1	George and Davidson (1966)
Cherry	Peach necrotic ringspot	—	George and Davidson (1966)
Cherry	Sour cherry yellows	77.7	George and Davidson (1966)
Cherry	Sour cherry yellows	17	Gilmer and Way (1963)
Peach	Latent viruses	2.4	Fridlund (1966)
Peach	Necrotic ringspot	b	Williams *et al.* (1963)
Peach	Prune dwarf	b	Williams *et al.* (1963)
Peach	Prune dwarf	15	Gilmer and Way (1960)
Prune	Prune dwarf (sour cherry yellows)	b	George and Davidson (1964)
Raspberry	Tomato blackring	1	Cadman (1963)
Raspberry	Raspberry ringspot	10	Cadman (1963)
Raspberry, black	Black raspberry latent	b	Converse and Lister (1969)
Solanaceae			
Tomato	Potato spindle tuber	a	Bensen and Sing (1964)
Potato	Potato spindle tuber	c	Singh (1967)
Rutaceae			
Citrus	Psorosis	19	Childs and Johnson (1966)
Vitaceae			
V. labrusca	TMV	20	Gilman and Kelts (1968)
Agaricaceae	Die-back (through		
Mushroom	spores)	—	Schiller *et al.* (1963)

a By pollen also.
b By pollen only.
c By pollen and ovule.

the same virus in cucumber; and TMV, 22% seed-transmitted in pungent pepper (McKinney, 1952), compared with none in tomato. In only one of these cases (common bean mosaic) was virus found in the embryos, as is shown in Crowley's figure (Fig. 11.1), but embryo infection may occur during germination if the testa is infected, as Crowley proved experimentally with TMV in pungent pepper. No satisfactory explanation for the negative results obtained by Crowley with TSW in *Cineraria* is available.

Bennett's theory (1940) is that most viruses are unable to survive in the micro- or macrospores or in the embryo sacs, and viruses are unable to infect developing embryos because of the lack of plasmodesmatal connections with the surrounding tissues.

Bennett (1969) reviewed the subject of seed transmission at length and should be referred to for a detailed account of the hypotheses involved. Viruses may be carried on the surfaces of seeds but this requires that the virus remain active on the seed until germination; a court of entry would then be required, however. Viruses that occur in the seed but not in the embryo result in very few infections in the growing seedling; one exception is barley stripe mosaic virus which can have as high a titer in the endosperm as in the leaf tissue; the virus invades the embryo and is seed transmitted. According to Timian (1970), infection occurs through the ovules only if the parental plants are infected with the virus for at least 12 days before fertilization and a longer time is necessary for transmission through pollen. Some strains of the virus are readily transmitted through the seed of some barley varieties; some not through the seed of any cultivar. The manner in which seed-borne viruses are transmitted to seedlings, even though they do not occur in the embryo, is not entirely clear. Tobacco mosaic is not found in the embryo of *M. platycarpa* although 38% of seed transmission has been reported.

Seed transmission occurs in cantaloupe infected with squash mosaic. In addition, the infection in the parent plant results in lower melon weight, seed number and weight, and percent germination. 12% of the seeds contained detectable amounts of virus in the embryos and 22% in the seedlings; in 30 day-old plants, however, this percentage was down to 13. Storage of seed for 2 yr reduced infection. The results were obtained by assay (Powell and Schlegel, 1970).

The longevity of a virus transmitted through pollen was strikingly illustrated by Williams and Smith (1967) who recovered pear decline from pollen stored for more than 5½ years. With the recent finding that pear decline is now included in the diseases associated with mycoplasma, this report is of especial interest. Infection with stubborn disease of citrus results in excessive abortion of seeds (Carpenter *et al.*, 1965). The infield sources of infection with cucumber mosaic was determined by Tomlinson

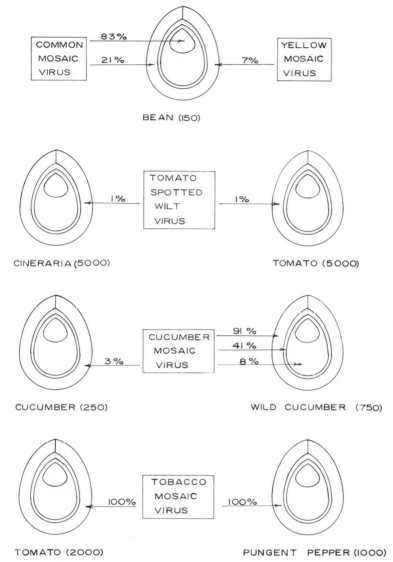

FIGURE 11.1. Diagrammatic representation of the distribution of viruses in the seeds of various plants. Numbers in parentheses indicate the number of seeds dissected (After Crowley, 1957).

and Carter (1966) to be due to seed transmission of the virus through *Stellaria media* (chickweed). Seed infection with alfalfa mosaic virus is responsible for most initial AMV infections in alfalfa stands according to Frosheiser (1970). 93% infection was found in alfalfa seedlings. An unusual consequence of seed infection was reported by Wilson and Dean (1964), who found that bean flour is an efficient source of inoculum of common bean mosaic.

Genetic Aspects

Genetic aspects were considered by Couch (1955), who found that the genotype of the host is the main factor in determining percentage of seed transmission of lettuce mosaic virus. There is some evidence that resistance to infection is a normal characteristic of haploid gemetephytic cells, which would explain the (relative) rarity of seed transmission.

INTERNATIONAL ASPECTS OF SEED TRANSMISSION. These are of importance to many newly developing countries in tropical areas, where huge volumes of seed crops are imported, primarily for direct use as food but frequently also for seed. It is manifestly impossible to control this international transport of viruses (as well as bacterial and fungal infected seed) by any possible quarantine system, so the only hope is for the development of tolerant and resistant strains for each country's conditions.

Evidence for the absence of seed transmission is not often reported, either because negative evidence is expected and is taken for granted, or it is not tested. An exception can be noted. McClean (1957) tested hundreds of citrus seedlings over an 8-yr period and found no evidence that tristeza virus was ever seed transmitted. Evidence on sugar beet yellows is somewhat equivocal. Ernould (1950), working with Belgian material, found no transmission through seed of beet mosaic and beet yellows, but found 33% transmission of yellows from progeny of the Irish virus line, the latter presumably being a different but unidentified strain. Ernould's figure for the Irish line corresponds closely with the original determination of 30% by Clinch and Loughnane (1948); on the other hand, Nikolić (1956) found positive evidence of seed transmission in Yugoslavia only to the extent of 0.01%. It is easy to visualize differences in seed transmission due to specific hosts, but not so easy to account for different results on the basis of virus strains.

Transmission through pollen to seed has been recorded for elm mosaic even though the healthy mother trees were not infected from this source (Callahan, 1957). The virus remained viable after 3 yr storage of the pollen at 0° C. Pollen transmission of necrotic ringspot virus of cherry to the seed, but not back from the seed to the mother tree, also occurs (Way and Gilmer, 1958).

TRANSMISSION THROUGH SOIL

Most of the problems involved in soil transmission of viruses appear to have been resolved during the last decade by the discovery of nematode transmitted viruses and of those transmitted by soil fungi. Seven viruses are known to be fungus-transmitted; lettuce big vein, tobacco stunt, and tobacco necrosis transmitted by *Olpidium brassicae;* cucumber necrosis by *Olpidium cucurbitaceae;* wheat mosaic by *Polymyxa graminis;* potato mop top by *Sporangospora subterranea;* and potato virus X by *Synchrytrium endobicticum* (Cadman, 1963; Campbell and Fry, 1966; Hiruki, 1968; Jones, 1968; Jones and Harrison, 1969; Teakle, 1969).

The subject has recently been reviewed by Grogan and Campbell (1966) who suggest that the presumptive evidence is that soil-borne mosaics are fungus transmitted. Teakle's paper (1969) is also a review. He lists four of the fungi just referred to and does not recognize *O. cucurbitaceae* as a vector of cucumber necrosis. A useful feature of his review is the inclusion of a long table of nonvector species of fungi. Jones and Harrison (1969) have explored in detail the transmission of potato mop top by *S. subterranea* (Walbr.). A mushroom die-back virus is transmitted through spores (Schisler *et al.,* 1963). Nematodes involved in soil transmission are included in Table 12.3.

Much remains to be done on the vector-virus relationships but from the examples now known, there is clear evidence of lack of vector specificity. Tobacco ringspot, for example, is transmitted by grasshoppers (Walters, 1952; Dunleavy, 1957), but it is also nematode transmitted. Tobacco ringspot is transmitted also to leaves by *Thrips tabaci* (Bergeson *et al.,* 1964; Messieha, 1969). Lettuce big vein is transmitted by aphids as well as by fungi. There are still some problems in soil transmission. Chlorotic streak of sugarcane is transmitted in soil water, its release being through the medium of root exudates (Sturgess, 1962). Any virus transmitted by soil organisms to the roots could be a candidate for other vectors if the infection becomes systemic and can be acquired through the leaves. The question of failure of viruses to be found in leachate through the soil has largely been settled by the discovery of the soil-borne vectors. There may be an exception in the case of wheat yellow mosaic and barley yellow mosaic which, according to Miyamoto (1959, 1959a), exist adsorbed in (on) soil particles of the clay fraction which have colloidal characteristics and the ability to adsorb proteins.

TRANSMISSION THROUGH DODDER

The use of parasitic plants as a vehicle for the transmission of plant viruses was first described by Bennett (1940, 1944, 1944a); a total of ten species of *Cuscuta* have since been used. These are *C. californica* Choisya, *C. campestris* Yunck, *C. americana* L, *C. gronovii* Willd., *C. europae* L, *C. epithymum* Murr., *C. subinclusa* D. and H., *C. lupuliformis* Krocker, *C. japonica* Choisya, and *C. chinensis* Lam. In order for this type of transmission to take place, the parasitic plant must first establish vascular connnection with the virus-infected plant. This is accomplished by the development of a suckerlike process arising from the epidermis of the parasite, which adheres to the host plant and from which develops the haustorium. This organ is morphologically an adventitious root, and when it contacts the cells of the host, it penetrates deep into the host tissues through the medium of elongate, hyphalike strands which fuse into a more or less compact tissue. Some of the strands make contact with the xylem of the host; others with the phloem. An unbroken connection between the host and parasite is therefore maintained through the mature haustorium.

Lackey (1949) found considerable differences between species of *Cuscuta* in their reaction on a number of host plants. Three species did equally well on sugar beet and tobacco but not so well on cucumber; on the stems of squash there is a layer of large cells which appears to act as a barrier to vascular contact with the haustoria. The haustoria of *C. americana* induce some cell hypertrophy in the tomato stem which appears to prevent haustorial hyphae from contacting vascular tissue.

These reactions are illustrated in Fig. 11.2.

Food transport from the host plant to the dodder occurs through the haustorium, and the acquisition of virus by the dodder is accomplished in the same manner. It will follow that viruses of the yellows type, having high concentrations of virus in the phloem, will be more effectively transmitted to the dodder than in other types where concentration of virus in the host-plant phloem is low.

The transmission of viruses from dodder to the host plant, according to Bennett (1944a), might take place (1) by movement of virus from the phloem of dodder through the haustorium into the phloem of the host, counter to the prevailing direction of food movement, and (2) by passage from the parenchyma of the haustorium into that of the host through plasmodesmatal strands or from naked protoplasm of invading hyphalike cells. Bennett also suggests that viruses of the yellows type may be transmitted by the first method, and the mosaiclike viruses by the second.

Dodder may acquire viruses but be unable to transmit them. Canova

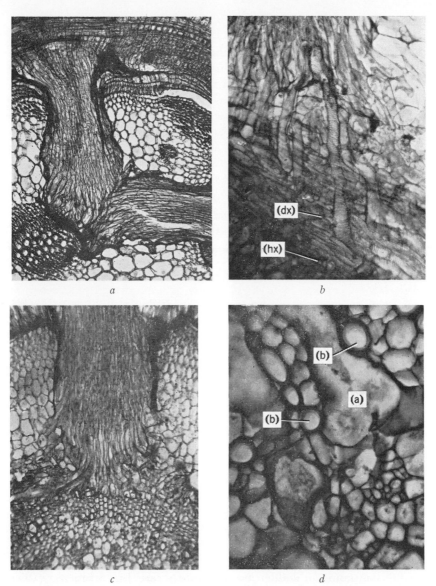

FIGURE 11.2. Dodder haustoria in beet and tobacco. *a,* Two haustoria of *Cuscuta subinclusa* contacting vascular bundle of sugar-beet petiole; *b,* longitudinal section of beet petiole showing how xylem vessels of dodder haustorium (DX) blend with xylem of host plant (hx); *c,* same dodder on *Nicotiana tabacum* with haustorial hyphae of haustorium contacting external phloem; and *d,* haustorial hypha (a) that has penetrated beyond xylem (b) and is growing in direction of internal phloem of tobacco stem (Photographs: Lackey, 1949).

(1955) was unable to get transmission of beet yellows virus by *Cuscuta epithymum,* but the dodder growing on the diseased beet had acquired the virus. This was shown by successful transmission by *M. periscae* (Sulz.) from the dodder to beet.

Schmelzer (1956) used nine *Cuscuta* species as virus carriers and found differences between them in their transmitting ability. Table 11.3 (from Bennett, 1944) illustrates this ability for several viruses and three species of *Cuscuta.*

TABLE 11.3.   Comparison of Ability of Cuscuta subinclusa,
C. campestris, and C. californica
to Transmit Viruses [a]

| | | Number of plants inoculated and infected | | | | | |
| | | *Cuscuta subinclusa* | | *Cuscuta campestris* | | *Cuscuta californica* | |
Disease induced by tested virus	Plant to which transfer was made	Inoc	Inf	Inoc	Inf	Inoc	Inf
Sugar beet curly top	Sugar beet	80	8	80	39	80	7
Sugar beet curly top	Tobacco var. Turkish	80	12	80	6	80	11
Cucumber mosaic	Sugar beet	80	8	80	32	80	19
Cucumber mosaic	Tobacco var. Turkish	60	60	60	60	60	60
Mustard mosaic	*Brassica adpressa*	100	12	100	13	100	97
Tobacco mosaic	Tobacco var. Turkish	80	0	80	0	80	0
Dodder latent mosaic	Pokeweed	100	100	100	100	100	100
Sugar beet mosaic	Sugar beet	80	0	80	0	80	0
Tobacco etch	Tobacco var. Turkish	40	0	40	0	40	40
Tomato spotted wilt	*Nicotiana glutinosa*	20	0	20	14	20	1

[a] After Bennett (1944).

Visible symptoms on dodder are apparently rare, but Schmelzer (1955, 1956) found that potato-stem-mottle virus induced symptoms on *C. campestris, C. subinclusa,* and *C. lupuliformis. C. campestris* also showed symptoms following acquisition of cucumber mosaic.

Dodder transmission of viruses, tabulated in useful detail by Fulton (1964), shows 17 species of *Cuscuta* as established vectors. Since then four citrus viruses have been successfully transmitted (Weathers and Hartung, 1964; Weathers, 1965) as well as barley yellow dwarf (Timian, 1964). The latest review is by Hosford (1967).

TRANSMISSION BY GRAFTING

According to Smith (1951), all viruses which are systemic in their hosts can be transmitted by grafting. Bawden (1950) considers the method so characteristic that it has become the criterion of a virus disease. There is no good reason why vascular, bacterial, and fungal parasites should not be transmitted by grafting, but this is so rare that the possibility is usually ignored. The standard grafting methods used in horticulture are often sufficient.

Cleft Grafts

The cleft graft is one in which the top of a healthy plant is cut off, a slit is cut in the stem, and into this a shoot from a diseased plant is inserted after being cut into a wedge shape. Union between scion and stock is hastened if the plant is kept in damp, warm conditions. Transmission of the virus from scion to stock is evident in the succulent side shoots produced in the stock.

Natural Root Grafts

Root grafts are essential for transmission of certain viruses which occur only in the root in phony peach. Natural root grafts occur, and viruses may be transmitted under field conditions in this manner (Hunter *et al.*, 1958) (Fig. 11.3).

Tongue Grafts

Two other types of graft, the "tongue graft" and the "approach graft," have been used for indexing strawberries by grafting to *Fragaria vesca* (Miller, 1952). For tongue grafting, a thin cut is made about ¾ in. long and $\frac{1}{32}$ in. deep parallel to the surface of the runner of the donor, and near its tip. A thin piece of tissue is sliced off and an oblique cut, ½ in. long, is then made starting at one end of the parallel incision. The receptor plant is similarly treated except that the oblique cut is made in the opposite direction. The tongue or wedge-shaped cut of the donor is then inserted in the cut of the receptor and the whole bound together with Scotch tape, a piece of electrician's rubber tape, and a coating of asphalt emulsion. The approach graft is advisable when the runners are small in cross-section. The same type of initial cutting and slicing is done as for the tongue graft, but the two surfaces are then bound together without tonguing, and the bond tightened as before (see Fig. 11.4).

FIGURE 11.3.   Transmission of viruses through natural root grafts. Two apple trees showing root graft through which apple mosaic was transmitted (Photograph: S. A. Runsey, after Hunter *et al.*, 1958).

Tongue grafting is actually a variant of inarch grafting (Harris, 1932; Harris and King, 1942), since in both cases the donor and receptor runners are not detached from the growing plants. With "bottle grafting," the scion is cut off and grafted to the stock, but its lower free end is kept in a nutrient solution until the union is firm.

Excised Leaf Grafts

The excised leaf can also be grafted by scraping the petiole and then inserting it into a receptor plant stem (Bringhurst and Voth, 1956) (Fig. 11.5). The method has an advantage in that since the donor tissue is a single leaf, many replications can be made from a single donor plant.

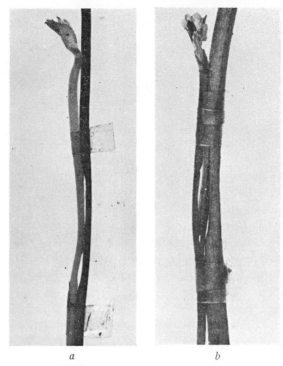

a   *b*

FIGURE 11.4. Types of grafts used in grafting strawberries. *a*, approach graft; and *b*, a tongue graft. The grafts are held together with Scotch tape, the rubber tape and asphalt emulsion have not been applied (Photographs: Miller, 1952).

The successful union of leaf and petiole, however, does not always ensure transmission (Fulton, 1957; Cropley, 1958). Jorgensen (1957) simplified the method of Bringhurst and Voth by using only small boat-shaped pieces of petiole from donor plants and inserting these singly or in pairs into slits in the petioles of receptor plants.

Miller (1958) compared the efficiency of excised leaf grafts and stolon grafts in transmitting strawberry viruses and found that the former produced symptoms earlier; sometimes a higher percentage of the excised leaf grafts established union; and symptoms were more severe than when stolon grafts were used.

Budding

Budding is extensively used in the transmission of viruses affecting woody plants and trees. A bud is cut with cortical and phloem tissue attached,

FIGURE 11.5.   Steps in grafting excised leaves by a modification of the Bringhurst and Voth technique. *a,* Prepared leaf scion from plant being indexed; *b,* stock leaf from indicator plant, showing slit in petiole prior to insertion of scion; *c,* scion and stock in place; and *d,* final step, showing scion and stock tightly wrapped with latex tape and coated with water-asphalt emulsion (Photographs: Miller, 1952).

and this is then inserted in slits which reach the cambium of the healthy plant.

Bark Grafts

Bark grafts, the insertion of diseased stem tissue between bark and wood, and even the insertion of cut, diseased leaves under the bark of healthy trees, have all been used successfully (Sreenivasaya, 1930; Cochran and La Rue, 1944; Wallace, 1947; Blattny and Limberk, 1956).

Core Grafts

With tubers, core grafting is the favored method. A core is removed from the receptor tuber by means of a cork borer; a similar but slightly larger diameter core is removed from the donor tuber and pressed into the hole made in the receptor tuber. Davidson (1955) reports the use of tuber grafting in studies on potato leaf roll transmission in which healthy and diseased sets were bound together; the method was no more effective than was core grafting.

MECHANICAL TRANSMISSIONS

Methods of Transmission

ACCIDENTAL TRANSMISSION.   This a fairly broad category involving many variants and many techniques. Mechanical transmission is of paramount importance in the laboratory and of considerable importance in the field. Tobacco mosaic and potato X, while not insect-transmitted (the records of mealybug transmittal of tobacco mosaic definitely need confirmation), are nevertheless so easily transmitted mechanically that transmittal from rubbing together diseased and healthy leaves, from the handling of the tobacco plants by workers, or from infected plant debris in the soil, can be expected.

Potato viruses X and F can spread between plants that are in contact, and wind movement is a factor in increasing the rate of spread (Loughnane and Murphy, 1938). Merriam and Bonde (1954) showed that transmission of potato spindle tuber virus is effected by brushing healthy plants with diseased plant foliage, or by driving contaminated tractor wheels over the leaves of healthy plants. Potato spindle tuber is transmitted also by the seed cutting knife, by contact of freshly cut seed-pieces with each other, and by piercing seed-pieces with contaminated picker-planter knives (Bonde and Merriam, 1951). One strain of peach ringspot virus was mechanically inoculated to beans, and leaf friction was sufficient to transmit the virus from inoculated to healthy plants in the same pot (Yarwood, 1957a).

Transmission of exocortis virus of citrus by contaminated knives was demonstrated by Garnsey and Widden (1970). Etrog citrus was used as the source plant and various citrus cultivars as the test plants. Some of these were not susceptible; others were intermediate.

PLANNED FRICTION TRANSMISSION.   A planned friction method was used by Celino and Martinez (1956) in transmitting *abaca* mosaic to healthy *abaca* (*Musa textilis* Neé), arrowroot, maize, and *Canna indica* L seedlings. A partly emerged, infected leaf is cut with a safety razor blade, and the cut edge of the leaf is drawn gently against healthy leaves which have been dusted with carborundum 300. A modification of this method was developed by Yarwood (1953), who rubbed the cut edges of leaf disks from the source plant over test leaves which had been previously dusted with carborundum. He found the method particularly suitable with unstable viruses such as tomato spotted wilt and apple mosaic.

SAP TRANSMISSION.   The most commonly used method is the rubbing of sap extracted from diseased plants onto healthy leaves. The term "sap-transmissible," applied to viruses, usually involves this method, and it is not surprising that it has become standard practice to test any new virus in this manner. An important advantage is that sap-transmissible viruses can be easily studied *in vitro*. The literature of plant virology would be shorn of some of its most exciting chapters if tobacco mosaic had not been so amenable to this technique.

The early use of sap inoculation was somewhat primitive, the method being either to inject sap or to scratch through drops of infective sap placed on a healthy leaf. Severin (1924) succeeded in transmitting curly top to sugar beet in a few cases by needle puncture through a blob of juice dropped into the heart of the healthy test plant. Crude as this method was, it has not been improved upon since, and curly top is still considered as essentially not mechanically transmissible.

The disadvantage of these early methods was that they caused severe injury to the tissues; it is now recognized that the most effective methods depend on those that admit the virus into the host cell with the least accompanying injury. Holmes (1931) used a piece of cheesecloth dipped in the infected sap and rubbed lightly over the test leaf. With a virus as stable and infectious as tobacco mosaic, this method was successful.

A most sophisticated method of using infected plant sap or homogenates from viruliferous vectors is to inject these into uninfected vectors by means of an extremely finely drawn glass capillary. The method is now routine for studies with leafhopper and aphid vectors.

USE OF ABRASIVES.   Rubbing gently with the finger or a glass spatula became standard practice, although, when working with tobacco mosaic, the latter was to be preferred for tobacco smokers. Improvement in the efficiency of sap inoculation came with the discovery by Rawlins and Tompkins (1936, 1940) that the addition of finely divided carborundum powder of 400–500 mesh greatly increased the successful transmission of viruses difficult to transmit with previously used methods.

More easily transmitted viruses, such as tobacco mosaic, produce many more local lesions when rubbed in with abrasive-loaded sap than without the abrasive. Kalmus and Kassanis (1945) estimated the increase as equivalent to a virus concentration increase of 100 times. The necessity for standardization of technique when determining virus concentration by the local lesion method is apparent.

Dilution end points are used to aid in characterizing viruses but the addition of abrasives can make a radical change in the dilution end point, and the particle size of the carborundum is a factor (Costa, 1944). The use of abrasives, therefore, merely facilitates entry. Host plants differ in their susceptibility to this treatment, and viruses differ in their capacity to take advantage of it.

The method ordinarily used for inoculating sugarcane seedlings with mosaic virus is to appress a diseased leaf tightly to a healthy one and make pinpricks through it to the healthy leaf (Liu, 1949). Harpaz (1959) transmitted a new maize virus by needle. This was an injection method which used a 1-cc tuberculin syringe to introduce the inoculum into the stems of 3-wk-old seedlings. Abbott (1949) compared this method with one using washed 80-mesh sea-island sand as an abrasive. One very susceptible and two moderately susceptible clones of sugarcane were used as test plants. The results are shown in Table 11.4.

TABLE 11.4. *Mosaic Infection in Sugarcane Clones Inoculated by Abrasion or by Needle Prick* [a]

	Percentage of plants infected [b]	
Sugarcane clone	Abrasion method	Needle-prick method
Louisiana purple	100	89
Co. 281	89	60
Co. 290	86	44

[a] After Abbott (1949).
[b] Differences between methods for each variety significant at the 5% level.

Some clones known to be resistant either to the pinprick method or under field conditions were sometimes infected by the abrasion method. This might simply mean that the number of virus particles in infected sap, and the number of inoculation points provided by the abrasive, were sufficient to find the occasional receptive cell in the resistant plant. The statistical probability of this happening is greater with the artificial technique than it would be with a natural vector.

Hutton and Peak (1951) called attention to the necessity for standardization of inoculation techniques when resistant or susceptible phenotypes are being sought. Abrasives to provide the maximum number of virus entry points can best be achieved by the use of diatomaceous silica which can be added to the inoculum in measured quantities, or to the stock solution of phosphate buffer. Table 11.5 gives the results of a test with spotted wilt virus.

TABLE 11.5.   Carborundum Dusting Compared with the Use of Bulk Buffer Containing 1.5% Hyflo Super-Cel for Spotted Wilt Inoculation of N. glutinosa at 10-Day Intervals [a]

Treatment	Mean number of lesions at 10-day intervals			
	0	1	2	3
Carborundum dusting	53.5	28.5	37.0	40.2
Hyflo Super-Cel	46.5	22.0	43.0	45.7

[a] After Hutton and Peak (1951).

Beraha *et al.* (1955) confirmed Yarwood's findings (1952) that carborundum increased the amount of infection beyond that due to the addition of buffer. They also found that the number of particles, not their size, determined their effectiveness. The mechanism of action of abrasives, stated by Bawden (1950) to be a reduction in the resistance of the host, is shown to be solely a matter of providing wounds for entry of the virus which is independent of the abrasive particle at optimum conditions.

If virus is adsorbed on the abrasive particles, which probably rarely occurs, it may be carried into the wounded tissues, according to Beraha *et al.* (1955). An adsorbed particle, however, is not a free agent and the bond would have to be broken by action of the cell contents before the virus particle would be free to multiply or even to act on the cell into which it was introduced.

Factors Affecting Effectiveness of Mechanical Transmission

CHEMICALS.   There have been many attempts to improve mechanical transmission, particularly of those viruses which are unstable and are soon inactivated *in vitro*. Hildebrand (1956) added reducing substances to sap of sweet potato infected with the virus of internal cork disease and found that cysteine (0.08M) apparently makes it readily transmissible. Caffeine at 0.5%, added to the suspending medium, increased the rate of transmission of necrotic ringspot virus from cherry to cucumber, and it extended the period during the spring when transmission is possible; it also increased the longevity of the virus *in vitro* and made transmission from *Prunus tomentosa* Thunb. possible (Diener and Weaver, 1959). Hougas (1951) had already extended the longevity *in vitro* of potato yellow dwarf virus by the addition of 0.01–0.04M cysteine hydrochloride.

Substances which inactivate viruses in expressed sap will, of course, prevent or reduce the effectiveness of mechanical transmission. Tannins in the juice of dahlia mosaic and virus-infected strawberry plants are cases in point. When tannins are removed by extraction with ethanol after lyophylization, a high percent of transmission is obtained (Cornuet, 1952). Vaughn (1956) used a modified Soxhlet apparatus to remove the tannins.

Effectiveness of mechanical transmission can be increased by techniques designed to increase susceptibility of the host plant. Yarwood (1954) used zinc sulfate at 0.001–0.03% in which to dip bean leaves after inoculation with TMV, and greatly increased the number of lesions by so doing. Detached bean leaves inoculated in their upper surface and floated on a substrate of 0.01–0.1% zinc sulfate were injured by the chemical, but both size of lesions and amount of virus present were increased. This technique involved, therefore, not only a transmission method but also a chemical breakdown of resistance favoring virus multiplications.

Heat treatment of beans, cowpeas, and cucumber, by dipping the plants in water at 50° C for 20 sec after inoculation with a number of viruses, delayed the appearance of necrotic lesions but increased their size and number. "Heating peach plants before inoculation with peach yellow-bud mosaic virus, by dipping the tops of the plants in water at 45° or 50° C for various intervals of time, resulted in the first juice transmission of a virus infection to fruit trees" (Yarwood, 1956, 1958).

Susceptibility of beans to several viruses was increased by immersing them in water for several hours before inoculation, although the addition of phosphate buffer to the inoculum had the same effect with considerably less trouble (Yarwood, 1959). Susceptibility can also be increased by rust infection. Beans are more susceptible to TMV, TNV, tobacco ringspot, alfalfa mosaic, cucumber mosaic, and apple mosaic viruses when first infected with rust (Yarwood, 1951, 1955). The mechanism of this increase in susceptibility is obscure.

The most generally used chemical supplement to inoculum is K_2HOP_4. Thornberry (1935) used 1.7% of this chemical to 10^{-3} TMV inoculum and increased the number of lesions on beans 5-fold. Carborundum will reduce the amount of inoculum necessary to induce one lesion of TMV bean leaf, but even then the amount necessary is 100 times more if phosphate is not used (Yarwood, 1957).

The effect of phosphate has been generally ascribed to pH, and it is true that buffers normally used contain phosphate. However, it is unlikely that the effect is entirely due to pH, for buffers at the same pH but containing no phosphate give a lower infection rate. Yarwood (1952) advanced the hypothesis that "phosphate increases susceptibility by its chemical action in increasing the size and number of cell perforations started by the act of inoculation by rubbing the leaf surface," but Hildebrand (1956) states that the effectiveness of the basic method with the very highly unstable sweet potato cork virus suggests rather "that the effect of the potassium phosphate buffer is to lubricate, to prevent chemically deleterious action of the host tissue sap on the virus, and to protect the wounds and exposed protoplasm by filming over the leaf surface."

Yarwood (1957) lists a large number of other chemicals which have been tested. In general, those that have shown an inhibiting effect are those which either kill or severely injure the inoculated cells; the effect of those which allow increased virus development is due to a milder type of injury that permits entry and increase. Besides phosphate, only sulfite (Na_2SO_2) has increased infection materially. Best and Samuel (1936) ascribed the effect of sulfite to its reducing property in preventing the rapid oxidation of spotted wilt virus, but the same effect of sulfite was observed by Yarwood on TMV infection on beans.

EXPOSURE TO DARKNESS.   Exposure to light is a factor in conditioning plants to susceptibility to infection. If plants are left in darkness prior to inoculation, the number of local lesions formed following inoculation is greatly increased (Samuel and Bald, 1933). Hougas (1951) used 18–24 hr of darkness and increased incidence of infection with potato yellow dwarf 2–4 times. Bawden and Roberts (1948) suggested that the preinoculation darkness affected susceptibility through the slowing down of the photosynthetic process which, under normal light conditions, would interfere with the passage of viruses from cell to cell or with their initial establishment.

Costa and Bennett (1955) achieved transmission of sugar beet yellows virus and found that a preconditioning period of 4 or 5 days in the dark or reduced light increased the percentage of transmission. The variety of plant inoculated, plant source of inoculum, the strain of the virus used, the treatment of the inoculum, and even the differential susceptibility of leaves on a single plant were all additional factors.

Mundry and Rohmer (1958) used a 4-day period to precondition sugar beets to infection with the European sugar beet yellows virus and reported successful results. The tissue selected for inoculation may be important. For some difficult virus isolations, the use of flowers from infected plants has proved useful. McWhorter (1953) used flower inocula from cucumber mosaic virus, from the white streak disease of gladiolus, and components of the latent virus complex of cherry. Inoculum from gladiolus was only four times as effective as leaf inoculum, but from cherry, the flower inoculum gave up to 175 lesions, as compared with none for leaf inoculum, when twin primary leaves of *Vigna sinensis* Endl. were used. When test plants were inoculated separately with flowers and leaf inocula, the flowers induced 440 lesions, and the leaf none. Milbrath (1953) used cherry flower petals infected with stone fruit latent virus complex to inoculate cowpea and cucumber. The leaf inoculum gave the expected reaction on cucumber but none on cowpea; the flower inoculum produced symptoms on both hosts.

NEW TECHNIQUES. New techniques are continually being developed for mechanical transmission. Some of these are stimulated by the necessity for large-scale inoculation of seedlings for resistance tests. McKinney and Fellows (1951) used a power-driven atomizer to spray wheat seedlings with inoculum from wheat streak mosaic virus. As these were sprayed, the leaves were drawn quickly between thumb and forefinger. When Celite abrasive was added, 100% infection was obtained with susceptible varieties. This technique was a modification of those of previous workers (Duggar and Johnson, 1933; Richards and Munger, 1944).

A more refined type of atomizer, the commercial artist's airbrush, was used by Lindner and Kirkpatrick (1959) to inoculate cucumber seedlings with TMV. The optimum conditions were: (1) 1 cm distance from airbrush nozzle to leaf surface, (2) 4 atm of air pressure, (3) inoculum delivery rate of 10 ml/min, (4) inoculation time of about 4 sec to cover cotyledon completely, and (5) 1% by weight of 600-mesh carborundum in the inoculum.

With viruses that are predominantly in the epidermis, Yarwood (1957b) found that a stiff poster brush stroked in turn over the donor and receptor plant leaves was an efficient transmission method.

An ingenious method of sap transference was used by Nagaraj *et al.* (1954) in studies of the possible virus nature of a wilt (root) disease of coconuts in Travancore and Cochin (southwest India). A severed root tip from a diseased tree was joined to the cut end of a root from a nearby healthy one by means of a rubber tube filled with sterile water. Sap was taken up from the diseased tip into the healthy tree. No results were reported and the success of the transmission, in the event that a virus is involved, might depend on the short distance of *in vitro* movement that would be required of the virus through the (no longer sterile) water connection.

THE INDICATOR PLANT AS AN AID IN VIRUS TRANSMISSION

The host range of many viruses is extensive, but the host plants of any one virus are not equally susceptible, and all are not handled with equal facility. Furthermore, some plants have the capacity to react differentially to several viruses. When a plant is used as a test plant to determine the presence of a virus in another plant, it is known as an indicator plant. Many indicator plants are not normally in the host range of viruses which they are used to indicate, but have been discovered either by chance or by deliberate screening of a number of candidates. The ideal indicator plant is one which is easily grown from seed, is usable when quite small, and which produces distinctive symptoms rapidly, without much variation between individual plants. An indicator plant may serve when the principal host of the virus being tested is dormant, when the period for symptoms in the principal host is very long, or when a complex of viruses can be resolved through the use of indicator plants which react differentially. Much of the earlier work of Johnson and Smith with indicators was done with potato virus complexes.

Physalis angulata L is propagated from the true seed; it is a rather small plant, easily grown in small pots. It is a useful indicator plant for potato leaf roll virus because symptoms appear in about 10 days after inoculation (Hovey and Bonde, 1948). Kirkpatrick (1948) used *Physalis floridana* Rydb. and *Datura stramonium* L to indicate leaf roll; Maramorosch (1955) preferred potato seedlings.

Nicandra physaloides (L) Gaertn. serves to indicate potato virus Y and is differential for at least one of its strains (Silberschmidt and Rostom, 1955). Other differential hosts for virus Y have been described by Hutton and Peak (1952). They used *Lycium halimifolium* Mill., *L. rhombifolium* Moench, *P. floridana,* and *Capsicum annuum* L. In 1953 Ross also found *P. floridana* suitable for virus Y, if it is preconditioned by means of extra light during the winter and has postinoculation shading and temperatures of 65–70° F.

Potato virus S induces local lesions on guar (*Cyamopsis tetragonoloba* L), but seven other plant species tested were negative (Yarwood and Gold, 1955). The use of *Solanum demissum* Lindl. as an indicator of virus A in potato eliminated the necessity for detection of the virus in potato seedlings by grafting. Instead, juice inoculation of the indicator plant is made with the sap of the seedlings and symptoms appear in 3–4 days (Webb and Buck, 1955). An F_1 hybrid between *S. demissum* and the Aquila potato variety was used by Köhler (1953) to indicate potato viruses A, X, and Y, stem mottle, bouquet ringspot, and various strains of tobacco mosaic virus.

Some plants are such ubiquitous indicators that their value in specific cases will depend on a very detailed knowledge of symptom reactions. An example is *Chenopodium amaranticolor* Costa and Reyn., which Hollings (1956, 1957) has shown to be susceptible to 40 viruses with symptoms of diagnostic value in many cases.

The study of stone fruit viruses has been greatly facilitated by the discovery of indicator plants. Moore *et al.* (1948) first described the transmission of a virus from sour cherry to cucumber. They ground up very young cherry leaves which were just beginning to show the initial symptoms of necrotic ringspot and used the juice to inoculate mechnically young cucumber cotyledons. Percent transmission was low from cherry to cucumber, but once in cucumber, the virus was transmitted readily to other cucumber. Mulder (1954) confirmed this result in Europe. Boyle *et al.* (1954) improved the technique by standardizing conditions of inoculation and environment. A large series of herbaceous plants was tested but only cucumber and squash were found to be hosts. Milbrath (1956) used squash as a differential host for strains of stone fruit ringspot viruses.

The cherry-to-cucumber transmission is particularly fortunate since the symptoms on cucumber are unlike those of any other known virus on that plant. It has not been possible to establish the identity of the virus or viruses from sour cherries since a method for reciprocal transmission from cucumber or squash to cherry has yet to be worked out (Gilmer, 1957).

Cherry infection with little cherry virus is normally only revealed in bearing trees, but Wilks and Welsh (1955) found that the foliage of certain sweet cherry varieties could be used as an indicator. The symptom is a reddening of the leaf lamina, and is specific for little cherry; three other viruses on cherry—sour cherry yellows, mora, and rusty mottle—failed to induce leaf symptoms. Gilmer and Brase (1956) tested six species of *Prunus* as indexing hosts for stone fruit viruses. In all instances, none was successful. Cucumber seedlings were superior to any of the *Prunus* species in revealing the presence of viruses, although they were not reliable for determining virus content. *P. tomentosa* was found by Fink (1955) to be a reliable index plant for determining virus infection in cherry nursery stock, but there were only unconfirmed indications that distinct types of symptoms on *P. tomentosa* might be used to distinguish viruses, complexes, or strains.

Four fruit tree virus strains, peach yellow leaf roll, Napa buckskin, Green Valley buckskin, and a related strain, were transmitted from peach to celery and back from celery to peach trees (Jensen, 1956). This is apparently the first case recorded where reciprocal transmissions to and from fruit trees and herbaceous hosts have been successful.

Williams *et al.* (1971) describe the transmission of a virus from pear trees to cucumber, apparently the first such from pear to a herbaceous host.

Many suitable indicators for fruit tree viruses are too sensitive to cold injury to survive in areas of cold winters, and cold-tolerant plants are necessary. Fridlund (1958) found the Krikon plum and Ember plum were extremely cold-resistant and suitable as indicators for prune dwarf virus and line pattern virus respectively.

Knorr (1956) investigated tristeza virus in Argentina; the symptoms produced on the eight indicator plants he used are shown in Table 11.6.

TABLE 11.6.   *Indicator Plants of the Tristeza Virus*
as Determined in Argentina in Tests
in Which the Vector Aphis citricidus *Kirk.*
Was Used as Inoculating Agent [a]

Species	Symptoms produced[b]
Aeglopsis chevalieri Swing.	VC, ST, SP
Afraegle paniculata (Schum.) Engl.	VC
Citrus aurantifolia (Christm.) Swing.	VC, ST, SP
Citrus combava Raf.	VC
Citrus hystrix DC.	VC, SP
Citrus paradisi Macf.	SP
Citrus sinensis (L) Osbeck	SP[c]
Pamburus missiones (Wt.) Swing.	VC

[a] After Knorr (1956).
[b] VC: vein-clearing, similar to the reaction caused by tristeza virus in the immature leaves of Key lime indicator plants; ST: stunting of over-all growth of plant; SP: stem-pitting in the wood of decorticated stems.
[c] In occasional plants only.

Scarlett O'Hara morning glory is an excellent indicator plant for sweet potato internal cork disease virus. Flat-tipped tweezers were used to press virus extract out onto the buffer-carborundum-coated leaf surface. The period for expression of symptoms is very short, from 7–15 days, as compared with a period of 1 yr in sweet potato inoculated by grafting (Hildebrand, 1956a). Sometimes a horticultural variety serves as an indicator plant if it is more susceptible and amenable to inoculation. The Blazing Glory chrysanthemum variety is the best test variety for determining the presence of flower distortion virus in other varieties of chrysanthemum. It shows rosetting of shoot tips 2 mo after graft inoculation (Brierley, 1955).

Suitable indicator plants for many viruses affecting monocotyledonous plants are not easily found, but Costa and Penteado (1950) found 4 out of 33 inbred lines of maize susceptible to sugarcane mosaic.

Some indicator plants are particularly useful in quantitative assay studies; many of the older studies were concerned with assay of tobacco mosaic. Ramamurthi *et al.* (1958) used *Gomphrena globosa* L to assay vein mosaic of red clover. *G. globosa* is the only known local lesion host of this virus. The cowpea has been similarly used to assay cucumber virus 1 (Sill and Walker, 1952). Crowley (1954) elaborated on the technique but reported that considerable variation in the reaction of individual plants had not been eliminated.

Indicator plants are of especial significance in revealing latent virus infection. There are two recognized "truly latent" viruses (Smith, 1952), the paracrinkle virus of King Edward potato (Salaman and LePelley, 1930), and the latent virus of dodder, *Cuscuta* spp. (Bennett, 1944). The latent virus of sugar beets and mangolds (Smith, 1951) has since been shown to be a bacterium, by Yarwood *et al.* (1961). Paracrinkle still remains the classic case. It was for many years only detectable by grafting on to other varieties of potato, thus leading to the hypothesis that the virus was artificially created by the act of grafting (Darlington, 1957). However, Bawden *et al.* (1950) succeeded in transmitting it mechanically to other varieties of potato and also to tomato plants. Tomatoes show no external disease symptoms, but internally manifest the fibrous particles characteristic of paracrinkle in King Edward potato. The latent virus of dodder produces mottling or necrosis on sugar beet, cantaloupe, tomato, potato, and a number of other plants, although completely symptomless on dodder itself (Bennett, 1944).

The symptomless carrier of a virus is more commonly recognized, perhaps because the definition is not quite so restrictive as that of the latent infection. As with latent viruses, however, apparently healthy plants are sources of viruses pathogenic to other plants. Again, the indicator plant is an essential tool.

Johnson (1925) showed that the sap from apparently healthy potatoes

always produced virus diseases when inoculated to tobacco and other solanaceous plants. The virus so revealed is now known as potato virus X,
and it is also known to exist in many strains, materially affecting symptom expression in those plants that react to infection with symptoms. In
the United States, commercial varieties are universally infected by it but
are symptomless.

Capsicum annuum serves as an indicator plant for potato virus X and
also separates that virus from the virus of potato *Aucuba* mosaic (Maris
and Rozendaal, 1956). Roberts (1953) reported that *D. stramonium* was
the principal indicator plant used to test for virus X in potatoes in Scotland (See also Chapter 13). *D. stramonium* is considered the most sensitive indicator host (Smith, 1952), but White Burley tobacco and *N. glutinosa* are also useful indicators. Wilkinson and Blodgett (1948) found that
Gomphrena globosa develops local lesions when infected with virus X,
but there is no systemic invasion. Some indicator plants are limited in
their reaction to infection. Examples are *Amaranthus retroflexus* L and
Digitalis lanata L, which develop local lesions only when infected with a
particular strain of potato virus X (Larson, 1944).

Indicator plants for strawberry viruses have been used for many years,
and the wild species, *Fragaria vesca* L and *F. virginiana* Dutch., were
shown to be suitable (Harris, 1937). Hildebrand (1941) found clonal differences in *F. virginiana* used as an indicator of viruses which were symptomless in two commercial varieties, Premier and Glen Mary. Some
clones were highly susceptible; others, robust in vegetative growth, were
susceptible but showed a capacity for partial recovery. Miller (1951,
1951a) found only one of many wild strawberry plants tested suitable as
an indicator; it was a triploid. Frazier (1951) used *F. bracteata* Heller.
Plakidas (1955) recognized seven distinct strawberry viruses. With one exception, the multiplier plant virus, the individual viruses are symptomless in commercial varieties. When symptom expression is encountered, it
is the result of multiple virus infection, which is frequently encountered
in nature. The resolving of such combinations depends on the availability of suitable indicator plants and on differential methods of transmission. Symptom severity must be rather limited in its usefulness, because
Miller (1958) reported that this can be materially affected by the method
of graft transmission used.

Hildebrand (1958) gives a good many examples of the use of indicator
plants to reveal "masked" viruses, but he uses so many variants of terms
defined in his introduction that it is not always clear just what the role of
the indicator plant is.

A latent virus of Norfolk Giant raspberry was demonstrated when Cadman (1951) found that a selection of this variety was a symptomless car-

rier of a virus, which he named leaf mottle. The entire selection is apparently infected. The virus was revealed by transfer to *Rubus saxatilis* L, the American black raspberry (*R. occidentalis* L var. Cumberland), and red raspberry varieties, Chartham and St. Walfried.

A latent virus of the lily species *Lilium tigrinum* Andr., *L. candidum* L, and *L. longiflorum* Thunb. can be demonstrated by transfer to tulips, where it causes the well-known symptom of flower color break (McWhorter, 1937). Frazier (1953) described a latent virus of *Fragaria vesca* and *F. bracteata,* the former the East Malling clone of that species, and the latter from California. It causes no symptoms in strawberry, but, as with other strawberry viruses already described, it induces more severe symptoms when other viruses are involved.

A virus latent in carnation and potato plants has been transmitted by sap inoculation and the aphid *Myzus persicae* to sweet william. The infected sweet william plants show no external symptoms; demonstration of transmission was through serological tests and electron microscopy. The particles revealed by the latter method closely approximate those found in King Edward potatoes carrying paracrinkle virus, but they were also found in many other potato varieties (Kassanis, 1954).

That other latent viruses of potatoes may exist is also suggested by Bagnall and Bradley (1955) on the basis of serological evidence. The sap of an apparently healthy plant which had previously reacted to antisera was inoculated into healthy seedlings. All reacted positively later, although uninoculated seedlings gave negative results.

A latent virus in dodder, when transmitted from dodder to crop plants, induced destructive disease of cantaloupe and buckwheat, and less critical symptoms on sugar beet, potato, and celery; it is symptomless in dodder (Bennett, 1944a). *Myzus persicae* transmitted a virus from turnip to *P. floridana, N. physaloides,* and several *Brassica* species with varying degrees of symptom expression, although it was symptomless on turnip and rutabaga (MacKinnon, 1956). Chrysanthemum latent virus was carried without symptoms by 11 varieties of chrysanthemum; Hollings (1957) lists the indicator plants, most of which are local lesion hosts, without systemic symptoms. Petunia is an efficient indicator.

REFERENCES

Abbott, E. V., "Comparison of methods for artificially inoculating sugar-cane seedlings with the mosaic virus," *Phytopathology,* **39,** 668–669 (1949).

Ainsworth, G. C., and Ogilvie, L., "Lettuce mosaic," *Ann. Appl. Biol.,* **26,** 279–297 (1939).

Anderson, C. W., "Seed transmission of three viruses in cowpea," *Phytopathology*, 47, 515 (1957).

Athow, K. L., and Bancroft, J. B., "Development and transmission of tobacco ringspot virus in soybean," *Phytopathology*, 49, 697–701 (1959).

Bagnall, R. H., and Bradley, R. H. E., "Note on a virus latent in potato plants," *Am. Potato J.*, 32, 252–253 (1955).

Bawden, F. C., *Plant Viruses and Virus Diseases*, 3rd ed., Chronica Botanica Co., Waltham, Mass., 1950.

Bawden, F. C., and Roberts, F. M., "Photosynthesis and predisposition of plants to infection with certain viruses," *Ann. Appl. Biol.*, 35, 418–428 (1948).

Bawden, F. C., Kassanis, B., and Nixon, H. L., "The mechanical transmission and some properties of potato paracrinkle virus," *J. Gen. Microbiol.*, 4, 210–219 (1950).

Bennett, C. W., "Acquisition and transmission of viruses by dodder (*Cuscuta subinclusa*)," *Phytopathology*, 30, 2 (1940).

Bennett, C. W., "The relations of viruses to plant tissues," *Botan. Rev.*, 6, 427–473 (1940a).

Bennett, C. W., "Latent virus of dodder and its effect on sugar beet and other plants," *Phytopathology*, 34, 77–91 (1944).

Bennett, C. W., "Studies of dodder transmission of plant viruses," *Phytopathology*, 34, 905–932 (1944a).

Bennett, C. W., "Lychnis ringspot," *Phytopathology*, 49, 706 (1959).

Bennett, C. W., "Seed transmission of plant viruses," *Advances in Virus Research*, 13, 221–262 (1969).

Bennett, C. W., and Esau, K., "Further studies on the relation of the curly top virus to plant tissues," *J. Agr. Research*, 53, 595–620 (1936).

Beraha, L., Varzandeh, M., and Thornberry, H. H., "Mechanism of the action of abrasives on infection by tobacco mosaic virus," *Virology*, 1, 141–151 (1955).

Bergeson, G. B., Athow, K. L., Laviolette, F. A., and Sister Mary Thomasine, "Transmission, movement and vector relationships of tobacco ringspot virus in soybean," *Phytopathology*, 54, 723–728 (1964).

Berkeley, G. H., and Madden, G. O., "Transmission of streak and mosaic diseases of tomato through seed," *Sci. Agr.*, 13, 194–197 (1932).

Best, R. J., and Samuel, G., "The reaction of the viruses of tomato spotted wilt and tobacco mosaic to the pH value of media containing them," *Ann. Appl. Biol.*, 23, 509–537 (1936).

Blakeslee, A. F., "A graft infectious disease of datura resembling a vegetative mutation," *J. Genet.*, 11, 17–36 (1921).

Blattny, C., and Limberk, J., "Implantace ulomku pletiv jako zpusob pro prenos stolburu (The suitability of implantation of tissue fragments for the transmission of stolbur.)," *Preslia*, 28, 413–414 (1956).

Bonde, R., and Merriam, D., "Studies on the dissemination of the potato spindle tuber virus by mechanical inoculation," *Am. Potato J.*, 28, 558–560 (1951).

Boyle, J. S., Moore, J. D., and Keitt, G. W., "Cucumber as a plant host in stone fruit virus research," *Phytopathology*, 44, 303–312 (1954).

Brierley, P., "Blazing Gold as a test variety for the chrysanthemum flower-distortion virus," *Plant Disease Reptr.*, 39, 899–901 (1955).

Brierley, P., and Travis, R. V., "Soil-borne viruses from chrysanthemum and begonia," *Plant Disease Reptr.*, 42, 1030–1033 (1958).

Bringhurst, R. S., and Voth, V., "Strawberry virus transmission by grafting excised leaves," *Plant Disease Reptr.*, 40, 596–600 (1956).

Cadman, C. H., "Studies in rubus virus diseases. I. A latent virus of Norfolk giant raspberry," *Ann. Appl. Biol.*, 38, 801–811 (1951).

Cadman, C. H., "Biology of soil-borne viruses," *Ann. Rev. Phytopathology,* 1, 143–172 (1965).

Caldwell, J., "Some effects of a plant virus on nuclear division," *Ann Appl. Biol.*, 39, 98–102 (1952).

Caldwell, J., "Seed transmission of viruses," *Nature,* 193, 457–459 (1962).

Callahan, K. L., "Prunus host range and pollen transmission of elm mosaic virus," *Dissertation Abstr.,* 17, 1861 (1957).

Campbell, R. N., and Fry, P. R., "The nature of the associations between *Olpidium brassicae* and lettuce big vein and tobacco necrosis," *Virology,* 29, 222–233 (1966).

Canova, A., "Rapporti tra *Cuscuta* e virus del giallume della Barbabietola," *Ann. Sper. Agrar. (Rome),* 9, 549–552 (1955).

Carpenter, J. B., Calavan, E. C., and Christiansen, D. W., "Occurrence of excessive seed absorption in citrus fruits affected with stubborn disease," *Plant Disease Reptr.,* 49, 668–672 (1965).

Cation, D., "Transmission of cherry yellows virus complex through seeds," *Phytopathology*, 39, 37–40 (1949).

Cation, D., "Further studies on transmission of ringspot and cherry yellows viruses through seeds," *Phytopathology*, 42, 4 (1952).

Celino, M. S., and Martinez, A. L., "Mechanical transmission of the abaca mosaic virus," *Philippine Agriculturist,* 40, 120–128 (1956).

Chamberlain, E. E., "Cucumber mosaic (*Cucumis* virus 1 of Smith, 1937)," *New Zealand Sci. Technol.,* A21, 74–90 (1939).

Clinch, P. E. M., and Loughnane, J. B., "Seed transmission of virus yellows of sugar beet (*Beta vulgaris* L.) and the existence of strains of this virus in Eire," *Sci. Proc. Roy. Dublin Soc.,* 24, 307–318 (1948).

Cochran, L. C., "Passage of the ring spot virus through Mazzard Cherry seeds," *Science,* 104, 269–270 (1946).

Cochran, L. C., and La Rue, J. L., "Some host-tissue relationships of the peach mosaic virus," *Phytopathology,* 34, 934 (1944).

Corbett, M. K., "A virus disease of lupines caused by bean yellow mosaic virus," *Phytopathology,* 48, 86–91 (1958).

Cornuet, P., "Sur l'extraction et l'inoculation par voie mecanique de certain virus affectant les Fraisiers," *Compt. Rend.*, **235**, 271–273 (1952).

Costa, A. S., "Multiplication of viruses in the dodder, *Cuscuta campestris*," *Phytopathology*, **34**, 151–162 (1944).

Costa, A. S., and Bennett, C. W., "Studies on mechanical transmission of the yellows virus of sugar beet," *Phytopathology*, **45**, 233–238 (1955).

Costa, A. S., and Penteado, M. P., "O milho como planta-teste para o virus do mosaico da cana de acúcar," *Bragantia*, **10**, 93–94 (1950).

Couch, H. B., "Studies on seed transmission of lettuce mosaic virus," *Phytopathology*, **45**, 63–70 (1955).

Cropley, R., "Erratic transmission of strawberry viruses by grafting excised leaves," *Ann. Rept. East Malling Research Sta., Kent*, **1957** 124–125 (1958).

Crowley, N. C., "Some variables affecting the use of cowpea as an assay host for cucumber mosaic virus," *Australian J. Biol. Sci.*, **7**, 141–150 (1954).

Crowley, N. C., "The effect of seed extracts on the infectivity of plant viruses and its bearing on seed transmission," *Australian J. Biol. Sci.*, **8**, 56–67 (1955).

Crowley, N. C., "Studies on the seed transmission of plant virus diseases," *Australian J. Biol. Sci.*, **10**, 449–464 (1957).

Darlington, C. D., "Gene and Virus," *Listener*, **57**, 473–474 (1957).

Davidson, T. R., "Tuber graft transmission of potato leaf roll," *Can. J. Agr. Sci.*, **35**, 238–241 (1955).

Desjardins, P. R., Latterell, R. L., and Mitchell, J. E., "Seed transmission of tobacco-ringspot virus in Lincoln variety of soybean," *Phytopathology*, **44**, 86 (1954).

Dickson, B. T., "Studies concerning mosaic disease," *MacDonald Coll. J. Ser., Tech. Bull.*, No. 2 (1922).

Dickson, B. T., and McRostie, G. P., "Further studies on mosaic," *Phytopathology*, **12**, 42 (1922).

Diener, T. O., and Weaver, M. L., "A caffeine additive to aid mechanical transmission of necrotic ring spot from fruit trees to cucumber," *Phytopathology*, **49**, 321–322 (1959).

Doolittle, S. P., and Gilbert, W. W., "Seed transmission of cucurbit mosaic by the wild cucumber," *Phytopathology*, **9**, 326–327 (1919).

Doolittle, S. P., and Thompson, R. C., "Artificial transmission of the virus of big vein of lettuce," *Phytopathology*, **35**, 484 (1945).

Duggar, B. M., and Johnson, B., "Stomatal infection with the virus of typical tobacco mosaic," *Phytopathology*, **23**, 934–948 (1933).

Dunleavy, J. M., "The grasshopper as a vector of tobacco ringspot in soybean," *Phytopathology*, **47**, 681–682 (1957).

Ernould, L., "La graine de betterave transmet-elle la jaunisse et la mosaique?" *Publs. Inst. Belge Amelioration Betterave*, **18**, 89–94 (1950).

Fajardo, T. G., "Studies on the mosaic disease of the bean (*Phaseolus vulgaris* L.)," *Phytopathology*, **20**, 469–494 (1930).

Fink, H. C., "*Prunus tomentosa* as an index plant for sour cherry viruses," *Phytopathology*, 45, 320–323 (1955).

Frazier, N. W., "*Fragaria bracteata* Heller as an indicator plant of strawberry viruses," *Plant Disease Reptr.*, 35, 127–128 (1951).

Frazier, N. W., "A latent virus of *Fragaria vesca*," *Plant Disease Reptr.*, 37, 606–608 (1953).

Fridlund, P. R., "Two promising cold-tolerant index hosts for the prune dwarf and line pattern viruses," *Plant Disease Reptr.*, 42, 1051–1053 (1958).

Frosheiser, F. I., "Virus infected seeds in alfalfa seed lots," *Plant Disease Reptr.*, 54, 591 (1970).

Fry, P. R., "Note on occurrence of a tobacco-necrosis virus in roots of lettuce showing big vein," *New Zealand J. Sci. Technol.*, A34, 224–225 (1952).

Fulton, J. P., "An evaluation of the use of excised leaf grafts in the strawberry virus studies," *Phytopathology*, 47, 521 (1957).

Fulton, R. W., "Transmission of plant viruses by grafting, dodder, seed and mechanical inoculation," *Plant Virology*, M. H. Corbett and H. D. Sisler, Eds., University of Florida Press, Gainesville, Fla., 1964.

Gardner, M. W., "Indiana plant diseases, 1926," *Proc. Indiana Acad. Sci., 1926*, 37, 417 (1927).

Garnsey, G. M., and Widden, R., "Transmission of exocortis virus to various citrus plants by knife-cut inoculations," *Phytopathology*, **60**, 1092 (Abstr.) (1970).

Gilmer, R. M., "Imported mahaleb seeds as carriers of necrotic ring spot virus," *Plant Disease Reptr.*, 39, 727–728 (1955).

Gilmer, R. M., "The behavior of some stone fruit virus isolates in cucumber and a new differential cucurbit host for a stone fruit virus," *Plant Disease Reptr.*, 41, 11–16 (1957).

Gilmer, R. M., and Brase, K. D., "The comparative value of various indexing hosts in detecting stone fruit viruses," *Plant Disease Reptr.*, 40, 767–777 (1956).

Gold, A. H., Suneson, C. A., Houston, B. R., and Oswald, J. W., "Electron microscopy and seed and pollen transmission of rod shaped particles associated with the false stripe virus disease of barley," *Phytopathology*, 44, 115–117 (1954).

Grogan, R. G., and Bardin, R., "Some aspects concerning seed transmission of lettuce mosaic virus," *Phytopathology*, 40, 965 (1950).

Grogan, R. G., and Campbell, R. N., "Fungi as vectors and hosts of viruses, *Ann. Rev. Phytopathology*, 4, 29–52 (1966).

Grogan, R. G., Hall, D. H., Kimble, K. A., and Kortsen, R. A., "Cucurbit virus diseases in the Imperial Valley of California," *Phytopathology*, 48, 393 (1958).

Hagborg, W. A. F., "Dwarfing of wheat and barley by the barley stripe-mosaic (false stripe) virus," *Can. J. Botan.*, 32, 24–37 (1954).

Harpaz, I., "Needle transmission of a new maize virus," *Nature*, 184, 77–78 (1959).

Harris, R. V., "Grafting as a method for investigating a possible virus disease of the strawberry," *J. Pomol. Hort. Sci.,* **10**, 35–41 (1932).

Harris, R. V., "Virus diseases in relation to strawberry cultivation in Great Britain—a synopsis of recent experiments at East Malling," *Ann. Rept. East Malling Research Sta., Kent,* **1936**, 201–211 (1937).

Harris, R. V., and King, M. E., "Studies on strawberry virus diseases. V. The use of *Fragaria vesca* L. as an indicator of yellow-edge and crinkle," *J. Pomol. Hort. Sci.,* **19**, 227–242 (1942).

Harrison, A. L., "Transmission of bean mosaic," *N. Y. State Agr. Expt. Sta. Bull.,* No. 236 (1935).

Harrison, B. D., "Soil transmission of Scottish raspberry leaf-curl disease," *Nature,* **178**, 553 (1956).

Harrison, B. D., "Beet ringspot—a soil-borne virus infecting potatoes in Scotland," *Proc. 3rd Conf. Potato Virus Diseases, Lisse-Wageningen,* **1957**, 24–28.

Harrison, B. D., "Soil transmission of beet ringspot virus to peach (*Prunus persica*)," *Nature,* **180**, 1055–1056 (1957a).

Harrison, B. D., "Raspberry yellow dwarf, a soil-borne virus," *Ann. Appl. Biol.,* **46**, 221–229 (1958).

Harrison, B. D., "The biology of soil-borne plant viruses," *Advances in Virus Research,* Academic Press, New York, 1960, volume 7, pp. 131–161.

Henderson, R. G., "Transmission of tobacco ringspot by seed of petunia," *Phytopathology,* **21**, 225–229 (1931).

Hendrix, J. W., "Soil transmission of tobacco ringspot virus," *Phytopathology,* **51**, 194 (1961).

Hildebrand, A. A., "The reaction of *Fragaria virginiana* to the virus of yellow-edge," *Can. J. Research, C19, 225–233 (1941).*

Hildebrand, A. A., "Mechanical transmission of sweet potato internal cork virus aided by cysteine," *Phytopathology,* **46**, 233 (1956).

Hildebrand, A. A., "Sweetpotato internal cork virus is indexed on Scarlett O'Hara morning glory," *Science,* **123**, 506–507 (1956a).

Hildebrand, A. A., "Masked virus infection in plants," *Ann. Rev. Microbiol.,* **12**, 441–467 (1958).

Hiruki, Chuji, "Persistence of plant virus in fungi and their transmission with reference to tobacco stunt virus," *First Intern. Phytopathology Congress, London,* (Abstract, p. 88) (1968).

Hobart, O. F., "Introduction and spread of necrotic ringspot virus in sour cherry nursery trees," *Iowa State Coll. J. Sci.,* **30**, 381–382 (1956).

Hollings, M., "*Chenopodium amaranticolor* as a test plant for plant viruses," *Plant Pathol.,* **5**, 57–60 (1956).

Hollings, M., "Reactions of some additional plant viruses on *Chenopodium amaranticolor*," *Plant Pathol.,* **6**, 133–135 (1957).

Hollings, M., "Investigations of chrysanthemum viruses. II. Virus B (mild mosaic) and chrysanthemum latent virus," *Ann. Appl. Biol.,* **45**, 589–602 (1957a).

Holmes, F. O., "Local lesions of mosaic in *Nicotiana tabacum* L.," *Contribs. Boyce Thompson Inst.,* 3, 163–172 (1931).

Hosford, R. M., "Transmission of plant viruses by dodder," *Botanical Rev.,* 33, 387–406 (1967).

Hougas, R. W., "Factors affecting sap transmission of the potato yellow-dwarf virus," *Phytopathology,* 41, 483–493 (1951).

Hovey, C., and Bonde, R., "*Physalis ingulata,* a test plant for the potato leafroll virus," *Phytopathology,* 38, 505–507 (1948).

Hunter, J. A., Chamberlain, E. E., and Atkinson, J. D., "Note on transmission of apple mosaic by natural root grafting," *New Zealand J. Agr. Research,* 1, 80–82 (1958).

Hutton, E. M., and Peak, A. R., "Inoculation techniques for studying the genetics of virus resistance in plants," *J. Australian Inst. Agr. Sci.,* 17, 193–198 (1951).

Hutton, E. M., and Peak, J. W., "Definition of potato virus Y strains by some solonaceous species," *Australian J. Agr. Research,* 3, 1–6, (1952).

Ikeno, S., "Studien über einen eigentümlichen Fall der infektiösen Buntblätterigkeit bei *Capsicum annuum,*" *Planta,* 11, 359–367 (1930).

Jagger, I. C., and Chandler, N., "Big vein, a disease of lettuce," *Phytopathology,* 24, 1253–1256 (1934).

Jensen, D. D., "Insect transmission of virus between tree and herbaceous plants," *Virology,* 2, 249–260 (1956).

Johnson, J., "Transmission of viruses from apparently healthy potatoes," *Wisconsin Univ. Agr. Expt. Sta. Research Bull.,* 63, 1–12 (1925).

Jones, L. K., "Streak and mosaic of *Cineraria,*" *Phytopathology,* 34, 941–953 (1944).

Jones, R. A. C., "Potato mop top and evidence for its transmission by *Spongosporasubterranea,*" *First Inter. Phytopathology Congress, London,* (Abstract, p. 100) (1968).

Jones, R. A. C., and Harrison, B. D., "Potato mop top. Transmission by *Spongospora subterranea* (Walbr.) Lagerh," *Ann. Appl. Biol.,* 63, 1–7 (1969).

Jorgensen, P. S., "Strawberry virus transmission by insert graft," *Plant Disease Reptr.,* 41, 1009–1010 (1957).

Kahn, R. P., "Seed transmission of the tomato-ringspot virus in the Lincoln variety of soybeans," *Phytopathology,* 46, 295 (1956).

Kalmus, H., and Kassanis, B., "The use of abrasives in the transmission of plant viruses," *Ann. Appl. Biol.,* 32, 230–234 (1945).

Kassanis, B., "A virus latent in carnation and potato plants," *Nature,* 173, 1097–1098 (1954).

Kendrick, J. B., "Cucurbit mosaic transmitted by muskmelon seed," *Phytopathology,* 24, 820–823 (1934).

Kendrick, J. B., and Gardner, M. W., "Soybean mosaic," *J. Agr. Research,* 22, 111–114 (1921).

Keur, J. Y., "Seed transmission of the virus causing variegation of *Abutilon,*" *Phytopathology,* 23, 20 (1933).

Kirkpatrick, H. C., "Indicator plants for studies with the leaf roll virus of potatoes," *Am. Potato J.*, 25, 283–290 (1948).

Knorr, L. C., "Suscepts, indicators, and filters of tristeza virus, and some differences between tristeza in Argentina and in Florida," *Phytopathology*, 46, 557–560 (1956).

Koehler, B., Bever, W. M., and Bonnett, O. T., "Soil-borne wheat mosaic," *Illinois Univ. Agr. Expt. Sta. Bull.*, 556, 567–599 (1952).

Köhler, E., "Der Solanum demissum-Bastard 'A 6' als Testpflanze verschiedener Mosaikviren," *Züchter*, 23, 173–176 (1953).

Lackey, C. F., "Reactions of three species of dodder to host plants used in virus disease investigations," *Phytopathology*, 39, 562–567 (1949).

Larson, R. H., "The identity of the virus causing punctate necrosis and mottle in potatoes," *Phytopathology*, 34, 1006 (1944).

Lindner, R. C., and Kirkpatrick, H. C., "The airbrush as a tool in virus inoculations," *Phytopathology*, 49, 507–509 (1959).

Lister, R. M., "Soil-borne virus diseases in strawberry," *Plant Pathol.*, 7, 92–94 (1958).

Lister, R. M., "Transmission of soil-borne viruses through seed," *Virology*, 10, 547–549 (1960).

Lister, R. M., and Murant, A. F., "Seed transmission of nematode-borne viruses," *Ann. Appl. Biol.*, 59, 49–62 (1967).

Liu, H. P., "A comparative study of inducing mosaic infection in sugar cane by various inoculating methods," *Rept. Taiwan Sugar Expt. Sta.*, 4, 210–220 (1949).

Loughnane, J. B., and Murphy, P. A., "Dissemination of potato viruses X and F by leaf contact," *Sci. Proc. Roy. Dublin Soc.*, 22, 1–15 (1938).

MacKinnon, J. P., "A virus latent in turnips," *Can. J. Botany*, 34, 131 (1956).

Mahoney, C. H., "Seed transmission of mosaic in inbred lines of muskmelons (*Cucumis melo* L.)," *Proc. Am. Soc. Hort. Sci.*, 32, 477–480 (1935).

Maramorosch, K., "Seedlings of *Solanum tuberosum* L. as indicator plants for potato leafroll virus," *Am. Potato J.*, 32, 49–50 (1955).

Marcelli, E., "Osservazioni su di una nuova virosi del Tobacco Transmissibile per seme," *Il Tabacco*, 59, 404–409, 676–677 (1955).

Maris, B., and Rozendaal, A., "Enkele proeven met stammen van het X-en het aucubabontvirus van de Aardappel," *Plantenziekten*, 62, 12–18 (1956).

Mastenbroek, C., "Enkele veldwaarnemigen over viruszlekten van Lupine en een onderzoek over haar mosaickziekte," *Tijdschr. Plantenziekten*, 48, 97–118 (1942).

Mather, J. C., "Trials on soil transmission of tobacco necrosis viruses in tulips," *Plant Pathol.*, 4, 96–97 (1955).

McClean, A. P. D., "Bunchy-top disease of the tomato: additional host plants, and the transmission of the virus through the seed of infected plants," *Sci. Bull. Dept. Agr. Union S. Africa*, No. 256 (1948).

McClean, A. P. D., "Tristeza virus of citrus; evidence for absence of seed transmission," *Plant Disease Reptr.,* 41, 821 (1957).

McClintock, J. A., "Is cucumber mosaic carried by seed?" *Science,* 44, 786 (1916).

McClintock, J. A., "Lima bean mosaic," *Phytopathology,* 7, 60–61 (1917).

McKinney, H. H., "Soil factors in relation to incidence and symptom expression of virus diseases," *Soil Sci.,* 61, 93–100 (1946).

McKinney, H. H., "Mosaics of winter oats induced by soil-borne viruses," *Phytopathology,* 36, 359–369 (1946a).

McKinney, H. H., "A seed-borne virus causing false-stripe symptoms in barley," *Plant Disease Reptr.,* 35, 48 (1951).

McKinney, H. H., "Two strains of tobacco mosaic virus, one of which is seed-borne in an etch-immune pungent pepper," *Plant Disease Reptr.,* 36, 184–187 (1952).

McKinney, H. H., and Fellows, H., "A method for inoculating varietal test nurseries with the wheat streak-mosaic virus," *Plant Disease Reptr.,* 35, 264–266 (1951).

McKinney, H. H., Eckerson, S. H., and Webb, R. W., "A mosaic disease of winter wheat and winter rye," *U. S. Dept. Agr. Bull.,* No. 1361 (1925).

McLean, D. M., "Studies on mosaic of cowpeas *Vigna sinensis,*" *Phytopathology,* 31, 420–430 (1941).

McNeal, F. H., and Afanasiev, M. M., "Transmission of barley stripe mosaic through the seed in 11 varieties of spring wheat," *Plant Disease Reptr.,* 39, 460–462 (1955).

McWhorter, F. P., "A latent virus of lily," *Science,* 86, 179 (1937).

McWhorter, F. P., "The utility of flower tissues for making inocula for difficult virus isolations," *Phytopathology,* 43, 479 (1953).

Merriam, D., and Bonde, R., "Dissemination of spindle tuber by contaminated tractor wheels and by foliage contact with diseased potato plants," *Phytopathology,* 44, 111 (1954).

Messieha, M., "Transmission of tobacco ringspot virus by thrips," *Phytopathology,* 59, 943 (1969).

Middleton, J. T., "Seed transmission of squash-mosaic virus," *Phytopathology,* 34, 405–410 (1944).

Milbrath, J. A., "Transmission of components of the stonefruit latent virus complex to cowpea and cucumber from cherry flower petals," *Phytopathology,* 43, 479–480 (1953).

Milbrath, J. A., "Squash as a differential host for strains of stone fruit ring-spot viruses," *Phytopathology,* 46, 638–639 (1956).

Miller, P. W., "An eastern Oregon *Fragaria vesca* suitable as an indicator for some strawberry virus diseases," *Plant Disease Reptr.,* 35, 61–62 (1951).

Miller, P. W., "Sensitivity of some species of *Fragaria* to strawberry yellows and crinkle virus diseases," *Plant Disease Reptr.,* 35, 259–261 (1951a).

Miller, P. W., "Technique for indexing strawberries for viruses by grafting to *Fragaria vesca,*" *Plant Disease Reptr.,* 36, 94–96 (1952).

Miller, P. W., "Comparative efficiency of excised leaf-petiole grafts and stolon grafts for transmitting certain strawberry viruses," *Plant Disease Reptr.,* 42, 1043–1047 (1958).

Miyamoto, Y., "The nature of soil transmission in soil-borne plant viruses," *Virology,* 7, 250–251 (1959).

Miyamoto, Y., "Further evidence for the longevity of soil-borne plant viruses adsorbed on soil particles," *Virology,* 9, 290–291 (1959a).

Moore, J. D., Boyle, J. S., and Keitt, G. W., "Mechanical transmission of a virus disease to cucumber from sour cherry," *Science,* 108, 623–624 (1948).

Mulder, D., "De overbrenging van een virusziekte van zure kers op Komkommer," *Tijdschr. Plantenziekten,* 60, 265–266 (1955).

Mundry, K. W., and Rohmer, I., "Über die mechanische Übertragung des vergilbungsvirus der Rüben," *Phytopathol. Z.,* 31, 305–318 (1958).

Nagaraj, A. N., Davis, T. A., and Menon, K. P. V., "Sap transfusion, a new device for virus transmission trials in palms," *Indian Coconut J.,* 7, 91–98 (1954).

Newhall, A. G., "Seed transmission of lettuce mosaic," *Phytopathology,* 13, 104–106 (1923).

Nikolić, V. J., "Prenošenje virusa žutice šećerne repe semenom (Transmission of sugar beet yellows virus by seed)," *Zashtita Bilja,* 35, 79–82 (1956).

Ogilvie, L., Mulligan, B. O., and Brain, P. W., "Progress report on vegetable diseases, VI," *Ann. Rept. Agr. Hort. Research Sta., Bristol,* 1934, 175–194 (1935).

Oswald, J. W., and Bowman, T., "Studies on a soil-borne potato virus disease in California," *Phytopathology,* 48, 396 (1958).

Plakidas, A. G., "Virus diseases of strawberry, a review," *Plant Disease Reptr.,* 39, 525–541 (1955).

Powell, C. C., and Schlegel, D. E., "Factors influencing seed transmission of squash mosaic in cantaloupe," *Phytopathology,* 60, 1466 (1970).

Pryor, D. E., "Exploratory experiments with the big-vein disease of lettuce," *Phytopathology,* 36, 264–272 (1946).

Quantz, L., "Untersuchungen uber ein samenubertragbares Mosaikvirus der Ackerbohne (*Vicia faba*)," *Phytopathol. Z.,* 20, 421–448 (1953).

Rader, W. E., Fitzpatrick, H. F., and Hildebrand, E. M., "A seed-borne virus of musk melon," *Phytopathology,* 37, 809–816 (1947).

Ramamurthi, C. S., Ross, A. F., and Roberts, D. A., "Use of *Gomphrena globosa* in quantitative work with red clover vein mosaic virus," *Phytopathology,* 48, 336–337 (1958).

Rawlins, T. E., and Tompkins, C. M., "Studies on the effect of carborundum as an abrasive in plant virus inoculations," *Phytopathology,* 26, 578–587 (1936).

Rawlins, T. E., and Tompkins, C. M., "Carborundum for plant-virus inoculations," *Phytopathmlogy,* 30, 185b,x6186 (1940).

Reddick, E., "Seed transmission of potato virus diseases," *Am. Potato J.,* 13, 118–124 (1936).

Reddick, E., and Stewart, V. B., "Transmission of the virus of bean mosaic in seed and observations on the thermal death-point of seed and virus, *Phytopathology,* 9, 445–450 (1919).

Reyes, T. T., "Seed transmission of coffee ring spot by Excelsa coffee (*Coffea excelsa*)," *Plant Disease Reptr.,* 45, 185 (1961).

Richards, B. L., Jr., and Munger, H. M., "A rapid method for mechanically transmitting plant viruses," *Phytopathology,* 34, 1010 (1944).

Roberts, F. M., "The infection of plants by viruses through roots," *Ann. Appl. Biol.,* 37, 385–396 (1950).

Roberts, J. D., "Virus tested potatoes in Scotland," *Am. Potato J.,* 30, 197–204 (1953).

Ross, A. F., "*Physalis floridana* as a local lesion test plant for potato virus Y," *Phytopathology,* 43, 1–8 (1953).

Salaman, R. N., and LePelley, R. H., "Paracrinkle, a potato disease of the virus type," *Proc. Roy. Soc. (London),* B106, 140 (1930).

Salamon, E. S., and Ware, W. M., "The chlorotic disease of the hop. IV. Transmission by seed," *Ann. Appl. Biol.,* 22, 728–730 (1935).

Samuel, G., and Bald, J. G., "On the use of the primary lesions in quantitative work with two plant viruses," *Ann. Appl. Biol.,* 20, 70–99 (1933).

Schisler, L. C., Sinden, J. W., and Sigel, E. M., "Transmission of a virus disease of mushroom by infected spores," *Phytopathology,* 53, 888 (Abstr.) (1963).

Schmeizer, K., "Die übertragbarkeit des Tabakmauche-virus durche *Cuscutaarten,*" "*Naturwissenschaften,* 42, 19 (1955).

Schmelzer, K., "Beitrage zur Kenntnis der übertragbarkeit von viran durch *Cuscutaarien,*" *Phytopathol. Z.,* 28, 1–56 (1956).

Severin, H. H. P., "Curly leaf transmission experiments," *Phytopathology,* 14, 80–93 (1924).

Silberschmidt, K., and Rostom, E., "A valuable indicator plant for a strain of potato virus Y," *Am. Potato J.,* 32, 222–227 (1955).

Sill, W. H., and Walker, J. C., "Cowpea as an assay host for cucumber virus 1," *Phytopathology,* 42, 328–330 (1952).

Smith, K. M., "A latent virus in sugar-beets and mangolds," *Nature,* 167, 1061 (1951).

Smith, K. M., *Recent Advances in the Study of Plant Viruses,* 2nd ed., Blakiston, Philadelphia, 1952.

Smith, K. M., and Short, M. E., "Lettuce ringspot: A soil-borne virus disease," *Plant Pathol.,* 8, 54–56 (1959).

Snyder, W. C., "A seed-borne mosaic of asparagus bean, *Vigna sesquipedalis,*" *Phytopathology,* 32, 518–523 (1942).

Sprau, F., "Zur frage der übertragung des Y-virus der Kartoffel durch Samen," *Pflanzenschutz Ber.,* 3, 128–129 (1951).

Sreenivasaya, M., "Contributions to the study of spike disease of sandal. Part XI. New methods of disease transmission and their significance," *J. Indian Inst. Sci.,* 13A, 113–117 (1930).

Sturgess, O. W., "Studies with chlorotic streak disease of sugarcane. V. Factors affecting transmission," *Tech. Commun. Bur. Sugar Exp. Sta. Queensland,* 1, 1–10 (1962).

Sutic, D., "Die Rolle des Paprikasamens bei der Virusübertragung," *Phytopathol. Z.,* 36, 84–93 (1959).

Teakle, D. S., "Fungi as vectors and hosts of viruses," *Viruses, Vectors and Vegetation,* Wiley-Interscience, New York, 1969.

Thomas, W. D., Jr., and Graham, R. W., "Seed transmission of red node virus in pinto beans," *Phytopathology,* 41, 959–962 (1951).

Thompson, R. C., Doolittle, S. P., and Smith, F. F., "Investigations on the transmission of big vein of lettuce," *Phytopathology,* 34, 900–904 (1944).

Thornberry, H. H., "Effect of phosphate buffers on infectivity of tobaccomosaic virus," *Phytopathology,* 25, 618–627 (1935).

Timian, R. G., "Seed transmission of barley stripe mosaic plants in relation to the length of time parental plants are infected," *Phytopathology,* 60, 1316 (Abstr.) (1970).

Tomlinson, J. A., and Carter, A. L, "Seed transmission of CMV virus in chickweed (*Stellaria media*) in relation to the ecology of the virus," *Ann. Appl. Biol.,* 66, 381 (1966).

Traversi, B. A., "Estudio inicial sobro una enfirmedad del girasol (*Helianthus annuus* L.) en Argentina," *Rev. Invest. Agr. (Buenos Aires),* 3, 345–351 (1949).

Troll, H. J., "Zur Frage der Braunevirusübertragung durch das saatgut bei Lupinus lutous," *Nachrbl. Dent. Pflanzensdutzdienst (Berlin),* 11, 218–222 (1957).

Uppal, B. N., "India, a new virus disease of *Dolichos biflorus,*" *Intern. Bull. Plant Protect.,* 5, 163 (1931).

Valleau, W. D., "Seed transmission and sterility studies of two strains of tobacco ringspot," *Kentucky Agr. Expt. Sta. Bull.,* No. 327 (1941).

van Koot, Y., "Enkele nieuwe gezichtspunten betreffende het virus van het tomatenmozaiek," *Tijdschr. Plantenziekten,* 55, 152–166 (1949).

Vasudeva, R. S., Raychaudhuri, S. P., and Pathanian, P. S., "Yellow mosaic of lettuce," *Current Sci. (India),* 17, 244–245 (1948).

Vaughan, E. K., "A device for the rapid removal of tannins from virus infected plant tissues before extraction of inoculum," *Tijdschr. Plantenziekten,* 62, 266–270 (1956).

Wagnon, H. K., Traylor, J. A., Williams, H.E., and Weinberger, J. H., "Observations on the passage of peach necrotic leaf spot and peach ringspot," *Plant Disease Reptr.,* 44, 117–119 (1960).

Walkinshaw, C. H., and Larson, R. H., "A soil-borne virus associated with the corky ringspot disease of potato," *Nature,* 181, 1146 (1958).

Wallace, J. M., "The use of leaf tissue in graft-transmission of psorosis virus," *Phytopathology*, 37, 149–152 (1947).

Wallace, J. M., and Drake, R. J., "Seed transmission of the avocado sunblotch virus," *Citrus Leaves*, (Dec. 1953).

Walters, H. J., "Some relationships of three plant viruses to the differential grasshopper, *Melanoplus differentialis* (Thos.)," *Phytopathology*, 42, 355–362 (1952).

Way, R. D., and Gilmer, R. M., "Pollen transmission of necrotic ringspot virus in cherry," *Plant Disease Reptr.*, 42, 1222–1224 (1958).

Weathers, L. G., "Transmission of exocortis virus of citrus by *Cuscuta subinclusa*," *Plant Disease Reptr.*, 49, 189–190 (1965).

Weathers, L. G., and Hartung, M. K., "Transmission of citrus viruses by dodder, *Cuscuta subinclusa*," *Plant Disease Reptr.*, 48, 102–103 (1964).

Webb, R. E., and Buck, R. W., "A diagnostic host for potato virus A," *Am. Potato J.*, 32, 248–253 (1955).

Wilkinson, R. E., and Blodgett, F. M., "*Gomphrena globosa*, a useful plant for qualitative and quantitative work with potato virus X," *Phytopathology*, 38, 28 (1948).

Wilks, J. M., and Welsh, M. F., "Sweet cherry foliage indicator hosts for the virus that causes little cherry," *Can. J. Agr. Sci.*, 35, 595–600 (1955).

Williams, H. E., and Smith, S. H., "Recovery of virus from stored pear pollen," *Phytopathology*, 57, 1011 (1967).

Wilson, V. E., and Dean, L. L., "Flour of infected bean seed as a source of virus," *Phytopathology*, 54, 489 (1964).

Yarwood, C. E., "Associations of rust and virus infections," *Science*, 114, 127–128 (1951).

Yarwood, C. E., "The phosphate effect in plant virus inoculations," *Phytopathology*, 42, 137–143 (1952).

Yarwood, C. E., "Quick virus inoculation by rubbing with fresh leaf discs," *Plant Disease Reptr.*, 37, 501–502 (1953).

Yarwood, C. E., "Zinc increases susceptibility of bean leaves to tobacco mosaic virus," *Phytopathology*, 44, 230–233 (1954).

Yarwood, C. E., "Mechanical transmission of an apple mosaic virus," *Hilgardia*, 23, 613–638 (1955).

Yarwood, C. E., "Heat induced susceptibility of beans to some viruses and fungi," *Phytopathology*, 46, 523–525 (1956).

Yarwood, C. E., "Mechanical transmission of plant viruses," *Advances in Virus Research*, Academic Press, New York, 1957, volume 4, pp. 243–278.

Yarwood, C. E., "Contact transmission of peach ringspot virus," *Phytopathology*, 47, 539 (1957a).

Yarwood, C. E., "A brush extraction method for transmission of viruses," *Phytopathology*, 47, 613–614 (1957b).

Yarwood, C. E., "Heat activation of virus infections," *Phytopathology*, 48, 39–46 (1958).

Yarwood, C. E., "Virus susceptibility increased by soaking bean leaves in water," *Plant Disease Reptr.*, 43, 841–844 (1959).

Yarwood, C. E., and Gold, A. H., "Guar as a local lesion host of potato virus S.," *Plant Disease Reptr.*, 39, 622 (1955).

Yarwood, C. E., Resonich, E. C., Ark, P. A., Schlegel, D. E., and Smith, K. M., "So-called beet latent virus is a bacterium," *Plant Disease Reptr.*, 45, 85–89 (1961).

Yu, T. F., "A mosaic disease of cowpea (*Vigna sinensis* Endl.)," *Ann. Appl. Biol.*, 33, 450–454 (1946).

CHAPTER **12** CHAPTER

Plant Virus Transmission
by Arthropods and
Other Animals

TRANSMISSION BY ARTHROPODS

Historical

About the same time that Meyer, Iwanowski, and Beijerinck were working on tobacco mosaic in Europe, and insect transmission had not yet been thought of, a Japanese worker (Takata, 1895, 1896) confirmed an observation by a rice farmer that a disease of rice was associated with the leafhopper *Deltocephalus dorsalis* Motsch. This work was published in Japanese and was not available to most workers; so it was not surprising that when Ball (1906) observed the relationship between *Circulifer tenellus* (Baker) and curly top of sugar beet, he thought of it as the first demonstration of its kind. Even then, he did not make the virus etiology clear, suggesting instead that the disease was due to the feeding effect of the insect in inducing the galllike protuberances typical of the disease in sugar beet. The virus etiology was established by Shaw (1910) and R. E. Smith and Boncquet (1915).

479

Storey's work on the transmission of maize streak virus (1925) by *Cicadulina mbila* (Naude) and Kunkel's on aster yellows virus transmission by *Cicadula sexnotata* Fall (1926) were major contributions in this early period. There are now more than 120 species of leafhoppers implicated in plant virus transmission.

In the early period of work on insect transmission, when the now classic examples of curly top, aster yellows, and maize streak were being studied intensively, a broad categorization into "yellows" diseases and "mosaics" placed the leafhopper-transmitted viruses in the former category. This generalization is still useful but is no longer inclusive, there being many other symptom expressions of virus infection transmitted by leafhoppers.

Leafhopper transmitted viruses are entirely, with one exception, characterized by persistence in the insect, and in some cases they are propagative. The exception is tungo disease of rice which is nonpersistent in its vector, *Nephotettix impicticeps* Ish. (Ling, 1966).

Aphididae

The largest group of insect vectors from the standpoint both of numbers of viruses, as well as of species of insect involved, is the Aphididae. Allard (1914) experimented with aphid transmission of tobacco mosaic virus and concluded that *Myzus persicae* (Sulz.) was a vector, but the aphid-vector relationships with tobacco mosaic virus are still obscure. According to K. M. Smith (1957), no aphid has been convicted of transmitting this virus, but this opinion disregards the report by Hoggan (1934) that it was readily transmitted from tomato-to-tomato by several species of aphids. The lack of confirmation in the intervening years suggests that perhaps Hoggan was working with some highly specific strain of the virus which has not been recognized since (See Lojek, 1969).

In the early twenties, the aphid as a vector of viruses was firmly established by the work of Doolittle (1920), who used *M. persicae* to transmit cucumber mosaic; Brandes (1920) transmitted sugarcane mosaic with *Aphid maidis* Fitch; Jagger (1921) transmitted lettuce mosaic with *M. persicae;* and Doolittle and Jones (1925) transmitted pea mosaic with *Macrosiphum pisi* Harris. Perhaps more significant because of its universal applicability was the discovery that the so-called potato-degeneration diseases were aphid-transmitted (Quanjer, 1920; Murphy, 1922; Schultz and Folsom, 1923).

M. persicae is known as the transmitter of more than 100 viruses. This insect's unique position may stem from the fact that not only is it a very efficient vector, but also that it has probably been tested more thoroughly than any other species, and, when it has been proved to be a vector, the search for additional vector species has not been pushed.

Most of the aphid-borne viruses induce mosaic symptoms in the host plants, but, as with leafhoppers and yellows diseases, there are some notable exceptions, e.g., yellow dwarf of onion, sugar beet yellows, and various vein-banding and vein-chlorosis-inducing viruses.

Miridae; Piesmidae

The plant bugs, Hemiptera and Heteroptera, have not proved to be efficient vectors. Heinze (1959) lists 13 species, mostly Miridae, which have been implicated from time to time, but only two species of Piesmidae can be considered authentic vectors. It is worth noting that both these species belong to the same genus, *Piesma;* both attack the same crop plant, sugar beet, one in Europe and Asia (*P. quadrata* Fieb.), the other in the United States (*P. cinereum* (Say)).

All the records of Miridae as vectors may not be in error, although Gibbs (1969) reference to *Lygus pratense* as the vector of rape savoy virus is questionable in view of the number of aphid species recorded as vectors. The observation that mirid saliva was extremely toxic and therefore likely to interfere with virus transmission by killing cells at the puncture point was made by Smith (1957), and the phytotoxicity of the oral secretions of this group of insects has been fully confirmed (see Chapter 4). These oral secretions, however, may undergo temporary changes either in quality or quantity; during such periods, the vector capacity of the insect may change also.

Aleyrodidae

The Aleyrodidae, or white flies, are insidious contaminants of many greenhouse and field plants. They are difficult insects to manipulate; for this reason, perhaps, some of the records of transmission by these insects are inadequate. K. M. Smith (1957a) records only 9 viruses transmitted by white flies, but Heinze (1959) lists 14 species of white fly transmitting over 20 viruses, but with frequent questioning of the record.

The best-known diseases caused by white fly-transmitted viruses are: cassava mosaic which, apparently, occurs wherever cassava is grown, except perhaps Malaysia; cotton leaf curl, which is best known from the Sudan; tobacco leaf curl, from the Transvaal in South Africa and many other tropical localities; and of special, but not economic, interest, *Abutilon* mosaic. This last disease was long known as a graft-transmissible variegation, highly prized as a horticultural variety. Orlando and Silberschmidt

(1945, 1946), however, showed that the disease was due to a virus transmitted by *Bemisia tabaci* (Genn.). *Sida rhombifolia* L is also a host according to these authors. The writer has observed *Sida* mosaic in fairly remote areas in Brazil, which indicates that it is widespread in nature.

Costa (1969) has critically and inclusively reviewed the role of white flies as vectors. As a group, white fly-transmitted viruses are not mechanically transmissible, and, as that would indicate, are persistent in the vector. There are exceptions to both these characteristics. Costa and Bennett as early as 1950 mechanically transmitted *Euphorbia mosaic,* and Sheffield (1958) successfully inoculated a white fly-vectored virus of sweet potato into indicator plants. Persistence of the viruses in white flies is usual, but an exception is cucumber vein-yellowing virus, transmitted by *Bemisia tabaci* (Genn.). This virus is nonpersistent in the vector and is not readily mechanically transmissible (Harpaz and Cohen, 1965). Costa's review calls attention to the need for clarification, not only of the vector species synonymy but also of the plant diseases involved.

Coccoidea

The scale insects (Coccoidea) are poorly represented in the list of insect vectors; in fact, the only members of the family involved are several species of mealybugs. With the exception of *Pseudococcus sacharifolii,* transmitting sugarcane spike virus, and the pineapple latent virus, transmitted by the pineapple mealybugs, the only viruses transmitted by mealybugs are those responsible for swollen shoot disease of cacao or diseases closely related to it. Although some 20 species of mealybug in West Africa have been recorded as transmitting one or more strains of swollen shoot (Posnette, 1950; Leston, 1970), this species of economic importance in West Africa is *Pseudococcus njalensis* Laing.

In Trinidad, where diseases of cacao related to swollen shoot occur, the recorded vectors are other species of Pseudococcidae (Kirkpatrick, 1950); in Ceylon, *Planococcus lilacinus* (Ckll.) is the vector (Carter, 1956); in Java, *Planococcus citri* (Risso).

Thysanoptera

The Thysanoptera, or thrips, are a peculiarly limited group of vectors in that only one plant virus is transmitted by them. This is known variously as tomato spotted wilt, pineapple yellow spot, vira-cabeca or carcova of tomato and tobacco, tobacco kromneck, etc. The distribution of the disease in Australia, Hawaii, South Africa, Great Britain, and central Europe is well authenticated. The illustration by Smith (1957) of yellow spot of pineapple in the Philippines (after Serrano, 1935) is clearly not that of yellow spot. The only authentic illustrations of yellow spot of pineapple thus far published are those of Linford (1932) and Carter (1939). The fact that the thrips vector in South Africa is a flower feeder (*Frankliniella schultzei* (Tribom)) has naturally limited symptom expression to fruit, although Carter (1939) found a few prefruiting plants with typical symptoms. The Queen variety of pineapple grown in the Bathurst area in South Africa is particularly susceptible to yellow spot in fruit.

Heinze (1959) faithfully cites many other records of thrips-transmitted viruses, but the adequacy of many of the records can be questioned (Sakimura, 1962). Tomato spotted wilt can no longer be considered the only thrips-transmitted virus, however. Pistacio rosette virus is reported as thrips-transmitted, and this record has not, apparently, been challenged (Kreutzberg, 1940). The most recent record is that of the transmission of tobacco ringspot virus, nematode transmitted to roots but by *Thrips tabaci* into leaves (Bergeson *et al.*, 1964; Messieha, 1969).

Mandibulate Insects

The mandibulate insects as potential vectors have received comparatively little attention, although one of the earliest records of insects as vectors included two cucumber beetle species (along with *Aphis gossypii* (Glov.)) as vectors of cucumber mosaic (Doolittle, 1920). The early concept that insects were merely inoculation needles has perhaps had its influence, since, obviously, biting mouth parts are but clumsy needles.

Nevertheless, there are now more examples known. Two species of grasshopper, the long-horned (*Leptophyes punctatissima* (Bosc.)) and the shorthorned (*Chorthippus bicolor* (Charp.)), transmit the virus of turnip yellow mosaic, as also do several species of flea beetles of the genus *Phyllotreta*. Markham and Smith (1949) reported that other beetles such as weevils, and even the larvae as well as the adults of the mustard beetle, *Phaedon cochliariae* F., also serve as vectors. In addition to turnip yellow mosaic, grasshoppers have also been known to transmit potato spindle tuber virus (Goss, 1931). *Melanoplus differentialis* (Thos.) transmitted potato virus X, tobacco ringspot virus, and TMV (Walters, 1921).

A mosaic of cowpea, *Vigna sinensis* Endl., is transmitted by a beetle, *Ceratoma trifurcata* Forster (C. E. Smith, 1924; McLean, 1941), while Dale (1949) reported another species of the same genus (*C. ruficornia* Oliv.) as a vector of a similar and perhaps the same virus in Trinidad. Two western cucumber beetles, *Acalymma melanocephala* (F.) and *A. trivittata* (F.), transmit the virus of squash mosaic (Freitag, 1941). A number of vectors belonging to other orders have been recorded, many of them as vectors of tobacco mosaic. These insects include butterflies, moths, leatherjackets (Tipulidae), and leafminers (Agromyzidae).

There seems to be no good reason to suppose, however, that mandibulate insects should be inefficient vectors of those plant viruses which are easily mechanically transmitted. This last factor may well be the reason for the increasing number of references to the transmission of tobacco mosaic which include biting larvae of Lepidoptera and some Diptera. Thus far, turnip yellow mosaic, potato spindle tuber, potato virus X, tobacco ringspot, tobacco mosaic, cowpea mosaic, squash mosaic, radish mosaic, bean pod mosaic, pumpkin mosaic, and soybean mosaic are all mandibulate-insect-transmitted. *Sesselia* (*Malacasoma*) *pusilla* (Gerst.), a chrysomelid beetle, transmits rice yellow mottle in Kenya for five days after an acquisition. Avoidance of overlapping growing seasons is needed to prevent rapid extension of the affected area (Bakker, 1970). The transmission of radish mosaic is normally by aphids; its transmission by flea beetles appears to be incidental (Campbell and Colt, 1967). It can be noted in passing that Hansen (1970) has rejected mandibulate vectors as not important enough to be included in his classification of plant viruses.

Mites

The recent discovery that the Acarina include virus vectors was actually a rediscovery. Amos *et al.* (1927) reported the transmission of black currant reversion virus by *Eriophyes ribi* Nalepa. Almost 30 years later, Slykhuis (1955) showed that wheat streak virus was transmitted by *Aceria tulipae* Keifer; Flock and Wallace (1955) showed that fig mosaic virus was transmitted by *Acerius ficus* (Cotte). Wilson *et al.* (1955) reported that an undetermined species of eriophyid transmitted the virus of peach mosaic; the mite was later described by Keifer and Wilson (1956) as *E. insidiosus* K. and W. In 1958 Mulligan associated rye-grass mosaic virus with *Abacarus hystrix* (Nalepa). There are a number of records of mite transmission of viruses (Heinze, 1959) which perhaps need confirmation. Mites are extremely difficult to handle, but once established on a plant they are even more difficult to remove; extreme care must be taken to avoid confusion between symptoms due to toxic feeding and those due to virus infection. Confusion on this point may explain the failure to appreciate the early record of Amos *et al.* (1927).

It has proved difficult to maintain colonies of some eriophyid mites for use in transmission tests, but Oldfield and Wilson (1970) achieved this by using ornamental peach varieties, on which the mite *Eriophyes insidiosus* Keifer and Wilson, develops larger populations than it does on commercial varieties.

Transmission by mites has been reviewed by Slykhuis (1965, 1969) and Oldfield (1970). The validity of mite transmission of reversion disease of black currant appears to have been firmly established (Thresh, 1964). Citrus leprosis has been transmitted by mites, and its former status as a phytotoxemia denied, primarily on the basis of regional occurrence of the disease and general distribution of the mite. Citrus leprosis is found in one area; and, an as yet undescribed ringspot disease, also associated with mites, in another. In a third area there is no evidence of either disease although the mite, *Brevipalpus phoenicis* (Geij), is present (Rosillo *et al.,* 1964).

On the contrary, Knorr's critical analysis (1968) shows clearly that the problem is by no means settled. Considering the three hypotheses proposed, fungus, mite, or virus etiology, the fungus hypothesis can apparently be discarded. The evidence for a mite-induced phytotoxemia includes, (1) nymphs reared from eggs in a petrie dish and adults taken directly from *Bidens pilosa* induced leprosis on citrus seedlings, and (2) control by the use of sulphur as a miticide which could not be expected to completely eliminate the mites. Evidence for a virus is primarily on a single datum, the transmission by grafting; but this was only achieved when leprosis affected patches were inserted into the immature green shoots. This, however, suggests the possibility of phytotoxin-induced mutagenic effects, which are not unknown. A major problem is that of the taxonomy of the mites, although all appear to be in the genus *Brevipalpus*. Knorr concludes that, as the taxonomy is currently understood, two species, *B. californicus* (Banks) and *B. obovatus* Donnadieu, are associated with leprosis in Florida, and these two species have been added to the virus-vector list, even though the final word on the etiology of the disease has not yet been said. *B. phoenicis* is not associated with leprosis in Florida; the phytotoxic effects of the mite's feeding have been reviewed in Chapters 6 and 7.

TRANSMISSION BY ANIMALS OTHER THAN ARTHROPODS

Heinze (1959) records the transmission of tobacco mosaic by several species of slugs and snails. With one exception, these determinations were all made by Heinze himself (1958). TMV can be transmitted by sparrows according to Broadbent (1965), but his is probably a crude mechanical transmission by frightened sparrows enclosed in a cage. The most important group of animals in this section are the nematodes, and since Hewitt *et al.* (1958) reported that *Xiphinema index* Thorne and Allen is the vector of fanleaf of grape, there has been intense activity concentrated on this subject, with the result that there is now a total of 11 viruses transmitted by nematodes involving 19 species of the vectors (Pitcher, 1965; Taylor and Cadman, 1969). The separate table (Table 12.3) in which the nematode vectors are listed is a combination of the lists provided by these two reviews (See also Cadman, 1963).

The Virus-Vector Lists

The virus-vector lists have been changed drastically, and the vector-virus list eliminated; newly described vectors have been placed in a new combined table. The experience of workers has been that reference is first made to the virus, and in the virus-vector list the vectors are listed but without the vector citation.

The most difficult problem encountered in the preparation of lists such as these is that of synonomy. When these were first compiled, the taxonomic arrangements followed were those used by Heinze (1959), but for the aphids, the taxonomy he followed has not proved acceptable to aphid taxonomists in many specific instances. Kennedy *et al.* provided a conspectus of the aphid vectors embodying the current (1962) view of the taxonomy of the species involved, but there is not complete unanimity in acceptance.

The serious error in the aphid species listed in Table 12.1 was the use of *Myzodes* for *Myzus,* and the error was confusing to students using the table, particularly in the case of *Myzus persicae.*

A complete conspectus of aphid vectors which will agree with modern concepts and with complete synonomy is urgently needed. The same can be said for the Aleyrodidae, but in this case synonyms for many of the white fly viruses as described would also be essential.

Nielson has followed up his 1962 paper on the leafhoppers with a monographic study of the Cicadellidae and the viruses they transmit (1968). This is an essential reference.

Table 10.2 of the first edition, including the viruses for which no vector is known, has been omitted; occasionally an item from this list has a vector reported, but if so it will appear in the newer lists. Newly recorded viruses are in a separate table (Table 12.2) arranged in what is hoped will be an acceptable manner. A new and separate table has been written (Table 12.3) to include all the nematode-transmitted viruses now known.

TABLE 12.1. A List of Plant Viruses and Their Insect Vectors

Virus	Virus citation[a]	Vector
Abaca bunchy top	Ocfemia (1930)	Pentalonia caladii v.d. Goot
		Pentalonia nigronervosa Coq.
Abutilon infectious variegation	Morren (1869), Brierley (1944)	Bemisia inconspicua (Q)
		Bemisia tabaci (Gen.)
Ajuga	Heinze (1957)	Dysaulacorthum pseudosolani (Theob.)
		Dysaulacorthum vincae (Walk.)
		Macrosiphon solanifolii (Ashm.)
		Myzus persicae (Sulz.)
Anemone latent	Smith (1957)	Dysaulacorthum pseudosolani (Theob.)
		Dysaulacorthum vincae (Walk.)
		Myzus persicae (Sulz.)
Anemone mosaic	Hollings (1957)	Dysaulacorthum pseudosolani (Theob.)
		Macrosiphon solanifolii (Ashm.)
Arabis mosaic	Smith and Markham (1944)	Xiphinema diversi-caudatum (Micoletsky)
Artichoke (Cynarus) latent	Costa et al. (1959)	Myzus persicae (Sulz.)
Aster yellows (California)	Jensen and Tate (1947)	Acinopterus angulatus Lawson
		Chlorotettix similis De Long
		Chlorotettix viridius Van Duzee
		Colladonus commissus (Van Duzee)
		Colladonus flavocapitatus (Van Duzee)
		Colladonus geminatus (Van Duzee)
		Colladonus intricatus (Ball)
		Colladonus kirkaldyi (Ball)
		Colladonus montanus (Van Duzee)
		Colladonus rupinatus (Ball)

TABLE 12.1. *Continued*

Virus	Virus citation[a]	Vector
		Colladonus rupinatus (Ball) var. Drunneus (De Long and Severin)
		Euscelidius variegatus (Kirschbaum)
		Fieberiella flori (Ståhl)
		Gyponana hasta De Long
		Idiodonus heidmanni (Ball)
		Macrosteles cristatus (Rib.)
		Macrosteles fascifrons (Ståhl)
		Macrosteles laevis (Rib.)
		Macrosteles masatonis (Mats.)
		Macrosteles quadripunctulatus (Kirschb.)
		Paraphlepsius apertinus (Osborn and Lathrop)
		Scaphytopius dubius (Van Duzee)
		Scaphytopius irroratus (Van Duzee)
		Texananus incurvatus (Osborn)
		Texananus lathropi (Baker)
		Texananus latipex De Long
		Texananus oregonus (Ball)
		Texananus pergradus De Long
		Texananus spatulatus (Van Duzee)
Aster yellows	Kunkel (1926, 1936)	
Aster yellows (Japan)	Fukushi (1930)	*Macrosteles fascifrons* (Ståhl)
Aster yellows (Russia)	Shuchov and Vovk (1945)	*Macrosteles masatonis* (Mats.)
Banana bunchy top	Magee (1927)	*Macrosteles quadripunctulatus* (Kirschb.)
Barley yellow dwarf	Oswald and Houston (1951)	*Pentalonia nigronervosa* Coq.
		Metopolophium dirhodum (Walk.)
		Neomyzus circumflexus (Buckt.)

Disease	Vector	Reference
Bean common mosaic	*Rhopalosiphon annuae* (Ostl.) *Rhopalosiphon maidis* (Fitch) *Rhopalsiphon padi* (L) *Rhopalosiphon prunifoliae* (Fitch) *Schizaphis graminum* (Rond.) *Sitobium avenae* (F.) *Sitobium granarium* (Kirby) *Acyrthosiphon destructor* (Johnson) *Acyrthosiphon onobrychis* (B.d.F.) *Aphidula spiraecola* (Patch) *Aphis fabae* Scop. *Brevicoryne brassicae* (L) *Bruchidius* (Acanthoscelides) *obtectus* Say *Cerosipha gossypii* (Glov.) *Dactynotus ambrosiae* (Thomas) *Hayhurstia atriplicis* (L) *Lipaphis pseudobrassicae* (Davis) *Macrosiphon solanifolii* (Ashm.) *Mcgoura viciae* (Buckt.) *Myzus persicae* (Sulz.) *Pergandeida craccivora* (Koch)	Stewart and Reddick (1917)
Bean double yellow mosaic	*Bemisia tabaci* (Gen.)	Smith (1957a)
Bean leaf wilt	*Myzus persicae* (Sulz.)	Johnson (1942)
Bean yellow mosaic	*Acyrthosiphon destructor* (Johnson) *Acrthosiphon onobrychis* (B.d.F.) *Aphis* (Doralis) *fabae* Scop. *Aulacorthum barri* (Essig) *Brachycaudus cardui* (L) *Brevicoryne brassicae* (L) *Cerosipha gossypii* (Glov.)	Pierce (1934)

TABLE 12.1. Continued

Virus	Virus citation[a]	Vector
		Comaphis helianthi (Monell)
		Dysaulacorthum pseudosolani (Theob.)
		Dysaulacorthum vincae (Walk.)
		Hayhurstia atriplicis (L)
		Hyperomyzus sonchi (Ostl.)
		Macrosiphon rosae (L)
		Macrosiphon solanifolii (Ashm.)
		Myzus cerasifolii (Fitch)
		Myzus certus (Walk.)
		Myzus persicae (Sulz.)
		Pentatrichopus fragaefolii (Cock.)
		Pergandeida craccae (L)
		Sitobium granarium (Kirby)
		Yezabura tulipae (B.d.F.)
Broad bean leaf roll	Quantz and Völk (1959)	*Acyrthosiphon onobrychis* (B.d.F.)
Broad bean mottle	Bawden et al. (1951)	*Acyrthosiphon onobrychis* (B.d.F.)
		Aphis fabae Scop.
Broad bean wilt	Stubbs (1947)	*Myzus persicae* (Sulz.)
Beet curly top	Boncquet and Hartung (1915)	*Circulifer tenellus* (Baker)
Beet curly top (Argentina)	Bennett et al. (1946)	*Agalliana ensigera* Oman
Beet curly top (Brazil)	Bennett and Costa (1949)	*Agallia albidula* Uhl.
Beet curly top (Turkey)	Bennett and Tanrisever (1958)	*Circulifer opacipennis* (Lth.)
Beet leaf curl	Wille (1928)	*Piesma quadratum* (Fieb.)
Beet mosaic	Robbins (1921)	*Acyrthosiphon destructor* (Johnson)
		Acyrthosiphon onobrychis (B.d.F.)
		Aphidula (?) *glycines* (Mats.)

		Aphidula pomi (De Geer)
		Aphis apii Theob.
		Aphis fabae Scop.
		Brevicoryne brassicae (L)
		Cavariella archangelicae (Scop.)
		Cerosipha apigraveolens (Essig)
		Cerosipha gossypii (Glov.)
		Dactynotus cichoricola H.R.L.
		Dysaulacorthum pseudosolani (Theob.)
		Hayhurstia atriplicis (L)
		Hyadaphis conii (Davids.)
		Lipaphis pseudobrassicae (Davis)
		Macrosiphon solanifolii (Ashm.)
		Myzotoxoptera tulipaella (Theob.)
		Myzus persicae (Sulz.)
		Neomyzus circumflexus (Buckt.)
		Paczoskia oblonga Mordv.
		Pergandeida craccivora (Koch)
		Rhopalomyzus ascalonicus (Doncaster)
		Yezabura inculta (Walk.)
		Yezabura middletoni (Thomas)
Beet pseudo-curly top	Simons and Coe (1958)	*Micrutalis* sp. Membracidae
Beet savoy	Coon, Kotilla, and Stewart (1937)	*Piesma cinereum* (Say)
Beet yellow net	Sylvester (1948)	*Aphis fabae* Scop.
		Dysaulacorthum pseudosolani (Theob.)
		Macrosiphon solanifolii (Ashm.)
		Myzus persicae (Sulz.)
Beet yellow wilt	Bennett and Munck (1946)	*Atanus exitiosus* Beamer
Beet yellows	Watson (1942)	*Acyrthosiphon onobrychis* (B.d.F.)
		Aphidula nasturtii (Kalt.)

TABLE 12.1. Continued

Virus	Virus citation[a]	Vector
		Aphis fabae Scop.
		Aulacorthum pelargonii (Kalt.)
		Dysaulacorthum pseudosolani (Theob.)
		Dysaulacorthum vincae (Walk.)
		Hayhurstia atriplicis (L)
		Macrosiphon solanifolii (Ashm.)
		Metopolophium primulae (Theob.)
		Myzotoxoptera tulipaella (Theob.)
		Myzus persicae (Sulz.)
		Myzus portulacae (Macch.)
		Neomyzus circumflexus (Buckt.)
		Rhopalomyzus ascalonicus (Doncaster)
Beet 41 yellows	Watson (1952)	*Myzus persicae* (Sulz.)
Beet (wild) water mottle	Watson (1958)	*Rhopalomyzus ascalonicus* (Donc.)
Blackberry dwarf	Zeller (1927)	*Pentatrichopus fragaefolii* (Cock.)
Black currant reversion	Lees (1920)	*Phytoptus ribis* (Nalepa)
Cabbage black ringspot	Smith (1935)	*Acyrthosiphon onobrychis* (B.d.F.)
		Aphis fabae Scop.
		Brevicoryne brassicae (L)
		Dysaphis radicola (Mordv.)
		Dysaulacorthum pseudosolani (Theob.)
		Dysaulacorthum vincae (Walk.)
		Macrosiphon rosae (L)
		Macrosiphon solanifolii (Ashm.)
		Megoura viciae (Buckt.)
		Myzus persicae (Sulz.)

Cabbage ring necrosis	Larson and Walker (1941)	*Neomyzus circumflexus* (Buckt.)
		Rhopalomyzus ascalonicus (Doncaster)
		Brevicoryne brassicae (L.)
		Myzus persicae (Sulz.)
Cacao mosaic disease (Trinidad)	Kirkpatrick (1950)	*Dysmicoccus brevipes* (Ckll.)
		Ferrisiana virgata (Ckll.)
		Planococcus citri (Risso)
		Pseudococcus comstocki (Kuw.)
Cacao swollen shoot	Stevens (1937)	*Dysmicoccus brevipes* (Ckll.)
		Farinococcus loranthi Strickland
		Ferrisiana virgata (Ckll.)
		Formicoccus tafoensis Strickland
		Paraputo ritchiei Laing
		Planococcus citri (Risso)
		Pseudococcus bukobensis Laing
		Pseudococcus celtis Strickland
		Pseudococcus concavocerarii James
		Pseudococcus gahani Green
		Pseudococcus hargreavesi Laing
		Pseudococcus longgispinus T.-T.
		Pseudococcus masakensis James
		Pseudococcus njalensis Laing
		Tylococcus westwoodi Strickland
		Planococcus citri (Risso)
Cacao swollen shoot (Ceylon)	Joachim (1954)	*Planococcus lilacinus* (Ckll.)
Canna mosaic	Fukushi (1932)	*Cerosipha gossypii* (Glov.)
		Macrosiphon solanifolii (Ashm.)
		Myzus persicae (Sulz.)
		Neomyzus circumflexus (Buckt.)
		Phopalosiphon maidis (Fitch)

TABLE 12.1. Continued

Virus	Virus citation[a]	Vector
Cantaloupe mosaic	Dickson et al. (1949)	Acyrthosiphon destructor (Johnson)
		Cerosipha gossypii (Glov.)
		Macrosiphon solanifolii (Ashm.)
		Myzus persicae (Sulz.)
		Pergandeida craccivora (Koch)
		Rhopalosiphon maidis (Fitch)
Carnation mosaic	Creager (1943)	Myzus persicae (Sulz.)
Carnation yellows streak	Brierley and Smith (1957)	Myzus persicae (Sulz.)
Carnation vein mottle	Kassanis (1955)	Myzus persicae (Sulz.)
Carrot motley dwarf	Stubbs (1948)	Cavarella aegopodii (Scop.)
		Semiaphis heraclei (Takah.)
Cassava brown streak	Storey (1936)	Bemisia vayssierei Frappa
Cassava mosaic	Lefevre (1935)	Bemisia nigeriensis Corb.
		Bemisia tabaci (Gen.)
		Bemisia vayssierei Frappa
Cauliflower mosaic	Tompkins (1934)	Acyrthosiphon destructor (Johnson)
		Acyrthosiphon onobrychis (B.d.F.)
		Aphis apii Theob.
		Aphis fabae Scop.
		Brevicoryne brassicae (L)
		Cavariella archangelicae (Scop.)
		Cerosipha apigraveolens (Essig)
		Cerosipha gossypii (Glov.)
		Cryptomyzus ribis (L)
		Dysaphis radicola (Mordv.)
		Dysaulacorthum pseudosolani (Theob.)

Celery calico

Hyadaphis conii (Davids.)
Hyadaphis mellifera (Hottes)
Hyperomyzus lactucae (L)
Lipaphis pseudobrassicae (Davis)
Macrosiphon rosae (L.)
Macrosiphon solanifolii (Ashm.)
Megoura viciae (Buckt.)
Myzotoxoptera tulipaella (Theob.)
Myzus persicae (Sulz.)
Myzus portulacae (Macch.)
Nasonovia ribisnigri (Mosley)
Neomyzus circumflexus (Buckt.)
Rhopalomyzus ascalonicus (Doncaster)
Sappaphis mali (Ferr.)
Yezabura inculta (Walk.)
Yezabura middletoni (Thomas)
Aphis apii Theob.

Severin and Freitag (1935)

Celery mosaic

Cerosipha apigraveolens (Essig)
Cerosipha gossypii (Glov.)
Dysaulacorthum pseudosolani (Theob.)
Hyadaphis mellifera (Hottes)
Micromyzus violae (Perg.)
Myzus persicae (Sulz.)
Neomyzus circumflexus (Buckt.)
Yezabura inculta (Walk.)
Yezabura middletoni (Thomas)
Acyrthosiphon destructor (Johnson)
Aphidula heraclella (Davis)
Aphis apii Theob.
Aphis fabae Scop.

Severin and Freitag (1935)

TABLE 12.1. Continued

Virus	Virus citation[a]	Vector
		Brevicoryne brassicae (L)
		Cavariella archangelicae (Scop.)
		Cerosipha apigraveolens (Essig)
		Cerosipha gossypii (Glov.)
		Dysaulacorthum pseudosolani (Theob.)
		Hyadaphis conii (Davids.)
		Hyadaphis mellifera (Hottes)
		Hyatopterus pruni (Geoffr.)
		Lipaphis pseudobrassicae (Davis)
		Macrosiphon rosae (L)
		Macrosiphon solanifolii (Ashm.)
		Myzus cerasi (F.)
		Myzus persicae (Sulz.)
		Neomyzus circumflexus (Buckt.)
		Yezabura inculta (Walk.)
		Yezabura middletoni (Thomas)
Celery yellow spot	Freitag and Severin (1945)	*Hyadaphis conii* (Davids.)
		Hyadaphis mellifera (Hottes)
Centrosema mosaic	Van Velsen and Crowley (1961)	*Brachycaudus helichrysi* (Kalt.)
		Cerosipha gossypii (Glov.)
		Nysius spp. (brown)
		Nysius spp. (green)
Cherry little cherry	Foster and Lott (1947)	*Macrosteles fascifrons* (Stål)
		Psammotettix lividellus (Zett.)
		Scaphytopius acutus (Say)
Cherry ringspot	Christoff (1958)	*Myzus pruniavium* C.B.

Chilli leaf curl	Joachim (1954)	*Bemisia inconspicua* (Q)
Chilli (pepper) mosaic	Ferguson (1951)	*Cerosipha gossypii* (Glov.)
Chilli (pepper) vein-banding	Dale (1954)	*Cerosipha gossypii* (Glov.)
		Macrosiphon solanifolii (Ashm.)
		Myzus persicae (Sulz.)
Chrysanthemum Noordam's B	Hollings (1957)	*Brachycaudus helichrysi* (Kalt.)
		Coloradoa rufomaculata (Wilson)
		Dysaulacorthum pseudosolani (Theob.)
		Dysaulacorthum vincae (Walk.)
		Macrosiphon solanifolii (Ashm.)
		Myzus persicae (Sulz.)
		Pyrethromyzus sanborni (Gill.)
Chrysanthemum stunt	Brierley and Smith (1949)	*Brachycaudus helichrysi* (Kalt.)
		Coloradoa rufomaculata (Wilson)
		Myzus persicae (Sulz.)
		Neomyzus circumflexus (Buckt.)
		Pyrethromyzus sanborni (Gill.)
Cineraria mosaic	Jones (1942)	*Brachycaudus marutae* (Ostl.)
Citrus dwarf disease	Yamada and Sawamura (1952)	*Geisha distinctissima* (Walk.)
Citrus infectious mottling	Fawcett (1936)	*Toxoptera aurantii* (B.d.F.)
Citrus tristeza	Meneghini (1946)	*Aphidula spiraecola* (Patch)
		Cerosipha gossypii (Glov.)
		Toxoptera aurantii (B.d.F.)
		Toxoptera citricida (Kirk.)
Citrus vein enation	Wallace and Drake (1953)	*Myzus persicae* (Sulz.)
		Toxoptera citricida (Kirk.)
Clover big vein	Black (1944)	*Agallia constricta* Van Duzee
		Agallia quadripunctata Provancher
		Agalliopsis novella (Say)
Clover club leaf	Black (1944)	*Agalliopsis novella* (Say)

497

TABLE 12.1. *Continued*

Virus	Virus citation [a]	Vector
Clover (Alsike) mosaic	Zanumeyer and Wade (1935)	*Acyrthosiphon destructor* (Johnson)
Clover (Subterranean) mosaic	Aitken and Grieve (1943)	*Macrosiphon solanifolii* (Ashm.)
		Myzus persicae (Sulz.)
Clover phyllody	Evenhuis (1958)	*Aphrodes albifrons* (L)
		Euscelis bilobatus Wagner
		Euscelis galiberti Ribaut
		Euscelis lineolatus Brullé
		Macrosteles cristatus Rib.
		Macrosteles viridigriseus (Edwards)
Clover chlorotic stunt	de Fluiter, Evenhuis, and van der Meer (1955)	*Euscelis plebejus plebejus* (Fall.)
Clover stunt (Australia)	Grylls and Butler (1956)	*Pergandeida craccivora* (Koch)
Clover (red) vein mosaic	Osborn (1937)	*Acyrthosiphon destructor* (Johnson)
		Acyrthosiphon onobrychis (B.d.F.)
		Myzus persicae (Sulz.)
Clover (white) (Japan)	Koshimizu and Iizuka (1957)	*Myzus persicae* (Sulz.)
Clover witch's broom	Frazier and Posnette (1957)	*Euscelis bilobatus* Wagner
		Euscelis lineolatus Brullé
Cocksfoot streak	Smith (1957)	*Macrosiphon solanifolii* (Ashm.)
		Myzus persicae (Sulz.)
Coffee arabica blister spot	Wellman (1957)	*Toxoptera aurantii* (B.d.F.)
Cotton anthocyanosis	Costa (1956)	*Cerosipha gossypii* (Glov.)
		Myzus persicae (Sulz.)
Cotton leaf crumple	Dickson *et al.* (1954)	*Bemisia tabaci* (Gen.)
		Trialeurodes abutilonea Hald.
Cotton leaf curl	Kirkpatrick (1931)	*Bemisia goldingi* Corb.

Cowpea mosaic	McLean (1941)	*Bemisia tabaci* (Gen.)
		Acyrthosiphon destructor (Johnson)
		Aphis fabae Scop.
		Ceratoma trifurcata Forst.
		Cerosipha gossypii (Glov.)
		Macrosiphon solanifolii (Ashm.)
		Myzus persicae (Sulz.)
Cowpea mosaic	Dale (1949)	*Andrector ruficornis* (Ol.)
Crotalaria mosaic	Johnson and Lefebvre (1938)	*Cerosipha gossypii* (Glov.)
		Myzus persicae (Sulz.)
Cucumber mosaic	Doolittle (1916)	*Acyrthosiphon destructor* (Johnson)
		Acyrthosiphon onobrychis (B.d.F.)
		Anuraphis subterranea (Walk.)
		Aphidula mordvilkiana (Dobz.)
		Aphidula nasturtii (Kalt.)
		Aphidula urticata (F.)
		Aphis (*Doralis*) *cirsii-acanthoides* Scop.
		Aphis (*Doralis*) *fabae* Scop.
		Brachycaudus helichrysi (Kalt.)
		Brachycaudus lychnidis (L)
		Brevicoryne brassicae (L)
		Cavariella aegopodii (Scop.)
		Cavariella pastinacae (L)
		Cerosipha affinis (Del Gu)
		Cerosipha confusa (Walk.)
		Cerosipha epilobiina (Walk.)
		Cerosipha gossypii (Glov.)
		Cerosipha serpylli (Koch)
		Cerosipha verbasci (Schrk.)
		Chaitophorus betulinus v.d.G.

TABLE 12.1. Continued

Virus	Virus citation [a]	Vector
		Coloradoa tanacetina (Walk.)
		Cryptomyzus korschelti C.B.
		Dactynotus cichoricola (H.R.L.)
		Dactynotus erigeronensis (Thomas)
		Dactynotus henrichi C.B.
		Dactynotus jaceae (L)
		Dactynotus obscurus (Koch)
		Dactynotus tanaceti (L)
		Delphiniobium junakkianum (Karsch)
		Dysaulacorthum pseudosolani (Theob.)
		Dysaulacorthum vincae (Walk.)
		Hyalopterus pruni (Geoffr.)
		Hyperomyzus lactucae (L)
		Lipaphis erysimi (Kalt.)
		Macrosiphon solanifolii (Ashm.)
		Metopeurum fuscoviride Stroyan
		Metopolophium occidentale H.R.L.
		Mysella galeopsidis (Kalt.)
		Myzotoxoptera tulipaella (Theob.)
		Myzus auctus (Wlk.)
		Myzus ligustri (Mosley)
		Myzus persicae (Sulz.)
		Myzus portulacae (Macch.)
		Nasonovia ribisnigri (Mosley)
		Nectarosiphon idaei C.B.
		Neomyzus circumflexus (Buckt.)

Disease	Reference	Vector
		Pergandeida craccae (L)
		Rhopalomyzus ascalonicus (Donc.)
		Rhopalosiphon nymphaeae (L)
		Rhopalosiphoninus latysiphon (Davids.)
		Rungsia maydis Pass.
Dahlia mosaic	Brierley (1933)	*Toxopterina plantiginis* (Goeze)
		Aphis (Doralis) fabae Scop.
		Brachycaudus helichrysi (Kalt.)
		Cerosipha gossypii (Glov.)
		Dysaulacorthum pseudosolani (Theob.)
		Macrosiphon solanifolii (Ashm.)
		Myzus persicae (Sulz.)
		Myzus portulacae (Macch.)
		Neomyzus circumflexus (Buckt.)
		Pergandeida cytisorum (Htg.)
		Rhopalomyzus ascalonicus (Donc.)
Dandelion yellow mosaic	Kassanis (1944)	*Dysaulacorthum pseudosolani* (Theob.)
		Dysaulacorthum vincae (Walk.)
		Myzus portulacae (Macch.)
		Nasonovia ribisnigri (Mosley)
		Rhopalomyzus ascalonicus (Donc.)
Datura distortion mosaic	Capoor and Varma (1952)	*Myzus persicae* (Sulz.)
Datura rugose leaf curl	Grylls (1954)	*Austroagallia torrida* Evans
Dolichos lablab yellow mosaic	Capoor and Varma (1950)	*Bemisia tabaci* (Gen.)
Dock mosaic (New Zealand)	Chamberlain and Matthews (1949)	*Macrosiphon solanifolii* (Ashm.)
		Myzus persicae (Sulz.)
Elm phloem necrosis	Swingle (1938)	*Scaphoideus luteolus* Van Duzee
Euphorbia mosaic	Costa and Bennett (1950)	*Bemisia tabaci* (Gen.)
Fig mosaic	Condit and Horne (1933)	*Aceria ficus* Cotte
Filaree red leaf	Frazier (1951)	*Aulacorthum geranii* (Kalt.)

TABLE 12.1. Continued

Virus	Virus citation[a]	Vector
		Aulacorthum pelargonii (Kalt.)
		Aphis fabae Scop.
		Dysaulacorthum pseudosolani (Theob.)
		Macrosiphon solanifolii (Ashm.)
		Rhopalosiphon prunifoliae (Fitch)
Flax crinkle	Frederiksen and Goth (1959)	*Macrosteles fascifrons* (Stål)
Freesia mosaic	Woodward, in Smith (1957a)	*Dysaulacorthum vincae* (Walk.)
		Macrosiphon solanifolii (Ashm.)
Gooseberry vein-banding	Posnette (1952)	*Aphidula grossulariae* (Kalt.)
		Aphidula schneideri (C.B.)
		Nasonovia ribisnigri (Mosley)
Gramineae mosaic	Celino and Martinez (1956)	*Cerosipha gossypii* (Glov.)
		Rhopalosiphon maidis (Fitch)
Grape fanleaf (strain of *Arabis* mosaic)	Hewitt (1950)	*Xiphinema index* (Thorne and Allen)
Groundnut mosaic	Thung (1947)	*Nesophrosyne argentatus* (Berg.)
Groundnut rosette	Storey and Bottomley (1928)	*Pergandeida craccivora* (Koch)
		Pergandeida robiniae Macch.
Groundnut witch's broom	Thung (1947)	*Nesophrosyne argentatus* (Ev.)
Henbane mosaic	Hamilton (1932)	*Macrosiphon solanifolii* (Ashm.)
		Myzus persicae (Sulz.)
		Neomyzus circumflexus (Buckt.)
		Rhopalomyzus ascalonicus (Donc.)
Hibiscus rosa-sinensis leaf curl	Vasudeva, Varma, and Capoor (1955)	*Bemisia tabaci* (Gen.)
Hibiscus yellow vein mosaic	Capoor and Varma (1950)	*Bemisia tabaci* (Gen.)
Holodiscus witch's broom	Zeller (1931)	*Brachycaudus spiraeae* (Ostl.)

Disease	Reference	Vectors
Hop line pattern mosaic	Paine (1953)	*Macrosiphon solanifolii* (Ashm.) *Phorodon humuli* (Schrk.)
Hop mosaic	Salmon (1923)	*Macrosiphon solanifolii* (Ashm.) *Phorodon humuli* (Schrk.)
Hop split leaf blotch	Keyworth (1951)	*Macrosiphon solanifolii* (Ashm.) *Phorodon humuli* (Schrk.)
Iris beardless mosaic	Travis (1957)	*Aphis fabae* Scop. *Macrosiphon solanifolii* (Ashm.) *Myzus persicae* (Sulz.)
Iris mosaic	Brierley and McWhorter (1936)	*Aphis fabae* Scop. *Macrosiphon solanifolii* (Ashm.) *Myzus persicae* (Sulz.)
Laburnum infectious variegation	Brierley (1944)	*Pergandeida cytisorum* (Htg.)
Lettuce mosaic	Jagger (1921)	*Aulacorthum barri* (Essig) *Aulacorthum pelargonii* (Kalt.) *Aulacorthum scariolae* (Nevs.) *Cerosipha gossypii* (Glov.) *Dactynotus sonchi* (L) *Dysaulacorthum pseudosolani* (Mosley) *Macrosiphon solanifolii* (Ashm.) *Myzus persicae* (Sulz.) *Nasonovia ribisnigri* (Mosley)
Lily coarse mottle	Brierley and Smith (1944)	*Cerosipha gossypii* (Glov.) *Macrosiphon solanifolii* (Ashm.) *Myzus persicae* (Sulz.)
Lily latent mosaic	Brierley and Smith (1944)	*Macrosiphon solanifolii* (Ashm.) *Myzus persicae* (Sulz.) *Myzus persicae* (Sulz.)
Lily ringspot	Smith (1950)	*Cerosipha gossypii* (Glov.)
Lily rosette	Ogilvie (1928)	*Sitobium* (?) *lilii* (Monell)

TABLE 12.1. *Continued*

Virus	Virus citation[a]	Vector
Lily symptomless	Brierley and Smith (1944)	*Cerosipha gossypii* (Glov.)
Lucerne dwarf	Weimer (1931, 1936)	*Aphrophora angulata* Ball
		Aphrophora permutata Uhler
		Carneocephala flaviceps (Riley)
		Carneocephala fulgida Nott.
		Carneocephala trigutata Nott.
		Clastoptera brunnea Ball
		Cuerna occidentalis Oman and Beamer
		Cuerna yuccae Oman and Beamer
		Draeculacephala antica (Walk.)
		Draeculacephala californica Davids and Frazier
		Draeculacephala crassicornis Van Duzee
		Draeculacephala inscripta Van Duzee
		Draeculacephala minerva Ball
		Draeculacephala noveboracensis (Fitch)
		Draeculacephala portola Ball
		Friscanus friscanus Ball
		Graphocephala cythura (Baker)
		Helochara delta Oman
		Homalodisca insolita (Wlk.)
		Homalodisca liturata Ball
		Homalodisca triquétra (F.)
		Hordnia circellata (Baker)
		Keonolla confluens (Uhl.)
		Keonolla dolobrata (Ball)
		Neokolla hieroglyphica (Say)

Host / Disease	Reference	Vector species
		Neokolla severini DeLong
		Oncometopia undata (F.)
		Paragonia confusa Oman
		Paragonia furcata Oman
		Paragonia 13-punctata (Ball)
		Paragonia triunata (Ball)
		Philaenus spumarius L var. *fabricii* Van Duzee
		Philaenus spumarius L. var. *impressus* DeLong and Severin
		Philaenus spumarius L var. *marginellus* F.
		Philaenus spumarius L var. *pallidus* Zetterstedt
		Philaenus spumarius spumarius L
Lucerne mosaic	Weimer (1931)	*Acyrthosiphon destructor* (Johnson)
		Acyrthosiphon onobrychis (B.d.F.)
		Aphis fabae Scop.
		Myzus persicae (Sulz.)
Lucerne witch's broom	Edwards (1936)	*Hyalesthes obsoletus* Sign.
Lycium		*Nesophrosyme argentatus* (Berg.)
		Scaphytopius acutus (Say)
		Myzus persicae (Sulz.)
Maize leaf fleck	Heinze (1957)	*Myzus persicae* (Sulz.)
	Stoner (1952)	*Rhopalosiphon maidis* (Fitch)
		Rhopalosiphon prunifoliae (Fitch)
		Cerosipha gossypii (Glov.)
Maize mosaic	Kunkel (1923)	*Myzus persicae* (Sulz.)
		Peregrinus maidis Ashm.
		Rhopalosiphon maidis (Fitch)
Maize rough dwarfing	Klinowski and Kreutsberg (1958)	*Byrsocrypta ulmi* (L)
Maize streak	Storey (1925)	*Cicadulina bimaculata* (Evans)
		Cicadulina mbila (Naude)

TABLE 12.1. *Continued*

Virus	Virus citation[a]	Vector
Maize stunt	Kunkel (1946)	*Cicadulina Storeyi* China
		Cicadulina zeae China
		Dalbulus elimatus (Ball)
		Dalbulus maidis (De L. and W.)
Maize Wallaby ear	Schindler (1942)	*Cicadulina bimaculata* (Evans)
Malva	Ryschkow (1946)	*Myzus persicae* (Sulz.)
Mulberry mosaic	Ho and Li (1936)	*Eutettix disciguttus* Walk.
Muskmelon mosaic	Rader *et al.* (1947)	*Diabrotica undecimpunctata* Mann.
Narcissus mosaic	McWhorter and Weiss (1932)	*Acyrthosiphon destructor* (Johnson)
		Aphis (Doralis) fabae Scop.
		Aphis (Doralis) rumicis L
		Dysaulacorthum pseudosolani (Theob.)
		Macrosiphon rosae (L)
		Macrosiphon solanifolii (Ashm.)
		Myzus cerasi (F.)
		Nearctaphis crataegifoliae (Fitch)
Narcissus yellow stripe	van Slogteren (1955)	*Aphidula ageratoidis* (Ostl.)
		Aphis (Doralis) fabae Scop.
Oat pseudo-rosette	Grebennikov (1941)	*Calligypona marginata* (F.)
Oat sterile dwarfing	Heikinheimo (1957)	*Calligypona pellucida* (F.)
Oat yellow leaf	Farrar (1958)	*Rhopalosiphon prunifoliae* (Fitch)
Onion yellow dwarf	Melhus *et al.* (1929)	*Acyrthosiphon destructor* (Johnson)
		Acyrthosiphon onobrychis (B.d.F.)
		Acyrthosiphon porosum (Sanderson)
		Aphidula ageratoidis (Ostl.)
		Aphidula pomi (De Geer)

Aphis (Doralis) fabae Scop.
Aphis sombucifoliae Fitch
Aulacorthum purpurascens (Oestlund)
Brachycaudus cardui (L)
Brevicoryne brassicae (L)
Calaphis betulella (Walsh)
Capitophorus flaveolus (Walk.)
Cerosipha (?) *decepta* (Hottess and Frison)
Cerosipha forbesi (Weed)
Cerosipha gossypii (Glov.)
Cerosipha oenotherae (Ostl.)
Ceruraphis viburnicola (Gillette)
Comaphis helianthi (Monell)
Cryptomyzus ribis (L)
Dactynotus ambrosiae (Thomas)
Dactynotus gravicornis (Patch)
Dactynotus (?) *oder* (Patch)
Dactynotus taraxaci (Kalt.)
Drepanaphis acerifolii (Thomas)
Dysaulacorthum pseudosolani (Theob.)
Dysaulacorthum vincae (Walk.)
Hayhurstia atriplicis (L)
Hoplochaitophorus quercicola (Monell)
Hyalopterus pruni (Geoffr.)
Hyperomyzus (?) *monardae* (Williams)
Hysteroneura setariae (Thomas)
Lipaphis pseudobrassicae (Davis)
Microsiphon artemisiae (Gill.)
Macrosiphon rosae (L)
Macrosiphon solanifolii (Ashm.)

TABLE 12.1. *Continued*

Virus	Virus citation [a]	Vector
		Monellia caryae (Monell)
		Monellia caryella (Fitch)
		Myzocallis alhambra (Davidson)
		Myzocallis asclepiadis (Monell)
		Myzus cerasi (F.)
		Myzus persicae (Sulz.)
		Myzus portulacae (Macch.)
		Nectarosiphon rossi (Hottes and Frison)
		Nectarosiphon rubi (Kalt.)
		Neomyzus circumflexus (Buckt.)
		Neothomasia populicola (Thomas)
		Paczoskia rudbeckiae (Fitch)
		Pergandeida craccivora (Koch)
		Periphyllus negundinis (Thomas)
		Prociphilus fraxinifolii (Riley)
		Pseudomicrella viminalis (Monell)
		Pterocomma salicis (L)
		Rhopalomyzus ascalonicus (Donc.)
		Rhopalosiphon maidis (Fitch)
		Rhopalosiphon prunifoliae (Fitch)
		Rhopalosiphon rhois (Monell)
		Stagona corrugatans (Sirrine)
		Thripsaphis ballii (Gillette)
Orchid (*Cattleya*) mosaic	Jensen (1949)	*Myzus persicae* (Sulz.)
Ornithogalum mosaic	Smith and Brierley (1944)	*Cerosipha gossypii* (Glov.)
		Macrosiphon solanifolii (Ashm.)

Disease	Reference	Insect vectors
Papaw bunchy top	Bird and Adsuar (1952)	*Myzus persicae* (Sulz.) *Neomyzus circumflexus* (Buckt.) *Sitobium (?) lilii* (Monell)
Papaw mosaic	Capoor and Varma (1948)	*Empoasca dilitaria* DeLong and Davids. *Empoasca papayae* Oman *Aphidula umbrella* (C.B.) *Cerosipha gossypii* (Glov.) *Myzus persicae* (Sulz.)
Papaw ringspot	Jensen (1947)	*Aphis fabae* Scop. *Cerosipha gossypii* (Glov.) *Macrosiphon solanifolii* (Ashm.) *Micromyzus formosanus* (Tak.) *Myzus persicae* (Sulz.) *Pergandeida craccivora* (Koch)
Pawpaw mosaic (Puerto Rico)	Adsuar (1946)	*Aphidula spiraecola* (Patch) *Acyrthosiphon destructor* (Johnson)
Pea enation mosaic	Pierce (1935)	*Acyrthosiphon onobrychis* (B.d.F.) *Cerosipha gossypii* (Glov.) *Macrosiphon solanifolii* (Ashm.) *Myzus persicae* (Sulz.) *Myzus portulacae* (Macch.)
Pea leaf roll	Quantz and Volk (1954)	*Myzus persicae* (Sulz.)
Pea leaf rolling	Schultz and Folsom (1923)	*Macrosiphon solanifolii* (Ashm.)
Pea mosaic	Doolittle and Jones (1925)	*Aphis (Doralis) fabae* Scop. *Cavariella aegopodii* (Scop.) *Cerosipha gossypii* (Glov.) *Hayhurstia atriplicis* (L) *Lipaphis pseudobrassicae* (Davis) *Macrosiphon rosae* (L) *Pentatrichopus fragaefolii* (Cock.)

TABLE 12.1. *Continued*

Virus	Virus citation[a]	Vector
Pea mottle	F. Johnson (1942)	*Pergandeida craccivora* (Koch)
		Toxoptera citricida (Kirk.)
		Acyrthosiphon destructor (Johnson)
		Acyrthosiphon onobrychis (B.d.F.)
		Aphis fabae Scop.
		Myzus persicae (Sulz.)
Pea streak	Zaumeyer (1938),	*Acyrthosiphon destructor* (Johnson)
Pea wilt	F. Johnson (1942)	*Acyrthosiphon destructor* (Johnson)
		Acyrthosiphon onobrychis (B.d.F.)
		Aphidula nasturtii (Kalt.)
		Aphis fabae Scop.
		Myzus persicae (Sulz.)
Peach little peach	E. F. Smith (1891)	*Macropsis trimaculata* Fitch
		Philaenus spumarius L
Peach mosaic	Hutchins (1932)	*Eriophyes insidiosus* Keifer and Wilson
		Myzus persicae (Sulz.)
		Rhopalosiphon padi (L)
Peach ringspot	Cochran and Hutchins (1941)	*Myzus pruniavium* C.B.
Peach Eastern X disease	Stoddard (1938)	*Colladonus clitellarius* (Say)
		Scaphytopius acutus (Say)
Peach Western X disease	Reeves and Hutchins (1941)	*Colladonus geminatus* (Van Duzee)
		Colladonus montanus (Van Duzee)
		Fieberiella flori (Stål)
		Gyponana striata Burm.
		Keonolla confluens Uhl.
		Osbornellus borealis De L. and M.

Disease	Reference	Vector
Peach yellow bud mosaic strain	Samson and Imle (1942)	*Scaphytopius acutus* (Say)
Peach yellows	E. F. Smith (1888), Kunkel (1933)	*Xiphinema americanum* (Cobb)
		Macropsis trimaculatá (Fitch)
		Philaenus spumarius L
Phony peach	Neal (1921)	*Cuerna costalis* (F.)
		Graphocephala versuta (Say)
		Homalodisca coagulata (Say)
		Homalodisca insolita (Wlk.)
		Homalodisca triquétra (F.)
		Oncometopia orbona (F.)
Pelargonium leaf curl	Pape (1927)	*Bemisia vayssierei* Frappa
Phormium yellow leaf	Boyce *et al.* (1951)	*Oliarius atkinsoni* Myers
Physalis floridana yellow net	Webb (1955)	*Myzus persicae* (Sulz.)
Plum leaf margin scorch	Turica (1956)	*Myzus persicae* (Sulz.)
Plum narrow-striped variegation	Christoff (1958)	*Brachycaudus helichrysi* (Kalt.)
Plum (myrobalan) panash mosaic	Blattny (1958)	*Brachycaudus helichrysi* (Kalt.)
		Myzus persicae (Sulz.)
		Phorodon humuli (Schrk.)
Plum pox	Atanasoff (1932)	*Brachycaudus helichrysi* (Kalt.)
		Phorodon humuli (Schrk.)
Poison hemlock ringspot	Freitag and Severin (1945)	*Aphis fabae* Scop.
		Cavariella aegopodii (Scop.)
		Cerosipha gossypii (Glov.)
		Dysaulacorthum pseudosolani (Theob.)
		Hyadaphis conii (Davids.)
		Lipaphis pseudobrassicae (Davis)
		Myzus persicae (Sulz.)
		Neomyzus circumflexus (Buckt.)
		Yezabura incluta (Walk.)
		Yezabura middletoni (Thomas)

511

TABLE 12.1. *Continued*

Virus	Virus citation [a]	Vector
Poppy mosaic	Heinze (1952)	*Aphis fabae* Scop.
		Myzus persicae (Sulz.)
Potato aucuba mosaic	Quanjer (1931)	*Aphidula nasturtii* (Kalt.)
		Dysaulacorthum pseudosolani (Theob.)
		Dysaulacorthum vincae (Walk.)
		Myzus persicae (Sulz.)
		Neomyzus circumflexus (Buckt.)
Potato leaf roll	Appel (1907)	*Aphidula nasturtii* (Kalt.)
		Dysaulacorthum pseudosolani (Theob.)
		Myzotoxoptera tulipaella (Theob.)
		Myzus persicae (Sulz.)
		Myzus portulacae (Macch.)
		Neomyzus circumflexus (Buckt.)
		Rhopalomyzus ascalonicus (Donc.)
		Rhopalosiphoninus latysiphon (Davids.)
Potato leaf-rolling mosaic	Schultz and Folsom (1923)	*Dysaulacorthum pseudosolani* (Theob.)
		Macrosiphon solanifolii (Ashm.)
		Myzus persicae (Sulz.)
Potato paracrinkle	Salaman and LePelley (1930)	*Myzus persicae* (Sulz.)
Potato spindle tuber	Folsom (1923)	*Disonycha triangularis* Say
		Epithrix cucumeris Harr.
		Exitianus exitiosus (Uhl.)
		Leptinotarsa decemlineata (Say)
		Macrosiphon solanifolii (Ashm.)
		Melanoplus angustipennis Dodge
		Melanoplus bivittatus Say

Potato stem mottle	Rozendaal (1947)	*Melanoplus plumbeus* Dodge
		Myzus persicae (Sulz.)
		Systena elongata (F.)
		Systena taeniata Say
Potato virus A	Murphy and McKay (1932)	*Tettigonia viridissima* (L)
		Aphidula nasturtii (Kalt.)
		Myzus persicae (Sulz.)
		Neomyzus circumflexus (Buckt.)
Potato virus X	K. M. Smith (1931)	*Leptinotarsa decemlineata* (Say)
		Melanoplus differentialis (Thos.)
		Tettigonia viridissima (L)
Potato virus Y	K. M. Smith (1931)	*Aphidula nasturtii* (Kalt.)
		Aphis fabae Scop.
		Carvariella pastinacae (L)
		Cerosipha gossypii (Glov.)
		Dysaulacorthum pseudosolani (Theob.)
		Macrosiphon solanifolii (Ashm.)
		Myzus auctus (Wlk.)
		Myzus persicae (Sulz.)
		Myzus portulacae (Macch.)
		Neomyzus circumflexus (Buckt.)
		Pyrethromyzus sanborni (Gill.)
		Tettigonia viridissima (L)
Potato witch's broom	Hungerford and Dana (1924)	*Hyalesthes obsoletus* Sign.
		Ophiola flavopicta (Ishihara)
		Peragallia sinuata (Mulsant and Rey)
Potato yellow dwarf	Barrus and Chupp (1922)	*Aceratagallia curvata* Oman
		Aceratagallia longula (Van Duzee)
		Aceratagallia obscura Oman
		Aceratagallia sanguinolenta (Provancher)

TABLE 12.1. *Continued*

Virus	Virus citation [a]	Vector
Primula mosaic	Tompkins and Middleton (1941)	*Agallia constricta* Van Duzee
		Agallia quadripunctata Provancher
		Agalliopsis novella (Say)
		Acyrthosiphon destructor (Johnson)
		Aphis apii Theob.
		Aphis (*Doralis*) *fabae* Scop.
		Brevicoryne brassicae (L)
		Cerosipha gossypii (Glov.)
		Dysaulacorthum pseudosolani (Theob.)
		Macrosiphon solanifolii (Ashm.)
		Myzus portulacae (Macch.)
		Neomyzus circumflexus (Buckt.)
		Yezabura inculta (Walk.)
Radish mosaic	Tompkins (1939)	*Acyrthosiphon destructor* (Johnson)
		Aphis apii Theob.
		Aphis fabae Scop.
		Brevicoryne brassicae (L)
		Cavariella aegopodii (Scop.)
		Cerosipha gossypii (Glov.)
		Dysaulacorthum pseudosolani (Theob.)
		Lipaphis pseudobrassicae (Davis)
		Myzus persicae (Sulz.)
		Myzus portulacae (Macch.)
		Neomyzus circumflexus (Buckt.)
		Yezabura inculta (Walk.)
Rape savoy	Kaufmann (1936)	*Brevicoryne brassicae* (L)

Disease	Reference	Vector(s)
Raspberry alpha leaf curl	Bennett (1930)	*Dysaulacorthum pseudosolani* (Theob.), *Dysaulacorthum vincae* (Walk.), *Myzus persicae* (Sulz.), *Neomyzus circumflexus* (Buckt.)
Raspberry curly dwarf	Prentice and Harris (1950)	*Aphidula idaei* (v. d. Goot), *Aphidula rubicola* (Oestlund), *Aphidula rubiphila* (Patch), *Nectarosiphon idaei* C.B., *Nectarosiphon rubicola* (Oestl.)
Raspberry leaf spot	Cadman (1952)	*Aphidula idaei* (v. d. Goot), *Nectarosiphon idaei* C.B.
Raspberry (black) necrosis	Stace-Smith (1955)	*Nectarosiphon idaei* C.B., *Nectarosiphon rubi* (Kalt.)
Raspberry (moderate) vein chlorosis	Cadman (1952)	*Nectarosiphon idaei* C.B., *Nectarosiphon rubi* (Kalt.)
Raspberry vein-banding	Cadman (1952)	*Aphidula idaei* (v. d. Goot), *Nectarosiphon idaei* C.B.
Raspberry yellow blotch curl	Chamberlain (1938)	*Nectarosiphon idaei* C.B.
Raspberry yellow mosaic	Bennett (1927)	*Aphidula idaei* (v. d. Goot), *Nectarosiphon idaei* C.B., *Nectarosiphon rubicola* (Oestl.)
Rhubarb mosaic	Chamberlain and Matthews (1954)	*Macrosiphon solanifolii* (Ashm.), *Myzus persicae* (Sulz.)
Rice dwarf	Fukushi (1931)	*Inazuma dorsalis* (Motsch.), *Nephotettix cincticeps* (Uhl.)
Rice stripe	Kuribayashi (1931)	*Calligypona marginata* (F.)
Rice white leaf (hoja blanca)	Lasaga (1957, Acuna (1958)	*Sogata cubana* Crawf., *Sogata oryzicola* Muir
Rose wilt	Grieve (1931)	*Macrosiphon rosae* (L)
Rubus stunt	Prentice (1950)	*Macropsis fuscula* (Zett.)

TABLE 12.1. *Continued*

Virus	Virus citation [a]	Vector
Rubus yellow net	Stace-Smith (1955)	*Nectarosiphon idaei* C.B.
		Nectarosiphon rubi (Kalt.)
Rye grass mosaic	Mulligan (1958)	*Abaracus hystrix* (Nalepa)
Salvia virus I	Roland (1950)	*Myzus persicae* (Sulz.)
Sandal spike	Venkata, Rao, and Iyengar (1934)	*Coelidia (Jassus) indica* (Walk.)
Southern stolbur	Musil and Valenta (1958)	*Aphrodes bicinctus* (Schrk.)
		Macrosteles laevis Rib.
Sowbane mosaic	Costa, de Silva, and Duffus (1958)	*Lyriomyza langei* Frick
Soybean mosaic	Gardner and Kendrick (1921)	*Acyrthosiphon destructor* (Johnson)
		Acyrthosiphon onobrychis (B.d.F.)
		Aphidula nasturtii (Kalt.)
		Aphis fabae Scop.
		Cerosipha gossypii (Glov.)
		Dysaulacorthum pseudosolani (Theob.)
		Dysaulacorthum vincae (Walk.)
		Macrosiphon solanifolii (Ashm.)
		Myzus persicae (Sulz.)
		Myzus portulacae (Macch.)
		Neomyzus circumflexus (Buckt.)
Spinach yellow dwarf	Severn and Little (1947)	*Myzus persicae* (Sulz.)
Squash mosaic	Middleton (1949)	*Acalymma duodecimpunctata* (Oliv.)
		Acalymma melanocephala (F.)
		Acalymma trivittata (Mann.)
		Diabrotica balteata Lec.
		Diabrotica undecimpunctata (Mann.)
Stock mosaic	Tompkins (1939)	*Brevicoryne brassicae* (L)

Disease	Reference	Vector
Strawberry crinkle	Zeller and Vaughan (1932)	*Capitophorus elaeagni* (Del Gu.)
		Lipaphis pseudobrassicae (Davis)
		Myzus persicae (Sulz.)
		Pergandeida craccivora (Koch)
		Acythosiphon porosum (Sanderson)
		Aulacorthum pelargonii (Kalt.)
		Aulacorthum rogersii (Theob.)
		Dysaulacorthum vincae (Walk.)
		Myzaphis rosarum (Kalt.)
		Myzus portulacae (Macch.)
		Nectarosiphon idaei C.B.
		Nectarosiphon rubi (Kalt.)
		Pentatrichopus fragaefolii (Cock.)
		Pentatrichopus (?) minor (Forbes)
		Pentatrichopus tetrarhodus (Walk.)
		Rhopalomyzus ascalonicus (Doncaster)
Strawberry green petal	Posnette (1953)	*Euscelis bilobatus* Wagner
		Euscelis galiberti Ribaut
		Euscelis lineolatus Brullé
		Macrosteles viridigriseus (Edwards)
Strawberry latent	Frazier (1953)	*Pentatrichopus fragaefolii* (Cock.)
Strawberry mild yellow edge	Prentice (1948)	*Pentatrichopus fragaefolii* (Cock.)
Strawberry mottle	Prentice (1952)	*Pentatrichopus fragaefolii* (Cock.)
Strawberry necrosis	Schoniger (1958)	*Pentatrichopus fragaefolii* (Cock.)
Strawberry stunt	Zeller and Weaver (1941)	*Pentatrichopus fragaefolii* (Cock.)
Strawberry vein-banding	Frazier (1955)	*Aulacorthum pelargonii* (Kalt.)
		Dysaulacorthum pseudosolani (Theob.)
		Dysaulacorthum vincae (Walk.)
		Myzus portulacae (Macch.)
		Nectarosiphon idaei C.B.

TABLE 12.1. *Continued*

Virus	Virus citation [a]	Vector
Strawberry witch's broom	Zeller (1927)	*Pentatrichopus fragaefolii* (Cock.)
		Pentatrichopus thomasi H.R.L.
Sugarcane chlorotic streak	Abbott and Ingram (1942)	*Pentatrichopus fragaefolii* (Cock.)
		Draeculacephala portola Ball
		Tomaspis liturata Lep. and Serv. var. *ruforivulata* Stål
Sugarcane Fiji disease	Kunkel (1924)	*Perkinsiella saccharicida* Kirk.
		Perkinsiella vastatrix Breddin
		Perkinsiella vitiensis Kirk.
Sugarcane mosaic	Brandes (1920)	*Cerosipha gossypii* (Glov.)
		Hysteroneura cyperi (v.d.G.) Ainslie
		Hysteroneura setariae (Thomas)
		Rhopalosiphon maidis (Fitch)
		Schizaphis graminum (Rond.)
Sugarcane streak	Storey (1925)	*Cicadulina mbila* (Naude)
Sunflower mosaic	Traversi (1949)	*Myzus persicae* (Sulz.)
		Thrips tabaci Lind.
Sweet potato feathery mottle	Doolittle and Harter (1945)	*Aphis apii* Theob.
		Cerosipha gossypii (Glov.)
		Myzus persicae (Sulz.)
Sweet potato internal cork	Nusbaum (1947)	*Cerosipha gossypii* (Glov.)
		Macrosiphon solanifolii (Ashm.)
		Myzus persicae (**Sulz.**)
Sweet potato mild mottle	Sheffield (1957)	*Myzus persicae* (Sulz.)
Sweet potato stunt	Sheffield (1957)	*Bemisia tabaci* (Gen.)
Teasel mosaic	Stoner (1951)	*Macrosiphon rosae* (L.)
		Myzus persicae (Sulz.)

Disease	Reference	Organisms
Thimbleberry ringspot	Stace-Smith (1958)	*Nectarosiphon davidsoni* (Mason) *Nectarosiphon maximum* (Mason) *Nectaropiphon parviflori* (Hill)
Tigridia mosaic	Smith and Brierley (1944)	*Cerosipha gossypii* (Glov.) *Neomyzus circumflexus* (Buckt.) *Sitobium* (?) *lilii* (Monell)
Tobacco browning of midribs	Völk (1957)	*Aphis fabae* Scop. *Aphidula nasturtii* (Kalt.)
Tobacco etch	E. M. Johnson (1930)	*Aphidula nasturtii* (Kalt.) *Aphis fabae* Scop. *Cerosipha gossypii* (Glov.) *Macrosiphon solanifolii* (Ashm.) *Myzus persicae* (Sulz.) *Neomyzus circumflexus* (Buckt.) *Rhopalomyzus ascalonicus* (Donc.)
Tobacco leaf curl	Storey (1932)	*Aleurotrachelus socialis* Bondar *Bemisia tabaci* (Gen.) *Bemisia tuberculata* Bondar *Trialeurodes natalensis* Corb.
Tobacco mosaic	Allard (1914)	*Decticus verrucivorus* (L) *Deroceras agreste* (L) *Deroceras reticulatum* (O.F.Müll.) *Goniodiscus rotundatus* (O.F.Müll.) *Heliothis assulta* (Gn.) *Lehmannia marginata* (O.F.Müll.) *Liriomyza langei* Frick *Melanoplus differentialis* (Thos.) *Oxychilus draparnaldi* (Beck.) *Phytometra gamma* (L) *Protoparce sexta* Johan.

TABLE 12.1. *Continued*

Virus	Virus citation [a]	Vector
		Tettigonia cantans (Fuessly)
		Tettigonia viridissima (L.)
Tobacco rattle virus (Dutch culture)	Quanjer (1943)	*Trichodorus pachydermus* Seinhorst
Tobacco rattle virus (English culture)	Quanjer (1943)	*Trichodorus primitivus* (DeMan) Micol.
Tobacco ringspot	Fromme et al. (1927)	*Melanoplus differentialis* (Thos.)
		Tettigonia viridissima (L)
Tobacco yellow dwarf	Hill (1937)	*Nesophrosyne argentatus* (Ev.)
Tomato aspermy	Blencowe and Caldwell (1949)	*Brachycaudus helichrysi* (Kalt.)
		Coloradoa rufomaculata (Wilson)
		Dysaulacorthum pseudosolani (Theob.)
		Dysaulacorthum vincae (Walk.)
		Macrosiphon solanifolii (Ashm.)
		Myzotoxoptera tulipaella (Theob.)
		Myzus persicae (Sulz.)
		Pyrethoromyzus sanborni (Gill.)
Tomato big bud (Stolbur)	Samuel et al. (1933)	*Agallia sinuata* (Mulsant and Rey)
		Aphrodes bicinctus (Schrk.)
		Hyalesthes mlokosiewiczii (Sign.)
		Hyalesthes obsoletus (Sign.)
		Macrosteles cristatus (Rib.)
		Macrosteles laevis (Rib.)
		Nesophrosyne argantatus (Ev.)
Tomato black ring, English strain (lettuce ringspot)	Smith (1945)	*Longidorus attenuatus* Hooper
Tomato black ring, Scottish strain (beet ringspot)	Smith (1945)	*Longidorus elongatus* (de Man)

Disease	Reference	Vector species
Tomato dwarfing	Heinze (1959)	*Aphidula nasturtii* (Kalt.)
		Aphis fabae Scop.
		Cerosipha gossypii (Glov.)
		Coloradoa rufomaculata (Wilson)
		Dysaulacorthum vincae (Walk.)
		Metopolophium occidentale H.R.L.
		Myzus persicae (Sulz.)
		Myzus portulacae (Macch.)
		Rhopalomyzus ascalonicus (Donc.)
Tomato ringspot	Thomas *et al.* (1944)	*Xiphinema americanum* (Cobb)
Tomato spotted wilt	Samuel *et al.* (1930)	*Frankliniella fusca* (Hinds)
		Frankliniella moultoni Hood
		Frankliniella occidentalis Perg.
		Frankliniella schultzei (Trybom)
		Frankliniella tenuicornis (Uzel)
		Liothrips pistaciae Kreutzberg
		Thrips tabaci Lind.
Tomato yellow net	Sylvester (1954)	*Myzus persicae* (Sulz.)
Tomato yellow top	Costa (1949)	*Macrosiphon solanifolii* (Ashm.)
		Myzus persicae (Sulz.)
Tropaeolum mosaic	Jensen (1950)	*Myzus persicae* (Sulz.)
		Yezabura inculta (Walk.)
Tropaeolum ringspot	K. M. Smith (1957a)	*Aphis fabae* Scop.
		Macrosiphon solanifolii (Ashm.)
		Myzus persicae (Sulz.)
Tulip breaking	McKenny and Hughes (1931)	*Aphis (Doralis) fabae* Scop.
		Cerosipha gossypii (Glov.)
		Dysaulacorthum pseudosolani (Theob.)
		Dysaulacorthum vincae (Walk.)
		Macrosiphon solanifolii (Ashm.)

TABLE 12.1. Continued

Virus	Virus citation[a]	Vector
		Myzus persicae (Sulz.)
		Neomyzus circumflexus (Buckt.)
		Yezabura (B.d.F.)
Turnip crinkle	Broadbent and Blencowe (1955)	Locusta migratoria L
		Macrosiphon solanifolii (Ashm.)
		Myzus persicae (Sulz.)
		Neomyzus circumflexus (Buckt.)
		Phyllotreta aerea All.
		Phyllotreta atra (F.)
		Phyllotreta consobrina Curt.
		Phyllotreta cruciferae Goeze
		Phyllotreta diademata Foud.
		Phyllotreta nemorum (L)
		Phyllotreta nigripes F.
		Phyllotreta undulata Kutsch.
		Phyllotreta vittula Redt.
		Phytomyza rufipes Mg.
		Pieris brassicae L
		Psylliodes chrysocephala (L)
		Psylliodes cuprea Koch
		Rhopalomyzus ascalonicus (Doncaster)
Turnip mosaic	Gardner and Kendrick (1921)	Acyrthosiphon onobrychis (B.d.F.)
		Aphidula nasturtii (Kalt.)
		Aphidula pomi (De Geer)
		Aphis (Doralis) acetosae L
		Aphis fabae Scop.

Brachycaudus cardui (L)
Brevicoryne brassicae (L)
Capitophorus elaeagni (Del Gu.)
Cavariella pastinacae (L)
Cerosipha epilobiina (Wlk.)
Cerosipha gossypii (Glov.)
Cerosipha helianthemi (Ferr.)
Coloradoa rufomaculata (Wilson)
Cryptomyzus ribis (L)
Dactynotus rapunculoidis C.B.
Dysaulacorthum aegopodii C.B.
Dysaulacorthum vincae (Walk.)
Hyperozymus lactucae (L)
Lipaphis erysimi (Kalt.)
Lipaphis fritzmulleri C.B.
Lipaphis pseudobrassicae (Davis)
Macrosiphon daphnidis C.B.
Macrosiphoniella artemisiae (B.d.F.)
Macrosiphoniella millefolii (Deg.)
Macrosiphoniella tanacetaria (Kalt.)
Metopolophium primulae (Theob.)
Myzotoxoptera tulipaella (Theob.)
Myzus certus (Walk.)
Myzus persicae (Sulz.)
Neomyzus circumflexus (Buckt.)
Pergandeida craccivora (Koch)
Pergandeida loti (Kalt.)
Pergandeida medicaginis (Koch)
Pyrethromyzus sanborni (Gill.)
Rhopalomyzus ascalonicus (Donc.)

TABLE 12.1. *Continued*

Virus	Virus citation[a]	Vector
Turnip yellow mosaic	Markham and Smith (1946)	*Rhopalomyzus lonicerae* (Sieb.)
		Rhopalosiphon nymphaeae (L)
		Ceuthorrhynchus assimilis Payk.
		Chortippus bicolor (Charp.)
		Forficula auricularia L
		Leptophyes punctatissima (Bosc.)
		Locusta migratoria L
		Myzus persicae (Sulz.)
		Neomyzus circumflexus (Buckt.)
		Phaedon cochleariae F.
		Phyllotreta aerea All.
		Phyllotreta atra (F.)
		Phyllotreta consobrina Curt.
		Phyllotreta cruciferae Goeze
		Phyllotreta diademata Foud.
		Phyllotreta nemorum (L)
		Phyllotreta nigripes F.
		Phyllotreta undulata Kutsch.
		Phyllotreta vittula Redt.
		Pieris brassicae L.
		Psylliodes chrysocephala (L)
		Psylliodes cuprea Koch
Turnip yellows	Vanderwalle and Roland (1951)	*Myzus persicae* (Sulz.)
Vaccinium false blossom	Shear (1916)	*Cicadella aurata* (L)
		Scleroracus vaccinii (Van Duzee)
Vaccinium stunt	Wilcox (1942)	*Scaphytopius magdalensis* (Prov.)

Vine infectious degeneration	Branas (1948)	*Viteus vitifolii* (Shim.) Fitch *Xiphinema index* Thorne and Allen
Watercress mosaic	Roland (1952)	*Aphis fabae* Scop. *Brevicoryne brassicae* (L) *Myzus persicae* (Sulz.)
Watermelon mosaic	Freitag (1952), Anderson (1954)	*Cerosipha gossypii* (Glov.) *Myzus persicae* (Sulz.)
Wheat spot mosaic	Slykhuis (1956)	*Aceria tulipae* Keifer
Wheat streak mosaic	McKinney (1944, 1953)	*Aceria tulipae* Keifer
Wheat striate mosaic	Slykhuis (1952)	*Endria (Polyamia) inimica* (Say)
Wheat (European) striate mosaic	Slykhuis, Watson, and Mulligan (1957)	*Calligypona pellucida* (F.)
Winter wheat mosaic	Zazhurilo and Sitnikova (1939)	*Psammotettix striatus* (L)
Yellow mosaic of stone fruits	Blattny (1958)	*Brachycaudus helichrysi* (Kalt.) *Phorodon humuli* (Schrk.)

[a] Authors and dates are as shown in the *Review of Applied Mycology, Supplement,* No. 35, issued in August, 1957. Later references follow the same pattern, including those listed in Phytopathological Paper No. 9 Commonwealth Mycological Institute, Kew, Surrey, England, 1968.

TABLE 12.2 New Records of Viruses and Their Vectors

Virus	Vector	Vector reference
	Leafhopper vectors	
Aster yellows (Japan)	*Scleroracus flavipictus* (Ishihara)	Fukushi and Mehani (1964)
Aster yellows (USA)	*Endria inimica* (Say)	Chiyowski (1963)
Aster yellows	*Scaphytopius acutus* (Say)	Chiyowski (1963)
Ballota split leaf	*Philaenus spumaris* (L)	Harpaz (1962)
Beet yellow wilt	*Paratanus exitiosus* Beamer	Bennett et al. (1967)
Cereal striate mosaic	*Nesoclutha obscura* (Evans)	Grylls (1963)
Clover leaf chlorosis	*Macrosteles viridigriseus* (Edw.)	Harpaz (1962)
(median flavescence)	*Anaceratagallia* (M) *levis* (Rib.)	Harpaz (1962)
Clover phyllody [a]	*Euscelidius variegatus* (Kirschbaum)	Gianetti (1969)
Clover phyllody (Europe)	*Euscelis plebeja* (Fallén)	Musil (1961)
Clover phyllody (virescence)	*Euscelis lineolata* (Brullé)	Harpaz (1962)
Clover phyllody (Strawberry)	*Aphrodes bicincta* (Shrank)	Frazier and Posnette (1957)
Clover phyllody	*Paraphlepsius irroratus* (Say)	Chiykowski (1965)
Corn stunt [a]	*Graminella nigrifrons* (Forbes)	Chiykowski (1965)
Corn stunt	*Deltocephalus sonorus* Ball	Grenados et al. (1966)
Gladiolus "germes fin"	*Euscelis plebeja* (Fallén)	Albany (1966)
Grass witch's broom	*Maloxodes farinosus* Fennah	Kulkarni (1969)
Pierce's disease of grape	*Cuerna costalis* (F)	Kaloostian et al. (1962)
(Lucerne dwarf)	*Homalodisca coagulata* (Say)	Kaloostian et al. (1962)
(Lucerne dwarf)	*Oncometopia arbona* (Fabricius)	Kaloostian et al. (1962)
Maize rough dwarf	*Calligypona pellucida* (F)	Harpaz (1962)
Maize stripe	*Calligypone pellucida* (F)	Blattny et al. (1965)
Mulberry dwarf [a]	*Hishimonoides sellatiformis* Ish.	Ishihara (1965)
Peach X Eastern	*Norvellina seminuda* (Say)	Gilmer et al. (1966)

Peach X (Eastern)	*Paraphlepsius irroratus* (Say)	Gilmer *et al.* (1966)
Oat blue dwarf	*Macrosteles fascifrons* (Stal.)	Banteri and Moore (1962)
Oat sterile dwarf [a]	*Calligypona obscurella* (Boh)	Ikaeimo and Raathikäinen (1961)
Pangola grass stunting	*Sogatella furcifera* (Horv.)	Dirven and Van Hoof (1961)
Rice grassy stunt [a]	*Nilaparvata lugens* Stal.	Rivers *et al.* (1966)
Rice red leaf	*Nephotettix impicticeps* Ish.	IRRI Report (1965)
Rice stripe [a]	*Calligypona apicalis* (Mast.)	Hirai (1968)
Rice transitory yellowing	*Nephotettix apicalis* (Motsch.)	Chiu *et al.* (1965)
Rice tungro	*Nephotettix impicticeps* Ish.	Rivers and Ou (1965)
Sowbane mosaic	*Circulifer tenellus* (Baker)	Bennett and Costa (1961)
Soybean rosette	*Nesophrosyne orientalis* Matsumura	Lo (1966)
Sweet potato witch's broom	*Nesophrosyne ryukyuensis* Ish.	Shinkai (1964)
Tomato pseudo-curly top	*Micrutalis malleifera* Fowler	Simons (1962)
Vine golden flavescence	*Scaphoideus littoralis* Ball	Schvester *et al.* (1962)
Wheat striate mosaic	*Calligypona obscurella* (Boh.)	Ikaheimo and Raathikäinen (1961)

Aphid vectors

Artichoke latent	*Myzus persicae* (Sulz.)	Marrau and Mehani (1964)
Abaca mosaic	*Toxoptera citricidus* (Kirk)	Gavarra and Eleja (1966)
Barley yellow dwarf	*Rhopalosiphon padi* (L)	CSIRO Report (1961)
	Macrosiphon avenae miscanthi (F)	CSIRO Report (1961)
	Myzus poae (Gill)	Orlob *et al.* (1961)
	Myzus persicae (Sulz.)	Smith (1963)
	Sipha agropyrella Hrl.	Smith (1963)
Beet yellow mosaic	*Brachycaudus helichrysi* (Kalt)	Sole and Swenson (1964)
	Caveriella aegopodii (Scop.)	Sole and Swenson (1964)
	Thericaphis rishmi (Born)	Sole and Swenson (1964)
Beet mosaic	*Myzus ajugae* Schout	Heinze (1962)

TABLE 12.2 Continued

Virus	Vector	Vector reference
	Aphid vectors	
Beet yellows	*Myzus ajugae* Schout	Heinze (1962)
	Hyperomyzus staphyleae Koch	Tahon (1964)
Beet yellow stunt	*Myzus persicae* (Sulz)	Duffus (1964)
Brassia junca mosaic	*Aphis rumicis* (L)	Azad *et al.* (1963)
Cardamon mosaic	*Pentalonia nigronervosa* (Coq.)	Varma (1962)
Cardamon mosaic streak	*Rhopalosiphon maidis* (Fitch)	Raychaudhuri and Chatterjee (1964)
Clover yellow vein	*Myzus persicae* (Sulz)	Hollings and Mariani (1964)
	Acyrthosiphon pisum (Harris)	Hollings and Mariani (1965)
Clover (white) mosaic USA	*Acyrthosiphon pisum* (Harris)	Goth (1962)
Cherry latent	*Myzus persicae* (Sulz)	Milbrath and Swenson (1964)
Citrus reticulate yellows	*Aphid* spp.	Schwarz (1965)
Citrus tristea	*Aphis craccivora* (Koch)	Varma *et al.* (1965)
	Dactynotus jaceae (L)	Varma *et al.* (1965)
Cucumber mosaic	*Myzocallis asclepiadis* (Monell)	McClanahan and Guyer (1964)
Dahlia mosaic	*Idiopterus nephrelepdidae* (Davis)	Iglison (1963)
Gooseberry vein-banding	*Hyperomyzus pallidus*	Posnette (1964)
Groundnut mosaic	*Aphis craccivora* (Koch)	Hull (1964)
Groundnut rosette	*Aphis gossypii* (Glov.)	Adams (1966)
Lettuce mosaic	*Pemphigus bursarium* (L)	McClean (1962)
Lettuce necrotic yellows	*Hyperomyzus lactucae* (L)	Stubbs and Grogan (1963)
Lettuce mosaic (Japan)	*Macrosiphon matsumuracum* Hori	Koshimuzu and Iisuka (1963)
Lucerne mosaic	*Myzus ajugae* Schout	Heinze (1962)
	Cerosipha thalictri	Heinze (1962)
Maize dwarf mosaic	*Schizaphis graminum* (Rond.)	Daniels and Toler (1969)

Disease	Vector	Reference
Mulberry mosaic	*Myzus persicae* (Sulz.)	Chatterjee and Raychaudhuri (1963)
Nasturtium mosaic	*Rhopalosiphon maidis* (Fitch)	Raychaudhuri et al. (1962)
	Aphis gossypii Glover	Bisht (1962)
	Brevicoryne brassicae (L)	Bisht (1961)
	Myzus persicae (Sulz.)	Bisht (1961)
	Rhopalosiphon padi (L)	Bisht (1961)
Passiflora chlorotic spot	*Aphis gossypii* Glover	Van Velsen (1961)
Passion fruit mosaic	*Aphis gossypii* Glover	Martini (1962)
Peanut stunt	*Myzus persicae* (Sulz.)	Miller and Troutman (1966)
	Aphis craccivora (Koch)	Hebert (1967)
	Aphis spiraecola Patch	Hebert (1967)
Pea seed-borne; fizzle top	*Acyrthosiphon pisum* (Harris)	Gonzales and Hagedorn (1970)
	Macrosiphon euphorbiae Thomas	Gonzales and Hagedorn (1970)
	Myzus persicae (Sulz.)	Gonzales and Hagedorn (1970)
Pea mosaic	*Cryptomyzus ribis* (L)	Kvicala (1965)
	Appelia schwartzii (Borner)	Kvicala (1965)
P. floridana mild chlorosis	*Myzus persicae* (Sulz.)	McKinnon (1965)
Potato veinal necrosis	*Brachycaudus helichyrsi* (Kalt.)	Edwards (1965)
Radish mosaic (Taiwan)	*Brevicoryne brassicae* (L)	Kou (1961)
	Myzus persicae (Sulz.)	Kou (1961)
	Rhopalosiphon pseudobrassicae (Davis)	Kou (1961)
Raspberry beta leaf curl	*Aphidula rubicola* (Oest.)	Converse (1962)
Sowthistle yellow vein	*Hyperomyzus lactucae* (L)	Duffus (1960)
Soybean stunt	*Aphis glycine* (Mats.)	Koshimizu and Iisuka (1963)
	Myzus persicae (Sulz.)	Koshimizu and Iisuka (1963)
	Rhopalosiphon prunifoliae (Fitch)	Koshimizu and Iisuka (1963)
Sugarbeet yellows	*Idiopterus nephrelepidae* (Davis)	Iglisch (1963)
Sugarcane mosaic	*Acyrthosiphon pisum* (Harris)	Abbot and Carpentier (1963)
	Dactynotus (M) *ambrosiae* (Thom.)	Abbot and Carpentier (1963)

TABLE 12.2. *Continued*

Virus	Vector	Vector reference
	Aphid vectors	
Strawberry pseudo-mild yellow edge	*Hyperomyzus lactucae* (L)	Abbot and Carpentier (1963)
	Chaetosiphon jacobi (Hrl)	Frazier (1966)
Sweet potato russet crack	*Myzus persicae* (Sulz.)	Martin and Daines (1964)
Tomato yellow top	*Macrosiphon euphorbiae* (Thomas)	Braithwaite and Blake (1961)
Turnip mosaic	*Myzus ajugae* Schout	Heinze (1962)
	Cerosipha thalictri	Heinze (1962)
	Idiopterus nephrelepidae (Davis)	Iglisch (1963)
	White fly vectors	
Acalypha indica mosaic	*Bemisia tabaci* (Gennadius)	Chenulu and Pastek (1965)
Bean crumpling	*Bemisia tabaci* (Gennadius)	Costa (1965)
Bean golden mosaic	*Bemisia tabaci* (Gennadius)	Costa (1965)
Bean mottled dwarf	*Bemisia tabaci* (Gennadius)	Costa (1965)
Beet pseudo-yellows	*Trialeurodes vaporariorum* (West)	Duffus (1965)
Cotton main vein thickening	*Bemisia tabaci* (Gennadius)	Dour and Dour (1964)
Cotton small vein thickening	*Bemisia tabaci* (Gennadius)	Dour and Dour (1964)
Cucumber vein yellowing (bottle gourd mosaic)	*Bemisia tabaci* (Gennadius)	Cohen and Bohn (1960)
Malvaiscus arboreus Cav. leaf curl		Mikherjie and Raychaudhuri (1964)

Beetle vectors

	Ceratoma ruficornis	Walters and Henry (1970)
Southern bean mosaic (cowpea strain)		
Bean pod mosaic	*Ceratoma trifurcata* Forst.	Ross (1963)
Cowpea chlorotic mottle	*Ceratoma trifurcata* Forst.	Walters and Dodd (1969)
	Diabrotica undecimpunctata Mann.	Walters and Dodd (1969)
Cowpea mosaic	*Ceratoma trifurcata* Forst.	Smith (1924)
	Ceratoma trifurcata Forst.	Van Hoof (1962)
	Ceratoma trifurcata Forst.	Walters and Barnett (1964)
	Ootheca mutabilis (Sahlbergi)	Chant (1959)
	Diabrotica (?) *laeta* (F)	Van Hoof (1962)
	Epilachna varivestis Mulsant	Jensen and Staples (1970)
Pumpkin mosaic	*Diabrotica longicornis* (Say)	Stoner (1963)
	Diabrotica undecimpunctata Mann.	Stoner (1963)
	Diabrotica virgifera (LeConte)	Stoner (1963)
	Acalymma vittata (F)	Stoner (1963)
Radish enation mosaic	*Diabrotica undecimpunctata* Mann.	Campbell and Colt (1967)
Radish mosaic	*Phyllotreta* spp.	Campbell and Colt (1967)
	Phyllotreta striolata (F)	Tochihara (1968)
Soybean mosaic	*Halictus* spp.	Bennett and Costa (1961)
Rice yellow mottle	*Sesselia pusilla* (*Malacasoma*) Gerst.	Bakker (1970)

Grasshopper vector

Pumpkin mosaic	*Melanoplus differentialis* (Thom.)	Stoner (1963)

531

TABLE 12.2. *Continued*

Virus	Vector	Vector reference
	Psyllid vectors	
Citrus greening [a]	*Trioza erytreae* (De G.)	McClean and Oberholzer (1965)
Citrus leaf mottle yellows	*Diaphorina citri* Kuwauama	Martinez and Wallace (1967)
Pea red leaf mottling	*Psylla piri* (L)	Hardtl (1967)
Pear decline [a]	*Psylla pyricola* Forst.	Shalla *et al.* (1963)
Pear leaf curl (decline)	*Psylla pyricola* Forst.	Kaloostian and Jones (1968)
	Mite vectors	
Agropyron mosaic	*Abacarus hystrix* (Nal.)	Slykhuis (1965)
Citrus leprosis (Brazil)	*Brevipalpus phoenicis* (Geij.)	Rosille *et al.* (1964)
(Florida)	*Brevipalpus californicus* (Banks)	Knorr (1968)
	Brevipalpus obovatus Donnadieu	Knorr and Denmark (1970)
Pigeon pea sterility	*Aceria cajani* Ch. Bozovanna	Seth (1962)
Potato Y	*Tetranychus telarius* (L)	Schultz (1963)
Plum latent	*Vasates fockeui* (Nal.)	Proesler and Keglar (1966)
Rose rosette	*Phyllocoptes fructiphilus* Keifer	Arlington *et al.* (1968)
	Mirid vector	
Potato gothic (spindle tuber)	*Lygus pratensis* (L)	Gerasimov (1965)
	Coccid vector	
Sugarcane spike	*Pseudococcus sacharifolii*	Ali (1962, 1963)

[a] Diseases marked thus are now associated with mycoplasma.
NOTE—See footnote a, Table 12.1 for references.

TABLE 12.3. Nematode Vectors of Polyhedral (NEPO) Viruses [a]

Virus	Strain	Vector	Reference
		Genus Xiphinema	
Ash ringspot	Type	*X. americanum*	Hibben and Walter (1971)
Arabis mosaic	Type	*X. diversicaudatum*	Harrison and Cadman (1959)
Arabis mosaic	Grapevine fanleaf	*X. index*	Hewitt *et al.* (1958)
Arabis mosaic	Type and Rhubarb mosaic isolate	*X. coxi*	Fritsche (1964)
Arabis mosaic	Grapevine fanleaf	*X. italiae*	Cogen *et al.* (1970)
Cherry leafroll	Type	*X. coxi*	Fritsche (1964)
Cherry rasp leaf	Type	*X. americanum*	Nyland *et al.* (1969)
Strawberry latent ringspot	Type	*X. diversicaudatum*	Lister (1964)
Tobacco ringspot	Arkansas isolate	*X. americanum*	Fulton (1962)
Romato ringspot	Peach yellow bud isolate	*X. americanum*	Breece and Hart (1959)
Tomato ringspot	Grape yellow vein isolate	*X. americanum*	Teliz *et al.* (1966)
Weidelgras mosaic	Type	*X. diversicaudatum*	Schmidt, Fritsche, and Lehmann (1963)
		X. coxi	
		Genus Longidorus	
Raspberry ringspot	Type	*L. elongatus*	Taylor (1962)
Raspberry ringspot	English	*L. macrosoma*	Harrison (1962)
Tomato black ring	English	*L. attenuatus*	Harrision *et al.* (1961)
Tomato black ring	Scottish	*L. elongatus*	Harrison *et al.* (1961)

TABLE 12.3. *Continued*

Nematode Vectors of Tubular (NETU) Viruses

Genus Trichodorus

Virus	Strain	Vector	Reference
Pea early browning	Dutch isolate	*T. pachydermus* *T. teres*	Van Hoof (1962)
Pea early browning	English isolate	*T. anemones* *T. pachydermus* *T. viruliferous*	Harrison (1967) Gibbs and Harrison (1964) Gibbs and Harrison (1963)
Tobacco rattle	Dutch isolate	*T. pachydermus*	Sol and Seinhorst (1961)
Tobacco rattle	Dutch (Gladiolus) notchleaf isolate	*T. similis*	Cremer and Kooistra (1964)
Tobacco rattle	English isolate	*T. primitivus*	Harrison (1961)
Tobacco rattle	German isolate	*T. primitivus*	Sänger (1961)
Tobacco rattle	Scottish isolate	*T. primitivus*	Mowat and Taylor (1962)
Tobacco rattle	U.S.A. California isolate	*T. allius* *T. christiei* } *T. porosus*	Ayala and Allen (1966)
Tobacco rattle	U.S.A. Oregon isolate	*T. allius*	Jensen and Allen (1964)
Tobacco rattle	U.S.A. Wisconsin isolate	*Y. christiei*	Walkinshaw *et al.* (1961)

[a]From Pitcher (1965) and Taylor and Cadman (1969) with additions.

REFERENCES

Allard, H. A., "The mosaic disease of tobacco," *U.S. Dept. Agr. Bull.,* No. 40 (1914).

Amos, J., Hatton, R. G., Knight, R. C., and Massee, A. M., "Experiments in the transmission of reversion in black currants," *Ann. Rept. East Malling Research Sta., Kent,* 1925, 126–150 (1927).

Bakker, W., "Rice yellow mottle, a mechanically transmissible disease of rice in Kenya," *Neth. J. Plant Pathology,* 76, 53–63 (1970).

Ball, E. D., "The sugar beet leafhopper (*Eutettix tenella*)," *Utah Agr. Expt. Sta., 16th Ann. Rept.,* 1904–1905 (1906).

Bergeson, G. B., Athow, K. L., Laviolette, F. A., and Sister Mary Thomasine, "Transmission, movement and vector relationships of tobacco ringspot virus in soybean," *Phytopathology,* 54, 723–728 (1964).

Brandes, E. W., "Artificial and insect transmission of sugar cane mosaic," *J. Agr. Research,* 19, 131–138 (1920).

Breece, J. R., and Hart, W. H., "A possible association of nematodes with the spread of peach yellow bud mosaic virus," *Plant Disease Reptr.,* 43, 989–990 (1959).

Broadbent, L., "The epidemiology of tomato mosaic, VIII Virus infection through tomato roots. IX Transmission by birds," *Ann. Appl. Biol.,* 55, 57–66 (1965).

Cadman, C. H., "Biology of soil borne viruses," *Ann. Rev. Phytopathology,* 1, 143–172 (1963).

Campbell, R. N., and Colt, W. M., "Transmission of radish mosaic virus," *Phytopathology,* 57, 504–506 (1967).

Carter, W., "Geographical distribution of yellow spot of pineapples," *Phytopathology,* 29, 285–287 (1939).

Carter, W., "Notes on some mealybugs (Coccidae) of economic importance in Ceylon," *F.A.O. Plant Protect. Bull.,* 4, 49–52 (1956).

Costa, A. S., "White flies as virus vectors," *Viruses, Vectors and Vegetation,* Wiley-Interscience, New York, 1969.

Costa, A. S., and Bennett, C. W., "White fly transmitted mosaic of *Euphorbia prunifolia,*" *Phytopathology,* 40, 266–283 (1950).

Dale, W. T., "Observations on a virus disease of certain crucifers in Trinidad," *Ann. Appl. Biol.,* 35, 598–604 (1948).

Dickson, R. C., "A working list of the names of aphid vectors," *Plant Disease Reptr.,* 39, 445–452 (1955).

Doolittle, S. P., "The mosaic disease of cicurbits," *U.S. Dept. Agr. Bull.,* No. 879 (1920).

Doolittle, S. P., and Jones, F. R., "The mosaic disease in the garden pea and other legumes," *Phytopathology,* 15, 763–772 (1925).

Flock, R. A., and Wallace, J. M., "Transmission of fig mosaic by the eriophyid mite *Aceria ficus*," *Phytopathology*, 45, 52–54 (1955).

Freitag, J., "Insect transmission, host range, and properties of squash mosaic virus," *Phytopathology*, 31, 8 (1941).

Goss, R. W., "Infection experiments with spindle tuber and unmottled curly dwarf of the potato," *Nebraska Univ. Expt. Sta. Research Bull.*, No. 53 (1931).

Gibbs, A. J., "Plant Virus Classification," *Advances in Virus Research*, 14, 263–328 (1969).

Hansen, H. P., "Contribution to the systematic plant virology," *D. S. R. Forlag, Royal Vet. Agric. Univ., Copenhagen*, pp. 1–110 (1970).

Harpaz, I., and Cohen, S., "Semipersistent relationship between cucumber vein yellowing virus (CVYV) and its vector, the tobacco white fly (*Bemisia tabaci* Gennadius)," *Phytopath. Zeit.*, 54, 240–248 (1965).

Harrison, B. D., and Cadman, C. H., "Role of a dagger nematode (*Xiphinema* sp.) in outbreaks of plant diseases caused by *Arabis* mosaic virus," *Nature*, 184, 1624–1626 (1959).

Harrison, B. D., Mowat, W. P., and Taylor, C. E., "Transmission of a strain of tobacco black ring by *Longidorus elongatus*," *Virology*, 14, 480–485 (1961).

Heinze, K., "Können Schnecken pflanzliche Virosen übertragen?" *Z. Pflanzenkrkh. u. Pflanzenschutz*, 65, 193–198 (1958).

Heinze, K., *Phytopathogene Viren und ihre Ubertrager*, Duncker & Humblot, Berlin, 1959.

Hewitt, W. B., Raski, D. J., and Goheen, A. C., "Transmission of fan-leaf virus by *Xiphinema index* Thorne and Allen," *Phytopathology*, 48, 393–394 (1958).

Hewitt, W. B., Raski, D. T., and Goheen, A. C., "Nematode vector of soil-borne fan leaf virus of grapevines," *Phytopathology*, 48, 586–595 (1958a).

Hoggan, I. A., "Transmissibility by aphids of the tobacco mosaic virus from different hosts," *J. Agr. Research*, 49, 1135–1142 (1934).

Jagger, I. C., "A transmissible mosaic disease of lettuce," *J. Agr. Research*, 20, 737–740 (1921).

Jha, A., and Posnette, A. F., "Transmission of a virus to strawberry plants by a nematode (*Xiphinema* sp.)," *Nature*, 184, 962–963 (1959).

Kennedy, J. S., Day, M. F., and Eastop, V. F., "A conspectus of aphids as vectors of plant viruses," *Commonwealth Inst. Entomology*, London, 1962, 114 pages.

Keifer, H. H., and Wilson, N. S., *Calif. Dept. Agr. Bull.*, No. 44 (1956).

Kirkpatrick, T. W., "Insect transmission of cacao virus disease in Trinidad," *Bull. Entomol. Res.*, 41, 99–117 (1950).

Knorr, L. C., "Studies on the etiology of leprosis in citrus," *Proc. 4th Conf. Inter. Orgn. Citrus Virologists*, University of Florida Press, Gainesville, Fla., 1968.

Kunkel, L. O., "Studies on aster yellows," *Am. J. Botan.*, 13, 646–705 (1926).

Leston, D., "Entomology of the cocoa farm," *Ann. Rev. Entomology*, 15, 273–294 (1970).

Linford, M. B., "Transmission of the pineapple yellow spot virus by *Thrips tabaci*," *Phytopathology*, **22**, 301–324 (1932).

Ling, K. C., "Non-persistence of the tungro virus of rice in its leafhopper vector, *Nephotettix impicticeps*," *Phytopathology*, **56**, 1252–1256 (1966).

Lojek, J., "Transmission of purified TMV by aphids," *Proc. Can. Phytopathological Soc.*, **36**, 13–26 (Abstr.) (1969).

Markham, R., and Smith, K. M., "Studies on the virus of turnip yellow mosaic," *Parasitology*, **39**, 330–342 (1949).

McLean, D. M., "Studies on mosaic of cowpea *Vigna sinensis*," *Phytopathology*, **31**, 420–430 (1941).

Messieha, M., "Transmission of tobacco ringspot virus by thrips," *Phytopathology*, **59**, 943 (1969).

Mulligan, T., "Transmission of rye grass mosaic virus," *Rept. Rothamsted Expt. Sta.*, **1957**, 110–111 (1958).

Murphy, P. A., "Leaf roll and mosaic, two important diseases of the potato," *Irish J. Dept. Agr. Technol. Instr.*, **22**, 281–284 (1922).

Nielson, M. W., "A synonymical list of leafhopper vectors of plant viruses (Homoptera, Cicadellidae)," *U.S. Dept. Agr.*, ARS-33-74 (1962).

Nielson, M. W., "The leafhopper vectors of phytopathogenic viruses (Homoptera, Cicadellidae). Taxonomy, biology and transmission," *Tech. Bull.*, 1382, 1–386 *USDA, ARS*, (1968).

Ochs, G., "Untersuchengen uber Verbreitung der Rebenviren durch Vektoren," *Naturwissenschaften*, **45**, 193 (1958).

Oldfield, G. N., "Mite transmission of plant viruses," *Ann. Rev. Entomology*, **15**, 343–380 (1970).

Oldfield, G. N., and Wilson, N. S., "Establishing colonies of *Eriophyes insidiosus*, the vector of peach mosaic virus," *J. Econ. Entomol.*, **63**, 1006 (1970).

Orlando, A., and Silberschmidt, K., "The vector of the 'infectious chlorosis' of malvaceous plants," *Biologico*, **11**, 139–140 (1945).

Orlando, A., and Silberschmidt, K., "Estudos sobre a disseminacao natural do virus da 'clorose infecciosa' das Malvaceas (*Abutilon* virus I Baur) e a sau relacão com o inseto-vetor, *Bemisia tabaci* (Genn.) (Homoptera-Aleyrodidae)," *Agr. Inst. Biol. (São Paulo)*, **17**, 1–36 (1946).

Pitcher, R. S., "Interrelationships of nematodes and other pathogens of plants," *Helminth. Abstracts*, **34**, 1 (1965).

Posnette, A. F., "Virus diseases of cacao in West Africa VII. Virus transmission by different vector species," *Ann. Appl. Biol.*, **37**, 378–384 (1950).

Quanjer, H. M., "The mosaic disease of the Solanaceae, its relation to the phloem necrosis, and its effect upon potato culture," *Phytopathology*, **10**, 35–47 (1920).

Rosillo, M. A., Puzzi, D., and Stamato, W., "Experiencia de campe visando e controle de acaro *Brevipalpus pheonicis* (Geijskes) em cultura de citrus," *Arq. Inst. Biol. (Sao Paulo)*, **31**, 41–43 (1964).

Sakimura, K., "The present status of thrips-borne viruses," *Biological Aspects of Virus Transmission,* Academic Press, New York, 1962.

Schultz, E. S., and Folsom, D., "Transmission, variation and control of certain degeneration diseases of Irish potato," *J. Agr. Research,* 25, 43–117 (1923).

Serrano, F. B., "Pineapple yellow spot in the Philippines," *Philippine J. Sci.,* 58, 481–493 (1935).

Shaw, H. B., "The curlytop of beets," *U.S. Dept. Agr. Bur. Plant Ind. Bull.,* No. 181 (1910).

Sheffield, F. M. L., "Virus diseases of sweet potatoes in East Africa II. Transmission to alternate hosts," *Phytopathology,* 48, 582–590 (1958).

Slykhuis, J. T., "*Aceria tulipae* Keifer (Acarina, Eriophyidae) in relation to the spread of wheat streak mosaic," *Phytopathology,* 45, 116–128 (1955).

Slykhuis, J. T., "Mite transmission of plant viruses," *Advances in Virus Research,* 11, 97–137 (1965).

Slykhuis, J. T., "Mites as vectors of plant viruses," *Viruses, Vectors and Vegetation,* Wiley-Interscience, New York, 1969, pp. 121–142.

Smith, K. M., "Arthropods as vectors and reservoirs of phytopathogenic viruses," *Handbuch der Virusforschung,* Springer, Wien, 1957, pp. 143–176.

Smith, K. M., *A Textbook of Plant Virus Diseases,* 2nd ed., Little, Brown, Boston, 1957a.

Smith, R. E., and Boncquet, A., "New light on curlytop of the sugar beet," *Phytopathology,* 5, 103–107 (1915).

Sol, H. H., van Heuven, J. C., and Seinhorst, J. W., "Transmission of rattle virus and *Atropa belladonna* mosaic by nematodes," *Tijdschr. Plantenziekten,* 66, 228–231 (1960).

Storey, H. H., "Streak disease of sugar cane," *S. African Dept. Agr. Scil Bull.,* 39 (1925).

Takata, K., "Results of experiments with stunt disease of rice," *J. Japan Agr. Soc.,* 171, 1–4 (1895); 172 13–22 (1896).

Taylor, C. E., and Cadman, C. H., "Nematode vectors," *Viruses, Vectors and Vegetation,* Wiley-Interscience, New York, 1969, pp. 55–94.

Thresh, J. M., "Increased susceptibility to the mite vector (*Phytopus ribis* Nal.) caused by infection with black currant reversion virus. Association between black currant reversion virus and its gallmite vector," *Nature,* **202,** 1028, 1085–1087 (1964).

Walters, H. J., "Some relationships of three plant viruses to the differential grasshopper, *Melanoplus differentialis* (Thos.)," *Phytopathology,* 42, 355–362 (1952).

Wilson, N. S., Jones, L. S., and Cochran, L. C., "An eriophyid mite vector of the peach mosaic virus," *Plant Disease Reptr.,* 39, 889–892 (1955).

13

Virus-Vector

Relationships

THE FEEDING PROCESS OF VECTOR INSECTS

The feeding process of vector insects has been studied specifically in connection with virus transmission in a few cases; it will be dealt with here as an addendum to the treatment already given insect feeding processes in Chapter 4. The earlier work on leafhopper feeding happened to be restricted to vectors which are primarily phloem feeders. There are many references to sucking insects as phloem feeders, mesophyll feeders, or xylem feeders, but it should not be taken for granted that the objective tissue is the only tissue tapped. This is particularly true of the intracellular feeders, but even an intercellular feeder must necessarily pass through intercellular sap en route to its objective tissue.

The Mechanism of Feeding

Storey (1938) reported that *Cicadulina mbila* Naude must reach the phloem before it can infect a plant; it can nevertheless acquire the virus from the mesophyll. Since it must penetrate the mesophyll before it can reach the phloem, acquisition is easily possible if, in passing through the mesophyll, the insect ingests.

The case of *Circulifer tenellus* (Baker) is in another category. In curly-top-infected plants, the virus is limited to the phloem, and this tissue is the objective of the insect when it feeds. It can neither acquire nor transmit the virus unless it reaches the phloem (Bennett, 1934). Fife and Frampton (1936) attempted to determine the mechanism whereby the insect was able to reach the phloem, and they concluded that it depended on a pH gradient of increasing alkalinity from the cortex to the vascular bundle, the difference being about 1.6 pH units. These experiments have been widely quoted and accepted as a generalization for leafhoppers, but without confirmatory evidence.

IN LEAFHOPPERS.   Day *et al.* (1952) studied the feeding of the leafhopper *Orosius argentatus* (Evans) in comparison with that of a number of other leafhoppers. Study of the feeding tracks deposited in a single host (*Malva parviflora* L) revealed major differences between species, both in the percentage of the tracks reaching specific tissues and in the tissue damage that resulted. The authors' diagram is shown as Fig. 13.1, and their summarized data are listed in Table 13.1. There were, however, even greater differences when a single species of insect was fed on other

TABLE 13.1.   Tissues in Which Feeding Tracks of Jassids Terminate[a]

Jassid	Host plant and tissue	Total tracks counted	Percentage of tracks in		
			Phloem	Xylem	Parenchyma or mesenchyma
Orosius argentatus adults	*Beta vulgaris* petiole	114	48	4	48
O. argentatus adults	Large *Datura stramonium* petiole	53	2	1	97
O. argentatus 1st–3rd instar	*B. vulgaris* petiole	112	23[b]		77
O. argentatus 4th instar	*B. vulgaris* petiole	106	18[b]		82
O. argentatus adults	*B. vulgaris* leaf	55	29	2	69
Euscelis punctatus adults	*Medicago sativa* stems	54	50	24	26

[a] Data from Day *et. al.* (1952).
[b] Most of these tracks terminated in the very small bundles in the "wings" of these petioles, which are unusually large; in these bundles it was impossible to determine precisely where the insect was feeding.

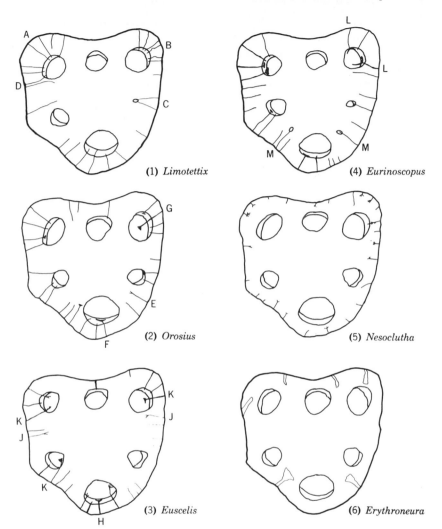

FIGURE 13.1.   Diagrams of transverse sections of petioles of *Malva* showing feeding tracks of (1) *Limotettix,* (2) *Orosius,* (3) *Euscelis,* (4) *Eurinoscopus,* (5) *Nesoclutha,* and (6) *Erythroneura* (After Day *et al.,* 1952).

host plants. *O. argentatus,* for example, feeding on *Beta vulgaris* L, reached the phloem in 48%, the xylem in 4%, and the parenchyma in 48% of the cases examined. On a large *Datura stramonium* L, however, only 2% reached the phloem, 1% the xylem, and 97% the parenchyma. The curving of tracks toward a particular tissue is taken as evidence that the leafhopper is actually showing a preference for that tissue. The differences between tissues reached by a single vector, feeding on different host plants, might well account for the different infection rates for a given virus that occur in susceptible crops.

These same authors attempted to repeat Fife and Frampton's experiments on pH gradient, but came to the conclusion that no gradient of any kind assists the leafhopper in locating its preferred feeding tissue. Instead, the conclusion was reached that the feeding behavior of the species studied could be assigned to three categories: (1) primarily parenchyma feeders, (2) species whose tracks reach the phloem in 30–50% of cases but which apparently also feed on parenchyma, and (3) species that reach the phloem more frequently than would occur by chance. All three groups feed on whatever tissues are reached by preliminary probing, but they feed longer when they reach preferred tissues.

Two conditions must be met with respect to the feeding behavior of a leafhopper if it is to be an efficient vector of a phloem-restricted virus: (1) it must reach the phloem in a high percentage of its feeds, and (2) it must not cause undue damage to the phloem which might prevent infection by the virus. With these conditions in mind, it is possible to estimate the relative efficiency of known vector species with respect to transmission of the phloem-restricted viruses by comparison of their feeding habits. There are many other factors besides tissue preference which determine whether a leafhopper can be a virus vector.

Day *et al.* (1952) have described the mechanism of feeding, as determined by observation of insects feeding through a membrane, as follows: "The insect braces its prothoracic and mesothoracic legs, rotates the head so as to bring the rostrum (which normally is directed posteriorly between the prothoracic legs) perpendicular to the surface to be penetrated. The surface of a plastic membrane is actually dented by the pressure of the mouth-parts and almost immediately the feeding stylets, comprising the mandibles and maxillae, appear through the membrane, and extend about 70μ into the fluid. The insect may then either withdraw the stylets and move off, leaving no trace of its presence, or settle to feed. In the latter event it proceeds to form a salivary sheath. The stylets perform a fairly rapid oscillating motion. As they are withdrawn a small particle appears at the tip. This is shaped into a tube-like sheath by the next forward movement. Each movement adds to the length of the sheath at its distal end (although sometimes the sheath forks at the tip). It reaches a length of about 70μ in a minute or so, but may be added to at any time subsequently during feeding. The stylets then extend beyond the sheath (sometimes for brief periods to more than three times its length) and are moved to and fro in the fluid. This probably occurs during actual ingestion. Withdrawal of the mouth-parts from the sheath occurs after feeding is complete or if the insect is disturbed. It may be affected with considerable speed and without difficulty. The hind legs are placed on the membrane and the insect withdraws the stylets and moves away. If the disturbance is caused by another insect or change of light intensity, the stylets may be reinserted in a position close to the previous one, often without any other movement of the insect."

With the exception of some vectors in the family Cercopidae, all the vectors of Pierce's disease of grapes (alfalfa dwarf) are in the subfamily Cicadellinae. These insects feed for long periods of time and give off large quantities of excrement, amounting to as much as 2.5 cc in a 24-hr period in the case of *Draeculacephala minerva* Ball, without withdrawing the mouth parts from the plant tissue. This is a clear indication that the insect is feeding in some part of the vascular system. Four species were studied by Houston *et al.* (1947) with respect to the method of feeding and the tissues involved. All four species were found to be xylem feeders; the virus can multiply and cause the disease only when introduced into the xylem.

IN WHITE FLIES.   A comparison of the feeding behavior of two insect species of white flies on cotton is possible through the work of Pollard (1955, 1958). *Bemisia tabaci* (Genn), a white fly, is an important vector of leaf-curl virus. It is also known for severe direct effects of feeding on the cotton plant such as leaf shedding, although visible effects on the cellular structure are slight. The entry of the stylets is intercellular for the most part, although occasional intracellular and stomatal entry takes place. The primary objective is the phloem, although a few end in parenchyma, and, very rarely, in xylem. Stylet tracks are scarce, and when they do occur are poorly developed.

The life cycle of a white fly and its feeding habits as described by Hildebrand (1959) are complicated. There are four stages: the egg, three larval stages, pupal stage, and adult. Only the first larval "crawler" stage is mobile and only then for less than 24 hours, after which its stylet, arising in the rostrum, penetrates the leaf in a feeding position where it remains permanently. The larval stages and emerging adult leave this original stylet *in situ,* and the adult develops a proboscis with a new stylet. This feeding habit clearly indicates that unless it can be shown that white fly-transmitted viruses are transmitted through the egg, only the adult can carry a virus from one host to another. The virus can be acquired by the larva and survives pupation.

IN APHIS.   On the other hand, *Aphis gossypii* (Glover) a vector of many viruses of plants other than cotton, while pursuing an intercellular course, deposits well-developed stylet tracks. Pollard describes the feeding process of this insect; the most significant finding is the length of time required to attain the final feeding site, 20 min. Penetration of the stylets through the epidermis is very rapid and is accomplished in 42 sec. In 2 min the stylets are halfway through the mesophyll, but penetration rate decreases then, and the entire mesophyll is penetrated in not less than 9 min. This brings the stylets to the vascular tissue, when the rate of penetration decreases very rapidly, an additional 10 min being required to pass through the first 6μ of the phloem cells. Insertion then accelerates and about 26μ of phloem is traversed in the next 8 min. There is then a 2nd drop in penetration rate, and the stylet apex comes to rest finally in about 30 min. A curious part of the feeding behavior is the rotational twisting of the stylets after they are inserted in the leaf. Of 20 examined, 14 showed either clockwise or counterclockwise rotation.

Aphis fabae scop. is another well-known vector of virus diseases whose feeding process has been studied without regard to its virus-vector relationship, but purely as a contribution to the biology of the insect. Banks and Nixon (1959) studied the excretion and feeding rates of this insect using host plants grown in water culture and made radioactive with P^{32}. Their conclusion was that the amounts of sap ingested at first were small, but that the rate of ingestion increased rapidly between 12 and 16 hr. The maximum rate of feeding was estimated at 0.2 mg of sap/hr, an uptake of 59% of the mean body weight of the insects/hr. This is much more than the 10% of the insect's body weight/hr recorded for *M. persicae* (Sulz.) (Watson and Nixon, 1953). The amount of sap ingested plainly bears no relationship to virus acquisition or transmission, except perhaps to rapidly dilute virus concentration when the insect moves from an infected to a healthy plant. In the case of viruses retained for long periods in the vectors, reservoirs for the viruses must be available that are relatively unafffected by the ingested food stream.

Wisconsin pea streak virus is transmitted by *Macrosiphum pisi* (Johnson), but the insect is an inefficient vector. Skotland and Hagedorn (1955) attempted to relate this inefficiency to the feeding behavior of the insect. The insect produces stylet sheaths, and, of these, 122 terminated in the parenchyma, 152 in the phloem, and 4 in xylem. The path of the stylets was intercellular except in rare instances. The number of aphids reaching the phloem increased with increasing feeding time; but even with 60 min feeding time only 3 aphids acquired the virus, as determined by test feedings. The authors suggest that the intercellular feeding habit is the reason for the inefficiency as a vector, but this is no barrier to the efficiency of other virus vectors, and there is nothing in the data on the objective tissues to explain the inefficiency. The reasons for vector inefficiency must be sought elsewhere.

Roberts (1940) investigated the feeding and penetration rates of three species of aphid, *M. persicae, Myzus (Neomyzus) circumflexus* (Buckt.), and *Macrosiphum gei* (Koch), and found no relationship between infectivity and either the tissue tapped or the method of penetration. *M. persicae* stylets followed an intracellular course in the majority of cases with feedings up to 5 min, but tracks left by the insect after 1 hr feeding were equally divided between an intra- and an intercellular course. The phloem was reached only after the 1 hr feedings. *M. circumflexus* was predominantly intercellular in its penetration and reached the phloem freely after 1 hr feeding. *M. gei* had an entirely intracellular method of penetration, and the phloem was reached freely after 1 hr feeding. All these records were from feeding on tobacco. *M. persicae* on sugar beet, while following the same intracellular course as in tobacco, rarely reached the phloem in 15 min, and even with 24 hr feeding did not always reach it. All three aphids share the characteristic of being most efficient vectors with a 2-min feeding on the virus source plant in the case of *Hyoscyamus* virus III, potato virus Y, and cucumber virus 1.

It is clear, therefore, that both time and method of penetration may vary greatly in the case of the first three viruses named, but in the case of sugar beet yellows, the length of time to reach the phloem and increased infectivity with increased feeding time may not be merely coincidental, and suggest that acquisition of the sugar beet yellows virus from the phloem may be necessary for infectivity. For all three aphids, infectivity decreases rapidly with increased feeding times on the infected plant. With *M. persicae*, however, increased feeding periods on sugar-beet-yellows-infected plants increases infectivity with the yellows virus.

USE OF RADIOACTIVE TECHNIQUES.   There have been some attempts made to determine the feeding behavior of aphids by using radioactive tracer techniques. Hamilton (1935) used radioactive polonium and fed *M. persicae* on solutions by using a piece of epidermis stripped from a cabbage leaf as the feeding membrane. Insects differed in the amount imbibed, but a constant proportion of the imbibed material was transmitted to a leaf when the insect was transferred from the feeding solution. This was taken as evidence that the solution was not merely picked up by adherence to the stylets, but took the route considered to be normal for ingested material, namely, through the alimentary canal into the blood stream and thence to the salivary glands from whence it was ejected down through the stylets.

In the meantime, techniques for handling and recording radioactive isotopes have greatly improved. Day and Irzykiewicz (1953) used radioactive phosphorus ($Na_2HP^{32}O_4$ or $NaH_2P^{32}O_4$) incorporated into sucrose solutions, or into plant tissues, by permitting the petiole of an isolated leaf of Chinese cabbage (*Brassica chinensis* L) to stand in Knop's solution, containing the isotope, until a measured volume of solution had been absorbed by the leaf. *M. persicae* and *Brevicoryne brassicae* (L) were used as the test insects and artificial feeding was accomplished through a thin plastic membrane. No uptake could be measured during the first few minutes of penetration; but *Myzus* ingested approximately 0.07 mg in 1 hr, *B. brassicae* less than 1/20 of that amount. Both species reinjected in a subsequent feed a fraction of 1% of the amount ingested. While this was a very much smaller quantity than reported by Hamilton, the latter did not account for the possibility that excreted isotope might have been absorbed by the plant. Day and Irzykiewicz also reported the mechanical carriage of P^{32} solution on the feeding stylets of *M. persicae,* but the effects of adding a wetting agent to the solution, which were to separate the stylets and materially reduce the amount of P^{32} carried thereon, suggest that there was capillary flow into the closed stylets.

FEEDING BEHAVIOR.   Bradley and Rideout (1953) found no evidence that the feeding behavior of four species of aphids could be related to their efficiency as vectors of potato virus Y (PVY). Acquisition feeding punctures were frequently very brief and over 70% lasted less than 30 sec. Feeding on the first test plants was slightly longer, 50% lasting less than 30 sec and 20% over a minute; 74% of *M. persicae* transmitted PVY after acquisition feedings of 5–15 sec. Efficiency of transmission was highest with *M. persicae,* followed by *Aphis abbreviata* Patch, *Macrosiphum solanifolii* (Ashm.), and *Myzus solani* (Kalt.), in that order. It seems clear that neither objective tissues nor the manner of penetration to them play any significant part in either acquisition or transmission of PVY by these aphid species, and the explanation appears to be that in the case of PVY, the virus is carried on the stylet tip of the aphid (Bradley and Ganong, 1955, 1955a). Bradley (1956) went further and reported that when the stylets did penetrate beyond the first layer of plant cells, both acquisition and transmission by *M. persicae* were rarely achieved. The same appears to be true with three other viruses transmitted by *M. persicae:* cucumber mosaic, henbane mosaic, and tobacco severe etch (Bradley and Ganong, 1957).

The feeding of Thysanoptera does not appear to have been studied with respect to the role of these insects as vectors, but there is one detailed account of the feeding of *Thrips tabaci* Lind. which describes the feeding process and the tissues affected (Wardle and Simpson, 1927). The mouth parts of the thrips form a short conical proboscis, or mouth cone, the rear wall of which is provided by the triangular labium (Fig. 13.2).

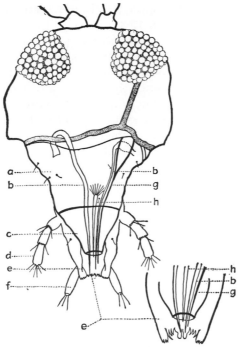

FIGURE 13.2. The asymmetrical mouth parts of *Thrips tabaci.*
Key: a, clypeus; b, maxilla; c, labrum; d, maxillary palp; e, labium;
f, labial palp; g, mandible; h, hypopharynx (After Leach, 1940).

The tips of the maxillae and the labium are provided with hooks, and it
is suggested that these hooks are for the purpose of gripping the surface
of the leaf closely.

There are four actual piercing organs: the paired stylets, the median
stylet, and the left stylet. The paired stylets are considered to be maxil-
liary laciniae; the median one, the hypopharynx; and the left one, the
mandible. In the feeding process, the mouth cone is applied firmly to the
leaf surface which is gripped by the hooked tips of the maxillae and labium. The head of the thrips moves slightly upward and downward; at
each swing upward the mandible is protruded, at each swing downward
the mandible is withdrawn. The first step in the feeding process then, is a
gashing action, designed to break through the cuticle and outer epidermal wall. When this is accomplished, the longer and weaker maxillae
function, and the boundary between the epidermis and the outermost
mesophyll layer is broken down. In the closely applied mouth cone is a
partial vacuum which aids in the passage of the chloroplasts into the
pharynx. Usually, after gashing one epidermal cell and absorbing the

contents of the underlying mesophyll cells, the insect moves on and gashes another cell farther away (Fig. 13.3).

This feeding process has not been considered in relation to virus transmission by this insect, but it poses some interesting problems. The mandibular gash in the epidermis would not be expected to have any significance in virus transmission since there is no direct connection between that organ and the pharynx. The function of the maxillae would also appear to be one of breaking down cell walls so that the contents of the mesophyll cells would be available to the insect's suction apparatus. There is some evidence that tomato spotted wilt is primarily located in the mesophyll, and the insect's feeding habit makes it almost imperative that that is the tissue into which the virus is introduced. The virus is, however, not too dependent for its introduction into a relatively uninjured cell, because feeding actually does not occur until severe cell damage has been done.

There were no significant additions to the data just presented on the feeding of thrips until just recently. Mound (1971), with the aid of the stereo-electron microscope, has shown that the stylets of *T. tabaci* actually form a tube and that the single mandible is a stout structure with a solid apex. Mound suggests that the deep feeding reported by Sakimura (1962) involves sucking through the long stylets as in the Hemiptera, and the shallow feeding involves applying the mouth cone directly onto the leaf surface after the leaf surface has been macerated. Mound's superb stereo-electron micrographs bring an entirely new light on what has been a very difficult problem. The significance of the findings with respect to virus transmission by thrips is obvious (Fig. 13.4).

Regurgitation by biting insects seems to account for the transmission of turnip yellow mosaic virus by flea beetles and grasshoppers (Markham and K. M. Smith, 1949). Walters (1952) described the dark brown fluid which is emitted from grasshopper mouths as buccal fluid; he considers it certain that it is made up of regurgitated material from the insect's crop as well as salivary secretions. Walters found that the buccal fluid, while having some inhibitive effects on the viruses of tobacco mosaic, potato ringspot, and tobacco ringspot, nevertheless was the vehicle for all three viruses named, transmitted by the differential grasshopper, *Melanoplus differentialis* (Thos.).

These have been the subject of several recent reviews (Sylvester, 1962, 1969; Kennedy *et al.,* 1962; Rochow, 1963; Bradley, 1964; Smith, 1965; Maramorosch, 1964; Sinha, 1968). The most recent and inclusive is to be found in *Virus, Vectors, and Vegetation* (K. Maramorosch, Ed., 1969), comprising contributions of 33 authors. Most of the 29 papers related to virus-vector relationships and previous reviews have been brought up to

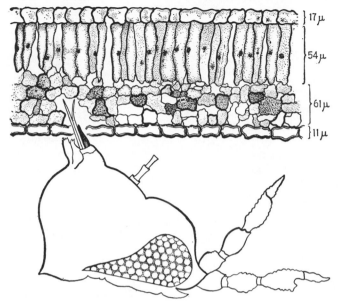

FIGURE 13.3. A diagram of the head of a feeding thrips, indicating its size in relation to the leaf tissues and the depth of the feeding wounds (After Leach, 1940).

date with a wealth of detail. The volume provides an inclusive treatment of this specialized section.

Sinha (1968) reviewed recent work on the leafhopper-transmitted viruses. Serology has proved of limited value in determining relationships; potato yellow dwarf and wound tumor viruses are exceptions. The use of fluorescent antibody technique has facilitated detection of viruliferous individuals in a colony. Virus has been detected in many insect vectors' organs and the sequence of these infections determined. Viruses have been shown to multiply in nonvector insects, and puncturing the abdomen, the method used with inactive races of a vector species, does not result in the ability to transmit.

Injections of leafhoppers with suspensions of viruliferous insects is now a well established technique, but with wheat striate mosaic virus and *Endria inimica* (Say), success of the method depends on concentration, infectivity being reduced drastically from a concentration of 10^{-7} to 10^{-4} (Lee and Bell, 1963). A curious datum is that the suspensions which induce infectivity in the insect were negative in mechanical transmission tests. The injection technique used for leafhoppers has also been developed for use with aphids (Mueller and Rochow, 1961; Muller, 1965). It is now routine

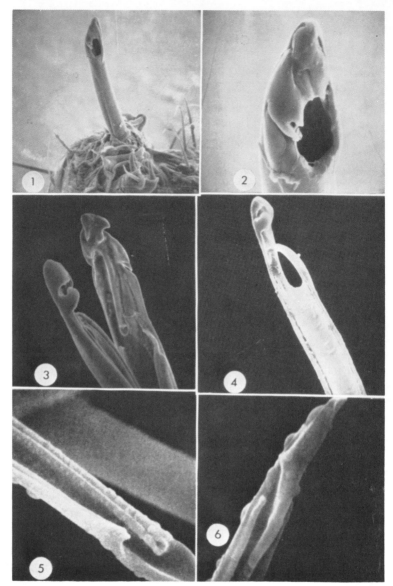

FIGURE 13.4. The feeding apparatus of thrips. 1–2, *Idolothrips spectrum* Haliday; 1, paired stylets forming feeding tube protruding from mouth cone, right labial palp in foreground, X500; 2, detail of apex of stylets with feeding aperture X2100; 3–4, *Macrothrips papuensis* Bagnall; 3, stylet apices separated showing internal ridge of right stylet with feeding channel opening to left, X700; 4, inner surface of left stylet with grooves on lateral margins, X700; 5–6, *Thrips tabaci;* 5, detail of broken stylet showing grooves on lateral margins, X10,000; 6, apex of right stylet with internal ridge and feeding channel opening to upper left, X7300 (Photograph: L. A. Mound, by permission of the Commonwealth Agricultural Bureaux).

and has proven a potent technique for the study of virus-vector relationships of aphid-transmitted viruses.

VECTOR SPECIFICITY IN TRANSMISSION OF PLANT VIRUSES

In the early days of work with transmission of plant viruses, specificity presented fewer problems than it does now. This was largely because the oldest known leafhopper vectors all displayed specificity to a high degree. With both viruses and vectors becoming known in increasing numbers, it has become clear that the concept of specificity is a relative one. Day and Bennetts (1954) reviewed the evidence for both plant and animal viruses; their review forms the basis for this section.

Stage or Species Specificity

The most specific type of vector relationship occurs when only one stage of a single species of insect can acquire or transmit a virus. One such example is found in sugar beet leaf curl, which is transmitted only by the adult but is acquired by either nymph or adult of *Piesma quadratum* (Fieb.). There are no data available on whether *Piesma cinereum* (Say), transmitting beet savoy, belongs in the same category. More than one species of thrips is involved in the transmission of spotted wilt virus, but extreme specificity exists in the vector-virus relationship, since only the larvae can acquire the virus, although both larvae and adults can transmit it.

There are two other examples of the ability of nymphs, but not adults, to acquire viruses (Zazhurilo and Sitnikova, 1941): *Psammotettix striatus* (L), vector of winter wheat mosaic; and two vectors of Fiji disease of sugarcane, *Perkinsiella saccharicida* Kirk, and *Perkinsiella vestatrix* Breddin (Montgomery and Bell, 1933; Ocfemia, 1933). In the case of winter wheat mosaic, all stages can transmit but only the first three larval stages can acquire.

Less marked, but still highly specific, are those cases wherein a virus is known to be transmitted by only a single vector species. The best known examples of this type are leafhopper-transmitted viruses of beet curly top, aster yellows, and cranberry false blossom. Clover club-leaf virus can only be transmitted by *Agalliopsis novella* (Say). In all these cases, both nymphal and adult stages can both acquire and transmit.

Western-X-disease virus can be acquired and transmitted by all five nymphal instars of *Colladonus montanus* (Van D.), but not by nymphs of

Keonolla confluens (Uhl.) or *Scaphytopius acutus* (Say). Transmission was effected, however, by adults of *S. acutus* that acquired the virus in the fifth nymphal instar (Wolfe, 1955).

Sinha (1960) found that nymphs of *Delphacodes pellucida* (F.) were more effective vectors than adults of the same species and of the same race; 34 of 71 adult insects that had fed on infected plants as nymphs transmitted the virus, as compared with 10 out of 72 that had fed first as adults.

Piesma quadrata (Fieb.), vector of sugar beet leaf curl, can only acquire the virus in the nymphal stage, and cannot transmit until the adult stage is reached. The virus is propagative in the insect (Proeseler, 1964).

Barley yellow dwarf virus was transmitted with greater facility by nymphs of the aphid *Rhopalosiphon maidis* (Fitch) than by adults. The ratio was higher for acquisition feeding periods of 1 and 2 days than when longer periods were used. Apterous adults were more efficient than were alates. When acquisition was by 12 hr-old nymphs, their increased efficiency was carried over to both apterous and alate adults; their efficiency was greater than when acquisition was by the adult stages. The principal BYDV isolate used was 6417, but the greater efficiency of the nymphs in transmission was true also of three other isolates, all *R. maidis*-specific; one other isolate did not show greater nymphal efficiency (Gill, 1970).

Specificity may be lost if 2 strains of a virus (BYDV) are used to doubly infect oat seedlings, one of the strains being normally transmitted by *R. fitchii* (Sand.) and the other by *M. avenae*. With double infections, *R. fitchii* transmitted both strains of BYDV; the specific transmission of the *M. avenae* strain was still maintained (Rochow, 1965).

Variations between clones of a single aphid species in specific transmission is well known (Rochow, 1963; Smith, 1965), but physiological variations may also be a factor in transmission of a specific virus strain. *R. padi* from Kansas regularly transmitted one strain of BYDV which was only rarely transmitted by the New York clone of the same aphid.

Taxonomic consideration also enters into this complex since morphological differences were observed in the Kansas clone of *R. padi*. Actually the two clones of *R. padi* differed morphologically, physiologically, and presumably genetically, but the sum of the differences was not considered adequate to separate the two clones. To complete the complication, differences between the two clones of *R. padi* with respect to the transmission of BYDV strains were pronounced at 15–20°C, but the only consistant transmissions of the virus strain by the New York clone occurred at 30°C (Rochow and Eastop, 1966).

Rochow's recent review (1969) on barley yellow dwarf and its strains encompasses 20 text pages and 69 references, making it perhaps the most complete treatment of the problems of specificity of transmission of a single virus yet published.

Clones of *Myzus persicae* varied in their ability to transmit potato-leaf-roll virus even when the acquisition was by means of injection; this variation was also expressed as a difference in reproductive capacity and ability to survive injection (Clark and Ross, 1964).

One color form of *M. pisum* out of four used failed to transmit pea enation mosaic, but each of the four forms was transferred to the test plant from a different host plant; the test plant was *Vicia faba* (Hinz, 1969). Whether or not the presence of virus inhibitors in the saliva of aphids is affected by their specific food plants would provide a useful research project.

Potato virus Y, in two strains, MPVY and RPVY, was used to compare the comparative transmission of two aphid species, the green peach aphid (*M. persicae*) and the potato aphid (*M. euphorbiae*). The test plant was "California Wonder" pepper. RPVY inoculated by *M. persicae* caused both severe and mild symptoms, but the potato aphids caused only severe symptoms. Symptoms of MPVY were always severe. The mild symptoms of RPVY could be restored to the severe syndrome by serial sap inoculations in pepper. The variability of RPVY transmission raises an interesting question, e.g., why does a plant inoculated by aphids require many more weeks to become a virus source plant than when sap inoculation is used? There was no difference in titers of sap-transmissible virus and aphid-inoculated RPVY. The author's speculations conclude that differences in probing between aphid species, in availability of several viruses to the aphids, and / or differences in virulence between the mild and severe symptoms, are probably responsible for variation in vector efficiencies (Simons, 1969).

Acquisition and transmittal of tomato blackring virus strains by the nematode *Longidorus elongatus* (de Mann) is limited to the larvae (Harrison *et al.,* 1961), but *Arabis* mosaic virus is transmitted more freely by adults of *X. diversicaudatum* Micol. Both larval and adult forms of *L. elongatus* transmit raspberry mosaic. Severe and yellow strains of cowpea mosaic virus are transmitted by five species of chrysomelid beetle, all within the subfamily Galerucinae. Jansen and Staples (1971) tested these species for vector specificity but none was shown. The banded cucumber beetle *Diabrotica balteata* Le Conte and the bean leaf beetle, *Ceratoma trifurcata* (Forst.) were more efficient transmitters and retained the virus longer than three other species; they fed more extensively on both source and test plants.

Simons and Eastop (1970) studied temperature effects on transmission following the work of Sylvester (1964). They used three strains of potato virus Y and one of cucumber mosaic; the aphids used were *M. persicae* and *M. euphorbiae*. All virus-vector combinations showed positive correlations between increased temperatures and transmission efficiency. With the potato aphid, all three strains transmitted equally well. *M. persicae* transmitted each strain with a different efficiency. DPYV was not temperature labile; RPVY was; both were inefficiently transmitted by *M. persicae*. The authors suggest the possibility that the diphenol-polyphenoloxidase described by Miles (1968) was involved.

Area Specificity

Area (regional) specificity is best exemplified by the case of spotted wilt which is transmitted by different species in different parts of the world. Three curly top viruses, or related strains, found in the United States, Brazil, and Argentina, are each transmitted by a different species of leafhopper; it is now known that one of these, *Circulifer tenellus* (Baker), is a native of the Mediterranean region and is an introduced species into the United States (Ohman, 1949). Curly top in Turkey appears to be the same as in the United States (Bennett and Aziz, 1957).

Group Specificity

It can be presumed that if it were possible to bring all these vectors together, their vector capacity would fall into the group specificity category of Day and Bennetts. The term is somewhat ambiguous. If it refers to taxonomic groups of insects, then Pierce's disease of grape (alfalfa dwarf), with its 20 known tettigellinid leafhopper vectors and 4 cercopid vectors, would present an ideal example, even though two separate families are involved. Onion yellow dwarf, however, with more than 50 aphid vectors, represents group specificity only to the extent that all the vectors are in the Aphididae, and most workers would regard this case as one of complete lack of specificity.

Both the New Jersey and New York varieties of potato yellow dwarf virus can be considered examples of group specificity since both are transmitted by specific groups of leafhoppers. The groups overlap only in the cases of *Agalliopsis novella* (Say), which is an extremely poor vector of either virus, and *Agallia quadripunctata* (Prov.), which is an extremely poor vector of the New York variety.

Regional Specificity

The fact that groups of vectors are frequently closely related left Day and Bennetts with the problem of accounting for some examples of transmission of a single virus by unrelated insects. Tomato big bud in Australia is transmitted by a deltocephalinid, whereas in the U.S.S.R. the vector is a cixiid; for this reason it is doubted that the viruses can be the same. Similarly, Australian witch's broom, with a vector in the Deltocephalini, may be distinct from lucerne witch's broom, which is transmitted by a vector in the Scaphytopiini. These last two cases surely could represent examples of regional specificity.

If the aster yellows virus from the type localities in the eastern United States and the California aster yellows virus are truly closely related strains of the same virus, then the concept of group specificity seems inapplicable. In the one case, that of the eastern aster yellows, only one vector is known, but the California aster yellows has many leafhopper vectors in the Deltocephalini and one in a different tribe, Iassine (*Gyponana hasta* DeLong).

If a single virus can be transmitted by two vectors completely separated taxonomically, as was thought to be the case with feathery mottle virus of sweet potato, such an example could best be characterized as a case of limited specificity. In the instance cited, however, it was shown that three viruses are involved. Hildebrand (1960) makes it clear that feathery mottle virus is a complex of yellow dwarf that is white fly transmitted; internal cork and leafspot are aphid transmitted.

Regional specificity in the transmission of virus strains has been reported by Toko and Bruehl (1956, 1957) in connection with barley yellow dwarf virus (BYDV) isolates from Washington, and for the same virus from New York by Rochow (1958). In Washington, both apple grain and English grain aphids transmitted from most of the field samples, but one isolate was transmitted only by apple grain aphids and another one only by English grain aphids. In New York, however, virus was transmitted usually from field samples only by English grain aphids. On the basis of this evidence, it was concluded that the strains of BYDV could be differentiated by means of specific vectors. When, however, the two species of aphids from both regions were compared in both places with the local virus strains, it was found that virus transmission by apple grain aphids from New York and Washington was identical in all but one of 16 tests, and transmission by the English grain aphid was identical in 17 out of 20 tests. These tests were conducted in New York by Rochow (1958). The comparable and coordinated tests in Washington (Bruehl,

1958) showed that the Washington isolates were transmitted equally well by both species of aphids. The conclusion, shared by both workers, was that the observed vector specificity was actually due to a regional difference in the virus complex.

Vector Nonspecificity

Nonspecificity of vectors is reserved by Day and Bennetts for those cases in which the virus is transmitted by a wide variety of vectors. An example is that of potato spindle tuber, which has been transmitted by aphids and a variety of leaf-feeding insects, including grasshoppers and larvae and adults of beetles. Such viruses are characteristically mechanically transmitted.

Nonspecificity is evidently not equated with lack of specificity by Day and Bennetts, for they refer to the latter as characteristic of certain aphid-borne viruses, including, presumably, those involved in group specificity. Perhaps the status of the relationship could be indicated by the term "virus-vector" when the virus is determinative (as in onion yellow dwarf), and by "vector virus" when marked specificity or limited group specificity of vectors occurs (as in some leafhopper-transmitted viruses or spotted wilt).

Day and Bennetts list some possible factors in specificity of vectors in a suggested sequence in which they might be encountered: (1) ecological factors, (2) feeding behavior, (3) ingestion, (4) blood, (5) salivary glands, and (6) immunity in insects.

Factor of Ingestion

Feeding behavior, except in a few cases of tissue-restricted viruses, does not appear to be the obvious factor in specificity as Day and Bennetts describe it. The amount of sap ingested may be important if virus concentration in the plant is low, but vigorous feeders, such as *Lygus* (Flemion *et al.,* 1952) may not be vectors of any virus disease. As Day and Bennetts remark, data on the amount of material ingested become of value only when the concentration of the virus in the ingested material is known; to date, this information is not known for any plant virus, and getting it would obviously present formidable technical difficulties.

Ingestion is apparently essential for some types of transmission but not for others. Day and Bennetts have shown this by diagrams (Fig. 13.5 and 13.6) of types of transmission in which ingestion is not a precursor of transmission, and have already been shown to be independent of feeding behavior. Where ingestion is necessary, as in those cases in which long re-

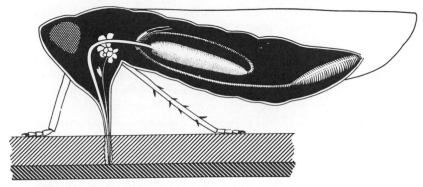

FIGURE 13.5.   Diagrammatic sagittal section of leafhopper, showing the route taken by viruses during the transmission cycle (After
Day and Bennetts, 1954).

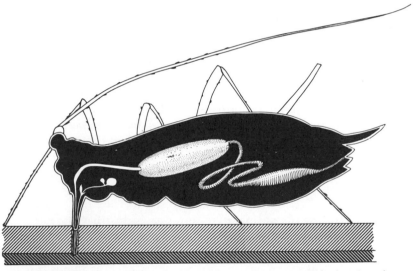

FIGURE 13.6.   Diagrammatic sagittal section of aphid, showing the
route taken by viruses during the transmission cycle (After Day and
Bennetts, 1954).

tention of the virus in the insect's body occurs, the fate of the virus after
ingestion is significant with respect to specificity. Ingestion *per se* is not
enough to ensure transmission. There are well-authenticated cases in
which virus can be found in the bodies of nonvector insects; for example,
Bennett and Wallace (1938) showed that curly top virus could be acquired and carried in the bodies of aphids, thrips, mites, and nonvector
species of leafhoppers for long periods of time, but it is not transmitted

by these insects. K. M. Smith (1941) recovered curly top virus from flea beetles and tobacco mosaic from the feces of caterpillars. K. M. Smith (1957) also records that the virus of turnip yellows mosaic is resistant to inactivation by digestive juices; it has been recovered and recrystalized from the feces of the cabbage white butterfly (*Pieris brassicae* L) and is not inactivated by passage through the alimentary canal of snails.

It has been generally considered, with the exception of a few cases where virus is regurgitated, that ingested viruses must pass through the gut wall into the hemolymph. Virus inactivators present in the gut, gut wall, or hemolymph must therefore either not operate in these cases or be saturated. Inactivators are not always present or effective, since it is possible to demonstrate viruses in homogenates obtained from macerated vectors (McClintock and L. B. Smith, 1918; C. E. Smith, 1924; Freitag, 1941). It is possible, however, that the process of maceration resulted in the inactivation of the inactivators, either chemically or by dilution.

There is evidence that inability to pass through the gut wall constitutes a barrier to rendering an insect infective. Storey's now classic experiments with *Cicadulina mbila* (Naude, 1933) demonstrated that if the gut of an individual of an inactive race was punctured, it became infective (Fig. 13.7). Sinha (1960) found that the capacity of adults to transmit was greatly improved if the adults, even those of an active race, were punctured through the anterior portion of the abdomen either immediately before or immediately after an acquisition feed, but not seven days after the acquisition feed.

Puncturing the mid-gut of adult *T. tabaci* at the midpoint of a 48-hr feeding on infected tomato plants did not render them capable of acquiring the virus of spotted wilt; the reasons for the growth-stage specificity exhibited by thrips vectors still remains obscure (Day and Irzykiewicz, 1954). Shortcircuiting the normal ingestion process by injecting virus into the bodies of insects has been achieved by several workers, as discussed in connection with multiplication of viruses in their vectors.

Insect Blood as a Virus Reservoir

The blood of insects is generally considered to be the virus reservoir in the insect vector. Bennett and Wallace (1938) found that the blood of *Circulifer tenellus* (Baker) contained a relatively strong concentration of the virus of curly top, but the highest concentration was found in the alimentary canal and a considerable amount in the feces. This indicates that only a portion of the virus reaching the gut passes through into the hemolymph, even in effective vectors. There is no evidence that virus has ever been found in the blood of nonvectors. It is possible that specificity

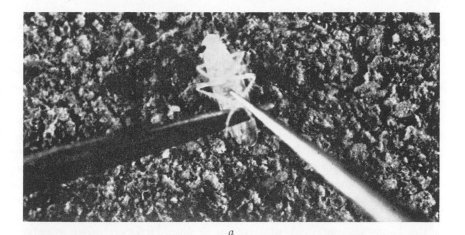

a

b   c

FIGURE 13.7. The method of inoculation of *Cicadulina mbila*. *a*, The insect is steadied by means of the "holding needle" on the left and punctured in the abdomen with the "inoculating needle" on the right; *b*, the method of deep inoculation of *C. mbila*. The point of the needle is thrust right through the anterior part of the insect's abdomen. *c*, The method of inoculation of *C. mbila* by means of a micropipet (Photographs: Storey, 1933).

may be influenced by differences in hemolymph between one insect species and another, but such differences are unlikely. An insect species may, however, carry specific inactivators in its blood which would mitigate against its success as a vector; but until virus can be shown to pass into the blood of a nonvector through the epithelium of the mid-gut, the question of inactivation in the blood is not moot.

Storey (1933) did, however, inject virus into the body of *Peregrinus maidis* Ashm., and presumably into the hemocoele, but without rendering the insect infective. In this case, an experimental condition was set up that meets the requirement just referred to, namely, that virus was presumably introduced into the blood of an insect which remained a nonvector. Tobacco mosaic virus was inactivated when injected into the blood of living grasshoppers, but inactivation was not very rapid since 6 of the 12 insects remained infective for 12 days. However, when grasshoppers were fed on virus-infected plants, no virus could be found in the blood (Walters, 1952).

It is presumed that all viruses which are ingested and pass through the gut into the blood find their way to the host plant through the salivary glands. This suggests another possible barrier to virus transmission, but the permeability of the salivary glands surely does not depend on the *quantity* of virus available for penetration, as Day and Bennetts suggest in accounting for the long latent periods in some insects. Intermittent transmission as described by Bennett and Wallace (1938) may mean, as Bawden *et al.* (1954) suggest, that the virus only penetrates the salivary glands in adequate quantity at intervals. But the virus is not in solution in the hemolymph, and there is no reason to suppose that the distribution of virus particles in the hemolymph is uniform, especially if the sites for virus multiplication are not limited to the hemolymph. Regularity of transmission would seem to be more remarkable than intermittent transmission, on these grounds.

Richardson and Sylvester (1965) found that honeydew from viruliferous aphids was superior as a source of inoculum to that from plant extracts or hemolymph from viruliferous aphids. This finding has more than passing interest. It is well known that many insect parasites feed on honeydew. If this feeding is extensive enough it might well result in the virus being transmitted occasionally by the parasite, either through contamination of the mouth parts or through contamination of the egg surfaces; the latter is more likely since the eggs are deposited in the insect's body. Ants are also heavy feeders on honeydew, and if the ants happen to be species which lacerate tissue, they also might be vectors on occasion. Another possibility is that honeydew, being an excretion, provides a mechanism for the rejection of virus by the insect.

Complex Transmissions

There are other examples of transmission, which involve specificity, that were not considered by Day and Bennetts. These cases have been noted by Smith (1957), who speaks first of selective transmission, in which one species of aphid selects one out of a complex of two viruses. Both of these viruses may be aphid-borne or only one may be. Kvicala (1945) showed that when *M. Persicae* and *B. brassicae* are colonized on cauliflower seedlings infected with cabbage black ringspot and cauliflower mosaic viruses, they are able to transmit both, but when *Myzus ornatus* Laing is similarly colonized, only cauliflower mosaic is transmitted; the black ringspot virus is left behind. If cucumber mosaic virus and potato virus Y are present together in an infected plant, *Myzus ascolonicus* (Donc.) will select out cucumber mosaic virus but will not transmit PVY. The same species, when fed on plants infected with a mixture of henbane mosaic and severe etch viruses, will transmit only the henbane mosaic virus (Doncaster and Kassanis, 1946).

When two viruses are present, only one of which is aphid-transmitted, it is not surprising that only that virus, i.e. the aphid-transmitted one, is picked out of the combination. An example of this is found when cucumber mosaic and tobacco mosaic are together in the same plant. Only the former is transmitted. When two viruses together result in a disease symptom characteristic of infection with the combination, nevertheless the two viruses are still separate entities; if one is aphid-transmitted and the other is not, then the complex can be separated by using the aphid vector to extract the one virus and leave the other. An example is the combination of potato viruses X and Y which together result in the symptoms known as rugose mosaic disease, but only one (virus Y) is aphid-transmitted.

A more complex form of specificity is found when one virus cannot be transmitted unless it is accompanied in the plant by another. K. M. Smith has shown (1946) that tobacco rosette is a complex resulting from the presence of two viruses, the mottle and vein-distorting viruses. When together, they are both aphid-borne, but when separated, only the vein-distorting virus can be transmitted by aphids.

A complex situation is described by Watson and Sergeant (1962) in which transmission of one virus is affected by the presence of a second one. Red leaf and mottle viruses in carrot cause the syndrome known as carrot motley dwarf. Red leaf is transmitted only by the aphid *Cavariella aegopodiae* (Scop.) and is not mechanically transmitted; its host range is limited to the Umbelliferae. Mottle virus is mechanically transmitted and has a wider host range, but it can only be acquired and transmitted by aphids from double-infected plants. This transmission can occur to plants not susceptible to red leaf; the complex can therefore be broken down in this manner.

Hull and Adams (1968) describe a complex which includes an assistor virus. Groundnut rosette disease is caused by the groundnut rosette virus plus a symptomless virus. The assistor virus (GRAV) must be present for aphid transmission to occur.

Parsnip yellow fleck (PYFV) is a semipersistent virus transmitted by *Cavariella aegopodii* (Schop.), but only in the presence in the source plants (*Anthriscus sylvestris* and others) of the second virus *Anthriscus* yellows (AYV) (Murant and Goold, 1968).

The mechanism of transmission of potato virus C and *Aucuba* mosaic has been described by Kassanis and Govier (1971). Transmission by insects occurs from plants also infected with a helper virus that is insect-transmitted and is nonpersistent. The help must be provided either at the stage of acquisition of the helper virus by *M. persicae* or during injection of the virus into the test plant.

Freitag (1969) calls such cases examples of dependent transmission and the designation seems appropriate. The experimental data are adequate, but the explanations, different in each case, provide some interesting hypotheses for further work.

Cross Protection

The protection afforded a plant against a second strain of a virus when the plant is infected with another strain of the same virus is well established and is a basis for establishing strain relationships. A possibly analogous situation obtains among leafhopper vectors, provided that the viruses are propagative in the insect, but not otherwise. The phenomenon was first established by Kunkel (1955, 1957) and Freitag (1967) for the aster yellows agent (*Macrosteles fascifrons* Stål.) Maramorosch (1958) demonstrated it for the corn stunt agent and *Dalbulus maidis* (DeL and W.). Freitag (1969) reported a detailed study of strains of aster yellows and interference in the leafhopper between strains, which is the most complete study of the phenomenon available. Should the mycoplasma hypothesis for the etiology of aster yellows be finally confirmed and strains of the organism successfully cultured, some exciting prospects lie ahead.

It is difficult to account for these complex situations on the basis of the specificity categories of Day and Bennetts, but they are, nevertheless, examples of extreme specificity, which suggest that perhaps specificity is a function of the virus as well as of the vector.

GENETIC VARIABILITY AS A
FACTOR IN SPECIFICITY

Genetic variability among individuals of an insect vector species in their ability to transmit plant viruses was first proved by Storey (1932), although the possibility that such might be the case had been considered by K. M. Smith (1932a) and Hogan (1929, 1931). Storey, working with *Cicadulina mbila* Naude, an efficient vector of streak disease virus of maize, occasionally encountered individuals of the species which were unable to transmit the virus. In breeding experiments Storey first took a mixed culture, and from this he isolated pure lines which he designated as "active" and "inactive." Each individual of these pure lines was tested for ability to transmit. When a colony was derived from the mating of a pair, both of which were from an active colony, the progeny were active. Inactive progeny resulted from the mating of two inactive individuals.

When the two pure races were crossed, using an active female crossed with an inactive male, the F_1 generation was entirely active; the F_2 generation was comprised of active females and both active and inactive males. When an inactive female was crossed with an active male, the F_1 generation was comprised of active females and inactive males, while both active and inactive males and females appeared in the F_2 generation. The ability to transmit is inherited as a single, dominant, sex-linked Mendelian factor.

Fukushi (1934) observed a wide variation in the ability of *Nephotettix apicalis* var. *cincticeps* Uhl. to transmit the virus of stunt disease of rice; many individuals apparently were unable to transmit at all even though fed throughout their nymphal stages on infected plants. Fukushi recorded no breeding experiments, however. Bennett and Wallace (1938) found similar variation among individuals of *C. tenellus* in their ability to transmit curly top, but breeding experiments failed to isolate a pure inactive strain such as Storey described. It was possible, however, to breed strains which differed in efficiency, and these differences were heritable.

Black (1943) studied the same problem in connection with the transmittal of potato yellow dwarf virus by *Aceratagallia sanguinolenta* (Prov.). As in the case of curly top, it was not possible to isolate a race, of which no member would transmit, but colonies were developed by selection which showed significant differences in transmitting ability. For that reason, Black's explanation of the genetics of these races is more complicated than Storey's. After ten generations, however, 80% of the active race, 2% of the inactive race, and 30% of the hybrids proved infective when tested under the same conditions. Males resulting from crosses between inactive males and active females were more active than the females derived from such crosses, and they were more active than the males and females resulting from reciprocal crosses.

Bjorling and Ossiannilsson (1958) found that different strains of *M. persicae* showed statistically significant differences in transmitting ability for both beet yellows and potato leaf roll. Differences between strains appeared to be genetically determined. When strains of *M. persicae* were crossed, the descendents significantly differed from the parent strains in virus-transmitting ability. *A. fabae* also showed strain differences, those between the most and the least effective being statistically significant. Among 85 strains of *M. persicae,* transmitting abilities could be grouped in a continuous series showing from 10–80% average frequency of transmission, but there were no relationships between frequencies and the host plant species from which the strain was derived. These data suggest that the strains were actually genetically derived and not merely collections from separate host plants.

Active and inactive races of *M. persicae* have also been described by Stubbs (1955a), who used a yellows virus of spinach as his test material. The transmitting ability of viviparously produced progeny of individual apterae selected at random from a stock colony was tested, and the results made clear that there were wide differences in infectivity between the groups of five aphids used in each test. Some were completely negative. Four clonal cultures developed from individual mother aphids showed a high percent of transmission by two of these and completely negative results with the others.

Races of *Delphacodes pellucida* Fabr., described as efficient and inefficient, are referred to under transovarial transmission (Watson and Sinha, 1959). Galvez *et al.* (1960) found that only 9% of 500 individuals of *Sogota orizicola* Muir were capable of transmitting hoja blanca disease of rice in Colombia, although efficient individuals are capable of transmitting for their entire life. No breeding experiments are recorded, but if the low percent of individuals capable of transmitting represents a genetic segregation, the inactive group would appear to be the dominant. However, *Sogatodes oryzicola* (Muir) collected from the field population shows only a small percentage of individuals capable of transmitting, but with selective mating, this percentage of active progeny can be brought up to 80–90% (Hendrick *et al.,* 1965).

The case of *Aceratagallia calcaris* Oman and yellow vein of sugarbeets has been described by Staples *et al.* (1970). The insect transmits with low efficiency, which was not influenced in five hybrids and two inbreds of sugarbeet. There is an incubation period of 27–29 days in the insect, and symptoms appear from 33 days. There is evidently a low percent of individuals able to transmit, and there is no consistency of transmission from infective colonies, which in the case of other leafhopper vectors would be clearly positive. In these respects the case appears to be unique; it may however be a case similar to that of *S. oryzicola* and hoja blanca.

The vector of groundnut mosaic virus, *Aphis craccivora* (Koch), varies in inherent efficiency in transmission with different races. One race failed to transmit altogether (Storey and Ryland, 1955).

A genetic complex was shown to occur (Nagaraj and Black, 1962) in *Agallia constricta* Van D., which transmits two viruses independently. There are four races involved in the transmission of potato yellow dwarf virus, of which two are efficient and two are not. The efficiency of transovarial transmission of wound tumor virus is also modified by race; two races are more efficient vectors than the third.

Specificity in transmission of virus serotypes by different nematode species is indicated (Dias and Harrison, 1963) and may be correlated with the antigenic constitution of the virus (Harrison, 1964).

McEwan and Kawanishi (1967) failed to find the occasional *Peregrinus maidis* Ashm. with unusually short incubation periods of corn mosaic virus reported by Carter (1941) as occurring in a few cases with genetically related insects, nor did they get the transmission efficiency of the planthopper which Carter reported. With 25 years separating the two studies, however, and the certainty that the biotypes used were completely unrelated, this is not surprising. The finding in both studies of occasional aphid transmission has yet to be explained; in retrospect it might be ascribed to contamination of the aphid colonies with sugarcane mosaic virus.

TRANSOVARIAL TRANSMISSION

The acquisition of plant viruses by insect vectors can occur transovarially, although only a few viruses were known to be so transmitted by 1960. Three Agalliinae: *Agalliopsis novella* (Say), *Agallia constricta* Van D., and *Austroagallia torrida* (Evans), one Deltocephalinae: *Nephotettix apicalis* Motsch., and two *Delphacodes* species: *D. striatella* Fallén and *D. pellucida* Fabr., have been shown to acquire viruses by transmission through the egg.

Proving transovarial transmission presents some technical difficulties. Fukushi (1933, 1935), who was the first to demonstrate that a virus could pass through the egg of the vector, did so by breeding a single infective female with an infective male of *N. apicalis* and moving the progeny at the moment of hatching from the egg over to healthy rice plants. By retaining single newly hatched nymphs on the first rice plant for 5 days, and then for 1 day in serial transfer, Fukushi was able to show that no virus had been introduced into the first (5-day feeding) rice plant fed upon. None of the insects became infective earlier than 9 days after hatching. Since the incubation period for symptoms in the plant is a minimum of 6 days and usually between 9 and 13 days, none of the plants fed on for 1 day could have been a source of virus, for generally it is only the day or so before symptoms appear in the plant that an insect can recover virus from the plant. There is a rare exception to this generalization; Storey (1939) showed that maize streak virus could be recovered in 1 or 2 days from the same leaf as had been fed upon by infective insects. This is not true, however, of rice plants infected by stunt virus.

Yamada and Yamamoto (1955, 1956) conducted extensive experiments on the transmission of stripe disease of rice by *Delphacodes striatellus* Fall. Using essentially the same techniques as Fukushi's, they demonstrated transovarial transmission when the female parents were infected. By feeding the freshly hatched nymphs on fresh rice plants serially, they were able to show that the virus was transmitted through 24 generations.

Fukushi's experiment was carried on for more than a year, during which time 6 generations of leafhoppers were derived from the original female. From these, 82 infective leafhoppers infected 1200 rice plants. Fig. 13.8 is a diagram of Fukushi's definitive experiment.

Black (1953a) analyzed the controversy which followed Fukushi's publication, but his paper was concerned primarily with the problem of multiplication of the virus in the insect. The transovarial passage of the virus was not challenged.

Black (1948) showed that *Agalliopsis novella* (Say), transmitting clover-club-leaf virus, acquired the virus through the egg. Transmission by this method was effective, with 24 out of 27 insects transmitting. Black used the same method as Fukushi except that he transferred the insects at weekly intervals. However, none of the insects infected the test plants until at least 3 wk had elapsed; the best transmission occurred during the 7–11 wk, inclusive. Later, Black (1950, 1953) reported similar results in an experiment wherein 21 generations of insects were developed over a period of 5 yr, an experiment that was so conclusive that it is now considered to be the model of its kind. The experiment was developed to test multiplication of the virus in the insect since transovarial acquisition had already been proved, but the additional evidence for the latter is overwhelming.

The same species of leafhopper, *A. novella,* was also shown to acquire the wound tumor virus transovarially, but only about 1.8% of the prog-

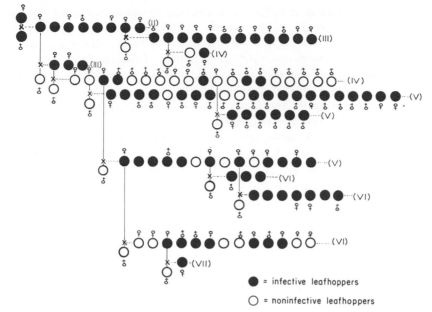

FIGURE 13.8. Diagram showing transovarial transmission of rice dwarf virus by *Nephotettix apicalis cincticeps* (After Fukushi, 1939).

eny were infective. An occasional female showed a much higher rate of transovarial passage. Females of *A. constricta,* reared on crimson clover infected with New Jersey potato yellow dwarf virus, oviposited eggs which produced about 0.8% of infective progeny (Black, 1953).

Grylls (1954) recorded the transovarial transmission of rugose leaf curl of alfalfa (lucerne) in *Austroagallia torrida.* The disease has a long incubation period in the plant, 3 or 4 mo in the indicator plant used (*Malva parviflora* L), which is perhaps why Grylls did not think it necessary to transfer the single, newly emerged nymphs to a succession of test plants. Approximately 3 mo after the insects were removed, 42% of the test plants showed symptoms.

Yamada and Yamamoto (1955) showed that *Delphacodes striatella* Fallén transmitted the virus of rice stripe disease and that transovarial passage occurred. Watson and Sinha (1959) found some interesting aspects of transovarial transmission in a study of *Delphacodes pellucida* Fabr., the vector of European wheat striate mosaic. Transmission through the egg is a characteristic of efficient transmitter races of the vector, the nymphs of inefficient races inheriting little or no virus. Even the efficient races transmit to the eggs only if they have had a long feeding period on the infected plants, and females fed on infected plants only during mating usually did not transmit to their progeny, although they themselves were able to infect after a latent period of 15 days after the acquisition feeding.

The transovarial transmission of hoja blanca of rice has been shown to occur in the progeny of nontransmitting females of *Sogatodes oryzicola* (Muir), indicating at least that the insect might be viruliferous but non-infective even though the mechanism for transovarial transmission is not impaired (Showers and Everett, 1967).

Wheat streak mosaic virus, transmitted by *Calligypona pellucida* (F.) after transovarial acquisition, nevertheless increases in infectivity by 61.6% after an incubation period of 20–34 days and 98.2% after 40–51 days (Prusa and Vacke, 1960). If infectivity increases after the initial transmission from the infected mother to the egg, then the number of virus particles transmitted was too low for maximum infectivity; the latent period elapsing before this was reached was so long as to definitely place the virus as propagative.

Although transovarial transmission has been known in the past only for the leafhopper vectors, the phenomenon is now known to occur in two aphid vectors, *Myzus persicae* (Sulz.) transmitting potato leaf roll (Myamoto and Myamoto, 1966), and *Hyperomyzus lactucae* (L) transmitting yellow vein of sowthistle (Sylvester, 1969). A third candidate is turnip mosaic propagative in its vector (Duffus, 1963) and lettuce necrotic yellows (Stubbs, 1968).

As far as it is known there are no cases of transovarial transmission among white fly-transmitted viruses, but negative results have been reported (Costa, 1969).

The role of the symbionts in the vector-virus relationships of a leafhopper was possibly considered for the first time by Carter (1939) who reported negative results from studies of active and inactive races of *Cicadulina mbila* (Naude), but with respect to viruses and their relationship to the mycetomes and symbionts of vectors, there have been some important contributions made during the last decade. Vidano (1966) showed that virus particles of maize rough dwarf occurred in the salivary glands and mycetomes of *Laodephax striatellus* Fallén. In the mycetomes they were found in high concentration in lumen cells usually containing *Blastomycetes* as symbionts.

The role of the mycetomes and symbionts of vectors in the mechanism of transovarial transmission has been reported in elegant detail by Nazu, whose earlier work (1963, 1965) was followed by a complete treatment in 1969. This last report clearly establishes the presence of virus particles in the mycetome and the attachment of these to the symbionts. In this manner the virus could be carried along in the hemolymph and enter the egg along with the symbiont (Nazu, 1969). Perhaps if titer of the virus particles in the hemolymph were high, some might become attached to the symbiont while it was en route through the hemolymph without having gone through this process in the mycetome.

There must, however, be major limiting factors involved with respect to hemipterous vectors in general, because otherwise, transovarial transmission would be a much more common occurrence than it is now known to be. Carter, in the first edition of this book, suggested the possibility of involvement of the mycetomes and symbionts in transovarial transmission, and noted then that the conditions for entry must be quite specific and limited in time. The known cases of transovarial transmission are shown in Table 13.2.

TABLE 13.2. Viruses Transovarially Transmitted in Their Vectors

Virus	Vector	Reference
Alfalfa rugose leaf curl	*Asteragallia torrida* (Evans)	Grylls (1954)
Clover club-leaf	*Agalliopsis novella* (Say)	Black (1948)
Clover witch's broom	*Euscelis plebeja* (Fallén)	Posnette and Ellenberger (1963)
European wheat striate mosaic	*Calligypona pellucida* (F.)	Watson and Sinha (1959)
Oat dwarf tillering	*Calligypona pellucida* (F.)	Lindsten (1961)
Oat striate mosaic	*Calligypona pellucida* (F.)	Lindsten (1961)
Potato leafroll	*Myzus persicae* (Sulz.)	Miyamoto and Miyamoto (1966)
Potato yellow dwarf (N.J.)	*Agallia constricta* Van D.	Black (1950)
Rice dwarf (stunt)	*Nephotettix cincticeps* Ühler	Fukushi (1934)
	Inazuma dorsalis Motsch.	Shinkai (1954)
Rice hoja blanca	*Sogatodes oryzicola* (Muir)	Galvez *et al.* (1960)
Rice stripe	*Delphacodes striatella* Fallén	Kuribayashi (1931)
Sowthistle yellow vein	*Hyperomyzus lactucae* (L)	Sylvester (1969)
Strawberry green petal	*Euscelis plebejus* (Fallén)	Posnette and Ellenberger (1963)
Winter wheat mosaic	*Psammotettix striatus* (L)	Shaskol Skaya (1962)
Wound tumor	*Agallia novella* (Say)	Black (1963)

MULTIPLICATION OF VIRUSES IN INSECT VECTORS

Fukushi was the first to interpret his data on transmission through the egg as demonstrating the multiplication of virus in the vector (1939, 1940). The hub of the controversy that followed was whether the amount of virus in the female leafhopper with which Fukushi began his test was sufficient to produce all the infections recorded through succeeding generations, without multiplication taking place. It was perhaps true that there was enough virus in a single leafhopper egg to infect all the 1200 plants reported by Fukushi, but the virus charge still had to be distributed through six generations during a period of over one year. Black (1953a) calculated that the virus present in the original leafhopper underwent an estimated minimum dilution of 1:563,000 and quite properly points out that this minimum was greatly underestimated. Only some of the eggs of each female were used, and there was probably a greater amount of virus left in each female than entered the eggs.

Black's classic study of clover-club-leaf virus was made with full recognition of the questions raised in the discussion of Fukushi's experiments. Black mated a virus-bearing female leafhopper with a virus-free male, and the pair were then caged on a Grimm alfalfa plant (*Medicago sativa* L). This plant is immune to the virus. Extreme precautions were taken to prevent the germination of any weed seeds in the pots in which the alfalfa was grown in order to avoid any possibility of contamination with other plants that might have been symptomless carriers of the virus. The female produced 42 nymphs, of which 21 were each tested on a series of four crimson clover plants (*Trifolium incarnatum* L) and then discarded. Of the 21 nymphs, 15 produced infections, which meant that the virus in the original female had been diluted approximately 1:30 in each of her progeny, provided that all the virus in the female had been available for distribution to the eggs, which was extremely unlikely. The 21 remaining progeny, when adult, were mated with virus-free males. This procedure was continued for more than 5 yr through 21 generations of insects grown only on immune alfalfa, without loss of infectivity. The main line of descent is shown in Table 13.3.

It should be noted that the dilution of the original virus, if no multiplication in the insect occurred, exceeded $1:2.8 \times 10^{-26}$. Such a dilution, even with viruses easily mechanically transmitted (and therefore amenable to dilution studies), is completely out of the range of that known for any virus. The conclusion that multiplication of the virus occurred in the insect seems, therefore, to be inescapable.

TABLE 13.3. *Calculated Minimum Dilution of Clover-Club-Leaf Virus during Passage Through the Egg of Its Insect Vector for 21 Generations* [a]

Generation	Number of progeny	Infectivity tests	Reciprocal of dilution	
			Each generation	Total
1	42	15/21	30	30
2	101	6/15	40	12×10^2
3	35	3/15	7	84×10^2
4	54	9/15	32	$2,688 \times 10^2$
5	60	3/15	12	$32,256 \times 10^2$
6	89	9/15	53	$1,709,568 \times 10^2$
7	104	6/15	42	$71,801,856 \times 10^2$
8	106	15/20	80	$574,414,848 \times 10^3$
9	64	3/20	10	$574,414,848 \times 10^4$
10	173	5/11	79	$453,787,729 \times 10^6$
11	88	6/21	25	$113,446,932 \times 10^8$
12	31	2/11	6	$680,681,594 \times 10^8$
13	52	0/15	0	$680,681,594 \times 10^8$
14	59	1/10	6	$408,408,956 \times 10^9$
15	129	1/5	26	$106,186,328 \times 10^{11}$
16	60	11/20	33	$350,414,885 \times 10^{12}$
17	114	2/15	15	$525,622,327 \times 10^{13}$
18	50	2/15	7	$367,935,629 \times 10^{14}$
19	107	1/10	11	$404,729,192 \times 10^{15}$
20	128	5/10	64	$259,026,683 \times 10^{17}$
21	105	1/10	11	$284,929,351 \times 10^{18}$
				or 2.8×10^{26}

[a] Data from Black (1953b).

There have been other approaches to the problem of multiplication of virus in insect vectors. Kunkel (1926), on the basis of an incubation (latent) period in the insect of 9 days, and the life-long retention of the virus in the insect, suggested that perhaps the virus multiplied in the insect.

Kunkel (1937) followed this suggestion by experimenting with the effect of heat on the virus in the insect. Ineffective leafhoppers, kept at approximately 35°C for varying lengths of time, were unable to transmit the virus immediately when they were returned to more normal temperatures of around 24°C, but eventually regained the ability to transmit without access to fresh sources of virus. On holding at the higher temperatures for

12 days or longer, the ability to transmit was completely lost, and the insects could only be rendered infective again by feeding on a fresh source of virus. Kunkel interpreted the effect of the shorter heat treatments as one which lowered the virus concentration in proportion to the temperature and duration of the treatment, and the time for regaining infectivity as a period comparable to the original latent period in the insect. His term "heat-induced incubation period" is not appropriate; the heat treatment induced the virus loss, not its restoration.

Kunkel reported that the leafhoppers could be grown at 35°C in what appeared to be a normal manner, but at this temperature the normal sigmoid curve for the development of insects has already turned down. This slowing down of the rate of development of the insect would be difficult to observe, especially with adults in varying stages of maturation. This observation does not negate Kunkel's findings in any manner, but it does suggest that the insect was not merely a convenient test tube in which to subject the virus to high temperatures. On the contrary, Kunkel's basic thesis, which went beyond the mere subjection of a virus to heat, and which envisioned the virus as a living system, would gain support from the concept that the insect's physiology, as expressed in rate of development, coincided with that of the virus. The complete loss of the virus at the longest exposure periods could mean one of two things; that the virus was actually "killed," having a lower thermal death point than the insect, or that the virus was exhausted by excretion during the abnormal metabolism of the insect at the high temperatures used. A leafhopper kept at 35°C feeds and excretes at a high rate. The data from Kunkel's experiments have been summarized by Black (1953) and are shown in Table 13.4.

Black (1941) used an entirely different approach. Starting with a large colony of leafhoppers previously checked and found virus-free, Black then transferred the colony to infected aster plants for one day and subsequently transferred the colony to a succession of aster plants. On the second, fourth, eighth, twelfth, and sixteenth days after feeding on the diseased plants, samples of 50 insects were withdrawn from the colony and tested for virus at different dilutions. Black's procedure is repeated here verbatim: "In order to test for virus, the following procedure was carried out at 0°C.: The sample of source insects was ground. Sufficient neutral 0.85 percent NaCl sollution was added to dilute the weighted leafhoppers the desired amount. The suspension was cleared by low-speed centrifugation, and minute quantities from each dilution were injected by glass capillaries into 120 young, virus-free adults. These were then placed on virus immune rye plants, and maintained in the greenhouse for three weeks, while the virus passed through the incubation period. The surviv-

TABLE 13.4. Effects of Heat Treatments on Aster Yellows Virus in Macrosteles divisus (Uhl.)[a]

Experiment	Temperature of treatment, °C	Days of heat treatment followed by days elapsing before infectivity was regained
A1[b]	31	2–0[c]
A2	31	13–00[d]
A3	31	7–3, 10–3
A5	32	6–16, 12–00
A6	32	1–0, 2–7, 3–8
A7	32	4–2, 5–5, 10–6
A8	32	7–2, 7–6, 8–13, 8–17
B[b]	36	1–1, 2–3, 3–3, 4–6, 5–8, 6–13, 7–23, 8–00

[a] Data from Kunkel and summarized by Black (1953b).
[b] Kunkel reported experiments labeled A in 1937, B in 1941.
[c] The first figure indicates the number of days the insects were kept at the temperature indicated in Col. 2; the second figure indicates the days that elapsed after the heat treatment before the insects regained their ability to infect plants.
[d] The symbol 00 indicates that the insects never regained their ability to infect plants.

ing injected insects were then tested in colonies of five on aster plants; each colony fed on one aster plant for one week, and on a second aster plant for two weeks." The results are shown in Table 13.5. It seems clear from this evidence that multiplication of the virus had been demonstrated.

Maramorosch (1952) succeeded in applying the mechanical inoculation technique to achieve serial passage of the aster yellows virus through 10 groups of leafhoppers. It was calculated that the virus would be diluted approximately 10^{-4} at each passage and would become progressively less concentrated unless it multiplied to a like extent. During each of the first 6 passages, the insects were kept on rye for 30 days, rye being immune to infection by the virus. The survivors were then tested individually on aster plants for 2 days before being used as a source of virus for subsequent passages. During the last 4 passages, the insects were kept for long periods on asters, being transferred to fresh plants 3 times each week or with 5-day intervals. The results of this experiment are in Table 13.5A.

The dilution of the original virus used for injection in the tenth passage, as shown in the table, would have reached 10^{-40}, but Maramorosch (1952a) showed that the virus in the tenth passage was just as concen-

TABLE 13.5. Transmission of Aster Yellows Virus [a,b]

Ratio of number of colonies infective to number of colonies tested

Day of injection[c]	Dilution of juice injected			Control insects	
	10^{-1}	10^{-2}	10^{-3}	Uninoculated	Inoculated[d]
2	0/19	0/20		0/17	17/19
4	0/19	1/19		0/18	18/18
8		3/20	1/20	0/19	19/19
12		13/14	18/18	0/18	20/20
16		7/24[e]	9/19	0/17	17/17

[a] After Black (1953b).

[b] By colonies of 5 leafhoppers inoculated with various dilutions of juice from insects fed on diseased aster plants for 1 day.

[c] During these days, juice from source insects was injected into test insects.

[d] Inoculated with infective insect juice diluted 10^{-2}.

[e] In this case, only 24 insects survived, and they were tested individually on asters.

TABLE 13.5A. Serial Passage of Aster Yellows Virus Through the Aster Leafhopper [a]

Passage	No. of injected insects	Days after injection on		Survivors	Infective survivors	Control insects	Calculated dilution
		Rye	Asters				
1	200	30	2	A 36	8	100	$10^{-3,8}$
2	300	30	2	16	3	100	10^{-8}
3	150	30	2	2	0	100	10^{-12}
4[b,c]	300	30	2	34	5	100	10^{-16}
5	100	30	2	5	0	100	10^{-20}
6[c]	200	30	2	22	5	100	10^{-24}
7[c]	100	6	54[d]	B 8	2	20	10^{-28}
8	240	40	25[e]	12	4	20	10^{-32}
9	50	19	13[e]	16	6	20	10^{-36}
10	50	30	20	4	1	20	10^{-40}

[a] Data from Maramorosch and summarized by Black (1953b).

[b] The diluted insect juice in the fourth passage was filtered through sintered glass.

[c] Penicillin was added in the fourth, fifth, sixth, and seventh passages.

[d] Individual insects were transferred to individual aster plants three times weekly.

[e] Individual insects were transferred to individual aster plants every 5 days.

trated as in the first. This was achieved by using the correlation that exists between concentration of the virus and the incubation period for symptom expression in the plant. Insects were injected with 1 / 8000 cc volume of inoculum derived from dilutions of 10^{-1} and 10^{-3} of 100 macerated leafhoppers. At the first passage with a dilution of 10^{-1}, the shortest incubation periods were 11, 14, and 15 days. With a dilution of 10^{-3}, the shortest periods were 24, 25, 28, and 38 days. Although there are some apparent fluctuations in concentration on the seventh passage, the minimum incubation periods of 23, 25, 27, and 29 days on the ninth passage indicated that there had been no reduction in virus concentration after eight passages.

The wound tumor virus has also been successfully passed serially through the insect vector *Agallia constricta* Van Duzee, by Black and Brakke (1952). In these experiments, the injected leafhoppers were kept on immune Grimm alfalfa for 2 wk and then tested individually on susceptible crimson clover plants for 1 mo. Other insects from the same lot, used as the source of the virus for succeeding passages, were kept on the alfalfa for a month. The dilution of the original source of virus at each transfer was calculated by using the weights of the leafhoppers, the total volumes of the dilutions injected, and the proportion of the injected insects used as a source for further passages. The results are shown in Table 13.6.

The logarithms of the percentages (Table 13.6) of infective leafhoppers inoculated by injection were directly proportional to the logarithms of the virus concentration; it is obvious that the original source virus was diluted to about 10^{-18}, although the concentration of virus in the insects at each passage was at approximately the same level.

All the experiments thus far discussed have, perforce, dealt with insects which were kept on plant tissue. Maramorosch (1956) attempted to eliminate this factor so as to test virus multiplication in the absence of plant juices. His method was to feed nymphs from a virus-free stock on infected plants for 2 days, to cut up these nymphs in small pieces, and then to culture them in an artificial medium for 10 days. The cultures were then crushed and centrifuged, and the cell-free supernatant fluid was used for the injection of virus-free adult leafhoppers. Eleven of these leafhoppers became infective after a latent period of a minimum of 29 days after injection. Three of the 11 transmitted only once.

While the tissue culture medium was completely plant-cell-free before the cut-up nymphs were introduced, the removal of the nymphs from infected plants just before this treatment must have introduced into the culture medium all the residual plant material present in the nymphs at the time of transfer. Perhaps the evidence for virus multiplication in the absence of plant cells would have been more complete if it had been pos-

TABLE 13.6. *Serial Transmission of Wound Tumor Virus Through an Insect Vector (Agallia constricta) Grown on Immune Alfalfa*[a]

| Passage | Infectivity of insects inoculated with insect juice at indicated dilutions | | | | | | | Fraction of original virus used as source for each dilution series (minus log) | Uninoculated control insects |
	10^{-7}	10^{-6}	10^{-5}	10^{-4}	10^{-3}	10^{-2}	$10^{-1.5}$		
1	0/24[b]	0/21	0/17	1/20	1/20	6/20	9/30	0.00	0/30
2			0/28	1/28	5/27		6/28	3.06	0/27
3			1/16	3/20	4/20		4/10	6.05	0/20
4				0/20	3/27	11/27		8.76	0/24
5				1/27	5/29	11/30		12.13	0/25
6				0/25	8/28	15/28		15.24	0/27
7				2/30	8/28	17/30		18.32	0/28
Total infective	0/24	0/21	1/61	8/170	34/179	60/135	19/68		0/181
Percentage infective	0	0	1.6	4.7	19.0	44.4	27.9		0

[a] After Black (1953b).
[b] Numerator is the number of insects infective; denominator is the number of insects tested.

sible to feed the nymphs artificially for as long as possible, short of the normal latent period, before maceration and introduction into the culture medium. "Culture medium" might also properly be used in quotes; it is possible that it was merely a suitable medium for *in vitro* survival of the virus.

There are conflicting views on whether the virus of potato leaf roll multiplies in its insect vector, just as there is a similar conflict as to whether there is a latent period of the virus in the insect. The two phenomena are, of course, related, for if transmission is affected without a latent period in the insect, there would be no question that multiplication had not occurred. Latent periods of up to 49 hr (Kassanis, 1952) and 54 hr (K. M. Smith, 1929, 1931) have been recorded; but, on the other hand, other workers have reported that the virus could be transmitted by aphids with a minimum latent period of 1½ hr (Kirkpatrick and Ross, 1952), while Klostermeyer (1953) reported transmission following a combined acquisition and inoculation feeding period of 20 min.

Day (1955) was able to recover virus from the hemolymph of aphids fed on an infected plant for 2 wk and inject this into virus-free aphids. Only 6 / 146 infections occurred, however, a frequency too small to permit using the technique for serial passaging. Day's conclusion that there is multiplication, at least to a limited extent in the aphid vectors, is based on three experimental results: (1) the ability of the aphid to transmit is not related to the duration of the acquisition, (2) the maximum efficiency of transmittal increases with time, and (3) the frequency distribution of the number of infections plotted against the duration of the infection feeding time follows an exponential curve.

Heinze (1955) reported a higher degree of success with transmission of leaf roll virus by injection than did Day, and, on the basis of this finding, Stegwee and Ponsen (1958) attempted to perform serial passage experiments of leaf roll virus through a succession of *M. persicae*. When aphids were injected, using the technique described by Worst (1954), either with juice obtained by macerating the insects or with hemolymph drawn from the insect with the injection needle, infections resulted with dilutions of 1:10 at 7–10 days after injection when the insects were serially transferred at 3–4 day intervals to *Physalis floridana* test plants. Using aphids injected with dilutions of 1:5, however, the largest number of infections occurred on the 1st series of test plants. The number dropped rapidly with the two succeeding transfers, this drop coinciding with the beginning of a high percentage of infections following injections with the greater dilutions. The authors noted this result, which is shown in Table 13.7, but made no attempt to relate it to their final conclusions. It should be noted, however, that Stegwee and Ponsen's table reveals no significant

TABLE 13.7.   Transmission of Potato Leaf Roll Virus by Injected Myzus persicae[a]

	Plants infected/total tested				
	Test feeding, in days after injection				
	0–3	4–6	7–10	11–13	14–17
Aphids injected with					
Juice 1 : 100[b]	0/9	0/5	0/3	0/3	0/2
Juice 1 : 10[b]	0/8	0/8	4/8	3/8	2/4
Juice 1 : 10[c]	0/30	0/20	7/13	6/8	4/8
Juice 1 : 5[d]	5/10	2/9	1/5	—	—
Hemolymph	4/10	2/5	0/2	—	—
Average mortality, %	45	62	73	82	100
Noninoculated controls					
Viruliferous	20/20	17/20	16/17	—	—
Virus-free	0/20	0/20	0/20	0/20	—
Average mortality, %	20	42	53	70	96
Insect-free control plants	0/25				

[a] After Stegwee and Ponsen (1958). Virus-free aphids were injected with approximately 0.004 liter of virus-containing solution and subsequently were transferred for periods of 3 to 4 days to successive *Physalis floridana* test plants.
[b] Five aphids per test plant.
[c] One aphid per test plant.
[d] Calculated from the data for insects injected with juice 1 : 100 and 1 : 10.

differences in percent of infection obtained for the three periods after injection with juice at 1:10 or for the two last periods at 1:5. The results from Table 13.7 support the validity of Day's hypothesis (1955) that, with a large amount of virus ingested, some virus may reach the salivary glands before multiplication has occurred, but they also suggest that, if this occurs, not only is there no multiplication, but also there is no latent period and very short retention.

The authors were, however, successful in showing serial passage of the virus through 15 transfers from insect to insect, involving a dilution of 10^{-21}, leading to the reasonable conclusion that multiplication in the insect vector did, in fact, occur.

Harrison (1958) succeeded in rendering *M. persicae* infective by injecting small volumes of extracts of aphids which had acquired potato leaf roll by feeding on infected plants. Fifty percent of the injected aphids transmitted the virus for periods up to 18 days. There was apparently a latent period of 20 hr. The infectivity of the aphid extracts increased with the length of the acquisition feeding. However, when insects fed 1–2 days on potato infected by leaf roll were transferred to turnip, their virus content decreased with time.

Harrison interprets his data as indicating lack of multiplication of the virus in the aphid, particularly: (1) the increase in infectivity of extracts with increasing acquisition feeding time, and (2) the decrease in infectivity of extracts with time, which is interpreted as a gradual loss of the original "charge" and which applies also to aphids that have acquired infectivity by feeding. This latter point is not considered conclusive by Stegwee and Ponsen (1958), who prefer to interpret the reduction in concentration as a "net" decrease, being the difference in the amount ingested and increased by multiplication within the vector, and the amount egested by the insect in feeding. To establish their point, however, would require quantitative measures of virus concentration ingested in a total acquisition feed, as well as the amount necessary to infect when egested. Clearly, the amount ingested, although affected by virus concentration at the acquisition sites, is more than enough to infect.

The best documented case against multiplication of a virus in its insect vector is that of curly top virus and *C. tenellus* (Freitag, 1936; Bennett and Wallace, 1938). Freitag assumed that if multiplication of the virus of curly top occurred, the insects should retain the infective capacity throughout their entire adult life. Furthermore, an insect fed for only a short time on a diseased beet should also be able to cause as many infections over a period of time as those fed for longer periods. Freitag's results appear to be conclusive in that he showed clearly that there was a gradual decrease in percentage of beets infected by the leafhopper when the insects were transferred at daily intervals to fresh beets over a 30-day period. The longer the acquistion feed, the longer the insects remained highly infective. When infective adults were caged on an immune plant (sweet corn) and tested every tenth day for their capacity to transmit, there was a gradual loss from 50% transmission on the first day to 3.1% on the ninetieth day. However, when insects, which had lost their capacity to transmit as they approached old age, were allowed to feed on a fresh source of virus, their transmitting capacity was equal to that of freshly molted adults.

Freitag's conclusion was that there was no evidence for multiplication. Bennett and Wallace (1938), agreeing that long retention of the curly top virus occurs in the vector, also agree with Freitag that multiplication does not occur, or at least does not occur in quantity sufficient to maintain the original virus content, which, they showed clearly, decreased with time; and the smaller the initial charge (i.e., the length of acquisition feeding), the more rapid the decrease. Leafhoppers rendered noninfective by exhaustion of their virus content were readily made infective again when fed on a fresh source of virus. Long retention of a virus in the insect vector does not necessarily imply that multiplication is taking place. As Bennett and Wallace point out, retention might be affected by the extent and distribution of virus reservoirs in the insect and the rate at which virus is diffused from, and passed out of, these reservoirs.

Maramorosch (1955) succeeded in mechanically transmitting the curly top virus to *C. tenellus* by needle injection, but he appears to ascribe more importance to the effect of dilution of the inoculum in influencing "incubation period" than his data support. His results are shown in Table 13.8.

Table 13.8 shows that the 3 cases of short incubation period in the insects injected with a 1:30 dilution were followed by long periods of failure to transmit; 2 insects out of 5 injected with the 1:300 dilution had shorter incubation periods than 2 out of 6 injected with the higher concentration.

A more significant comparison can be made between the record of transmissions for the 1:30 and 1:300 dilutions. Note that the frequency of transmission was considerably higher in the 1:30 series, which would be expected on the basis of Bennett and Wallace's interpretation. If serial transfer of the virus from insect to insect could be achieved, it would, as Maramorosch suggests, be much more convincing evidence for multiplication of the curly top virus in its vector. Table 13.9 summarizes the evidence on multiplication of plant viruses in their insect vectors.

PERSISTENCE OF VIRUSES IN INSECTS

It became apparent quite early in the study of vector-virus relationships that insect transmission of plant viruses fell into two rather broad categories according to events immediately following the virus acquisition feeding: (1) those in which the insect is unable to infect the first series of test plants on which it feeds for a short time, but after a lapse of time could begin a period of infectivity which might continue throughout its' lifetime; and (2) those in which the insect infects the first plant on which it feeds, but few or none after that (Storey, 1939).

The Latent Period

Clearly the difference thus far lies in the occurrence of a waiting or latent period in the insect in the first type of transmission, which does not occur in the second. This period has also been known as the incubation period, but that term is best restricted to the time between exposure to infection

TABLE 13.8. *Effect of Dosage on Incubation Period of Curly Top Virus in Mechanically Inoculated Beet Leafhoppers*[a]

Dilution	Insect No.	Transmission record (days of transfer)																											Incubation period (days)
		1	2	3	4	6	7	8	9	10	12	13	14	15	17	19	20	21	22	23	26	27	28	29	30	35	36	37	
1:30	1	−	+	+	+	−	−	−	P	−	−	+	−	+	+	−	−	+	−	D									1
	2	−	−	+	+	−	−	−	−	−	−	−	−	+	−	−	−	−	−	−	−	−	−	−	D				2
	3	−	−	+	−	−	−	−	−	−	−	−	P	−	−	+	−	+	+	+	−	+	−	+D					2
	4	−	−	−	−	−	+	+	+	+	+	+	+	+	+	+	+	−	+	+	+	+D							6
	5	−	−	P	P	−	−	+	−	+	+	+	−	+	+	−	−	+	+	+	+	−	+	+	+	−	+	−	7
	6	−	−	−	−	−	−	−	+	+	+	−	−	+	−	+	+	+	+	+	+	−	+	+	+	+	−	+	9
		(No. 7–15 did not transmit)																											
1:300	1	−	−	−	−	+	+	P	−	−	−	+	−	−	+	+	−	−	+	+	+	−	−	−	+	+	−	+	5
	2	−	−	−	−	+	+	−	D																				5
	3	−	−	−	−	−	−	−	−	−	−	−	−	+	−	+	+	−	−	−D	−	−	−	−	−				14
	4	−	−	−	−	−	−	−	−	−	−	−	−	−	+	−	−	−	−	+	−	+	+	−	−	−	−	−	16
	5	−	−	−	−	−	−	−	−	−	−	−	−	−	−	−	+	−	−	−	−	−	+	−D					20
		(No. 6–12 did not transmit)																											

[a] After Maramorosch (1955).

Key: − indicates healthy plant; + indicates diseased plant; P indicates that plant died prematurely, not due to curly top; D indicates that leafhopper was found dead.

TABLE 13.9. *Summary of Evidence on Multiplication of Plant Viruses in Their Insect Vectors*

Virus (common name)	Method of demonstrating virus multiplication	Reference
Rice stunt	Transovarial passage (6 generations)	Fukushi (1940)
Aster yellows	Effect of heat on incubation period	Kunkel (1937, 1941)
	Titration of virus, acquired by feeding, during incubation	Black (1941)
	Correlation of incubation period in plant and insect	Kunkel (1937)
	Titration of virus acquired by feeding as well as by injection	Maramorosch (1953)
	Serial passage (10 transfers)	Maramorosch (1952)
	Effect of volume on incubation	Maramorosch (1953)
Clover club-leaf	Transovarial passage (21 generations)	Black (1950)
Wound tumor	Arrested incubation period at 0° C	
	Effect of dosage	
	Correlation of incubation in plants and insects at various temperatures	
	Serial passages (7 transfers)	Black and Brakke (1952)
Corn stunt	Correlation of incubation in plant and insect	Kunkel (1948)
	Serial passage (3 transfers)	Maramorosch (1952)
Potato leaf roll	Transmitting ability unrelated to length of acquisition feed	Day (1955)
	Maximum efficiency increases with time, and frequency of infection follows exponential curve	
	Serial passage (15 transfers)	Stegwee and Ponsen (1958)
	Conclusion negative [a]	Harrison (1958)
Curly top	Conclusion negative [a]	Freitag (1936)
	Conclusion negative [a]	Bennett and Wallace (1938)
	Positive conclusion indicated	Maramorosch (1955)

[a] See text pp. 570–580.

and the development of symptoms in the infected plant. If the term "latent period" is applied to the plant it should be limited in its meaning to the period between exposure to infection and the time when the insect can acquire the virus from the plant. The latent period in the plant may be the same length as the incubation period, but is often shorter. The latent periods in the insect vectors of a number of viruses are shown in Table 13.10. It will be noted that most of the insects listed in Table 13.10 are leafhoppers, but all fall into the first of Storey's categories.

TABLE 13.10.   Viruses with Latent Periods in Their Vectors

Virus	Insect vector	Latent period
Abaca bunchy top	*Pentalonia nigronervosa* Coq.	24 hr
Aster yellows	*Cicadula sexnotata* Fall.	10 days
Argentine curly top	*Agalliana ensigera* (Oman)	24–72 hr
Brazilian curly top	*Agallia albidula* Uhl.	24 hr
California aster yellows	*Gyponana hasta* DeL.	19–35 days
Cereal mosaic	*Delphax striatella* Fall.	6 days
Corn stunt	*Baldulus maidis* (DeL. and W.)	16 days
Corn mosaic	*Perigrinus maidis* Ashm.	4 days
Cotton leaf curl	*Bemisia gossypiperda* M. and L.	6 hr
Clover big vein	*Agallia constricta* Van D. *Agallia quadripunctata* Prov. *Agalliopsis novella* (Say)	13–15 days at 25–32° C
Cranberry false blossom	*Euscelis striatulus* Fall.	Undetermined
Elm phloem necrosis	*Scaphoideus luteolus* Van D.	Several days
Maize streak	*Cicadulina mbila* (Naude)	6 hr
Pea enation mosaic	*Myzus persicae* (Sulz.)	12 hr
Peach yellows	*Macropsis trimaculata* (Fitch)	8 days
Peach X-disease	*Collodonus geminatus* (Van Duzee)	22–30 days
Pierce's grapevine	*Draeculacephala minerva* Ball.	Less than 4 days
Potato leaf roll	*Myzus persicae* (Sulz.)	24 hr
Potato yellow dwarf	*Aceratagallia sanguinolenta* Prov.	9 days
Rice stunt	*Nephotettix apicalis* Motsch	3 days
Strawberry mild yellow edge	*Pentatrichopus fragariae* (Theob.)	1 day
Strawberry virus 3	*Pentatrichopus fragaefolii* (Cock.)	10–19 days
Sugar beet curly top	*Circulifer tenellus* (Baker)	1–44 days; mean 9.6 days
Sugar beet yellows	*Myzus persicae* (Sulz.)	30 min
Tobacco rosette	*Myzus persicae* (Sulz.)	Less than 1 day
Tomato spotted wilt	*Frankliniella insularis* Frankl.	5 days
Turnip yellow mosaic	*Phaedon cochleariae* Fabr.	24 hr

The Persistent and Nonpersistent Virus

Watson and Roberts (1939, 1940), in a study practically parallel to that of Storey, developed the concept, now classic, of the persistent and nonpersistent viruses. The persistent virus is essentially the same as in Storey's first category but with some essential additional requirements. There is generally a long minimum acquisition feeding period from several hours to many days, followed by a long latent period before the insect can transmit. There is also a long retention period in the insect, often for its entire remaining life, and there is a high degree of vector-virus specificity. Since Watson and Roberts published, the long retention period has been equated with virus multiplication (see p. 570) and specificity has been recognized as a many-faceted complex. On the other hand, the nonpersistent virus is acquired by the vector during a short acquisition feeding. There is only a short retention period, little or no vector-virus specificity (unless the group specificity of Day and Bennetts is considered), and most of the viruses in this category are easily mechanically transmitted.

There is little room for difference of opinion on the first category of the persistent virus, especially as it relates to the leafhoppers, thrips, and white flies. Prescribing the limits of persistence, on the one hand, and of nonpersistence, on the other, however, has proved difficult; with the steady accumulation of case histories, the distinction between the two categories is no longer as clear cut as the original generalization promised. No other generalization in the field of vector-virus relationships, however, has proved more effective in stimulating original research and the development of ingenious techniques.

The first modification of the categories of persistence and nonpersistence in viruses came from Watson herself. Watson had already studied the effects of preacquisition feed-fasting on the efficiency of transmission (1938). Her conclusion was that fasting up to an hour prior to the acquisition feed increased efficiency greatly, but efficiency did not increase proportionately with longer starvation.

This increased efficiency was inversely correlated with the length of the acquisition feeding, there being no difference between fasted and continuously fed aphids if the aphids were fed for an hour on infected leaves. Watson explained this by postulating the action of an inactivating enzyme, secreted by the insect only when it was actively feeding and therefore absent in the starved insect.

The terms persistent and nonpersistent refer to the length of time for which the viruses remain active in the vectors, and it is true that, in general, the persistent viruses are retained longer in the vectors than are the

TABLE 13.11. *Inoculation Threshold Periods of Some Nonpersistent Aphid-Borne Viruses* [a]

Virus	Inoculation threshold period (sec) [b]
Spinach blight (cucumber mosaic)	300
Cucumber mosaic	300
Red clover mosaic	600
Pea virus 2 (pea mosaic)	300
Hyoscyamus virus III	120
Tobacco etch	120
Poison hemlock ringspot	300
Beet mosaic	300
Spinach yellow dwarf	300
Cauliflower mosaic	300
Cauliflower mosaic complex	
Turnip virus 1 component	300
Cauliflower virus 1 component	300
Mild stock mosaic	300
Beet mosaic	10 [c]
Cabbage mosaic	300
Papaya ringspot	300
Brassica nigra virus	5
Dahlia mosaic	300
Pea mosaic	120
Severe stock mosaic	600
Japanese radish stunt	300
Cucumber mosaic, strains of	60
Alfalfa mosaic, strain of	10–20
Henbane mosaic, strain of	5
Potato virus A	10 [c]
Cabbage black ringspot	5

[a] After Sylvester (1954).
[b] Figures in column represent lowest interval tried; that interval gave a positive result.
[c] In this case, a 5-second interval gave negative results.

nonpersistent. However, it has become evident that there is so much variation in the retention time that the terms persistent and nonpersistent are no longer adequate to demark clearly two categories. Watson (1946) demonstrated that there was actually an overlapping of the virus survival times in the vectors, by a study of *M. persicae* as the common vector of sugar beet mosaic and sugar beet yellows. Sugar beet mosaic, although a nonpersistent virus, can nevertheless be transmitted to a series of test plants before infectivity is lost. Beet yellows, on the other hand, has the characteristics of a persistent virus in that infectivity is increased with increasing length of acquisition and transmission feeding, and is not affected by preliminary starvation. The main distinction between the two viruses appears to be the difference in response to preliminary fasting by the vector; and Watson suggests that this response is a better criterion for the two types of viruses than the retention factor, which is the basis for the concept of the persistent and nonpersistent categories. Sylvester (1954) has reported serial transmission tests of the nonpersistent *Brassica nigra* (L) virus by *M. persicae*. In twenty consecutive feedings, positive transmissions occurred throughout the series.

Sylvester's paper (1954), which is the most complete study available, lists the acquisition threshold, inoculation threshold, and transmission

TABLE 13.12.   Acquisition Threshold Periods of Some Nonpersistent Aphid-Borne Viruses [a]

Virus	Acquisition threshold period (sec) [b]
Spinach blight (cucumber mosaic)	600
Cucumber mosaic	300
Hyoscyamus virus III	120
Red clover mosaic	300
Pea virus 2 (pea mosaic)	300
Cucumber mosaic (strains Y and G)	120
Mild broad bean mosaic	600
Potato virus Y	120
Onion yellow dwarf	1800
Tobacco etch	120
Poison hemlock ringspot	300
Beet mosaic	120
Beet mosaic	300
Lettuce mosaic	300
Spinach yellow dwarf	300
Cabbage mosaic	300
Canna mosaic	300
Cauliflower mosaic	300–600
Cauliflower mosaic complex	
Turnip virus 1 component	300
Cauliflower virus 1 component	300
Mild stock mosaic	300
Aspermy of tomato	180

threshold periods for a number of nonpersistent viruses. The acquisition and inoculation thresholds are defined as the periods necessary for successful acquisition and inoculation, respectively; the transmission threshold, as the sum of the first two. These data are given in Tables 13.11–13.13. The retention periods for some nonpersistent viruses are shown in Table 13.14.

It will be clear from the data of these tables that although there is considerable variation in the time factors involved, the viruses concerned are definitely grouped separately from those shown in Table 13.10.

Sylvester has proposed an intermediate grouping, semipersistent, to include viruses such as beet yellows, which appears to have some of the characteristics of both the persistent and the nonpersistent types. Following Watson (1946), Sylvester (1956) used *M. persicae* with beet mosaic and beet yellows, but compared the two viruses on the basis of acquisi-

Beet mosaic	10^{c}
Brussels sprouts necrosis virus	300
Cantaloupe mosaic	60
Nemesia virus (cucumber mosaic strain)	120
Papaya ringspot	120
Subterranean clover virus	10^{c}
Brassica nigra virus	60
Dahlia mosaic	60
Nasturtium mosaic	5
Pea mosaic	120
Potato acuba mosaic (virus F+G)	300
Primula mosaic	120
Radish mosaic	30
Severe stock mosaic	30
Japanese radish stunt	300
Cucumber mosaic	120^{d}
Watermelon mosaic	18–36
Alfalfa mosaic (strain of)	15–45
Henbane mosaic (strain of)	$5-10^{e}$
Lettuce mosaic	60–300
Western celery mosaic	10^{c}
Potato virus A	15^{c}
Cabbage black ringspot	10^{c}

[a] After Sylvester (1954).
[b] Figures in column represent lowest interval tried; that interval gave positive results.
[c] In these cases a 5-sec interval gave negative results.
[d] In this case a 30-sec interval gave negative results.
[e] In this case a 0–5-sec interval gave negative results.

TABLE 13.13.   *Transmission Threshold Periods of Some Nonpersistent Aphid-Borne Viruses* [a]

Virus	Transmission threshold periods (sec) [b]
Spinach blight (cucumber mosaic)	900
Cucumber mosaic	600
Pea virus 2 (pea mosaic)	300
Hyoscyamus virus III	240
Tobacco etch	240
Poison hemlock ringspot	600
Beet mosaic	480
Beet mosaic	600
Spinach yellow dwarf	600
Cauliflower mosaic	600
Cauliflower mosaic complex	
Turnip virus 1 component	600
Cauliflower virus 1 component	600
Mild stock basic	300
Beet mosaic	42 [c]
Cabbage mosaic	600
Papaya ringspot	420
Brassica nigra virus	39 [c]
Dahlia mosaic	360
Pea mosaic	420
Severe stock mosaic	900
Cucumber mosaic, strains of	180
Alfalfa mosaic, strain of	25–65 [c]
Henbane mosaic, strain of	300
Cabbage black ringspot	30 [c]

[a] After Sylvester (1954).
[b] Figure in column is lowest interval tested or published.
[c] In these instances specific trials were made to determine the minimum.

tion, inoculation, and retention rather than preliminary fasting. The acquisition curves are similar, but the time scale for beet mosaic is in seconds while that for beet yellows is in hours. The suggestion is made that the difference in acquisition lies in plant-tissue-virus relationships, beet mosaic being acquired best from the epidermis and beet yellows from the mesophyll, or mesophyll-phloem region.

Inoculation thresholds are of the same order, that of beet mosaic being chartable in seconds and beet yellows in hours. Virus retention curves are

TABLE 13.14.   Maximum Reported Retention of Some of the Nonpersistent Viruses by Feeding Aphids[a]

Virus	Vector	Retention
Hyoscyamus virus III	*Myzus persicae*	6 but not 12 hr
Red clover mosaic	*M. persicae*	30 min
Pea virus 2 (pea mosaic)	*Macrosiphum pisi*	25 min
Vein mosaic of red clover	*M. pisi*	More than 1, but less than 24 hr
Western celery mosaic	*Aphis ferruginea striata*	Up to 10 hr
Tobacco etch	*Myzus persicae*	15 but not 30 min
Poison hemlock ringspot	*Rhopalosiphum conii*	8 hr
Beet mosaic	*Myzus persicae*	3 hr
Spinach yellow dwarf	*M. persicae*	2 hr
Cauliflower mosaic	*Brevicoryne brassicae*	2 but not 3 hr
Mild stock mosaic	*Rhopalosiphum pseudo-brassicae*	10 min
Cabbage mosaic	*Myzus persicae*	70 min (10-min transfers)
Papaya ringspot	*M. persicae*	5 min
Dahlia mosaic	*M. persicae*	More than 2 but less than 3 hr
Radish mosaic	*Brevicoryne brassicae* and *Rhopalosiphum pseudo-brassicae*	3 hr
Severe stock mosaic	*Myzus persicae*	2 hr
Japanese radish stunt	*M. persicae*	1 hr
Potato A	*M. persicae*	20 min
Cabbage black ringspot	*M. persicae*	55–110 min

[a] After Sylvester (1954).

also similar, the difference again being the time scale.

All these differences are essentially quantitative rather than qualitative, a conclusion in agreement with that reached before by K. M. Smith and Lea (1946) for the differences between persistent and nonpersistent viruses. Both viruses are dependent on insects for their transmission, but the vector-virus relationship is more specific with beet yellows, and Sylvester generalized that the more dependent a virus is on a particular host plant and the more specific the tissue regions involved, the more tenacious the vector-virus relationships are likely to be. Sylvester later (1958) developed what he designated an "operational classification" which summarizes, in graphic tabular form, the vector-virus relationships found among many aphids. These are shown in Table 13.15 and Fig. 13.9.

TABLE 13.15. Operational Classification of the Vector–Virus Relationships Found Among Aphids [a],[b]

Test	Aphid-borne plant virus		
	Nonpersistent	Semipersistent	Persistent
Juice inoculation	+	±	−
Tissue of acquisition	epidermal	mesophyll-phloem	mesophyll-phloem
Preliminary fasting	+	0	0
Prolonged access time	−	+	+
Latent period	none	none	variable
Retention (feeding)	min to hr	hr to days	days to life
Retention (fasting)	>feeding	<(feeding)	=feeding
Retention (ecdysis)	−(?)	+(?)	±
Vector specificity	generally low	medium	medium to high

[a] After Sylvester (1958).
[b] The symbols are used in the following context: +: possible, or increased; −: not normally possible, or decreased; 0: no effect; >: more than; <: less than; ?: theoretical, not confirmed experimentally; =: the same as.

THEORIES ON THE MECHANISM OF TRANSMISSION OF THE NONPERSISTENT VIRUS

The mechanisms whereby nonpersistent viruses are transmitted have been the subject of continuous debate since Hoggan (1931, 1933, 1934) proposed that these viruses are transmitted mechanically on the insect's stylets. Watson and Roberts (1939) argued that when several plants could be infected in succession by a single aphid, the mechanical transmission hypothesis was untenable. Again (1940) these same authors summarized the available evidence against the hypothesis. In their view, the hypothesis of mechanical transmission rests on three arguments: (1) that the transmission threshold is too short to have permitted the virus to have entered the insect's body and passed into the plant via the blood and salivary secretion, (2) that nonpersistent viruses are only transmitted by aphids to a single plant or, rarely, a second, after a single acquisition feeding, the assumption being that the feeding process cleans the stylets, the loss of infectivity therefore being also a mechanical process, and (3) retention during fasting was longer than during feeding, indicating that the stylets were not being decontaminated as rapidly during fasting, but that the virus was being lost at rates comparable with those of inactivation *in vitro*. Watson and Roberts (1940) concluded, on the contrary,

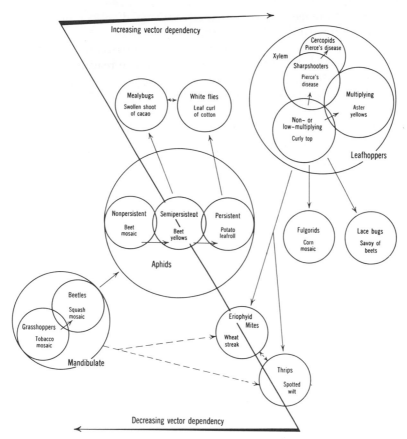

FIGURE 13.9. Diagram illustrating hypothetical relationships of the various groups of arthropod vectors of plant viruses (After Sylvester, 1958).

that the loss of infectivity was not due to mechanical cleansing of stylets but rather to the activity of a substance produced by the aphid while feeding but not while fasting. The effect of preliminary fasting in increasing the efficiency of nonpersistent virus vectors was ascribed to the absence of this inactivating substance in the fasted insect.

The Inactivator Hypothesis

The hypothesis that an inactivator is involved has found support from numerous other workers, although there are some anomalous situations reported. Kassanis (1947) showed that the vectors of dandelion mosaic virus do not improve efficiency with preliminary fasting; the length of the acquisition feeding is positively correlated with efficiency, but the virus is retained for only one hour. The first two characteristics would place the virus as persistent; the last, as nonpersistent. Prentice and Harris (1946) reported similar situations with certain strawberry viruses.

If the inactivator hypothesis includes as an essential part the concept of a generalized inactivator in aphids, then it fails to explain vector specificity and the lack of aphid transmission of many viruses, such as tobacco mosaic, potato X, turnip yellow mosaic, and others, which occur in relatively high concentrations in the plant hosts, are readily juice-transmissible, and are quite stable *in vitro*.

There appears to be no good reason, moreover, why a generalized inactivator should be postulated. There is ample evidence that specific insects produce specific oral secretions (See Chapters 4–6), and it should be expected that inactivators would also be specific to the insect producing them. Among strains of one insect species, differences in the inactivator might well be quantitative, while they are qualitative between species. With the concept of a specific inactivator, vector specificity should follow as a logical consequence.

The inability of aphids to transmit certain viruses is more difficult to explain by the specific inactivator hypothesis, but it is questionable that the hypothesis should be burdened with the necessity of offering an explanation in order to defend its validity otherwise.

The inhibitory action of insect saliva on viruses has been held responsible for the failure of aphids to transmit TMV, but Lojek showed (1969) that when TMV is purified, it can be transmitted by *Myzus persicae* and retained by the insect for five hours. On the other hand, Nishi (1962) found that potato virus X, while not aphid-transmitted, was nevertheless inhibited by aphid saliva. Later Nishi (1969) found that the inhibitory action was reversible. He appears to have successfully isolated and purified the inhibitory substance and described its properties. The saliva (buccal fluid) of the grasshopper *Tettigonia viridissima* (L) inactivates TMV on the mouth parts of the grasshopper in one hour, but both TMV and potato virus Y were transmitted before inactivation occurred (Schmetterer, 1961) (See also Orlob, 1963).

So-Called Biological Transmission

Another mechanism suggested for the aphid transmittal of the nonpersistent viruses is a so-called biological transmission, in which the virus is thought to be ingested, passing to the mid-gut, from which it moves into the hemocoele and reaches the salivary glands. This mechanism is generally agreed upon for the persistent viruses having a latent period in the insect, but there is no support for this suggestion in connection with the nonpersistent viruses. K. M. Smith and Lea (1946), referred to by Day and Irzykiewicz (1954) as suggesting this mechanism, actually make no such proposal. They simply attempt to show, by simple quantitative arguments, that the differences between the persistent and nonpersistent viruses are essentially quantitative, being differences in concentration in host plants and differences in the rate of inactivation in the insect vectors. Perhaps the reason for ascribing the suggestion of a biological transmission mechanism to these authors was their reference to the feeding of the aphid resulting in a full stomach after a 2-minute feed.

Hypotheses based on mechanical transmission, resulting from contamination of stylets only, actually do not or should not take into account ingestion, but rather only probing.

Role of Stylets

Since the salivary apparatus is clearly concerned in any proposed mechanism for transmission, it is not surprising that a hypothesis based on the feeding process and the role of the salivary sheath should be offered. Bradley (1952) found that where *M. persicae* feeds only on the epidermal cells, with brief punctures sufficing to acquire and transmit henbane mosaic, little or no saliva is injected. Bradley presumably referred solely to the gelling secretion (See Miles, 1959). However, even the gelling secretion is deposited earlier than Bradley recognized, for Van Hoof (1957) observed that the apex of the rostrum regularly makes a salivary imprint on the cuticula.

A well-defined sheath is taken as a barrier to acquisition or transmittal of viruses, the explanation being that the sheath serves as a filter for the insect's food, and only during its absence during brief feeding punctures can the insect acquire or transmit viruses. No one has yet shown, however, that the stylet sheath, composed as it is of a hardened gel, is even permeable, let alone permeable enough to function as a filter for the relatively huge quantity of sap imbibed by the insect.

Bradley and Ganong (1955a, 1957), however, have shown rather conclusively that some viruses at least are carried on the stylet tips of aphids; all those named are of the nonpersistent type. These findings lead the authors back to the hypothesis of Hoggan (1933), namely that the stylets carry the virus mechanically. The sites on the stylets where virus may be carried are a matter of conjecture, with as many opinions as possibilities. The feeding duct is favored as a site by Bawden *et al.* (1954) and Sylvester (1954). Van der Want (1954) thinks that viruses are carried externally on the stylets, and Day and Irzykiewicz (1954) tend to agree, with some reservations.

On the basis of the structure of the stylets (Fig. 13.10), Van Hoof (1957) suggests that, as the stylets are retracted, plant material is taken out and is retained behind the ridges on the tips of the mandibulary and/or the maxillary stylets; a healthy plant can later be infected with this material.

What most authors have failed to recognize is that the stylet sheath is actually an open-ended capillary tube. While it is being formed by the insect, it is quite possible that the extrusion of the gelling compound, as the insect seeks its favored tissue, might prevent capillary movement into the sheath tube. But once the favored tissue is reached, then both the pressure of the cell sap and capillary action in the sheath tube would function to bring the virus up into the feeding duct. It is not necessary, however, to postulate such action for short feeding periods in which a capillary sheath is not involved. Miles (1959) has at long last demonstrated what was logically obvious, namely, that, in addition to sheath material, the insect also secretes another component. It is independent of the gelling material, watery, and can be drawn back by capillarity (see Fig. 5.13). A single puncture with ejection of the watery component and subsequent capillary movement back into the stylet canal would suffice to "charge" the aphid with a dose of virus, infective in the next plant on which the aphid fed. At the same time, however, inactivating substances in the secretion, presumably in the nongelling component, would begin to function on a quantitative basis, depending on the relative quantities of plant sap, virus, and secretion components in the capillary-attracted column.

Sylvester (1954) and Day and Irzykiewicz (1954), in independent and parallel studies, proposed modifications of the mechanical transmission hypothesis to include the action of inhibitors. Sylvester made the following points: (1) transmission is mechanical in the sense that virus is carried in the food canal of aphids, (2) different species of aphids feed in a similar manner and in similar areas during initial stages of penetration, with variations depending on the tissue sought, (3) such aphids acquire a similar charge of virus when feeding but a short time on a given virus source plant, and (4) the action of inactivators which are present in the salivary secretions is not upon the virus, but rather upon the host plant cells into which the virus is injected; that is, the insect renders the host plant cell resistant or practically immune to infection. These considerations were used by Sylvester to present his "incompatibility hypothesis," which led to the conclusion that vector efficiency and specificity are due to compatibility factors which are dependent upon specific interactions

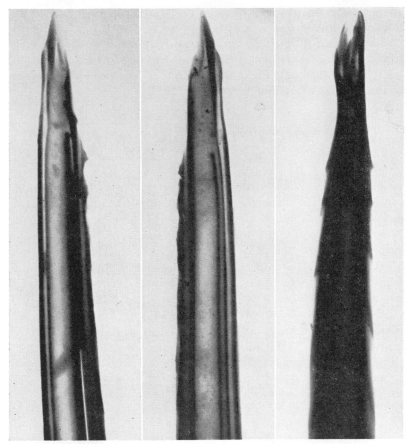

FIGURE 13.10. Electron micrographs of the apices of both of the maxillary stylets of *Myzus persicae* (Sulz.) and of the two together. Note that the apices do not form a closed needle's eye (Photograph: Van Hoof, 1957).

among the viruses, the saliva of the aphids, and the host plant cells being inoculated.

The working hypothesis of Day and Irzykiewicz (1954) is as follows: the stylets of the aphid become contaminated during probing in infected tissue; some of the virus on the stylets is inactivated by the salivary fluids, and viruses differ in their susceptibility to these inhibitors, and different species of aphids vary in the production of the inhibitor. Fasting decreases the activity of the inhibitors. The authors describe this as a "modified mechanical" hypothesis; used to reclassify the aphid-borne viruses, it led them to propose the terms "vector-direct" for those viruses previously known as nonpersistent and "vector-latent" for the persistent viruses. A useful test for which group a virus belongs to is to determine whether the vector can transmit following a molt. A positive result would place it in the persistent category. Stubbs (1955) found this to be true for carrot motley dwarf virus, although on other grounds it had been classified as a nonpersistent virus (Day and Irzykiewicz, 1954).

THE TRANSMISSION OF PERSISTENT VIRUSES BY APHIDS

Relatively few aphid-borne viruses are persistent viruses, but some few require a latent period between acquisition and transmission feedings. There is such a latent period, for example, in *M. persicae,* following feeding on an infected potato-leaf-roll plant, but workers have differed on its duration. Kirkpatrick and Ross (1952) reported a period of 1½ hours, but MacCarthy (1954) records periods of up to 120 hours. These differences have been thought to be due to differences between strains of the vector, and to differences in virus concentration in different host plants (Day, 1955a). This virus appears to be a model of its kind since (1) it can be isolated from the hemolymph of infected aphids, (2) the virus survives the molting process, (3) the ability to transmit is retained for at least 15 days, and (4) there is some evidence that the virus multiplies in the insect (Day, 1955a).

Another type of persistent virus is that of pea enation mosaic (Simons, 1954). This virus also has a latent period in its vector, *Macrosiphum pisi* (Kalt.), with nymphs showing a shorter mean latent period (30 hours) than the adults (56.8 hours). There is a positive correlation between the length of acquisition feeding and retention of the virus, but no effect on transmission due to postacquisition starvation. This is contrary to the experience of Watson (1946) with beet yellows, another persistent aphid-borne virus.

Day (1955) interprets Simons' results as indicating that no multiplication of the virus occurs in *M. pisi* and suggests that persistent viruses may include both types of transmission, that is, one involving multiplication of the virus in the vector and the other not. The latent period is an essential to the persistent virus category, but Simons makes it clear that the length of this period is extremely variable, and, in fact, may be so short as to be undemonstrable.

It has been shown earlier that great variation exists among the nonpersistent viruses with respect to their vector-virus relationships, that some viruses appear to have some characteristics of both persistence categories, and even in the small number of aphid-borne persistent viruses there are important anomalies. This all lends emphasis to the suggestion by Sylvester (1958) that the aphid-borne viruses can be considered as forming a continuum ranging in complexity from those which are transmissible mechanically on the stylets to those in which the virus is internally translocated among the various body tissues prior to injection via the salivary glands. Terms to describe categories among these, such as persistent, semipersistent, nonpersistent, or vector-direct and vector-latent, are terms of convenience and the dividing lines between them are not sharply defined. These examples will suffice to show that, with our present knowledge, it is not yet possible to wrap up vector-virus relationships in neat little packages of dogmatic classification.

There is steady progress, however, toward clarification. Day and Venables (1961) have set up rather precise definitions of the terms "persistent" and "nonpersistent" as follows: "A persistent virus is one in which: (a) the transmission time is long; (b) the virus is recoverable from the haemolymph of a vector; (c) the virus is transmitted following the moult of a vector; and (d) the vector is capable of becoming infective when purified virus is inoculated into the haemocoele. A nonpersistent virus is one in which (a) the transmission time is short; (b) the virus is not recoverable from the haemolymph; (c) the vector is not capable of transmitting following a moult; and (d) the vector does not become infective when purified virus is inoculated into the haemocoele."

More recently, Kennedy, Day, and Eastop (1962) have proposed the terms "stylet-borne" for those aphid-transmitted viruses carried inside the labial groove. The internally borne viruses are called "circulative" (after Black, 1953b). These terms refer to the inferred route of transport and are preferred by the authors to the terms persistent and nonpersistent, which simply refer to transmission characteristics.

There appears to be no place in the Kennedy *et al.* categories for the semipersistent virus (Frazier, 1966), nor can the differences be accounted for on the basis of mechanism of transport. The semipersistent viruses are not retained through molting, but they are clearly not stylet-borne, unless it can be shown that there is some mechanism whereby virus can be retained in the stylets for the duration of feedings over a period which may extend to hours; in the case of beet yellows, for two days (Sinha, 1968).

Bradley (1964) reviewed the extensive research on the transmission of the nonpersistent viruses and with the establishing of these as stylet-borne. Pirone more recently (1969) reviewed the same subject and concluded that while the evidence is that these viruses are carried at the distal end of the stylets, and that epidermal tissues are involved in both acquisition and transmittal, there is as yet no satisfactory explanation for the observed differences in efficiency or specificity of transmission.

The mechanisms of transmission which appear to be generally agreed upon place the aphid-borne viruses into two general categories: (1) depending on the route of transmission, or (2) depending on virus retention and transmission by the vector, i.e.:

(1) a. Virus carried on the stylet tips and lost on molting
     b. Virus ingested and transported in hemolymph, circulative, and retained through molting
(2) a. Nonpersistent, stylet-borne
     b. Semipersistent, not stylet-borne
     c. Persistent, circulative

Circulative is equated with propagative, but the terms are not necessarily synonymous; pea enation mosaic is circulative (Nault *et al.,* 1964) but whether it is propagative, that is, multiplies in the insect, is a matter of debate (Sylvester and Richardson, 1969). Sylvester has a critical review of these aspects (1966).

Examples can be quoted from other orders, e.g., *Bemisia tabaci* Gennadius and yellow leaf curl virus (Cohen, 1967), and of course the classical example of *Circulifer tenellus* (Baker) and curly top of sugarbeets.

Virus-vector relationships between oat-blue-dwarf virus and its one reported vector, the aster leafhopper (*Macrosteles fascifrons* Stål) have been described by Banttari and Zeyen (1970). A few nymphs and adults acquired the virus after a 15 min acquisition period, but the highest numbers became viruliferous with longer feedings, 1 wk for the nymphs and 2 days for adults. The shortest incubation period of the virus in leafhoppers was 7 days, but the majority transmitted between the 17th and the 34th day; with few exceptions, transmission ceased on the 66th day. Ground leafhoppers were a more efficient inoculum than were plant extracts when used for artificial inoculation.

The authors conclude that the indications are that the virus is propagative in the leafhopper, but from the erratic results with serial transmissions, it is possible that individual variation might result in circulative transmission, while in others, both circulative and propagative transmission occurs.

The multiplication of the aster-yellows agent following feeding on a diseased plant but not after injection was demonstrated for a nonvector insect, *Agallia quadripunctata* (Prov.) (Sinha and Chiykowski, 1967). Acquisition of viruses by nonvectors is known but this case appears to be unique.

The transmission of pea enation mosaic virus by the pea aphid is affected by the age of the vector, and by the length of acquisition and inoculation periods. Young nymphs of the pea aphid (*A. pisum*) are most efficient. Altering the acquisition access period can vary the virus dosage, and these variations can in turn affect the number of insects that transmit, the net transmission rate (which is defined as the average number of plants infected per vector), and, to a lesser extent, the length of time inoculativity is retained. With a circulative, but not propagative virus, such as PEMV, a decrease in the amount of virus in the vector will be expressed in a reduced probability of inoculating any given plant, rather than the period over which this probability will be distributed (Kyriaopoulou and Sylvester, 1969). Sylvester and Richardson (1966) had previously determined the effect of temperature on transmission and the biology of *A. pisum*.

CYCLICAL TRANSMISSION

Cyclical transmissibility of barley yellow dwarf by *M. avenae* occurred regularly in experiments described by Gill (1969). Increase to a major peak of transmission began on the 4th day from the start of the inoculation period into the plants and the 1st major peak occurred on the 6th day. From that point on, transmission occurred in a cyclical pattern. The emergence of successive leaves coincided with successive valleys in the transmission curves. The period for symptom development was dependent upon both temperature and light intensity, confirming Stimmann and Swenson (1967) in this respect. See also Tu and Ford (1969) with respect to maize dwarf mosaic in corn.

"Periodic" transmission of tomato yellow leaf curl by *B. tabaci* is reported by Cohen and Harpaz (1964) to describe a situation wherein the vector undergoes cycles of loss of infectivity and cannot be recharged until this occurs. Later, Cohen (1967) accounted for this by the presence of an inhibiting factor which develops in the insect at the beginning of the cycle of infectivity loss. *A. pisum* can partially restore the inoculative potential for pea enation mosaic virus by recharging, but whether the same phenomenon, as reported by Cohen, occurs is not known.

What is perhaps a unique case is that of *Scaphoideus litteralis* Ball transmitting the golden flavescence of vine. According to Caudwell (1964) and Schvester *et al.* (1962), insecticide treatments eliminate the insect and the vine recovers spontaneously from the disease. Reappearance of the disease at intervals is explained by reinfestation and reinfection by the vector. There is a strong implication here that a phytotoxemia is also involved, but a virus may perhaps also have cycles of high and low titer in the plant, with the lowest points failing to induce symptoms.

THE HOST PLANT IN RELATION TO TRANSMISSIBILITY AND VECTOR EFFICIENCY

It has been assumed for the most part that only the virus and the vector are involved in transmission, but Watson and Nixon (1953) and Sylvester and Simons (1951) have called attention to the role of the host plant in determining both transmissibility and vector efficiency.

One of the earliest references to the effect of the host plant was that by Hoggan (1934), who recorded aphid transmission of tobacco mosaic from tomato, but not from tobacco to tobacco. Severin (1947) found that two races of the aster-yellows-virus vector (*Macrosteles divisus* (Uhler) differed in their ability to infect celery but not asters. Sylvester and Simons (1951) found that the green peach aphid was able to inoculate *Brassica nigra* virus into mustard plants but not into pak choi. The false cabbage aphid had a lower efficiency rating but was not selective as to host plant species inoculated.

Young leaves of infected plants are better sources of virus inoculum than are old leaves (Kassanis, 1952; Cadman, 1954). Sometimes one host is a better source of inoculum than another, although the latter may be a better host for the insect and more susceptible to infection. This was shown by Simons (1954) in connection with southern cucumber mosaic virus in Swiss chard (the better inoculum source) and California Wonder chili, the better rearing host and the more susceptible.

Watson (1965) has studied the effect of different host plants of potato virus C in determining its transmission by aphids. Starting with a stock of the virus in Edgecote Purple potatoes, which in 1945 was not transmissible by aphids, the virus was maintained by mechanical inoculation in *N. glutinosa* L and *N. tabacum* L. In 1955 it was found to be transmissible by about 1 in 20 aphids compared with 1 in 2 for potato virus Y. When inoculated to Majestic potatoes and then returned to tobacco plants, virus C usually ceased to be aphid-transmissible and did not recover this property in any subsequent subcultures. Watson concluded that the ability of a virus to be transmitted by aphids seems to be genetically determined and linked with other inherent properties of the virus. On the basis of mutation, the potato strain changed to the *Nicotiana* strain when transferred from potato to *Nicotiana,* and the reverse mutation occurred when reinoculated to potato.

Jensen (1959) worked with an aphid-transmitted virus of nasturtium, and, after discontinuance of vector work, maintained the viruses by juice inoculation for a period of 2 yr. When vector work was resumed, it was found that transmission occurred only rarely.

Black (1953) found that potato yellow dwarf, after being transmitted mechanically for some years in *Nicotiana rustica* L, was incapable of being transmitted by leafhoppers. Similarly, tomato aspermy virus (Hollings, 1955) lost most of its ability to be transmitted by aphids after a period of sap inoculation through tobacco plants.

All the cases quoted above involved rather long periods of absence from insect transmission, but Swenson (1957) reported that 1 of 2 isolates of bean yellows virus lost its ability to be transmitted by three species of aphids within a period of 1–4 mo.

Even more suggestive of host plant-virus relationships is the evidence that the presence of a complex of viruses in a plant may effect the multiplication of one of them as well as its transmissibility by aphids (Badami and Kassanis, 1958). Equally indicative of the complexity of the factors affecting transmissibility is the case of tobacco rosette, which is a complex of mottle virus and vein-distorting virus. The mottle virus is easily mechanically transmitted, but it cannot be transmitted by *M. persicae* unless it occurs in the same plant that carries the vein-distorting virus. The effect of a large number of insects, whether infective or not, in decreasing the probability of transmission is probably due to modification of the test plant, according to Kirkpatrick and Ross (1952); but it could involve the gross effects of large quantities of inactivating substances which presumably are produced by the noninfective insect to the same extent as by the infective, but with qualitative differences.

The influence of the host plant on the vector capacity of *Myzus persicae* (Sulz.) is illustrated by the case of potato leaf roll. A higher rate of transmission occurs when the insects are fed on *P. floridana* or *B. oekinensis* than when they are fed on other hosts. Transmission rates for sugar beet mosaic and beet yellows are also highest following vector feeding on *P. floridana* (Hinz, 1969).

The bean leaf beetle, *Ceratoma ruficornis* Forst., transmits the cowpea strain of southern bean mosaic virus. Retention is a function of the length of the acquisition feeding period, and with 24 hr feeding, it lasts for 19 days. Daily transfers, however, do not result in regular transmissions. When the insects are fed on resistant plants (*Phaseolus vulgaris,* Great Northern var.), transmissions only occur by groups of insects that feed on resistant plants for 3 days. The authors interpret this finding as indicating a virus-inhibiting factor in the resistant bean plants (Walters and Henry, 1970).

The effect of source and test plants of cowpea mosaic virus on the efficiency and retention of infectivity of the bean leaf beetle and the spotted cucumber beetle was recorded by Jansen and Staples (1970). With the bean leaf beetle (*C. trifurcta*), cowpea-to-cowpea and soybeanto-cowpea transfers resulted in 100% and 95% transmission respectively. From cowpea-to-soybean and soybean-to-soybean, the results were 24% and 20% respectively. However, in the case of the cucumber beetle (*D. undecimpunctata*), the efficiency of transmission was much lower, i.e., cowpea 72%, soybean-to-cowpea 32%, while cowpea-to-soybean and soybean-to-soybean transfers resulted in no transmission whatever.

Swenson (1969) has analyzed the problem of host plant susceptibility to infection by insect transmission. Choosing the most susceptible host plants would increase efficiency of transmission; the most resistant host plants might well be sources of field resistance. Results of evaluation of susceptibility by mechanical inoculation tests show little similarity to those obtained by inoculation by insects. Temperature and plant age are the only factors thus far showing an appreciable effect on plant susceptibility to the aphid stylet-borne viruses; concerning the circulative viruses, the data are too sparse to allow of any conclusion.

EFFECTS OF VIRUSES ON INSECT VECTORS

Beneficial effects of plant viruses on their insect vectors have been reported, but it would be more appropriate to relate these effects to the nutrition provided by virus-infected plants. Carter (1939) attributed the presence of higher populations of thrips on *Emilia sonchifolia* L to the curling of the leaves and the delayed maturity of the plant. Kennedy (1951) found that sugar beet mosaic virus so changed the physiology of sugar beet plants that *A. fabae* reproduced more freely on diseased than on healthy leaves. The same conclusions had been reached by Severin (1946) to explain the survival of nine species of leafhoppers on celery or aster yellows plants infected with California aster yellows virus when none survived on healthy plants.

The mite, *Phytopus ribis* (Nal.), vector of reversion disease of black currant, produces galls on the axillary buds and malformations on the apical buds. The mites increase much more rapidly on diseased plants, and this is associated with the reduced hairiness which results from virus infection (Thresh, 1964). The increased fecundity of the mite may not, therefore, be a direct result of feeding on virus-infected tissue, since, if the explanation offered is correct, any variety of black currant with reduced hairiness would also be a more favorable host for the mite, irrespective of whether the plant was virus-infected or not. Aphids on barley yellow-dwarf-infected oats showed a reduction in oxygen consumption of 13.8%, but they had greater longevity and an increased nymphal production (Miller and Coons, 1964).

Maramorosch (1958) used a different approach. *Dalbulus maidis* (De-Long and Wolc.) and *D. eliminatus* (Ball) are both limited in the normal host range to corn and teosinte (*Euchlaena mexicana* Schrad.), the latter a wild grass. Neither insect species is a vector of aster yellows, but both were able to survive on diseased aster yellows plants for long periods, even though no progeny developed, although neither species survived the transfer from their normal hosts to healthy aster plants. The transfer from one host to another, even within the normal host range of an insect, is not always easy, and it is quite in order to postulate that the diseased aster yellows plant served as a suitable transitional host between the corn and healthy asters. Even then, the host transfer was not complete since no progeny resulted. In none of the cases quoted is there any evidence that the results were due to the virus *per se*.

M. persicae, M. ascolonicus, A. fabae, and *M. solani* preferred infected sugar beet plants to healthy ones and severely infected plants to those less affected; they bred more rapidly and lived longer on diseased than on healthy plants. The multiplication rate on healthy plants was increased if the plants were sprayed either with sugar solutions or casein hydrolysate, although the effect was less than that induced by the virus infection (Baker, 1960).

M. persicae failed to colonize on healthy China aster plants but did so when the plants were infected with aster-yellows agent. The length of time the aphids lived on the healthy plant was influenced by the host plant sequence. Those reared on infected asters lived longer on healthy aster plants than did aphids from a colony reared on cabbage. *R. padi* maintained a colony on infected asters, but survived for only 3–7 days on healthy ones (Saini and Peterson, 1965).

M. persicae greatly prefers healthy Chinese cabbage to that infected with turnip mosaic virus. The infected seedlings had a higher sugar content but a lower nitrogen content; the nonproteinaceous nitrogen was reduced by 86.7%. They also showed increased osmotic pressure and absorption pressure with a consequent reduction in imbibition pressure (Kuan Chih-Ohu and Wang Shu, 1965).

M. persicae colonizes very poorly on plants infected with cucumber mosaic. The suggestion is made that the reduced amino acid content of the CMV-infected plants is the factor involved; with viruses that increase the amino acid content, the reverse could be expected (Lowe and Strong, 1963). However, increases in amino acids occur with oat sterile dwarf virus and with European wheat striate mosaic virus transmitted by the leafhopper *Javesella pellucida* (F.), but fecundity of the aphid *R. padi* was decreased on oats; in another aphid (*M. avenae*), fecundity was increased in the case of OSDV, but impaired with EWSMV (Laurema *et al.*, 1966).

Euscelis plebejus, the vector of witch's broom virus of clover, suffers retarded development and reduced survival when feeding on infected plants (Posnette and Ellenberger, 1963). The rice delphacid (*S. oryzicola*) shows a reduced longevity when it has acquired hoja blanca transovarially (Showers and Everett, 1967). The hoja blanca virus of rice affects the vector, *S. oryzicola,* by reducing the number of eggs laid so that fewer nymphs emerge. Fewer nymphs develop to maturation, and fertility of both sexes of adults is reduced, along with longevity. The authors (Jennings and Pinada, 1971) postulate that the greatly reduced numbers of actual and potential vectors in the field populations (less than 1%) have occurred through the agency of the pathogenic effects of the virus on the insects.

Effects of virus infection on the metabolism of insects have been reported from Japan (Yoshii and Kiso, 1957). Changes in oxygen consumption and total phosphorus were reduced in both host plant and insect vector. The virus was the one causing dwarf disease of orange and the vector was *Geisha distinctissima* Wal. Further work with this combination by Yoshii *et al.* in 1959 led to the conclusion that the infective insect showed abnormal metabolism in respiration, protein, and carbohydrates. Their data on sugars, however, are not too convincing.

Yoshii *et al.* (1959a), working with *Nephotettix apicalis* var. *cincticeps* (Uhl.) and rice stunt disease, again found that abnormal phosphorus metabolism was common to both infected plant and infected vector. The incorporation of P^{32} into RNA and DNA showed differences, however, between plant and insect. RNA-P did not differ between healthy and diseased plants, but DNA-P was much higher in the diseased plant. However, the radioactivity in RNA-P and DNA-P in the infective insect was always found at a higher level than in the noninfective insect. Yoshii and Kiso (1959), continuing their work with the dwarf disease of orange, studied the rate of incorporation of P^{32} into the phosphorus fractions of the plant and leafhopper, and they found that both were at a relatively higher level when virus-infected. In none of these papers of Yoshii and his co-authors is there any suggestion that the abnormal (different) metabolism of the virus-infected insect was in any way related to longevity, reproduction, or the maintenance of infective capacity.

Deleterious effects of viruses on insect vectors have been looked for. Dobroscky (1929) compared infective and noninfective leafhopper vectors of aster yellows but could find no histological differences. Hartzell (1937) could demonstrate what he considered to be intracellular inclusions in the infective vector of peach yellows (*Macropsis trimaculata* Fitch) that were similar to those found in the infected peach. This study was made at the time that research was concentrated on finding "organisms"; with the passing of that phase, no further interest was taken. From the standpoint of the possible pathogenicity of the virus to its insect vector, however, studies such as this and that of Blattny (1931), who reported changes in the salivary glands of insects infected with potato leaf roll virus, would bear repeating. The longevity of *Macrosteles fascifrons* (Stål) on healthy and diseased aster plants was compared, but no difference was found (Severin, 1947). On the other hand, Watson and Sinha (1959) reported that infective adult *D. pellucida* females that had fed on infected plants as nymphs had 40% fewer progeny than did those that fed on healthy plants, and some embryos died in the egg at a late stage of development. Adults that had fed on diseased plants as nymphs, but which did not transmit the virus as adults, laid healthy eggs. This was interpreted by Sinha (1960) as evidence that egg mortality was independent of the nutrition of the mother, but due rather to the pathogenicity of the virus to the eggs.

Cytological evidence of pathogenicity of a virus to the insect vector has been reported by Littau and Maramorosch (1956), from a study of the fat body of infective and noninfective aster leafhoppers (*Macrosteles fascifrons* Stål). In the noninfective insects the nuclei tended to be rounded or to have smooth contours; only a few were stellate and these occurred in 36% of the leafhoppers. The cytoplasm was generally homogeneous with a large number of vacuoles of varying size, and the cells were intact. On the other hand, in the infective insects almost all nuclei of the fat body cells were stellate, and this condition held for 95% of the leafhoppers; the cytoplasm was reticulate and many cells appeared broken in the sections.

Although these differences between the infective and noninfective insects seem to be clear, evidence for pathogenicity of the virus to the insect does not appear to be available. On the contrary, infectivity in this insect does not have any observable effects on longevity or reproductive vigor (Severin, 1947a). The fat body is presumably a storage organ and could well be modified by the insect's nutrition. Furthermore, the difference between the infective and noninfective insect is a quantitative one, since some stellate nuclei were found in 36% of the noninfectives.

The most convincing evidence for the deleterious effect of virus infection on an insect vector is that provided by Jensen (1950a), who showed that the peach-yellow-leaf-roll strain of Western X disease virus causes the early death of its insect vector, *Colladonus montanus* (Van Duzee). Jensen used celery as both virus source plant and test plant. The comparative longevity of insects fed on diseased and healthy celery differed materially, those from the diseased plant having a mean longevity of 38 days as compared with 82 days for the noninfective controls. However, Jensen's experimental technique avoided the question of the effect being one of the changed nutrition available to the insect feeding on a diseased plant.

When individuals from a colony from a diseased plant were tested separately for virus transmission and longevity, the mean longevity of 116 leafhoppers that transmitted the virus was 20 days compared to 51 days for the 64 nontransmitting individuals. Since, in a given experiment, all the insects came from the same colony, all fed on the diseased plant for the same length of time, and each insect was tested separately for virus transmission and longevity, the results seem to establish that the virus did, in fact, exert a lethal effect on the vector.

Cytological changes in viruliferous insects are now known to be frequent (Jensen, 1963; Maramorosch and Jensen, 1963) although not encountered universally. The two references are to reviews, that by Jensen being expanded somewhat in the joint review. There is little to indicate that these changes adversely affect the insect, although some of them are so drastic that it could be expected.

For example, *N. cincticeps* infected with rice yellow dwarf virus shows extreme changes in cell nuclei, and these are followed by vacuolation of the cytoplasm to the point of complete reticulation (Takahashi, 1963). Lomakina *et al.* (1963) described both cytological and histochemical changes in *P. striatum* infected with winter wheat mosaic. Nymphs, transovarially infected, had impaired structure and development. Schmidt (1959, 1960) reported changes in *M. persicae* which were drastic enough to seriously affect the insect, but possibly only through an acceleration of normal cell degeneration.

Cytopathic changes occur in most tissues of leafhoppers infested with Western X disease; also nuclear divisions within cellular divisions occur in adipose tissue and salivary glands, with irregular cell divisions in the acini of the anterior lobe of the salivary glands. Proliferative symptoms are rare and are not found in healthy leafhoppers. There are proliferative outgrowths from the aesophageal valve, spermtheca, and rectal bud. With these several foci, systemic stimulus is indicated (Whitcomb and Jensen, 1968).

The latest inclusive review in this field is that by Jensen (1969), who discussed inclusion bodies and the cytopathology and histopathology in leafhoppers at length, since it is with that group of insects that most of the work has been done. There are, however, two species of aphids in which cytological studies have been noted, *Myzus persicae* (Sulz.) and *Macrosiphon granarium* (Kirby), but in neither case was the evidence adequately positive.

The fairly frequent evidence that many vectors thrive better on diseased plants, even though there are exceptions and the converse holds true, suggests again that nutrition is the primary factor. When internal changes occur in infective vectors, and when reduced fecundity, sterility, and a shortened life span occur, the presumption is that these changes are deleterious and the result of the virus infection in the insect.

Sylvester and Richardson (1971) have shown that virus strains will differ in their effect on longevity of infected insects. They selected strains of the sowthistle yellow vein virus acquired from sowthistle by the aphid *Hyperomyzus lactucae* (L) by serially transferring the virus by injection 20 times. By the fifth passage, survival of serially injected aphids was less than survival of those which had had none, or but one, injection, but even these latter had a poor survival rate compared with controls which had been injected with hemolymph from healthy aphids. In one isolate, however, the survival rate continued to decline after the 17 passages, and plants infected with this isolate rarely developed symptoms. Another isolate which did not show further survival decline after the fifth passage could still produce symptoms in plants. This study clearly showed that virus strains could be isolated by using serial passages of injected inoculum and that these isolates varied in their deleterious effects on the insect's survival rate.

REFERENCES

Badami, R. S., and Kassanis, B., "Potato viruses," *Rept. Rothamsted Expt. Sta.,* **1957,** 107–109 (1958).

Baker, P. F., "Aphid behaviour on healthy and on yellows infected sugarbeet plants," *Ann. Appl. Biol.,* **48,** 384–391 (1960).

Banks, C. J., and Noxon, H. L., "The feeding and excretion rates of *Aphis fabae* Scop. on *Vicia faba* L.," *Entomol. Exp. Appl.,* **2,** 77–81 (1959).

Bantarri, E. E., and Zeyen, R. J., "Transmission of oat blue dwarf virus by the aster leafhopper following natural acquisition or inoculation," *Phytopathology,* **60,** 399 (1970).

Bawden, F. C., Hamlyn, B. M. G., and Watson, M. A., "The distribution of viruses in different leaf tissues and its influence on virus transmission by aphids," *Ann. Appl. Biol.,* **41,** 229–239 (1954).

Bennett, C. W., "Plant tissue relations of the sugar beet curly top virus," *J. Agr. Research,* **48,** 665–701 (1934).

Bennett, C. W., and Tanrisever, A., "Sugar beet curly top disease in Turkey," *Plant Disease Reptr.,* **41,** 721–725 (1957).

Bennett, C. W., and Wallace, H. E., "Relation of the curly top virus to the vector *Eutettix tenellus*," *J. Agr. Research,* **56,** 31–51 (1938).

Bjorling, K., and Ossiannilsson, F., "Investigations on individual variations in the virus transmission ability of different aphid species," *Socker Hand II,* 14, 115–116 (1958).

Black, L. M., "Further evidence for multiplication of the aster yellows virus in the aster leafhopper," *Phytopathology,* 31, 120–135 (1941).

Black, L. M., "Genetic variation in the clover leafhopper's ability to transmit potato yellow dwarf virus," *Genetics,* 28, 200–209 (1943).

Black, L. M., "Transmission of clover club-leaf virus through the egg of its insect vector," *Phytopathology,* 38, 2 (1948).

Black, L. M., "A plant virus that multiplies in its insect vector," *Nature,* **166,** 852–853 (1950).

Black, L. M., "Occasional transmission of some plant viruses through the eggs of their insect vectors," *Phytopathology,* 43, 9–10 (1953).

Black, L. M., "Loss of vector transmissibility by viruses normally insect transmitted," *Phytopathology,* 43, 466 (1953a).

Black, L. M., "Viruses that reproduce in plants and insects," *Ann. N. Y. Acad. Sci.,* 56, 398–413 (1953b).

Black, L. M., and Brakke, M. K., "Multiplication of wound-tumor virus in an insect vector," *Phytopathology,* 42, 269–273 (1952).

Blattny, C., "Ize zjistiti pritomnost viru pusobiciho nektere choroby Bramboru v jejich prenaseci, msicich? (Can the viruses that cause certain potato diseases be detected in their aphid vector?)," *Trans. Roy. Bohemian Univ. Sci.,* 2, (1931).

Bradley, R. H. E., "Studies on the aphid transmission of a strain of henbane mosaic virus," *Ann. Appl. Biol.,* 39, 78–97 (1952).

Bradley, R. H. E., "Effect of depth of stylet penetration on aphid transmission of potato virus Y," *Can. J. Microbiol.,* 2, 539–547 (1956).

Bradley, R. H. E., "Aphid transmission of stylet-borne viruses," *Plant Virology* M. K. Corbett, and H. D. Sisler, Eds., University of Florida Press, Gainesville, Fla., (1964).

Bradley, R. H. E., and Rideout, D. W., "Comparative transmission of potato virus Y by four aphid species that infest potato," *Can. J. Zool.,* 31, 333–341 (1953).

Bradley, R. H. E., and Ganong, R. Y., "Evidence that potato virus Y is carried near the tip of the stylets of the aphid vector *Myzus persicae* (Sulz.)," *Can. J. Microbiol.,* 1, 775–782 (1955).

Bradley, R. H. E., and Ganong, R. Y., "Some effects of formaldehyde on potato virus Y *in vitro,* and ability of aphids to transmit the virus when their stylets are treated with formaldehyde," *Can. J. Microbiol.,* 1, 783–793 (1955a).

Bradley, R. H. E., and Ganong, R. Y., "Three more viruses borne at the stylet tips of the aphid *Myzus persicae* (Sulz.)," *Can. J. Microbiol.,* 3, 669–670 (1957).

Bruehl, G. W., "Comparison of eastern and western aphids in the transmission of barley yellow dwarf virus," *Plant Disease Reptr.,* 42, 909–911 (1958).

Cadman, C. H., "Studies in *Rubus* virus diseases, VI. Aphid transmission of raspberry leaf mottle virus," *Ann. Appl. Biol.*, 41, 207–214 (1954).

Carter, W., "Populations of *Thrips tabaci* with special reference to virus transmission," *J. Animal Ecol.*, 8, 261–276 (1939).

Carter, W., "Studies on the relationship between insect vector and virus with special reference to the insect's symbionts," *Proc. 3rd Intern. Congress for Microbiology*, 1939 (1940).

Carter, W. "*Peregrinus maidis* (Ash.) and the transmission of corn mosaic," *Ann. Entomol. Soc. Amer.*, 34, 551 (1941).

Caudwell, A., "Identification d'une nouvelle maladie a virus de la vigne 'La Flavescence Doree'. Etude des Phenomenes de localization des symptomes et de retablissement," *Annales des Epiphyties 15 Horse-serie I*, 1–193 (1964).

Chalfant, R. B., and Chapman, R. K., "Transmission of cabbage viruses A and B by the cabbage aphid and the green peach aphid," *J. Econ. Entomol.*, 55, 584–590 (1962).

Clark, M. F., and Ross, A. F., "Variation among clons of *Myzus persicae* in ability to transmit potato leafroll virus acquired by injection," *Phytopathology*, 54, 199–204 (1964).

Cohen, S., "The occurrence in the body of *Bemisia tabaci* of a factor apparently related to the phenomenon of a 'periodic acquisition' of tomato yellow leaf curl virus," *Virology*, 31, 180–183 (1967).

Cohen, S., and Harpaz, I., "Periodic, rather than continual acquisition of a new tomato virus by its vector, the tobacco white fly (*Bemisia tabaci* Genn.)," *Entomol. Exp. Appl.*, 7, 155–166 (1964).

Costa, A. S., "White flies as virus vectors," *Viruses, Vectors and Vegetation"*, K. Maramorosch, Ed., Wiley-Interscience, New York, 1969, pp. 95–120.

Day, M. D., "Mechanism of transmission of viruses by arthropods," *Exptl. Parasitol.*, 4, 387–418 (1955).

Day, M. F., "The mechanism of the transmission of potato leafroll virus by aphids," *Australian J. Biol. Sci.*, 8, 498–513 (1955a)

Day, M. F., and Bennetts, M. J., "A review of problems of specificity in arthropod vectors of plants and animal diseases," *Council Scientific Industrial Res. Orgn., Canberra*, 1–172 (1954).

Day, M. F., and Irzykiewicz, H., "Feeding behaviour of the aphids *Myzus persicae* and *Breviocoryne brassicae,* studied with radio-phosphorus," *Australian J. Biol. Sci.*, 6, 98–108 (1953).

Day, M. F., and Irzykiewicz, H., "On the mechanism of transmission of non-persistent phytopathogenic viruses by aphids," *Australian J. Biol. Sci.*, 7, 251–273 (1954).

Day, M. F., and Irzykiewicz, H., "Physiological studies on thrips in relation to transmission of tomato spotted wilt virus," *Australian J. Biol. Sci.*, 7, 274–281 (1954a).

Day, M. F., and Venables, D. G., "The transmission of cauliflower mosaic virus by aphids," *Australian J. Biol. Sci.*, 14, 187–197 (1961).

Day, M. F., Irzykiewicz, H. and McKinnon, A., "Observations on the feeding of the virus vector *Orosius argentatus* (Evans) and comparisons with certain other jassids," *Australian J. Sci. Research Ser. B.,* 5, 128–142 (1952).

Dias, H. F., and Harrison, B. D., "The relationship between grapevine fanleaf, grapevine yellow mosaic and *Arabis* mosaic viruses," *Ann. App. Biol.,* 51, 97–105 (1963).

Dobroscky, I. B., "Is the aster yellows virus detectable in its insect vector?" *Phytopathology,* 19, 1009–1015 (1929).

Doncaster, J. P., and Kassanis, B., "The shallot aphis, *Myzus ascalonicus* Doncaster, and its behaviour as a vector of plant viruses," *Ann. Appl. Biol.,* 33, 66–68 (1946).

Duffus, J. E., "Possible multiplication in the aphid vector of sowthistle yellow vein virus," *Virology,* 21, 194–202 (1963).

Fife, J. M., and Frampton, V. L., "The pH gradient extending from the phloem into the parenchyma of the sugar beet and its relation to the feeding behaviour of *Eutettix tenellus,*" *J. Agr. Research,* 53, 581–593 (1936),..

Flemion, F., Miller, L. P., and Weed, R. M., "An estimate of the quantity of oral secretion deposited by *Lygus* when feeding on bean tissue," *Contribs. Boyce Thompson Inst.,* 16, 429–433 (1952).

Frazier, N. W., "Nonretention of two semi-persistent strawberry viruses through ecdysis by their aphid vectors," *Phytopathology,* 56, 1318 (1966).

Freitag, J. H., "Negative evidence on multiplication of curly top virus in the beet leafhopper *Eutettix tenellus,*" *Hilgardia,* 10, 305–342 (1936).

Freitag, J. H., "Insect transmission, host range and properties of squash mosaic virus," *Phytopathology,* 31, 8 (1941).

Freitag, J. H., "Interaction between strains of aster yellows virus in six-spotted leafhopper *Macrosteles fascifrons,*" *Phytopathology,* 57, 535–537 (1967).

Freitag, J. H., "Interactions of plant viruses and virus strains in their insect vectors," *Viruses, Vectors and Vegetation,* Wiley-Interscience, New York, 1969, pp. 303–326.

Fukushi, T., "Transmission of the virus through the eggs of an insect vector," *Proc. Imp. Acad. (Tokyo),* 9, 457–460 (1933).

Fukushi, T., "Studies on the dwarf disease of rice plant," *J. Fac. Agr. Hokkaido Imp. Univ.,* 37, 41–164 (1934).

Fukushi, T., "Multiplication of virus in its insect vector," *Proc. Imp. Acad. (Tokyo),* 11, 301–303 (1935).

Fukushi, T., "Retention of virus by its insect vectors through several generations," *Proc. Imp. Acad. (Tokyo),* 15, 142–145 (1939).

Fukushi, T., "Further studies on the dwarf disease of rice plant," *J. Fac. Agr. Hokkaido Imp. Univ.,* 45, 83–154 (1940).

Galvez, G. E., Jennings, P. R., and Thurston, H. D., "Transmission studies of hoja blanca of rice in Colombia," *Plant Disease Reptr.,* 44, 80–81 (1960).

Gill, C. C., "Cyclical transmissibility of barley yellow dwarf virus from oats with increasing age of infection," *Phytopathology*, 59, 23–28 (1969).

Gill, C. C., "Aphid nymphs transmit an isolate of barley yellow-dwarf more efficiently than do adults," *Phytopathology*, 60, 1747 (1970).

Grylls, N. E., "Rugose leaf curl—A new virus disease transovarially transmitted by the leafhopper *Austroagallia torrida*," *Australian J. Biol. Sci.*, 7, 47–58 (1954).

Hamilton, M. A., "Further experiments on the artificial feeding of *Myzus persicae* (Sulz.)," *Ann. Appl. Biol.*, 22, 243–258 (1935).

Harrison, B. D., "Studies on the behaviour of potato leaf roll and other viruses in the body of their aphid vector, *Myzus persicae* Sulz.)," *Virology*, 6, 265–277 (1958).

Harrison, B. D., "Specific nematode vectors for aerologically distinctive forms of raspberry ringspot and tomato black-ring viruses," *Virology*, 22, 544–550 (1964).

Harrison, B. D., and Winslow, E. D., "Laboratory and field studies on the relation of *Arabis* mosaic virus to its nematode vector *Xiphinema diversicaudatum* (Micol.)," *Ann. Appl. Biol.*, 49, 621–633 (1961).

Hartzell, A., "Movement of intracellular bodies associated with peach yellows," *Contribs. Boyce Thompson Inst.*, 8, 375–388 (1937).

Heinze, K., "Versuche zur uebertragung des blattrolivirus der Kartoffel in den Ubertrager *(Myzodes persicae* (Sulz.) mit Injdktionsverfahren," *Phytopathol. Z.*, 25, 103–108 (1955).

Hendrick, R. D., Everett, T. R., Lamey, H. A., and Showers, W. B., "An improved method of selecting and breeding for active vectors of hoja blanca virus," *J. Econ. Entomol.*, 58, 539–542 (1965).

Hildebrand, E. M., "A white fly, *Trialeurodes abutilonia*, an insect vector of sweetpotato feathery mottle in Maryland," *Plant Disease Reptr.*, 43, 712–714 (1959).

Hildebrand, E. M., "The feathery mottle virus complex of sweetpotato," *Phytopathology*, 50, 751–757 (1960).

Hinz, B., "Influence of food plants on vector capacity of aphids *(Myzus persicae)*," *Zentralbl. Bakteriol. Infectionskrankh Hyg. Zw. Abt.*, 123, 230–236 (1969).

Hoggan, I. A., "The peach aphid *(Myzus persicae* (Sulz.)) as an agent in virus transmission," *Phytopathology*, 19, 109–123 (1929).

Hoggan, I. A., "Further studies on aphid transmission of plant viruses," *Phytopathology*, 21, 199–212 (1931).

Hoggan, I. A., "Some factors involved in aphid transmission of the cucumber-mosaic virus to tobacco," *J. Agr. Research*, 47, 689–704 (1933).

Hoggan, I. A., "Transmissibility by aphids of the tobacco mosaic virus from different hosts," *J. Agr. Research*, 49, 1135–1142 (1934).

Hollings, M., "Investigation of chrysanthemum viruses. I. Aspermy flower distortion," *Ann. Appl. Biol.*, 43, 86–102 (1955).

Hood, L. R., and McLean, D. L., "Correlation of transmission of barley yellow-dwarf with salivation of *Acyrthosiphon pisum,*" *Ann. Entomol. Soc.,* **62,** 1398–1401 (1969).

Houston, B. R., Esau, K., and Hewitt, W. B., "The mode of vector feeding and the tissues involved in the transmission of Pierce's disease virus in grape and alfalfa," *Phytopathology,* 37, 247–253 (1947).

Hull, R., and Adams, A. N., "Groundnut rosette and its assistor virus," *Ann. Appl. Biol.,* **62,** 139–145 (1968).

Jansen, W. P., and Staples, R., "Effect of cowpeas and soybeans as source of test plants of cowpea mosaic virus on vector efficiency and retention of infectivity of the bean leaf beetle and the spotted cucumber beetle," *Plant Disease Reptr.,* **54,** 1053 (1970).

Jansen, W. P., and Staples, R., "Specificity of transmission of cowpea mosaic virus by species within the subfamily Galerucinae, family Chrysomelidae," *J. Econ. Entomol.,* **64,** 364 (1971).

Jennings, P. R., and Pineda, A. T., "The effect of the hoja blanca virus on its insect vector," *Phytopathology,* **61,** 142 (1971).

Jensen, D. D., "Insects, both hosts and vectors of plant diseases," *Pan-Pacific Entomologist,* **35,** 65–82 (1959).

Jensen, D. D., "A plant virus lethal to its insect vector," *Virology,* **8,** 164–175 (1959a).

Jensen, D. D., "Effect of plant viruses on insects," *Annals. N. Y. Acad. Sci.,* **105,** 685–712 (1963).

Jensen, D. D., "Insect diseases induced by plant pathogenic viruses," *Viruses, Vectors and Vegetation,* K. Maramorosch, K. Ed., Wiley-Interscience, New York, 1969, pp. 505–526.

Kassanis, B., "Studies on dandelion yellow mosaic and other virus diseases of lettuce," *Ann. Appl. Biol.,* 34, 412–421 (1947).

Kassanis, B., "Some factors affecting the transmission of leaf-roll virus by aphids," *Ann. Appl. Biol.,* **39,** 157–167 (1952).

Kassanis, B., and Govier, D. A., "New evidence on the mechanism of aphid transmission of potato virus C and potato *Aucuba* mosaic viruses," *J. Gen. Virology,* **10,** 99 (1971).

Kennedy, J. S., "Benefits to aphids from feeding on galled and virus-infected leaves," *Nature,* **168,** 825–826 (1951).

Kennedy, J. S., Day, M. F., and Eastop, V. F., "A conspectus of aphids as vectors of plant viruses," *Commonwealth Inst. Entomol. (London),* 114 (1962).

Kirkpatrick, H. C., and Ross, A. F., "Aphid-transmission of potato leafroll virus to solanaceous species," *Phytopathology,* **42,** 540–546 (1952).

Klostermeyer, E. C., "Entomological aspects of the potato leaf roll problem in central Washington," *Wash. State Coll. Agr. Expt. Sta. Tech. Bull.,* No. 9 (1953).

Kuan, C. and Wang, S., "On some physiological changes of Chinese cabbage infected by the Kwuting strain of turnip mosaic virus in relation to the development of *Myzus persicae* (Sulzer)," *Acta Phytophyl. Sinica,* 4, 27–33 (1965).

Kunkel, L. O., "Studies on aster yellows," *Am. J. Botan.*, 13, 646–705 (1926).

Kunkel, L. O., "Effect of heat on ability of *Cicadula sexnotata* (Fall.) to transmit aster yellows," *Am. J. Botan.*, 24, 316–327 (1937).

Kunkel, L. O., "Heat cure of aster yellows in periwinkles," *Am. J. Botan.*, 28, 761–769 (1941).

Kunkel, L. O., "Cross protection between strains of yellows-type viruses," *Advances in Virus Research*, 3, 251–273 (1955).

Kunkel, L. O., "Acquired immunity from infection by strains of aster yellows virus in the aster leafhopper," *Science*, 126, 1233 (Abstr.) (1957).

Kunkel, L. O. "Studies on a new corn virus disease," *Arch. Ges. Virusforsch*, 4, 24–46 (1948).

Kvicala, B. A., "Selective power in virus transmission exhibited by an aphis," *Nature*, 155, 174–175 (1945).

Kyriakopoulou, P. E., and Sylvester, E. S., "Vector age and length of acquisition and inoculation access periods as factors in transmission of pea enation mosaic virus by the pea aphid," *J. Econ. Entomol.*, 62, 1423 (1969).

Laurema, S., Markkula, M., and Raatikainen, M., "The effect of virus diseases transmitted by the leafhopper *Javsella pellucida* (F.) on the concentration of free amino acids in oats and on the reproduction of aphids," *Annales Agric. Fenniae*, 5, 94–99 (1966).

Lee, P. E., and Bell, W., "Some properties of wheat striate mosaic virus," *Can. J. Botany*, 41, 767–771 (1963).

Littau, V. C., and Maramorosch, K., "Cytological effects of aster-yellows virus on its insect vector," *Virology*, 2, 128–130 (1956).

Lomakina, L. Y., Razvyazkina, G. M., and Shubnikova, E. A., "Cytological and histochemical changes in fat body of *Psammotettix striatus* Fallen infected with winter wheat mosaic virus," *Vaprosy Virusologii*, 2, 168–172 (in Russian) (1963).

Lowe, S., and Strong, F. A., "The unsuitability of some viruliferous plants as hosts for the green peach aphid, *Myzus persicae*," *j. econ. Entomol.*, 56, 307–309 (1963).

MacCarthy, H. R., "Aphid transmission of potato leafroll virus," *Phytopathology*, 44, 167–174 (1954).

Maramorosch, K., "Direct evidence for the multiplication of aster-yellows virus in its insect vector," *Phytopathology*, 42, 59–64 (1952).

Maramorosch, K., "Multiplication of aster-yellows virus in its vector," *Nature*, 169, 4292 (1952a).

Maramorosch, K., "Incubation period of aster yellows virus," *Am. J. Botan.*, 40, 797–808 (1953).

Maramorosch, K., "Mechanical transmission of curly top virus to its insect vector by needle inoculation," *Virology*, 1, 286–300 (1955).

Maramorosch, K., "Multiplication of aster yellows virus *in vitro* preparations of insect tissues," *Virology*, 2, 369–376 (1956).

Maramorosch, K., "Beneficial effect of virus diseased plants on non-vector insects," *Tijdschr. Plantenziekten*, 64, 383–391 (1958).

Maramorosch, K. "Cross protection between two strains of corn stunt in an insect vector," *Virology,* 6, 411–420 (1958).

Maramorosch, K., "Virus vector; vectors of circulative, propagative viruses," *Plant Virology,* M. K. Corbett and M. D. Sisler, Eds., University of Florida Press, Gainesville, Fla., 1964, pp. 175–194.

Maramorosch, K., "The fate of plant-pathogenic viruses in insect vectors," *Viruses, Vectors and Vegetation,* Wiley-Interscience, New York, 1969.

Maramorosch, K., and Jensen, D. D., "Harmful and beneficial effects of plant viruses in insects," *Ann. Rev. Microbiol.,* 17, 495–530 (1963).

Markham, R., and Smith, K. M., "Studies on the virus of turnip yellow mosaic," *Parasitology,* 39, 330–342 (1949).

McClintock, J. A., and Smith, L. B., "True nature of spinach blight and the relation of insects to its transmission," *J. Agr. Research,* 14, 1–60 (1918).

McEwen, F. L., and Kawanishi, C. Y., "Insect transmission of corn mosaic; laboratory studies in Hawaii," *J. Econ. Entomol.,* 60, 1413 (1967).

Miles, P. W., "Secretion of two types of saliva by an aphid," *Nature,* 183, 756 (1959).

Miles, P. W., "Insect secretions in plants," *Ann. Rev. Phytopathology,* 6, 137–164 (1968).

Miller, J., and Coon, B. F., "The effect of barley yellow-dwarf virus on the biology of its vector, the English grain aphid, *Macrosiphon graminum,*" *J. Econ. Entomol.,* 57, 970–974 (1964).

Miyamoto, S., and Miyamoto, Y., "Notes on the transmission of potato leafroll virus," *Sci. Rept. Hyogo Univ. Agric. Ser. Plant Protection,* 7, 51–66 (1966).

Montogomery, R. W., and Bell, A. F., "Fiji disease of sugar cane and its transmission," *Queensland Bur. Sugar Expt. Sta. Bull.,* No. 4 (1933).

Mound, L. A., "The feeding apparatus of thrips," *Bull. Entomol. Res.,* 60, 547–548 (1971).

Mueller, W. C., and Rochow, W. F., "An aphid injection method for the transmission of barley yellow dwarf virus," *Virology,* 14, 253–258 (1961).

Muller, I., "The Cornell Plantations," *Winter,* 1964–65.

Murant, A. F., and Goold, R. A., "Purification, properties and transmission of parsnip yellow fleck, a semi-persistent aphid borne virus," *Ann. Appl. Biol.,* 68, 123–137 (1968).

Nagaraj, A. N., and Black, L. M., "Hereditary variation in the ability of a leafhopper to transmit two unrelated viruses," *Virology,* 16, 152–162 (1962).

Namba, R., "Aphid transmission of plant viruses from the epidermis and subepidermal tissues; *Myzus persicae* (Sulzer)—cucumber mosaic virus," *Virology,* 16, 267 (1962).

Nasu, S., "Some studies on some leafhoppers and planthoppers which transmit virus diseases of rice plant in Japan," *Bull. Kyushu Agric. Exp. Sta.,* 8, 153–349 (1963).

Nasu, S., "Electron microscopic studies on transovarial passage of rice dwarf virus," *Japan J. Appl. Entomol. Zool.,* 9, 225–237 (1965).

Nasu, S., "Electron microscopy of the transovarial passage of rice dwarf virus," *Viruses, Vectors and Vegetation,* Wiley-Interscience, New York, 1969, pp. 433–448.

Nault, L. R., Gyrisco, G. G., and Rochow, W. F., "Biological relationship between pea enation mosaic virus and its vector, the pea aphid," *Phytopathology,* 54, 1269 (1964).

Nishi, Y., "On the effect of substances secreted by aphid on the infectivity of potato virus X," (In Japanese), *Proc. Assoc. Plant Protection Kyushu,* 8, 56–57 (1962).

Nishi, Y., "Inhibition of viruses by vector saliva," *Viruses, Vectors and Vegetation,* Wiley-Interscience, New York, 1969.

Ocfemia, G. O., "An insect vector of the Fiji disease of sugar cane," *Am. J. Botan.,* 21, 113–120 (1933).

Ohman, P. W., "The nearctic leafhoppers (Homoptera: Cicadellidae). A generic classification and check list," *Mem. Entomol. Soc. Wash.,* No. 3 (1949).

Orlob, G. B., "Reappraisal of transmission of tobacco mosaic virus in insects," *Phytopathology,* 53, 822–830 (1963).

Pirone, T., "Mechanism of transmission of stylet-borne viruses," *Viruses, Vectors and Vegetation,* Wiley-Interscience, New York, 1969, pp. 199–210.

Pollard, D. G., "Feeding habits of the cotton whitefly *Bemisia tabaci* (Genn.) (Homoptera; Aleyrodidae)," *Ann. Appl. Biol.,* 43, 664–671 (1955).

Pollard, D. G., "Feeding of the cotton aphid (*Aphis gossypii* Glover)," *Empire Cotton Growing Rev.,* 35, 244–253 (1958).

Posnette, A. F., and Ellenberger, C. E., "Further studies of green petal and other leafhopper-transmitted viruses infecting strawberry and clover," *Ann. Appl. Biol.,* 51, 69–83 (1963).

Prentice, I. W., and Harris, R. V., "Resolution of strawberry complexes by means of aphis vector *Capitophorus fragariae* Theob.," *Ann. Appl. Biol.,* 33, 50–53 (1946).

Proeseler, G., "Die Übertragung des Virus der Rübenkräuselkrankeit mit Hilfe der Injektionsmethode. Der Nachweiss der Vermehrung des Rübenkräuselkrankeit-Virus in *Piesma quadrata* (Fieb.) mit Hilfe der Injektionstechnik," *Naturwissenschaften,* 50, 553 (1963); 51, 150–51 (1964).

Prusa, V., and Vacke, J., "Wheat striate mosaic virus. Transmission of wheat striate virus by the eggs of the vector *Calligypona pellucida* Fabr.," *Biol. Plant. Acad. Sci. Bohemosl.,* 2, 277–289 (1960).

Richardson, J., and Sylvester, E. S., "Aphid honeydew as inoculum for the injection of pea aphids (*Macrosiphon, Acyrthosiphon*) pisum (Harris) with pea-enation mosaic virus," *Virology,* 25, 472–475 (1965).

Roberts, F. M., "Studies on the feeding methods and penetration rates of *Myzus persicae* (Sulz.), *Myzus circumflexus* Buckt., and *Macrosiphum gei* Koch," *Ann. Appl. Biol.,* 27, 348–358 (1940).

Rochow, W. F., "The role of aphids in vector specificity of barley yellow dwarf virus," *Plant Disease Reptr.,* 42, 905–908 (1958).

Rochow, W. F., "Variation within and among aphid vectors of plant viruses," *Ann. N. Y. Acad. Sci.,* 105, 713–729 (1963).

Rochow, W. F., "Apparent loss of vector specificity following double infection by the two strains of barley yellow dwarf virus," *Phytopathology,* 55, 62 (1965).

Rochow, W. F., "Specificity in aphid transmission of a circulative plant virus," *Viruses, Vectors and Vegetation,* Wiley-Interscience, New York, 1969, pp. 175–198.

Rochow, W. F., and Eastop, V. F., "Variation between *Rhopalosiphon padi* and transmission of barley yellow-dwarf virus by clons of four aphid species," *Virology,* 30, 286–296 (1966).

Saini, R. S., and Peterson, A. G., "Colonization of yellows-infested asters by two species of aphids," *J. Econ. Entomol.,* 58, 537–539 (1965).

Sakimura, K. "The present status of thrips-borne viruses," *Biological Transmission of Disease Agents,* Academic Press, New York, 1962.

Schmidt, H. B., "Beitrage zur Kenntnis der Übertragung pflanzlicher Viren durch Aphiden," *Biol. Zentr.,* 78, 889–936 (1959).

Schmidt, H. B., "Die Übertragung pflanzlicher Viren durch Insekten," *Deut. Acad. Wiss. Berlin,* 2, 214–223 (1960).

Schmutterer, H., "Zur Kentnis der Beissinsekten als Übertrager pflanzlicher Viren Teil 1. "On biting insects as vectors of plant viruses," *Z. Angew. Ent.,* 47, 277–301; 416–439 (1961).

Schvester, D., Carle, P., and Moutous, G., "*Scaphoideus littoralis* Ball (Homopt. Jassidae) cicadelle vectrice de la flavescence doree de la vigne," *Rev. Zool. Agric.,* 61, 117–144 (1962).

Severin, H. H. P., "Longevity, or life histories, of leafhopper species on virus-infected and on healthy plants," *Hilgardia,* 17, 121–137 (1946).

Severin, H. H. P., "Newly discovered leafhopper vectors of California aster-yellows virus," *Hilgardia,* 17, 155–523 (1947).

Severin, H. H. P., "Longevity of noninfective and infective leafhoppers on a plant nonsusceptible to a virus," *Hilgardia,* 17, 541–543 (1947a).

Showers, W. B., and Everett, T. R., "Transovarial acquisition of hoja blanca virus by the rice delphacid," *J. Econ. Entomol.,* 60, 547 (1967).

Simons, J. N., "Vector-virus relationships of pea-enation mosaic and the pea aphid *Macrosiphum pisi* (Kalt.)," *Phytopathology,* 44, 283–289 (1954).

Simons, J. N., "The pseudo-curly top disease in South Florida," *J. Econ. Entomol.,* 55, 358–363 (1962).

Simons, J. N., "Differential transmission of closely related strains of potato virus Y by the green peach aphid and the potato aphid," *J. Econ. Entomol.,* 62, 1088 (1969).

Simons, J. N., and Eastop, V. F., "Temperature effects on aphid transmission of non-persistent viruses with notes on morphological variations within clones of aphid with different vector efficiencies," *J. Econ. Entomol.,* 63, 484 (1970).

Sinha, R. C., "Comparison of the ability of nymph and adult *Delphacodes pellucida* (Fabricius) to transmit European wheat striate mosaic virus," *Virology,* **10,** 344–352 (1960).

Sinha, R. C., "Recent work on leafhopper transmitted viruses," *Adv. Virus Res.,* **13,** 181–223 (1968).

Sinha, R. C., and Chiykowski, L. N., "Multiplication of aster yellows virus in a non-vector leafhopper," *Virology,* **31,** 461–466 (1967).

Skotland, C. B., and Hagedorn, D. J., "Vector-feeding and plant-tissue relationships in the transmission of the Wisconsin pea streak virus," *Phytopathology,* **45,** 665–666 (1955).

Smith, C. E., "Transmission of cowpea mosaic by the bean leaf beetle," *Science,* **60,** 268 (1924).

Smith, K. M., "Studies on potato virus diseases. V. Insect transmission of potato leaf-roll," *Ann. Appl. Biol.,* **16,** 209–229 (1929).

Smith, K. M., "Studies on potato virus diseases. IX. Some further experiments on the insect transmission of potato leaf-roll," *Ann. Appl. Biol.,* **18,** 141–157 (1931).

Smith, K. M., "Virus diseases of plants and their relationship with insect vectors," *Biol. Revs.,* **6,** 302–344 (1931a).

Smith, K. M., "Some notes on the relationship of plant viruses with vector and non-vector insects," *Parasitology,* **33,** 110–116 (1941).

Smith, K. M., "The transmission of a plant virus complex by aphids," *Parasitology,* **37,** 131–134 (1946).

Smith, K. M., "Arthropods as vectors and reservoirs of phytopathogenic viruses," *Handbuch der Virusforschung,* Springer, Wien, 1957, pp. 143–176.

Smith, K. M., "Plant virus-vector relationships," *Advances in Virus Research,* **9,** 61–96 (1965).

Smith, K. M., and Lea, D. E., "The transmission of plant viruses by aphids," *Parasitology,* **37,** 25–37 (1946).

Staples, R., Jansen, W. P., and Andersen, L. W., "Biology and relationship of the leafhopper *Aceratagallia calcaris* Ohman to yellow vein disease of sugarbeets," *J. Econ. Entomol.,* **63,** 1460–1463 (1970).

Stegwee, D., and Ponsen, M. B., "Multiplication of potato leaf-roll virus in the aphid *Myzus persicae* (Sulz.)" *Ent. Exptl. Appl.,* **1,** 291–300 (1958).

Stimmann, M. W., and Swenson, K. G., "Aphid transmission of cucumber mosaic virus affected by temperature and age of infection in diseased plants," *Phytopathology,* **57,** 1074 (1967).

Storey, H. H., "The inheritance by an insect vector of the ability to transmit a plant virus," *Proc. Roy. Soc. (London),* **b112,** 46–60 (1932).

Storey, H. H., "Investigations of the mechanism of the transmission of plant viruses by insect vectors. I.," *Proc. Roy. Soc. (London),* **b113,** 463–485 (1933).

Storey, H. H., "Investigations of the mechanism of the transmission of plant viruses by insect vectors. II. The part played by puncture in transmission," *Proc. Roy. Soc. (London),* **B125,** 455–477 (1938).

Storey, H. H., "Transmission of plant viruses by insects," *Botan. Rev.,* 5, 240–272 (1939).

Storey, H. H., "Investigations of the mechanism of the transmission of plant viruses by insect vectors. III. The insect's saliva," *Proc. Roy. Soc. (London),* B127, 526–543 (1939a).

Storey, H. H., and Ryland, A. K., "Transmission of groundnut rosette virus," *Ann. Appl. Biol.,* 43, 423–432 (1955).

Stubbs, L. L., "Retention of carrot motley dwarf virus in *Cavariella aegopodii* (Scop.) following a moult," *J. Australian Inst. Agr. Sci.,* 21, 267–268 (1955).

Stubbs, L. L., "Strains of *Myzus persicae* (Sulz.) active and inactive with respect to virus transmission," *Australian J. Biol. Sci.,* 8, 68–74 (1955a).

Stubbs, L. L., "Lettuce necrotic yellows virus," *First Intern. Phytopath. Congress,* 195 (abstract) (1968).

Swenson, K. G., "Transmission of bean yellow mosaic by aphids," *J. Econ. Entomol.,* 50, 727–731 (1957).

Swenson, K. G., "Plant susceptibility to virus infection by insect transmission," *Viruses, Vectors and Vegetation,* Wiley-Interscience, New York, 1969, pp. 143–158.

Sylvester, E. S., "Aphid transmission of nonpersistent plant viruses with special reference to the *Brassica nigra* virus," *Hilgardia,* 23, 53–98 (1954).

Sylvester, E. S., "Beet mosaic and beet yellows transmission by the green peach aphid," *J. Am. Soc. Sugar Beet Technologists,* 9, 56–61 (1956).

Sylvester, E. S., "Aphid transmission of plant viruses," *Proc. 10th Intern. Congr. Entomol., Montreal 1956,* 195–200 (1958).

Sylvester, E. S., "Mechanisms of plant virus transmission by aphids," *Biological Transmission of Disease Agents,* Academic Press, New York, 1962.

Sylvester, E. S., "Some effects of temperature on the transmission of cabbage mosaic virus by *Myzus persicae,*" *J. Econ. Entomol.,* 57, 536 (1964).

Sylvester, E. S., "Evidence of transovarial transmission of the sow thistle yellow vein virus in the aphid *Hyperomyzus lactucae* (L.)," *Virology,* 38, 440 (1969).

Sylvester, E. S., "Virus transmission by aphids—a viewpoint," *Viruses, Vectors and Vegetation,*" Wiley-Interscience, New York, 1969.

Sylvester, E. S., and Richardson, J., "Some effects of temperature on the transmission of pea enation mosaic and on the biology of the pea aphid vector," *J. Econ. Entomol,* 59, 255 (1966).

Sylvester, E. S., and Richardson, J., " 'Recharging' pea aphids with pea enation mosaic virus," *Virology,* 30, 592–597 (1966).

Sylvester, E. S., and Richardson, J., "Decreased survival of *Hyperomyzus lactucae* inoculated with serially passed sowthistle yellow vein virus," *Virology* (1971).

Sylvester, E. S., and Simons, J. N., "Relation of plant species inoculated to efficiency of aphids in the transmission of *Brassica nigra* virus," *Phytopathology,* 41, 908–910 (1951).

Takahashi, Y., "Detecting method of the viruliferous green rice leafhopper, *Nephotoettix cincticeps* Uhler, a vector of the yellow dwarf disease of the rice plant," *Japanese J. Appl. Ent. Zool.,* 7, 200–206 (1936).

Thresh, J. M., "Increased susceptibility to the mite vector (*Phytopus ribis* Nal.) caused by infection with black currant reversion virus," *Nature,* **202,** 1028 (1964).

Toko, H. V., and Bruehl, G. W., "Apple-grain and English grain aphids as vectors of the Washington strain of the cereal yellow-dwarf virus," *Plant Disease Reptr.,* 40, 284–288 (1956).

Toko, H. V., and Bruehl, G. W., "Strains of the cereal yellow dwarf virus differentiated by means of the apple-grain and the English grain aphids," *Phytopathology,* 47, 536 (1957).

Tu, J. C., and Ford, R. E., "Effect of temperature on maize dwarf virus infection, incubation and multiplication in corn," *Phytopathology,* 59, 699 (1969).

van der Want, J. P. H., "Ondersoekingen over virusziekten van de boon (*Phaseolus vulgaris* L.)," Dissertation, Wageningen, 1954.

Van Hoof, H. A., "On the mechanism of transmission of some plant viruses," *Verhendel Akad. Wetenschap.,* **60,** 314–317 (1957).

Vidano, C., "Maize rough dwarf virus in salivary glands and mycetomes of *Laodephax striatellus* Fallén," *Atti. Accad. Sci. Torino,* **100,** 731–748 (1966).

Walters, H. J., "Some relationships of three plant viruses to the differential grasshopper, *Melanoplus differentialis* (Thos.)," *Phytopathology,* 42, 355–362 (1952).

Walters, H. J., and Henry, D. G., "Bean leaf beetle as a vector of cowpea strain of southern bean mosaic virus," *Phytopathology,* **60,** 177 (1970).

Wardle, R. A., and Simpson, R., "The biology of *Thysanoptera* with reference to the cotton plant. III. The relation between feeding habits and plant lesions," *Ann. Appl. Biol.,* 14, 513–528 (1927).

Watson, M. A., "Further studies on the relationship between *Hyoscyamus* virus 3 and aphid *Myzus persicae* (Sulz.) with special reference to the effects of fasting," *Proc. Roy. Soc. (London),* B125 144–170 (1938).

Watson, M. A., "The transmission of beet mosaic and beet yellows viruses by aphids; a comparative study of a non-persistent and a persistent virus having host plants and vectors in common," *Proc. Roy. Soc. (London),* B133, 200–219 (1946).

Watson, M. A., "The effect of different host plants of potato virus C in determining its transmission by aphids," *Ann. Appl. Biol.,* 44, 599–607 (1956).

Watson, M. A., and Nixon, H. L., "Studies on the feeding of *Myzus persicae* (Sulz.) on radioactive plants," *Ann. Appl. Biol.,* 40, 537–545 (1953).

Watson, M. A., and Roberts, F. M., "A comparative study of the transmission of *Hyoscyamus* virus 3, potato virus Y and cucumber virus 1 by the vectors *Myzus persicae* (Sulz.), *M. circumflexus* (Buckt.), and *Macrosiphum gei* (Koch)," *Proc. Roy. Soc. (London),* **B127,** 543–576 (1939).

Watson, M. A., and Roberts, F. M., "Evidence against the hypothesis that certain plant viruses are transmitted mechanically by aphids," *Ann. Appl. Biol.,* **27**, 227–233 (1940).

Watson, M. A., and Sinha, R. C., "Studies on the transmission of European wheat striate mosaic by *Delphacodes pellucida* Fabr.," *Virology,* **8**, 139–163 (1959).

Watson, M. W., and Serjeant, E. P., "Carrot motley dwarf," *Rept. Rothamsted Exp. Stn.,* 1961, (1962).

Whitcomb, R. F., and Jensen, D. D., "Proliferative symptoms in leafhoppers infected with Western-X virus disease," *Virology,* 35, 68 (1968).

Wolfe, H. R., "Relation of leafhopper nymphs to the western-X disease virus," *J. Econ. Entomol., 48, 588–590 (1955).*

Worst, J., "A micro-injection system," *Quart. J. Microscop. Sci.,* **95**, 469–472 (1954).

Yamada, W., and Yamamoto, H., "Studies on the stripe disease of rice plant. I. On the virus transmission by an insect, *Delphacodes striatella* Fallén," *Spec. Bull. Okayama Pref. Agr. Expt. Sta.,* 52, 93–112 (1955).

Yamada, W., and Yamamoto, H., "Studies on the stripe disease of rice plant. III. Host plants, incubation period in the rice plant, and retention and overwintering of the virus in the insect, *Delphacodes striatella* Fallén," *Spec. Bull. Okayama Pref. Agr. Expt. Sta.,* 55, 35–56 (1956).

Yoshii, H. and Kiso, A., "Studies on the nature of insect-transmission in plant viruses (II). Some researches on the unhealthy metabolism in the viruliferous plant hopper, *Geisha distinctissima* Wal., which is the insect vector of the dwarf disease of Satsuma orange," *Virus (Osaka),* 7, 315–320 (1957).

Yoshii, H., and Kiso, A., "Studies on the nature of insect-transmission in plant viruses (VIII). On the incorporation of P[32] into virus-transmitting green broad winged planthopper and virus-infected Satsuma orange," *Ann. Phytopathol. Soc.,* 24, 175–181 (1959).

Yoshii, H., Kiso, A., and Kikumoto, T., "Studies on the nature of insect-transmission in plant viruses (VII). On the metabolic components in the orange virus-transmitting green broad winged planthopper," *Virus (Osaka),* 9, 462–467 (1959).

Yoshii, H., Kiso, A., Yamaguchi, T., and Miyauchi, S., "Studies on the nature of insect-transmission in plant virus diseases (VI). On the incorporation of P[32] into stunt virus-infected rice plant and virus-transmitting green rice leafhopper," *Virus (Osaka),* 9, 453–462 (1959a).

Zazhurilo, V. K., and Sitnikova, G. M., "The relation of the virus of winter wheat mosaic to its vector (*Deltocephalus striatus* L.)," *Lenin Acad. Agr. Sci.,* 6, 27–29 (1941).

CHAPTER 14 CHAPTER

Ecological Aspects of
Plant Virus Transmission

The movement of insects from one host plant to another is the commonplace activity upon which all transmission of plant virus diseases by arthropods depends (Carter, 1961). This movement is conditioned by many factors: the normal life history of the insect, its host range and host preferences, the availability and condition of these hosts, and their status as virus reservoirs. Superimposing their determinative influence on all these biotic factors arc the physical factors of the environment. This chapter is concerned with the interrelationships of these factors.

VECTOR MOVEMENT,
DISPERSAL, AND MIGRATION

Movement, dispersal, and migration are terms whose definition varies with the user. Movement should perhaps be restricted to changes of position intramurally. That is, an apterous aphid moving from one part of the plant to another, or from plant to plant in the same location, would be considered in movement. Alate aphids rising above a potato field and settling down again would fall in the same category.

Dispersal has been defined as flights within natural breeding areas

(Severin, 1933; Linn, 1940) and migration as flights out of the natural breeding areas. Perhaps "permanent" would be a better word than "natural." Dispersal would include such phenomena as flight of aphids from a primary to a secondary host, i.e., the short flight of *Piesma quadratum* (Fieber) from the woodlands where it overwinters to young sugar beets, and the back-and-forth movements from weeds to cultivated crops of many vector species. Dispersal is no doubt conditioned by many instinctive reflexes and the local pressures of the environment, but it still involves considerable choice, *in situ,* and in host plant selection.

Migration, the flight out of permanent breeding grounds, usually involves long distances, sometimes hundreds of miles, and infers an obligatory movement imposed by climate or other factors. Dispersal could be a prelude to migration, because at that time, the insect, moving in short flights, is vulnerable to strong convection currents and high winds. Within the limits imposed by season, dispersal of a migrating insect can occur at the end of the migration when the insect is occupying its temporary breeding grounds.

Simpson (1940) restricted the use of the term migration to movements between primary and secondary hosts occurring in the spring and fall, and dispersal to movements occurring within or between plantings of secondary host plants.

Migration, according to Davidson (1927), has a specialized meaning with respect to aphids, namely, the flight of sexual and asexual forms between winter and summer hosts. Aphids are called migratory or nonmigratory according to whether or not they so alternate. C. G. Johnson's usage (1954) differs somewhat in that he includes some aphid flights as migrations where there is no alternation of summer and winter hosts.

Lawson *et al.* (1951) preferred the word dissemination to dispersal or migration, the objection to the word dispersal being that it, by definition, connotes no directional movement; and to the word migration because it has certain specialized (subjective) meanings when used in connection with birds, mammals, and fishes. Neither objection seems valid; in fact, Lawson *et al.* found it convenient to use the term flight to describe movement from one place to another. It is not at all certain that the migration stimuli of insects differ fundamentally from those of any other animals; whether it be crowding, food supply, or just the urge to get up and go, associated with maturation.

A distinction between active and passive migration has been made; active migration is held to be purposeful flight, taken only in still dry air for short distances, while passive migration has been taken to involve long-distance flights (Doncaster, 1943). Whether passive flights are, in fact, flights, and not merely involuntary wind-controlled movement of buoyant particles would be difficult, if not impossible, to determine.

Mass migrations of insects are defined by C. G. Johnson (1960) as "the exodus flight by new adults from the breeding site followed by one or more flights, varying in character from species to species, to a new breeding site." Johnson considers that the primary function of the adult is dispersal and that adversity, in the shape of crowding, lack of food supply, or space, coincides with mass exodus but is not necessarily the cause of it.

The beet leafhopper weed-host complex has, according to Douglass (1954), changed in favor of the insect in some areas during the last 30 years, but unless the insect is becoming acclimatized, as Douglass suggests, the end point of the long-distance migration will be in an area climatically unsuitable for the insect's survival (Carter, 1930). Even the permanence of extension of the breeding grounds north and east into the southern Great Plains is dependent on subnormal rainfall (Douglass *et al.*, 1956).

The leafhopper that transmits aster yellows virus, *Macrosteles fascifrons* (Stål), has not been considered a long-distance migrant, but recent data indicate that this insect migrates into Canada from the northern United States. Extensive damage to lettuce and other susceptible crops occurred in Manitoba, and, only slightly less, in Saskatchewan; 14% of the immigrants were infective on arrival. Evidence of the extension of the migration as far north as Edmonton has been recorded (Sackston, 1958; Westdal *et al.*, 1959).

Migration of *M. fascifrons,* vector of aster yellows agent from the south, has been followed in Wisconsin. The leafhopper travels several hundred miles, and the stimulus for migration appears to be the maturation of host grain crops in the source areas (Drake and Chapman, 1965). A similar study with essentially the same conclusions was made in Minnesota (Meade and Peterson, 1964). The potato leafhopper, *E. fabae,* is also a long-range migrant from the extreme southern United States, moving under the influence of warm south winds up the Mississippi Valley (Pienowski and Medler, 1964). The insect cannot survive the winter in the Midwest (Decker and Cunningham, 1968). Seasonal aggregations of the insect occur at the margins of fields (due to accumulation of migrants) or in elevated areas (coinciding with maximum population density) (Kieckheker and Medler, 1966).

The insignificance of *D. maidis* as a vector of corn stunt disease in Mississippi has been established by a study of the migration northward from the southwestern U. S. and possibly Mexico. The insect does not occur early in the season in the southeastern states, but the disease incidence is already high. The insect cannot survive the winters in Mississippi (Pitre *et al.*, 1967) (See also Pitre and Hepner, 1967).

The development of a migratory form of *Aceria tulipae* Keifer may account for the occasional rapid spread of winter wheat mosaic. The type of food available to the mites appears to be a factor in determining whether the mites are migratory or not, the proportion of those migrating being much higher on wheat than on fruiting culms. Survival of the migratory forms is significantly greater (82 vs. 55%) than that of the non-migratory forms, although the fecundity is the same in both forms (Somsen, 1966). This same mite colonizes wheat in the spring and early summer; in the late summer and autumn it appears on corn. On wheat it infests secluded areas under leaf ligules and sheaths under glumes on wheat heads; on corn, under the husks on the corn ears. Other grass-infesting eriophyids readily colonize leaf blades (Nault and Steyr, 1969). Proeseler (1967) considers that active migration of vector species of eriophyids of less importance than passive dissemination through the agencies of insects, wind, and rain.

The virus of groundnut mosaic is not carried over in groundnut. Other food plants of the vector were investigated but none were found capable of maintaining the virus. Adams (1967) concluded that the virus was reintroduced each year by long distance migration of the vector aphid, *A. craccivora*.

An epidemic of BYDV occurred in 1969 in Manitoba. Two strains of the virus were involved, one *R. maidis*-specific, the other not, being transmitted by three other species of aphid. The aphids bred more rapidly on barley than on wheat and oats. The loss was estimated at 1,380,860 bushels of barley, and losses on other cereal crops were high. Early migration of the aphids' vectors and late-seeded crops combined to bring about the epidemic. Distribution of the disease indicated that the infestation was from aphids carried long distances (Gill, 1970).

Lettuce mosaic yellows is a serious virus disease in South Australia, and the virus is transmitted by *Hyperomyzus lactucae*. Randalls and Crowley (1970) studied the relationship between the incidence of the disease and the activity of the vector. The seasonal incidence was high only in certain periods and negligible at other times. There was a positive correlation between disease incidence and the migratory activities of the vector. Disease incidence peaked 4–5 weeks after peak numbers of the aphids were trapped, and the high trap catches occurred when the weekly temperature means were in the range of 60–72°F (15.6–22.2°C). Meteorological data could well be used to predict flights and disease incidence.

Altitude definitely affects both initial infestation and the summer migration by *M. persicae*. As altitude increases, both of these factors decrease. Although plants growing at altitudes of 3000 feet or over became infested, slopes and hills exposed to winds in prealpine regions were less liable to infection and were, therefore, more suitable for production of seed potatoes (Meier, 1958).

METHODS OF MEASURING VECTOR POPULATIONS AND THEIR MOVEMENTS

Quantitative Methods

Dispersal and migration are quantitative, and, with the exception of the sugar beet leafhopper, have been studied most in connection with aphid vectors.

Much of the earlier work was done by Davies (1932), who periodically examined 100 leaves taken at random from varying positions on the plants during the season. Initial infestation, its growth, and movement throughout the crop could therefore be estimated. Any sampling method is suspect, but any such method consistently followed will provide comparable data. For tracing movement of aphids within fields, an extension of the counting method was necessary. Davies developed a method for locating the aphids on the sampled leaves, and weekly records disclosed that at least 84% of each of two species moved from leaf to leaf during the weekly period.

To avoid variation in size and position of leaves, some workers have attempted estimates of aphids per plant (Bald *et al.*, 1946). When the leaf position is considered, the sampling method becomes complicated, but Bald *et al.* (1950) took the total count per plant and then divided the lower leaves into two groups: the basal leaves, which are the lowest two or three, and the "ground canopy," the next three leaves that form a canopy over the ground without touching it. Location of nymphs differed with species, and adults were more randomly distributed than nymphs. Broadbent (1953) estimated aphid density by multiplying the number of aphids on upper, middle, and lower leaf by the average number of leaves for each zone.

Shands *et al.* (1954) counted wingless aphids by using two subunits: one consisted of the terminal and two opposite basal leaflets from three leaves, taken from the top, middle, and bottom of the plant; the other, one half of each of those leaflets. The effect of using the second was greatly to reduce labor, and the results were comparable.

Strickland (1954) developed an aphid-counting grid after testing volumetric methods. The latter method was shown to be inadequate when it was found that 2500 small nymphs, or 600 third-instar nymphs, occupied the same volume (1 cc) as did 250 alatae. Combinations of the 3-leaf method and the grid count were used by Church and Strickland (1954) in sampling aphids on Brussels sprouts.

The standard unit for determining populations of sugar beet leafhoppers on wild and cultivated hosts has been, for many years, 50 sweeps of a net (Carter, 1930; Fox, 1938). The method has obvious disadvantages (DeLong, 1932; Gray and Treloar, 1933), but it has universal applicability within its limits. It is practically useless to determine the early movements of leafhoppers into beet fields when the plants are very small. The appearance of males is the best criterion of the spring movement (Douglass *et al.*, 1946), and the first quantitative counts are made by the "hand and knee method" wherein the observer moves on hands and knees along a row of sugar beets, gently disturbing the plants as he moves along, and making an actual count of the leafhoppers disturbed. Hills (1933) developed a sampling cage, 1 ft sq, which gives a somewhat more accurate count, and Shands *et al.* (1942, 1955) have described wind-vane aphid traps.

Survey methods in the United States have been brought to a fair degree of standardization and published by the Plant Pest Control Branch, ARS (Anon., 1955, 1958). Vectors included in these are the beet leafhopper and the major vector species of aphids. The onion thrips is also included, but on onions only, not on any other virus host. Some survey methods are necessarily drastic. Strickland (1951a), in order to determine the distribution and density of mealybug populations on cacao trees, felled 12 trees on each of ten 1-ac plots each month and counted all the mealybugs found on trunk, branches, twigs, and leaf canopy.

Aphid Traps

Many methods have been devised for estimating the numbers of insects in flight. For aphids, the sticky trap has been widely used (Doncaster and Gregory, 1948; Broadbent *et al.*, 1948); a series of graphs comparing population counts by the leaf method and by trapping are shown in Fig. 14.1 and 14.2.

The value of aphid traps is recognized by C. G. Johnson (1952), who nevertheless suggests caution in using them as exclusive instruments for the analysis of population and activity factors. Johnson's data indicate that the greatest number of aphids migrate in windy weather, but alighting requires calm conditions. Host plants offering little shelter might then be colonized during infrequent calm periods. This type of aphid distribution might be associated with the transmission of the nonpersistent viruses. On the other hand, crops which provide considerable shelter may be colonized continuously by longer range flights of wind-borne aphids,

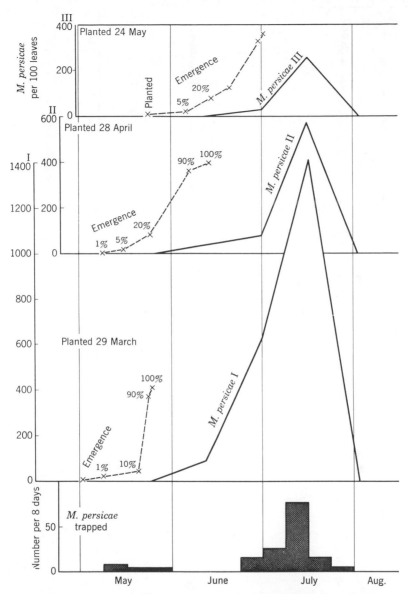

FIGURE 14.1. Number of *Myzus persicae*/100 leaves and percentage emergence of plants. Also numbers of *M. persicae* caught on trap in 8-day period (After Doncaster and Gregory, 1948).

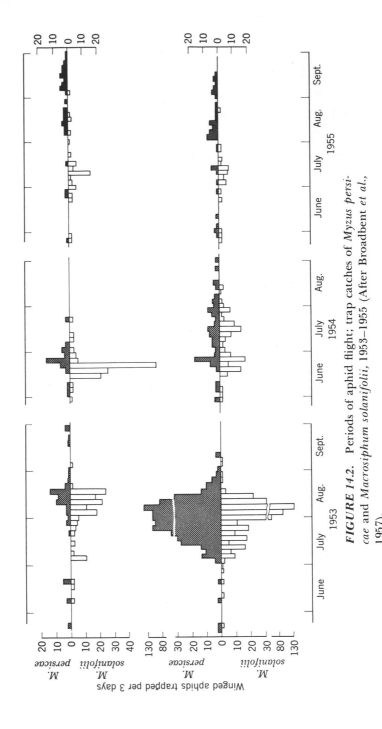

FIGURE 14.2. Periods of aphid flight; trap catches of *Myzus persicae* and *Macrosiphum solanifolii*, 1953–1955 (After Broadbent et al., 1957).

favoring the transmission of the persistent viruses. For leafhoppers, Lawson *et al.* (1951) reviewed the use of various mechanical devices, depending on air movement for efficiency. For sampling air-borne mite populations, Staples and Allington (1959) used grease-coated 1 by 3 in. slides and concluded that good estimates of mite populations in volunteer wheat could be obtained by use of a single trap. The white flies have, thus far, eluded any attempt to determine their movements on a quantitative basis, as have the mandibulate vectors, beetles and grasshoppers.

Height of Insect Vector Flights

The height of insect vector flights has received attention. Broadbent (1948) found that 68% of *Myzus persicae* (Sulz.) were taken at a mean height of 68 in., 25–28% at 38 in., and 5–8% at 7 in. By means of sticky drum traps, Nowak (1951) caught this same species up to 45 m, but Glick (1939), reporting on the results of airplane trapping, recorded *M. persicae* taken at 3000 ft and *Macrosiphum pisi* Harris at 200, 1000, 2000, and 4000 ft.

C. G. Johnson and Penman (1951) found a linear relationship between the log. of aphid density and the log. of ht with a range of 10–2000 ft. Shands *et al.* (1956) studied the low-elevation movement of aphids for several years, using a combination of wind-vane traps and tall tanglefoot screens (Fig. 14.3 and 14.4). The three species of aphid studied, buckthorn, green peach, and potato aphids, all had their peak populations around the 25-ft level. Heathcote (1958) used yellow water traps (after Moericke did so in 1951) and yellow cylindrical traps (after Broadbent *et al.*, 1948) to determine flight at different "heights," that is, from bare ground to 152 cm. These studies obviously pertain to movement and dispersal; a migrating insect would not get very far traveling 6 ft from the ground. Dickson *et al.* (1956) found that a 5-ft high trap gave the best relative count. Traps at 1, 5, and 12 ft revealed differences between species, *Aphis spiraecola* Patch showing somewhat higher counts at the higher level, while the reverse was true of *A. gossypii* Glover. In all cases, more aphids were caught on the leeward side of the citrus trees than on the windward side.

Broadbent (1948) found a relationship between height of trap and species caught, different species flying at different heights; but again, the highest trap was 5–6 ft from the ground; voluntary flight decreased with increasing height. Perhaps both movement and dispersal are primarily voluntary, and the closer to the ground the aphid flies the less the possibility of its being swept into obligatory migration. Leafhoppers have been trapped in various types of mechanical devices (Lawson *et al.*, 1951)

FIGURE 14.3. Wind-vane trap for measuring aphid flight at low altitudes (Photograph: Shands, U.S. Department of Agriculture).

at heights up to 127 ft (Hills, 1937).

Migration is not necessarily achieved in a single flight. Dorst and Davis (1937) traced long-distance movements of the beet leafhopper by planting a series of 100-sq-ft plots to favored weed hosts in the desert areas over which the insect migrates. The nearest winter breeding area to the sugar beet fields is about 200 mi distant. Daily collections on these plots defi-

FIGURE 14.4. Adhesive screen for trapping aphids (Photograph: Shands *et al.*, 1956a).

nitely placed the origin of the leafhoppers as well as the time and magnitude of dispersal. Douglass *et al.* (1946) determined the time and magnitude of dispersal of leafhoppers into beet fields by the use of the Hills' sampler (Hills, 1933), which traps the insects in circular areas containing 1 sq ft. Strickland (1950) caught a few small nymphs of mealybugs on sticky traps in the cocoa-growing area of Ghana. The numbers were not as significant as the indication that wind could be a factor in the movement of the insects.

PHYSICAL FACTORS OF THE ENVIRONMENT IN RELATION TO VECTOR MOVEMENTS

Factors affecting movement of vectors, migration, dispersal, and movement, are a complex of biotic and physical factors; the latter are more easily measured. For aphids, Davies (1935) demonstrated with controlled laboratory experiments, that high humidities inhibited flight of *M. persi-*

cae, and that temperatures from 70–90°F were the most favorable. Thomas and Vevai (1940) extended these studies and suggested some arbitrary standards for the flight of aphids in general. These were: (1) a temperature above 70°F, (2) a relative humidity below 80%, (3) a wind velocity below 5 mph, and (4) a big difference in maximum and minimum temperatures (more than 12°F) for the day. The most important factor limiting flight appeared to be wind velocity. Davies' laboratory experiments were thus confirmed in general by the field data of Thomas and Vevai, but later experiments under controlled conditions by Broadbent (1949) indicated that reactions to relative humidity are complex.

Humidity & Light Intensity

Increases in relative humidity to a higher level retarded flight activity; changes to a lower level increased activity, but aphids adjusted to humidities between 50 and 80% and flew readily at these humidities with temperatures below 80°F. High humidity and high temperature (90°F) sometimes inhibited flight. Light intensity between 100–1000 fc made little difference to flight, but below 100 fc flight activity declined rapidly. These light intensities were from artificial light. Markkula (1953) records differences in aphid takeoff in varying degrees of sunlight; 43, 20, and 11 takeoffs / min of *B. brassicae* (L) were observed with full sunshine, thin clouds, and dense clouds, respectively.

Wind Velocity

Cornwell (1960) conducted a detailed study of the aerial dispersal of the pseudococcid vectors of swollen shoot of cacao. Wind speeds in the cocoa canopy were low, being only 1 / 10 to 1 / 20 of those in the open, where the maximum recorded was 7 mph. Eight vector species became established on cacao after dispersal by air currents. This is apparently accomplished when mealybugs fall from the branches, are carried by air currents, and become laterally dispersed at levels a few feet from the ground.

Aerial dispersal is more pronounced during dry conditions, whether during the main dry season from December to February or during the shorter dry period in July and August. The maximum recorded distance of mealybug aerial dispersal from surrounding vegetation to cacao seedlings was 340 feet. The relationship between numbers of mealybugs caught on sticky traps and infestation of trap seedlings is shown in Fig. 14.5. Watson and Heathcote (1965) considered that the sticky trap is the most reliable tool for developing data on insect populations with regard to disease spread.

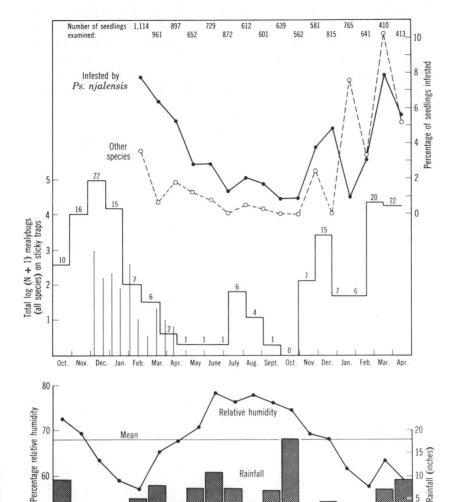

FIGURE 14.5. Seasonal changes in the catch of airborne mealybugs on sticky traps and on cacao seedlings within farmer's cacao. The open histograms are data from sticky traps at 2–30 ft above ground; the thin solid lines from traps at 6 ft above the ground. Numbers of air-borne mealybugs are highest during months of low rainfall and periods of low relative humidity (After Cornwell, 1960).

Myzus persicae is blown in from the coastal or mountainous areas of southern California to the interior desert agricultural areas. This was determined by using sticky traps in the mountain passes and near mountain slopes. The aphids arrive in September and October when favored crops and weeds are germinating. Low populations of parasites and predators at that time permit an explosive increase in aphid populations (Dickson and Laird, 1967).

An area comprising some 10,000 square miles to the northwest of the central coastal districts of New South Wales serves as a breeding area for *Aphis craccivora* Koch. After a period of suitable weather and host conditions for breeding, the weather becomes dry, the host plants dry up, and the exodus occurs. Migration follows periods of northwesterly upper winds which bring the aphids to the coastal regions, 200–300 miles away, in less than 24 hours. One instance is recorded of the arrival in Sydney of a dense swarm of aphids which had been apparently carried out to sea by the upper winds and brought in to the coast again on a sea breeze (B. Johnson, 1957).

This would appear to be a situation favorable for the transmission of persistent viruses, but not for the nonpersistent. It might depend, however, on the temperature conditions of the upper winds. An aphid carried in the upper air currents might well have its body temperature reduced below the threshold of development while in involuntary flight. Experiments on the longevity of the nonpersistent viruses in vectors held at such low temperatures might be well worthwhile.

If all the possible physical factors in the microenvironment are considered, the varying humidities, temperatures, and light exposures that could be found within a few feet, generalizations based on broad field studies might be modified, but only in a very limited degree.

Leafhopper movements have been studied extensively. The physical factors most closely associated with their movements are those of temperature and wind. Lawson *et al.* (1951) found that *Circulifer tenellus* (Baker) flew when temperatures were above 60–64°F, especially in the crepuscular periods near sunrise and sunset. Since temperatures are more likely to be favorable above the threshold in the evening, flights at that time are more common. The number of leafhoppers moving on any particular day is proportional to the temperature. The sexes react differently to winds of low velocity, the males tending to swarm around conspicuous objects. With winds of higher velocity, both sexes are carried along in the wind stream, not necessarily the prevailing winds, but in those that are blowing at the time when light and temperature conditions are most favorable for flight. A low correlation between catch of leafhoppers and humidity was shown to be attributable to the fact that both catch and humidity were correlated with temperature. Lawson *et al.* believed that most of the flying population traveled below 10,000 ft on "surface" winds; in fact, the bulk of their catches were from 5 to 50 ft, with only a few caught at the 130-ft level. Ball (1917), however, reported finding beet leafhoppers on the top of Pike's Peak at 14,000 ft; although flights at this altitude are probably rare, they would account for long-distance migrations of the insect into areas far east of the permanent breeding grounds. Douglass (1954) summarized the records of the occurrence of curly top in these areas, and since then there have been additional records from Maryland in 1958 and 1959 (Schneider, 1959; Heggestad and Moore, 1959). Bennett (1971) has included these considerations in his recent monograph on curly top and in his review (1967) on the epidemiology of leafhopper-transmitted viruses. It is perhaps significant that these last records are for the virus disease, not the leafhopper; the latter could easily migrate beyond its survival range but live long enough to feed and transmit the virus. Douglass quotes Glick's (1939) findings on the influence of air currents on insect migration; the rougher the air, the greater the proportion of insects found at the higher elevations. Leafhoppers at active temperatures live only a short time if unfed. This would require that insects in obligatory long-distance migration be carried passively in air currents at temperatures lower than their threshold of development, as was suggested by Carter (1927) and Severin (1933).

Temperature and Rainfall

Severe winters and cold spring weather are a major deterrent to many species. Carter (1930) showed the correlation between severe winters and relative freedom from beet leafhopper outbreaks the following year. Later workers extended this to include the influence of spring weather. Annand *et al.* (1932) reported the necessity of taking the spring weather into account since this determines the time of the spring movement into cultivated fields. Fox (1945) considered the time of the spring movement of more importance than the peak population.

Cook (1945) used temperature summation to determine the spring movement of the leafhopper in the San Joaquin Valley. Although Cook interpreted the correlation between degree days and spring flights solely on the development of the insect and on the assumption that maturity was the principal stimulus, it is probable that the temperature summation related to plant development and maturity would coincide in most years. Cook's data showing expected and actual dates are shown in Table 14.1.

When the hibernating insect is the vector, as is the case with the two

TABLE 14.1. *Comparisons of Actual Dates of Spring Migrations of the Beet Leafhopper in the San Joaquin Valley with Dates Corresponding to the Summation of 825 Day-Degrees above 45° F after January 1* [a]

Year and item	Southern		Central		Northern foothills
	Foothills	Plains	Foothills	Plains	
1936					
Observed			4/9–12	4/17	4/17
Calculated	3/18	3/30	4/9	4/16	4/18
Difference [b]			+1.5	+1.0	−1.0
1937					
Observed	4/20–25	4/25–5/2	5/2	5/2	
Calculated	4/18	4/22	4/28	5/6	5/15
Difference [b]	+4.5	+6.5	+4.0	−4.0	
1938					
Observed	4/20–25	4/20–25	5/3–7	5/7–12	5/12–17
Calculated	4/19	4/26	4/29	5/7	5/14
Difference [b]	+3.5	−3.5	+6.0	+2.5	+0.5
1939					
Observed	4/4–5	4/15–19	4/15–19	4/21–27	4/27–5/5
Calculated	4/11	4/13	4/16	4/24	4/26
Difference [b]	−6.5	+4.0	+1.0	0.0	+5.0
1940					
Observed	4/1–2	4/1–8	4/15–19	4/15–19	4/27
Calculated	3/27	4/1	4/9	4/12	4/24
Difference [b]	+5.5	+3.5	+8.0	+5.0	+3.0

[a] After Cook (1945).

[b] A minus sign (−) before the difference indicates that the observed migration occurred before the calculated date, and a plus sign (+) that the observed migration occurred after the calculated date. Differences given are between the calculated date and the midpoint of the observed migration period when this is more than 1 day.

species of *Piesma, P. quadratum,* the vector of beet leaf curl, and *P. cinerum,* the vector of savoy, winter conditions in the unploughed areas, headlands, and woodlands where the insect hibernates will determine the extent of survival and the degree of infection in beets the following spring (Wille, 1929; Coons *et al.*, 1958).

Low temperatures in spring were shown by Shands *et al.* (1958) to markedly affect the survival of stem-mother aphids. This was not as a result of sudden drops in temperature below survival levels, but rather the general temperature levels during the developmental period. These same temperature conditions also reduce the amount of favorable foliage.

The severity of winter weather has a marked effect on the spring populations and migrations of *M. persicae.* During mild winters or in well-protected locations, the viviparous females can overwinter successfully on crops other than the true winter hosts. During severe winters, the over-wintering females are eliminated, so that the population in the spring must develop from eggs laid on the winter hosts (peach, apricot, etc.). Populations of aphid species which overwinter only in the egg stage are not as susceptible to variations in the severity of winters (Watson *et al.,* 1951).

Hull and Watson (1945) reported that sugar beet yellows develops seriously only after a succession of mild winters. The first mild winter leaves infestations confined largely to seed crop areas, but if the following winter is also mild, then some plants of the heavily infested root crop remain in the ground, together with infected mangel, spinach, and spinach beet. These serve as widespread sources of infection for the succeeding year.

Curly top incidence in tomatoes in the Yakima Valley in the state of Washington depends on daytime temperatures affecting dispersal during the early growing period. Disease severity is more highly correlated with winter rainfall than with spring leafhopper populations (Clark, 1968).

The vector of sweet potato mosaic in Georgia (*B. tabaci*) normally overwinters in Florida and migrates north. The late spring in Florida is thought to be responsible for the absence of the vector in Georgia in 1958. Rainfall was also high in Georgia that year, a negative factor, for the insect thrives best under relatively dry conditions. Roguing is practised, and dieldrin application for the sweet potato weevil has no doubt contributed to a complex of factors which has resulted in the apparent eradication of sweet potato mosaic from Georgia (Giradeau and Ratcliffe, 1963).

Slykhuis (1963) considers *Endria inimica,* vector of wheat striate mosaic, to be a real threat to the western wheat grower because of its breeding on many alternate host plants and (Slykhuis and Sherwood, 1964) the increased activity due to high temperatures which may result in epidemics in early spring wheat. Hagborg (1963) questions this on the grounds that the insect has a very low fecundity and a slow rate of reproduction. Hagborg has a point, but if the insect is migratory, movement from southern breeding grounds affected by high temperatures might possibly result in infections of epidemic proportions.

Whereas climate will determine the geographic range of an insect, weather is responsible for the fluctuations that determine its seasonal activity and numbers. Even in tropical West Africa, where weather seems to be extremely uniform, seasonal differences in temperature associated with rainfall conditions affect mealybug vector populations materially. The decline of populations during the first part of the year is associated with periods of increasing rainfall. Maximum populations occur in October and November, but changes during the year are of a 5- or 6-fold order, and the activities of parasites and predators, again related to season, are held to be important (Cornwell, 1957).

The leafhoppers and cercopids that transmit Pierce's disease of grapevine in California breed on wild host plants and alfalfa. Wet seasons are favorable for the development of large populations of the vectors, and outbreaks of the disease are positively correlated with rainfall (Winkler, 1949).

Cataclysmic Weather Conditions

Cataclysmic weather conditions sometimes affect vector populations. Shands *et al.* (1956a) reported that the two hurricanes in northeastern Maine so reduced aphid populations that feeding damage was not economic; fall populations of *M. persicae* were so reduced on both potato and winter hosts that reduction of populations the following spring occurred. The hurricanes occurred when the fall migrants were beginning to mature and disperse to primary hosts. The effect on the buckthorn aphid was not so pronounced, because, on the primary host (*Rhamnus alnifolia* L'Her), populations were already large, egg deposition had started, and the plant is low-growing and in sheltered environments.

Staples and Allington (1956) reported that hail, occurring just before wheat harvest, forced a volunteer crop of wheat. A wheat mosaic epidemic followed because of heavy infestations of *Aceria tulipae* Keifer.

Little cherry disease in British Columbia, known for its capacity to reach epidemic proportions, has declined in the Kootenay Lakes area following the most severe winter in the history of the area (Wilks and Welsh, 1964). Such an event could bring vector populations down to an extremely low level and could have left many virus sources vectorless. If continued long enough, transmissibility of the virus through insects could be lost.

HOST PLANT SEQUENCE

It is conceivable that a monophytophagous insect, living on a host plant with a seed-transmitted virus, could exist, but only if the host and virus had reached a mutually tolerant equilibrium. Otherwise, the result would be the elimination of all three. All insect vectors of virus diseases, as far as is known, are polyphytophagous. Their movement between plants of the same host species can be conditioned by the insect's behavior or the activities of man, while their movement between host plants of different species can be brought about by obligate facets of the insect's biology, or by all the interrelated factors of their host plant ecology.

Intrahost Movement

Intrahost movement has been studied in connection with aphid biology. Davies (1932) found that 84% of *Macrosiphum euphorbiae* (Thomas) and 73% of *M. persicae* changed site within 24 hours. There is considerable movement from leaf to leaf of adjoining plants when these are in contact (Doncaster and Gregory, 1948; Czerwinski, 1943). There is a difference between species. *M. persicae* is generally considered to be restless, particularly when temperatures are high. The apterae of *M. euphorbiae* move more freely from plant to plant than either *M. persicae* or *Aphis rhamni* (Fonscolombe) (Broadbent, 1953). Short, frequent flights are common on calm, warm evenings, on calm, cloudy days, or in the early morning, and such flights will invariably result in alighting on other plants (Broadbent, 1949).

Cornwell (1956) studied the behavior of the mealybug vectors of swollen shoot disease on cacao. When diseased trees are cut down and slash piled, the mealybug populations are usually low, since many have left the infected tree early in the infection stage. The cut branches wilt, but wilting is retarded in the pile, and many of the mealybugs remain on the cut wood, continuing to reproduce for nine weeks or more. Nymphs from this reproduction migrate but, when starved, their movement is greatly retarded and they will die within a week.

Feeding preferences of the green peach aphid on sugarbeet leaves infected with beet mosaic, western yellows, beet yellows and curlytop viruses differ. Apterous aphids preferred yellows-infected plants to symptomless mosaic-infested or curlytop-infected leaves. Movement of the apterous aphids was at random while on plants. Alates showed short-lived preferences for yellows-infected leaves but differences were lost after 48–96 hours. Curlytop-infected plants were avoided on both leaves and whole plants. The authors suggest that preference was not made in advance of feeding but only after selective test feedings (Macias and Mink, 1969). Swenson (1968) has reviewed the role of aphids in the ecology of plant viruses and given consideration to the aphids host plant, the dispersal of the insects from these, and the spread within cultivated fields.

Interhost Movement

Interhost movement is often forced on the insect by the activities of man. As long ago as 1928, Wolcott related the spread of sugarcane mosaic to the practice of weeding out small grass species among the cane plants. This resulted in the alatae flying off, but the apterae had no alternative but to move on to the nearest cane plant. Martin (1938) reported the same sequence of events in Hawaii.

Emilia sonchifolia L is a common weed in old pineapple fields, and it is frequently infected with TSW virus and heavily infested with *Thrips tabaci* Lind. Knocking down of these fields exposes fields to the leeward to infection (Carter, 1939). Hand weeding of *Emilia* in young fields will also expose nearby pineapple plants to infection, especially if the weeds are left in the field.

Weeds in many cultivated crops serve as reservoirs of both viruses and insect vectors, but the economic problem is to prevent the growth of these weeds altogether, since their control after establishment of vector populations again forces interhost movement and consequent virus spread. Chemical methods of weed control, especially the preemergence type, have greatly improved in recent years.

The insect's biology may impose an obligatory change of host plant; the prime examples of this are to be found among the aphids. Kring (1959) states that true migratory (dispersing) aphids have three characteristics in common: (1) eggs are produced only on a few plant species, the primary hosts, and only viviparous females occur on other host plants, the secondary hosts; (2) primary host plants usually are members of older plant families than the secondary hosts; and (3) male aphids are produced only on secondary hosts, while the oviparous females are produced only on the primary hosts. According to this definition, *A. gossypii* is not a true migrant because both males and females are produced, but there is a formidable list of viruses of which it is a vector.

The primary hosts of *M. persicae,* the most ubiquitous vector of viruses known, vary with geographical location and climate. In Maine, wild plum is the only recognized primary host and the aphid hibernates on this host in the egg stage (Simpson and Shands, 1949). In Holland, the insect may overwinter in glasshouses or in mangold clamps; but this is rare compared with hibernation on *Prunus serotina* Ehrhart, the American bird cherry, which has been planted in vast numbers as a protection for young coniferous forests, or on peach trees. According to Hille Ris Lambers (1955), 97% of migrants from winter hosts to summer hosts come from peach or *P. serotina*. The mangold clamp as a source of infection of sugar beet fields is illustrated in Fig. 14.6 (Broadbent *et al.,* 1949).

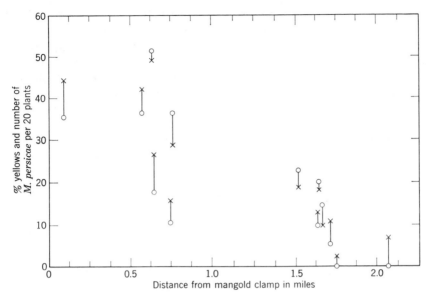

FIGURE 14.6. The relation between numbers of *Myzus persicae* and yellows infection in sugar beet fields and the distance from a mangold clamp. X=yellows; o=*M. persicae* (After Broadbent *et al.,* 1949).

In the northeast of Scotland, the insect overwinters mainly as apterous viviparae on various Cruciferae (savoy cabbage, cabbage, and Brussels sprouts) (Shaw, 1955). This is true also in England, particularly near towns and villages where many other cultivated crops also serve as overwintering hosts. In some parts of Europe, peach orchards are the main source of this aphid, which overwinters there in the egg stage, the winters being too severe for the viviparae to survive (Broadbent, 1953).

The generalized seasonal history of aphids with alternating hosts is as follows: the eggs hatch in early spring, and most of the winged-migrant forms mature in the third generation, moving to summer hosts—potatoes —and to many weed species. Reproduction continues on the secondary hosts until the end of the summer, when winged-fall migrants fly to the primary hosts and initiate colonies.

Since the magnitude of both spring flights outward from the primary hosts and the fall return flights depend on the production of alatae, the factors affecting this production are of some interest. Lal (1955) related production of alatae to decrease in the water content of the aphids, which was brought about by combinations of temperature and relative humidity which also diminished the water content of their host plants.

Crowding is associated with production of alatae, but this is probably due to inadequacy of the water supply. Dry atmospheric conditions, a reduced water content in the host plants, and crowding of the aphids will be followed by an increase in the production of winged migrants. Kennedy *et al.* (1958) confirmed the factor of water strain in the host plant in inducing increased production of alatae. The problem is complex. Sometimes a change of host plant will induce alatae production in the absence of water stress. *A. gossypii,* growing on *Commelina nudiflora* L in cages, produced no winged forms in Hawaii, but when *Bidens pilosa* L grew as a weed in the cages, alatae developed profusely. Host transfer can be forced by the maturing of a weed host, either according to normal seasonal development or through drought.

Shands and Simpson reported on the production of alatae under greenhouse conditions (1948). In this case, water stress was not a factor. They concluded that winged-form production was influenced by the size of the plant and the size of the aphid population.

Most authors agree that production of alatae is influenced by the factors of light, temperature, crowding, starvation, wilting of the host plant, and even parentage. Davidson (1929) reported that winged migrants of *Aphis rumicis* (L) developed as a result of an inherent tendency, but that the actual number at any particular time in the development of the colony was influenced by overcrowding and the condition of the host plant. It seems clear that the production of alatae in aphids, so essential to the spread of plant viruses, is brought about by this inherent tendency, essential to the survival of the species, and conditioned by physical and biotic factors of the environment.

Kennedy *et al.* (1958) explained the effect of water shortage in the host as an effect of concentration of solutes in the sap, increasing at the expense of quantity available. This increased sap concentration increased the production of winged forms. A somewhat analogous situation is found in the case of *C. tenellus.* Although there is an inherent tendency to migrate coincident with maturation, the major factor is the condition of the host plants related to sap concentration (Carter, 1927a; Lawson *et al.*, 1951). Thomas and Vevai (1940) listed sunny weather, low relative humidity, large fluctuations in temperature, and a long, dry period resulting in starvation as factors in the production of alatae. According to Wadley (1931), temperatures are critical in the case of *Toxoptera graminum* (Rondani).

Tomatoes grown adjacent to sugar beet fields, compared with those grown more remotely, indicated very little movement of *Circulifer tenellus* (Baker) within a given tomato field once the insect was established in the beet field (Clark, 1968a). It might be mentioned that this would be the expected occurrence since tomato is a poor host for the leafhopper, and infestation of tomatoes, with resultant curlytop, is incident to the insect's migration; it will die out on tomatoes, particularly when plant density is high, but it will flourish in beet fields.

Plant Succession

Host plant sequence is in many cases governed by natural plant associations. The factors of plant ecology must, therefore, be integrated with those of the insect's ecology, and the best documented case of its kind is that of the sugar beet leafhopper and its weed hosts. As Cook (1941) pointed out, the breeding grounds of the leafhopper are characteristically in arid country with a mean annual rainfall of 10–12 inches. All the breeding grounds but one (along the Rio Grande in Texas and New Mexico) are west of the Continental Divide (Fig. 14.7). Since the early work of Carter (1930) recognized the importance of weed-host succession in determining and maintaining populations of the leafhopper, there has been a continuous study made by many workers whose extensive findings can be referred to only in summary (Piemeisel, 1932, 1938, 1945; Piemeisel and Chamberlin, 1936; Piemeisel and Lawson, 1937; Lawson and Piemeisel, 1943; Fox, 1938, 1942, 1945; Piemeisel and Carsner, 1951; Piemeisel *et al.*, 1951; Douglass and Hallock, 1957). When the original vegetation of the San Joaquin Valley in California was mapped, five vegetation types were recognized. All have been greatly modified by cultivation and overgrazing. Plant species replacing the original vegetation are determined by land use. Overgrazing results in a stand of winter annuals, and, if too heavy, in bare soil. Intermittently-farmed lands produce large areas of summer annuals. The beet leafhopper in this area utilizes both winter and summer annuals, and it is on these hosts that the huge populations of the insect develop.

Piemeisel (1954) and his coworkers have shown conclusively that, with proper protection, the original vegetation can be restored. The accompanying diagram is, in essence, a summary of their conclusions (Fig.14.8).

In southern Idaho, a seven year study by Piemeisel (1938) showed essentially the same type of plant succession, but with special emphasis on fluctuations in dominance of the annual weeds. Lawson and Piemeisel (1943) followed with a similar study conducted in the San Joaquin Valley (Fig. 14.9).

Changes in leafhopper populations following these fluctuations were followed in a parallel study by Fox (1938), who found that when the plant succession had progressed to the point where downy chess (*Bromus*

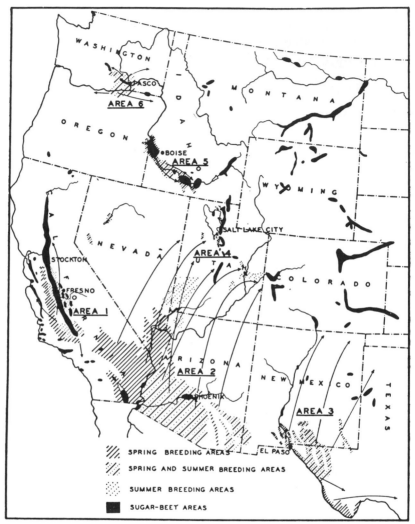

FIGURE 14.7. Major breeding areas of the beet leafhopper and sugar beet areas affected by them. Arrows indicate the general direction of spring movements of the insects (Map: Courtesy of the Agricultural Research Service, U.S. Department of Agriculture).

tectorum L) and sagebrush were dominant, the area lost its status as a leafhopper breeding ground.

Forty-three host plant species were studied by Douglass and Hallock (1957) to determine their relative importance. Host plants varied in their suitability for egg laying, and there were differences in the nymphal populations on various species as the summer progressed.

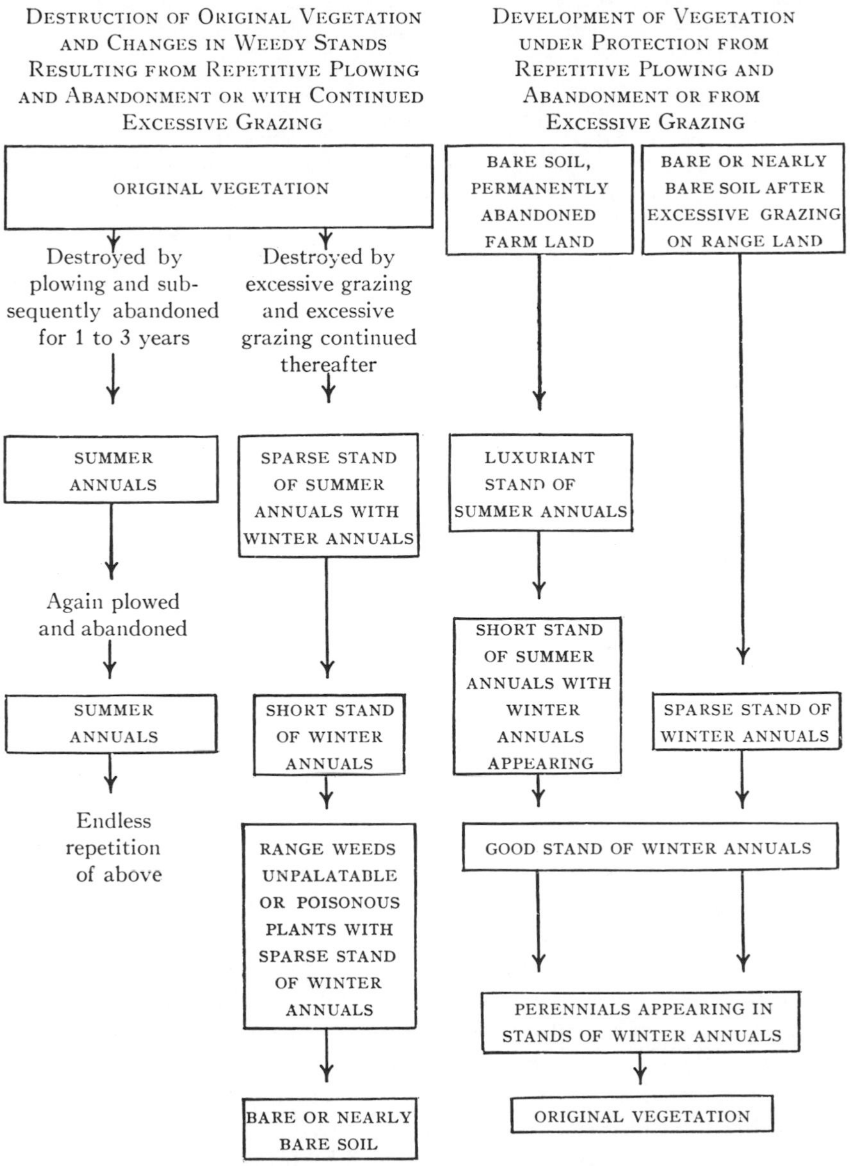

DESTRUCTION OF ORIGINAL VEGETATION AND CHANGES IN WEEDY STANDS RESULTING FROM REPETITIVE PLOWING AND ABANDONMENT OR WITH CONTINUED EXCESSIVE GRAZING

DEVELOPMENT OF VEGETATION UNDER PROTECTION FROM REPETITIVE PLOWING AND ABANDONMENT OR FROM EXCESSIVE GRAZING

ORIGINAL VEGETATION

BARE SOIL, PERMANENTLY ABANDONED FARM LAND

BARE OR NEARLY BARE SOIL AFTER EXCESSIVE GRAZING ON RANGE LAND

Destroyed by plowing and subsequently abandoned for 1 to 3 years

Destroyed by excessive grazing and excessive grazing continued thereafter

SUMMER ANNUALS

SPARSE STAND OF SUMMER ANNUALS WITH WINTER ANNUALS

LUXURIANT STAND OF SUMMER ANNUALS

Again plowed and abandoned

SUMMER ANNUALS

SHORT STAND OF WINTER ANNUALS

SHORT STAND OF SUMMER ANNUALS WITH WINTER ANNUALS APPEARING

SPARSE STAND OF WINTER ANNUALS

Endless repetition of above

RANGE WEEDS UNPALATABLE OR POISONOUS PLANTS WITH SPARSE STAND OF WINTER ANNUALS

GOOD STAND OF WINTER ANNUALS

PERENNIALS APPEARING IN STANDS OF WINTER ANNUALS

BARE OR NEARLY BARE SOIL

ORIGINAL VEGETATION

FIGURE 14.8. Diagrammatic representation of the changes in the stands that follow the destruction of the original vegetation. The left half of the figure deals with the changes as a result of continuous disturbance, and the right half with the changes taking place under protection from disturbances and leading to a reestablishment of the original cover (After Piemeisel and Lawson, 1937).

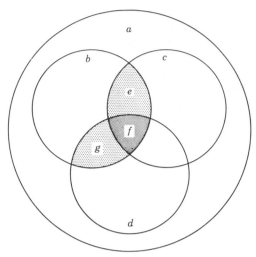

FIGURE 14.9. Diagram of the effects of rainfall, seed supply, competition, and the interaction of these factors on the growth of Russian thistle: *a,* covered by winter rains and with a uniformly dense cover of winter annuals, except within circle *c; b,* Russian thistle seed present; *c,* winter annuals sparse or absent; *d,* covered by late spring rains; *e,* where Russian thistle will grow but will die prematurely if its stand is dense; *f,* where Russian thistle will almost certainly grow and mature; and *g,* where Russian thistle will not survive except in years when late spring rains are unusually heavy (After Lawson and Piemeisel, 1943).

Host plant succession is not necessarily one of breeding hosts. Many are transient feeding hosts, although the insect may feed long enough to transmit viruses. Swenson and Nelson (1959) reported no colonization by 18 species of aphid, yet cucumber mosaic (CMV) of gladiolus was freely transmitted, and 6 new aphid vectors of CMV were discovered in the process. Few, if any, of the 50 species of aphids transmitting yellow dwarf on onion actually breed on the plant. Dickson *et al.* (1949) demonstrated that *M. persicae* did not colonize on the melon crop but was nevertheless responsible for the spread of cantaloupe mosaic in the desert valleys of California. Of the economic crops attacked by tomato spotted wilt virus, none are favorable hosts of the thrips vectors. The coincident geographic range of western yellow blight of tomatoes and curlytop of sugar beets was known for years before the coidentity of the two diseases was recognized (Shapovalov, 1928). The reason for this is that *C. tenellus* does not breed on tomato, although the feeding of transient leafhoppers can be economically devastating. Severin (1933) reported that many hosts in which the leafhopper oviposited were unsuitable for development of the nymphs to the winged stage. Many other plants, recorded as hosts, served only as hold-over hosts in times of stress, particularly during the short period in the fall between the drying up of the summer hosts and the germination of the winter annuals (Piemeisel and Lawson, 1937).

The transfer of insects from endemic to introduced hosts, when the endemic hosts are virus infected, can create a new host-vector-virus relationship, extremely favorable to the insect and critical for the introduced economic crop plant. Native hosts of the swollen shoot virus of cacao in West Africa are the endemic forest trees. Three families of the Tiliales, and the Malvacae, have thus far been implicated (Posnette *et al.*, 1950; Dale and Attafuah, 1957; Tinsley and Wharton, 1958). These forest trees stand as solitary giants in the forest; the undergrowing cacao trees form an almost continuous canopy through which the mealybug vectors can move unhindered, except for barriers imposed by occasional breaks in the continuity of the canopy.

Study of the host plant and environmental preferences can be used for vector separation. Hutchinson (1955) worked with two leafhopper species, *Scaphytopius magdalensis* (Prov.) and *S. verecundus* (Van D.). These insects had been reported to be covectors of blueberry stunt virus. Externally, the insects are similar, and they are separated morphologically by the male genitalia. Surveys, using sticky traps and sweep nets, showed that *S. magdalensis* was abundant in blueberry fields but absent from cranberry bogs. The converse was true of *S. verecundus*, which also prefers open, sunny environments, in preference to the dense growth of blueberry fields. The conclusion was reached that the *S. magdalensis* is the vector of blueberry stunt.

A somewhat similar situation, not yet completely clarified, is described by de Fluiter (1954) and de Fluiter and van der Meer (1958) in connection with *Rubus* stunt. The known vector is *Macropsis fuscula* (Zett.). Interhost transfer of the insect between raspberry and blackberry failed, suggesting the existence either of two host-limited strains of the leafhopper or a closely related species to account for the failure of *M. fuscula* to infect cultivated blackberries, although it breeds on several species of wild blackberries. A related species, *M. scotti* Edwards, is common on cultivated and some wild blackberries, but it cannot complete development on raspberries; its status as a vector is not yet reported.

Before *Jassus indicus* (Walk.) was determined to be the vector of spike disease of sandal, Rao (1943) recognized three area types coinciding with incidence of the disease: (1) the dry scrub jungle containing more or less dense stands of sandal and heavy undergrowth; (2) pole forests with only sparsely growing sandal and very variable undergrowth; and (3) culti-

vated lands, private estates, and village sites where large and healthy isolated trees of sandal are frequent, with no undergrowth. The disease is
most severe in the dry scrub; less so in the pole forests where it spreads
very slowly; and it is almost entirely absent in the cultivated lands, private holdings, and village sites.

THE VIRUS SOURCE

The virus source may be the hibernating insect itself. Curlytop is known
to overwinter in its insect host, *C. tenellus* (Wallace and Murphy, 1938),
although the availability of winter hosts of the insect might affect this. If
the hosts are immune to the virus, then the leafhopper may lose its infective charge. Severin (1934) suggested that infective power was not retained by adult female leafhoppers unless they could derive new sources
during the winter. When the spring annuals germinate in the fall, there
is ample opportunity for the leafhopper to feed intermittently on infected
plants. The two beet viruses transmitted by species of *Piesma* presumably
also overwinter in the insects.

Susceptible Plant Hosts

The most important sources of virus, however, are infected plants, and
Severin (1934) listed 14 species of annual plants and 3 perennials in
which the virus of curlytop overwintered in California, and 75 annuals
and 4 perennials that were susceptible (Severin, 1939). Wallace and Murphy (1938) studied the epidemiology of the disease in southern Idaho,
where 2,000,000 acres of primary wild host area are supplemented by
9,000,000 acres of secondary overgrazed sagebrush area. Infection of the
spring brood of leafhoppers, and the maintenance of virus reservoirs
throughout the summer and fall, are assured by the susceptibility of the
principal hosts, mustards and Russian thistle. Beets left in the field after
harvest survive the winter and are important sources of virus if the field
is replanted to beets the following year. In the Imperial Valley in California, with its extremely high summer temperature, Flock and Deal (1959)
found that sugar beets and weeds in the cultivated areas were the only
important sources. Aster yellows and Pierce's disease of grapes, both
leafhopper-transmitted, also have a long list of wild hosts which serve as
reservoirs (Freitag, 1951; Kunkel, 1954).

Anderson (1959) sampled 150 species of plants as possible field sources
of five pepper viruses. Seventy-nine of these were positive for one or more
viruses, and the five viruses were recovered from about 21 plant species.

Choke-cherries are a reservoir of X-disease of peach (Hildebrand and Palmiter, 1942). Seed crops of biennials are important sources, such as sugar beet yellows and beet mosaic. With some viruses, however, the most important sources are diseased plants within the crop itself (Doncaster and Gregory, 1948; Broadbent *et al.,* 1951; Broadbent, 1957).

In Hibernating Insects

Overwintering of vectors, viruses, and their alternate host plants play an important economic role. With rice dwarf, the overwintering generation of *N. cincticeps* is the most important, being responsible for primary infection in rice. The long latent period of the virus in the insect mitigates against second generation transmission (Sameshima and Nagai, 1962).

H. coagulata overwinters in the woodlands near peach orchards. Spread of phony peach begins at the edge of the field, the first row infection being equal to that in the next two rows, and so on. (KenKnight, 1961).

Aphid populations on potato crops in Idaho were positively correlated with closeness to towns, presumably because aphids could find many more suitable and protected niches and host plants for hibernation and spring emergence (Bishop, 1965).

Man sometimes provides suitable hibernation quarters for aphids. Smith and Brierley (1948) report that bulb storage houses permit aphids to survive and transmit viruses to sprouting bulbs during that period. Mangold and fodder beets serve as hibernation quarters for aphids when the roots are kept over the winter in clamps (piles covered with straw and soil). Glasshouses are also a means whereby vectors are protected during the winter.

Distribution and prevalence of virus source plants are more related to the prevalence of CMV and watermelon mosaic virus 2 of cantaloupe than to abundance of the aphid vectors in the arid lands of southwest Arizona. Neither virus depends on native plants. CMV persists in introduced-perennial ornamentals, while WMV 2 depends on annual plants, cultivated and uncultivated, which are not native to the area. The problem is brought about by man's conversion of a desert area to agriculture through irrigation. Each year there is a great spread of virus infection in cantaloupe fields which is favored by the low density of planting (Nelson and Tuttle, 1969).

Wild hosts of NEPO viruses are an essential factor in the spread of the viruses. These viruses, transmitted by *Longidorus elongatus,* infect a high proportion of weed seeds and dissemination of these appears to be the principal method of spread. Persistence in seed is essential for survival in soil since field nematode populations lose infectivity after a winter fallow. *Xiphinema* transmitted viruses, on the other hand, persist for many months in their vectors, but they occur only rarely in weed seeds in soils from outbreak areas (Lister and Murant, 1961; Murant and Lister, 1967).

Wild host plants are significant in the spread and maintenance of the aster yellows viruses in Europe, since infected crop plants play little if any role in further spread. Incidence in these is variable, sometimes very low or virtually absent. During these latter periods, the viruses circulate among wild plants with leafhoppers the most important vectors.

Sources of viruses in the field may be weeds or crop residues, but two diseases affecting the same crop, both aphid-transmitted, may present different aspects of the problem. Beet mosaic virus in volunteer beets from the previous crop appear to be the main source of the virus; western beet yellows was also found in volunteer beets in growing fields, but apart from these providing early season inoculum for WBYV, the main source of WBYV appears to be nonvolunteer sources for mid- and late-season infections (Howell and Mink, 1969). An example of one crop, winter wheat, serving as a reservoir for a virus affecting another crop, is seen in Boothroyd and Romaine's report on maize dwarf mosaic (1971).

In the epidemiology of virus diseases, the role of weed hosts of the vectors varies, depending on the relative attraction of the commercial crop to the vector and its value to the insect as a breeding host. Pitre and Boyd (1970) reported on experiments in which weed and weeded plots in corn fields were compared with respect to leafhopper (*Graminella nigrifrons* Forbes) populations and incidence of corn stunt. The insect was more abundant in the weed plots, but the incidence of the disease was higher in the weeded plots. The weeds are evidently preferred to corn as hosts, but since the plots were small, drifting over to the contiguous corn plots would no doubt occur; this would account for increased incidence in the weeded plots. New arrivals would land on corn in the weeded plots but would have a choice in those not weeded. Although it is not mentioned in the report, it is presumed that the weeds are virus hosts.

Lettuce necrotic yellows presents an interesting, possibly unique, situation. The vector, *Hypermyzus lactucae* (L), is only a transient feeder on lettuce; the wild host of the insect and of the virus is *Sonchus* spp. but *Sonchus* is not visibly affected by the virus (Stubbs and Grogan, 1963, 1963a). It is customary, when seeking a vector of a virus, to look for wild hosts showing symptoms of virus disease; the solution in this case indicates a certain amount of luck, but also a considerable amount of acumen on the part of the investigators.

Duffus (1968) summarized data from several studies which succeeded in separating the complex of five beet yellowing diseases into components:

(1) semipersistent aphid-transmitted viruses with rather restricted natural weed host ranges and localized distribution (beet yellows virus, beet yellows stunt virus), (2) persistent aphid-transmitted viruses with extremely wide host ranges and widespread distribution (beet western yellows, malva yellows virus), and (3) a white fly-transmitted virus (beet pseudo-yellows virus). The relative abundance and incidence of these viruses are directly related to their transmission characteristics and wild host plant range. The importance of wild plants in the ecology of beet viruses is dependent on a number of factors: (1) the distribution of the weed hosts in relation to the crop, (2) the incidence of virus in the weed hosts, (3) the occurrence, distribution, and prevalence of vectors in the crop and weed hosts, (4) the behavior and migration of the vectors as governed by their specificity, periodicity, and availability of food plants, and (5) the nature of the vector-virus relationship.

PHENOLOGY

The use of phenology, that is, the study of recurring natural phenomena in relation to climatic conditions, deserves more attention than it gets. Its great advantage is that the phenomena it measures are summaries of many individual factors which are difficult, if not impossible, to interrelate in any other way. The disadvantage is that, although the value of phenological data improves as the years of observation are increased, very few individuals are in a position to maintain these observations long enough. Research projects involving phenology are best organized on the institutional level so that continuity can be planned.

Virus-disease incidence is the phenological climax of phenomena of virus sources and their availability to the insect, the susceptibility of the virus host plant, vector-insect numbers, their movements into and distribution within cultivated crops, their vector efficiency, and their relationships with the viruses they carry. It is reasonably certain that for no plant virus disease are all the contributing factors and their interrelationships fully understood.

The strawberry aphid *Pentatrichopus fragaefolii* (Cock.) has two rhythms in its annual cycle, depending on climatic conditions during the four seasons. De Fluiter (1954) found that with a combination of severe winter, bad spring, and early summer weather, populations of the aphid reached only a low peak by autumn, and these were soon overtaken by frost. Under the converse conditions, i.e., mild winter and favorable spring and early summer weather, the first peak of population occurs in late June and early July. This then declines rapidly, but rises again in late September and October. Periods of a late production occur in May to

June and October to early December, and the population increases coincide with the production of young strawberry leaves.

Gibbs and Leston (1970) sampled the seasonal changes in population of some 36 species of insects on a cocoa farm and related these to weather, feeding habits, and food supply. Their paper is a model contribution.

The phenological approach has its greatest practical value as an aid to prediction of vectors and virus diseases. Carter (1930) used winter temperatures and populations of the vector surviving hibernation to predict outbreaks of curlytop of sugar beets.

Variations in weather affect the annual incidence of beet yellowing virus diseases. Incidence of these diseases can be related to the number of days in February and March when temperatures fall below 0°C and the mean weekly temperatures in April. The first item refers to survival of overwintering forms, the second to spring activity of survivors. Changes in climate (weather) and cultivation methods may reduce the accuracy of predictions based on an older prediction equation (Watson, 1966, 1967). Gibbs (1966) relates sunspot activity to beet yellows disease. His explanations correlate with Watson's emphasis on low temperature. Heavy snowfall during periods of low temperature may have to be considered as a factor in these prediction formulae (See also Hurst, 1965).

Two important studies, originally reported in Japanese journals in 1964, are now available in English in a new publication from the Japanese Institute of Agricultural Science in Hishigahara. Both refer to development of prediction methods for the control of the green rice leafhopper (*Nephotettix cincticeps* Ühler) and rice dwarf disease. Hashizume (1968) used, for forecasting, the period of activity of adults from hibernating populations, using net counts; there was a good correlation between these data and the size of the first generation. Peaks could be forecast by comparing the condition of the ovaries at regular intervals and comparing the results with those of previous years. The steady drift in from neighboring weeds flying at low altitude could be forestalled by using barriers about 1 meter high and combining this with Malathion sprays.

Murumatsu *et al.* in the same review journal reported on forecasting methods for the rice dwarf disease. It is possible to forecast the incidence of the disease by using as indices, the population density, or the density of viruliferous individuals of the overwintered generation, times the diseased-hill rate of the year before. This may be amended to take into account the population trend of the next generation. The diseased-hill rate is used as an indicator for the economic use of chemical control. A diagram summarizing this study is shown in Fig. 14.10.

Planting time in relation to migration and susceptibility to infection can be an important parameter in phenology. The age of sugar beets at

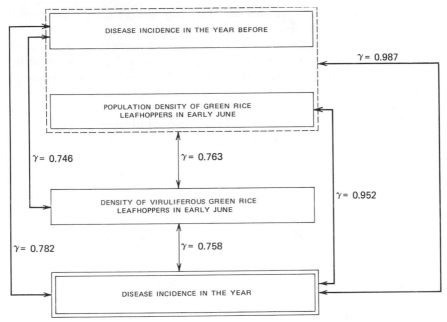

FIGURE 14.10. Correlations among the disease incidence, population density of green rice leafhoppers, and population density of viruliferous individuals (From Murumatsu *et al.,* 1968).

the time of infection is the dominant factor in disease epidemiology (Wallace and Murphy, 1938). Hills and Taylor (1954) showed that cucurbits infected with curlytop in the cotyledon or two-leaf growth stage were killed, but later infections resulted in stunting or growth retardation only. Shands *et al.* (1957) recognized the need for protecting potato plants in the early stage of leaf emergence, that being the critical time for establishment of early aphid migrants. Unfortunately, the phenology of the crop is close to that of the vector, and cultural modifications have their definite limitations. Broadbent and Doncaster (1949) recognized that migration time coincided with development of the host plants. Sometimes, however, a small difference is important as Hansen (1950) showed with sugar beet yellows. April 1st, 15th, and May 1st plantings were 38, 53, and 70% infected, respectively.

THE SIZE OF VECTOR POPULATIONS AND THEIR MOVEMENTS

The size of vector populations and their movements are of critical concern in disease incidence. Davies (1934) used an index figure in the evaluation of areas suitable for potato seed growing. Where there was a rapid increase of virus disease among potato stocks, index counts of *M. persicae* per 100 leaves always exceeded 100; where there had been no increase of virus infection, the figure did not exceed 25. For such an index to be valid, the index should reflect the magnitude of initial infestations of alates from outside sources and their subsequent spread, and not merely an index of the rate of reproduction of aphids within a field.

Bemisia tabaci (Gennadius) was first noted as injurious to cotton in 1960–61, but since then, each year there have been high populations occurring progressively earlier. The phenomenon is accounted for by the long growing season (ten months), insufficient destruction of crop residues, abundance of *Sida* (a malvaceous weed occurring along field edges), and a growing tolerance for the insecticide applied to control cotton bollworm (Kraemer, 1966). In an area where single cropping of rice is practised, population development of *S. oryzicola* in the southern United States is limited. When two crops are grown, populations increase (Van Hoof *et al.,* 1962). This development is to be expected, and it may well affect many rice areas and insect vectors of many other virus diseases of rice where double cropping is resorted to as a means of increasing rice production.

Incidence of Disease

Henner and Schreiner (1952) found some correlation between total numbers of *M. persicae* on potato leaves and the incidence of virus disease, but some low counts were obtained (less than 20 per leaf) in some areas of high disease incidence. Broadbent (1950) found correlation between potato-leaf-roll incidence and maximum counts per 100 leaves of *M. persicae* in England and Wales. Spread, however, was better correlated with the number of alate *M. persicae* caught on sticky traps than with total numbers from leaf counts.

Watson and Healy (1953) derived a mathematical formula to account for the observed incidence of beet yellows virus in sugar beets in England. Assuming that the crop is visited by N aphids at a time when the proportion of infected plants is k ; the predicted proportion of infection for a time 3–4 wk later is:

where
$$k_1 = k_0 + 100 \, (1 - k_0) \, (1 - e^{-n1})$$
$$1 = p \, [(1 - k_0) \, {}^t + k_0{}^t - 1] \, / k_0.$$

This formula adequately accounts for the observed spread of infection when $N = 1/10$, the sticky trap count for 3–4 wk preceding the time k_0

infection is observed; p, the probability of infection by a single aphid $=0.5$; and t, the number of movements per aphid effective for spreading beet yellows virus $=5$. *M. persicae* populations were not, however, correlated with the incidence of beet mosaic virus. Similar differences have been shown by Doncaster and Gregory (1948) and Broadbent (1950) in the correlations of incidence of potato leaf roll and potato virus Y with *M. persicae* populations. Leaf roll can be correlated but not virus Y. While virus-vector relationships are not in review here, it can be mentioned in passing that the explanation for the differences in correlation are to be found in the fact that beet mosaic virus and that of potato Y are both nonpersistent in the vector, and, therefore, quickly dissipated. *Dickson et al.* (1956) measured populations of flying aphids in southern California citrus groves and related these to the spread of tristeza (quick decline) virus. The rate of spread of the virus was about 2 new infections each year for each diseased tree. The green citrus aphid (*Aphis spiraecola* (Patch)) constituted 85% of the aphid catch; the melon aphid (*A. gossypii*), 3% of the catch. Although *A. spiraecola* is reported to be a vector of the virus in Florida, it is not known to be a vector in California. There, the melon aphid is the vector, although its efficiency is much below that reported for *A. spiraecola*. However, with 35,000 individuals flying to or from a single citrus tree in a single year, natural spread of the virus can be fully accounted for, even though an average of 5600 melon aphids was required to produce each experimental infection.

ESTIMATES OF VECTOR POPULATIONS

Advisory Services

Estimates of vector populations, coupled with phenological observations, have become of considerable importance in the development of advisory services to growers. Shands and his associates (1958a) have, for many years, followed the development of *M. persicae* populations on wild plum thickets and their migrations and development on potatoes. This information has been relayed to the farmers through the county agricultural agents. Timing and spray frequency have in this way been coordinated and adjusted to seasonal differences.

Wet and stormy springs may so reduce aphids on sugar beets that spraying may not be necessary, but this is determined by the arrival of winged aphids in the beet fields. In areas where beet yellows is prevalent, one aphid to four plants calls for a spray warning; in other areas one aphid to one plant is the standard, but no warning may be necessary if this level is not reached by mid-July (Hull, 1958, 1959).

Hille Ris Lambers (1955) bases an advisory service to potato seed growers, first, on the knowledge that 97% of the migration of *M. persicae* from winter hosts to summer hosts comes from peach and *P. serotina*. Estimates of population size are based on the number of fundatrices found annually, together with counts of predators. The exact time of migration is known, so that the degree of early spread of leaf roll can, to some extent, be predicted by multiplying the number of primary colonies by the number of alatae per colony. Confirmation is obtained by direct counts of alatae arriving on potato during the migration period. This applies to early spread caused by the alatae flying from the winter hosts. The main spread, occurring in July, is by alatae developed on potatoes and other herbaceous plants. Like Broadbent (1950), Hille Ris Lambers found no correlation between the number of aphids on the plants at that time and virus spread in the fields where the counts were made. The correlation is with trap catches, and the data from a large number of these are used to prescribe dates for early lifting of the seed potato crop. Since a few days' time can involve serious losses, trap catches are dealt with the day of arrival. Munster (1958) has described methods of determining aphid development with a similar objective in view.

Broadbent (1950) found correlation between numbers of alate aphids caught on sticky traps and the spread of leaf roll. Rugose mosaic spread was not so clearly correlated because of a number of vectors involved, but the average health of potato stocks for the following year was predictable on the basis of the average trap catch data. Hauschild (1947) has derived a theoretical formula, based on the initial health of the crop and the course of the aphid infestation, for use in a similar prediction.

In the case of curlytop of beets and other susceptible economic crops, the importance of the spring movement of leafhoppers into the cultivated fields depends on the date and duration of the movement and on the size of the incoming population. The size of the population depends on the size of the population dispersing from Russian thistle in the fall to winter and spring hosts, the winter mortality, and the condition of the surviving females. The numbers of spring generation leafhoppers entering cultivated fields depend on the location of the cultivated area vis-à-vis the breeding grounds, dispersal as determined by wind movement, and time of dispersal as determined by temperature and host plant abundance.

Statistical Correlations

Fisken (1959), first determining that the principal overwintering area for *M. persicae* was in the market garden area around Edinburgh, showed by trap counts that the initial infestation in crops within different areas did not differ materially between seasons, but that a hot, dry summer brought earlier infestation and earlier maximum populations. Distance from the dispersal area and the lateness of season are perhaps more important than actual numbers in determining relative freedom from leaf roll and Y viruses.

Fox (1945) posed the question as to what constitutes an economic population of leafhoppers. This is determined by several factors: the weather during the spring, the percentage of leafhoppers carrying virus (Wallace and Murphy, 1938, report variations from 4 to 67% from one year to another), and the age and size of the sugar beets at the time. With very early movements, severe damage occurs; later movements cause little or negligible damage, and the time of the spring movement is more important than the population peaks. In the writer's experience, however, the most damaging years have been those in which early spring movement and large populations coincided. Wallace and Murphy's data on the infectivity of beet leafhoppers are shown in Fig. 14.11. The infectivity of the vectors of Pierce's disease under natural conditions in a number of habitats was determined by Freitag and Frazier (1954). The disease occurs wherever the three more important vectors are found; the habitats included the seashore, the high mountains, the desert, and cultivated valleys.

When the invaded host plant is not a favorable host for the vector, disease incidence depends entirely on the size of the invading population. Percentage of disease incidence will, therefore, depend on plant density, and within the limits imposed by good agronomy, the percentage of diseased plants can be reduced by increasing the density of planting. Van der Plank and Anderssen (1944) showed this connection with tomato spotted wilt on tobacco. Increasing the number per acre n times reduced the proportion of diseased plants from $1 - q$ to $1 - n\sqrt{q}$, where q is the proportion of healthy plants at standard density. Linford (1943) reported a similar conclusion with the same virus in pineapple, thus confirming Carter's conclusion (1939) that infield populations on favorable weed hosts were not a factor in the disease incidence in pineapple plants in that field. Curlytop in tomatoes is a similar case of density of planting determining the percentage of plants infected (Shapovalov *et al.,* 1941).

Peasant cultivators in East Africa used poor agronomy in their growing of groundnuts (peanuts). Spacing was much too close, and weeding was delayed, but the method had pragmatic sanction. When more open spacing and clean culture were tried, groundnut rosette increased to serious proportions (Storey, 1935).

There are many other examples of the effect of plant spacing on disease incidence: Broadbent (1948a, 1957), Blencowe and Tinsley (1951),

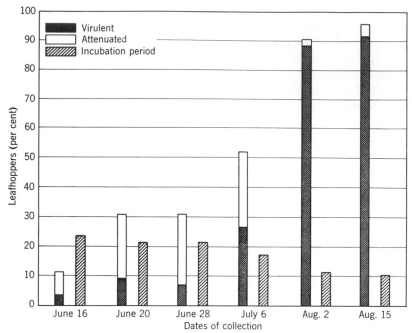

FIGURE 14.11. Percentage of leafhoppers with virulent and attenuated virus, total percentage viruliferous, and average incubation
period of all infections produced in greenhouse tests of collections
made at intervals from a beet field near Castleford, Idaho, during
1932 (After Wallace and Murphy, 1938).

and Shirahama (1957) all reporting such effects with aphid-transmitted
viruses, even where the crop plant was a suitable vector host.

PHYSIOLOGICAL AND BEHAVIORAL FACTORS

Since physiological and behavioral factors are the handmaidens of auto-
ecology, they are appropriately referred to here. Cauliflower mosaic virus
(CMV) is in high concentration in new leaves produced by infected
plants, while cabbage black ringspot virus (CBRSV) occurs mainly in
older leaves, and even there it is localized to the symptom areas; only in
recently infected plants does the virus occur in young leaves. The two
aphid vectors, *M. persicae* and *B. brassicae,* alight and feed on the upper
portions of the plant and are less likely, therefore, to acquire CBRSV
than CMV. The older leaves of cabbage are in a more favorable position
for aphids alighting than are those of cauliflower, which may explain the

more frequent infection of cabbage with CBRSV (Broadbent, 1954). There are differences in the reports on the influence of plant growth on aphid development. Bald *et al.* (1946) reported that aphids avoid infected plants because of stunting due to infection. Hull and Watson (1945) reported that vigorous, well-fertilized plants were more severely infested. On the other hand, according to Arenz (1951), *M. persicae* grew faster on potatoes with leaf roll and severe mosaic. *Aphis fabae* Scop. reproduces poorly on active presenescent leaves (Kennedy, 1951; Ibbotsen and Kennedy, 1950; Kennedy *et al.,* 1950). *C. tenellus* develops larger populations on diseased plants than on vigorous, densely foliaged, healthy ones (Carter, 1930; Romney, 1943). *Thrips tabaci* populations on diseased *Emilia* are higher than on healthy *Emilia* (Carter, 1939), while Severin (1946) reported that leafhoppers developed faster, lived longer, and reproduced more on the diseased than on the healthy plants. No explanation has been offered for the beneficial effects of the diseased plant on the insect, but presumably they can be ascribed to a higher concentration of cell sap and probably a higher proportion of carbohydrates, providing a more efficient diet for growth and reproduction. Three species of aphid, *M. persicae, A. fabae,* and *Aulacorthum solani* (Kalt.), preferred leaves severely infected with yellows virus to healthy ones, breeding more rapidly and living longer on them (Baker, 1960). While this would ultimately result in large populations of infective aphids, the preference shown for the diseased plants would restrain movement away to some degree.

The physiological microenvironment has been reduced to its microlimits by Bawden (1961) in a review of the effects of changing environment on the behavior of viruses. Bawden's microenvironment is the plant cell itself. According to him, environmental changes affect the behavior of viruses by selecting genetically different variants that are favored by the different environments. If we begin with this concept and extend it through the environmental effects on the physiology of the whole plant as affected by physical factors of the environment and the parallel effects on the virus vectors, the potential for variants is enormous; it would behoove mankind to gird itself for the continuing struggle.

Flight Behaviors

This is a somewhat controversial subject. For many years, the "flight activity" hypothesis was generally accepted. Essentially, this hypothesis ascribes the governing factors of aphid flight to the physical factors of wind, temperature, and humidity. C. G. Johnson (1954), however, in a detailed analysis of this hypothesis and a reconsideration of the supporting data, has presented another hypothesis concerning flight from breeding sites which is based on physiological factors, behavior being secondary. Molting of the winged nymphs into alatae, usually most intense early in the morning, is followed by one or more additional peaks during the day. Flight maturity of these newly emerged alatae is usually reached in 24 hours, after which they fly away in "flushes" on the first migratory flight. Since aphids lose their ability to fly, owing to autolysis of the flight muscles after a few days (B. Johnson, 1953), their ability to spread virus diseases subsequent to the first migration is considerably impaired. Changes in the number of aphids in the air throughout the day and from day to day are, therefore, actually collective population changes. Differences in flight capacity between spring and summer in forms of *M. persicae* have been recorded by Woodford (1968). Dickson (1949) ascribed the high efficiency of *M. persicae* in the spread of cantaloupe mosaic to the flight behavior of the aphid. Although the insect does not colonize the plant, it swarms over the crop in short flights, with short periods of feeding. Swenson and Nelson (1959) consider that whether the numbers of different species which alight on a crop plant are proportional to the numbers of those species which fly over the crop is still a moot point.

The behavior of migrant aphids in finding their host plants has some significance in virus spread. Kennedy *et al.* (1959, 1959a) found an intensely dispersive type of behavior in finding food plants, and he suggested that this may be common among aphids. All three species studied, *M. persicae, A. fabae,* and *B. brassicae,* showed the same general behavior, i.e., alighting, probing, and taking off, from either favored or nonfavored host plants, although probing occurred more often on the favored food plant. This type of behavior would favor the dissemination of the nonpersistent viruses simply by virtue of alighting, probing, and taking off being achieved in a short space of time. Neitzel and Muller (1959) also reported on flight behavior as a factor in field infection. Percentage of infection was clearly related to the number of alates, and the edges of the plots had the highest infection, which decreased sharply toward the center of the field. The vectors flew close to the ground and stopped at the edge of the crop. Less difference from edge to center was observed when flight occurred to a late-planted field and the plants were small.

Finally, some reference should be made to the complexity of some biocoenoses involved in virus transmission. Perhaps the most elaborate of these is that concerned with swollen shoot of cacao. According to Strickland (1951, 1951a), no less than 17 species of pseudococcid, 75 species of ants, and 3 arachnoids are involved. There are three distinct ecological niches: (1) the association between mealybugs and myrmicine ants; (2) mealybugs of the genera *Paraputo* and *Formicoccus* and the wild host trees; and (3) a negative association between myrmicines and *Oecophylla* and *Macromischoides.*

Pseudococcus njalensis Laing is always attended by ants and *Cremato-gaster* density is closely correlated with mealybug density. Tree-to-tree variation in mealybug density is largely dependent on the identity of the dominant ant species. There is no direct correlation between mealybug density and incidence of swollen shoot. Mealybug populations in the devastated areas are very low compared with areas of active spread.

The latest contribution to this subject, and a major one, is that of Leston (1970). Leston has reviewed the role of mealybugs and ants in the transmission and incidence of swollen shoot virus. Although most of the 20 species of mealybug in the West African area are capable of transmitting swollen shoot strains, *P. njalensis* is the principal and most effective vector. Tree-to-tree variation in ant species and *P. njalensis* is maintained at a higher level than other *Pseudococcus* species which are only casually ant attended. Leston's paper has 150 references.

In conclusion, it can be said that these ecological aspects are an essential part of the biological approach to plant viruses.

REFERENCES

Adams, A. N., "The vectors and alternative hosts of groundnut rosette mosaic virus in Central Province, Malawi," *Rhodesia, Zambia, Malawi, J. Agric. Res.,* 5, 145–151 (1967).

Anderson, C. W., "A study of field sources and spread of five viruses of peppers in central Florida," *Phytopathology,* 49, 97–101 (1959).

Annand, P. N., Chamberlin, J. C., Henderson, C. F., and Waters, H. A., "Movements of the beet leafhopper in 1930 in southern Idaho," *U.S. Dept. Agr. Circ.,* No. 244 (1932).

Anon., "Survey methods," *U.S. Dept. Agr. ARS, Coop. Econ. Inst. Rept.,* (1955, 1958).

Arenz, B., "Der Einfluss verschiedener auf die Resistenz der Kartoffel gegen die Pfirischblattlaus. Weitere erhebnisse über die Resistenz der Kartoffel gegen die Pfirischblattlaus," *Z. Pflanzenbau Pflanzenschutz,* 2, 49–67 (1951).

Baker, P. F., "Aphid behaviour on healthy and yellows-infected sugar beet," *Ann. Appl. Biol.,* 48, 384–391 (1960).

Bald, J. G., Norris, D. O., and Helson, G. A., "Transmission of potato virus diseases. V. Aphid populations, resistance and tolerance of potato varieties to leaf roll," *Australia, Commonwealth Council Sci. Ind. Research Bull.,* No. 196 (1946).

Bald, J. G., Norris, D. O., and Helson, G. A., "Transmission of potato virus diaeases. VI. The distribution of the aphid vectors on sampled leaves and shoots," *Australian J. Agr. Research,* 1, 18–32 (1950).

Ball, E. D., "The beet leafhopper and the curly-leaf disease that it transmits," *Utah State Agr. Coll. Expt. Sta. Bull.,* No. 155 (1917).

Bawden, F. C., "Some effects of changing environment on the behaviour of plant viruses," In "Microbiological Reaction to Environment," *Symposium Soc. Gen. Microbiol.*, 11, 296–316 (1961).

Bennett, C. W., "The curly top of sugar beets and other plants," *Amer. Phytopathological Society,* Monograph No. 7 (1971).

Bennett, C. W., "Epidemiology of leafhopper-transmitted viruses," *Ann. Rev. Phytopathology*, 5, 87 (1967).

Bishop, G. W., "Green peach aphid distribution and potato leafroll virus occurrence in the seed-potato producing areas of Idaho," *J. Econ. Entomol.*, 58, 150–153 (1965).

Blencowe, J. W., and Tinsley, T. W., "The influence of density of plant population on the incidence of yellows in sugar beet crops," *Ann. Appl. Biol.*, 38, 395–401 (1951).

Boothroyd, C. W., and Romaine, C. P., "Winter wheat as a reservoir for maize dwarf mosaic," *Phytopathology*, 61, 885 (Abstr.) (1971).

Broadbent, L., "Aphis migration and the efficiency of the trapping method," *Ann. Appl. Biol.*, 35, 379–394 (1948).

Broadbent, L., "Methods of recording aphid populations for use in research on potato virus diseases," *Ann. Appl. Biol.*, 35, 551–566 (1948a).

Broadbent, L., "Factors affecting the activity of alatae of the aphids *Myzus persicae* (Sulz.) and *Brevicoryne brassicae* (L)," *Ann. Appl. Biol.*, 36, 40–62 (1949).

Broadbent, L., "The correlation of aphid numbers with the spread of leaf roll and rugose mosaic in potato crops," *Ann. Appl. Biol.*, 37, 58–65 (1950).

Broadbent, L., "Aphids and virus diseases in potato crops," *Biol. Revs. Cambridge Phil. Soc.*, 28, 350–380 (1953).

Broadbent, L., "The different distribution of two *Brassica* viruses in the plant and its influence on spread in the field," *Ann. Appl. Biol.*, 41, 174–182 (1954).

Broadbent, L., "Investigation of virus disease of *Brassica* crops," *Agr. Research Council Rept. Ser.*, No. 14 (1957).

Broadbent, L., and Doncaster, J. P., "Alate aphids trapped in the British Isles, 1942–1947," *Entomologist's Monthly Mag.*, 85, 174–182 (1949).

Broadbent, L., Doncaster, J. P., Hull, R., and Watson, M. A., "Equipment used for trapping and identifying alate aphids," *Proc. Roy. Entomol. Soc. (London),* A23, 57–58 (1948).

Broadbent, L., Cornford, C. E., Hull, R., and Tinsley, T. W., "Overwintering of aphids, especially *Myzus persicae* (Sulz.) in root clumps," *Ann. Appl. Biol.*, 36, 513–524 (1949).

Broadbent, L., Tinsley, T. W., Buddin, W., and Roberts, E. T., "The spread of lettuce mosaic in the field," *Ann. Appl. Biol.*, 38, 689–706 (1951).

Carter, W., "Extensions of the known range of *E. tenellus* Baker and curlytop of sugar beets," *J. Econ. Entomol.*, 20, 714–717 (1927).

Carter, W., "Population of *Eutettix tenellus* Baker and the osmotic concentrations of its host plants," *Ecology*, 8, 350–352 (1927a).

Carter, W., "Ecological studies of the beet leafhopper," *U.S. Dept. Agr. Tech. Bull.*, No. 206 (1930).

Carter, W., "Populations of *Thrips tabaci* with special reference to virus transmission," *J. Animal Ecol.*, 8, 261–276 (1939).

Carter, W., "Ecological aspects of plant virus transmissions," *Ann. Rev. Entomol.*, 6, 347–370 (1961).

Church, B. M., and Strickland, A. H., "Sampling aphid populations on Brussels sprout," *Plant Pathol.*, 3, 76–80 (1954).

Clark, R. L., "Epidemiology of tomato curlytop in the Yakima Valley," *Phytopathology*, 58, 811–813 (1968).

Clark, R. L., "Restricted local spread of curlytop in tomatoes," *Plant Disease Reptr.*, 52, 17–18 (1968a).

Cook, W. C., "The beet leafhopper," *U.S. Dept. Agr. Farmer's Bull.*, No. 1886 (1941).

Cook, W. C., "The relation of spring movements of the beet leafhopper (*Eutettix tenellus* Baker) in central California to temperature accumulations (Homoptera)," *Ann. Entomol. Soc. Amer.*, 37, 149–162 (1945).

Coons, G. H., Stewart, D., Rockstahler, H. W., and Schneider, C. L., "Incidence of savoy in relation to the variety of sugar beets and to the proximity of wintering habitat of the vector, *Piesma cinerea*," *Plant Desease Reptr.*, 42, 502–511 (1958).

Cornwell, P. B., "Some aspects of mealybug behaviour in relation to the efficiency of measures for the control of virus diseases of cacao in the Gold Coast," *Bull. Entomol. Res.*, 47, 137–166 (1956).

Cornwell, P. B., "An investigation into the effect of cultural conditions on populations of the vectors of virus diseases of cacao in Ghana with an evaluation of seasonal population trends," *Bull. Entomol. Res.*, 48, 375–396 (1957).

Cornwell, P. B., "Movements of the vectors of virus diseases of cacao in Ghana. II. Wind Movements and aerial dispersal," *Bull. Entomol. Res.*, 51, 175–201 (1960).

Czerwinski, H., "Untersuchugen und Beobachtungen uber die Blattlaus *Myzodes persicae* (Sulz.) als Verbreiter des Kartoffelabbaues auf dem Versuchsfeld des Instituts fur Acker-und Pflanzenbau Berlin-Dahlem und dem Versuchsgut Thyrow," *Angew. Botan.*, 25, 20–50 (1943).

Dale, W. T, and Attafuah, A., *W. African Cacao Research Inst., Tafo, Gold Coast, Ann. Rept.*, 1955–56, 28–30 (1957).

Davidson, J., "The biological and ecological aspects of migration in aphids," *Sci. Progr.*, 21, 641–658; 22, 57–69 (1927).

Davidson, J., "On the occurrence of the parthenogenetic and sexual forms in *Aphis rumicis* L., with special reference to the influence of environmental factors," *Ann. Appl. Biol.*, 16, 104–133 (1929).

Davies, W. M., "Ecological studies on aphids infesting the potato crop," *Bull. Entomol. Res.*, 23, 535–548 (1932).

Davies, W. M., "Studies on aphids infesting the potato crop. II. Aphis survey: Its bearing upon the selection of districts for seed potato production," *Ann. App. Biol.,* 21, 283–299 (1934).

Davies, W. M., "Studies on aphids infesting potato crop. III. Effect of variation in relative humidity on the flight of *Myzus persicae* Sulz.," *Ann. Appl. Biol.,* 22, 106–115 (1935).

Decker, G. C., and Cunningham, H. B., "Winter survival and overwintering area of the potato leafhopper," *J. Econ. Entomol.,* 61, 154 (1968).

de Fluiter, H. J., "Observations on the phenology of the strawberry aphid, *Pentatrichopus fragaefolii* Cock., in the Netherlands," *Entomol. Ber., (Amsterdam),* 15, 94–98 (1954).

de Fluiter, H. J., and van der Meer, F. A., "The biology and control of *Macropsis fuscula* Zett., the vector of the *Rubus* stunt virus," *Proc. Intern. Congr. Entomol., 10th Congr., Montreal, 1956,* 341–345 (1958).

De Long, D. M., "Some problems encountered in the estimation of insect populations by the sweeping methods," *Ann. Entomol. Soc. Amer.,* 25, 13–17 (1932).

Dickson, R. C., "Aphid flights in relation to cantaloupe mosaic," *Plant Disease Reptr., Suppl.,* No. 180, 7–8 (1949).

Dickson, R. C., and Laird, E. F., Jr., "Fall dispersal of green peach aphids to desert valleys," *Ann. Entomol. Soc.,* 60, 1088 (1967).

Dickson, R. C., Swift, J. E., Anderson, L. D., and Middleton, J. T., "Insect vectors of cantaloupe mosaic in California's desert valleys," *J. Econ. Entomol.,* 42, 770–774 (1949).

Dickson, R. C., Johnson, M. M., Flock, R. A., and Laird, E. F., Jr., "Flying aphid populations in southern California citrus groves and their relation to the transmission of the tristeza virus," *Phytopathology,* 46, 204–210 (1956).

Doncaster, J. P., "The life history of *Aphis (Doralis) rhamni* B.d.F. in eastern England," *Ann. Appl. Biol.,* 30, 101–104 (1943).

Doncaster, J. P., and Gregory, P. H., "The spread of virus disease in the potato crop," *Agr. Research Council Rept. Ser.,* No. 7 (1948). H.M.S.O.

Dorst, H. E., and Davis, E. W., "Tracing long-distance movements of beet leafhopper in the desert," *J. Econ. Entomol.,* 30, 948–954 (1937).

Douglass, J. R., "Outbreaks of beet leafhoppers north and east of the permanent breeding areas," *Proc. Am. Soc. Sugar Beet Technologists,* 8, 185–193 (1954).

Douglass, J. R., Hallock, H. C., Fox, D. E., and Hofmaster, R. N., "Movements of spring-generation beet leafhoppers into beet fields of south-central Idaho," *Proc. Am. Soc. Sugar Beet Technologists, 4th Gen. Meeting* (1946).

Douglass, J. R., Peay, W. E., and Cowger, J. I., "Beet leafhopper and curlytop conditions in the southern Great Plains and adjacent areas," *J. Econ. Entomol.,* 49, 95–99 (1956).

Douglass, J. R., and Hallock, H. C., "Relative importance of various host plants of the beet leafhopper in southern Idaho," *U.S. Dept. Agr. Tech. Bull.,* No. 1155 (1957).

Drake, D. C., and Chapman, R. K., "Migration of the six-spotted leafhopper, *Macrosteles fascifrons* (Stål). I. Evidence for long-distance migration of the six-spotted leafhopper in Wisconsin," *Res. Bull. Univ. Wisconsin,* 261 (1965).

Duffus, J. E., "Beet yellowing viruses in the U.S.A.," *First International Congress Phytopathology London* (Abstracts, p. 49) (1968).

Fisken, A. G., "Factors affecting the spread of aphid-borne viruses in potato in eastern Scotland. 1. Overwintering of potato aphids, particularly *Myzus persicae* (Sulz.). 2. Infestation of the potato crop by potato aphids, particularly *Myzus persicae* (Sulz.)," *Ann. Appl. Biol.,* 47, 264–273, 274–286 (1959).

Flock, R. A., and Deal, A. S., "A survey of beet leafhopper populations on sugar beets in the Imperial Valley, California, 1953–1958," *J. Econ. Entomol.,* 52, 470–473 (1959).

Fox, D. E., "Occurrence of the beet leafhopper and associated insects on secondary plant successions in southern Idaho," *U.S. Dept. Agr. Tech. Bull.,* No. 607 (1938).

Fox, D. E., "Beet leafhopper populations on various types of Russian thistle stands," *Proc. Am. Soc. Sugar Beet Technologists,* 3, 465 (1942).

Fox, D. E., "Factors affecting curlytop damage to sugar beets in southern Idaho," *U.S. Dept. Agr. Tech. Bull.,* No. 897 (1945).

Freitag, J. H., "Host range of the Pierce's disease virus of grapes as determined by insect transmission," *Phytopathology,* 41, 920–934 (1951).

Freitag, J. H., and Frazier, N. W., "Natural infectivity of leafhopper vectors of Pierce's disease of grape in California," *Phytopathology,* 44, 7–11 (1954).

Galvez, G. E., Thurston, H. D., and Jennings, P. R., "Host range and insect transmission of the hoja blanca disease of rice," *Plant Disease Reptr.,* 45, 949–953 (1961).

Gibbs, A. J., "A possible correlation between sugar beet yellows incidence and sun-spot activity," *Plant Pathology,* 15, 150–152 (1966).

Gibbs, D. G., and Leston, D., "Insect phenology in a forest cocoa farm locality in West Africa," *J. Appl. Ecology,* 7, 519–548 (1970).

Gill, C. C., "Epidemiology of barley yellow dwarf in Manitoba and effect of the virus on yield of cereals," *Phytopathology,* 60, 1826 (1970).

Girardeau, J. H., Jr. and Ratcliffe, T. J., "Apparent eradication of mosaic of sweet potatoes in Georgia aided by the disappearance of the insect vector," *Plant Disease Reptr.,* 47, 791–793 (1963).

Glick, P. A., "The distribution of insects, spiders and mites in the air," *U.S. Dept. Agr. Tech. Bull.,* No. 673 (1939).

Gray, H. E., and Treloar, A. E., "On the enumeration of insect populations by the method of net collection," *Ecology,* 14, 356–367 (1933).

Hagborg, W. A. F., "Wheat striate mosaic, a sword of Damocles hanging over the western wheat grower, or not?," *Canadian Plant Disease Survey,* 43, 194–195 (1963).

Hansen, H. P., "Investigations on virus yellows of beets in Denmark," *Trans. Danish Acad. Tech. Sci.,* 1, 1 (1950).

Hashizuma, B., "Studies on forecasting and control of the rice green leafhopper *Nephotettix cincticeps* Ühler with special reference to eradication of the rice dwarf disease," *Review Plant Protection Research (Japan),* 1, 78 (1968).

Hauschild, I., "Zur Beurteilung des Planzengutwertes von Saatkartoffelfeldern unter Berucksichtigung des Auftretens der Übertrager der Kartoffelvirosen," *Zuchter,* 17, 241 (1947).

Heathcote, G. D., "The comparison of yellow cylindrical, flat and water traps, and of Johnson suction traps, for sampling aphids," *Ann. Appl. Biol.,* 45, 133–139 (1957).

Heathcote, G. D., "Effect of height on catches of aphids in water and sticky traps," *Plant Pathol.,* 7, 32–36 (1958).

Heggestad, H. E., and Moore, E. L., "Occurrence of curlytop on tobacco in Maryland and North Carolina in 1958," *Plant Disease Reptr.,* 43, 682–684 (1959).

Henner, J., and Schreiner, O., "Untersuchunger uber das Auftreten von Blattlausen an Kartoffeln in Osterreich in den Jahren 1950–1951 in Zusammenhang mit virosem Kartoffelabbau," *Pflanzenschutz Ber.,* 8, 150–159 (1952).

Hildebrand, E. M., and Palmiter, D. H., "Control of X disease of peaches by killing chokecherry weed with ammonium sulfamate," *Agr. Newsletter,* 10, 73–75 (1942).

Hille Ris Lambers, D., "Potato aphids and virus diseases in the Netherlands," *Ann. Appl. Biol.,* 42, 355–360 (1955).

Hills, O. A., "A new method for collecting samples of insect populations," *J. Econ. Entomol.,* 26, 906–910 (1933).

Hills, O. A., "The beet leafhopper in the central Columbia River breeding area," *J. Agr. Research,* 55, 21–31 (1937).

Hills, O. A., and Taylor, E. A., "Effect of curlytop infective beet leafhoppers on cantaloupe plants in varying stages of developments," *J. Econ. Entomol.,* 47, 44–48 (1954).

Howell, W. E., and Mink, G. I., "Correlation of virus occurrence in volunteer and commercial sugar beets," *Phytopathology,* 59, 1032 (Abstr.) (1969).

Hull, R., "Spraying sugar beet to control virus yellows disease," *Brit. Sugar Beet Rev.,* 26, 112 (1958).

Hull, R., "Sugar beet yellows. The search for control," *Agriculture (London),* 65, 62–65 (1959).

Hull, R., and Watson, M. A., "Virus yellows of sugar beet," *Agriculture (London),* 52, 66–70 (1945).

Hurst, G. W., "Forecasting the severity of sugar beet yellows," *Plant Pathology,* 14, 47–53 (1965).

Hutchinson, M. T., "An ecological study of the leafhopper vectors of blueberry stunt," *J. Econ. Entomol.,* 48, 1–8 (1955).

Ibbotsen, A., and Kennedy, J. S., "The distribution of aphid infestation in relation to leaf age. II. The progress of *Aphis fabae* Scop. infestation on sugar beet in pots," *Ann. Appl. Biol.,* 37, 680–696 (1950).

Jensen, D. D., "A plant virus lethal to its insect vector," *Virology,* 8, 164–175 (1959).

Johnson, B., "Flight muscle autolysis and reproduction in aphids," *Nature,* 175, 813 (1953).

Johnson, B., "Studies on the dispersal by upper winds of *Aphis craccivora* Koch in New South Wales," *Proc. Linnean Soc. N. S. Wales,* 82, 191–198 (1957).

Johnson, C. G., "A new approach to the problems of the spread of aphids and insect trapping," *Nature,* 170, 147 (1952).

Johnson, C. G., "Aphid migration in relation to weather," *Biol. Revs. Cambridge Phil. Soc.,* 29, 87–118 (1954).

Johnson, C. G., "A basis for a general system of insect migration and dispersal by flight," *Nature,* 186, 348–350 (1960).

Johnson, C. G., and Penman, H. L., "Relationship of aphid density to altitude," *Nature,* 168, 337–338 (1951).

KenKnight, G., "Spread of phony disease into Georgia peach orchards," *Phytopathology,* 51, 345–349 (1961).

Kennedy, J. S., "A biological approach to plant viruses," *Nature,* 168, 890 (1951).

Kennedy, J. S., Ibbotsen, A., and Booth, C. O., "The distribution of aphid infection in relation to leaf age. I. *Myzus persicae* (Sulz.) and *Aphis fabae* Scop. on spindle trees and sugar beet plants," *Ann. Appl. Biol.,* 37, 651–679 (1950).

Kennedy, J. S., Lamb, K. P., and Booth, C. O., "Responses of *Aphis fabae* Scop. to water shortage in host plants in pots," *Entomol. Exp. Appl.,* 1, 274–291 (1958).

Kennedy, J. S., Booth, C. O., and Kershaw, W. J. S., "Host finding by aphids in the field. I. Gynoparae of *Myzus persicae* (Sulz.)," *Ann. Appl. Biol.,* 47, 410–423 (1959).

Kennedy, J. S., Booth, C. O., and Kershaw, W. J. S., "Host finding by aphids in the field. II. *Aphis fabae* Scop. (gynoparae) and *Brevicoryne brassicae* L.; with a reappraisal of the role of host finding behaviour in virus spread," *Ann. Appl. Biol.,* 47, 424–444 (1959a).

Kieckheker, R. W., and Medler, J. T., "Aggregations of the potato leafhopper in alfalfa fields in Wisconsin," *Ann. Entomol. Soc. Amer.,* 59, 180–182 (1966).

Kraemer, P., "Serious increase in cotton whitefly and virus transmission in Central America," *J. Econ. Entomol.,* 59, 1531, (1966).

Kring, J. B., "The life cycle of the melon aphid, *Aphis gossypii* Glover, an example of facultative migration," *Ann. Entomol. Soc. Amer.,* 52, 284–286 (1959).

Kunkel, L. O., "Maintenance of yellows-type viruses in plant and insect reservoirs," *Dynamics of Virus and Rickettsial Infections, International Symposium,* Blakiston, New York, 1954, pp. 209–221.

Lal, R., "Effect of the water content of aphids and their host plants on the appearance of alatae," *Indian J. Entomol.,* 17, 52–63 (1955).

Lawson, F. R., and Piemeisel, R. L., "The ecology of the principal summer weed hosts of the beet leafhopper in the San Joaquin Valley, California," *U.S. Dept. Agr. Tech. Bull.,* No. 848 (1943).

Lawson, F. R., Chamberlin, J. C., and York, G. T., "Dissemination of the beet leafhopper in California," *U.S. Dept. Agr. Tech. Bull.,* No. 1030, 1–59 (1951).

Leston, D., "Entomology of the Cocoa Farm," *Ann. Rev. Entomol.,* 15, 273–294. (1970).

Linford, M. B., "Influence of plant populations upon incidence of pineapple yellow spot," *Phytopathology,* 3, 408–410 (1943).

Linn, M. B., "The yellows disease of lettuce and endive," *Cornell Univ. Agr. Expt. Sta. Bull.,* No. 742 (1940).

Lister, R. M., and Murant, A. F., "Soil borne viruses," *Rept. Scottish Hort. Res. Inst., 1960–1961,* 56 (1961).

Macias, M., and Mink, G. I., "Preference of green peach aphids for virus-infected sugar beet leaves," *J. Econ. Entomol.,* 62, 28, (1969).

Markkula, M., "Biologisch-okologische Untersuchungen uber die Kohlblattlaus, *Brevicoryne brassicae* (L.) (Hem., Aphididae)," *Ann. Zool. Soc. Zool. Botan. Fennicae Vanamo,* 15, 1–113 (1953).

Martin, J. P., *Sugar cane diseases in Hawaii,* Honolulu Advertiser Co., Honolulu, 1938.

Meade, A. B., and Peterson, A. G., "Origin of populations of the six-spotted leafhopper, *Macrosteles fascifrons* (Stål) in Anoka County, Minnesota," *J. Econ. Entomol.,* 57, 878–881, (1964).

Meier, W., "Der einfluss der Hohenlage und gelandeklimatischer Faktoren aud das Auftreten Pfirsichblattlaus (*Myzus persicae* Sulz.) in Kartoffelfeldern der Schweizder grunen," *European Potato J.,* 1, 25–46 (1958).

Moericke, V., "Eine Farbfalle zur Kontrolle des Fluges von Blattlausen insbesondere der Pfirsichblattlaus *Myzodes persicae* (Sulz.)," *Nachrbl. Deut. Planzenschutzdienstes (Stuttgart),* 3, 23–24 (1951).

Munster, J., "Methods for the observation of the development of virus-transmitting aphids with a view to determining the date for early harvesting and the effects of this on the yield of seed potatoes," *European Potato J.,* 1, 31–41 (1958).

Murant, A. F., and Lister, R. M., "Seed transmission in the ecology of nematode-borne viruses," *Ann. Appl. Biol.,* 59, 63–76 (1967).

Murumatsu, Y., Furuki, I., Kawaguchi, K., Sawaki, T., and Takeshima, S., "Forecasting rice dwarf disease, II. Forecasting methods," *Review Plant Protection Research (Japan),* 1, 83, (1968).

Nault, L. R., and Steyr, W. E., "Dispersal of *Aceria tulipae* and three other grass-infesting eriophyid mites in Ohio," *Ann. Entomol. Soc. Amer.,* 62, 1446–1455 (1969).

Neitzel, K., and Muller, H. J., "Erhoehter Virusbefell in den Rundreihen von Kartoffel-bestaenden als Folge des Flugerhaltens der Vektoren," *Entomol. Exptl. Appl.,* 2, 27–37 (1959).

Nelson, M. R., and Tuttle, D. M., "The epidemiology of cucumber mosaic and watermelon mosaic 2 in cantaloups in an arid climate," *Phytopathology,* **59,** 849 (1969).

Nowak, W., "Das Vorkommen der grünen Pfirsich-Blattlaus (*Myzodes persicae* Sulz.) als Luftplankton in verscheidener Hohe über dem Eadboden," *Z. Pflanzenbau Pflanzenshutz,* **2,** 123–130 (1951).

Piemeisel, R. L., "Weedy abandoned lands and the weed hosts of the beet leafhopper," *U.S. Dept. Agr. Circ.,* No. 229 (1932).

Piemeisel, R. L., "Changes in weedy plant cover on cleared sagebrush land and their probable causes," *U.S. Dept. Agr. Tech. Bull.,* No. 654 (1938).

Piemeisel, R. L., "Natural replacement of weed hosts of the beet leafhopper as affected by rodents," *U.S. Dept. Agr. Circ.,* No. 739 (1945).

Piemeisel, R. L., "Replacement control; changes in vegetation in relation to control of pests and diseases," *Botan. Rev.,* **20,** 1–32 (1954).

Piemeisel, R. L., and Carsner, E., "Replacement control and biological control," *Science,* **113,** 2923 (1951).

Piemeisel, R. L., and Chamberlin, J. C., "Land improvement measures in relation to a possible control of the beet leafhopper and curly top," *U.S. Dept. Agr. Circ.,* No. 416 (1936).

Piemeisel, R. L., and Lawson, F. R., "Types of vegetation in the San Joaquin Valley of California and their relation to the beet leafhopper," *U.S. Dept. Agr. Tech. Bull.,* No. 557 (1937).

Piemeisel, R. L., Lawson, F. R., and Carsner, E., "Weeds, insects plant diseases, and dust storms," *Sci. Monthly,* **73,** 124–128 (1951).

Pienowski, R. L., and Medler, J. T., "Synoptic weather conditions associated with long range movement of the potato leafhopper, *Empoasca fabae* into Wisconsin," *Ann. Entomol. Soc. Amer.,* **57,** 588–591 (1964).

Pitre, H. N., and Boyd, F. J., "A study of the role of weeds in corn fields in the epidemiology of corn stunt diseases," *J. Econ. Entomol.,* **63,** 195 (1970).

Pitre, H. N., Douglas, W. A., Combs, R. L., Jr., and Hepner, L. W., "Annual movement of *Dalbulus maidis* into the Southeastern United States and its role as a vector of corn stunt virus," *J. Econ. Entomol.,* **60,** 616 (1967).

Pitre, H. N., and Hepner, L. W., "Seasonal incidence of indigeneous leafhoppers (Homoptera; Cicadellidae) on corn and several winter crops in Mississippi," *Ann. Entomol. Soc. Amer.,* **60,** 1044 (1967).

Posnette, A. F., Robertson, N. F., and Todd, J. M., "Virus diseases of cacao in West Africa. V. Alternative host plants," *Ann. Appl. Biol.,* **37,** 229–240 (1950).

Proeseler, G., "Transmission of phytopathogenic viruses by gall mites," (Trans.) *Arch. PflSchutz,* **3,** 163–175 (1967).

Randles, J. W., and Crowley, N. C., "Epidemiology of lettuce necrotic yellows virus in S. Australia. I. Relationship between disease incidence and activity of *Hyperomyzus lactucae*," *Aust. J. Agric. Res.,* **21,** 447 (1970).

Rao, M. G. V., "The role of undergrowth in the spread of spike disease of sandal," *Mysore Sandal Spike Invest. Ser. Bull.*, No. 4 (1934).

Romney, V. E., "The beet leafhopper and its control on beets grown for seed in Arizona and New Mexico," *U.S. Dept. Agr. Tech. Bull.*, No. 855 (1943).

Sackston, W. E., "Aster yellows: a virus that attacks many plants," *Research for Farmers*, 3, 15–16 (1958).

Sameshima, T., and Nagai, K., "Relation between the life-history of green rice leafhopper *Nephotettix cincticeps* Uhler and rice yellow dwarf propagation," (In Japanese), *Japan J. Appl. Ent. Zool.*, 6, 267–273 (1962).

Schneider, C. L., "Occurrence of curlytop of sugar beets in Maryland in 1958," *Plant Disease Reptr.*, 43, 681 (1959).

Severin, H. H. P., "Field observations on the beet leafhopper *Eutettix tenellus*, in California," *Hilgardia*, 7, 281–360 (1933).

Severin, H. H. P., "Weed host range and overwintering of curlytop virus," *Hilgardia*, 8, 263–280 (1934).

Severin, H. H. P., "Factors affecting curlytop infectivity of beet leafhopper *Eutettix tenellus*," *Hilgardia*, 12, 497–530 (1939).

Severin, H. H. P., "Longevity or life hostories of leafhopper species on virus infected and on healthy plants," *Hilgardia*, 17, 121–133 (1946).

Shands, W. A., and Simpson, G. W., "The production of alate forms of *Myzus persicae* on *Brassica campestris* in the greenhouse," *J. Agr. Research*, 76, 165–173 (1948).

Shands, W. A., Simpson, G. W., and Covell, M., "Aphids caught in wind-vane traps with openings of different sizes," *J. Econ. Entomol.*, 48, 624–625 (1955).

Shands, W. A., Simpson, G. W., and Dudley, J. E., Jr., "Low-elevation movement of some species of aphids," *J. Econ. Entomol.*, 49, 771–776 (1956).

Shands, W. A., Simpson, G. W., and Lathrop, F. H., "An aphid trap," *U.S. Dept. Agr., U.S. Bur. Entomol. Plant Quarantine*, ET-196 (1942).

Shands, W. A., Simpson, G. W., and Reed, L. B., "Subunits of sample for estimating aphid abundance on potatoes," *J. Econ. Entomol.*, 47, 1024–1027 (1954).

Shands, W. A., Simpson, G. W., and Wave, H. E., "Some effects of two hurricanes upon populations of potato-infesting aphids in northeastern Maine," *J. Econ. Entomol.*, 49, 252–253 (1956a).

Shands, W. A., Simpson, G. W., and Wave, H. E., "Aphid studies stress need for early spraying," *Maine Farm Research*, 5, 1 (1957).

Shands, W. A., Simpson, G. W., and Wave, H. E., "Effect of low temperatures on survival of stem-mother aphids in northeastern Maine," *J. Econ. Entomol.*, 51, 184–186 (1958).

Shands, W. A., Simpson, G. W., and Wave, H. E., "This spring watch for green peach aphids and prepare to control them," *Potato Councillor*, 4, 16 (1958a).

Shapovalov, M., "Yellows, a serious disease of tomatoes," *U.S. Dept. Agr. Misc. Pub.*, 13, 4 (1928).

Shapovalov, M., Blood, H. L, and Christiansen, R. M., "Response of the tomato plant to spacing," *Utah Acad. Sci. Proc.,* **18,** 91–94 (1941).

Shaw, M. W., "Overwintering of *Myzus persicae* (Sulz.) in northeast Scotland," *Plant Pathol.,* 4, 137–138 (1955).

Shirahama, K., "Studies on radish mosaic disease and its controlling," (Eng. summary), *Agricultural Improvement Extension Work Conference, Tokyo* (1957).

Simpson, G. W., "Aphids and their relation to the field transmission of potato virus diseases in northeastern Maine," *Maine Agr. Expt. Sta. Bull.,* No. 403 (1940).

Simpson, G. W., and Shands, W. A., "Progress on some important insect and disease problems of Irish potato production in Maine," *Maine Agr. Expt. Sta. Bull.,* No. 470 (1949).

Slykhuis, J. T., "Factors related to the potential destructiveness of wheat striate mosaic on the prairies," *29th Session Can. Phytopath. Soc. Abstr. Proceedings,* **30,** 10–21 (1963).

Slykhuis, J. T., and Sherwood, P. L., "Temperature in relation to the transmission and pathogenicity of wheat striate mosaic virus," *Can. J. Botany,* **42,** 1123–1133 (1964).

Smith, F. F., and Brierley, P., "Aphid transmission of lily viruses during storage of the bulbs," *Phytopathology,* **28,** 841–844 (1948).

Somsen, H. W., "Development of migratory form of wheat curl mite," *J. Econ. Entomol.,* **59,** 1283–1284 (1966).

Staples, R., and Allington, W. B., "Streak mosaic of wheat in Nebraska and its control," *Nebraska Univ. Agr. Expt. Sta. Research Bull.,* No. 178 (1956).

Staples, R., and Allington, W. B., "The efficiency of sticky traps in sampling epidemic populations of the eriophyid mite *Aceria tulipae* (K.), Vector of wheat streak mosaic virus," *Ann. Entomol. Soc. Amer.,* **52,** 159–164 (1959).

Storey, H. H., "Virus diseases of East African plants. III. Rosette disease of ground-nuts," *E. African Agr. J.,* 1, 3 (1935).

Strickland, A. H., "The dispersal of Pseudococcidae (Hemiptera-Hmoptera) by air currents in the Gold Coast," *Proc. Roy Entomol. Soc. (London),* **A25,** 1–9 (1950).

Strickland, A. H., "The entomology of swollen shoot of cacao. I. The insect species involved, with notes on their biology," *Bull. Entomol. Res.,* 41, 725–748 (1951).

Strickland, A. H., "The entomology of swollen shoot of cacao. II. The bionomics and ecology of the species involved," *Bull. Entomol. Res.,* 42, 65–103 (1951a).

Strickland, A. H., "An aphid counting grid," *Plant Pathol.,* 3, 73–75 (1954).

Stubbs, L. L., and Grogan, R. G., "Lettuce necrotic yellows virus," *Nature,* **197,** 1229 (1963).

Stubbs, L. L., and Grogan, R. G., "Necrotic yellows, a newly recognized virus disease of lettuce," *Aust. J. Agric. Res.,* 14, 439–459 (1963a).

Swenson, K. G., "Role of aphids in the ecology of plant viruses," *Ann. Rev. Phytopathology,* 7, 351–374 (1968).

Swenson, K. G., and Nelson, R. L., "Relation of aphids to the spread of cucumber mosaic virus in gladiolus," *J. Econ. Entomol.,* 52, 421–425 (1959).

Thomas, I., and Vevai, E. J., "Aphis migration. An analysis of the results of five seasons' trapping in north Wales," *Ann. Appl. Biol.,* 27, 393–405 (1940).

Tinsley, T. W., and Wharton, A. L., "Studies on the host ranges of viruses from *Theobroma cacao* L.," *Ann. Appl. Biol.,* 46, 1–6 (1958).

Valenta, A., "Investigations on stolbur and related viruses," (*Ge*) *Conf. Potato Virus Disease Proc.,* 141–145 (1960–1961).

Van der Plank, J. E., and Andersen, E. E., "Kromnek disease of tobacco; a mathematical solution to a problem of disease," *Union S. Africa Dept. Agr. Sci. Bull.,* No. 240 (1944).

Van Hoof, H. A., Stubbs, R. W., and Wouters, L., "Beschouwingen over hoja blanca en zijn overbrenger *Sogata oryziocola* Muir. (Observations on hoja blanca and its vector *S. oryzicola*)," *Surinam Landb.,* 10, 1–18 (1962).

Wadley, F. M., "Ecology of *Toxoptera graminum,* especially as to factors affecting importance in the northern United States," *Ann. Entomol. Soc. Amer.,* 24, 325–395 (1931).

Wallace, J. M., and Murphy, A. M., "Studies on the epidemiology of curly top in southern Idaho, with special reference to sugar beets and weed hosts of the vector *Eutettix tenellus,*" *U.S. Dept. Agr. Tech. Bull.,* No. 624 (1938).

Watson, M. A., "The relation of annual incidence of beet yellowing viruses in sugar beet to variations in weather," *Plant Pathology,* 15, 145–149 (1966).

Watson, M. A., "Epidemiology of aphid-transmitted plant viruses," *Outl. Agric.,* 5, 155 (1967) (Rothamsted Expn. Stn., Harpenden, England).

Watson, M. A., and Heathcote, G. D., *Rept. Rothamsted Exp. Stn.,* 1965 (1966).

Watson, M. A., Hull, R., Blencowe, J. W., and Hamlyn, B. M. G., "The spread of beet yellows and beet mosaic viruses in the sugar beet root crop. I Field observations on the virus diseases of sugar beet and their vectors *Myzus persicae* Sulz. and *Aphis fabae* Koch," *Ann. Appl. Biol.,* 38, 743–764 (1951).

Watson, M. A., and Healy, M. J. R., "The spread of beet yellows and beet mosaic viruses in the sugar beet root crop. II. The effects of aphid numbers on disease incidence," *Ann. Appl. Biol.,* 40, 38–59 (1953).

Watson, M. A., and Sinha, R. C., "Studies on the transmission of European wheat striate mosaic virus by *Delphacodes pellucida* Fabr.," *Virology,* 8, 130–163 (1959).

Westdal, P. H., Richardson, H. P., and Barrett, C. F., "The six-spotted leafhopper, *Macrosteles fascifrons* Stål, in relation to aster yellows virus in Manitoba in 1959," *Bull. Entomol. Soc. Amer.,* 5, Abstr. No. 83 (1959).

Wilks, J. M., and Walsh, M. F., "Apparent reduction of little cherry disease spread in British Columbia," *Can. Plant. Dis. Survey,* 44, 2 (1964).

Wille, J., "The beet leaf bug (*Piesma quadrata* Fieb.)," *Ann. Appl. Biol.,* 16, 645 (1929).

Winkler, A. J., "Pierce's disease investigations," *Hilgardia,* 19, 207–264 (1949).

Wolcott, G. N., "Increase of insect transmitted plant disease and insect damage through weed destruction in tropical agriculture," *Ecology,* 9, 461–466 (1928).

Woodford, J. A. T., "Difference in flight capacity between naturally occurring spring and summer forms of *M. persicae* (Sulz.) (Hemip. Aphididae)," *Nature,* 217, 583–584 (1968).

15

The Control of Viruses and Virus Diseases of Plants

The control of virus diseases has many aspects—the very number and kinds of approaches to control indicate the complexity of the problem and the absence of any one effective method. It is possible to go further and say that for no one virus disease is there a single method of control that is completely effective.

CULTURAL METHODS OF CONTROL

Cultural methods of control are the simplest that the grower can use, but they have limitations imposed by season and crop requirements.

Isolation

Spatial isolation of the site will reduce the risk of infection of beets by yellows but cannot be expected to eliminate it when infective aphids are carried by the ebb and flow of wind currents in the general area. In spite of this, however, if steckling bed sites are chosen in areas where neither root nor seed crops are grown, but where the climate and soil are suitable, the production of stecklings has been successful.

Cauliflower mosaic in broccoli can be reduced if seedlings are raised half a mile from old infected plants. Internal cork disease of sweet potato can be avoided by planting 100 yd or more from cork-infected plants (Martin and Kantack, 1960).

Vector dispersal from an infected source is of prime importance in virus infection of cultivated crops. Shepard and Ellis (1970) concluded that to provide protection for sugar beets against beet yellows and beet mosaic, both transmitted by *Myzus persicae,* the distance from a large area of cultivated and infected beets should not be less than 15–20 mi and 12–15 mi respectively. Sole spring plantings in inland valleys are practically free from these diseases. However, isolation of the scale necessary would require a highly organized system of discrete planting and harvested areas, to avoid overlapping and decreasing the effectiveness of a host-free period.

Lettuce mosaic has been practically eliminated from seed crops by growing the crop in an area where climatic factors eliminated the vector aphids during a critical period (Stubbs and O'Laughlin, 1962) (See also Gonzales and Rawlins, 1969).

Barrier Crops

Barrier crops as a method of achieving isolation have been used. Screen cultures are used in France to separate beet steckling beds 20 yd long and 5 yd wide. These bands are about 1 yd wide and are sown to oats, maize, hemp, or sunflower (R. Hull, 1952) (Table 15.1).

Surrounding the seed bed with a barrier crop of kale or barley will decrease the incidence of cauliflower mosaic, even if diseased plants are only 5 yd away (Jenkinson, 1955).

TABLE 15.1.   Percentage of Virus Yellows in Stecklings and the Yield of Seed [a,b]

	Virus yellows, %	Yield of seed, cwt. ac
Stecklings sown in barley, Apr. 7	12.2	26.8
Stecklings sown in open beds, July 12	100.0	14.4
Stecklings sown in open beds, Sept. 2	86.7	19.2
Stecklings sown along with barley, Sept. 2	43.6	22.0
Standard error ± 0.728		

[a] After Hull (1952).
[b] From experimental plots comparing undersowing with open beds at Terrington, 1948–1949.

Density of Stand

Density of stand is correlated with reduced infection by yellows in both seed and root crops of sugar beet because the denser foliage provides an environment unsuitable for aphids to develop, and, on a percentage basis, a given number of infective aphids will infect a smaller percentage in a denser planting. Much the same microenvironment is afforded if weeds are allowed to grow. A practical variation of this is the use of a cover crop such as barley. The beets are sown as an undercrop at the same time as the barley, in April; they make little growth before the barley is cut but grow rapidly in the autumn. Table 15.1 gives the results obtained in an experimental confirmation of this method.

Close planting of tobacco is a practical way of providing against losses due to tomato spotted wilt (Kromneck disease) in South Africa (van der Plank and Anderssen, 1944). Since the thrips vector does not breed on the tobacco and does not move from plant to plant, losses are reduced on a percentage basis by close planting. Linford (1943) reported a similar relationship between percentage infection and planting density in pineapple fields, when the same virus and vector relationships were involved. Shapovalov (1941) noted that dense tomato plantings were less affected by curlytop than were more open plantings. The tomato plant is a poor host for the leafhopper vector, but the dense growth is also an unfavorable factor in this case.

Plant density can affect disease incidence. Native methods of groundnut planting used high densities. When for agronomic reasons, lower densities were advocated, rosette disease increased. New explanations are that the optometer landing response of alate aphids (*A. craccivora* (Koch)) is affected by density of ground cover (A'Brook, 1964), or that the aphids are attracted more to the increased yellowing of plants grown in low density (Hull, 1964) (See also Muller, 1964). Further studies on the effect of plant spacing on groundnut mosaic incidence were reported by A'Brook (1968). Trapped aphids at the 3 ft level were more numerous over open- than over close-spacing. The contrast between soil and plants is a factor in widely-spaced planting. At crop height, an alighting response was shown only by those species that fed on groundnut. Control of rosette disease by early planting and close spacing appears to be due to the inhibition of landing response of the aphid by the continuous ground cover resulting from close spacing.

Planting and Harvesting Dates

Planting and harvesting dates can be manipulated to some extent as control measures. Schultz *et al.* (1944) found that early harvest of seed potatoes in Maine greatly reduced leaf roll, mosaic, and spindle-tuber infection even in seed produced under conditions of poor isolation. Bagnall (1953), however, found that early harvest was not an economical control method for virus Y infection; Hansing (1943) had come to the same conclusion in the case of potato yellow dwarf.

Early planting of potatoes destined for the market actually increases the incidence of virus disease, but when coupled with appropriate fertilizing, yields are greatly increased; the incidence of virus diseases in the year of infection causes inappreciable loss of yield (Broadbent *et al.*, 1952).

Although early-grown stecklings are more liable to heavy infection with beet yellows, later sowing has its disadvantages: the season is shortened and growth is too small to produce a good seed crop the following year (Table 15.2) (R. Hull, 1952).

TABLE 15.2. *Mean Percentage of Virus Yellows in Seed Crops Planted with Stecklings from Various Centers* [a]

	Sowing, 1946		Sowing, 1945	
Center	July	August	July	August
Roxwell, Essex	77.4	20.6	36.8	21.2
Maldon, Essex	5.9	8.9	14.7	7.7
Harpenden, Herts.			10.0	26.5
St. Neott's, Beds.	70.0	8.0	5.7	5.0
Biggleswade, Beds.	8.0	5.8	4.6	4.0
Kettering, Northants.			51.8	44.6
Terrington, Lincs.	92.0	44.0	45.2	43.5
Spalding, Lincs.	74.9	33.3	100.0	100.0
Stainton-le-Vale, Lincs.			29.8	5.0
Market Weighton, Yorks.			39.2	6.4
Sutton Bonnington, Leics.			48.0	20.5
Shottle, Derbyshire			21.1	9.8
Little Leigh, Cheshire	19.4	0.8		

[a] After Hull (1952).

Broadbent and Doncaster (1949) recognized that migration time coincided with development of the host plants. Sometimes, however, a very small time difference is important, as Hansen (1950) found with sugar beet yellows. April 1st, 15th, and May 1st plantings were 38, 53, and 70% infected, respectively. Whether it was practical for planting to be finished by April 1, and whether the farmer knew any good reason why it should have been, are quite different matters.

Sometimes the early attractiveness of crop plants is taken advantage of. Border rows of sugar beets are planted ahead of the normal planting season in the areas likely to be affected by *P. quadrata*. These serve as a trap for the insects coming in from their hibernation quarters, and they can be destroyed, either by crushing the young infested beets, or, more recently, by the use of insecticides before the normal planting is made.

There are many other factors affecting susceptibility, but they are physiological rather than ecological. It is of interest, for example, to know that artificially modifying day length can affect susceptibility, but insolation is rather regular for any one latitude.

Avoidance of Infection

Avoidance of infection is practiced extensively in areas where environmental conditions in the production of seed potatoes do not favor aphid establishment and virus spread. In Great Britain these conditions are found in general at altitudes of over 500 ft and with rainfall of more than 35 in., although tradition and experience have been determining factors in the establishment of seed-growing areas. For example, there are good seed-producing areas in eastern Scotland, but Fisken (1959) concluded that freedom from leaf roll and virus Y depended on the lateness of aphid infestation rather than freedom from it. Attempts have been made to designate suitable areas on the basis of potential virus spread, based on a study of aphid movements rather than on virus spread within a crop (Hollings, 1955). The "angle of colonization" was arrived at by plotting the percentage of sample plants infested at each count against time, and then by drawing a line from an arbitrary zero point of June 1 to the peak of the curve. This index of aphid movement provides a more suitable comparison between areas than does virus spread, which for any particular location is influenced strongly by the incidence of infection in the seed used (Hollings, 1960). The destruction of potato haulms in seed potato crops before maturity stops aphid development and avoids any further spread of virus within the field. This practice is essential for certification in Switzerland (Salzmann, 1953).

BY VIRUS-FREE SEED. Seed-transmitted virus control must begin at the source, and developing a stock of virus-free seed is a major control measure. Grogan *et al.* (1951) achieved this by growing lettuce seedlings in an insect-free greenhouse, roguing all infected plants, and then planting out in a mosaic-free area. Zink *et al.* (1956, 1957) and Broadbent *et al.* (1951) confirmed that the use of mosaic-free seed was the most effective control measure available, but the percent of seed transmission must be below 0.1

if control is to be satisfactory. Tobacco mosaic in tomato seeds can be reduced materially by fermenting the seeds for 10 days (Sinclair, 1959).

BY REMOVAL OF VIRUS SOURCES.   Removal of virus sources is probably the oldest method of control and usually the first to be tried. It includes field sanitation—the removal of old diseased plants and volunteers —and roguing of diseased plants in the growing field. The value of the latter procedure depends upon the objective. For the production of virus-free seed or planting stock, roguing is essential to meet certification requirements. For the production of table (ware) potatoes, however, it is not considered worthwhile (Broadbent *et al.,* 1950).

Groundnut mosaic is epidemic in some South African districts and control requires that all volunteer plants, whether diseased or not, be removed. Roguing of diseased plants in the new planting is also necessary for 2 mo after planting (Klesser and LeRoux, 1957).

Roguing is obviously not called for if there is no spread within the crop, as in kromneck in tobacco and yellow spot in pineapple.

The most extensive roguing is done in attempts to suppress spread of the viruses infecting trees. A 21-yr record of the Colorado peach mosaic suppression program was issued by List *et al.* (1956), during which time no vector was known. The first 2 yr were devoted to cutting out the accumulated infections; this amounted to 36.5% of the total number of trees destroyed; current infection leveled off, but for 17 yr there was no significant reduction in the total annual number of current infections. Once an orchard was infected, the probability of current infection was ten times as great as it was in previously uninfected orchards. The opinion was expressed that eradication is impossible, but a loss of 6.5 trees/1000 is considered cheap insurance.

Phony peach disease was known in 1952 in the United States in 17 out of 26 of the states surveyed. In 6 of these states, the disease was eliminated by roguing, and it was greatly reduced in 8 others.

Reappearance of the disease in a number of counties is ascribed to the presence of wild plums in the vicinity; wild plum eradication could reduce disease incidence. Roguing of diseased trees is considered an effective method of control in commercial plantings (Persons, 1952). In Missouri, where phony peach disease is the most serious stone fruit disease known, it is believed that eradication has provided an economical control. This is in spite of the fact that three of the four leafhopper vectors of the disease occur in the state. The wild plum does not appear to be an important virus source in that state (Millikan, 1955). Banana bunchy top is being successfully controlled by roguing, the incidence being reduced to 1 plant in 2000 in 1948, or only 1/12 the number found in 1937

(Cann, 1949). In 1952, only 15 infected plants were found per 100 ac. The cost of the control is about £A 28,000 per year, *cira* $62,000 (Cann, 1952).

The most extensive programs of diseased tree removal are those in Ghana and Nigeria, where swollen shoot disease of cacao has devastated large areas of cocoa plantings. Procedures have differed materially between the two countries. In Ghana, only the infected trees have been cut out; recently, a voluntary contact-tree removal program was initiated. All outbreaks discovered are treated within a month or 6 wk, and all treated outbreaks are subject to regular monthly reinspection until deemed "controlled"; that is, until no infected trees have been found for 2 yr. Hammond (1957), from whose paper this summary of the Ghana program is largely drawn, stated that only 2% of the outbreaks (on 124,098 farms) were controlled without contact-tree removal.

The campaign started in 1946; it was interrupted as a result of violent repercussions from the farmers, but it has continued at a steady pace since 1949. Hammond reported that 63 million trees had been cut out up to June, 1957. The 100-million mark was passed by the end of 1960. In the Eastern Province of Ghana there has been evidence of a leveling off at about 215,000 trees a month. The total cost of the operation countrywide is huge; approximately 20,000 men were involved. In spite of the fact that in both Ghana and Nigeria there are areas where control has been abandoned, techniques have improved over the years, and education of farmers has had its effect on growers' cooperation. If contact-tree removal becomes general practice in Ghana, spread of the disease may be appreciably slowed down, although there will be an enormous increase in the number of trees cut out.

In Nigeria, until 1950, only diseased trees were removed. This policy was drastically changed, and trees were removed for a distance of 30 yd in every direction. This involved the destruction of large numbers of healthy trees, and the procedure has since been modified to provide for the removal of apparently healthy trees to distances of 5, 10, and 15 yd around outbreaks of 0–6, 6–50, and 50–200 infected trees, respectively (Thresh, 1959). The basic difficulties involved are essentially the same for both countries, but perhaps they are more acute in Ghana where the forest is much more dense, the virus strains more virulent, and the infection in forest trees more widespread. At present, the only method of detecting field infection is by routine inspection for symptoms. Even after obviously diseased trees are removed, the symptoms on some trees are difficult to spot; the long period for expression of symptoms leaves infected trees symptomless and dangerous sources of virus. Cornwell (1956) has shown that such trees will support larger populations of mealybugs than will severely affected trees, and that as the trees deteriorate the mealybugs tend to disperse. Expression of symptoms can be hastened if the contact trees, which often include those still in the latent period, are coppiced; that is, cut down to a foot or so from the ground. The new flush will usually show symptoms if the tree is infected; however, coppicing needs to be done at the advent of the dry season if regeneration is to be successful, so its use is limited.

In spite of the fact that millions of trees are still being cut out and new outbreaks are adding to the burden of constant reinspection, no acceptable substitute control method is as yet available. It is clear, however, that cutting out will not eradicate swollen shoot, and some alternative or subsidiary measure must be agreed upon.

HEAT THERAPY OF VIRUS-DISEASED PLANTS

Inactivation of Viruses by Heat

Temperature may affect susceptibility to infection, the time required for the expression of symptoms (incubation period), the degree of symptom expression, the rate of multiplication of the virus *in vivo,* and the emergence of attenuated strains (Kassanis, 1957). This discussion, however, is concerned only with the use of heat as a practical control measure, particularly in those cases where the majority or all of the plants in a vegetatively propagated clone are infected. Treatment with hot water at 52–55°C for 30 min improved the growth of the sugarcane seed-pieces affected with sereh disease (Wilbrink, 1923; Houtman, 1925). Martin (1933) successfully applied the hot-water treatment to control chlorotic streak of sugarcane. The use of this method in sugarcane culture now appears to be standard practice for all those virus disease infections which are susceptible to treatment. There is varietal susceptibility to heat in Puerto Rico, where the treatment is used to control chlorotic streak (Adsuar, 1956). In Australia, large acreages of sugarcane are planted with seed-pieces treated with hot water to control ratoon stunting disease. The recommended treatment is 50°C for 20 min; longer periods may adversely affect germination, while post-treatment handling must avoid heating in the closed bags used (Mungomery, 1953; Bieske, 1954). There is varietal difference in the ability of the cane to withstand temperature, a difference of 1°C (from 53–52°C) being significant (Knust, 1953).

The early work with sugarcane was done with diseases whose virus nature was not well understood at the time. It remained for Kunkel (1936) to establish definitely that peach trees infected with yellows disease virus were cured by incubating potted trees in a hot room at from 34.4–

35.3°C, or by immersing the plants in water held at 50°C.

Three other virus diseases of peach, little peach, red suture, and rosette, were also cured by exposure to heat, the last-named being somewhat more refractory than the others. On the other hand, these same treatments had no effect on peach mosaic (Kunkel, 1936a). Kunkel continued his studies on heat treatment, and in a succession of papers (1941, 1942, 1943, 1952) reported the cure of aster yellows in periwinkle by heat treatment at 38–42°C for 2 wk. It was possible to reinfect and cure as often as desired. Cranberry false blossom in periwinkle (*Vinca rosea* L) was cured by treatment for 2 wk at 40°C; potato witch's broom was cured in the same host by 13 days' treatment, as were potato tubers up to ¾ in. in diameter. In the case of alfalfa witch's broom in *V. rosea,* however, both temperature and time were critical, 7 days at 42°C affecting cure while a 6-day exposure did not.

At first, the indications were that curing by heat was limited to viruses with a low thermal death point *in vitro*. As Kassanis (1954, 1957) has pointed out, however, heat therapy does not appear to be correlated with the thermal inactivation point of the virus *in vitro*. For example, the end points for tomato bushy stunt and tomato spotted wilt are 80 and 45°C, respectively; yet the former is cured by exposure to 36°C for 3–4 weeks and the latter is not. In general, viruses with spherical particles can be inactivated by heat; those with rod-shaped particles are less sensitive. An exception is the treatment of watermelon seed to eliminate the pimple virus by treatment with infrared at 158°F for 30 min (Rosberg, 1957). Some strawberry viruses can be heat-inactivated. Posnette *et al.* (1953) maintained plants at 37°C for 7–11 days and inactivated viruses of the crinkle type. Later, Posnette and Cropley (1958) tested the reactions of 20 varieties of strawberry to heat treatment. There was considerable variation between varieties; Perle de Prague was freed from mottle virus in 6 days at 37°C. Crinkle virus required 30–50 days; yellow edge virus, 26 days (Fig. 15.1 and 15.2).

Strawberry plants can be conditioned to withstand temperatures at which virus inactivation occurs by growing them at 27°C for several weeks before subjecting them to the higher inactivating temperatures, which varied diurnally from 31–48°C. The effect of the treatment was to reduce symptom expression and improve growth, but one plant was obtained which was completely symptom-free (Fulton, 1954). Bovey (1954) reported 5.4% of 698 plants cured by exposure to 37°C for 10 days.

There is a distinction made between permanent cure and one that is only temporary. In the latter case, heat may so reduce the concentration of virus that the plant may remain symptomless and without transmissible virus for a time, but nevertheless is virus-infected. Posnette and Crop-

FIGURE 15.1. Cucumber plants freed from cucumber mosaic by heat. The plants on the right were infected with different strains of cucumber mosaic virus and were photographed 40 days after infection. The plants on the left were placed at 36°C 17 days after infection and photographed after being at 36°C for 23 days. (Photograph Kassanis 1957)

ley (1958) suggest the possibility that heat may affect a mixture of strains differentially, inactivating the more virulent strains and leaving the avirulent strains behind to provide a measure of cross protection.

Hunter *et al.* (1959) applied heat therapy to seedlings budded with strains of apple mosaic, having found that wholly infected plants and scion wood were more difficult to treat, or somewhat refractory. Exposures for 28–40 days at 37°C resulted in symptom-free growth during the following summer when the check trees were all visibly infected. This result is interesting as an example of heat therapy of a newly infected plant, but there are better ways to practical control.

Leaf roll virus can be inactivated in potato tubers by 25 days' exposure to 37°C, but this treatment did not affect either virus X or Y (Kassanis, 1949, 1950). The temperatures reached in the Indian plains inactivate potato leaf roll in potatoes stored there (Thirumalachar, 1954).

Neither tristeza nor psorosis virus of citrus can be inactivated *in vivo* by hot-water treatments of 10 min at 52°C, 180 min at 45°C, or 3 days at 35°C (Price and Knorr, 1956). Table 15.3 lists viruses that have been inactivated in plants by heat treatment.

a

b

FIGURE 15.2 (*a*) and (*b*), *Abutilon striatum* (Dicks.) plants of the same age; (*a*) kept at ordinary greenhouse temperatures; (*b*) after 28 days at 36°C. (Photograph Kassanis 1957)

TABLE 15.3. *List of Viruses That Have Been Inactivated in Plants by Heat* [a]

Virus	Plant	Treatment	Temperature (°C)	Length of treatment
Abutilon variegation	*Abutilon striatum* (plants)	Hot air	36	3–4 wk
Apple mosaic	Budded seedlings	Hot air	37	28–40 days
Aster yellows	*Vinca rosea* and *Nicotiana rustica* (plants)	Hot air	38–42	2–3 wk
	Vinca rosea (plants)	Hot water	40–45	2½–24 hr
Carnation ringspot	Carnation (plants)	Hot air	36	3–4 wk
Cherry necrotic rusty mottle	Cherry budsticks	Hot water	50	10 min
Cherry ringspot	Cherry budsticks	Hot air	100	17–24 days
Citrus tristeza	Potted plants	Hot air	95° F ± 3°	121–360 days
Cranberry false blossom	Cranberry and *Vinca rosea* (plants)	Hot air	42	8 days
Cucumber mosaic	Cucumber, tobacco, *Datura stramonium* (plants)	Hot air	36	3–4 wk
Little peach	Peach (budsticks)	Hot water	50	3 min
Peach red suture	Peach (budsticks)	Hot water	50	3 min
Peach rosette	Peach (budsticks)	Hot water	50	8 min
Peach X-disease (yellow-red virosis)	Peach (budwood)	Hot water	50	6–15 min
Peach yellows	Peach (trees)	Hot air	35	24 days
	Peach (dormant trees)	Hot water	50	10 min
Phony peach	Peach (dormant trees)	Hot water	48	40 min
Potato leaf roll	Potato (tubers)	Hot air	37	15–30 days

Potato witch's broom	*Vinca rosea* (plants)	Hot air	42	13 days
	Potato (tubers)	Hot air	36	6 days
Raspberry leaf mottle	Raspberry (plants)	Hot air	32–35	1–4 wk
Raspberry leaf spot	Raspberry (plants)	Hot air	32–35	1–4 wk
Raspberry unidentified latent virus	Raspberry (plants)	Hot air	32–35	1–4 wk
Raspberry *Rubus* stunt	Raspberry (canes)	Hot water	45	1½–2 hr
Strawberry leaf burn or X	Strawberry (plants)	Hot air	37	7–11 days
Strawberry virus 1 (mottle)	Strawberry (plants)	Hot air	37	7–11 days
Strawberry virus 3 (crinkle)	Strawberry (plants)	Hot air	37	7–11 days
Strawberry virus 4 (vein chlorosis)	Strawberry (plants)	Hot air	37	7–11 days
Strawberry virus 2 (mild yellow edge)	Strawberry (plants)	Hot air	37	16 days
Strawberry nonpersistent viruses	Strawberry (plants)	Hot water	43–48	½–7 hr
Strawberry nonpersistent viruses	Strawberry (plants)	Hot air	36–38	8–12 days
Strawberry type 2	Strawberry (plants)	Hot air	38	8 days
Strawberry viruses (unidentified)	Strawberry (plants)	Hot air	37	10 days
Sugarcane chlorotic streak	Sugarcane (cuttings)	Hot water	52	20 min
Sugarcane ratoon stunt	Sugarcane (setts)	Hot water	50	2 hours
Sugarcane ratoon stunt	Sugarcane (cuttings)	Hot water	50	20 min
Sugarcane sereh disease	Sugarcane (cuttings)	Hot water	52–55	30 min
Tobacco ringspot	Tobacco (plants)	Hot air	37	3–4 wk
Tomato aspermy	Tomato and tobacco (plants)	Hot air	36	3–4 wk
Tomato aspermy	Chrysanthemum (plants)	Hot air	36	3–4 wk
Tomato bushy stunt	*Datura stramonium* (plants)	Hot air	36	3–4 wk

[a] After Kassanis (1957).

While there is no doubt that heat therapy has its place, it is not a practical control method for large-scale field use except in those cases in which short exposures are successful, as with sugarcane. Another useful practical application is the sterilization of tobacco waste, a source of infection by tobacco mosaic when it is used as a fertilizer or as an additive in mixed fertilizers. Treatment at 100°C in a moist atmosphere for 15 min or longer suffices to sterilize (Beinhart and Morgan, 1952). Apart from instances of this kind, heat therapy finds its usefulness principally as a technique for developing healthy propagation stocks (Roland, 1957).

Development of virus-free propagation stocks by heat treatment depends largely on the treatment differentially affecting the virus in old and new tissues, so that even if the plant is not entirely free from virus, it is possible to grow plants from cuttings that are virus-free. Kassanis (1954) obtained a high proportion of healthy cuttings from treated virus-infected plants, even though symptoms reappeared in the treated plants after intervals of 10 days to 5 wk. Similar results have been reported by Chant (1959) for cassava mosaic. Grant (1957) found that new growth produced on heat-treated citrus budsticks was free of virus and could be used as a source of healthy budwood. Tip cuttings from heat-treated hydrangea plants infected with ringspot virus and heated in a chamber held at 35°C for from 4 to 19 wk showed a fair proportion of virus-free plants after 12-, 15-, and 19-day exposure periods. None of the treated source plants was at any time virus-free, but the conditions for virus multiplication were so unfavorable that tip cuttings could be removed before virus invasion had occurred (Brierley, 1957). Chrysanthemum plants infected with aspermy, stunt, and ring pattern viruses had their growing tips pinched off before treatment at an air temperature of 97°F for 3–4 wk, and they threw lateral shoots which grew into virus-free plants. Viruses B, D, vein mottle, and an unidentified virus were not affected by the treatment, but Hollings and Kassanis (1957) consider that the results are of practical use; the most destructive viruses are among those controlled, and even when only part of a virus mixture is eliminated, flower quality is often improved.

Heat treatment for control of ratoon stunting disease of sugarcane is still used (Liu *et al.,* 1963), but this can render the plants more susceptible to sugarcane mosaic (Zumme, 1967). Heat treatment of sugarcane mosaic infections is not new, but Bandy (1970) modified the treatment and cured the infection by repeating the hot water treatments at daily intervals at 54.8, 56.5, and 57.3°C for 7 min each time. Two or more vegetative generations later, the plants were still free of symptoms; symptoms followed reinoculation.

Natural high temperatures in the Imperial Valley inactivated one out of five strawberry viruses (crinkle); it is likely that the vectors would also find the environment difficult (Frazier *et al.*, 1965). Another use of heat is the "hot knife" method used by Freitag for removing orchid flower spikes (1965). Heat treatment freed a variety of carnation (Joker) from carnation mottle virus, but it took 138 days at 38°C. The cost and economic value of this treatment were not discussed (Hakkaat and Gordonova, 1968).

INDEXING: A METHOD FOR PRODUCTION OF VIRUS-FREE PLANTS

Indexing means the testing of plants for the detection of virus infection. The potato has probably received more attention than any other crop plant because of the importance of planting virus-free seed-pieces. Indexing tubers is accomplished by removing one eye-piece from the stem end of each tuber, growing it in a pot or greenhouse flat, and, when the plant is 6–8 in. high, recording its condition with respect to virus disease symptoms. Growers provide the indexing authority with sample lots of tubers; the healthy ones are returned to them and are then planted in a tuber-unit seed plot as far removed from sources of reinfection as possible. During the growing season, the plantings are inspected and rogued. The progeny are used for increase and the following year marketed as seed potatoes. Indexing of one or more tubers from selected hills is also practiced, and, while not as accurate as the single tuber method, it is satisfactory when there is a low mosaic incidence.

The tuber-unit planting is actually a form of indexing, but instead of actually testing each tuber in the greenhouse, plots are planted in the field with tubers usually divided into four pieces. The success of this method depends on skill in determining field symptoms and care in roguing, but as a seed growers' method it is practical even though it cannot be compared with the tuber index method for accuracy. These methods are so well established that detailed descriptions go back 30 years (Schultz *et al.*, 1934).

A very practical method for eliminating purple-top hair sprout from potato seed is to move the tubers to warm storage for 10–14 days before planting to induce slight sprouting; this permits roguing of diseased tubers (Larson, 1959).

Certification Programs

Certification programs are based on indexing. In the United States these are individual state functions. Strawberry certification standards in Oregon are set up by a Certification Board at Oregon State College, which also operates a Strawberry Plant Propagation Center. Indexing of each mother plant is accomplished by making three leaf-petiole grafts, one known to be infected with mild mottle, one infected with vein-banding, and the third one virus-free. This method reveals not only virus infection but also the components of virus complexes (Dobie *et al.*, 1958). Plants passing this test are propagated and runners distributed to growers and cooperating institutions. No virus disease tolerance is allowed in the propagation center. For nurseries growing certified plants, an extremely low tolerance is allowed, as shown in Table 15.4.

TABLE 15.4. Tolerances Allowed in Growing Certified Strawberry Plants in Oregon [a]

	Tolerance		
Factors	First and second inspections (%)	Third and subsequent inspections (%)	Spring inspection (%)
Total diseases	2	1	1
Virus diseases	1	0.5	0
Red stele	0	0	0
Variety mixture	0	0	0
Nematodes			
Root knot			
Foliar	0	0	0
Bulb and stem			
Cyclamen mite	0	0	0

[a] From Anon., Oregon State College Certification Board (1955).

The Arkansas and Michigan programs are very similar, indexing being accomplished by grafting to various indicators and plants of the wild strawberry *Fragaria vesca* L (Fulton and Seymour, 1957; Fulton and Lovitt, 1958).

A sugar beet steckling certification scheme is in effect in Great Britain. Farmers are provided with stecklings grown in suitable isolation but inspected throughout the season. An evaluation is made at the end of October. The plant population is estimated by counting the plants in 1-yd lengths of row at 20 random positions; counts of diseased plants are made from 20 10-yd lengths of row and the percentage of infection is calculated. More than 10% results in condemnation; less than 1% in certification. Cases falling between 1 and 10% are subject to arbitration (and probably considerable difference of opinion as to whether the yellow plants are, in fact, suffering from virus yellows) (R. Hull, 1952).

Difficulty in diagnosis because of heat spot, resulting from growing the plants at high temperatures to induce runner formation, has been reported by Smeets and Wassenaar (1956). A clone of *F. vesca* which does not develop heat spot would be most useful.

Leaf roll and mosaic are two viruses which have been greatly reduced in importance by seed-certification schemes, based on field inspections (Bawden *et al.*, 1948). Virus X presents another problem. First virus-free plants must be found and then multiplied by the tuber-unit method, or multiplied in areas particularly suitable for seed growing. Testing for virus freedom can be done either by sap inoculation from individual tubers or from composite samples to *D. stramonium* L.

Serological Methods

Serological methods can also be used. Virus X is a good antigen; if purified virus X is used, difficulties with antigenicity of normal plant constituents are avoided.

The most extensive use of serological methods on a practical scale appears to have been done in Holland. D. H. M. van Slogteren (1959) reports success in developing a polyvalent antiserum that can be used to test in one reaction on the potato viruses X, Y, S, and *Aucuba* mosaic. The scale of this operation can be judged by the use of horses for the production of the antisera. E. van Slogteren (1955) lists 46 varieties of potato seed stocks cleared of viruses by the use of serological diagnosis; and antisera for some 25 different viruses infecting flower bulbs and field crops are used to test seed growers' propagation stocks.

Apical and Meristem Cultures

The production of virus-free stocks can also be achieved by taking advantage of the fact that some plants grow and elongate faster than the virus can occupy the new tissue.

Holmes (1955) eliminated spotted wilt from a clone of dahlias by rooting small tip cuttings from diseased plants, improved a stock of chrysanthemums by freeing them of one of two contaminating viruses (Holmes, 1956a), and cleared a stock of sweet potato of a virus-induced chlorotic spotting of foliage by the same method (Holmes, 1956). Hildebrand (1957) freed seven varieties of sweet potato from infection with cork virus by growing the plants in a warm greenhouse to induce flushing of new growth and then growing new plants from 3-node tip cuttings. Baxter and McGlohon (1959) freed white clover of bean yellow mosaic by rooting 1-cm-long terminal cuttings at 10°C.

An extreme development of the use of apical cuttings involved growing excised tips in culture solution (Kassanis, 1957a; Nielsen, 1960). Thomson (1956a, 1957) heat treated apices after establishing them in culture. The game is presumably worth the candle if the clone freed of virus is very valuable and the desired result cannot be achieved in a more direct manner.

Kassanis and Varma (1967) produced virus-free clones of potatoes by meristem culture. Out of 20 developing from 196, 19 were free of viruses A, X, and S. Meristem culture resulted in the elimination of four carnation and sweet william viruses, of 10% in the case of carnation mosaic virus, of 71% of carnation latent virus, and of 83 and 88% respectively of CRSV and CVNV (Stone, 1968). Rhubarb was freed of four viruses by apical tip culture, but *Arabis* mosaic sometimes survived (Walkey, 1968). In horseradish, out of 30 clones containing sap-transmissible virus, one was freed of cabbage black ring and one of cauliflower mosaic by apical culture (Hickman and Varma, 1968). Vine (1968) freed strawberry plants from gooseberry vein-banding mosaic and a small number (1–3) from four other varieties. This was accomplished by an improved technique for apical tissue culture.

The commercial meristem culture of *Cattleya* and *Cymbidium* orchid plants, popular in the United States, has frequently failed to prevent passage of viruses present in the mother plant, and if this occurs it is an effective means of virus dissemination. Hollings (1965), in his review of disease control through virus-free stocks, stresses the necessity to maintain the health of foundation stocks, whether these are derived from indexing techniques, chemical control, heat treatment, or meristem and apical tissue culture. The possibilities for elimination of viruses from plants by meristem culture or other means have been fairly well demonstrated, but maintaining these stocks in the hands of growers and nursery men is quite another matter.

CHEMICAL CONTROL OF VIRUSES

Chemotherapy for diseased plants has considerable attraction, not only for the sophisticated research worker, but also for the untutored but intelligent farmer. Members of the Swollen Shoot Commission to the Gold Coast in 1948 were repeatedly told by cocoa farmers that when a man became sick, he was not cut down, but was treated and made well again;

why not do the same with diseased cacao trees? This is a consummation devoutly wished for.

The basis for chemotherapy of plant virus diseases will perhaps be laid by the study of inhibitors of virus infection and multiplication. The literature is replete with studies of inhibitors *in vitro,* but *in vivo* inhibition, which is essential from the standpoint of control, is not so well documented. Inhibitors of infection would fall into the class of prophylactics, and prophylactic control measures would obviously be limited to special situations, perhaps greenhouse and nursery areas where the cost of repeated treatments in the absence of disease would not appear redundant to growers. Inhibitors of virus multiplication, on the other hand, might find many uses. Lowered virus concentration could be reflected in reduced availability to vectors, and therefore to reduced virus spread. Tolerance might be less burdensome to the plant if virus invasion were halted or the intensity of its attack reduced.

Bawden (1954) has reviewed this field of inquiry, so the few cases considered here represent only a sampling of those reported. Inhibitors of infection are found in plant extracts, but these same substances do not prevent the multiplication of viruses in plants that contain them, nor do they prevent insects from acquiring viruses from such plants; in other words, they only inhibit when released from the discipline of the systems of which they are normal constituents.

Material sedimented from the sap of tobacco leaves inactivated tobacco ringspot but not tobacco mosaic virus. The material sedimented from healthy tobacco, or tobacco infected with tobacco mosaic virus, is less effective as an inactivator than that from leaves infected with tobacco necrosis or tobacco ringspot viruses (Bawden and Pirie, 1957).

Takahashi (1942) reported on a virus inactivator derived from yeast. When the concentration of the inactivator was doubled, the virus activity was halved, leading Takahashi to postulate that the virus particle combined with the inactivator. There was no *in vivo* inactivation.

Sap from ten species of plants high in tannin were shown by Yoshii *et al.* (1954) to inactivate tobacco mosaic virus. Inhibition of infection, rather than of virus increase, by sap from *Capsicum frutescens* L, was demonstrated by McKeen (1956). The effect was only observable when the sap was either applied before inoculation or mixed with the inoculum. There was some evidence of translocation of the inactivating substance. As with other inhibitors from plant sap, the decrease in number of local lesions was independent of the virus used and depended only on the species of plant to which inoculations were made (Gendron and Kassanis, 1954). McKeen (1956) speculates on the possibility that inhibitors in the cells of virus source plants, presumably in the same cells that contain virus particles, are picked up by insect vectors and injected into the test plant with the virus. For those plants in which the inhibitor functions, the vector might be innocently guilty of reduced efficiency. Simons' (1955) data on the transmission of southern cucumber mosaic by *Myzus persicae* (Sulz.), showing that chard was a better virus source than pepper, provided support for McKeen's speculation.

Milk is a most interesting natural inhibitor in the control of virus infection in greenhouse tomatoes (J. G. Cook, 1957; Crowley, 1958). Hare and Lucas (1959) provided experimental data on the use of milk to prevent contact transmission of tobacco mosaic virus. Control was obtained in the greenhouse on pepper, tomatoes, and tobacco; in the field, on tobacco.

Antibiotics

Antibiotics as virus inhibitors have been tested, and many reduce virus multiplication *in vitro*. J. Johnson (1938) cultured *Aerobacter aerogenes* V. Tieg and *Aspergillus niger* (Kruse) Beijerinck, and from the cultures obtained substances which inhibited tobacco mosaic virus; action was instantaneous. The substances were very stable to alcohol, chloroform, mercuric chloride, and charcoal, and less sensitive to heat than *Phytolacca* juice or trypsin. Beale and Jones' (1951) results with commercial preparations of antibiotics against TMV and potato yellow dwarf were negative. More recent studies have dealt with the effects of more purified antibiotics. The limitations are clearly indicated by Shimomura and Hirai (1959), who tested 30 such compounds and found 10 that showed 20% or more inhibition of tobacco mosaic virus. The floating-leaf-disk method was used, and determinations were made by the chemical method of Bancroft and Curtis (1957). Table 15.5 summarizes the results.

Gray (1955) had reported that an antiviral agent from *Nocardia,* Noformicin, when sprayed on to intact plants infected with bean mosaic or tobacco mosaic viruses, reduced both local lesions and systemic infection; but the effect was not lasting, although there was translocation of the antibiotic from the base of a bean leaf to the tip and in the reverse direction. *In vitro* treatment, curiously enough, had no effects, although Schlegel and Rawlins (1954) found it effective when used on floating leaf disks. As with other inhibitors, it was necessary for the applications to be made close to the inoculation time.

Cytovirin, isolated from a *Streptomyces* species, was also reported on by Gray (1957) and Lucas and Winstead (1958). According to Gray, local lesion formation by southern bean mosaic and tobacco mosaic viruses was completely prevented by spray levels of 0.5–1 ppm. of crystalline Cytovirin. A spray level of 100 ppm. of the crude material protected plants

TABLE 15.5. Comparison of the Chemical Analysis of TMV Multiplication That Is Inhibited by Some of the Antiviral Substances [a]

Antiviral substances	Conc. of chemicals in $\mu g/ml$	Percentage of increase (+) or decrease (−) in TMV multiplication and average percentages of deviation from mean		Effect of substances on leaves
		$(NH_4)_2SO_4$ method [b]	TCA method [c]	
No. 213 [d]	50	-14 ± 3.5	-27 ± 3.5	Slight yellowing
No. 217 [e]	50	-17 ± 1.5	-30 ± 0.0	None
Dextromycin sulfate	50	-29 ± 3.0	-39 ± 8.0	None

[a] After Shimomura and Hirai (1959).
[b] By the ammonium sulfate precipitation method.
[c] By the method of Bancroft and Curtis (1957).
[d] No. 213: benzaldehyde thiosemicarbazone.
[e] No. 217: benzalacetone thiosemicarbazone.

from systemic infections of tomato spotted wilt and tobacco mosaic. Stunting of infected plants was markedly reduced by a postinfection treatment. While these results appear promising, the time factors are the same as for other inhibitors; treatment and inoculation must be fairly close together. Once a virus is fully established systemically, Cytovirin is ineffective.

Blasticidin-S (BeS) taken up through the roots of rice plants inhibited the transmission of rice stripe virus by the leafhopper *Laedelphax striatellus,* reduced transmission by progeny and the rate of transovarial transmission. The material was soon degraded into compounds inert against rice viruses, but 2 days after spray or root treatment, BeS benzylaminobenzene (BAS-H3) was still in effective concentration in both rice shoots and leafhoppers. The material is widely used in Japan against blast disease of rice caused by *Pyricularia oryzae* (Hirai *et al.,* 1968).

Inhibitors

Most of the inhibitors affecting only inocula or leaves treated close to inoculation time are substances with large molecules. In contrast, the analogs of purines and pyrimidines are small molecules. They can diffuse into leaves and inhibit infection as well as reduce the rate of multiplication (Bawden, 1961). Thiouracil has been used extensively in experiments (Holmes, 1955; Kassanis and Tinsley, 1957). This compound will check multiplication, which resumes, however, when the treatment is stopped. Unfortunately, the compound is phytotoxic (Bawden and Kassanis, 1954). Porter and Weinstein (1957, 1960) showed that disturbances in protein and nucleic acid metabolism resulted from thiouracil treatment. However, the resulting interference with virus multiplication was accompanied by too much phytotoxicity to be worthwhile from the control standpoint. Kurtzman *et al.* (1960) tested 14 different purines for their effect on multiplication of tobacco mosaic virus in tissue cultures. 6-Methylpurine was the first compound observed to inhibit virus production without phytotoxicity. It reduced the number of local lesions by 40% at 0.1 mg/liter; 0.25 mg/liter caused tissue damage. At 10 mg/liter symptoms on whole plants were delayed for 7 days. The authors consider that 6-methylpurine has possibilities for practical control.

Hirai *et al.* (1958) tested a series of thiosemicarbazones, but only benzylacetone, unsubstituted, reduced virus multiplication as much as 20%. Kooistra (1959) tested a series of nitrosohydroxyaryl compounds, first in the greenhouse and then in the field. The search was for inhibitors of infection to be inoculated simultaneously with the virus. In field trials with sugar beet, 1-nitroso-2-naphthol gave an average yield improvement of 7% ± 1.5; results with the other compounds were insignificant. On tobacco, nitrosothiouracil mixtures were useful, but on other plants they were too phytotoxic. The beneficial effects of the nitroso compounds were found only at concentrations of 0.1%, with 2 or at most 3 sprayings permissible. Beyond that, phytotoxicity outweighted any beneficial results. Kooistra tabulated all the chemicals reported to have antiviral properties (Table 15.6).

Action of inhibitors on host susceptibility was suggested by Simons *et al.* (1963) to account for the action of sap from succulents on TMV production. An unusual use of sap as an inhibitor was reported from Venezuela. Sap from *Carioca cauliflora* inhibited papaya mosaic, but this was also achieved by grafting *C. cauliflora* on to infected papaya which later gave rise to healthy leaves (Micheletti, 1962).

Inhibitors of virus infection are most commonly reported from laboratory and greenhouse studies, but there are a few practical applications. The use of milk, fat, or corn or mineral oil reduced lettuce mosaic in seedlings (Hein, 1965); TMV was reduced in tomatoes with milk sprays (32% increase in yield) (Denby and Wilks, 1965); fresh skim milk delayed severe potato X and TMV in the field for a few weeks (Jaeger,

TABLE 15.6. Antivirus Compounds [a]

Compound	Virus	Host plant	Method
Thiouracil	TMV	Tobacco	4
			3
			1, 4
			4
			4
			1
	Lucerne mosaic	Tobacco	1, 3
		Beans	
	TMV	Tobacco spp.	3, 4
	Several other		
	viruses	Several other plants	
	Stone fruit	Cucumber	2
	TMV	Tomato	
	Ringspot	Plum tree	5
	Cucumber mosaic	Tobacco	1
Guanazolo (8-azaguanine)	Lucerne mosaic	Clover	1, 3
	TMV	Tobacco (also	
		N. glutinosa)	
		Tomato	
	Stone fruit	Cucumber	2
	TMV	Tomato	
	Virus yellows	Sugar beet	1, 3
	Ringspot	Plum tree	5
Guanazolo (8-azaguanine)	TMV	Tobacco	3
2-Thiocytosine			
2-Thiothymine			
2,6-Diaminopurine			
6-Methylpurine	TMV	Tobacco	7
6-Chloropurine			
Pyrimidines	TMV	Tobacco	4
Thiazole			
Malonic acid			
Nicotinic acid			
Diazouracil	TMV	Tobacco	4
dl- and l-isoleucine			
Several other chemicals			
Malachite green	TMV	Tobacco	4
	TMV	Tomato	3
	Potato virus X		
	Potato virus X	Potato var.	7
		Green mountain	
		(Early Carmen)	

TABLE 15.6. **Continued**

Compound	Virus	Host plant	Method
Malachite green	TMV	Tobacco	4?
Other triphenylmethane dyes, as victoria blue, night blue			
Methylene blue	Tissue tumor	*Rumex acetosa*	7
Crystal violet			
Malachite green			
Neutral red			
Chloramphenicol	Stone fruit	Cucumber	2
	TMV	Tomato	
Naphthalene acetic acid	Tissue tumor	*Rumex acetosa*	7
	TMV	Tobacco	7
	Lettuce big vein	Lettuce seedlings	1
Indolyl acetic acid	Tissue tumor	*Rumex acetosa*	7
Colchicine			
Coconut milk	TMV	Tobacco	7
Naphthalene acetic acid			
Coconut milk	TMV	Tobacco	7
Aspartic acid			
Glutamic acid			
2,4-Dichlorophenoxyacetic acid	TMV	Tobacco	7
Methoxine			
Naphthalene acetic acid			
2,4-Dichlorophenoxyacetic acid	Chrysanthemum mosaic	Chrysanthemum	1
2,4,6-Trichlorophenoxyacetic acid	Lettuce big vein	Lettuce seedlings	1
Indole acetic acid			
Ammonium salt of 2,4-dichlorophenoxyacetic acid	TMV	Tobacco	1, 3, 6
	Tomato aspermy	Chrysanthemum	
3-Chlorophenoxyacetic acid			
Cortisone	TMV	Tobacco	4
Calcium chloride	Carnation mosaic	Carnation	5
			1, 3
	Lettuce big vein	Lettuce seedlings	1
Potassium bromide	TMV	Tobacco	1
Sodium bromide			1
Ammonium bromide			
Sodium chloride	Potato virus X	Seed potatoes	6
Potassium chloride			
Trisodium phosphate	TMV	Tomato seed	6
Sodium hydroxide	TMV	Tomato seed	6
	Cucumber mosaic		

Compound	Virus	Host plant	Method
Mineral salts	TMV	Pinto bean	1, 6
Zinc	TMV	Tobacco	1, 4
Zinc sulfate	Peach X	Peach tree	5
	Carnation mosaic	Carnation	1, 3
	Lettuce big vein	Lettuce seedlings	1
Zinc chloride	TMV	Tobacco	4
Zinc sulfate	Carnation mosaic	*Dianthus barbatus*	
	Strawberry virus 2	Strawberry	1
Nitrous acid (HNO_2)	TMV	Tobacco	5
	Tissue tumor	Clover	
Milk	TMV	Tomato	3
Trichothecin	TMV	Tobacco	3, 4
Ribonuclease	TMV	Tobacco	2
4-Chloro-3,5-dimethyl-phenoxyethanol	TMV	Tobacco	1
Tannic acid	TMV	Tomato	3
Streptomycin	TMV	Tobacco	4
	Necrosis ringspot	Cowpea	
Vitamin B_1	TMV	Tobacco	4
		N. glutinosa	
MK 61 (antibiotic from *Nocordia* species)	TMV	Tobacco	4
Noformicin (antibiotic)	TMV	Tobacco	1, 3
	Southern bean mosaic	Pinto bean	
Cytovirin (antibiotic)	TMV	Tobacco	1, 3
	Southern bean mosaic	Pinto bean	
Quinhydrones	Peach X	Peach tree	5
8-Hydroxyquinoline			
Calcium hydroxyquinoline			
Magnesium hydroxyquinoline			
o-Nitrophenol			
Urea			
Sodium thiosulfate			
p-Aminobenzene sulfanilamide			
p-Aminobenzene sulfanilamide mixed with maltose or dextrose	Peach X	Peach tree	5
p-Toluenesulfanilamide			
Hydroquinone			

TABLE 15.6. Continued

Compound	Virus	Host plant	Method
Maltose			
Dextrose			
Sulfaguanidine	Carnation mosaic	Carnation	5
Sulfasuxidine			
Sulfathalidine			
Sulfamerizine			
p-Aminobenzene- sulfoxylamide			
Sodium diethyldithio- carbamate			
Hydroxyquinoline sulfate	Carnation mosaic	Carnation	1, 3
Sulfathiazole			
Sulfaguanidine			
Sulfasuxidine			
Sulfathalicidine			
Sulfamerizine			

Ethanol containing crotonal-
    dehyde and possibly small
    quantities of other alde-
    hydes, higher alcohols,
    and/or methanol, and/or
    pyridine bases
	Several viruses	Several plants	1, 3

Sodium salt of 4-sulfon-
    amidephenyl
Azo-7-acetylamino-
    1-oxynaphthaline
	TMV	Tobacco	1

1966). On the other hand, negative results with milk and rice polish against potato Y, and a significant increase in infection when they were used, was reported by Shands and Simpson (1964); the results are ascribed to inhibition or stimulation of vector feeding, not to effects on the virus itself.

Miscellaneous Chemical Treatments

The empirical treatment of virus diseased plants with various chemicals could be expected, as also could the great variability in results reported.

The earliest reported attempt seems to be that of Stoddard (1942) who records cure of X-disease in peach buds by soaking in water solutions of quinhydrone, urea, and sodium thiosulfate.

Compound	Virus	Host plant	Method
3,6-disulfonic acid (Prontosil)			
Radiophosphorus (P[32])	TMV	Tobacco	4
Furfuryl-5-chloro-2-methyl carbanilate	Southern bean mosaic	Pinto bean	5
Furfuryl carbanilate			
Nemagon	Peach yellow bud mosaic	Peach tree	1
Carbon bisulfide			
Methyl bromide			
DD mixture			
Vapam			

[a] From Kooistra (1959).
Method: 1. Absorption of the chemotherapeutic substances through the root system. 2. Vacuum infiltration, mostly of the leaves. 3. Spraying. 4. Floating the detached leaves or leaf disks on the solution of the chemical under test, whether or not solidified with agar. 5. Wrapping a cotton pad drenched with the compound round the shoots; applying the chemical at a certain spot of the plant in a lanoline paste, or through a woollen thread passed into the stem; dipping an intact bud or leaf in a solution of the compound for a certain time; injecting the plants with the chemical, etc. 6. Treatment of seeds, seed potatoes, or beet stecklings with a solution of the compound. 7. Tissue cultures, the compound added to the tissue nutrient solution.

Treatment of leaf roll-infected potato tubers with sodium sulfide, potassium hydroxide, potassium permanganate, and ascorbic acid were uniformly unsuccessful (Köhler and Hauschild, 1950).

Treatment of infected, unrooted, carnation cuttings with 0.2% calcium chloride and 0.25% zinc sulfate eliminated carnation mosaic from over 50% of them. Phytotoxicity was such that the treatment was deemed impractical, but pre- and post-treatment conditioning could possibly have reduced this (Rumley and W. D. Thomas, 1951). Highly significant reduction in activity of the same carnation virus was obtained by injection of infected plants with a number of sulfanilamide compounds (W. D. Thomas and Baker, 1949). Another sulfa drug (Prontosil) had some effect in culture solution in which infected tobacco plants (tobacco mosaic and potato virus X) were grown (Rodriguez, 1950). This, however, was more likely an effect on the host rather than on the virus. Carter (1939) fed Prontosil to infective leafhoppers (*Peregrinus maidis* (Ashm.)) until the red color was diffused throughout the insect's body and was ejected in large quantity in the feces, but without affecting the infectiveness of the insect. Couranjou's experiments with citrus psorosis (1951) would seem to be another case of treatment affecting either plant susceptibility or tolerance. According to Couranjou, only trees suffering nutrient deficiency were attacked in North Africa; when the deficiency was corrected by the applicaton of potassium permanganate applied around the roots three times a year for 4 yr at the rate of 3 kg per tree, "a complete and permanent cure" was effected.

Similar results were reported using Trucidor, a polynitrocycloaralcoyloxbenzene, applied at the rate of 200–500 g per tree; 418 badly affected 18-yr old Thompson navel orange trees were treated with 500 g Trucidor per tree. Six mo later, 70% were "completely cured," with evidence of psorosis still remaining on only the most severely affected trees. It is surprising that this apparently unequivocal result has gone unnoticed.

This situation may be analogous to that obtaining in Nigeria, where capsid (Miridae) control restored the leafy canopy of trees infected with swollen shoot virus. Unsprayed trees were moribund (Fig. 15.3 and 15.4). The virus strain is a mild one by comparison with those in Ghana, but it is clear that in the absence of the debilitating toxic effect of capsid feeding, the trees can tolerate the virus and produce an economic yield even though the trees remain infected (Thresh, 1960).

Dyes have been used in an attempt to control virus diseases, but the results lack uniformity. Nickell (1951) reported transient inhibition of growth of virus tumor tissue in culture by methylene blue, crystal violet, and malachite green at concentrations above 0.01 ppm. Takahashi (1948) showed that tobacco mosaic synthesis was inhibited by malachite green. Norris' results (1953, 1954) with malachite green and potato virus X were questioned by Thomson (1956), who suggested that the stem apices found healthy by Norris after treatment with the dye had not been reached by the virus in the first place. Spectrophotometric titer of tobacco mosaic was not affected by malachite green (Kirkpatrick and Lindner, 1954).

Growth substances may find a use in control, but the results from experiments with gibberellic acid suggest that the treatment resupplies a deficiency of growth elongation hormones brought about as one aspect of the infection. The effect finally wears off but is favorable as long as it lasts (Chessin, 1957; Maramorosch, 1957; J. Hull and Klos, 1958).

A number of growth regulators were tried by Rich (1956) in an attempt to control lettuce big vein. Seedlings were watered with solutions of the chemicals before being transplanted. Naphthaleneacetic acid at 50 and 100 ppm, indoleacetic acid at 50 ppm, 2, 4, 6-trichlorophenoxyacetic acid at 50 ppm, calcium chloride at 1000 ppm, and zinc sulfate at 250 ppm reduced disease significantly. In addition, the zinc sulfate increased

FIGURE 15.3. Effect of insecticidal control of capsides (Miridae) on regeneration of cacao trees infected with swollen shoot virus. Unsprayed plot (Photograph: West African Cacao Research Institute, Ibadan, Nigeria).

fresh weight significantly over the check.

The nonpersistent and some semipersistent viruses continue to be the object of oil treatments for control of the diseases. A successful use of oil for protecting against stylet-borne viruses (potato X) was achieved by sprays of oil (viscosity 90–105 S.U. sec at $100°F$). Six applications reduced yield in 1 yr, presumably due to phytotoxicity, but not at all in another. With 3 applications, no reduction occurred in 3 yr testing (Bradley *et al.,* 1966). Later, Bradley (1968) summarized results over a period of years. With about 100 stylet-borne viruses known, many causing serious losses, practical control by the use of oil sprays could generally be important. Results showing reduction in virus spread from 50 to 90% have been recorded, a considerable improvement over conventional aphicides. Weekly oil sprays appear to be necessary; no cost data are included.

Sugar beet yellows symptoms are suppressed by oil, but the effect is local, not systemic. Neither the movement of virus nor its transmission by vectors was interfered with (Crane and Calpouzos, 1969). Suppression of symptoms for three weeks is encouraging but of no apparent economic importance.

FIGURE 15.4. Effect of insecticidal control of capsids (Miridae) on regeneration of cacao trees infected with swollen shoot virus. Sprayed plots (Photograph: West African Cacao Research Institute, Ibadan, Nigeria).

Control of primary infecton of nonpersistent and semipersistent viruses was accomplished with oil; mixtures with insecticides were ideal combinations (Vanderveken *et al.,* 1968). Vanderveken and Vilain (1967) found that inhibition of transmission was greatest when both source and indicator plants were sprayed.

Source plant spraying alone was more effective with the nonpersistent viruses; with severe beet yellows, transmission was also inhibited when the aphids were allowed to feed on healthy oil-sprayed plants for 1–24 hr before the acquisition feed or between acquisition and inoculation feedings.

Uptake of lettuce and beet mosaics was impeded when oil was sprayed on the lower leaf surfaces and *M. persicae* placed on the upper surfaces. In sugar beets, oil penetrated subepidermal intercellular spaces and into substantial cavities of the untreated upper surfaces within 15 min (Kulp, 1968). Watson (1967) has pointd out that nonpersistent viruses, acquired and transmitted by aphids within a very short time, are very difficult to control by insecticides, and disease may actually be increased because of aphid irritation. However, oil sprays, while having some insecticidal effect, having been used for years for that purpose, apparently serve as prophylactics against infection and for that reason occupy a unique position.

CHEMICAL CONTROL OF VECTORS

Broadbent (1957) reviewed progress in this field up to February 1956, but was able to encompass the subject, introduction and all, in 15 pages. This was a good measure of the slow progress made up to that time with insecticidal control of vectors as a means of controlling virus diseases of plants. His latest review (1969) makes only passing reference to use of chemicals for vector control. Accumulated experience during the last decade, however, indicates that there is considerable promise of satisfactory results from chemical control of vectors in some cases, and also that the organophosphorus compounds have been widely used for this purpose. A selection of reported results is shown in Table 15.7.

There are certain basic difficulties which are not related to the effectiveness of insecticides *per se*. There is probably no virus vector known for which an effective insecticide is not available. When nonvector insects or noninfective vectors infest a crop, damage is usually proportional to the number of insects and the length of time they feed, unless there are virus sources within the field. In this latter case, the spread of such viruses will be determined by the vector's biology, its intracrop movements, and the characteristics of the virus or viruses available for acquisition. When infective vectors fly in from some outside virus source, the problem is rendered much more difficult.

Timing of spray applications becomes important, but proper timing is of little use if flights continue over a long period of time. Since aphids are the most ubiquitous vectors of plant viruses, it will be useful to consider in what manner their biology and vector-virus relationships affect control measures. With a few exceptions (beet stecklings and stored flower bulbs), crops are infested by winged forms flying in from outside. Their apterous progeny move but little, but when new alatae are developed, spread within the field can be widespread.

This again is influenced by the type of virus involved. If the virus is nonpersistent, it can be acquired by the aphid in a minute's feeding and can be transmitted in an equally short time to a healthy plant. The fact that many of the nonpersistent viruses occur in the epidermal cells of the infected host (Bawden *et al.*, 1954) particularly adapts them to easy access by the aphid. To prevent the spread of such viruses requires an insecticide that will kill incoming alatae quickly and yet be persistent enough in its residual action to remain effective over a relatively long flight pe-

TABLE 15.7. *Control of Virus Disease by Chemical Control of Their Vectors*

Disease	Vector	Treatment	Notes	Reference
Barley yellow dwarf	*Rhopalosiphon padi*	Metasystox 2.5 oz/ac	13 bu/ac increase in wheat yld.	Smith (1964)
	Rhopalosiphon padi	Systemic phosphates 1 spray	5–24 bu/ac increase in wheat yld.	Smith (1963)
Bean yellow mosaic	*Myzus persicae* *Aphis craccivora*	Systemic phosphates	Dimethoate reduced infection from 81% to 13%	Leuck *et al.* (1962)
Beet mosaic	Aphids	Metasystox-R; 12 oz/ac 40 gal/ac 3 times	9.5 tons/ac increase in May plantings; none earlier	Hills *et al.* (1964)
Beet western yellows	Aphids	Metasystox-R; 12 oz/ac 40 gal/ac 3 times	9.5 tons/ac increase in May plantings; none earlier	Hills *et al.* (1964)
Beet yellows	Aphids	Preplanting Disyston 2 lb/ac+Disyston 8 oz/ac/44 gal once	Effective	Sylvester *et al.* (1961)
Beet yellows	Aphids	Metasystox	20% increase in yield	Thielemann (1962)
Beet yellows	*Aphis fabae*	3–4 sprays Thimeton	80–90% disease reduction	Smruz (1962)
Citrus tristeza	*Toxoptera citricidus* and *T. aurantica*	Metasystox 0.1% Thimeton 0.05%	Infected trees should be sprayed in early flush	Tao and Tan (1961)
Curlytop cantaloupe snap beans	*Circulifer tenellus*	Phorate+4–8% sugarbeet juice	Reduced incidence sometimes	Hills *et al.* (1964)
	Circulifer tenellus	Dimethoate 0.7 lb/ac +5% beet juice or sugar	Six-fold increase	Peay and Oliver (1964)
Lettuce yellows	*Macrosteles fascifrons*	Systemic phosphates at planting	Effective for 4–5 weeks	Thompson (1961)

Disease	Vector	Treatment	Result	Reference
Potato leaf roll	*Myzus persicae*	Systemic phosphates	2 applications after plants emerged stopped field spread	Van der Wolfe (1964)
Potato leaf roll and potato Y	*Aphis gossypii* *Myzus persicae*	DDT; Endrin; Systemic phosphates Weekly sprays	Incidence reduced	Nirula (1962)
Phony peach	*Homalodisca coagulata* *Cuerna costalis*	Disyston granules 150 gm 10%/tree	100% mortality in 1 day 100% mortality in 3 days	Kaloostian and Pollard (1962) Kaloostian and Pollard (1962)
Peach mosaic	*Eriophyes insidiosus*	Diazinon 1 lb/50% W.P. 100/gal/ac once yearly	Spread restricted	Jones *et al.* (1970)
Reversion of black currant	*Phytopus ribis*	Endrin 6 sprays 0.4%	Mites eliminated; disease reduced	Thresh (1965)
Rice virus disease	*N. impicticeps*	Carbaryl phosphamidon Phorate 2.75 lb/ac 15 day intervals	6-fold increase	Pathak *et al.* (1967)
Rubus stunt	*Macropsis fuscula*	Tar distillate ovacide Parathion in spring	40–80% infection reduced to 1% in 3 yr	Van der Meer and de Fluiter (1962)
Sugarcane mosaic	*Hysteroneura setariae*	Insecticides	Complete control of vector No reduction in disease	Zumme and Charpentier (1965)
Tomato spotted wilt	*Thrips tabaci*	Autumn ploughing BHC 45 lb/ac/12%	Incidence reduced	Lulov (1961)
Tomato yellow top	*Bemisia tabaci*	Endrin. 6 sprays at 5-day intervals	About 90% reduction in disease incidence	Cohen *et al.* (1963)
Arabis mosaic Raspberry ring-spot Tomato blackring	*Xiphinema diversicaudatum*	DD; MB at 2 lb 100 sq. feet	Complete disease control	Harrison *et al.* (1963)

riod. Such a chemical has yet to be discovered.

On the other hand, with the persistent viruses, time factors impose certain limitations on spread by vectors which have a less casual vector-virus relationship. The process of acquisition takes time, for the virus, often predominantly in the phloem, must be picked up by the vector, pass through the gut into the hemocoele, and thence to the salivary glands before the insect is infective. Once this process has been completed, the vector is doubly dangerous, for it can infect a series of plants in succession without further access to a fresh virus source. Again, suitable insecticides are those with long residual activity, and even if they were available, there are legal limitations to their use, at least in the United States.

Leafhoppers, thrips, and other virus vectors are involved in the transmission of persistent viruses and long-residual insecticides are essential. It is quite clear that each disease presents a separate and distinct problem, and the extent to which success has been achieved can be judged from consideration of some of the reported cases.

Potato Diseases

The nonpersistent aphid-borne diseases of potato appear to be poor subjects for insecticidal control. In spite of weekly sprays with parathion, Emilsson and Castberg (1952) failed to reduce the spread of virus Y. Even when the frequency of parathion spraying was increased to once every 2 days, neither virus Y nor virus A was prevented from spreading (Hey, 1952). Munster and Murbach (1952) reduced aphid populations by 90% by mid-July by using 0.2% Pestox 3H, a systemic insecticide, at 14-, 21-, and 28-day intervals, beginning June 5. When the haulms were removed August 1, viruses X, Y, and leaf roll were reduced slightly. When the haulms were not removed, the sprayed plots showed an increase in infection. Spraying was ineffective in the critical period between June 1 and 20. Actually, spraying disturbed the insects and caused more than normal movements between plants. Spraying was not recommended. Potato leaf roll, being a persistent virus and requiring some 56 hr after acquisition before it can be transmitted, has proved to be more amenable to insecticidal control.

DDT, when used by growers, generally in an area with cooperative dusting programs, reduced leaf roll in Washington state to very low percentages; in one variety, White Rose, to zero (King, 1952). This was an accumulated result achieved with potato seed stocks over a period of 5 yr. Gibson *et al.* (1951) reported excellent control of aphids using DDT, sulfur dusts, or Systox, but incidence of leaf roll was high because of incoming infective migrants (Fernow and Kerr, 1953).

Reduction of the spread of leaf roll in central Washington state through control of *M. persicae* has varied from considerable to none at all. In some cases, aphid control measures have resulted in an increase of disease spread. Klostermeyer (1959) explains this on the basis of the size of the aphid migration and the reduced attractiveness to winged migrants of plants heavily infested with apterous aphids. When the aphid migration is low and of short duration, apterous colonies are small and winged aphids will continue to alight on the same plants. Under these circumstances, aphid control results in reduced leaf roll spread. When, however, the aphid migration is large, plants become heavily populated with apterae and are not attractive to winged migrants. Aphid control reduces the populations of apterae, and the plants are attractive to winged migrants which may be already infective from feeding on diseased plants in early potato fields. Leaf roll may, therefore, spread rapidly, and the net result is an increase in leaf roll even though the spray controlled the aphid population. There was an inverse relationship between aphid populations and percent net necrosis (which was used by Klostermeyer as an index of leaf roll) for all the insecticides used—Schradan, Diazinon, and Parathion.

Spread of leaf roll within the field, but not from outside, was greatly reduced by Systox applications, which did not, however, affect the spread of virus Y. Because of the operator hazards involved, it was recommended that spraying be limited to individual infected or suspected plants, with dye added to the spray to identify the plants for later roguing (Lambers *et al.,* 1953). It is difficult to follow the logic of this recommendation.

Broadbent *et al.* (1960) showed that spraying with DDT (4 sprays of emulsion at 2 lb active DDT/ac at 14-day intervals) was enough to prevent secondary spread of leaf roll and virus Y, but it did not prevent infection by flights from outside. This risk will remain until all potato crops are freed from virus, but if spraying could be done generally before the summer dispersal of aphids, the risk would be materially reduced.

The incorporation of Thimet granular in a fertilizer mixture ($3\frac{1}{3}$ lb–250 lb of 12-12-12) failed to reduce virus Y infection, but it significantly reduced purple-top wilt (aster yellows virus) (Hoyman, 1958).

The use of soil insecticides to control potato aphids was tried by Burt *et al.* (1960). There might be a number of advantages to such a practice: avoiding mechanical damage to growing plants, avoiding compaction due to machinery, and preventing infection of plants when they first emerge. Thimet and Rogor, both systemic organophosphorus compounds, were used either in furrow application with the fertilizer or as individual doses beneath each tuber. All treatments prevented or decreased leaf roll but had only minor effects on virus Y. A remarkable observation was that the harvested tubers contained some residues, although very small; shoots from these tubers carried fewer aphids than did those on the controls, although the controls grew more rapidly than did the treated ones. Residue hazards make the method undesirable for table potatoes, but for seed crops it has promise.

Sugar Beets

Disease control in sugar beets by the use of insecticides has probably received as much or more attention than it has in potatoes, although one major disease, virus yellows, has been known for a much shorter time. Much of this work has been done in Europe; the American point of view is somewhat negative, since in all the spray tests recorded by Bennett (1960), the cost exceeded the value of the increased yields obtained. McLean (1957) showed in greenhouse experiments that Thimet and Demeton, both systemic organophosphorus compounds, had residual action for 18–24 days. Aphids, however, were infective for 80–100 min after feeding on an insecticide-treated plant; since such feeding causes irritation before death ensues, interplant movement is stimulated, with consequent virus spread.

In Great Britain, Hull has published an annual summary of the sugar beet yellows situation since 1953 (R. Hull, 1954–1959). It is clear that the value of spraying is determined largely by the size of the migrating aphid populations, the time of the spring flight, and the recognition by farmers of the necessity for spraying at the right time, that is, when the aphid infestations become noticeable. When, however, spring flights are unusually heavy and early, heavy losses can occur in spite of the spray, as in 1947, when one million tons of roots were lost.

With the advent of methyl demeton, results have improved, and it is estimated that a single spray will increase yields 7–8% in crops infected with yellows. Sugar content and juice purity are also improved, but R. Hull (1958) concludes that complete control is not yet possible. However, in 1959 in Great Britain, from 75 to 90% of the acreage in sugar beets was sprayed, and the practice is routine on many farms (Hull, 1959).

In the province of Limberg, Netherlands, 2 sprays of Systox reduced incidence of yellows by 50%, and 1 spray is considered profitable (Rietberg and Hijner, 1956).

Sugar Beet Curlytop

On sugar beets curlytop has been controlled by means of resistant varieties, but for some other crops for which no resistant varieties are available, vector control has been attempted on a large scale. The method was to attack the leafhopper populations at their source, that is, in the breeding areas in the desert foothills in California (Wallace, 1948). There was some controversy on the effectiveness of this program, but it has been continued. The original spray formulation, pyrethrum in diesel oil, was changed to an oil DDT with the advent of DDT. W. C. Cook (1943) attempted to evaluate the program, but in the absence of untreated areas this was not possible. Cook suggested that a maximum safe population could be set for each breeding area, and that control, to be effective, must reduce the population below that level. However, the spraying of large areas has its logistics to contend with, and the elaborate quantitative surveys necessary to meet Cook's suggestion, even if adequately interpreted, would actually increase complication. Armitage (1952) has described the program in more detail. There are three spray periods during the period from November 1 to April 20. The spray is a solution of DDT in oil and is applied at the rate of 2 gal /ac, including 1 lb actual DDT, either by plane or ground machine of a turbine blower type. The first spraying starts when the leafhoppers move from the matured and dry agricultural crops on the valley floor and accumulate on perennials (*Atriplex*) while awaiting the germination of the winter hosts. This phase may last for 6 weeks. When the winter hosts spring up on the floors of the canyons following the first winter rains, there is a second concentration and a second spraying. The third spraying is applied when the remaining females concentrate on sparse growth on low knolls where the spring or flight generation is developed. This area is also sprayed. The survivors of this generation move into the cultivated area, sometime between April 15 and April 20, en masse. Planting time for many crops can be delayed until after the flight has occurred, but early tomatoes are an exception, and growers depend on a high leafhopper kill for protection.

The problem can be intensified by drought cycles which permit the development of thousands of areas of Russian thistle on the grazing land on the valley floor. If the huge populations developed on these plants move to the foothills, the problem is out of control. Additional spraying is, therefore, necessary during a limited period in October. In 1951, 150,000 acres of such lands were sprayed. In wet years, the extent of spray operation is much reduced. The $300,000 spent per year is balanced by the fact that it amounts to less than ½ of 1% of the average return to growers.

On the smaller scale, Douglass *et al.* (1955) controlled leafhoppers in Idaho in areas contiguous to snap bean growing. It was estimated that curlytop losses were reduced from 11 to 1.9%. Treatment of the bean crop was not successful (Deen and Hallock, 1954), although there was some decrease in curlytop, and a yield increase. Hills *et al.* (1960) used systemic phosphates to protect beet plants grown for seed. Both seed and soil treatments were used, and protection was afforded up to the 4- to 6-leaf stage.

Curlytop and beet western yellows control were attempted by Landis *et al.* (1970) by seed treatment. Phorate and disulfoton were too toxic at effective doses, but disulfoton gave some protection against curlytop for 3 mo but none with *M. persicae* and beet western yellows. A stimulating effect on aphid populations was noted and ascribed to the homoligosis theory, that is, that stimulation could result from a stressing factor.

Control of the beet yellows viruses in sugar beet in the Davis area of California (BYV, BWYV) has been tested, using 4 variables, (1) delayed planting, (2) insecticides, (3) resistant variety, and (4) a combination of resistant variety and insecticides. The results, expressed in terms of yield, were as follows: delayed planting 0%; insecticide 13%; resistant variety 14%; and a combination of the previous two treatments, 39% increase. Beet mosaic virus, also serious in the same growing area, was apparently not affected by the treatments. The resistant variety used was USDAH9B, which is resistant to the yellows diseases but not to BMV.

Fewer numbers of aphids were found on the resistant variety, but the lack of any difference in the incidence of BMV was ascribed to the ability of *M. persicae* to transmit that virus with shorter penetration times than is necessary for the two yellows viruses. The failure of delayed planting to result in any increased yield is accounted for on the grounds that the sugar beet is a plant which requires rapid and continued vegetative growth, and delay in planting cuts down the growing period.

Some 30,000 acres of sugar beet are overwintered in the area to provide spring harvest. In spite of the huge reservoir of infection thus provided, the authors believe that sugar beet production in northern California could be improved by use of the USDA yellows resistant variety USDAH9B, planting not later than April 1, and using an insecticide to prevent aphid feeding through May (Hills *et al.,* 1969). Thiabendazole sprays affect the susceptibility of sugar beets to yellowing viruses and also affect the vector, *M. persicae*. According to Russell (1968), however, the reduced infection following foliar sprays is not due to insecticidal action. Aphids transferred to sprayed plants, or spraying aphids before transfer to unsprayed plants, showed reduced fecundity and much reduced nymphal numbers; the adults were not apparently otherwise affected. But these events are surely an example of a rather sophisticated insecticidal effect and one which might well fit into integrated control programs.

Miscellaneous Crops

Vector control attempts are reported against thrips transmitting tomato spotted wilt virus (Costa *et al.,* 1950). Parathion was the most promising.

Michelbacher *et al.* (1950) used DDT dust five or six times, but felt that disease reduction was not proportional to reduction in thrips numbers, and they quite rightly concluded that migratory thrips were responsible. Attempts to control yellow spot in pineapple by insecticides have failed completely thus far and for the same reason, viz., that infection is by migratory thrips, and infective populations do not build up in the crop itself.

Strawberry yellows spread was controlled better by dusting insecticides to control the aphid vector than by certifying stock and roguing, according to Breakey and Campbell (1951).

X-disease of peach transmitted by *Scaphytopius acutus* (Say) was not controlled by vector control. The presence of vectors over a long period during the year precludes successful control by sprays (Palmiter and Adams, 1957). Ornamental stock could not be protected against mosaic if migrating populations were high, despite effectiveness of the aphicides used (Jefferson and Eads, 1951). Swollen shoot of cacao, mealybug-transmitted, presents another widespread and difficult problem in vector control. Contact insecticides are of little use because most of the mealybugs in the canopy are under ant-built tents. Attention has, therefore, been directed to the use of systemic insecticides. Hanna *et al.* (1955) used Dimefox as soil application and trunk implantation, and judged by reductions in mealybug populations that the treatments were satisfactory. However, soil applications would not be likely to be effective in the dry season, and the trunk implantations actually caused considerable mechanical damage to the trees. Furthermore, taste tests of chocolate made from beans from treated trees indicated that off-flavors had developed.

There is, perhaps, a place for vector control, but my opinion is that it will be best used as a supplement to the cutting-out program, as a peripheral treatment around swollen shoot outbreaks.

The mealybugs can be indirectly controlled by controlling ants. This was attempted on a fairly large scale, using Dieldrin emulsion at 1 lb actual per ac, and was eminently successful. The side effects—increase of other pests—were serious, however; some formicide with a narrow spectrum of effectiveness would need to be found. This control also would be used only in the same limited peripheral areas as proposed for the systemic insecticides in order to be practical. Tolerant varieties might well eliminate the necessity for any insecticidal control.

Systox and American Cyanamid 12008 were used in experiments on barley yellow dwarf disease on oats as drenches, foliar sprays, and seed treatment. Good control of aphids was obtained, but there was no protection against barley yellow dwarf (Pizarro and Arny, 1958). Dickason *et al.* (1960) reported essentially the same results when barley was treated.

Insecticide control can be indirect. Ant attended colonies of *Toxoptera graminum* are prolific spreaders of barley yellow dwarf beginning at field edges (Orlob, 1963). Controlling the ants as a measure of control for disease spread does not appear to have been tried, but it should have possibilities.

Sometimes, control measures are complex. A case in point is that of banana mosaic (cucumber mosaic strain). Plantains are removed from the vicinity; infected mats of bananas are eradicated along with 80 adjoining ones. The areas so treated and 25 surrounding ac are sprayed with nicotine sulphate (1 pt/40% 100 gal at 200 gal/ac) three times at 20-day intervals. No banana plants are allowed to grow for 6 mo after any neighboring infections occur. Disease incidence was reduced on a large plantation from 21.5 infections per 100 mats to 0.12 (Adam, 1962). This is obviously a method acceptable only to large corporation agriculture, and even then it seems unnecessarily heroic; another plantation in the same geographical area achieved control by roguing young infected plants and replanting after a short interval. The basis for this latter control measure is the fact that only very young plants are susceptible.

SOIL TREATMENT. Soil treatment for control of soil-borne viruses goes back many years. McKinney (1923) saturated infested soil with a 2% solution of 40% formaldehyde, let the soil aerate for 5 wk before seeding, and found that wheat mosaic did not occur. F. Johnson (1945) used a number of volatile compounds as soil treatments: e.g., granulated calcium cyanide, carbon disulfide, chloropicrin, methyl bromide, rotenone, and naphthalene, all of which prevented infection with wheat mosaic. McKinney *et al.* (1957) found that soil was easily rendered wheat-mosaic-free by treatment with formaldehyde, chloropicrin, carbon disulfide, dichloropropene-dichloropropane (DD), or ethyl alcohol. Toluene was less effective. These authors came to the logical conclusion that all their evidence suggested the existence of a vector in which the virus overseasons. Peach yellow bud mosaic virus was eliminated from soil by the use of volatile soil fumigants (Wagnon and Traylor, 1957). These authors used soil from two depths, the top 4 in. and from a lower layer, 14 in. in depth. Seedlings grown in the untreated lower soil developed the disease, but those in the treated lower soil or the untreated upper soil did not. Again these results suggest a vector rather than a direct effect on the virus itself.

Heat treatment of soil inactivates wheat mosaic virus (F. Johnson, 1942). Heating soil to 40 or 50°C had no effect, but at 60, 70, and 100°C all the plants subsequently grown in the soil were healthy.

Nematode Vector Control

During the last decade, the role of nematodes in the transmission of soilborne viruses has been fully established, and success with control of these viruses by soil fumigation can be clearly ascribed to reduction in the vector populations. *Xiphenema diversicaudatum* Micol. populations were so greatly reduced that it was anticipated that buildup to damaging levels would be long delayed (Harrison *et al.*, 1963). The fumigants used were DD, or methyl bromide. The same fumigants, as well as quintozene, successfully controlled tomato blackring and raspberry ringspot in strawberries (Murant and Taylor, 1965) (See also Taylor and Murant, 1968). Fumigation, applied preplanting, controlled transmission of peach rosette mosaic by the nematodes, *X. americanum* and *Criconemoides* spp. (Klos *et al.*, 1967).

The search for new nematocides is currently a very active field; some of the newer ones are dry materials. A unique example of the volatile fumigant is DBCP (Nemagon) which can be applied to the soil around growing plants. Its use to maintain freedom from nematode activity after preplanting fumigation is well established in Hawaii (Schmidt, 1960), and extension to other tropical crops can be expected (Schmidt, 1969).

VECTOR-HOST-PLANT CONTROL

Control of vector hosts by chemicals has, no doubt, occurred frequently as a result of chemical weed sprays. One unique case, however, is that of the use of mangolds and beets when they are clamped for the winter. During this period of storage, the roots normally sprout rosettes of etiolated leaves, and *M persicae* will not only survive the winter on the leaves but will develop large colonies by the time the clamps are opened in the spring.

Spraying the growing crop with up to 1% maleic hydrazide in a triethanolamine formulation, 1 or 2 mo before the roots were dug, practically stopped the formulation of sprouts after the roots were clamped. October sprays of 0.25% were the most effective (Cornford, 1955) (Fig. 15.5).

Attacking virus vectors at their source, that is on wild vegetation, can sometimes bring a measure of control, and it has been proposed as a possible solution in other cases. Lettuce necrotic vein virus depends for its spread on vector populations developed on *Sonchus olericus*, sowthistle, growing in the vicinity. Eradication of the weed for a distance of 170–470 yards surrounding a lettuce field reduced infection from 75 to 6% (Stubbs *et al.*, 1963). Applying this method to a large area where many proverbially individualistic growers operate is difficult, but successful demonstration is usually convincing.

Aster yellows on celery was not controlled by measures applied to the crop, but elimination of the vector in adjoining natural breeding

FIGURE 15.5. The use of maleic hydrazide to depress shoot growth on clamped mangolds reduces the leaf area available to aphids overwintering in the clamps (Photograph: Cornford, 1955; Rothamsted Experiment Station).

grounds before migration occurred was proposed by Freitag *et al.* (1962). A cooperative pilot scheme over 6 years reduced loss from 30% to less than 5%. This control measure will remain valid even if the causal agent proves to be a mycoplasma.

Another successful case of control by alternate host destruction was reported by Wallis and Turner (1969) who burned drainage ditch banks harboring host plants of the vector of beet western yellows virus *M. persicae*. Aphid numbers were reduced by 51–91% and the number of diseased beets by 77–84%. Increase in yield was calculated to be 1.5–2.3 tons/ac at a cost of $2.50–$6.85/ac of beets protected. Even at the higher cost figure, the operation was profitable.

When the source is a crop being grown for seed, as in the case of sugar beet, the root crop nearby can be seriously affected, since control of the vectors on the seed crop is difficult. One suggestion is to grow seed crops every other year; this would presumably cut the danger in half (Rib-bands, 1963), but it also would cut the profit on the seed crop in half. In view of the latter control measure, however, stecklings are now grown without site restrictions, although application of insecticides is still neces-sary if the stecklings are grown on the same site as root crops.

Replacement Control

Replacement control has had limited opportunity, but it is an ecologi-cally sound method of controlling vector populations on wild hosts where the latter are predominant through overgrazing or the abandonment of marginal farm land. The beginning is the deterioration of the plant cover. This comes from misuse of land, overgrazing, and damage by ro-dents until the land is laid bare. Weeds fill in, and following these weeds come the insects and plant diseases. Piemeisel *et al.* (1951) and Piemeisel (1954) have described the situation as it has developed in the intermoun-tain states and California, where the sugar beet leafhopper has found its principal breeding grounds in just such situations. The remedy, accord-ing to these authors, is to permit the natural replacement of the original vegetation and to hasten it by seeding the appropriate grasses. The prob-lem is, of course, a complex one, but intelligent range management will do much to bring about a solution. If launched on the scale that is needed, it could bring vast areas back into productivity and at the same time practically eliminate the breeding grounds of many noxious insects and diseases, of which the sugar beet leafhopper and curlytop virus are the prime examples in the areas studied (Fig. 15.6) (See Chapter 14).

RESISTANT AND TOLERANT VARIETIES

The use of resistant varieties is an ideal way of effecting a measure, often considerable, of virus disease control. Complete immunity would, of course, be desirable, but there is little evidence that it occurs very often. The seedling potato variety USDA 41956 has long been considered im-mune from virus X, but Benson and Hooker (1960) recovered virus X from the below-ground portions of this seedling as well as from Tawa and Saco varieties, also considered immune. This finding does not negate the usefulness of these varieties for breeding purposes; it does mean that the term "immunity" has been used to designate what is actually a high degree of resistance or tolerance.

Definitions are necessary at this point. Immunity infers complete ab-sence of infection; resistance is the ability to ward off infection; the toler-

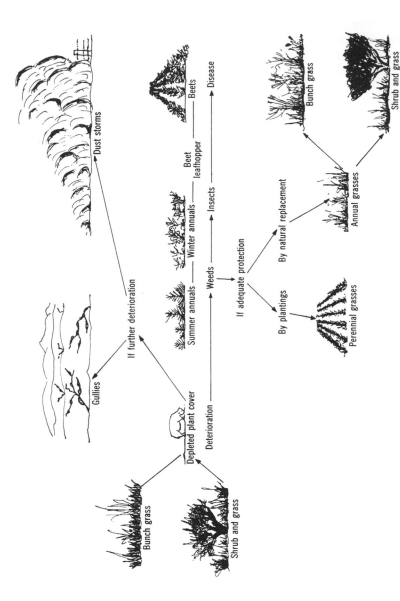

FIGURE 15.6. Replacement control of virus-vector host plants. A pictorial diagram illustrating the sequence of events when land is misused and when it is given adequate protection (From Piemeisel *et al.*, 1951).

ant variety is one that can survive and produce economic yields in spite of infection. The symptomless carrier of a virus represents tolerance in its most extreme form. There are some subdivisions of these definitions. Hypersensitivity is a defense mechanism which is considered as an aspect of resistance. The entry of the virus into the cell induces necrosis, which seals off the infection and prevents its spread. An extremely susceptible condition, therefore, becomes a factor in resistance (Bagnall, 1952).

Hooker *et al.* (1954) defined three types of resistance to virus X found among potatoes: "1) tolerant types or carriers of virus X with symptoms ranging from complete masking to a strong mottle which may be accompanied by necrosis depending upon the strains of virus X present and environmental conditions; 2) field immune or hypersensitive types responding to graft inoculation by well-developed top necrosis; and 3) immune types in which the virus fails to survive" (Fig. 15.7).

Resistance of potato to leaf roll appears to be dependent on a virus-inactivating mechanism which reduces field incidence and prevents the transmission of the virus to a proportion of the tuber progeny of infected plants.

This mechanism serves to reduce the virus concentration to a point at which some resistant but infected plants of the Bismarck variety are symptomless. The procedure used experimentally is to inoculate tuber sprouts of hybrid selections and varieties and retain those with leaf roll-free progeny (Hutton and Brock, 1953). Although these authors make no mention of it, it seems that a good case could be made for considering virus inactivation as a mechanism of tolerance rather than resistance. If we define resistance as the ability to repel infection, then the inactivation, if it functioned very rapidly, could prevent further infection and be considered a resistance mechanism. But, in a systemically infected plant, reduction of virus concentration would clearly be a tolerance mechanism. Hutton (1952) found evidence for virus inactivation as a resistance factor to virus X. When two viruses singly show practically no symptoms, but in combination, such as virus A with virus X, a distinct mosaic symptom, it is necessary to combine resistance to both viruses. This actually occurs in some commercial varieties: Epicure and Craig's Defiance are hypersensitive to both viruses A and X, and, therefore, extremely resistant under field conditions (Bagnall, 1953).

True resistance and field resistance are not synonymous, and the difference is important to the practical plant breeder. True resistance can be defined as resistance to infection when the pathogen is introduced from pure culture under ideal conditions for infection. Most laboratory and greenhouse tests measure true resistance. Field resistance is conditioned by many external environmental factors including the biotic status of the

FIGURE 15.7. Virus X resistance in potato. Symptoms of virus X infection on seedling plants 22 days after inoculating the leaves marked with india ink. *a,* Necrotic spotting of inoculated leaf and very early systemic necrosis in upper right leaf; *b,* inoculated leaves, severely necrotic; *c,* systemic top necrosis of the plant; *d,* no symptoms, either local or systemic, on an apparently immune plant (Photographs: Hooker *et al.,* 1954).

pathogen. These may be either favorable, if the environmental resistance to the entry and development of the pathogen is high; or unfavorable, if the converse holds true. The term "field immune" was used by Webb and Schultz (1959) to designate potato seedling varieties that had passed true resistance tests for virus A in the greenhouse, but were susceptible to infection by tuber grafting or inarch grafting with diseased plants; further, they developed top necrosis after grafting but had no recoverable virus in grafted plants or their progeny. Bagnall and Bradley (1958) found that the same relationship between the necrotic type reaction and survival in the tubers occurred in the field, but they called it "field potato resistance."

A high degree of tolerance to both turnip yellows (*Fusarium oxysporum* f. spp. *conglutans*), turnip mosaic, and cauliflower mosaic is found in inbred lines (Badger 1–10) of cabbages. This triple tolerance is obviously of great value if it can be entirely retained in commercially acceptable varieties (Pound *et al.*, 1965). For the brassicas, field selection of tolerance is the only method available for obtaining breeding material (Brusse, 1965).

Varieties may vary greatly in their resistance to hoja blanca disease of rice (Lamey *et al.*, 1964), and this is probably true of the resistance of rice to most, if not all, pathogens. Field resistance is essential, and fortunately it occurs more frequently than does true resistance, for the conditions for the latter are rarely encountered in the field. Incorporation of disease resistance in the rice crops of the world is of paramount importance if famine is to be averted among the millions of people for whom rice is a staple diet.

Resistance to leafhopper vectors of rice viruses by rice varieties was reported by Pathak *et al.* (1969). The vectors were *Nephotettix impicticeps* Ishi and *Nilaparvata lugens* (Stål). Only 3 out of 1000 varieties were highly resistant; Mungo was undamaged, nymphs dying within 10 days and none becoming adult. IRRI 8 and Pankara 202 were resistant to *Nephotettix*. Mungo crossed with IRRI 8 was highly resistant; resistance was dominant. While the proportion of varieties found resistant was extremely low, the possibility of combining this resistance with high yielding varieties will certainly not be overlooked.

What is called "general resistance" to 2 viruses, as well as resistance to the aphid vector of raspberry viruses, is found in the Norfolk Giant variety (Jennings, 1963). Lloyd George variety was found to be immune to the vector (*A. rubi*) (Daubeny and Stace-Smith, 1963).

Resistance to tobacco mosaic in many crops can be indicated by the local lesion response (Holmes, 1965a). Resistance is dominant in pepper (Cook, 1963), tomato, (Alexander, 1963) and beans (Thompson *et al.*, 1962). Pelham (1966) has reviewed work on resistance to TMV and has stressed the need for standardized techniques and a conservative attitude to release of so-called resistant varieties.

Fiji disease of sugarcane has been almost completely eliminated by the use of resistant varieties in commercial plantings. It is still a problem with small growers (Baudin and Antoine, 1962). Hughes has tabulated 50 sugarcane varieties for reaction to sugarcane mosaic, chlorotic streak, Fiji disease and ratoon stunting (Hughes, 1966).

Breeding for resistance to beet yellows and beet mild yellowing viruses is receiving much attention (Bell, 1963). Immunity has not been found, but host tolerance exists, unfortunately in combination with low sugar and high potassium content. This tolerance, however, extends to different strains, so field breakdown of the tolerance is not expected (Russell, 1964). The tolerant varieties lose 30% in yield when inoculated, compared with 60% loss by commercial varieties. Later, Russell (1966), working with virus-tolerant breeding material, found differences in resistance to beet yellows and mild yellowing viruses. The possibility of breeding sugar beet varieties in which the two forms of resistance, tolerance and resistance to infection, are combined, is pointed out.

Tolerance to curlytop of tomatoes of dwarf and woolly varieties, held up in severe curlytop years. One woolly dwarf variety showed tolerance in only one year (1962) (Oshima and Foskett, 1966). This could have been related to a specific virus strain prevalent that year. Where many strains of a virus occur naturally in the field, it presents an almost impossible problem for any resistance program, either immunity at one end of the scale, or extreme tolerance at the other. Tolerance no doubt does increase the difficulty of roguing practices (Longworth and Thresh, 1963), but as a practical matter, tolerance is often all that the plant breeder has available.

Maize dwarf virus was recently described. Of 600 hybrids tested, only nine showed significant resistance; two are probably immune (Wernham and Mackenzie, 1965). Tolerance is believed by Nault *et al.* (1971) to be more important than resistance to aphid and mechanical inoculations of maize dwarf in corn. One inbred line of corn was more susceptible to aphid inoculations; two others were less so. In five inbred lines, no evidence for immunity was found. There was no correlation between susceptibility to inoculation and severity of symptoms. When incidence of disease in the field is high, all the plants will be infected unless resistance is very strong; in these circumstances, tolerance is of economic importance.

There is only one report of immunity to yellow bean mosaic in cowpea according to Kuhn *et al.* (1965), but Robertson (1965) found that 79 varieties of cowpea inoculated with cowpea yellow mosaic showed local lesion reactions that indicated a range from susceptibility to resistance to immunity. Cowpea chlorotic mottle can attack all the commercial varieties and cultivars of cowpea tested, but eight introductions were resistant (Sewell *et al.*, 1965).

The length of time elapsing between inoculation by the aphid vector and symptom expression in 50% of the groundnut populations can be used as a guide to resistance to rosette disease (Nutman *et al.*, 1964).

There is tolerance to barley yellow dwarf, but different virus strains in different areas are a complication (Arny and Kedlinski, 1966). Of 6689 varieties of barley, only 117 proved to be resistant; most of these were of Ethiopean origin (Schaller *et al.*, 1963). The resistance is conditioned by a single gene, incompletely dominant (Damsteeght and Bruehl, 1964; Damsteeght, 1963).

Genetics of Resistance

Holmes (1954) has reviewed the subject of inheritance of resistance to plant viruses, but the number of viruses considered worthy of inclusion is surprisingly small and the number of crops involved not much greater. Resistance to virus may actually be associated with resistance to aphid vectors. Lamb (1960) showed this to be true in the case of some *Brassica* field crops in New Zealand.

Huber and Schwartze much earlier (1938) had shown varietal and heritable resistance to the attack of *Amphorophora rubi* Kalt. on raspberry. Other examples are found in Ivanoff (1945) for muskmelon resistance to *Aphis gossypii* (Glov.) and in Wilcox (1951) for cranberry varietal resistance to the leafhopper *Scleroracus vaccinii* (Van D.). Nothing short of immunity to vector attack, however, could be expected to be of significance in virus transmission. Tomato is "resistant," i.e., a poor host of *C. tenellus;* pineapple a very poor host for thrips; yet virus transmission does not appear to be impeded in either case.

Another rather curious aspect of resistance is the tendency to escape infection. This tendency, known as klendusity, is heritable. It was found in the Chilean tomato *Lycopersicon chilense* Dim. and in some of its hybrids with the cultivated tomato *L. esculentum* Mill. (Holmes, 1943). The mechanism is probably through physiological factors, as suggested by Troutman and Fulton (1958), for klendusity in tobacco and cucumber mosaic.

Factors for resistance are heritable for virus Y (Schultz *et al.*, 1934). A. A. Cook and Anderson (1960) found that inheritance of resistance to virus Y in two strains of peppers (*Capsicum annuum* L) was inherited as a single recessive factor. Both resistant pepper strains were homozygous for the recessive allele y^a; susceptible varieties were homozygous for the dominant allele Y^a.

The inheritance of susceptibility (symptom expression) to bean mosaic virus 4 is governed by a single allelomorphic pair of Mendelian factors. A local lesion type of mottling occurs in plants carrying the dominant allelomorph, whereas the homozygous recessive plants are susceptible to leaf-mottling, stunting, and yield reduction. The former are considered commercially resistant and are being used as parent material for further hybridization (Zaumeyer and Harter, 1943).

Resistance to pod mottle virus in bean also appears to be governed by a single allelomorphic gene pair. Varieties carrying the dominant local lesion character are commercially resistant (H. R. Thomas and Zaumeyer, 1950). Inheritance of resistance to local lesions of tobacco ringspot virus in black cowpea is through a single gene dominant (de Zeeuw and Ballard, 1959), as is the case with cucumber mosaic (Sinclair and Walker, 1955). The one resistance appears to be independent of the other, but this is to be expected according to Holmes (1958), who states: "Genes conferring resistance to known viruses, or virus strains, have never been found to confer resistance to other viruses." However, A. A. Cook (1960), reporting on resistance to two viruses in *Capiscum annuum*, found that resistance to tobacco etch virus resulted from a single recessive gene, as did also resistance to virus Y. Because recombinant genotypes could not be demonstrated, it was concluded that the same single recessive pleiotropic gene functioned in both cases. Further evidence of multiple resistance in the same strain of *C. annuum* was shown when it was found hypersensitive to mosaic (Cook and Anderson, 1959). Five distinct genes appear to be necessary to provide resistance to tomato spotted wilt in western Australia, but only one is needed in Hawaii and only one in the eastern United States; this reflects the heterozygosity of the Australian strains.

Resistance of beans to bean yellow mosaic virus, on the other hand, appears to be governed by a single gene pair without dominance; symptom severity and the time of symptom appearance are associated with modifying genes (Baggett and Frazier, 1957). Resistance to infection by tobacco mosaic in tomato is controlled by a single dominant gene (Holmes, 1954a).

Yellow *Aucuba* strain (TMV-YA) of tobacco mosaic virus increased in virulence when four passages were made through the resistant tomato N-32 which was derived from Holmes' P.I. 235673. A substrain of the virus multiplied selectively in relation to the standard or parent virus to the end that the parent virus was excluded. The increased virulence, therefore, came about through the selection of a substrain in the parent virus population. There was no evidence that the substrain in the resistant host reverted to the parent virus (Zitter and Murakishi, 1969). This example is referred to here since it indicates that the genetics of the virus, i.e., its mutability, could offset genetic resistance to a parent strain.

Melon mosaic virus is symptomless when inoculated into the cucumber cultivar Kyoto (K.3.f.), and virus concentration was significantly lower than in the susceptible Bet-Alfa (BA). Resistance is governed by a single dominant gene. When this gene is absent, different degrees of susceptibility are also correlated with concentration of MMV particles as determined by local lesion tests on *Chenopodium amaranticolor* (Cohen *et al.,* 1971).

Tolerance to virus infection as a control measure appears to be suspect, since a tolerant variety may be a source of virus infection. Boyce (1958) reported resistance to yellow dwarf of *Phormium,* but since the resistance might be due to partial tolerance it was not recommended for distribution to growers. The probability is, however, that many so-called resistant varieties are actually tolerant to some degree, and tolerance is probably less rare in crop plants than is effective resistance. Even if a tolerant tree, after infection, is a somewhat lower yielder than a healthy tree, it may still be expected to produce an economic return with proper agronomy. The argument that such tolerant, infected trees provide virus sources loses its force in areas where severe virus infection is rampant.

Tolerance to virus infection, both singly and in complexes, is used perforce in the development of new strawberry viruses, which are systematically exposed to infection before being released to growers. If no symptoms appear, the variety is not considered immune or resistant but as showing a degree of tolerance (Miller and Waldo, 1959).

This is true of cacao in Ghana and Nigeria, where tolerant varieties are more firmly delineated than are resistant ones. The potential value of these tolerant varieties is enormous for those areas which are massinfected and where control, by cutting out diseased trees, is either a failure or is too expensive.

Holmes (1965) has reviewed the work on the genetics of pathogenicity in viruses and resistance in host plants. He breaks the subject down under five headings of which four concern resistance. Natural selection for increased resistance in hosts carries with it the problem of tolerance. The tolerant but infected variety, a criterion most widely used in practical applications, has potential as a virus reservoir from which more virulent mutants can arise. Holmes considers that breeding to produce plants which show the disease clearly, but which are only rarely infected, is to be preferred. The ideal combination is a field resistant type of the disease which combines low transmissibility from affected tissues, ease of recognition of the disease, and relatively high resistance to infection.

On the other hand, Posnette (1969) points out that genetic tolerance is more common than resistance. In areas of high infection rates in sensitive varieties, replacement by tolerants would not increase the virus population from which other plants could become infected. Virus mutants would be less likely to establish in tolerant than in sensitive varieties or resistant plants in which the selection pressure is greater. In many tree crops, commercial varieties are so tolerant that no disease is evident and the virus is not determinable except by transfer to indicator plants. Most epidemics have occurred in monocultures of sensitive plants, e.g., citrus tristeza, little cherry, cacao swollen shoot and plum pox.

Unconscious selection for tolerance has allowed the development of crops in areas where endemic viruses would have rendered sensitive varieties uneconomic. In plant breeding programs the choice may lie between continuing a fruitless research for extreme resistance, or using tolerance. Conditions favoring the use of tolerance are, (1) high levels of infection (rapid spread), (2) virus-free material available for planting, (3) virus reservoirs outside the crop which cannot be reached, and (4) low incidence of other viruses in the crop.

Sources of Resistant Genes

Wild species are sometimes good sources of resistant genes. Owen (1944) called attention to the domestic beet species, *Beta vulgaris* L, and the wild beet, *B. maritima* L, which crosses freely with it, as sources of genetic variability. Miller (1959) found the wild strawberry *Fragaria chiloensis* was tolerant to many viruses and virus combinations affecting domestic strawberry. Many species of *Solanum* have been used in potato breeding. Webb and Hougas (1959) reported tests on 39 *Solanum* species and 58 species hybrids. Resistance to viruses A, X, Y, and leaf roll was found in greater or lesser degree, but all the species hybrids were found susceptible to spindle tuber. Wild *Lycopersicon* species have great potential as a reservoir of resistant genes for tobacco mosaic, spotted wilt, curly top, and many other pathogens (Doolittle, 1954). Extreme resistance in wild forms of *Beta maritima* L collected in Europe was recovered after hybridizing with susceptible commercial sugar beets (Coons *et al.*, 1931). *Manihot melanobasis* Mull. Arg. was reported by Jennings (1959) to be a valuable source of new genes for cassava improvement.

Cassava is widely grown in Malaysia, but after a mission extending for five months and reaching from the southern to the northern borders of West Malaysia, not a single case of cassava mosaic was seen. This experience was also that of pathologists familiar with the African situation who were working in the Malaysian area. It may be, of course, that the virus has never reached that country, but if that is not the explanation, there should be some valuable genes in the West Malaysian cassava. The intensity of exposure may greatly affect resistance. Under mild or moderately severe exposure, curlytop resistance appears to be dominant, and F_1 hybrids between resistant and susceptible resemble the resistant parents. With more drastic exposure, that is, midsummer plantings in areas of high leafhopper populations, dominance is lacking and the F_1 hybrids resemble the susceptible parent much more than the resistant one (Owen, 1944).

Although not a wild species, the runner bean *Phaseolus coccineus* L is considered to have a broad potential as a source of genes resistant to many viruses. Baggett and Frazier (1959) tested *P. coccineus* against 27 virus inocula, but susceptibility was found to only two strains of alfalfa mosaic.

There are some encouraging examples of success in breeding resistant varieties of commercial importance, even though there are many efforts which have not yet shed their academic mantle. As Holmes (1954) correctly states: "The differences that militate most decisively against acceptance of newly developed disease-resistant varieties are conspicuously unfavorable differences in total yield or in quality."

Use of Resistance as a Commercial Control Measure

Successful use of resistance as a commercial control measure can be reported. Sugarcane has been bred for resistance to mosaic and other diseases successfully, and many varieties are available that are essentially immune to mosaic (Brandes, 1925) and resistant to sereh disease (Abbott *et al.,* 1936). In Taiwan, Liu and Li (1953) compared sugarcane varieties for resistance to the three strains of mosaic recognized there. In Table 15.8 susceptibility of these varieties to mosaic in the generic sense, as reported by Summers *et al.* (1948), is compared with susceptibility to the virus strains. It is clear that a variety may even be immune to one strain but very susceptible to another. The development of sugar beet varieties resistant to curlytop has saved the sugar beet industry from extinction in many areas in the western United States. The first variety was released as U. S. No. 1, grown from selections by Carsner, (1926). While this variety had only an intermediate degree of curlytop resistance, it marked the turning point in the survival of sugar beet as an economic crop in curly top areas. There have been major improvements since, based on selection of extremely resistant individuals from U. S. No. 1 (Owen *et al.,* 1939).

TABLE 15.8.   Resistance to Mosaic by Saccharum *Species* [a]

Variety	Degree of susceptibility to mosaic reported by			
		Liu and Li (1953)		
	Summers *et al.* (1948)	SS	YS	FS
S. officinarum	Susceptible	V.S.	V.S.	S.
S. barberi	Susceptible	V.S.	V.S.	I.
S. robustum	Susceptible	S.	V.S.	V.S.
S. sinense	Highly resistant	I.	I.	S.
S. spontaneum	Immune	I.	I.	I.
S. officinarum X *S. barberi*	Susceptible	V.S.	V.S.	R.
S. officinarum X *S. sinense*	Resistant	V.S.	V.S.	V.S.
S. officinarum X *S. spontaneum*	Resistant	H.R.	H.R.	R.

[a] From Liu and Li (1953).
Key: SS=short-stripe type mosaic; YS=yellow-stripe type mosaic; FS=fine-stripe type mosaic; V.S.=very susceptible; S.=susceptible; R.=resistant; H.R.=highly resistant; I.=immune.

Curlytop resistance in beans has also been developed to the point of commercial usefulness. The Red Mexican is a naturally resistant variety, as is Burtner, the latter a rather poor quality snap bean. These two varieties have been used as parents for thousands of crosses which have been tested under the severest curlytop exposures possible in the Twin Falls, Idaho, area. The Red Mexican was also crossed with the common Pinto bean, and from this cross have been developed the varieties of Pinto bean resistant to mosaic and curlytop which are now largely used in the Pacific Northwest and Rocky Mountain states (Hungerford, 1951) (Fig. 15.8). Resistance to curlytop has been dealt with in monographic detail by Bennett (1971).

Cassava resistant to mosaic and brown streak virus has been developed in East Africa. While no industry has been involved, and, therefore, no statistical evidence for great improvement is available, this is one case in which the staple diet of a native peasantry has been vastly improved. Nichols (1947) described the techniques and the species used for interspecific hybrids with cassava (*Manihot utilissima* Pohl.).

Jennings (1957, 1959, 1960) has continued Nichols' studies and reports extremely high yield increases obtained with resistant varieties. It is strange that no reports on the effectiveness of the varieties have been

FIGURE 15.8. Curlytop resistance in beans. The blank spaces are plots that were planted to seedling varieties which were completely destroyed (Photograph: Piemeisel, U.S. Department of Agriculture).

made from other cassava growing areas where virus disease is rampant.

Jameson (1964) has continued work with cassava mosaic resistance in East Africa and has outlined the testing procedure and the limitation of the method. He reports that the disease has finally been almost extinguished by the use of resistant varieties. It was interesting to learn, after personal enquiries, that in areas remote from East Africa, tests of the East African resistant varieties had not proved as satisfactory in those areas.

Internal cork disease of sweet potato should be amenable to control by resistant varieties. Some clones are resistant to symptom expression, although they are symptomless carriers; some are resistant to virus multiplication (i.e., tolerant) (Nielsen and Pope, 1960).

CROSS PROTECTION

Cross protection afforded by mild strains against virulent ones also offers some opportunity for control in special circumstances, although very little, if any, use of the method on a commercial scale has been reported.

Salaman (1933) protected tobacco plants against a virulent strain of virus X by first inoculating with a very mild strain. He reported that neither the protective inoculation nor the subsequent virulent one had any appreciable effect on the health of the plant.

Bennett (1951) reviewed the subject of interference between plant virus strains. A high degree of protection is afforded between strains of some viruses, as with tobacco mosaic, and little or no protection between any of the strains of curlytop virus (Bennett, 1953).

Protection is not, however, limited to strains of one virus or closely related viruses. Thomson (1958) reports that nonspecific interference occurs on occasion between quite unrelated viruses.

The principal objection to the use of protective inoculation is the possibility that virus complexes may develop as a result of infection with unrelated or nonprotected viruses. The validity of this objection depends on the virulence of the virus being protected against. If the effect on the host is lethal, as is the case with the New Juaben strain of swollen shoot in Ghana, the objection is not valid. The real difficulty in this case lies in making the protective inoculations in the first place since, currently, it can be done only by budding. If the current work on mechanical transmission of the swollen shoot viruses develops favorably, this limitation would disappear (Brunt and Kenten, 1960).

Woodiness virus of passion fruit in Queensland exists in several strains, the mild strains protecting against the severe. Owing to the necessity of using *Fusarium*-resistant root stocks, grafting is necessary in any case, so it only remains to use scion material from a parent vine bearing the appropriate mild strain. This procedure as a control measure against woodiness has reached the practical field demonstration stage (Simmonds, 1959).

Posnette no longer advocates the use of mild strain protection in the control of cacao swollen shoot disease (Posnette and Todd, 1955). The political and psychological aspects of "introducing disease" into a planting made it impossible even to test the method on a field scale.

THE FUTURE OF VIRUS DISEASE CONTROL

If the main headings under which control has been considered are reviewed, it is clear that the agronomic methods offer some measure of ameolioration which is useful to the average grower. With the tendency for the number of individual farms to be reduced and the areas incorporated into larger units, some advantages of comparative isolation might be lost.

The search for resistant genes could well be widely extended, since many virus diseases do not appear to be amenable to any other method of control. Progress in fundamental studies of the viruses themselves, in the chemical inhibition of virus multiplication, and in vector-virus relationships does not appear to have been translatable to any great extent into techniques for the control of the virus diseases involved. Such studies should, however, provide the basis on which new approaches to the problems of control will develop. The development of systemic viricides might well be one such approach, but these will be used rationally only if there is an understanding of the ecological complex in which virus, vector, and host plant are so firmly integrated.

REFERENCES

Abbott, E. V., Summers, E. M., and Rands, R. D., "Disease resistance tests and seedling selections in 1935 at the U. S. Sugar Plant Field Station, Houma, La.," *U. S. Dept. Agr. Bur. Plant Ind. Sugar Bull.*, 14, 3–7 (1936).

A'Brook, J., "The effect of planting date and spacing on the incidence of groundnut rosette disease and the vector, *Aphis craccivora* Koch at Mokwa, Northern Nigeria," *Ann. Appl. Biol.*, 54, 199–208 (1964).

A'Brook, J., "The effect of plant spacing on the numbers of aphids trapped over the groundnut crop," *Ann. Appl. Biol.*, 61, 289–294 (1968).

Adam, A. V., "An effective program for the control of banana mosaic," *Plant Disease Reptr.*, 46, 366–370 (1962).

Adsuar, J., "Susceptibility of some sugarcane varieties to the treatment used in the control of chlorotic streak," *J. Agr. Univ. Puerto Rico*, 40, 67–69 (1956).

Alexander, L. J., "Transfer of a dominant type of resistance to the four known Ohio pathogenic strains of tobacco mosaic virus TMV from *Lycopersicon peruvianum* to *L. esculentum*," *Phytopathology*, 53, 869 Abstr. (1963).

Anon, "Strawberry plant certification standards," *Oregon State College Certification Board*, 1955.

Armitage, H. M., "Controlling curlytop virus in agricultural crops by reducing populations of overwintering beet leafhoppers," *J. Econ. Entomol.*, 45, 432–435 (1952).

Arny, D. C., and Jedlinski, H., "Resistance to the yellow dwarf virus in selected barley varieties," *Plant Disease Reptr.*, 50, 380–381 (1966).

Atanasoff, D., "Interferon and plant viruses," *Phytopath. Zeit.*, 47, 204–214 (1963).

Baggett, J. R., and Frazier, W. A., "The inheritance of resistance to bean yellow mosaic virus in *Phaseolus vulgaris*," *Proc. Am. Soc. Hort. Sci.,* **70**, 325–333 (1957).

Baggett, J. R., and Frazier, W. A., "Disease resistance in the runner bean, *Phaseolus coccineus* L," *Plant Disease Reptr.,* 43, 137–143 (1959).

Bagnall, R. H., "Hypersensitivity, a form of resistance to plant viruses in potatoes," *Quebec Soc. Protect. Plants, 34th Rept.* (1952).

Bagnall, R. H., "Notes on potato virus A.," *Proc. Can. Phytopathol. Soc.,* **20**, 11 (1953).

Bagnall, R. H., "The spread of potato virus Y in seed potatoes in relation to the date of harvesting and the prevalence of aphids," *Can. J. Agr. Sci.,* 33, 509–519 (1953).

Bagnall, R. H., and Bradley, R. H. E., "Resistance to virus Y in the potato," *Phytopathology*, 48, 121–125 (1958).

Bancroft, J. B., and Curtis, R. W., "A simple method using trichloroacetic acid for estimating tobacco mosaic concentration," *Phytopathology,* 47, 79–82 (1957).

Bandy, G. T. A., "Heat cure of sugarcane infected with sugarcane mosaic," *Phytopathology,* **60**, 1284 Abstr. (1970).

Bae, S. H., and Pathak, M. D., "Common leafhopper-planthopper populations and incidence of tungro virus in diazinon-treated and untreated rice plots," *J. Econ. Entomol.,* **62**, 773 (1969).

Baudin, P., and Antoine, R., "Fiji disease in Madagascar," *Proc. Intern. Soc. Sugarcane Technologists,* 11, 760–767 (1962).

Bawden, F. C., "Inhibitors and plant viruses," *Adv. Virus Res.,* **2**, 31–57 (1954).

Bawden, F. C., "Some effects of changing environment on the behaviour of plant viruses," *Symposium Soc. Gen. Microbiol.,* 11, 296–316 (1961).

Bawden, F. C., and Kassanis, B., "Some effects of thiouracil on virus-infected plants," *J. Gen. Microbiol.,* **10**, 160–173 (1954).

Bawden, F. C., and Pirie, N. W., "A virus-inactivating system from tobacco leaves," *J. Gen. Microbiol.,* **16**, 696–710 (1957).

Bawden, F. C., Hamlyn, B. M. G., and Watson, M. A., "The distribution of viruses in different leaf tissues and its influence on virus transmission by aphids," *Ann. Appl. Biol.,* 41, 229–239 (1954).

Bawden, F. C., Kassanis, B., and Roberts, F. M., "Studies on the importance and control of potato virus X," *Ann. Appl. Biol.,* 35, 250–265 (1948).

Baxter, L. W., Jr., and McGlohon, N. E., "A method of freeing white clover plants of bean yellow mosaic virus," *Phytopathology,* 49, 810–811 (1959).

Beale, H. P., and Jones, C. R., "Virus diseases of tobacco mosaic and potato yellow dwarf not controlled by certain purified antibiotics," *Contribs. Boyce Thompson Inst.,* 16, 395–407 (1951).

Beinhart, E. G., and Morgan, O. D., "Preliminary study of sterilizing tobacco stems against mosaic diseases," *U. S. Dept. Agr. Publ.,* AIC-334 (1952).

Bell, G. D. H., "Developments in sugar beet breeding," *J. Natl. Inst. Agric. Botany,* 9, 435–444 (1963).

Bennett, C. W., "Interference phenomena between plant viruses," *Ann. Rev. Microbiol.*, 5, 295–308 (1951).

Bennett, C. W., "Evidence of lack of interference between strains of the curly top virus," *Intern. Congr. Microbiol., Rome, 1953*, 3, 387–388 (1953).

Bennett, C. W., "Sugar beet yellows disease in the United States," *U. S. Dept. Agr. Tech. Bull.*, No. 1218 (1960).

Bennett, C. W., "Curlytop of sugar beet and other plants," *Monograph No. 7, Amer. Phytopath. Soc.* (1971).

Benson, A. P., and Hooker, W. J., "Isolation of virus X from "immune" varieties of potato *Solanum tuberosum*," *Phytopathology*, 50, 231–234 (1960).

Bieske, G., "Hot water treatment of cane plants in the Mulgrave area," *Cane Growers' Quart. Bull.*, 17, 78–79 (1954).

Bovey, R., "Guerison de fraisiers, atteints de virus, par traitement thermique," *Sta. Fédérales Essais Agr., Lausanne, Publ.*, No. 462 (1954).

Boyce, W. R., "Resistance to yellow-leaf disease in *Phormium tenax* Forst. and the occurrence of delayed expression of symptoms," *New Zealand J. Agr. Research*, 1, 31–36 (1958).

Bradley, R. H. E., "Use of oil sprays to control viruses," *First. Intern. Cong. Plant Pathology Abstr.*, 20, (1968).

Bradley, R. H. E., Moore, C. A., and Pond, D. D., "Spread of potato virus Y curtailed by oil," *Nature*, 209, 1370 (1966) (and personal communication 1968).

Brandes, E. W., and Klaphaak, P. J., "Breeding of disease resisting sugar plants for America," in *The Reference Book of the Sugar Industry of the World*, Louisiana Planter and Sugar Manufacturer Co., New Orleans, 1925 Vol. 3, pp. 50–57.

Breakey, E. P., and Campbell, L., "Suppression of strawberry yellows by controlling the aphid vector *Capitophorus fragaefolii* (Ckll.)," *Plant Disease Reptr.*, 35, 63–69 (1951).

Brierley, P., "Virus-free hydrangeas from tip cuttings of heat-treated ring-spot affected stock plants," *Plant Disease Reptr.*, 41, 1005 (1957).

Broadbent, L., "Insecticidal control of the spread of plant viruses," *Ann. Rev. Entomol.* 2, 339–354 (1957).

Broadbent, L., "Disease control through vector control," *Viruses, Vectors and Vegetation*, 593–630 Wiley-Interscience, New York, 1969.

Broadbent, L., Gregory, P. H., and Tinsley, T. W., "Roguing potato crops for virus disease," *Ann. Appl. Biol.*, 37, 64–650 (1950).

Broadbent, L., Gregory, P. H., and Tinsley, T. W., "The influence of planting date and manuring on the incidence of virus diseases in potato crops," *Ann. Appl. Biol.*, 39, 509–524 (1952).

Broadbent, L., Heathcote, G. D., and Burt, P. E., "Field trials on the retention of potato stocks in England," *European Potato J.*, 3, 251–262 (1960).

Broadbent, L., Tinsley, T. W., Buddin, W., and Roberts, E. T., "The spread of lettuce mosaic in the field," *Ann. Appl. Biol.*, 38, 689–706 (1951).

Brunt, A. A., and Kenten, R. H., "Mechanical transmission of cacao swollen-shoot virus," *Virology*, 12, 328–330 (1960).

Brusse, M. J., "Glasshouse and field observations on cauliflower and turnip mosaic viruses in relation to the breeding for virus resistance," *N. Z. J. Agric. Res.*, 8, 672–680 (1965).

Burt, P. E., Broadbent, L., and Heathcote, G. D., "The use of soil insecticides to control potato aphids and virus diseases," *Ann. Appl. Biol.*, 48, 580–590 (1960).

Cann, H. J., "Big decline in incidence of bunchy top," *Banana Bull. (Sydney)*, 13, 11 (1949).

Cann, H. J., "Bunchy top virus of bananas 'at all time low' in New South Wales," *Agr. Gaz. (New South Wales)*, 63, 73–76 (1952).

Carsner, E., "Resistance in sugar beets to curlytop," *U. S. Dept. Agr. Circ.*, No. 388 (1926).

Carter, W., "The use of prontosil as a vital dye for insects and plants," *Science*, 90, 394 (1939).

Chant, S. R., "A note on the inactivation of mosaic virus in Cassava (*Manihot utilissima* Pohl.) by heat treatment," *Empire J. Exptl. Agr.*, 27, 55–58 (1959).

Chessin, M., "Growth substances and stunting in virus-infected plants," *Proc. 3rd. Conf. Potato Virus Diseases*, Lisse-Wageningen, 80–83 (1957).

Cohen, S., Gertman, E., and Kedar, N., "Inheritance of resistance to melon mosaic virus in cucumbers," *Phytopathology*, 61, 253 (1971).

Cook, A. A., "Genetics of resistance in *Capsicum annuum* to two virus diseases," *Phytopathology*, 50, 364–367 (1960).

Cook, A. A., "Dominant resistance to tobacco mosaic virus in an exotic pepper," *Plant Disease Reptr.*, 47, 783–786 (1963).

Cook, A. A., and Anderson, C. W., "Multiple virus disease resistance in a strain of Capsicum annuum," *Phytopathology*, 49, 198–201 (1959).

Cook, A. A., and Anderson, C. W., "Inheritance of resistance to potato virus Y derived from two strains of *Capsicum annuum*," *Phytopathology*, 50, 73–75 (1960).

Cook, J. G., *Virus in the Cell*, Dial Press, New York, 1957.

Cook, W. C., "Evaluation of a field-control program directed against the beet leafhopper," *J. Econ. Entomol.*, 36, 382–385 (1943).

Coons, G. H., Stewart, D., and Elcock, H. A., "Sugar-beet strains resistant to leaf spot and curlytop," *Yearbook Agr. U. S. Dept. Agr.*, 1931, 493–496.

Cornford, C. E., "Maleic hydrazide as a shoot depressant for clamped mangolds and fodder beets," *Plant Pathol.*, 4, 89–90 (1955).

Cornwell, P. B., "Some aspects of mealybug behaviour in relation to the efficiency of measures for the control of virus diseases of cacao in the Gold Coast," *Bull. Entomol. Res.*, 47, 137–166 (1956).

Costa, A. S., Forster, R., and Fraga, R., and Fraga, G., "Controle de viracabaca do Tomate pela destruicao do vector," *Bragantia*, 10, 1–9 (1950).

Couranjou, A., "Echee à la psorose," *Fruits et Primeurs,* **21**, 2–5 (1951).

Crane, G. L., and Calpouzos, L., "Suppression of symptoms of sugar beet virus yellows by mineral oil," *Phytopathology,* 59, 697–702 (1969).

Crowley, N. C., "The use of skim milk in preventing the infection of glasshouse tomatoes by tobacco mosaic virus," *J. Australian Inst. Agr. Sci.,* 24, 261–264 (1958).

Damsteeght, V. D., "The inheritance of tolerance in barley to barley yellow dwarf virus," *Diss. Abstr. Wash. State Univ.,* 23, 12, 1;4492 (1963).

Damsteeght, V. D., and Bruehl, G. W., "The inheritance of resistance in barley to barley yellow dwarf virus," *Phytopathology,* 54, 219–224 (1964).

Daubeny, H. A., and Stace-Smith, R., "Note on immunity to the North American strain of the red raspberry mosaic vector, the aphid *Amphophora rubi* (Kalb.)" *Canad. J. Pl. Science,* 43, 413–415 (1963).

Deen, O. T., and Hallock, H. C., "Beet leafhopper control experiments on snap beans grown for seed," *J. Econ. Entomol.,* 47, 122–126 (1954).

Denby, L. G., and Wilks, J. M., "Milk sprays reduce spread of tobacco mosaic in tomatoes," *Res. Farmers,* 10, 1 (1965).

de Zeeuw, D. J., and Ballard, J. C., "Inheritance in cowpea of resistance to tobacco ringspot virus," *Phytopathology,* 49, 332–334 (1959).

Dickason, E. A., Raymer, W. B., and Foote, W. H., "Insecticide treatments for aphid control in relation to spread of barley yellow dwarf virus," *Plant Disease Reptr.,* 44, 501–504 (1960).

Dobie, N. D., Vaughan, E. K., Miller, P. W., and Waldo, G. F., "New facilities and procedures used for growing disease-free strawberry plants in Oregon," *Plant Disease Reptr.,* 42, 1048–1050 (1958).

Doolittle, S. P., "The use of wild *Lycopersicon* species for tomato disease control," *Phytopathology,* 44, 409–414 (1954).

Douglass, J. R. Romney, V. E., and Jones, E. W., "Beet leafhopper (*Circulifer tenellus*) control in weed-host areas of Idaho to protect snap bean seed from curly top," *U. S. Dept. Agr. Circ.,* No. 960 (1955).

Emilsson, B., and Castberg, C., "Control of potato aphids by spraying with parathion and the effect on spread of virus Y," *Acta Agr. Scand.,* 2, 247–257 (1952).

Fernow, K. H., and Kerr, S. H., "Leaf-roll control by use of insecticides," *Am. Potato J.,* 30, 187–196 (1953).

Fisken, A. G., "Factors affecting the spread of aphid-borne viruses in potato in eastern Scotland. II. Infestation of the potato crop by potato aphids, particularly *Myzus persicae* (Sulz.)," *Ann. Appl. Biol.,* 47, 274–286 (1959).

Frazier, N. W., Veth, V., and Bringhurst, R. S., "Inactivation of two strawberry viruses in plants grown in a natural high-temperature environment," *Phytopathology,* 55, 1203 (1965).

Freitag, J. H., "A hot knife to prevent virus spread in orchids," *Bull. Amer. Orchid Soc.,* 34, 501–502 (1965).

Freitag, J. H., Aldrich, T. M., and Drake, R. M., "The control of the spread of aster yellows in celery," *14th. Internl. Symp. on Phytopathology and Phytiatry,* (1962).

Fulton, J. P., "Heat treatments of virus-infected strawberry plants," *Plant Disease Reptr.,* 38, 147–149 (1954).

Fulton, J. P., and Seymour, C., "The Arkansas strawberry certification program," *Plant Disease Reptr.,* 41, 749–754 (1957).

Fulton, R. H., and Lovitt, D. F., "The Michigan virus-free strawberry certification program," *Mich. State Univ. Agr. Expt. Sta. Quart. Bull.,* 40, 575–580 (1958).

Gendron, Y., and Kassanis, B., "The importance of host species in determining the action of virus inhibitors," *Ann. Appl. Biol.,* 41, 183–188 (1954).

Gibson, K. E., and Fallini, J. T., "Beet leafhopper control in southern Idaho by seeding breeding areas to range grass," *U.S.D.A. ARS,* 33–83 (1963).

Gibson, K. E., Landis, B. J., and Klostermeyer, E. C., "Effect of aphid control on the spread of leaf roll in potatoes," *Am. Potato J.,* 28, 658–666 (1951).

Gonzales, D., and Rawlins, W. A., "Relation of aphid populations to field spread of lettuce mosaic virus," *J. Econ. Entomol.,* 62, 1109 (1969).

Grant, T. J., "Effect of heat treatments on tristeza and psorosis viruses in citrus," *Plant Disease Reptr.,* 41, 232–234 (1957).

Gray, R. A., "Activity of an antiviral agent from *Nocardia* on two viruses in intact plants," *Phytopathology,* 45, 281–285 (1955).

Gray, R. A., "Combating plant virus diseases with a new antiviral agent, Cytovirin," *Plant Disease Reptr.,* 41, 576–578 (1957).

Grogan, R. G., Welch, J. E., and Bardin, R., "The use of mosaic-free seed in controlling lettuce mosaic," *Phytopathology,* 41, 939 (1951).

Hakhaat, F. A., and Jordonova, J., "Heat treatment experiments with carnations for the elimination of carnation mottle and etched ring viruses," *Neth. J. Pl. Pathology,* 74, 146–149 (1968).

Hammond, P. S., "Notes on the progress of pest and disease control in Ghana," *Cocoa Conf. Rept. Cocoa Choc. and Confect. Alliance,* London, 1957.

Hanna, A. D., Judenko, E., and Heatherington, W., "Systemic insecticides for the control of insects transmitting swollen-shoot virus disease of cacao in the Gold Coast," *Bull Entomol. Res.,* 46, 669–710 (1955).

Hansing, E. D., "A study of the control of the yellow-dwarf disease of potatoes," *Cornell Univ. Agr. Expt. Sta, Bull.,* No. 792 (1943).

Hare, W. W., and Lucas, G. B., "Control of contact transmission of tobacco mosaic virus with milk," *Plant Disease Reptr.,* 43, 152–154 (1959).

Harrison, B. D., Peachey, J. E., and Winslow, R. D., "The use of nematocides to control the spread of *Arabis* mosaic by *Xiphinema diversicaudatum* (Micol.)," *Ann. Appl. Biol.,* 52, 243–255 (1963).

Hein, A., "Die wirkung emulgieter Fette auf die Übertragung nicht-persistent Viren durch *Myzus persicae* (Sulz.)," *Phytopathology Zeit.,* 52, 29–36 (1965).

Hey, A., "Verbreitung und Bekampfung virusubertragender Blattlause in Beziehung zum Auftreten von Kartoffelvirosen in Nachbau," *Nachrbl. Deut. Pflanzenschutzdienstes (Stuttgart)*, **6**, 181–187 (1952).

Hickman, A., and Varma, A., "Viruses in horse-radish," *Plant Pathol.*, **17**, 26–30 (1968).

Hildebrand, E. M., "Freeing sweet potato varieties from cork virus by a flush of growth and propagation with tip cuttings," *Phytopathology*, **47**, 452 (1957).

Hills, F. J., Lange, W. H., and Kishiyama, J., "Varietal resistance to yellows, vector control and planting date in the suppression of yellows and mosaic of sugar beet," *Phytopathology*, **59**, 1792 (1969).

Hills, O. A., Valcarce, A. C., Jewell, H. K., and Coudriet, D. C., "Beef leafhopper control in sugar beets by seed or soil treatment," *J. Am. Soc. Sugar Beet Technologists*, **11**, 15–24 (1960).

Hirai, T., Shimomura, T., and Nishikawa, Y., "Effect of thiosemicarbazones on the multiplication of tobacco mosaic virus," *Nature*, **181**, 352–353 (1958).

Hirai, T., Saito, T., Onda, H., Kitani, K., and Kiso, A., "Inhibition by blasticidin-S of the ability of leafhoppers to transmit rice stripe virus," *Phytopathology*, **58**, 602 (1968).

Hollings, M., "Aphid movement and virus spread in seed potato areas of England and Wales, 1950–53," *Plant Pathol.*, **4**, 73–82 (1955).

Hollings, M., "Aphid movement and virus spread in seed potato areas of England and Wales," *Plant Pathol.*, **9**, 1–7 (1960).

Hollings, M., "Disease control through virus-free stock," *Ann. Rev. Phytopathology*, **3**, 367–396 (1965).

Hollings, M., and Kassanis, B., "The cure of chrysanthemums from virus diseases by heat," *J. Roy. Hort. Soc.*, **82**, 339–342 (1957).

Holmes, F. O., "A tendency to escape tobacco-mosaic disease in derivatives from a hybrid tomato," *Phytopathology*, **33**, 691–697 (1943).

Holmes, F. O., "Inheritance of resistance to viral diseases in plants," *Advances in Virus Research*, **2**, 1–30 (1954).

Holmes, F. O., "Inheritance of resistance to infection by tobacco-mosaic virus in tomato," *Phytopathology*, **44**, 640–642 (1954a).

Holmes, F. O., "Preventive and curative effects of thiouracil treatments in mosaic-hypersensitive tobacco," *Virology*, **1**, 1–9 (1955).

Holmes, F. O., "Elimination of spotted wilt from dahlias by propagation of tip cuttings," *Phytopathology*, **45**, 224–226 (1955a).

Holmes, F. O., "Elimination of foliage spotting from sweet-potato," *Phytopathology*, **46**, 502–504 (1956).

Holmes, F. O., "Elimination of aspermy virus from the Nightingale chrysanthemum," *Phytopathology*, **46**, 599–600 (1956a).

Holmes, F. O., "A single-gene resistance test for viral relationship as applied to strains of spotted-wilt virus," *Virology*, **5**, 382–390 (1958).

Holmes, F. O., "Genetics of pathogenicity in viruses and of resistance in host plants," *Advances in Virus Research*, 11, 139–161 (1965).

Holmes, F. O., "Resistance to viral disease in tobacco," *Phytopathology*, 55, 504 Abstr. (1965a).

Hooker, W. J., Peterson, C. E., and Timian, R. G., "Virus X resistance in potato," *Am. Potato J.*, 31, 199–212 (1954).

Houtman, P. W., "The hot water treatment on a large scale of DI-52 setts affected by sereh disease," *Arch. Suikerind. Ned.-Indie.*, 33, 631–642 (1925).

Hoyman, W. G., "Effect of Thimet on incidence of virus Y and purple-top wilt in potatoes," *Am. Potato J.*, 35, 708–710 (1958).

Huber, G. A., and Schwartze, C. D., "Resistance in the red raspberry to the mosaic vector *Amphorophora rubi* Kalt.," *J. Agr. Research*, 57, 623–633 (1938).

Hughes, C. G., "Disease resistance of approved varieties," *Cane Growers Bull.*, 29 (3), (1966).

Hull, J., Jr., and Klos, E. J., "Responses of healthy, ring spot, and yellows virus-infected Montmorency cherry trees to gibberellic acid," *Mich. State Univ. Agr. Expt. Sta. Quart. Bull.*, 41, 19–23 (1958).

Hull, R., "Control of virus yellows in sugar beet seed crops," *J. Roy. Agr. Soc.*, 113, 86–102 (1952).

Hull, R., "Sugar beet yellows in Great Britain, 1954 to 1959," *Plant Pathol.*, 3, 130 (1954); 4, 134 (1955); 5, 130 (1956); 6, 131 (1957); 7, 131 (1958); 8, 145 (1959).

Hull, R., "Sugar beet yellows. The search for control," *Agriculture (London)*, 65, 62–65 (1958).

Hull, R., "Virus yellows was decreased by spraying. Precautions for 1960," *Brit. Sugar Beet Rev.*, 28, 24 (1959).

Hull, R., "Spread of groundnut mosaic virus by *Aphis craccivora* (Koch)," *Nature*, 202, 214 (1964).

Hungerford, C. W., "Plant breeders develop beans resistant to curlytop," *Idaho Univ. Agr. Expt. Sta. Sci. Bull.*, 36, 5 (1951).

Hunger, J. A., Chamberlain, E. E., and Atkinson, J. D., "Note on a modification in technique for inactivating apple mosaic virus in apple wood by heat treatment," *New Zealand J. Agr. Research*, 2, 945–946 (1959).

Hutton, E. M., "Some aspects of virus Y resistance in the potato (*Solanum tuberosum*)," *Australian J. Agr. Research*, 4, 362–371 (1952).

Hutton, E. M., and Brock, R. D., "Reactions and field resistance of some potato varieties and hybrids to the leaf-roll virus," *Australian J. Agr. Research*, 4, 256–263 (1953).

Ivanoff, S. S., "A seedling method for testing aphid resistance," *Heredity*, 36, 357–361 (1945).

Jaeger, S., "Milchspritzungen zue Einschrankung von Virusinfektion en im Gemusebau," *Z. PflKrankh. Pfl. Path. Pflschutz.*, 73, 215 (1966).

Jameson, J. D., "Resistance to cassava mosaic," (not title) *East African Agric. For. J.*, **29**, 208 (1964).

Jefferson, R. N., and Eads, C. O., "Control of aphids transmitting stock mosaic," *J. Econ. Entomol.*, **44**, 878–882 (1951).

Jenkinson, J. G., "The incidence and control of cauliflower mosaic in broccoli in south-west England," *Ann. Appl. Biol.*, **43**, 409–422 (1955).

Jennings, D. L., "Further studies in breeding cassava for virus resistance," *E. African Agr. J.*, **22**, 213–219 (1957).

Jennings, D. L., "*Manihot melanobasis* Mull. Arg.—A useful parent for cassava breeding," *Euphytica,* **8**, 157–162 (1959).

Jennings, D. L., "Observations on virus diseases of cassava in resistant and susceptible varieties. I. Mosaic disease," *Empire J. Eptl. Agr.*, **28**, 23–34 (1960).

Jennings, D. L., "Preliminary studies on breeding raspberries for resistance to mosaic disease," *Hort. Res.*, **2**, 82–96 (1963).

Johnson, F., "Heat inactivation of wheat mosaic virus in soils," *Science*, **95**, 610 (1942).

Johnson, F., "The effect of chemical soil treatments on the development of wheat mosaic," *Ohio J. Sci.*, **45**, 125–128 (1945).

Johnson, J., "Plant virus inhibitors produced by micro-organisms," *Science*, **88**, 552–553 (1938).

Jones, L. S., Wilson, N. S., Burr, W., and Barnes, M. M., "Restriction of peach mosaic virus spread through control of the vector mite *Eriophyes insidiosus*," *J. Econ. Entomol.*, **63**, 1551 (1970).

Kassanis, B., "Potato tubers freed from leaf-roll virus by heat," *Nature*, **164**, 881 (1949).

Kassanis, B., "Heat inactivation of leaf-roll in potato tubers," *Ann. Appl. Biol.*, **37**, 339–341 (1950).

Kassanis, B., "Heat-therapy of virus-infected plants," *Ann. Appl. Biol.*, **41**, 470–474 (1954).

Kassanis, B., "Effects of changing temperature in plant virus diseases," *Adv. Virus Res.*, 4, 221–241)1957).

Kassanis, B., "The use of tissue cultures to produce virus-free clones from infected potato varieties," *Ann. Appl. Biol.*, **45**, 422–427 (1957a).

Kassanis, B., and Tinsley, T. W., "The freeing of tobacco tissue cultures from potato virus Y by 2-thiouracil," *Proc. 3rd Conf. Potato Virus Diseases*, Lisse-Wageningen, 1957, p. 24–28.

Kassanis, B., and Varma, A., "Production of virus-free clones of some British potato varieties," *Ann. Appl. Biol.*, **59**, 447–450 (1967).

King, L. W., "Washington certified seed potatoes," *Am. Potato J.*, **29**, 53–54 (1952).

Kirkpatrick, H. C., and Lindner, R. C., "Studies concerning chemotherapy of two plant viruses," *Phytopathology*, **44**, 529–533 (1954).

Klesser, P. J., and LeRoux, P. M., "Groundnut rosette prevention the best method of control," *Farming in S. Africa*, 33, 40–41 (1957).

Klos, E. J., Fronek, F., Knierim, J. A., and Cation, D., "Peach rosette mosaic transmission and control studies," *Qu. Bull. Michigan St. Agric. Exp. Sta.*, 49, 287–293 (1967).

Klostermeyer, E. C., "The relationship between insecticide-altered aphid populations and the spread of potato leaf roll," *J. Econ. Entomol.*, 52, 727–730 (1959).

Knust, H. G., "Notes on the heat treatment of cane plants at the Bundaberg Sugar Experiment Station," *Proc. Queensland Soc. Sugar Cane Technologists, 20th Conf., Mackay*, 1953, pp. 131–134.

Köhler, E., and Hauschild, I., "Versuche zur Beeinflussung blattrollkranker Kartoffel-knollen durch Chemikalien," *Nahrbl. Deut. Pflanzenschutzdienstes (Stuttgart)*, 2, 24–26 (1950).

Kooistra, G., "Investigations into the chemotherapeutic activity of nitroso hydroxyaryl compounds towards some virus diseases," *Acta Botan. Neerl.*, 8, 363–411 (1959).

Kuhn, C. W., Brantley, B. B., and Sewell, G., "Immunity to bean yellow mosaic virus in cowpea," *Plant Disease Reptr.*, 49, 879–881 (1965).

Külps, G., "Studies on the protective effect of oils on virus transmission by aphids," (Trans.), *Zeit. PflKrankh. Pfl. Path. Pflschutz.*, 75, 213–217 (1968).

Kunkel, L. O., "Peach mosaic not cured by heat treatments," *Am. J. Botan.*, 23, 683–686 (1936).

Kunkel, L. O., "Heat treatments for the cure of yellows and other virus diseases of peach," *Phytopathology*, 26, 809–830 (1936a).

Kunkel, L. O., "Heat cure of aster yellows in periwinkle," *Am. J. Botan.*, 28, 761–769 (1941).

Kunkel, L. O., "False blossom in periwinkles and its cure by heat," *Science*, 95, 252 (1942).

Kunkel, L. O., "Potato witches' broom transmission by dodder and cure by heat," *Proc. Am. Phil. Soc.*, 86, 470–475 (1943).

Kunkel, L. O., "Transmission of alfalfa witches' broom to nonleguminous plants by dodder, and cure in periwinkle by heat," *Phytopathology*, 42, 27–31 (1952).

Kurtzman, R. H., Jr., Hildebrandt, A. C., Burris, R. H., and Riker, A. J., "Inhibition and stimulation of tobacco mosaic virus by purines," *Virology*, 10, 432–448 (1960).

Lamb, K. P., "Field trial of eight varieties of *Brassica* field crops in the Auckland district. I. Susceptibility to aphids and virus diseases," *New Zealand J. Agr. Research*, 3, 320–331 (1960).

Lambers, H. R., Reestman, A. J., and Schepers, A., "Insecticides against aphid vectors of potato viruses," *Neth. J. Agr. Sci.*, 1, 188–201 (1953).

Lamey, H. A., McMillian, W. W., and Hendrick, R. D., "Host ranges of the hoja blanca virus and its insect vector," *Phytopathology*, 54, 536–541 (1964).

Landis, J. B., Powell, D. M., and Hagel, G. T., "Attempt to suppress curlytop and beet western yellows by control of the beet leafhopper and the green peach aphid with insecticide-treated seed," *J. Econ. Entomol.*, 63, 493 (1970).

Larson, R. H., "Purple-top hair sprout and low soil temperature in relation to secondary or sprout tuber formation," *Am. Potato J.*, 36, 29–31 (1959).

Linford, M. B., "Influence of plant populations upon incidence of pineapple yellow spot," *Phytopathology*, 33, 408–410 (1943).

List, G. M., Landblom, N., and Sisson, M. A., "A study of records from the Colorado Peach Mosaic Suppression Program," *Colo. Agr. Expt. Sta. Tech. Bull.*, 59 (1956).

Liu, H. P., and Li, H. W., "Studies on the sugarcane mosaic virus in Taiwan. Part II. The mode of resistance of cane varieties and the wild relatives of cane to strains of mosaic," *Rept. Taiwan Sugar Exp. Sta. (Taiwan)*, 10, 89–103 (1953).

Liu, H. P., Lee, S. M., and Teng, W. S., "Studies on the ratoon stunting disease of sugarcane. I. Effect of one heat treatment on the control of RSD in seed cane with various years of infection," *Rept. Taiwan Sugar Exp. Sta.*, 32, 131–141 (1963).

Longworth, J. F., and Thresh, J. M., "The reaction of different cacao types to infection with swollen-shoot virus," *Ann. Appl. Biol.*, 52, 117–124 (1963).

Lucas, G. B., and Winstead, N. N., "Control of tobacco mosaic with an antiviral agent," *Phytopathology*, 48, 344 (1958).

Maramorosch, K., "Reversal of virus-caused stunting in plants by gibberellic acid," *Science*, 126, 651–652 (1957).

Martin, J. P., "Chlorotic streak disease," *Hawaiian Sugar Planters' Assoc. Rept.*, 1932, 32 (1933).

Martin, W. J., and Kantack, E. J., "Control of internal cork of sweet potato by isolation," *Phytopathology*, 50, 150–152 (1960).

McKeen, C. D., "The inhibitory activity of extract of *Capsicum frutescens* on plant virus infections," *Can. J. Botan.*, 34, 891–903 (1956).

McKinney, H. H., "Investigations of the rosette disease of wheat and its control," *J. Agr. Research*, 23, 771–800 (1923).

McKinney, H. H., Paden, W. R., and Koehler, B., "Studies on chemical control and overseasoning of, and natural inoculation with, the soil-borne viruses of wheat and oats," *Plant Disease Reptr.*, 41, 256–266 (1957).

McLean, D. M., "Effect of insecticide treatments of beets on transmission of yellows virus by *Myzus persicae*, *Phytopathology*, 47, 557–559 (1957).

Michellbacher, A. E., Gardner, M. W., Middlekauff, W. W., and Walz, A. J., "Field dusting with DDT to control thrips and spotted wilt in tomatoes," *Plant Disease Reptr.*, 34, 307–309 (1950).

Micheletti, De Z. D., "Naturaleza de la resistencia al mosaico de papaya en *Carioca cauliflora* inferida de reaciones entre injertes de *C. papaya* y *C. cauliflora*," *Agron. Trop. Maracay*, 12, 3–11 (1962).

Miller, P. W., "An improved method of testing the tolerance of strawberry varieties and new selected seedlings to virus infection," *Plant Disease Reptr.,* **43,** 1247–1249 (1959).

Miller, P. W., and Waldo, G. F., "The virus tolerance of *Fragaria chiloensis* compared with the Marshall variety," *Plant Disease Reptr.,* **43,** 1130–1131 (1959).

Millikan, D. F., "The phony peach disease," *Missouri Univ. Agr. Expt. Sta. Bull.,* No. 661, (1955).

Muller, H. J., "Über die Anflugdichte von Aphiden auf farbige Salatpflanzen," ("On the landing frequency of aphids on colored lettuce plants") *Ent. Exp. Appl.,* **7,** 85–104 (1964).

Mungomery, R. W., "Obstacles to be avoided in hot-water treating cane setts against ratoon stunting disease," *Cane Growers' Quart. Bull.,* **17,** 54–56 (1953).

Munster, J., and Murbach, R., "L'application d'insecticides contre les pucernos vecteurs des viroses de la pomme de terre peut-elle garantir la production de plantes de qualité?," *Rev. Romande Agr. Viticult. Arboricult.,* **8,** 41–43 (1952).

Murant, A. F., and Taylor, C. E., "Treatment of soil with chemicals to prevent transmission of tomato black ring and raspberry ringspot viruses by *Longidorus elongatus* (de Mam)," *Ann. Appl. Biol.,* **55,** 227–237 (1965).

Nault, L. R., Harlan, H. J., and Findley, W. R., "Comparitive susceptibility of corn to aphid and mechanical inoculation of maize dwarf mosaic virus," *J. Econ. Entomol.,* **64,** 19 (1971).

Nichols, R. F. W., "Breeding cassava for virus resistance," *E. African Agr. J.,* **12,** 184–194 (1947).

Nickell, L. G., "Effect of certain dyes on the growth *in vitro* of virus tumor tissue from *Rumex acetosa," Botan. Gaz.,* **112,** 290–293 (1951).

Nielsen, L. W., "Elimination of the internal cork virus by culturing apical meristems of infected sweet potatoes," *Phytopathology,* **50,** 840–841 (1960).

Nielsen, L. W., and Pope, D. T., "Resistance in sweet potato to the internal cork virus," *Plant Disease Reptr.,* **44,** 342–347 (1960).

Norris, D. O., "Reconstitution of virus X-saturated potato varieties with malachite green," *Nature,* **172,** 816 (1953).

Norris, D. O., "Development of virus-free stock of green mountain potato by treatment with malachite green," *Australian J. Agr. Research,* **5,** 658–663 (1954).

Nutman, F. J., Roberts, F. M., and Williamson, J. G., "Studies on the varietal resistance in the goundnut (*Arachnis hypoges* L.) to rosette disease," *Rhod. J. Agric. Res.,* **2,** 63–77 (1964).

Nyland, G., "Heat inactivation of ringspot virus in some stone fruit hosts," *Phytopathology,* **47,** 530 (1957).

Nyland, G., "Hot-water treatment of Lambert cherry budsticks infected with necrotic rusty mottle virus," *Phytopathology,* **49,** 157–158 (1959).

Orlob, G. B., "The role of ants in the epidemiology of barley yellow dwarf virus," *Ent. Exp. Applic.,* **6**, 95–106 (1963a).

Oshima, N., and Foskett, R. L., "Tolerance of woolly and dwarf tomatoes to curlytop virus in Colorado," *Plant Disease Reptr.,* **50**, 309 (1966).

Owen, F. V., "Variability in the species *Beta vulgaris* L. in relation to breeding possibilities with sugar beets," *J. Am. Soc. Agron.,* **36**, 566–569 (1944).

Owen, F. V., Abegg, F. A., Murphy, A. M., Tolman, B., Price, C., Larmer, F. G., and Carsner, E., "Curlytop resistant sugar beet varieties in 1938," *U.S. Dept. Agr. Circ.,* No. 513 (1939).

Palmiter, D. H., and Adams, J. A., "Seasonal occurrence of leafhopper vectors of X-disease virus in sprayed and unsprayed peach blocks," *Phytopathology,* **47**, 531 (1957).

Pathak, M. D., Cheng, C. H., and Fortumo, M. E., "Resistance to *Nephotettix impicticeps* Ishi. and *Nilaparvata lugens* (Stål) in varieties of rice," *Nature,* **223**, 502–504 (1969).

Pelham, J., "Resistance in tomato mosaic virus," *Euphytica,* **15**, 258–267 (1966).

Persons, T. D., "Phony peach disease—a review of organized control from 1929 to 1951 and the effect of recent developments on future control programs," *Phytopathology,* **42**, 286–287 (1952).

Piemeisel, R. L., "Replacement control: Changes in vegetation in relation to control of pests and diseases," *Botan. Rev.,* **20**, 1–32 (1954).

Piemeisel, R. L., Lawson, F. R., and Carsner, E., "Weeds, insects, plant diseases, and dust storms," *Sci. Monthly,* **73**, 124–128 (1951).

Pizarro, A. C., and Arny, D. C., "The persistence of systox in oat plants and its effect on the transmission of the barley yellow-dwarf virus," *Plant Disease Reptr.,* **42**, 229–232 (1958).

Porter, C. A., and Weinstein, L. H., "Biochemical changes induced by thiouracil in cucumber mosaic virus-infected and non-infected tobacco plants," *Contribs. Boyce Thompson Inst.,* **19**, 87–106 (1957).

Porter, C. A., and Weinstein, L. H., "Altered biochemical patterns induced in tobacco by cucumber mosaic virus infection by thiouracil, and by their interaction," *Contribs. Boyce Thompson Inst.,* **20**, 307–316 (1960).

Posnette, A. F., "The spread and control of viruses in fruit trees," *Sci. Hort.,* **13**, 87–89 (1958).

Posnette, A. F., "Tolerance of virus infection in crop plants," *Rev. Appl. Mycology,* **48**, 113 (1969).

Posnette, A. F., and Cropley, R., "Heat treatment for the inactivation of strawberry viruses," *J. Hort. Sci.,* **33**, 282–288 (1958).

Posnette, A. F., and Todd, J. McA., "Virus diseases of cacao in West Africa. IX. Strain variation and interference in virus 1A," *Ann. Appl. Biol.,* **43**, 433–453 (1955).

Posnette, A. F., Cropley, R., and Ellenberger, C. E., "Progress in the heat treatment for strawberry virus diseases," *Ann. Rept. East Malling Research Sta., Kent,* 322, 128–130 (1953).

Pound, G. S., Williams, P. H., and Walker, J. C., "Mosaic and yellows resistant inbred cabbage varieties," *Res. Bull. Agric. Exp. Sta. Univ. Wisconsin,* 259 (1965).

Price, W. C., and Knorr, L. C., "Kinetics of thermal destruction of citrus tissues in relation to the virus disease problem," *Phytopathology,* 46, 657–661 (1956).

Ribbands, C. R., "The spread of apterae of *Myzus persicae* (Sulz.) and of yellows viruses within a sugarbeet crop," *Bull. Entomol. Res.,* 54, 267–283 (1963).

Rich, S., "Chemotherapy of lettuce big vein," *Plant Disease Reptr.,* 40, 414–416 (1956).

Rietberg, H., and Hijner, J. A., "Die bekampfung der vergilbungskrankheit der ruben in den Neiderlanden," *Zucker Susswarenwirtsch.,* 9, 483–485 (1956).

Robertson, D. G., "The local lesion reaction for recognizing cowpea mosaic varieties immune from and resistant to cowpea yellow mosaic virus," *Phytopathology,* 55, 923–925 (1965).

Rodriguez, S. J., "Examen de los medios de lucha contra los virus. Ensayos espanoles," *Bol. Patol. Vegetal Entomol. Agr.,* 18, 129–132 (1950).

Roland, G., "La thermotherapie de viroses végétales," *Mededeel. Landbouwhoge-school en Opzoekingssta. Staat Gent,* 22, 553–560 (1957).

Rosberg, D. W., "Treatment of watermelon seed for the control of pimples virus disease," *Texas Agr. Expt. Sta. Progr. Rept.* (1957).

Rumley, G. E., and Thomas, W. D., Jr., "The inactivation of the carnation-mosaic virus," *Phytopathology,* 41, 301–303 (1951).

Russell, G. E., "Breeding for tolerance to beet yellows virus and beet mild yellowing virus in sugarbeet. I. Selection and breeding methods. II. The response of breeding material to infection with different virus strains," *Ann. Appl. Biol.,* 53, 363–376 (1964).

Russell, G. E., "Breeding for resistance to infection with yellowing viruses in sugarbeet. I. Resistance in virus tolerant breeding material," *Ann. Appl. Biol.,* 57, 311–319 (1966).

Russell, G. E., "Some effects of spraying with thiabendazole on the susceptibility of sugar beet to yellowing viruses and on their vector *Myzus persicae* (Sulz.)," *Ann. Appl. Biol.,* 62, 265–273 (1968).

Salaman, R. N., "Protective inoculation against a plant virus," *Nature,* 131, 468 (1933).

Salzmann, R., "Über das Totspritzen der Kartoffelstauden als Manahme zur Verhinderung der Virusausbreitung," *Annuaire Agr. Suisse,* 52, 707–738 (1953).

Schaller, C. W., Rasmussun, D. C., and Qualset, C. O., "Sources of resistance to the yellow dwarf in barley," *Crop Science,* 3, 342–344 (1963).

Schlegel, D. E., and Rawlins, T. E., "Inhibition of tobacco mosaic virus by an antibiotic from an actinomycete, *Nocardia* species," *Phytopathology,* 44, 328–329 (1954).

Schmidt, C. T., "Soil fumigant comprising 1,' 2-Dibromo-3-chloropropane," *U.S. Patent 2937936* (1960).

Schmidt, C. T., "*Meloidogyne coffeicola* (Goeldi) a serious root nematode problem in Brazilian coffee," *F.A.O. Pl. Prot. Bull.,* **17,** 56 (1969).

Schultz, E. S., Bonde, R., and Raleigh, W. P., "Isolated tuber-unit seed plots for the control of potato virus diseases and blackleg in northern Maine," *Maine Agr. Expt. Sta. Bull.,* No. 370 (1934).

Schultz, E. S., Bonde, R., and Raleigh, W. P., "Early harvesting of healthy seed potatoes for the control of potato diseases in Maine," *Maine Agr. Expt. Sta. Bull.,* No. 427 (1944).

Schultz, E. S., Clark, C. F., Bonde, R., Raleigh, W. P., and Stevenson, F. J., "Resistance of potato to mosaic and other virus diseases," *Phytopathology,* **24,** 116–132 (1934).

Sewell, G., Kuhn, C. W., and Brantley, B. B., "Resistance of southern pea, *Vigna sinensis* cowpea chlorotic mottle virus," *Proc. Amer. Soc. Hort. Sci.,* **86,** 487–490 (1965).

Shands, W. A., and Simpson, G. W., "Exploratory tests to prevent leaf roll-infection transmitted by aphids," *Am. Potato J.,* **41,** 23–27 (1964).

Shands, W. A., Webb, R. E., and Schultz, E. S., "Tests with milk and rice polish to prevent infection of Irish potato with virus Y transmitted by aphids," *Am. Potato J.,* **39,** 36–39 (1962).

Shapovalov, M., Blood, H. L., and Christiansen, R. M., "Tomato plant populations in relation to curlytop control," *Phytopathology,* **31,** 864 (1941).

Shepherd, R. J., and Hills, F. J., "Dispersal of beet yellows and beet mosaic virus in the inland valleys of California," *Phytopathology,* **60,** 798 (1970).

Shimomoru, T., and Hirai, T., "Studies on the chemotherapy for plant virus diseases. IV. Effect of the antibiotics on the multiplication of tobacco mosaic virus," *Ann. Phytopathol. Soc. Japan,* **24,** 93–96 (1959).

Simmonds, J. H., "Mild strain protection as a means of reducing losses from the Queensland woodiness virus in the passion vine," *Queensland J. Agr.,* **16,** 371–380 (1959).

Simons, J. N., "Some plant-vector-virus relationships of Southern cucumber mosaic virus," *Phytopathology,* **45,** 217–219 (1955).

Simons, J. N., Swidler, R., and Moss, L. M., "Succulent-type plants as sources of plant virus," *Phytopathology,* **53,** 677–683 (1963).

Sinclair, J. B., "Reduction of tobacco mosaic virus on tomato seed by fermentation," *Phytopathology,* **49,** 551 (1959).

Sinclair, J. B., and Walker, J. C., "Inheritance of resistance to cucumber mosaic virus in cowpea," *Phytopathology,* **45,** 563–564 (1955).

Smeets, L., and Wassenaar, L. M., "Problems of heat spot in *Fragaria vesca* L. when indexing strawberry selections for viruses," *Euphytica,* **5,** 51–54 (1956).

Stoddard, E. M., "Inactivating *in vivo* the virus of X-disease of peach by chemotherapy," *Phytopathology,* **32,** 17 (1942).

Stone, O. M., "The elimination of four viruses from carnation and sweet william by meristem culture," *Ann. Appl. Biol.,* **62,** 119–122 (1968).

Stubbs, L. L., Guy, A. D., and Stubbs, K. J., "Control of lettuce necrotic virus disease by the destruction of the common sowthistle (*Sonchus oleracus,*)" *Aust. J. Exp. Agric. Animal Husb.,* **3,** 215–218 (1963).

Stubbs, L. L., and O'Loughlin, G. T., "Climatic elimination of mosaic spread in lettuce seed crops in the Swan Hill region of the Murray River," *Aust. J. Exp. Agric. Animal Husb.,* **2,** 16 (1962).

Summers, E. M., Brandes, E. W., and Rands, R. D., "Mosaic of sugar cane in the United States, with special reference to strains of the virus," *U.S. Dept. Agr. Tech. Bull.,* **955,** 44–87 (1948).

Takahashi, W. N., "A virus inactivator from yeast," *Science,* **95,** 586–587 (1942).

Takahashi, W. N., "The inhibition of virus increase by malachite green," *Science,* **107,** 226 (1948).

Taylor, C. E., and Murant, A. F., "Chemical control of raspberry ringspot and tomato black ring in strawberry," *Plant Pathol.,* **17,** 171–178 (1968).

Thirumalachar, M. J., "Inactivation of potato leaf roll by high temperature storage of seed tubers in Indian Plains," *Phytopathol. Z.,* **22,** 429–436 (1954).

Thomas, H. R., and Zaumeyer, W. J., "Inheritance of symptom expression of pod mottle virus," *Phytopathology,* **40,** 1007–1010 (1950).

Thomas, W. D., and Baker, R. R., "Chemical inactivation of the carnation mosaic virus *in vivo*," *J. Colo.-Wyo. Acad. Sci.,* **4,** 51 (1949).

Thompson, A. E., Lower, R. L., and Thornberry, H. H., "Inheritance in beans of the necrotic reaction to tobacco mosaic virus," *J. Heredity,* **53,** 89–91 (1962).

Thomson, A. D., "Studies on the effect of malachite green on potato viruses X and Y," *Australian J. Agr. Research,* **7,** 428–434 (1956).

Thomson, A. D., "Heat treatment and tissue culture as a means of freeing potatoes from virus Y," *Nature,* **177,** 709 (1956a).

Thomson, A. D., "Elimination of potato virus Y from a potato variety," *New Zealand J. Sci. Technol.,* **A38,** 482–490 (1957).

Thomson, A. D., "Interference between plant viruses," *Nature,* **181,** 1547–1548 (1958).

Thresh, J. M., "The control of cacao swollen shoot disease in Nigeria," *Trop. Agr. (London),* **36,** 35–44 (1959).

Thresh, J. M., "Capsids as a factor influencing the effect of swollen shoot disease on cacao in Nigeria," *Empire J. Exptl. Agr.,* **28,** 193–200 (1960).

Thresh, J. M., "Warm water treatments to eliminate the gall mite from black currant cuttings (*Phytopus ribis* Nal.)," *51st Rept. E. Malling Res. Sta.,* **1963,** 131–132 (1964).

Troutman, J. L., and Fulton, R. W., "Resistance in tobacco to cucumber mosaic virus," *Virology,* **6,** 303–316 (1958).

Vanderveken, J., and Vilain, N., "Inhibition de la transmission aphidienne de quelques phytovirus au moyen de pulvéristrons d'huile," *Annals Epiphyt. (Hors-Ser)*, 18, 125–132 (1967).

Vanderveken, J. *et al.,* "On the use of soils in the control of transmission of plant viruses," *Annals. Gembloux,* 74, 47–52 (1968).

van der Plank, J. E., and Anderssen, E. E., "Kromnek disease of tobacco," *Farming in S. Africa,* 19, 391–394 (1944).

van Slogteren, D. H. M., "Some recent developments in indexing virus infected potato plants in the Netherlands," *Am. Potato J.,* 36, 303–304 (1959).

van Slogteren, E., "Serological diagnosis of plant virus diseases," *Ann. Appl. Biol.,* 42, 122–128 (1955).

Vine, S. J., "Improved culture of apical tissues for production of virus-free strawberries," *J. Hort. Sci.,* 43, 289–292; 293–297 (1968).

Wagnon, H. K., and Traylor, J. A., "Results of some soil treatments for elimination of peach yellow bud mosaic virus from soil," *Phytopathology,* 47, 537 (1957).

Walkey, D. G. A., "Production of virus-free rhubarb by apical tip meristem," *J. Hort. Sci.,* 43, 283–287 (1968).

Wallace, H. E., " 'Russian thistle elimination as a supplement to the beet leafhopper control program in California," *Calif. Dept. Agr. Bull.,* 37, 8–11 (1948).

Wallis, R. L., and Turner, J. E., "Burning weeds in drainage ditches to suppress populations of green peach aphids and incidence of beet western yellows virus disease of sugar beet," *J. Econ. Entomol.,* 62, 307 (1969).

Webb, R. E., and Hougas, R. W., "Preliminary evaluation of *Solanum* species and species hybrids for resistance to disease," *Plant Disease Reptr.,* 43, 144–151 (1959).

Webb, R. E., and Schultz, E. S., "The evaluation of potato seedling varieties for field immunity from potato virus A," *Am. Potato J.,* 36, 275–283 (1959).

Wernham, C. C., and Mackenzie, D. R., "Sources of resistance to maize mosaic virus," *Phytopathology,* 55, 1285 Abstr. (1965).

Wilbrink, G., "Hot water treatment of setts as a remedy for the sereh disease of sugar cane," *Mededeel. Proefsta. Java-Suikerind.,* 1, 1–15 (1923).

Wilcox, R. B., "Tests of cranberry varieties and seedlings for resistance to the leafhopper vector of false-blossom disease," *Phytopathology,* 41, 722–735 (1951).

Yoshii, H., Tominaga, Y., and Morioka, T., "On the inactivating effect of some higher plant juices against tobacco mosaic virus," *Ann. Phytopathol. Soc. Japan,* 19, 25–28 (1954).

Zaumeyer, W. J., and Harter, L. L., "Inheritance of symptom expression of bean mosaic virus 4," *J. Agr. Research,* 67, 287–292 (1943).

Zink, F. W., Grogan, R. G., and Bardin, R., "The comparative effect of mosaic-free seed and roguing as a control for common lettuce mosaic," *Proc. Am. Soc. Hort. Sci.,* 70, 277–280 (1957).

Zink, F. W., Grogan, R. G., and Welch, J. E., "The effect of the percentage of seed transmission upon subsequent spread of lettuce mosaic virus," *Phytopathology*, 46, 662–664 (1956).

Zitter, T. A., and Murakishi, H. H., "Nature of increased virulence in tobacco mosaic after passage in resistant tomato plants," *Phytopathology*, 59, 1736 (1969).

Zumme, N., "Effect of treatment of seed cane on susceptibility of sugarcane to sugarcane mosaic virus," *Phytopathology*, 57, 83–85 (1967).

Index

Acarina, see mites
Aflatoxins, in cotton seed, 82
  groundnut, 83
Aleyrodidae, 481
  see also White flies
Amino acids, in aphid saliva, 159
  in ergot, 77
  inducing galls, 237
  in plant bugs, 160
  virus particles, 322
    strain differences, 418
Anasa tristis and wilt, 260
Anatonmy hemipterous, 140
  of Thysantoptera, 547, 550
Antibiotics, against fire blight, 16
  walnut blight, 18
  viruses, 694
Anti-viral compounds, 697
Aphids, amino acids in saliva, 159
  feeding process, 140, 543
  galls formed by, 239
  phytotoxins, 174, 209, 260
  salivary secretions, 150, 544
  stylet tracks, 139, 142, 155
  transmitting viruses, 480, 487, 527
    transovarially, 565
  traps, 624, 626, 631
Apical tissue culture, 691
Apple, bacterial rot, 23
  brown rot, 85
  canker, 88
  hopperburn, 221
  woolly aphid, 98
Australia, prickly pear, 27

Bacterial pathogens, diseases caused
    by, 36
  control by, antibiotics, 16, 18, 19, 23
    insecticides, 25, 32
    resistance, 16, 30, 33

Bacterial pathogens (*continued*)
  species involved, Bacillus
    cacticidus, 28
    Bacterium cactivorum, 28
    Corynebacterium sepedonicum, 22
    Erwinia, amylovora, 10, 38
      var rubri, 16
      carnegieana, 25
      carotovora, 18, 33, 43
      salicis, 38
      tracheiphila, 29
    Pseudomonas, flectans, 19
      melophora, 23
      phaseocola, 19
      savastoni, 39
      solanacearum, 27
      spp, 38
    Salmonella montevideo, 28
    Xanthomonas, citri, 28
      juglandis, 17
      pelargonii, 28
      stewartii, 30
  Transmission, generalized, 10
    specialized, 28
  Vector-pathogen relationships, 43
Banana, 27, 714
Birds, and entry of fungi, 90
  transmitting tobacco mosaic, 485
Blister mite of cotton, 200
Blood, insect as virus reservoir, 558
Buccal fluid inactivating TMV, 592
  as transmission medium, 545

Cacao, capsids, 182, 197
  cushion gall, 88
  die-back, 87
  swollen-shoot, 362, 482, 493, 532
Cactus, giant, 25
Canker, of apple (tree cricket), 88
  cacao, 88

Canker (*continued*)
  fire blight, 10
  gnarled stem of tea, 182
  perennial of apple, 98
  plane trees, 114
  stem, of mango, 184
Capsids (mirids) causing phytotoxemia,
    182, 194
  Adelphocorus, on alfalfa, 197
    on cacao, 182, 194
    capsid blast, 182
      control by gamma BHC, 193
        resistance to, 194
      esterases in saliva, 188
      pockets in forest, 190
      inter-relationships with ants, 192
      initiating *Calonectria* die-back, 189
      species involved, 185, 191
  Helopeltis, on cashew, 185
    cinchona, 183
    cotton, 185
    mango, 184, 185
      angular spot, 184
      fruit scab; rot, 184–185
      stem canker, 184
    tea, 182–184
  Lygus, on cotton, 196
    on developing carrot seed, 195
      gummosis of peach, 247
      lima beans, 195
      strawberry, 196
Castor bean hopperburn, 267
Cat-facing of peaches, 224
Cecidogen as virus, 247
Celery aphid yellows, 261
Chemical control, of vectors, 705
  of viruses, 692
Citrus, *Brevipalpus* gall, 239
  bud mite, 234
  canker, 28
  exanthema, false, 218
  Fetola disease, 246
  leprosis, 485
  mycoplasma, 55
  psyllid gall, 244
  psorosis, 417, 702
  sooty mold, 78
  viruses, lists, 487, 526, 533
Classification, of bacterial diseases, 10
  of fungal infections, 70
    mycoplasmas, 61
    phytotoxemias, 164
    vector-pathogen relations, 43

Classification (*continued*)
  symbiosis, 43
  viruses, 342
  see also Identification, 334
Coccids phytotoxic, 171, 174, 181,
    211, 274, 285
  transmitting fungus, 86, 101
  virus vectors, 362, 482, 493, 532
    latent, 301
Color break, of orchids, 365
  of tulips, 318
  test for viruses, 404
Control of bacterial disease, apple
    rot by chemicals, 25
  banana wilt by roguing, 27
  cucurbit wilt by chemicals, 29
  fire blight by antibiotics, 16
    resistance, 16
  maize (corn) wilt by chemicals, 32
    resistance, 33
  olive knot by chemicals, 41
  potato blackleg, antibiotics, 36
  potato ringrot by chemicals, 23
  walnut blight, antibiotics, 18
    Cu compounds, 18
  of fungal disease, azalea flower
      spot, 76
    cotton boll rot, 96
    Dutch elm disease, 123, 124
    fig endosepsis, 73
    oak wilt, 110
    rhododendron bud blast, 93
    sugarcane rot, by resistance, 83
  of phytotoxemia, alfalfa plant
      bug, 217, 226
    Empoasca tumors by resistance, 245
    froghopper blight, 279
    hopperburn, 224, 226
    mealybug wilt, 306
  of virus disease, Chap. 15
Coreids on cacao, 198
  coconut, 197
    method of predicting damage, 197
    relation to Oecophila ants, 198
  papaya, 218
  premature nut fall of coconut, 198
    Macadamia nut, 198
Cotton, *Aspergillus* infection, 82
  Dysdercus, 93
  fleahoppers, 196
  Helopeltis, 185
  internal boll disease, 93

Cotton (*continued*)
  mites, blister, 230
  Nysius, 218
  split lesions, 196
  viruses, 498, 530
Cross protection, 319, 414–417,
    562, 730
Curlytop, 363, 368, 544
  monograph on, 635, 728
  vector control, 710

Dahlia, hopperburn, 221
Dispersal, of bacteria, 10
  of fungi, 117
  by bark beetles, 121
  of spores, 69
  of virus vectors, 656
  see also Migration
Dodder latent virus, 330
  transmitting viruses, 446
Dutch elm disease, 115
Dysdercus, on cotton, 93
  spp. saliva, 153

Enzymes in saliva, 158
Ergot, 70
Eriophyidae, see Mites

Fetola disease of orange, 246
Fig, mosaic, 179
  rot, souring, 71
  wasp, 72
Fruit malformations, 246
  smuts and molds, 84
  spoilage, 84
Fumigation of soil, 714
Fungi, as virus vectors, 445
Fungi, diseases caused by, 69
  aflatoxin produced by, 82
  associated with coccids, 101
  in insect's stylet pouches, 94
  insects and spermatization, 77, 109
  species of fungi involved, *Alternaria
    brassicola*, 90
    spp., 91
    tenuis, 82
  Aspergillus niger, 80, 82, 84
    flavus, 80, 82
  Boletus tropicus, 101
  Botrodiplodia theobromae, 87, 88

Fungi (*continued*)
  Calonectria rigidiuscula, 87, 88
  Cephalosporium diospyri, 79
    diffenbachia, 92
  Ceratocystis fagacearum, 103,
    107, 109
    fimbriata, 103, 114
    ips, 111
    pini 112 pilifera, 103
  Ceratostomella ulmi, 119
  Chalara quercina, 102
  Claviceps purpurea, 76
  Colletotrichum falcatum, 83
    lagenarium, 82
  Dacromyces sp., 112
  Diplodia pinea, 92
    recifensis, 112
  Endoconidiophora coerulescens, 114
  Endophyllum sp., 77
  Fusarium moniliforme, 82
    var *fici*, 71
    spp., 80, 90
    oxysporum, 83
    poae, 93
  Gleosporium musarum, 82
    perennans, 98
  Helminthosporium sp., 81
  Hemeleia vastatrix, 124
  Lasiodiplodia theobromae, 111
  Leptosphaeria coniothyrium, 88
  Monilia sp., 112
  Monilinia fructifolia, 85
  Nectria coccinea, 86
    cucurbitula, 92
  Nematospora corylii, 97, 98
  Nematospora gossypii, 93, 97
  Nigrospora oryzae, 98, 102
  Ovulina azaleae, 73
  Phoma lingam, 91
  Phytophora sp., 90
  Polyporus coffeae, 101
  Powdery mildews, 77
  Pullularia pullulans, 86
  Pycnostyanus azaleae, 93
  Rhizopus nigricans, 85
  Saccharomyces sp., 112
  Schlerotina fructicola, 85
    fructigens, 85
    laxa, 85
  Sclerospora sorghi, 91

Fungi (*continued*)
　Sooty molds, 78
　Sphaerotheca phytoptophila, 77
　Stereum sanguinolentum, 124
　Thieleviopsis paradoxa, 90
　Trichosporium tingens, 112

Galls, cushion of cacao, 88
　chemically induced, by adenine, 238
　　by heterauxin, 235
　　IAA, 237
　　kineton, 238
　　NAA, 237
　　2-4-D, 237
　insect induced, stimulus for, 234
　　by coccids, 238
　　　aphids, 98, 239
　　　　balsam, woolly, 244
　　　　Phylloxera, 240
　　　　woolly apple, 98
　　　　resistance to, 239
　　　eriophyid mites, 238, 239
　　　leafhoppers, 245
　　　psyllids, 244
　virus induced Fiji disease, 361
　　swollen-shoot of cacao, 362
Genetic factors in symptomology, 382
　in resistance, 723
　variability in vectors, 563
Gigantism floral, 361
Glands salivary, 141, 151, 159, 560
Gnat, fungus and ergot, 76
Graft root in oak wilt, 103, 104
　transmission, types of, 449
Grasses, 32, 76, 80
Grasshoppers buccal fluid, 548, 592
　feeding process, 548
Grasshoppers, and ringrot of
　　potatoes, 23
　as virus vectors, 483, 487, 531
Growth substances, *see also*
　　Hormones, 160, 214, 216,
　　283, 702
Guinea corn, 213
Gumming disease of coconuts, 197
Gummosis of fruits, 246

Halo blight of beans, 19
　scab of citrus, 179
Heat therapy, 682
Helopeltis, see also Capsids, 183-185

Helopeltis (*continued*)
Hemiptera—Homoptera
　feeding processes, *see also*
　　Phytotexemia, 139
　　artificial, 152
　mouth parts, anatomy of, 140
　　route followed, 142
　　of vector species, 547-549,
　　　550, 595
　salivary glands, 141, 151, 159
　　secretions and components of,
　　　150-160
　　　enzymes, 158
　　　free amino acids, 160
　　　growth inhibitors, 160
　　　growth promoters, 160
　　　growth regulators, 160
　　　indole acetic acid, 160
　　tracks in tissue, 142
　　nature of, 150
Hemolymph, 558, 560
Heterauxin and gall formation, 235
Heteroplasia, 236
Heterothallism, 102
Honey dew and ergot, 77
　as virus innoculum, 560
Hopperburn, 219-229
　apple, 221
　cotton, 224
　dahlia, 221
　eggplant, 221
　peanuts, 222
　potatoes, 219-229
　raspberry, 221
　vines, 224
Hormones, synthesis by bugs, 161
　growth, 214, 216, 236, 283
　suppressing insect damage, 160
Host plants of vectors, associations,
　　successions, 638
　of aster yellows; curlytop, 648
　relation to vector efficiency, 600
　of NEPO viruses, 649
　preferences, for diseased, 602
　　for healthy, 603
　reproduction, effect on, aphids, 603
　　leafhoppers, 604
　　mites, 602
　sequence of, for vectors, 638

Host plants of vectors (*continued*)
  affecting insect's biology, 640
  suitability of terminal host, 643
Hyperplasia, 10, 172, 185, 238
Hypersensitivity to viruses, 719
Hypertrophy, 222, 238, 446

Immunity, definition, 717, 719
  to gall formation, 240
  potato ring rot, 22
Inactivator hypothesis, 594, 601
  in insects, 558, 601
Inclusions, intracellular, 392
Incompatibility hypothesis, 594
Indexing for virus-free plants, 689
Indicator plants, 460
  of latent viruses, 463
Idole–3–acetic acid, 160, 237, 238
Infection, avoidance of, *see also* Virus
  control, 679
Inhibitors of viruses chemical, 692–700
  in insects, 601
    milk, mineral oil, 694, 703, 704
    papaya, 696
    sap, 457
    seed, 436
    succulents, 696

Jack fruit, *P. citri* from, 216
Jerusalem cherry, psyllids on, 280
Job's tears, bacterial infection, 32

Kapok, savoying, 216
Kataplasia, 236
Klendusity, 723

Latency of viruses, 330
  components of complexes, 376
  definitions, 331
  indicator plants for, 463
  in mealybug wilt, 302
  terminology, Madison, 331
Leafhoppers and fire blight, 15
  enzymes in, *see also* Phytotoxemia,
    Viruses, 158
Lesions, local, as landmark, 318
  in phytotoxemia, 171
  split, of cotton, 196
  symptoms of, 377

Madison terminology; latency, 331
Maize (corn) bacterial wilt, *see also*
  Viruses, 31
  borer, 80
  dent var., 30
  earworm, 80
  Golden Cross Bantam var., 33, 273
  Guinea, 5, 213
  maggot, 33
  molds, 80
    resistance to, 80
  red kernal disease, 234
  resistance to bacterial wilt, 33
  Stewart's disease, 30
  wallaby ear, 246
Maleic hydrazide as growth
  depressant, 715
Mandibulate insects as virus vectors,
  see also bacterial and fungal
  disease, 483, 487, 531
  buccal fluid as transmission
    medium, 545
  as virus inhibitor, 592
Mango, *Helopeltis* injury, 184
  mite injury, 231
  Recife disease, 112
Mealybug wilt, 285–303
  control, 306
  epidemiology, 295
  etiology, 300
  function of latent virus, 303
  recovery from, 289, 300
  soil type and susceptibility, 295
  symptom sequence, 281
Mechanical innoculation of insects,
  549, 560, 577, 580
Migration of vectors, *Aceria tulipae*, 624
  Aphis craccivora, 624
  Circulifer tenellus, 623, 633
  Dalbulus maidis, 623
  Hyperomyzus lactucae, 624
  Rhopalosiphon maidis, 624
methods of measuring, 625
  aphids, leaf counts, 625
    traps, 626–631
  leafhoppers, net sweeps, 626
    mechanical devices, 626–629
  mealybugs, tree counts, 625
    sticky traps, 631
  physical factors affecting
    altitude, 635

Migration of vectors (*continued*)
  cataclysmic weather, 638
  drought, 633
  humidities, 632
  temperature and rainfall, 635
  wind velocities, 632
Mirids, *see* Capsids
Mites, transmitting bacteria, 17
  walnut blight, 17
Mites, transmitting fungi, *Aspergillus*
    to groundnut, 83
  cotton lint rot, 98
  Dutch elm disease, 115
  fig spoilage, 84
  maize rot; wheat, 102
  oak wilt, 106
  witch's broom of hackberry, 77
Mites, transmitting phytotoxins,
    alfalfa witch's broom, 230
  big bud of black currant, 231
  blister mite of cotton, 230
  causing galls, 238, 239
  cereals, 232
  citrus, 179, 234
    associated with fungus, 179
  Douglas fir, 232
  fig, confused with mosaic, 179
  garlic, resembling virus, 232
  leaf spotting by, 179
  maize red kernel disease, 234
  mango, 231
  Myrobalan plum, 179
  peaches, 179, 232
  phyllody, chrysanthemum, 232
  red berry of blackberry, 230
  red leaf of vines, 231
Mites, transmitting viruses, 338,
    484, 487, 532
Movement of vectors, 622
  measurement of, 625
Mutualism, *see* Symbiosis
Mycetomes, *see* Symbiosis
Mycoplasma, classification of, 61
  culture *in vitro*, 60
  in insects, 56
  in plants, 51
  remission by antibiotics, 52–56
  viruses of, attached to, 64
    evolution from, 329

Nematodes as virus vectors, 445, 533
  control by soil fumigation, 714
Nitidulid sp. in oak wilt, 104
Nucleic acid, 319–326, 418
Nucleocapsid, 335
Nucleotides; 322
Nutfall, coconut, 198
  macadamia nut, 198

Oats, capsid damage to, 158
  leafhopper, 267

Panel disease of rubber, 103
Papaya, coreid damage to, 218
  virus inhibition, 696
Phenology, 651–656
  in bark bettle control, 124
    prediction of outbreaks, 651
      curlytop, 653
      beet yellowing, 652
      rice dwarf, 652
      strawberry aphid, 651
Phytotoxemia, criteria for
    determination, 162
  definitions, toxiniferous, toxi-
    cogenic, 162
  developing secondary symptoms,
    by aphids, 181
    coccids, 181
    coreids, 197
    mirids (capsids), 182–196
    mites, 181
    stinkbugs, 181
  local lesions, caused by aphids, 177
    coccids, 171–174
    leafhoppers, 174–177
    mites, 179
    stinkbugs, 181
  origin, from salivary secretions,
    162–164
  plant malformations, caused by
    aphids, 209
    cicadellids, 219–229
    coccids, 211
    flies (Diptera), 229
    lygeids, coreids, cercopids, 218
    membracids, 219
    mirids (capsids), 214
    mites, 229–234
    thrips, 214
  plant malformations on fruit, 246
    see also Galls

Phytoxemia (*continued*)
  systemic phytotoxemia, by
    aphids, 260
    affecting frost damage, 264
    cercopids (spittle insects), on
      sugarcane (froghoppers), 277
      effect of insect's saliva, 278
      preference for unhealthy
        cane, 278
      resistant cane var., 280
      soil pH, 279
    leafhoppers, 264-274
      maize, 273
      clover confused with virus, 274
      on sugarbeet; China aster, 264
      oats, 267
        by females only, 271
        diet affecting, 273
    mealybugs, on pineapple,
      stripe, 274
      white streak, 276
      wilt, 285
    psyllids on potato; tomato, 280
      by nymphs only, 283
      mycetomes, 285
  symptoms of, categories, 164
Pineapple, gummosis, 246
  mealybug green-spotting, 172
    stripe, 274
    white streak, 276
    wilt, 285
  yellow spot virus, 482
  symptoms, 360
Plant succession, 638
Populations of vectors, estimates, 655
  indexes, 655
  measurement, 653
  prediction, 651
  see also Phenology
Potato, blackleg, 33
  certification, seed, 689
  disease control, viruses, 708
  psyllid yellows, 280
  purple top, 371
  resistance to, 22
  ringrot, 22
  spindle sprout, 283
  see also Viruses
Prickly pear, 27
Prosoplasia, *see* Galls, 236
Protection, cross against viruses, 730

Psyllid yellows, potato, tomato, 280
P^{32}, 152, 545
pH gradiant, 542

Races, of fungi, 78
  virus vectors, 562
Radioactive tracers, 152
  techniques, 545
Raspberry blight, 16
  hopperburn, 221
Recife disease, of mango, 112
Replacement control of host
    plants, 717
  of vegetation, 717
Resistance to bacterial wilt, 38
  Calonectria on cacao, 88
  cassava mosaic, 5, 728
  cereal rots, 81
  corn molds, 80
  definitions, 717
    hypersensitivity, 719
    immunity, 717
    resistance, 717-719
    tolerance, 719
  field, *see also* true, 16, 719
  fire blight, 16
  froghopper blight, 280
  gene sources, 726
  genetics of, 723
  hopperburn, 224
  klendusity, 723
  Lindane, by capsids, 193
  Lygus bugs, 196
  mealybug wilt, 304
    in virgin lands, 296
  olive knot, 41
  perennial canker, 101
  Phylloxera,
    related to phenol content, 242
  potato ringrot, 23
  rice leafhoppers, 721
  rust fungi, 78
  seed infection, 436
  Stewart's disease, 33
  true and field resistance, 16, 719
    types of, 719
  virus diseases, 717
  woolly apple aphid, 101, 239
Ribonucleic acid, 319, 322, 335
Rice, chlorotic spotting, 176
  resistance to leafhopper, 721

Rice (*continued*)
  viruses, 338, 515, 527
    prediction of outbreaks, 652
Rubber, *Alternaria* blight, 91
  panel disease, 103

Salmonella in wheat, 28
Salivary apparatus, 593
  glands, 141, 151, 159, 560
  secretions gelling, 152, 594
    enzymes in, *see* Enzymes, 158
    precursors of, 156
    watery, 153, 158, 595
  sheath, nature of, 150, 593
    in *Aphis craccivora*, 153, 158
    Circulifer tenellus, 152
    Dysdercus fasciatus, 153
    Dysmicoccus brevipes, 148,
      150, 152
    Empoasca fabae, 150
    Myzus persicae, 152
    Oncopeltus fasciatus, 151
    Phylloxera, 150
    Stictocephala festiva, 150
  syringe, 141
Seed certification, potato, 22, 689
  corn maggot, 33
  infection by bacteria, 32
  transmitted viruses, 436
Serology, basis for virus groups,
      318, 337
  in diagnosis, 408
  indexing for control, 691
Slugs and snails, vectors of fungi, 90
  vectors of viruses, 485
Soil-borne viruses, 445
  control by fumigation, 714
  types and froghopper blight, 279
    Helopeltis on tea, 183
Specificity, of vectors, 551
  of aphid phytotoxins, 211, 264
Sterility, virus induced, 436
Strains, of leafhoppers, 563
  mealybugs, 213
  vectors, 563–565
  viruses, 377, 410
Stylets, diagrammed, 142, 547
  of thrips, electron micrograph, 550
    M. periscae, 595
  pouches, of *Dysdercus*, 94
  role of, 593

Stylets (*continued*)
  route followed, Aleyrodidae, 150
    Cicadellidae, 150
    Dysmicoccus brevipes, 148
      neo-brevipes, 152
    Longistigma caryae, 143
    Lygus lineolaris, 150
    Macrosiphon gei, 142
    Myzus persicae, 142, 146, 152
    Olivarus atkinsoni, 149
    Oncopeltus fasciatus, 151
    Pseudococcus adonidum, 150
    Therioaphis maculata, 143
    Tuberolachnus salignus, 148
  sheath, *see also* Secretions, 150, 593
Sugarbeet curlytop, 317, 329, 363, 446
  leafhopper toxins, 268
  viruses, list, 338, 490–492, 526
Sugarcane, red rot, 83
  serah disease, 363
  viruses, lists, 339, 518, 526, 527
Symbiosis, 1
  classification of, 42–44
    as mutualism, 101
    ecto, 42
    endo, 42, 70
  in apple maggot, 23
  in bark beetles, 111
  in chloropid flies, 38
  in Dutch elm disease, 118
  in olive fly, 39
  in *Paratioza cockerelli*, 285
  in seed corn maggot, 33
  in virus transmission, 569
  in woolly apple aphid, 98

Tea, gnarled stem canker, 182
Terminology, of latency, 331
Thrips, feeding processes, 547
    toxic, 214
  gall formers, 242
  peanut pouts, 222
  powdery mildew, 77
  transmitting bacteria, 20
  transmitting fungi, 84
  transmitting viruses, 482, 518, 521
  virus vectors, 482
Tissue culture apical, 691
  malformations, 209
Tolerance, to virus infection, 702,
      719, 722, 726
Tomato bushy stunt crystals, 320
  psyllid yellows, 280

Tomato rosette complex, 374
  spotted wilt in pineapple, 362
  world distribution, 333
Transmission, *see* of bacteria,
    fungi, mycoplasms, phyto-
    toxins, and viruses
  complexes, 561
  cyclical, 599
  deficient; dependent, 562
  mechanical, 453–460
    inoculation of insects, 549, 571
  modes of, 435
    by arthropods, 479
    by other animals, 485
  periodic, 599
  transovarial, 565–569
    methods for proving, 565
    role of symbionts, 568
Traps, sticky for aphids, 626–631
Tulip color break, 318, 355
Tumors, amorphous, 245
  root, 362

Variegation, infectious, 481
Vectors, of viruses, behavior,
    feeding, 542
  flight, 659
  heights of, 629
  control by chemicals, 705–715
    of persistent, non-persistent
      virus vectors, 705, 708
    of potato diseases, 708
      by spraying vector hosts, 711
      sugarbeet viruses, 710
    nematode vectors, 714
      host plants, wild, 715
        by chemicals, 715
      eradication, 717
      replacement, 717
    seed treatments, 714
    vectors on various crops, 712
      strawberry yellows, 713
      swollen-shoot of cacao, 713
      thrips and TSW, 712
      X-disease of peach, 713
    volatile fumigants, 715
  lists, 487, 526
  movements, *see also* Migration,
    621–637
  populations and disease
    incidence, 651

Vectors (*continued*)
  aphids, *B. tabaci*, 654
  estimates by advisory services, 655
  physiological; behavior
    factors, 658
  statistical correlations, 659

Viruses, biological approach;
    ecological, 332
  classification of viruses, 342
  control, by chemicals, 700
    antibiotics, 694
    anti-viral compounds, 697
    inhibitors, 695
    other chemicals, 700–705
  control of diseases, apical tissue
    culture, 691
    avoidance of infection, 679
      removal of virus sources, 680
      virus-free seed, 679
    certification programs, 689
    cultural, isolation, 675
      barrier crops, 676
      density of planting, 677
      planting and harvesting dates, 677
    heat therapy, 682–689
    indexing for virus-free plants, 689
    resistance and tolerance, 717
      commercial control measures, 727
        beans, against curlytop, 728
        cassava mosaic, 728
        internal cork of sweet
          potato, 729
        sugarbeet curlytop, 727
        sugarcane, mosaic and
          sereh, 727
      cross protection, 730
        passion fruit woodiness, 730
        swollen shoot of cacao, 730
        tobacco and virus X, 730
      definitions, 719
        hypersensitivity, 719
        immunity, resistance, 717
        tolerance, 717–719
        true and field resistance, 16, 719
      genetics of, 717
        resistance vs. tolerance, 717, 723
        sources of resistant genes, 726
        in various groups, 721
        serological methods, 691

Viruses (*continued*)
  in various crops, 721
    serological methods, 691
  cross protection, 319, 614–618,
    562, 730
  crystalline *in vitro*, 320, 338
  definition of a virus, 326
  identification and naming, 334
    criteria proposed, 334
      cryptograms, 335
      Gibbs and Hansen compared, 336
    groups, cereals, 338
      corn (maize), 338
      collective species, 337
        with serological affinities, 339
        elongated particles, 339
      potato, 337
      soil-borne, 338
      sugarbeet, 338
      sugarcanᵉ, 339
      transmission by insects, 339
      approved by INCV, 341
  inclusions, intracellular, 393
  indicator plants, 460
  latency, 330
  novogenesis, 329
  origin, speculations on, 327
  replication, 326
  strain diagnosis, 410
    by interference phenomena, 414
    physical and chemical
      properties, 418
  structure *in vivo*, 320
    current concept, 319
  symptoms, external, 351–379
    acute and chronic, 383
    factors affecting, 379
      incubation period, 383
    internal, 384
    intracellular inclusions, 392
    nuclear changes induced, 390
    physical and chemical, 403
    serilogical evidence, 408
  symptomless carriers, 318–330
Virus transmission, arthropods, 479
  by other animals, 485
    birds, 438, 482
    nematodes, 445, 485, 533
    slugs and snails, 438
  modes of, 435
    by dodder, 446
      grafting, 449–453

Virus transmission (*continued*)
  indicator plants for, 460
  inoculation of vectors, 549, 560,
    572, 577, 580
  mechanical methods, 453
    factors affecting, 456
  transovarial, 565–569
  through seed, genetic aspects, 444
    lists of, 437–441
    reviews, 436, 442
  through soil, 445
    by fungi, 445
    nematodes, 445, 485, 533
Virus-vector relationships, 539–607
  acquisition by non-vectors, 557, 598
  biological, in aphids, 592
    role of the stylets, 593
    inhibitors, 594
  complex transmissions, 561
    dependent, 562
  cross protection, in vectors, 562
  cyclical transmission, 599
  definitions of categories, 598
    circulative, propagative, 597
    persistent, non-persistent, 597
  feeding processes of vectors, 539
    aphids, 543
      radioactive techniques, 545
      behavior, 546
    leafhoppers, 539, 549
    mandibulate insects, 548
    thrips, 546
    white flies, 543
  genetic variability in vectors, 563
  honeydew as inoculum, 560
  host plants, effect on
    transmission, 600
  inactivator hypothesis, 591, 595
  ingestion as a factor, 556
  insect blood as reservoir, 558
  mechanical inoculation of vectors,
    549, 560, 577, 580
  multiplication in insects, 570–580
  persistence of virus in insects,
    580–590
    latent period, 580
    persistent and non-persistent,
      584, 597
    semi-persistent, 590
    theories of mechanism of, 590
  reviews, 548

Virus-vector relationships (*continued*)
  serial passage through vectors, 573
  stylet-borne viruses, 597
  transmission of persistent viruses, 596
  transovarial transmission, 565, 569
    role of symbionts in, 568
  vector specificity, 551–558
    area, 554
    group, 554
    non-specific, 556
    regional, 555
    stage, or species, 551

Virus-vector relationships (*continued*)
  virus effect on vectors, 602

Walnut blight, 17
Weather, cataclysmic, 638
Wheat, bacterial infection, 24, 38
White flies (Aleyrodidae), 150, 481
  feeding processes, 543
  as virus vectors, lists, 487, 530
    relationships, 599
Woolly apple aphid, 98

Zoocecidia, *see also* Galls, 234

ERRATA

entence of the preface should read "It does not p
e subject matter of the highly specialized virology

The second
to impinge on t
and reviews..."